STATISTICS
The Craft of Data Collection, Description, and Inference

Third Edition

Ditlev Monrad
William F. Stout
E. James Harner
Barbara A. Bailey
Robert L. Gould
Xuming He
Louis A. Roussos

With contributions from
Jerome S. Colburn

Möbius Communications, Ltd.
Champaign, Illinois

Statistics: The Craft of Data Collection, Description, and Inference, Third Edition

This book is published by Möbius Communications, Ltd.
1802 South Duncan Road, Champaign, IL 61822
Telephone: (217) 398-9086
www.8-mobius.com

Design and composition: Publication Services, Inc.
Cover design: David W. Eynon

ISBN: 1-891304-91-7

Copyright © 2002 by Möbius Communications, Ltd. All rights reserved. No part of this publication may be reproduced, stored, or transmitted by any means without the prior written permission of the publisher.

Printed in the United States of America.

Library of Congress Cataloging-in-Publication Data

CONTENTS

Preface		vii
Acknowledgements		x
About the Authors		xii

Part I Describing Data — 1

1 Exploring Data: Tables and Graphs — 2
- 1.1 The Science of Statistics — 5
- 1.2 Displaying Small Sets of Numbers: Dotplots and Stem-and-Leaf Displays — 11
- 1.3 Graphing Categorical Data — 20
- 1.4 Frequency Histograms — 24
- 1.5 Density Histograms — 30
- 1.6 Misusing Statistics — 37
- Chapter Review Exercises — 45

2 Summarizing Data by Numerical Measures: Centers and Spreads — 50
- 2.1 Centers of Data — 53
- 2.2 Mean versus Median as a Measure of Center — 60
- 2.3 Measuring Spread: The Standard Deviation — 69
- 2.4 Normal Distributions — 82
- 2.5 Boxplots: The Five-Number Summary — 89
- 2.6 Descriptive Data Analysis — 98
- Chapter Review Exercises — 107

3 Linear Relationships: Regression and Correlation — 112
- 3.1 Scatterplots — 115
- 3.2 The Correlation Coefficient — 120
- 3.3 Regression — 132
- 3.4 The Question of Causation — 148
- Chapter Review Exercises — 151

Part II Probability Modeling and Obtaining Data — 155

4 Probabilities and Simulation — 156
- 4.1 Experimental Probability — 159
- 4.2 Probability Models — 166
- 4.3 Simulation: A Powerful Tool for Learning and Doing Statistics — 182
- 4.4 Simulating Random Sampling via a Box Model — 200

		4.5	Random Sampling with or without Replacement	206
			Chapter Review Exercises	212

5 Expected Values and Simulation — 218

	5.1	The Expected Value (Theoretical Mean) of a Random Variable	221
	5.2	Using Five-Step Simulation to Estimate Mean Values	232
	5.3	The Standard Deviation of a Random Variable	239
		Chapter Review Exercises	244

6 Probability Distributions: The Essentials — 248

	6.1	The Binomial Distribution	251
	6.2	The Geometric Distribution	259
	6.3	The Poisson Distribution	263
	6.4	The Normal Probability Distribution	273
		Chapter Review Exercises	288

7 Obtaining Data: Random Sampling and Randomized Controlled Experiments — 292

	7.1	Introduction	294
	7.2	Experimental Design: Observational Studies versus Randomized Controlled Experiments	298
	7.3	Sampling from a Real Population: Probability Sampling versus Non-Probability Sampling	312
	7.4	From Research Question to Data Collection: Some Examples	320
		Chapter Review Exercises	326

Part III Statistical Inference: Estimation and Hypothesis Testing — 329

8 Confidence Interval Estimation — 330

	8.1	An Introduction to the Statistical Estimation Problem	333
	8.2	Random Samples, Estimators, and Standard Errors (SEs) of Estimation	338
	8.3	Central Limit Theorem (CLT) Estimator for the Sample Mean $\bar{X}$	346
	8.4	Large Sample Confidence Interval for the Population Mean μ	353
	8.5	Large Sample Confidence Interval for the Population Proportion p	358
	8.6	Bootstrapping a Sample, Bootstrapped SEs, and SE-based Bootstrapped CIs	362
	8.7	Bootstrapped Percentile-based CIs: A Universally Applicable Method	369
	8.8	Confidence Interval for the Mean μ when the Population Is Normal and the Sample Size Is Small	374
	8.9	Confidence Interval for the Difference between Two Population Means $\mu_X - \mu_Y$	379
	8.10	Confidence Interval for the Difference between Two Population Proportions $p_1 - p_2$	384
	8.11	Confidence Interval for the Difference between Two Population Means in Matched Pairs Case: $\mu_D = \mu_X - \mu_Y$	385

	8.12	Point Estimate for the Population Variance σ^2 and SDσ: Unbiasedness of an Estimator	386
		Chapter Review Exercises	390

9 Hypothesis Testing — 394
- 9.1 The Null Hypothesis and the Alternative Hypothesis — 399
- 9.2 Tests for a Population Proportion — 402
- 9.3 Tests for Randomized Controlled Experiments Producing Sample Proportions — 409
- 9.4 Tests for a Population Mean — 417
- 9.5 Tests for Two Population Proportions or Means — 433
- 9.6 One-sided and Two-sided Hypothesis Testing — 441
- 9.7 Significance Testing versus Acceptance/Rejection Testing: Concepts and Methods — 445
- 9.8 Test for a Population Standard Deviation: Issues — 452
- Chapter Review Exercises — 453

10 Chi-Square Testing — 458
- 10.1 Is the Die Fair? — 460
- 10.2 How Big a Difference Makes a Difference? — 463
- 10.3 The Chi-Square Statistic — 467
- 10.4 Real-Life Chi-Square Examples — 473
- 10.5 The Chi-Square Density — 481
- 10.6 The Chi-Square Distribution and Its Use for Chi-Square Testing — 490
- 10.7 Unequal Expected Frequencies — 497
- Chapter Review Exercises — 505

11 Inference About Regression — 510
- 11.1 Inference about Slope — 512
- 11.2 Inference about Prediction of Y and Estimation of $E(Y \mid x)$ — 530
- 11.3 Transformation of Variables — 533
- Chapter Review Exercises — 537

12 Analysis of Variance and Multiple Regression (Provided on included CD-Rom) — 1
- 12.1 Introduction to the Analysis of Variance — 2
- 12.2 Using the Bootstrap to Test for Differences among Several Population Means — 11
- 12.3 Mean Squares — 16
- 12.4 One-Way Analysis of Variance — 18
- 12.5 Multiple Comparisons of Population Means for a One-Way ANOVA — 25
- 12.6 Two-Way Analysis of Variance — 31
- 12.7 Multiple Regression: Simulation-Based Testing and F-Distribution Testing — 42
- 12.8 Multiple Regression with Many Explanatory Variables — 52
- Chapter Review Exercises — 57

Glossary — 541

Appendixes:

A	Computationally Generated Random Digits	561
B	Random Number Tables	563
C	Chi-Square Probabilities	565
D	Linear Interpolation	567
E	Normal Probabilities	569
F	Student's t Probabilities	571
G	Binomial Probabilities	575
H	F-Distribution Probabilities	583
I	Boneferroni Confidence Intervals	586
J	Cumulative Poisson Probabilities	587
	Index	589

PREFACE

It is the authors' strongly held belief that the most effective way for beginning statistics students to learn statistics well is for them to directly experience its central concepts and standard procedures by closely observing and working with simulated and real data sets. The particular advantage of the use of simulated data is that the student knows precisely the underlying mechanism—that is, the probability model—producing the data. Thus the students can see how close the data come, when studied and analyzed with their evolving statistical expertise, to revealing their underlying model-based truths. Learning statistics by direct contact with data is facilitated in this textbook by the presentation and demonstration of statistical concepts using the five-step simulation method throughout. The same experiential pedagogical approach, with its emphasis on simulation-based demonstration of statistical concepts, was previously developed into a precollege textbook coauthored by William Stout. This book was used in a five-year National Science Foundation Educational Directorate national teacher training grant for high school teachers, with William Stout as a principal investigator.

A vital complement to the experiential learning of statistical concepts is a clear and understandable verbal and formula-based presentation of the formal, logical, and deductive aspects of the subject, using the minimum amount of mathematical formulation of concepts to make the concepts and procedures easy to understand. In most cases this theoretical component is presented closely after or concurrently with a thorough experiential immersion in and discussion of the topic of interest. For example, the basic aspects of probability—often a difficult set of principles for beginning students to master—are first presented mainly as empirical phenomena in Chapters 4 and 5 and then are abstracted and expanded into a body of probability tools and statistical models in Chapters 4 and 5 as well as Chapter 6. The textbook is deliberate in providing thorough explanations of statistical concepts and methods, because such verbal explanations are key to producing an in-depth understanding of statistics.

The textbook pedagogically stands on three legs: (i) verbal explanations of concepts and procedures; (ii) clear formula-based descriptions and explanations of statistical concepts and procedures; and (iii) immersion in the world of chance phenomena and their statistical analyses. Immersion is achieved through textbook-supplied and student-created simulation studies facilitated by fast, sophisticated, and yet easy-to-use instructional software and through examples rich in real-world data.

To be most effective, an introductory statistics textbook must strike a balance between, on the one hand, helping the student learn basic statistical concepts and, hence, to think statistically and, on the other hand, exposing the student to the evolving body of statistical procedures that are widely used in practice and are judged to constitute statistical literacy. This balance, a central goal of this book, equips a student with both real statistical discernment and the needed familiarity with the body of statistical procedures so widely used in science, business, and government, and so often reported in the media.

We have carefully crafted the textbook to function equally well for either of two rather different audiences: concept-based "liberal arts" statistics courses and "methods-based" introductory statistics courses for future statistics users. The concept-oriented course is improved by the inclusion of the standard core of statistical methods as well as the newer methods based on resampling, and a methods course is deepened by an emphasis on underlying concepts. As a basic principle of good statistics instruction, the authors assert that both kinds of courses are made better and more complete by this balancing synthesis of methods and concepts.

In today's technology-rich learning environments, however, an introductory statistics textbook must be designed to make profound and meaningful use of available computer-based technology.

We believe that instructional software must be platform-independent, easy to use (point-and-click being the ideal) for all students, and thoroughly integrated with the textbook. In a statistics course, the software should make the concepts come alive, be a rich source of data, and make complex statistical computations fast and easy to carry out. The accompanying simulation-based instructional software does all of these things, using the "five-step method" of constructing simulations as the presentation tool to facilitate learning.

The textbook breaks new ground in content through its inclusion of bootstrap-based and other resampling inferential procedures as important components. The bootstrap approach is one of the most important statistical advances of the past 25 years and is rapidly becoming an essential component of the practicing statistician's toolbox. When presented appropriately, the bootstrap approach is intuitive and, hence, easily understandable by the beginning statistics student. Consistent with the instructional approach of the book, the bootstrap method is simulation-based and meshes seamlessly with the five-step method used throughout. Most important, it provides the student with a powerful cutting-edge method of statistical inference that, together with the traditional small-sample, normal-population inferential approaches and large-sample inferential approaches based on the central limit theorem and the analogous large-sample result for the chi-square statistic provides the student with enormous inferential power.

The book is written to be accessible to all students having at least a modest exposure to algebra; "intermediate algebra" suffices. Although formulas and graphs appear frequently and students are encouraged to think deeply about what they are learning, the amount of formal mathematical background needed is minimal. The book should work very well in any non-calculus-based introductory statistics course at any two-year or four-year college or university. It can be effectively used at the precollege level, especially for Advanced Placement statistics courses. It can be taught to science and math majors and to humanities and social science majors with equal ease. It presumes no prior exposure to statistics.

Several influential nationally circulated reports have stressed the need for new emphases in the teaching of statistics. For example, the widely influential American Statistical Association/Mathematical Association of America Cobb report recommends that the teaching of statistics be heavily data-based and proposes that more emphasis be placed on statistical concepts than on abstract theory. Further, it stresses *active learning*. This textbook is tailor-made to address these valid recommendations because of the reasons stated above, and because of the numerous exercises provided after each chapter and accompanying the instructional software. Thus, students have ample opportunity to practice using the concepts learned in the text.

Part I of the book begins with three chapters describing how one explores and summarizes data—data being the focus of statistics—with tables and graphs and by means of statistical indices. This includes an introduction to linear regression in Chapter 3.

Part II then presents probability modeling and describes how statisticians collect good data. Chapters 4 and 5 provide a heavily empirical introduction to probability through emphasis on probability and probability distributions, and their means and standard deviations. This is appropriate because probability is the logical underpinning of inferential statistics. In particular, probability and expected value are introduced as empirical concepts, with large simulated samples often used to accurately approximate unknown theoretical probabilities and expected values. This is facilitated by the accompanying instructional software.

Chapter 6 supplements the basic introduction to probability in Chapters 4 and 5 with the three core probability distributions: normal, binomial, and Poisson, and their standard large n relationships. Chapter 7 tells the student how one obtains good data for statistical analyses through random sampling of real populations and well-designed randomized controlled experiments, including a section on how students can validly collect their own data for statistical analyses.

Part III then presents statistical inference, focusing on confidence intervals and hypothesis testing. Chapter 8 presents confidence interval estimation, stressing the traditional normal population and large sample approaches as well as the distribution-free and more widely applicable bootstrapping approaches. These latter flow seamlessly from the simulation approach of the textbook.

Chapter 9 presents hypothesis testing, including the increasingly important simulation-based resampling bootstrap approach as well as the traditional methods based on small samples from a normal population or on large samples via the central limit theorem.

Chapter 10 then presents chi-square testing, beginning with the more widely applicable simulation-based approach. Then the traditional large sample chi-square distribution approach is presented. Coverage includes both the usual multinomial distribution chi-square test (null hypothesis of a specified many-sided, possibly unfair die) and contingency table-based chi-square tests.

Chapter 11 complements the descriptive treatment of linear regression from Chapter 3 with an inference-based treatment of linear regression. Finally, Chapter 12, either downloadable or provided on CD when the course syllabus calls for it, provides a very accessible introduction to analysis of variance and multiple regression, which together provide the backbone to much of applied statistical practice.

The entire book can easily be covered in a two-semester (or three-quarter) course. If desired, the instructor can omit Chapter 12, or even Chapter 10 without loss of continuity. In a one-semester course, Chapters 1–9 and possibly 11 can be covered, in which case the instructor is encouraged to monitor the ongoing cognitive progress of the class and set the pace accordingly. In a one-semester or one- or two-quarter course, the optional sections and other sections can be omitted to avoid undesirable time pressure, based on instructor tastes. In particular, Chapter 7 can be covered quite briefly if desired.

The apparent length of the textbook is a biased indicator of the pace that can be comfortably maintained through the textbook as the semester or quarter proceeds. Put simply, because of the space required for the textbook's emphasis on learning through simulation and the space consumed by the thoroughness of its verbal explanations, it functions like a 500–600 page textbook would typically function.

This textbook has been heavily influenced by instructor and student input based on classroom use of an earlier textbook written by some of the authors. In particular, Chapter 7 on data collection and Chapter 12 on analysis of variance and multiple regression were developed in response to such input.

We are confident that you will find teaching statistics from this textbook to be a rewarding experience. We look forward to interacting with instructors by means of the Möbius Web site for this textbook. We welcome queries and are eager to assist in teaching statistics in this very exciting, effective, and experientially-based way.

It is worth noting that, in addition to the college setting it was designed for, the textbook is also ideal for high school Advanced Placement statistic courses. Its coverage of the College Board's outline of topics covered by the AP exam in statistics is remarkably complete, and an AP supplement is provided that covers those very few topics in the College Board AP outline that are not in the textbook proper. Moreover, the College Board's recommended emphasis on taking a simulation approach, heavily using technology, and having students construct their own statistics knowledge base (through individual and group projects, etc.) mesh seamlessly with the textbook and its integrated instructional software supplement. Further, the integrated TI-83 supplement facilitates the taking of the AP statistics exam.

ACKNOWLEDGMENTS

The authors wish to acknowledge their appreciation for the superb, in-depth intellectual and technical support provided by Möbius Communications and Publication Services, Inc. in the development and preparation of this book. Möbius Communications Product Manager Louise Toft developed the Professional Profiles that open each chapter and was responsible for many of the practical details involved in taking a manuscript and turning it into a book. At Publication Services, Editing Manager Rob Siedenburg and Production Managers Foti Kutil and Ted Young guided this complex and challenging endeavor. Jason Brown, Ben Coblentz, and Alysia Cooley worked as a team to turn the materials from the different authors into a unified textbook. Jerome Colburn provided valuable suggestions for content improvements and considerable editorial expertise on earlier versions of this textbook. His contributions remain highly valued in this, the third edition.

The authors wish to thank the publishers of the following materials for granting permission for their use:

Chapter 1: opening photos courtesy of Dr. Andrea Donnellan.

Chapter 2: opening photos courtesy of Dr. Andrew Harris.

Chapter 3: opening photos courtesy of the National Safety Council.

Chapter 4: opening photos courtesy of Dr. Paul W. Chodas and the Solar System Dynamic Group at the Jet Propulsion Laboratory.

Chapter 5: opening illustrations courtesy of Dr. J. Steven Landefeld and the Bureau of Economic Analysis.

Chapter 6: opening photos courtesy of the Jet Propulsion Laboratory.

Chapter 7: opening photos courtesy of Bruce W. Hoynoski and Nielsen Media Research.

Chapter 8: opening photos courtesy of Office of Migratory Bird Management and the Duck Stamp Office, U. S. Fish and Wildlife Service.

Chapter 9: opening photos courtesy of Dr. Sean Todd and the Allied Whale Marine Mammal Research Facility at the College of the Atlantic.

Chapter 10: opening photos courtesy of Dr. Harold Brooks. The map is from the National Severe Storms Laboratory Web site, www.nssl.noaa.gove/~brooks.concannon and was used with the permission of the National Severe Storms Laboratory.

Chapter 11: opening photos courtesy of Dr. Jian Zhang.

Chapter 12: opening illustrations courtesy of Dr. Peter J. Wagner and the Field Museum of Natural History.

ABOUT THE AUTHORS

Ditlev Monrad is an Associate Professor in the Department of Statistics and the Department of Mathematics at the University of Illinois at Urbana-Champaign (UIUC). His research is in the area of theoretical probability. A dedicated teacher, Professor Monrad has served as undergraduate and graduate advisor for statistics majors at the UIUC. He has extensive experience in teaching introductory statistics and probability at the 4-year college level.

William F. Stout is a Professor of Statistics at the University of Illinois at Urbana-Champaign (UIUC), where he has been on the faculty since 1967. He is an internationally acclaimed researcher in the application of statistics to the fields of educational and psychological measurement. As the founder of the University of Illinois Statistical Laboratory for Educational and Psychological Measurement, he has led the development of new and widely applicable theories and methods for improving standardized testing. Stout is an associate editor of two leading measurement journals: *Psychometrika* and *Journal of Educational and Behavioral Statistics*. He is president of the international Psychometric Society and CEO of an educational measurement start-up company that provides statistical tools to carry out formative assessments of students using standardized tests.

E. James Harner is a Professor and Chair of the Department of Statistics at West Virginia University. His principal research efforts are in environmental statistics, statistical computing environments, and dynamic graphics, though he is increasingly focusing on learning-based probability models and bioinformatics. He has over 40 published research papers and has co-developed several statistical software and Web-based learning environments. Currently, he is directing the development of IDEAL (Intelligent Distributed Environment for Adaptive Learning), which will incorporate advanced tutoring and assessment subsystems into college instruction.

Barbara A. Bailey is an Assistant Professor of Statistics at the University of Illinois at Urbana-Champaign. Dr. Bailey conducts research in the area of biomathematics and biostatistics with an emphasis on environmental statistics. From 1996 to 1998 she conducted postdoctoral research at the National Center for Atmospheric Research in Boulder, Colorado, concentrating on modeling the spatial and temporal distribution of cloud cover. She has taught mathematics and statistics at both the high school and college level.

Robert L. Gould is the Vice-Chair of Undergraduate studies and the Director of the Center for Teaching Statistics in the Department of Statistics at the University of California at Los Angeles (UCLA). Dr. Gould's research has been in nonlinear estimation of random effects in a mixed model. Much of this research has been done in connection with the Alzheimer Disease Research Center at the University of California at San Diego. Other recent professional work has included the development (in cooperation with Professor Jane Friedman of the University of San Diego) of a high-school primer on statistical and probabilistic issues in DNA "fingerprinting." Dr. Gould takes an active leadership interest in statistics instruction at all levels. Though professionally interested in numbers, Dr. Gould appreciates the arts in his spare time; he is an amateur cellist.

Xuming He is a Professor in the Department of Statistics at the University of Illinois at Urbana-Champaign. He is the Director of the department's statistical consulting service (Illinois Statistics

Office). In addition, Dr. He serves as a consultant to the Argonne National Laboratory and is a member of the editorial board for the *Journal of Multivariate Analysis*, *Stastistica Sinica*, and *Statistics and Probability Letters*. A highly ranked instructor, Dr. He has taught statistics courses at both undergraduate and graduate levels since 1989.

Dr. He's principal research areas are regression models of large dimensions, regression splines and constrained models, robust methods in linear models, and asymptotics of IRT model estimation. Additional research has been conducted under grants from the National Security Agency and the University of Illinois Research Board.

Louis Roussos is Chief Psychometrician at Prime Assessment, a highly innovative developer of test-based formative assessments, and Assistant Professor of Educational Measurement in the Department of Educational Psychology at the University of Illinois at Urbana-Champaign. Once an aerospace engineer who developed mathematical models for acoustics and vibrations research, his major research interest now centers on analyzing educational and psychological tests. Among his many accomplishments, Dr. Roussos received the 1997 National Council of Measurement in Education Outstanding Dissertation Award and the 1999 American Psychological Association's Division 5 Dissertation Award.

With over a decade of statistics and mathematics teaching experience, Dr. Roussos has both pedagogical expertise and test analysis competence. He has developed new statistical techniques and software for dimensionality analysis of tests and test bias, and has designed computerized adaptive standardized tests. Dr. Roussos has developed new theoretical insights and procedures that constitute a major advance in test-based formative assessment.

Describing Data

CHAPTER 1 EXPLORING DATA: TABLES AND GRAPHS   2

CHAPTER 2 SUMMARIZING DATA BY NUMERICAL MEASURES: CENTERS AND SPREADS   50

CHAPTER 3 LINEAR RELATIONSHIPS: REGRESSION AND CORRELATION   112

PROFESSIONAL PROFILE

Dr. Andrea Donnellan
Supervisor,
Data Understanding Systems Group
Jet Propulsion Laboratory
Pasadena, California

Dr. Andrea Donnellan on the ice in the Clark Mountains near a Global Positioning System station in Marie Byrd Land, Antarctica, November 1999

Dr. Andrea Donnellan

Earthquakes do more than destroy homes and businesses, endanger lives, and disrupt roads, communications, and power distribution. Earthquakes also change the surface of our planet, putting stresses in new locations. Dr. Andrea Donnellan studies what happens to the earth's crust before and after earthquakes. She helped spur the design and installation of the 250-station continuously recording Global Positioning System (GPS) network that monitors the San Andreas Fault and other faults in California. Dr. Donnellan has also studied the deformation of the earth's surface in Mongolia, Antarctica, Bolivia, and Alaska. She looks at earthquake systems and how faults in the Earth's surface interact with each other. Her immediate goal is to gain an understanding of earthquake fault systems and the style and size of earthquakes that may occur over the next half-century. "We do not as yet," explains Dr. Donnellan, "understand enough to do short-term predictions."

In her work, Dr. Donnellan uses many types of data, including seismic data from the California Institute of Technology, synthetic aperture radar data from Japanese and European satellites, and information from the GPS satellites. Statistical analysis techniques allow her to make sense of this information and see how the earth deforms following a quake.

Dr. Donnellan also notes that the public can benefit from her work. By knowing where faults are and where the greatest risks for future quakes are located, we can set priorities for projects to retrofit buildings, bridges, and other structures to reduce the severity of structural damage in future quakes and perhaps save lives.

1

Exploring Data: Tables and Graphs

Statistical thinking will one day be as necessary for efficient citizenship as the ability to read and write.
H.G. Wells

Objectives

After studying this chapter, you will understand the following:

- The role of statistics
- Types of statistical data
- Graphical methods of summarizing data, including dotplots, stem-and-leaf plots, pie charts, bar charts, frequency histograms, and density histograms
- How to use a density histogram to summarize the shape of a data set
- How to recognize misleading graphical descriptions of data

1.1	THE SCIENCE OF STATISTICS	5
1.2	DISPLAYING SMALL SETS OF NUMBERS: DOTPLOTS AND STEM-AND-LEAF DISPLAYS	11
1.3	GRAPHING CATEGORICAL DATA	20
1.4	FREQUENCY HISTOGRAMS	24
1.5	DENSITY HISTOGRAMS	30
1.6	MISUSING STATISTICS	37
	CHAPTER REVIEW EXERCISES	45

KEY PROBLEM

Will the Real Author Please Stand Up?

It can be very important to find out who actually wrote a book or document. After the billionaire Howard Hughes died, several different wills came to light, and the courts had to decide which one was actually written by Hughes. Other questions of authorship may not have as much money riding on them but are very interesting. For example, various scholars have suggested that some or all of the plays and poems attributed to William Shakespeare were actually written by Sir Walter Raleigh, Sir Francis Bacon, or the 4th Earl of Oxford.

Davy Crockett (1786–1836) is now a legend of the American frontier, but between his days as a pioneer and hunter in Tennessee and his death in battle at the Alamo during the Texas war of independence, he also served in the U.S. Congress. There he opposed President Andrew Jackson's plans to seize the lands of the Native Americans east of the Mississippi that had been granted by treaties and deport the people along the "Trail of Tears" to what is now Oklahoma. Two books were published in Crockett's name in the 1830s in support of his political activities, and another was published in his name after his death about his experiences in the Texas war. Scholars have doubted whether the frontiersman actually wrote the books, because he did not learn to read and write until his early thirties. They especially doubted that he wrote the Texas war memoir, which lacks the "plain-speaking" style that Crockett was admired for.

How can we identify the author of a text that was published more than a century ago? We cannot cross-examine possible authors in court. We do not have any evidence except the words of the text in question and other texts known to have been written by the various possible authors. By using statistical techniques, we may be able to determine whether the text in question was written by the same author as one or another of the known texts.

Statisticians Frederick Mosteller and David Wallace invented a method that uses a set of words called "contentless" because they can be used anywhere in a piece of writing (see Table 1.1). Researchers count the rates at which each of these words occurs in the text in question and in the writing samples for which authorship is known and compare the patterns, or **distributions,** of these word counts. If the distribution of the rates (called **relative frequencies**) of these words in the text in question is very different from the distribution in a sample known to be written by the author, the text in question was probably written by someone else. Mosteller and Wallace first used their method to conclude that 12 of the *Federalist Papers* (a set of 85 essays in support of the U.S. Constitution, published during 1787 and 1788 under the pen name "Publius") had been written by James Madison rather than Alexander Hamilton.

David Salsburg and Dena Salsburg applied the method of Mosteller and Wallace to the three books attributed to Davy Crockett: *A Narrative of the Life of David Crockett, Written by Himself* (called *Narrative* for short); *An Account of Col. Crockett's Tour of the North and Down East* (called *Tour*); and *Col. Crockett's Exploits and Adventures in Texas* (called *Texas*). As a comparison sample, Salsburg and Salsburg used Crockett's speeches in Congress. Only nine of the 30 "contentless" words were used consistently in each of the four texts. Their distributions are shown in Table 1.2.

At first glance the differences among texts do not look very great. But one notes that the frequencies of *to* and *this* are lower in *Texas* than in the *Narrative*, the *Tour*, and the speeches; the frequency of *an* is much higher in *Texas* than in the others. These differences suggest that Crockett could have written *Narrative* and *Tour* and may not have written *Texas*. But this is only a vague impression!

Table 1.1 "Contentless" Words

according	considerable	probability
also	direction	there
although	enough	this
always	innovation	though
an	kind	to
apt	language	upon
both	matter	vigor
by	of	while
commonly	on	whilst
consequently	particularly	works

Source: D. Salsburg and D. Salsburg, Searching for the "real" Davy Crockett, *Chance*, vol. 12, No. 2 (Spring 1999), p. 31.

Table 1.2 Frequencies of "Contentless" Words (per 1000 Total Words) in Texts Attributed to Davy Crockett

Word	Narrative	Tour	Texas	Speeches
also	0.75	0.32	0.41	0.51
always	0.13	0.79	0.55	0.00
an	2.63	2.21	4.93	2.54
both	1.00	0.47	0.14	0.00
there	3.75	4.90	3.43	1.52
this	4.75	5.22	2.60	7.10
though	0.25	0.00	1.51	0.00
to	36.79	33.21	28.78	30.93
while	1.13	0.63	0.68	2.54

Source: Chance, vol. 12, No. 2 (Spring 1999), p. 32.

Applying a statistical technique based on the chi-square probability law (which we will study later in the text) to these word distributions, Salsburg and Salsburg found that the *Narrative*, the *Tour*, and the speeches were very possibly written by the same person, but that *Texas* was most likely written by someone else.

Salsburg and Salsburg were careful *not* to say, "Davy Crockett wrote the *Narrative*, the *Tour*, and the speeches." Such an answer would have gone beyond what their statistical methods could tell. At most we can say that if Crockett composed the speeches, he could well have written the *Narrative* and the *Tour* but *probably did not* write *Texas*.

The underlying concepts and methods of such statistical inference are the central focus of this book. The science of statistics is essential: Statistical reasoning enables us to draw conclusions based on data patterns that are often hidden from the statistically uninformed observer of the data. It also leads us to recognize the limits on what we can determine from the data. Author identification problems usually involve both of these aspects.

1.1 THE SCIENCE OF STATISTICS

As a science, **statistics** is the discipline of gathering data, describing data, and drawing conclusions (inferences) from data. **Data** are pieces of information obtained by counting, measuring, or categorizing, using some kind of observational process. In this book we will study the basic ideas in statistics and learn how to use those ideas to help interpret data that we encounter in the world around us.

On the other hand, **statistics** also refers to numbers that somehow inform us about the world around us. For example, a baseball fan might consider Barry Bonds's home runs in 2001 an interesting statistic. The word *statistics* originally meant the kind of information that a *state* (that is, a government) can use in managing public affairs. Accordingly there are statistics—that is, pieces of numerical information—about taxes collected, prices of goods, and weather patterns.

In everyday conversation any interesting number may be called a statistic, but to a statistician a statistic is a summarizing number computed from a collection of data, such as the average midterm score for a statistical class.

The daily newspaper is filled with statistics, presented in many clever ways, about topics ranging from sports and entertainment to politics and health. Here are some examples of statistics in both senses of the word: as numerical information and as the science of reasoning from data to reach conclusions about the unknown character of things (this is known as statistical **inference**).

Rating Winter Olympics Coverage When CBS broadcast the 1994 Winter Olympic Games in Lillehammer, Norway, it posted an average A.C. Nielsen rating of 27.8% of U.S. television households for its prime-time telecasts. That was up 49% from CBS's ratings for the 1992 winter games in Albertville, France. Here the 27.8% and 49% are statistics of interest. They were computed from surveys of a carefully selected but relatively small number of households. They are intended to be a representative **sample** that reflects the television viewing preferences of the much larger **population** of all households in the United States. In this textbook, you will learn how a surprisingly small but carefully selected sample can accurately predict properties of a much larger population it was sampled from.

Is Coffee Bad for Your Heart? Some medical studies in the seventies found a connection between coffee drinking and heart disease: The rate of heart disease was higher for coffee drinkers than for nondrinkers. Did this mean that coffee drinking *causes* heart disease? Further studies revealed that the coffee drinkers smoked more than the nondrinkers. It was their smoking that gave the coffee drinkers heart disease, not the coffee.

Does More Math Mean More Money? One of the most extensive studies ever made of the careers of high school graduates confirms the conventional wisdom that those who take more mathematics courses earn considerably more in the marketplace (*Education Week*, Jan. 11, 1989, p. 9). On the surface, it seems plausible that taking more math courses should increase one's earning potential. On the other hand, other characteristics of those who take lots of math (such as having the motivation to take on a difficult challenge, or being of high intelligence) may be the major causes of their higher incomes. Therefore, taking more math courses might not be the *cause* of higher earning potential. The science of statistics can tell us when two quantities are statistically associated—that is, they are large or small together. It is dangerous and incorrect, though, to conclude automatically that an increase in one quantity *causes* an increase in the other (see also Section 3.7 and the lung cancer/smoking debate in Chapter 6).

Want to Avoid a Cold? Don't Get Stressed Out! Survey research carried out in England indicates that persons under stress have a greater tendency to catch a cold than those who are not stressed ("Psychological Stress and Susceptibility to the Common Cold," *New England Journal of Medicine*, Aug. 29, 1991, pp. 606–612). Provided the experimental work that produced these data was done properly, it may be an example of valid statistical inference to conclude that stress increases susceptibility to colds. For example, did the study balance out or adjust for other factors that make catching a cold more likely and that go together with being stressed? If not, these other factors (such as lack of exercise or poor diet) could be the *real* cause of the stressed people catching colds more often!

Do Headaches Run in the Family? If you suffer from migraine headaches, don't blame your stars—but perhaps blame your genes. Migraine headaches may be a hereditary condition. Data show that if both father and mother have had them, there is a 75% chance that a son or daughter does, too. If only one parent has had migraine headaches, data show that the chance of the condition occurring is 50%. Even having distant relatives with migraine headaches increases the risk of having them by about 20% over the general population rate. Of course, concluding from these data that heredity causes migraine headaches is risky. For example, stress in the family environment could really be causing migraine headaches to occur more frequently in all family members. For that matter, the parents' frequent migraines could put stress on the family that then contributes to the children having migraine headaches. (From *Headway*, vol. 3, Dec. 1993.)

All these news items deal with topics of general interest. Each contains numerical information (that is, statistics) or presents statistical inferences drawn (rightly or wrongly!) from numerical information. Here are some questions these statistically based discussions may help us with:

- *Entertainment:* The popularity of one TV program as contrasted to another, such as the 1994 Winter Olympics versus the 1992 Winter Olympics
- *Health:* The relationship between stress and susceptibility to colds
- *Economics:* Whether the study of mathematics may increase one's earning power

We've said that statistics is the discipline of gathering numerical data, describing these data, and drawing conclusions from the data. Statistics can help us answer questions like these:

- Does the future economic growth of the United States depend on continued immigration?
- Why do casinos make a profit at roulette?
- How accurately can a person's future income be predicted from his or her education?
- How do medical researchers determine whether a new drug works or not?
- How accurately can a poll of 1500 eligible voters predict the outcome of an election?

This book contains many examples of real data sets as well as the basic ideas and techniques of gathering, describing, and analyzing data. After studying this book, you will be able to draw sound conclusions from data that are not necessarily obvious to the casual observer of data. Further, you will be able to tell when the reported conclusions of a study cannot be trusted to be valid, in spite of a glitzy graph or compelling words to the contrary.

The techniques of graphically representing data and numerically summarizing them with appropriate statistics (indices) in ways that make it easy to recognize the essence of the information in the data are called **descriptive statistics.** Alternately, the techniques used to draw conclusions from data using the theory of probability and the methods of statistics are called **inferential statistics.** Good inferential statistics usually needs to be preceded by a good descriptive statistical analysis of a data set. This textbook presents both of these aspects of the discipline of statistics.

This chapter presents some commonly used graphical techniques of descriptive statistics, beginning with the dotplot. The last section will cover some of the common ways in which the techniques of descriptive statistics have been abused to make *misleading* presentations.

We shall first consider the different types of statistical data we are going to study.

Types of Statistical Data Eye color, height, and number of siblings (brothers and sisters) vary from person to person. Such characteristics that vary from one individual to another are called **variables.** Some variables like height and number of siblings take numerical values and are called **quantitative.** Quantitative variables may be **discrete** or **continuous.** A **continuous variable**, such as height, can take any value in a given range. A **discrete variable**, such as number of siblings, can only take certain distinct values like 0, 1, 2, 3, and so on. No value in between is possible. A variable like eye color cannot be expressed as a number. Such variables are called **qualitative** or **categorical.** We call a categorical variable **ordered categorical** or **ranked** if the categories imply some order or relative position. Military rank, for example, is an ordered categorical variable.

Finally, the terms **categorical**, **ordered categorical**, **discrete**, and **continuous** are also used to describe data.

Section 1.1 Exercises

1. Charles Darwin conducted an experiment to determine whether cross-pollinated plants (that is, another plant is involved in the breeding of the seedling) or self-pollinated plants (seedling produced without genetic influence of another plant) have greater vigor. Darwin took 15 pairs of seedlings, each pair having the same age. One seedling in each pair was produced by cross-pollination and the other was produced by self-pollination. The members of each pair were

grown in as nearly identical conditions as possible. The data are the final heights (in inches) of each plant after a fixed period of time.

Pair number	Cross-pollinated	Self-pollinated
1	23.5	17.4
2	12	20.4
3	21	20
4	22	20
5	19.1	18.4
6	21.5	18.6
7	22.1	18.6
8	20.4	15.3
9	18.3	16.5
10	21.6	18
11	23.3	16.3
12	21	18
13	22.1	12.8
14	23	15.5
15	12	18

Source: Charles Darwin, *The Effect of Cross- and Self-Fertilization in the Vegetable Kingdom*, 2nd ed., London: John Murray, 1876.

a. What is the fraction of the 15 pairs of seedlings for which the cross-pollinated seedlings exhibited better growth than self-pollinated seedlings?
b. What seems reasonable to statistically infer from the data?
c. Identify the three major components of the science of statistics in the statement of the problem and in your answers to parts **a** and **b**.
d. If you as a farmer had the option of purchasing peas guaranteed to have been produced by cross-pollination, what would your decision be as a result of your (informal) statistical analysis conducted in part **b**?
e. What general name does Chapter 1 use for the kind of number computed in part **a**?
f. Suppose the fraction had been 10/15 in part **a**. Is your decision obvious? The lesson: We need to understand how to do statistical inference when the data do not give us an obvious answer! That is one reason for studying this textbook.

2. A survey was conducted by the *British Medical Journal* to study the relationship between snoring and heart disease.

Heart disease	Non-snorers	Occasional snorers	Snore nearly every night	Snore every night	Total
Yes	24	35	21	30	110
No	1355	603	192	224	2374
Total	1379	638	213	254	2484

Source: P. G. Norton and E. V. Dunn (1985) Snoring as a risk factor for the disease: An epidemiological survey, *British Medical Journal*, 291, 630–632.

a. Which of the four categories of snorers is most prone to heart disease? *Hint:* Would it be right to say that occasional snorers are most prone to heart disease because the greatest number (35) of the 110 persons who have heart disease are occasional snorers and no other category has such a high number of snorers with heart disease? Explain.
b. Is there any kind of relationship between the degree of snoring and being prone to heart disease?

c. Suppose a behavioral psychologist tells a person who snores often that he can condition him not to snore. Is this likely to reduce that person's risk of heart disease? In other words, is it plausible that it is the snoring that causes heart disease? If you do not think so, give another explanation of why the frequent snorers have more heart disease than the infrequent snorers.

3. The following data show the brain and body weights of selected animals (current and extinct).

Species	Body weight (kg)	Brain weight (g)
Mountain beaver	1.35	8.1
Grey wolf	36.33	119.5
Guinea pig	1.04	5.5
Diplodocus	11,700	50
Asian elephant	2547	4603
Potar monkey	10	115
Giraffe	529	680
Gorilla	207	406
Human	62	1320
African elephant	6654	5712
Triceratops	9400	70
Rhesus monkey	6.8	179
Mouse	0.023	0.4
Rabbit	2.5	12.1
Jaguar	100	157
Chimpanzee	52.16	440
Brachiosaurus	87,000	154.5
Rat	0.28	1.9
Mole	0.122	3

Source: H. J. Jerison (1973), *Evolution of the Brain and Intelligence*, New York: Academic Press.

a. Rank the animals from largest to smallest according to the body weight and then do the same for the brain weight. (For example, the African elephant ranks fourth in body weight but first in brain weight.)
b. Do the two rankings approximately match?
c. Do you notice that a certain group of animals seems very different in the closeness of matching of their ranks? (Statisticians call data points that are very different from the main body of the data "outliers.")

4. The table below gives the road distance and the shortest distance between 15 different pairs of points in a city. Plot the data on graph paper and discuss the nature of the relationship between the road and the shortest distance in this city. Why do the points not all lie on a straight line?

Road distance	Shortest distance	Road distance	Shortest distance
5.1	4.3	10.4	7.9
5.8	5.2	9.5	6.4
11.1	9.5	10.3	7.7
6.6	4.1	4.2	3.5
5.9	4.3	3.8	3.4
2.3	1.7	10.5	8.2
9.2	7.8	7.0	4.2
5.9	4.3		

5. A survey interviewed a representative cross section of several hundred Canadians, asking about their level of education and their level of participation in physical activity. "Active" persons were defined as those who take part in three or more hours of physical activity per week for nine months or more during the year.

Amount of education	Percentage who are active
Elementary school only	41
Some secondary school	53
Secondary certificate or diploma	58
University degree	63

   a. What relationship do you see between educational level and physical activity?
   b. Between which two *consecutive* educational levels was the change in percentage of physical activity largest?

6. *Infant mortality* in the following table refers to the number of infants who die for every 1000 births. *Life expectancy* is the number of years, on average, that a person can expect to live. Here are the infant mortality and the life expectancy in the United States for people born in the years 1920 through 1990:

Year of birth	Infant mortality (per 1000)	Life expectancy
1920	86	54
1930	65	60
1940	47	63
1950	29	68
1960	26	70
1970	20	71
1980	13	74
1990	9	76

   a. Look at infant mortality through the years. What trend do you see?
   b. What trend do you see for life expectancy?
   c. Which improved more dramatically from 1920 to 1990, infant mortality or life expectancy?
   d. What may be a cause of both of the trends you see?
   e. Can we conclude that lowering infant mortality is the sole cause of the increase in life expectancy?

7. Classify each of the following variables as either categorical, discrete quantitative, or continuous quantitative.
   a. race
   b. weight
   c. sex
   d. religion
   e. age
   f. employment status (employed, unemployed, not in the labor force)
   g. family size
   h. marital status (married, never married, widowed, divorced)
   i. college degree (bachelor, master, doctor)
   j. Olympic medal (bronze, silver, gold)
   k. letter grade in a college course

1.2 DISPLAYING SMALL SETS OF NUMBERS: DOTPLOTS AND STEM-AND-LEAF DISPLAYS

The simplest way to display a small set of numbers is to construct a *dotplot*.

Example 1.1   Heights of Eleven-year-old Girls

Eleven-year-old Julie is short for her age. She is only 138 cm tall (about 4½ feet tall). Should her parents worry that she is not growing properly? Table 1.3 gives the heights of a representative sample of 40 eleven-year-old girls.

Looking at Table 1.3 it is difficult for us to tell whether Julie is unusually short or just short. We have to display these numbers graphically. The simplest way to represent 40 heights graphically is to construct a **dotplot**.

The data in Table 1.3 are displayed in Figure 1.1. Each observation is represented by a dot. The dotplot reveals that 3 of the 40 girls in the sample were shorter than Julie. So although Julie is short, her height is within the normal range.

The dotplot reveals a lot about the data in Table 1.3. For example, we see that half of the girls in the sample were shorter than 147.5 cm and half were taller than 147.5 cm. We express this fact by saying that the *median height* of the girls in the sample is 147.5 cm. (See Section 2.1.) In Figure 1.2 we

Table 1.3   Heights (cm) of 40 Eleven-year-old Girls

144	152	148	146	156	144	142	147
146	141	160	147	153	150	150	139
157	138	148	144	155	148	149	135
137	158	142	152	145	136	156	148
143	151	150	147	144	153	146	149

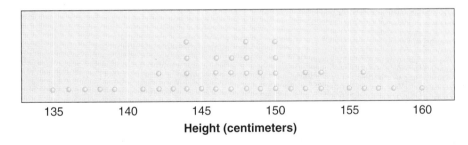

Figure 1.1   Dotplot for the Data in Table 1.3.

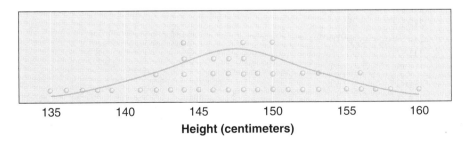

Figure 1.2   Heights of Eleven-year-old Girls.

have superimposed a smooth curve on the dotplot indicating the overall shape of the height distribution. The smooth curve makes the shape easier to see.

Dotplots work well for many small sets of numbers that, like in the data in Table 1.3, (a) have repeated values and (b) are not too spread out. The dotplot helps us discover and display the shape or distribution of the data.

Another graphical tool for describing the distribution of a small set of numbers is the **stem-and-leaf plot** (also called a **stemplot**). Such a plot is a mixture of a table and a graph.

Example 1.2   Luminosities (Brightnesses) of Stars

Table 1.4 lists 40 stars in a certain section of the sky in the constellation Taurus. (They were selected from the stars that can be seen without a telescope in that part of the sky; they are the ones whose distances from Earth are known most accurately.) Quite possibly this sample of stars can be viewed as representative of the populations of all visible stars. If so, properties of the sample can be inferred to be appropriately true for the population, too.

The numbers in the left column in the table were used by software on the satellite Hipparcos to identify the stars and have no meaning for us. The right column shows how much light each star puts out in comparison to our own star, the Sun. For example, star 18735 puts out seven times as much light as the Sun does. That is, if it were the same distance away from Earth as the Sun is, it would appear seven times as bright as the Sun does. Astronomers call this number the luminosity. These numbers are rounded to the nearest integer; to illustrate, a star with a measured luminosity of 0.79 would be listed with luminosity 1 in this table.

The stars are listed in order of their identification numbers. Thus, such a table is not designed to enable us to see important patterns in the data. We cannot easily see how the luminosities are grouped or any special characteristics of their distribution: Are they tightly clustered around one value, or are they spread out over a wide range of values? In other words, what is the distribution

Table 1.4   Star Luminosities

Star ID	Luminosity (Sun = 1)	Star ID	Luminosity (Sun = 1)
18735	7	20661	4
19038	43	20711	34
19076	1	20713	29
19261	7	20842	9
19877	5	20877	29
19990	7	20885	53
20087	13	20889	67
20205	60	20894	71
20219	9	20901	18
20255	6	20995	9
20261	14	21029	19
20284	6	21036	11
20400	8	21039	10
20455	53	21137	6
20484	10	21273	23
20542	19	21402	33
20614	8	21421	138
20635	35	21459	6
20641	12	21588	8
20648	30	21589	32

of the data? Are there any highly unusual values (that is, much different from the majority of the values)? We cannot easily tell from the table. We would have to reorder the data at least.

A stem-and-leaf plot is a way of listing a data set that displays the basic distribution or shape of the data. It is easy to construct and makes the data easier to interpret. This way of organizing data was developed by John Tukey, the founder of a modern approach to descriptive statistics called exploratory data analysis (EDA).

Here we present only the simplest method for constructing a stem-and-leaf diagram. Variations of this method enable us to decrease or increase the number of stems as is sometimes desirable (see for example *The ABCs of EDA* by Velleman and Hoaglin).

Constructing a Stem-and-Leaf Plot

The starting point for constructing a stem-and-leaf plot is a table of the numbers whose distribution (shape) we wish to display. Further, suppose we wish to display all the individual values as well, because those values are also interesting in addition to the general shape. This dual goal of wanting to display the general shape of the data while retaining all the individual values sometimes arises, for example, when one examines sports data. Some athletes have superb lifelong careers, performing at a high level for many years, whereas other athletes have brief moments in the sun, when they really shine, and then rapidly fade away. A stem-and-leaf plot can clearly contrast the two types of careers. We will consider Roger Maris's career as a Major League Baseball home run hitter and, in particular, construct a stem-and-leaf plot to see whether he falls clearly into either of these two categories. Consider the raw data, ordered from year to year for his 12 years as a Major League Baseball player:

14   28   16   39   61   33   23   26   8   13   9   5

The 61 home runs made Maris famous among baseball sports fans, because that broke one of baseball's most famous records, namely Babe Ruth's 60 home runs in one season—a record that would not be broken again until 1998. A casual glance at the data suggests that the 61 seems to be perhaps an extreme or outlying value for a home run career that was otherwise fairly ordinary, but clearly a display of these data that shows the distribution could be very informative! Also, for our purposes, this is a simple and straightforward example to illustrate the basic process of constructing a stem-and-leaf plot.

In a stem-and-leaf plot every value to be plotted is represented by its **stem** and its **leaf**. For the number 14, the stem is 1 and the leaf is 4; for 28 the stem is 2 and the leaf is 8. The number 5 is thought of as 05 and hence has a stem of 0 and a leaf of 5. When the stem and leaf are recombined, they reproduce the number. For example a stem of 44 and a leaf of 9 yields the number 449.

The basic idea is that the last digit (furthest to the right) of a number provides its leaf and the rest of the number (all numbers to the left of this last digit) supplies the stem. To illustrate:

138 yields a stem of 13 and a leaf of 8

33 yields a stem of 3 and a leaf of 3

9 yields a stem of 0 and a leaf of 9

To construct the simplest kind of stemplot, this is all you have to know about finding the stem and leaf of a number. Let us apply this knowledge and build the stem-and-leaf plot for the Maris career home run data. Clearly the stems run from 0, 1, ..., 6. We create a stem column that has as entries 0, 1, ..., 6. To the right of it we create a leaf column, where we will list the leaves of all numbers having the same stem, separated by commas.

Let us deal with the numbers with stem 1, namely the data points 13, 14, 16. We obtain

Stem	Leaf
0	
1	3, 4, 6
2	
3	
4	
5	
6	

This completes the tens row. Now we simply do the same for all the other stems to complete the plot

Stem	Leaf
0	5, 8, 9
1	3, 4, 6
2	3, 6, 8
3	3, 9
4	
5	
6	1

This graphical approach (in the sense that it displays the shape of the data) is often very useful. It shows at a glance the general shape of the data while retaining the values of each individual data point. For example, any visual representation of the Maris home run data that fails to identify the value 61 loses much of its interest. From the view of doing descriptive statistics, we see from the plot that over his career Maris fairly uniformly hit home runs in the 8-39 range (or 5-39 range, perhaps) and that 61 seems to be an extreme value, called an **outlier** by statisticians, that made Roger Maris famous. It would be interesting to compare the stem-and-leaf plot with that for Babe Ruth's home run career. We shall do so in Exercise 3.

Because each data value is represented by a one-leaf digit, the length of the string of leaves is proportional to the total number of data items that fall within that stem. This visual property, displayed in the Maris stem-and-leaf plot, is why stem-and-leaf plots show us the shape (distribution) of the data. Useful stem-and-leaf plots usually have between 5 and 15 stems, with larger data sets allowing the use of more stems.

Let's return to the star luminosity data of Table 1.4. Here is how the stem-and-leaf plot can be constructed for this data set. A brief inspection of Table 1.4 shows that the values range from 1 to the 70s, with one highly unusual value of 138. It is standard practice to list unusual values separately, putting unusually small values in a group labeled LO at the top and unusually high values in a group labeled HI at the bottom of the plot. So, grouping the data by tens (0–9, 10–19, etc.), we will have eight stems, which is a reasonable number. Now we write down the list of the stems. We start with a 0, which stands for the group of stars whose luminosities are less than 10 times the Sun's, namely in the range 0–9. In the next row we put a 1 for the group having luminosities from 10 to 19 times the Sun's, and so on.

Then we start making the leaves by writing down the data on the luminosity of each star. Let's start with the first two stars in the table:

Star 18735 luminosity 7 times the Sun's
Star 19038 luminosity 43 times the Sun's

We list those two data points as shown in the table below. That is, for star 19038 with its luminosity of 43, we separate the 4, the stem part (which in this table stands for 40), from 3, the leaf.

Stem	Leaf
0	7
1	
2	
3	
4	3
5	
6	
7	

To read stem-and-leaf plots, in general, we need a key. We write a key under the table:

Key: "4 3" stands for a luminosity 43 times that of the Sun.

The key tells us that in this table the stems are put in the tens place and the leaves are put in the units place. When you make a stem-and-leaf plot, you should always provide a key. For example, in another application "4 3" might stand for 0.043 or 4300: the key is needed to tell us which.

After the rest of the data are added, the completed stem-and-leaf plot is presented in Table 1.5. Notice that we have put the leaves of the table in order. That is a desirable step, and it is easy to do, especially by computer.

The unusual value, or outlier, is listed below the main portion of the table. If there were another outlying large value (say, 3100) we would have written "HI 138, 3100." If there were an outlying small value (which would not happen in this problem) we would have written it in a group labeled "LO" above the main portion of the table.

A stem-and-leaf plot has two very attractive properties. First, it shows the shape of the distribution of the data *at a glance*. We can see the **range** of the data (between 1 and 138, with all the values falling between 1 and 71 except outlier 138). We can see that the center of the data seems to lie in the 10–19 stem and that most of the data lie between 1 and 39. Moreover, we can distinguish several important aspects concerning the shape of the distribution of this data set:

- Is the distribution flat, or does it have a definite peak? The distribution is rather peaked, being high near 0 and steadily decreasing toward 70. By contrast, the Maris distribution is rather flat (or uniform) from 0 to 40.

Table 1.5 Stem-and-Leaf Plot of Star Luminosities

Stem	Leaf
0	1,4,5,6,6,6,6,7,7,7,8,8,8,9,9,9
1	0,0,1,2,3,4,8,9,9
2	3,9,9
3	0,2,3,4,5
4	3
5	3,3
6	0,7
7	1
HI	138

Key: "1 0" in this stem-and-leaf plot indicates a luminosity 10 times the Sun's.

- If the distribution has a peak, is the distribution symmetric, or is it skewed to one side or the other? The distribution of the luminosities is strongly skewed, which means much more stretched out from its center in one direction than the other, toward the high end of the range at the bottom of the plot. Imagine that the stemplot in Table 1.5 is turned counterclockwise 90 degrees. It will then have a long tail to the right and we say that the distribution is **skewed to the right** (following the usual convention based on other kinds of displays, such as the histograms introduced in Section 1.4, where data values increase from left to right). By contrast, if symmetric, the distribution would look about the same on either side of its center.
- Does the distribution contain one or more gaps? That is, sometimes a data set splits into two or more distinct chunks, often called clusters. (See Example 1.6.) In Table 1.5 there does not appear to be any large gaps in the 1–71 range.
- Does the distribution have any outliers (observations that do not follow the overall pattern of the data)? The star with luminosity 138 and, for the Maris data, the year with 61 home runs are outliers.

Second, the stem-and-leaf plot represents every point of the numerical data that was in the original table. That is, we can recover all of the data points from the stem-and-leaf diagram. In some other kinds of plots, such as the box-and-whisker plot introduced in the next chapter, the numbers that were used to make the plot are not recoverable from the plot—if you want to see the numbers, you must go back to the original list of data, such as a scientist's lab notebook or a computer file (which may not be possible).

For the technique of stem-and-leaf plotting to be widely applicable, several issues must be addressed and refinements of the standard approach developed, as remarked above. Going into detail about these is beyond the scope of this general introductory statistics textbook.

Comparing Populations: Back-to-Back Stem-and-Leaf Plots

One of the most common tasks in statistics is to compare two data sets that differ in some important characteristic, such as the annual incomes of male and female college graduates; the sizes of U.S. families and Brazilian families; or the cost of two-bedroom homes in San Francisco and in St. Louis, to name a few examples. The two data sets are usually regarded as representative samples from the two populations we want to compare. Statisticians look for differences between the samples and determine whether they can conclude that the populations differ. Trustworthy methods for comparing populations based on their representative samples will be presented later in this book, when statistical inference is studied.

One way to compare two data sets visually is to write their stem-and-leaf plots back to back. The stems are written in the middle of the table, the leaves for one data set are written to the left of the stem and extend leftward, and the leaves for the other data set are written to the right and extend rightward.

Example 1.3 Infant Mortality Rates in Europe

Tables 1.6 and 1.7 show the infant mortality rates per thousand births, measured over the years 1995 through 1999, for the countries in Europe that had not been under communist rule during the Cold War and those that had been, respectively. (Germany, listed among the noncommunist countries here, is a special case, because about a quarter of the country had been communist. But even then both parts had almost the same infant mortality rate.)

The completed back-to-back stem-and-leaf plots are shown in Table 1.8. The differences between the two data sets are clear. The noncommunist countries' infant mortality data from Table 1.6 are

Table 1.6 Infant Mortality Rates in Europe, 1995–1999: Noncommunist Countries

Country	Infant mortality (deaths per thousand live births)
Austria	6
Belgium	7
Cyprus	8
Denmark	7
Finland	5
France	7
Germany	6
Greece	8
Ireland	6
Italy	7
Luxembourg	9
Malta	10
Monaco	58
Netherlands	6
Norway	5
Portugal	8
Spain	7
Sweden	5
Switzerland	5
United Kingdom	6

Source: New York Times Almanac 2000, pp. 490–492.

Table 1.7 Infant Mortality Rates in Europe, 1995–1999: Former Communist Countries

Country	Infant mortality (deaths per thousand live births)
Albania	32
Belarus	15
Bosnia	13
Bulgaria	16
Croatia	10
Czech Republic	9
Estonia	12
Hungary	14
Latvia	16
Lithuania	13
Macedonia	23
Moldova	26
Poland	13
Romania	24
Russia	19
Slovakia	12
Slovenia	7
Yugoslavia	19

Source: New York Times Almanac 2000, pp. 490–492.

Table 1.8 Back-to-back Stem-and-Leaf Plot for Infant Mortality in Europe, 1995–2000

Leaf (Table 1.6 Data)	Stem	Leaf (Table 1.7 Data)
5,5,5,5,6,6,6,6,6,7,7,7,7,7,8,8,8,9	0	7,9
0	1	0,2,2,3,3,3,4,5,6,6,9,9
	2	3,4,6
	3	2
58	HI	

Key: "1 2" indicates 12 infant deaths per 1000 live births.

tightly concentrated, between 5 and 10, with a single outlier. The rates from the former communist countries in Table 1.6, on the other hand, are mainly distributed between 10 and 25.

The back-to-back stem-and-leaf plots reveal dramatically how much larger on average and how spread out the mortality rates in the former Communist countries are.

Also, the Monaco outlier deserves careful investigation. What is so unusual about Monaco (a very small country)? In this case, the outlier "58" is probably a misprint in the book from which the values were taken. The correct value is most likely 5.8, which we would round to six. Sometimes outliers result from mistakes like this; in other cases they are real and important values. They must be checked carefully in each case to determine whether they are mistakes or not.

CHAPTER 1 EXPLORING DATA: TABLES AND GRAPHS

Most standard statistical computer software packages for practitioners have a stem-and-leaf program that enables the operations we have described: constructing a standard stem-and-leaf plot, back-to-back plotting, representation of high- and low-valued outliers, and so on. Further refinements of the standard approach allow additional choices of stem intervals. The important thing for you as a student in an introductory statistics course is to be familiar with this powerful and often-used tool.

Section 1.2 Exercises

1. The origins of the Etruscan empire (centered in Italy around 800 B.C.) is a bit of a mystery to anthropologists (scholars who study human cultures). In a study, observations on the maximum head width (measured in millimeters) were taken on 84 skulls of Etruscan males and 70 modern Italian males. (*Source*: N. A. Barnicot and D. R. Brothwell (1959), The evaluation of metrical data in the comparison of ancient and modern bones. In *Medical Biology and Etruscan Origins*, G. E. W. Wolstenholme and C. M. O'Connor, eds., Little, Brown, p. 136.)

   Etruscan skulls

   ```
   141  148  132  138  154  142  150  146  155  158  150  140
   147  148  144  150  149  145  149  158  143  141  144  144
   126  140  144  142  141  140  145  135  147  146  141  136
   140  146  142  137  148  154  137  139  143  140  131  143
   141  149  148  135  148  152  143  144  141  143  147  146
   150  132  142  142  143  153  149  146  149  138  142  149
   142  137  134  144  146  147  140  142  140  137  152  145
   ```

   Italian skulls

   ```
   133  138  130  138  134  127  128  138  136  131  126  120
   124  132  132  125  139  127  133  136  121  131  125  130
   129  125  136  131  132  127  129  132  116  134  125  128
   139  132  130  132  128  139  135  133  128  130  130  143
   144  137  140  136  135  126  139  131  133  138  133  137
   140  130  137  134  130  148  135  138  135  138
   ```

   a. Compare a dotplot for the Etruscan skulls with a dotplot for the Italian skulls.
   b. Construct back-to-back stem-and-leaf plots for the Etruscan skulls and the Italian skulls.
   c. Is the hypothesis that Etruscans and modern Italians are of the same origins supported by the data (as it would be—somewhat—if the two plots seem quite similar)?

2. The ages of 34 Oscar-winning Best Actors and 34 Oscar-winning Best Actresses are as follows:

   Actors

   ```
   32  37  36  32  51  53  33  61  35  45  55  39  76  37  42  40  32  60  38
   56  48  48  40  43  62  43  42  44  41  56  39  46  31  47
   ```

   Actresses

   ```
   50  44  35  80  26  28  41  21  61  38  49  33  74  30  33  41  31  35  41
   42  37  26  34  34  35  26  61  60  34  24  30  37  31  27
   ```

   a. Construct back-to-back stem-and-leaf plots for actors and actresses.
   b. What things can you infer?

Section 1.2 Displaying Small Sets of Numbers

3. The number of home runs hit by Babe Ruth during his 15 years with the New York Yankees, from 1920 to 1934, is as follows:

   54, 59, 35, 41, 46, 25, 47, 60, 54, 46, 49, 46, 41, 34, 22

   The home run record of Roger Maris during his 10 years in the American League is as follows:

   14, 28, 16, 39, 61, 33, 23, 26, 8, 13

   a. Construct a back-to-back stem-and-leaf plot.
   b. Who was the better home run hitter? Explain.
   c. In any discussion of famous home run hitters, Hank Aaron certainly must be mentioned as well. Over 21 years with the Braves, Aaron's record was

   13, 27, 26, 44, 36, 39, 40,
   34, 45, 44, 24, 42, 44, 39,
   29, 44, 38, 47, 34, 40, 20

   Construct a back-to-back stem-and-leaf plot comparing Ruth and Aaron.

4. The following are the scores of students in a class for a test:

   77, 70, 65, 77, 60, 79, 71, 86, 82, 78, 76, 65, 81, 59, 100,
   62, 81, 67, 74, 71, 62, 84, 75, 70, 76, 72, 66, 73, 91, 49,
   66, 70, 81, 79, 72, 64, 95, 71, 66, 80, 78, 73, 69, 58, 89

   Construct a stem-and-leaf plot.

5. The following stem-and-leaf plot gives final scores on a statistics test for 20 students. From the table, write down each student's score.

Stem	Leaf
7	2,3,4,5,9
8	0,2,3,4,4,5,7,8
9	0,1,2,3,5,6,6

   Key: "8 2" stands for 82.

6. Using the stem-and-leaf plot of Exercise 4, answer the following questions:
   a. What was the lowest score obtained?
   b. What was the highest score obtained?
   c. How far apart were the lowest and highest scores? (This statistic is called the *range* of the data.)
   d. How would you describe the general shape of the distributions of the data?

7. The following table gives the high temperatures in degrees Fahrenheit for 20 cities on April 3, 1997.
   a. Construct a stem-and-leaf plot of the high temperatures.
   b. How far apart were the lowest and highest temperatures for these cities?

Asheville, North Carolina	72
Champaign, Illinois	68
Indianapolis, Indiana	69
Abilene, Texas	67
Los Cruces, New Mexico	72
Los Angeles, California	71
Billings, Montana	56
Chicago, Illinois	65
Albany, New York	53
Charleston, South Carolina	77
Miami, Florida	77
Birmingham, Alabama	74
Iowa City, Iowa	63
Detroit, Michigan	58
Molokai, Hawaii	77
St. Louis, Missouri	76
Lincoln, Nebraska	63
Boston, Massachusetts	53
Baltimore, Maryland	66
Boise, Idaho	60

8. The following table gives the unemployment rates for the so-called G7 nations (Group of Seven Leading Industrial Nations) for the years 1975 and 1996 (from *The World Almanac*, 1998).

	1975	1996
United States	8.5	5.4
Canada	6.9	9.7
Japan	1.9	3.4
France	4.2	12.6
Germany	3.4	7.2
Italy	3.4	12.1
Great Britain	4.6	8.2

a. Construct a stem-and-leaf plot for the employment rates in 1975.
b. What was the highest rate in 1975? Which country had it?
c. What was a typical unemployment rate in 1975?
d. Construct a stem-and-leaf plot for the employment rates in 1996.
e. What was the highest rate in 1996? Which country had it?
f. What was a typical unemployment rate in 1996?
g. Line up the stem-and-leaf plots for 1975 and 1996 back-to-back. Make a general comparison of the rates in 1975 to those in 1996.
h. For each of these nations, find the rate in 1996 minus the rate in 1975, and draw the stem-and-leaf plot of these differences.
i. How many of these differences are negative, if any? Which nations do they belong to?
j. What is the typical difference?

1.3 GRAPHING CATEGORICAL DATA

We cannot draw a stem-and-leaf plot for categorical data. We cannot say anything about the "shape" of a table of categorical data, because it would depend on the order of the categories, and we could put them in any order if we wanted to. For example, there would be no preferred order for the six continents from which a person's ancestors may have come. (Yes, they *could* be put in alphabetical order, but that order indicates nothing about the continents themselves and would not be the same when written in a different language, such as Russian.)

Two kinds of graphical displays are particularly useful for making comparisons involving categorical data: the **bar chart** and the **pie chart.**

Pareto Charts

A bar chart is constructed by drawing a vertical bar for each of the possible category values. All bars are of equal width. The height of each bar equals the number (or **frequency**) of persons (or objects) in that category. A vertical scale, usually at the left, helps the reader see the frequency. If the categories are arranged so that the tallest bar is to the left, then comes the second tallest, and so forth, the resulting graph is called a **Pareto Chart.** By arranging the bars in order of frequency, we focus attention on the more important categories. Thus a Pareto chart enables us to compare the numbers of objects in the various categories. Pareto charts are often used in engineering studies of causes of failure.

Example 1.4 Causes of Death

For Americans under 45 years of age, accidents are the leading cause of death. But for the entire U.S. population, accidents are only the fifth most common cause of death. Table 1.9 lists the number of deaths due to the eight leading causes of death in the United States in 1997. The cause of death is a categorical variable characterizing each death. Table 1.9 is a **frequency table.** The data in Table 1.9 are displayed graphically in Figure 1.3. Clearly, the vast majority of deaths are due to heart disease or cancer. Often in newspapers and magazines, the bars are separated for artistic reasons in bar charts.

Pie Charts

An important question that often arises in connection with categorical data is how much the frequency of each category contributes to the total frequency for all categories. For example, what

Table 1.9 Leading Causes of Death in the United States: 1997

Cause of Death	Number of Deaths
Heart disease	726,000
Cancer	537,000
Stroke	160,000
Pulmonary diseases	111,000
Accidents	92,000
Pneumonia and influenza	88,000
Diabetes	62,000
AIDS	30,000

Provisional data, estimated from a 10 percent sample of deaths. Source: *The New York Times Almanac 2000*, p. 392.

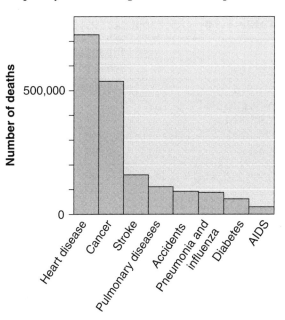

Figure 1.3 Pareto chart for leading causes of death in the United States.

proportion of cars sold in America are made in Asia? The pie chart is a graphical display commonly used to answer such questions. In a circle graph, each category value is represented by a wedge-shaped section of a circle between radiuses (lines drawn from the center of the circle to its edge). The size of each section is determined by the angle between the radiuses, which is equal to

$$\text{Angle between radiuses} = \frac{\text{Frequency of this category}}{\text{Total frequencies of all categories}} \times 360°$$

Just as the frequencies of all the categories add up to the total number of individuals being counted, the angles of the sections of the pie chart add up to 360 degrees, the number of degrees in a complete circle. Note that we cannot obtain or compare the total frequencies in the various categories from a pie chart, as contrasted with a bar graph.

Example 1.5 The U.S. Population by Marital Status

Classification by marital status provides categorical data that describe persons. Each adult is either *married, never married, divorced,* or *widowed.* The first two columns in Table 1.10 give the total number of occurrences (in millions) of each of these category values—in other words, the total number of persons having each marital status—in the United States in 1998.

The data in Table 1.10 could be displayed in a Pareto chart, but we shall instead use a **pie chart.**

The third and fourth columns in Table 1.10 give the proportions and the resulting sizes (in degrees) of the sections needed to draw a circle graph from these data. For example, the number of persons never married is 44.6 million. The proportion of the total number of adults is 44.6 million divided by the total, 195.5 million, or 0.228. The angle between the radiuses needed to draw the corresponding wedge is 0.228 × 360 degrees, rounded to 82 degrees.

The complete pie chart is shown in Figure 1.4. The number of adults who have never been married is nearly a quarter of the total; the number of adults who are currently married is somewhat over half the total.

Table 1.10 U.S. Population (Age 18 or Over) by Marital Status, 1998

Status	Millions of persons	Proportion of total	Angle (degrees)
Married	117.9	0.603	217
Never married	44.6	0.228	82
Widowed	13.6	0.070	25
Divorced	19.4	0.099	36
Total	195.5	1.000	360

Source: The New York Times Almanac 2000, p. 294.

Figure 1.4 Pie chart of the marital statuses of U.S. adults in 1998 from Table 1.10.

Section 1.3 Exercises

1. Classify each of the following variables as either quantitative, categorical, or ordered categorical.
   a. Blood type
   b. Grade-point average
   c. IQ
   d. Number of children
   e. Occupation
   f. State in which a person's legal residence is
   g. Systolic blood pressure

2. The following table gives the breakdown of total U.S. oil consumption (by percentage) by the purposes for which oil is used:

Use	Percentage
Gasoline	43.0
Heating oil	17.0
Industrial fuel oil	12.0
Jet fuel	7.0
Diesel fuel	5.5
Petrochemical	3.5
Other	12.0
Total	100

   a. Construct a pie chart from the data.
   b. Construct a Pareto chart.
   c. Which is more useful, or are they both useful for different purposes?
   d. What conclusions do you draw?

3. Randomly chosen customers were asked to name the brand of toothpaste (labeled A–E) they liked the most. The preferences showed are as follows:

   D, A, B, E, C, B, D, A, A, B, A, D, A, E, A, D, A, A, D, A,
   D, E, D, A, D, B, D, C, A, E, D, B, C, E, D, B, A, B, C, E

   a. Use an appropriate graphical tool that facilitates comparison among brands.
   b. Use an appropriate graphical tool that shows clearly the relative preference of each brand of toothpaste compared to the total for all brands.

4. A sample of employees at an organization start their work at the following times in the morning:

   9:00, 9:00, 8:30, 9:00, 8:00, 9:30, 9:00, 10:00, 8:30, 9:00, 9:30, 10:00, 8:00,
   8:00, 8:30, 9:00, 9:00, 8:00, 8:00, 9:00, 8:30, 8:30, 8:30, 9:00, 9:30, 10:00,
   9:30, 8:30, 9:00, 8:30, 9:30, 8:30, 9:30, 9:00, 9:00, 9:00, 9:00, 9:30, 9:00

   Use an appropriate graphical tool that facilitates an accurate comparison of the proportion of employees starting their work at 9:00 A.M. with the proportion starting their work at 9:30 A.M.

5. Bar graphs are sometimes used to compare categories even when we are looking at data other than frequencies. The table below shows the percentage of foreigners (from the perspective of the listed country) in the total labor force in different countries in 1991. (For more details on this table, see *Statistical Abstract of the United States, 1997.*)

Country	Percentage	Country	Percentage
United States	9.3	Germany	8.9
Australia	24.8	Japan	0.9
Austria	8.9	Luxembourg	33.3
Belgium	7.4	Netherlands	3.9
Canada	18.5	Norway	4.4
Denmark	2.5	Spain	0.4
France	6.2	Sweden	5.5

a. Depict these data using a bar graph. Then make comments on the distribution of the data.
b. Explain why the data cannot be displayed in a pie chart.

1.4 FREQUENCY HISTOGRAMS

In Section 1.3 we constructed bar graphs called Pareto charts for categorical (qualitative) data. We can do the same thing for quantitative (measured) data in order to study the distribution (shape) of the data. In this case, each bar represents the number of persons (things, observations, etc.) for which the measured value falls within a certain interval. Such a bar graph for measured data is called a **frequency histogram,** or just a histogram. (The word "histogram" comes from the Greek word *histos*, meaning pole, because the bars in the histogram look somewhat like a row of poles. Histograms appeared as early as 1786.)

The horizontal axis of such a graph indicates the range of the observed data values, and the vertical axis indicates the number of observations, or the **frequency,** within each interval in the range.

Example 1.6 The highest elevation in each of the 50 states

Consider the highest elevation for each state shown in Table 1.11. We shall construct a frequency histogram of these elevations in three steps.

Step 1. *Divide the range of the data into class intervals of equal width.* The data in Table 1.11 range from 345 (Florida) to 20,320 (Alaska). We choose as our class intervals 0 to 1500⁻, 1500 to 3000⁻,

Table 1.11 Highest Elevation in Feet

State	Elevation (feet)	State	Elevation (feet)	State	Elevation (feet)	State	Elevation (feet)
AL	2405	IN	1257	NE	5425	RI	812
AK	20320	IA	1670	NV	13140	SC	3560
AZ	12633	KS	4039	NH	6288	SD	7242
AR	2753	KY	4139	NO	1803	TE	6643
CA	14494	LA	535	NM	13161	TX	8749
CO	14433	ME	5267	NY	5344	UT	13528
CT	2380	MD	3360	NC	6684	VT	4393
DE	442	MA	3487	ND	3506	VA	5729
FL	345	MI	1979	OH	1549	WA	14410
GA	4784	MN	2301	OK	4973	WV	4861
HA	13796	MS	806	OR	11239	WI	1951
ID	12662	MO	1772	PA	3213	WY	13804
IL	1235	MT	12799				

Source: The Universal Almanac.

Section 1.4 Frequency Histograms

3000 to 4500⁻, and so on, each of width 1500. The minus signs indicate that the value 1500 is not included in the first interval but is included in the second; the value 3000 is not included in the second interval but is included in the third; and so forth.

Step 2. *Count the number of observations in each class interval.* The resulting table is called a *frequency table*. Here are the counts (frequencies):

Class interval	Tallies	Frequency
0 to 1,500⁻	⊦⊦⊦⊦ ∣∣	7
1,500 to 3,000⁻	⊦⊦⊦⊦ ⊦⊦⊦⊦	10
3,000 to 4,500⁻	⊦⊦⊦⊦ ∣∣∣	8
4,500 to 6,000⁻	⊦⊦⊦⊦ ∣∣	7
6,000 to 7,500⁻	∣∣∣∣	4
7,500 to 9,000⁻	∣	1
9,000 to 10,500⁻		0
10,500 to 12,000⁻	∣	1
12,000 to 13,500⁻	∣∣∣∣	5
13,500 to 15,000⁻	⊦⊦⊦⊦ ∣	6
15,000 to 16,500⁻		0
16,500 to 18,000⁻		0
18,000 to 19,500⁻		0
19,500 to 21,000⁻	∣	1

The highest elevation in Alabama is 2,405, so we put a tally mark next to the class interval 1,500 to 3,000⁻. The highest elevation in Alaska is 20,320, so we put a tally mark next to the class interval 19,500 to 21,000⁻, and so forth.

Step 3. *Draw the histogram.* In Figure 1.5 we mark elevation on the horizontal axis. The horizontal scale runs from 0 to 21,000. We mark frequency (number of states) on the vertical axis. For each class interval we draw a bar whose base is the class interval and whose height is the corresponding frequency.

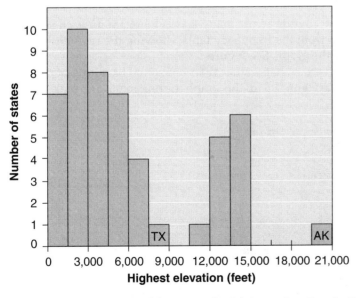

Figure 1.5 Frequency histogram for highest elevation in the 50 states.

26 CHAPTER 1 EXPLORING DATA: TABLES AND GRAPHS

We notice that the 50 states fall in three separate groups: Alaska with Mt. McKinley (elev. 20,320 feet) is in a group by itself. The middle group (with maximum elevations between 11,000 feet and 14,500 feet) consists of the Rocky Mountain states, the Pacific states, and Hawaii. Finally the main group (with maximum elevations under 9,000 feet) consists of the 37 states east of the Rocky Mountains. Within this group, only Texas ("TX" in Figure 1.3) which borders on the Rocky Mountains, has elevations above 7,500 feet. This is an example of a data set that has clusters with gaps between them.

You may ask how we chose our class intervals in Step 1. It is largely a matter of common sense. The number of class intervals should be large enough to display the important characteristics of the shape of data but small enough to provide an effective summary of the basic shape of the data. Not counting the four empty class intervals, we grouped the data in ten class intervals. With fewer observations we would have used fewer class intervals. As a general rule of thumb, 5 to 20 class intervals are usually appropriate. In this example we found the range of the data, 345 to 20,320, then we picked a slightly larger interval, 0 to 21,000, and divided it into 14 class intervals of equal width. We could just as well have used class intervals of slightly greater width, say 2,000 or 2,500. But if we use too few class intervals, then we throw away too much information about the shape of the distribution. Unlike a stem-and-leaf plot, the frequency table of a histogram only records the number of observations in each class interval, not the exact value of each data point. Figure 1.6 illustrates what happens if we use too few class intervals.

Figure 1.6 is not very informative. Important and interesting information about how the data is distributed is lost. We can still see that a majority of the states have maximum elevation under 5,000 feet. But we have blurred the distinction between the mountain states and the states east of the Rocky Mountains. The gap between the two groups has disappeared. Figure 1.6 simply has too few bars.

The next example illustrates the fact that every stem-and-leaf plot can be converted into a frequency histogram.

Example 1.7 Annual salaries

Look at the stem-and-leaf plot of annual salaries at a certain company to the nearest $1000 in Table 1.12. By simply enclosing each row (set of leaves) within a bar, we have a respectable histogram. See Figure 1.7(*a*).

We can also build a frequency table by counting how many leaves there are for each stem (see Table 1.12) and then draw the histogram from the frequency table, showing the frequencies (how many leaves there are) along one axis of the graph. See Figure 1.7(*b*).

Histograms are usually shown with vertical bars rather than horizontal bars. In that case, our histogram of the annual earnings looks like Figure 1.8. In this graph the horizontal axis scale (salaries

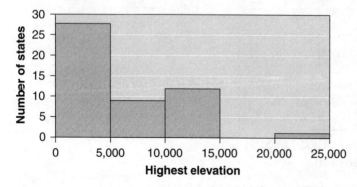

Figure 1.6 Frequency histogram for highest elevation in the 50 states.

Table 1.12 Stem-and-Leaf Plot of Annual Earnings (Thousands of Dollars)

Stem	Leaf
1	6,7
2	9,9,9,2,7,7,8,3,5
3	4,5,1,2,6,6,3,9
4	6,9,7,7,7,5,9,8,6,3,0,3,4,8,3
5	3,5,3,7,7,9
6	3,2,2,4,4,0,0
7	7,4

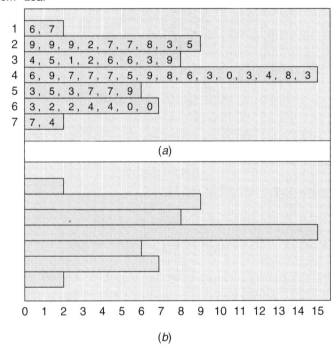

Figure 1.7 Histogram derived from stem-and-leaf plot of Table 1.12.

in $1000s) has been included to make it clear which interval in the range of incomes each bar represents, with the first interval from $10,000 to $20,000⁻, and so on.

Building a histogram from a frequency table rather than from a stem-and-leaf plot has a major advantage. We can freely choose the width of the class intervals in a frequency table. In this way we can choose a number of class intervals that does a nice job of showing interesting and important aspects of the shape of the data. Stem-and-leaf plots do not offer the same freedom of choice. For example, the frequency histogram in Figure 1.5 cannot be derived from any stem-and-leaf plot. Another limitation to the use of stem-and-leaf plots is that we can construct them only for small data sets.

Distribution Shapes

We have already seen that stem-and-leaf plots help reveal the underlying shape of a data set. Now we will see how useful histograms are for this purpose.

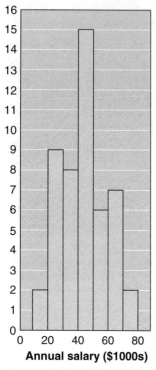

Figure 1.8 Histogram of Figure 1.7 put in usual position.

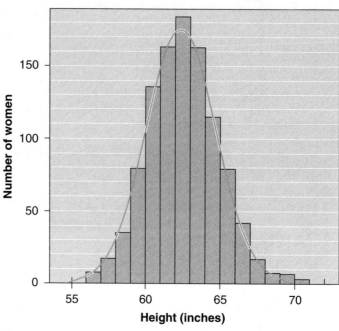

Figure 1.9 Height of mothers.

In 1903 Pearson and Lee published a paper on the laws of inheritance in humans. They studied over 1000 English families, relating the height, span of arms, and length of left forearm of the various family members. Figure 1.9 displays a frequency histogram for the heights of the 1052 mothers in the study. The heights were recorded to the nearest quarter of an inch. We have superimposed a smooth curve approximating the overall shape of the histogram. The smooth curve makes the shape easier to see.

The histogram in Figure 1.9 is "bell-shaped," a very commonly occurring distribution shape. Figure 1.10 displays some common distribution shapes: **bell-shaped**, **uniform**, **U-shaped**, **right-skewed** (or **skewed to the right**), and **left-skewed** (or **skewed to the left**). We say that figure (*d*) and (*e*) are "skewed" in the direction of the long tail.

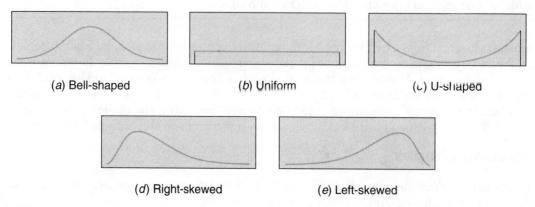

Figure 1.10 Common distribution shapes.

When we superimpose a smooth curve on a histogram we focus on the overall shape of the distribution and ignore minor irregularities.

Section 1.4 Exercises

1. One of the histograms sketched below shows the distribution of age at death from natural causes (heart disease, cancer, stroke, and so forth). The other shows the distribution of age at death from trauma (accident, suicide, murder). Which is which? Explain.

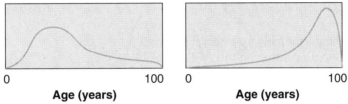

2. The following data give the per capita (per person) spending distribution in federal dollars for the 50 states. (For example, seven states receive an outlay in the $2,200–$2,399 interval).

Dollars per capita	Number of states
2000–2199	1
2200–2399	7
2400–2599	9
2600–2799	6
2800–2999	9
3000–3199	6
3200–3399	2
3400–3599	2
3600–3799	1
3800–3999	0
4000–4199	3
4200–4399	3
4400–4599	0
4600–4799	1
Total	50

Source: U.S. Bureau of Census, *Statistical Abstract of the United States*, 1997.

   a. Draw a histogram.
   b. What are your observations?

3. The following data show the time in days between 63 successive major earthquakes. An earthquake is included in this data set only if its magnitude was at least 7.5 on the Richter scale, or if over 1000 people were killed.

840	157	145	44	33	121	150	280	434	736	584	887	263
1901	695	294	562	721	76	710	46	402	194	759	319	460
40	1336	335	1354	454	36	667	40	556	99	304	375	567
139	780	203	436	30	384	129	9	209	599	83	832	328
246	1617	638	937	735	38	365	92	82	220			

a. Draw a histogram, using the class intervals 0 to 100⁻, 100 to 200⁻, and so on. List the values over 1000 as HI values.
b. What are your observations about the distribution?

4. Given below are measurements resulting from repeatedly carrying out an experiment.

75	95	100	10	16	50	26	84	29	80	35	57	27	50	12	51	12	80	29	68
27	84	28	52	53	61	40	94	88	52	54	14	36	64	40	87	28	52	65	31
101	40	93	33	65	33	64	69	44	102	46	95	34	66	99	39	84	44	79	47
37	70	55	72	59	66	44	49	70	56	62	32	62	41	78	49	61	37	80	59
63	42	92	76	67	67	70	73	74	85	77	86	90	76	103	103	90	90	100	

a. Construct a histogram using an interval size of 5.
b. Construct a histogram using an interval size of 10.
c. Which of the two interval sizes is more appropriate and why?

5. Given below are the observations from random sampling of a population.

65	77	31	54	50	73	55	43	30	58	64	39	77	53	63	46
78	65	52	34	52	76	33	62	43	33	51	51	66	48	71	47
72	35	67	61	75	66	49	30	68	50	80	75	70	76		

a. Construct a histogram using an interval size of 5.
b. Construct a histogram using an interval size of 10.
c. Which of the two interval sizes is more appropriate and why?

6. The National Earthquake Information Center has published the data shown below on magnitude (on the Richter scale) of earthquakes worldwide in 1998. Earthquakes of negligible magnitude are not included in the first category, indicating that the interval does not include 0.

Magnitude	Number of earthquakes
0⁺–1⁻	6
1–2⁻	580
2–3⁻	3,080
3–4⁻	4,518
4–5⁻	5,548
5–6⁻	674
6–7⁻	97
7–8⁻	9
8–9⁻	1

a. Construct a histogram.
b. Make your observations.

1.5 DENSITY HISTOGRAMS

Consider a variable such as the size of a family, the number of children in the family under 16 years of age, or the number of rooms in the family's home. The possible values for such **counting variables** are the numbers 0, 1, 2, 3, 4,… (called **integers**). When we plot histograms for such variables, we center the histogram columns at the possible integer values. To do so, we use the class intervals –0.5 to 0.5, 0.5 to 1.5, 1.5 to 2.5, and so forth, even though the variables cannot have a value like 0.5.

Example 1.8 Household Sizes

According to the U.S. Census Bureau, a household is any group of persons living together. In 1999 the Current Population Survey studied a representative cross section of a little more than 50,000 households, obtaining the following distribution.

Size	Class interval	Frequency	Percent
1	0.5 to 1.5	12,709	25.03
2	1.5 to 2.5	16,517	32.52
3	2.5 to 3.5	8,435	16.61
4	3.5 to 4.5	7,521	14.81
5	4.5 to 5.5	3,640	7.17
6	5.5 to 6.5	1,251	2.46
7	6.5 to 7.5	425	0.84
8	7.5 to 8.5	169	0.33
Over 8		118	0.23
Total		50,785	100.00

Source: U.S. Census Bureau

Because the **sample** of 50,000 is so large, the percentage of households of size 1 in the Current Population Survey (25.03%) is likely to be very close to the percentage of all 100 million U.S. households that are of size 1. The percentage of households of size 2 in the survey (32.52%) is likely to be very close to the percentage of all U.S. households that are of size 2, and so forth. We can therefore use the survey data to approximate very closely the distribution by size of all U.S. households. Figure 1.11 is the resulting **density histogram.** We note that household size data is right-skewed. We also note that the areas of the rectangles of a density histogram add to 100%.

In Section 1.4 we learned how to construct frequency histograms displaying the general shape, or distribution, of data sets. Often we want to compare two or more distributions. To illustrate, we compare the age distribution of the population of Germany with the age distribution of the population of Kenya. We want to know whether the German population is older than the population of

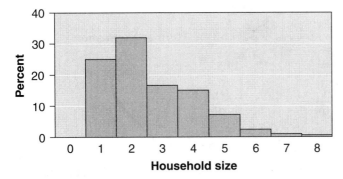

Figure 1.11 Distribution of U.S. households by size.

Table 1.13 Distribution of German and Kenyan Populations by Age (Year 2000)

Age group	Millions of Germans	Millions of Kenyans
0 to 15⁻	13.01	12.99
15 to 30⁻	14.41	9.59
30 to 50⁻	26.29	5.21
50 to 65⁻	15.64	1.72
65 to 80⁻	10.54	0.72
80 to 95⁻	2.91	0.11
Total	82.80	30.34

Kenya. We shall use the data in Table 1.13 taken from the U.S. Bureau of the Census, International Data Base.

We first note that, because the population of Germany is almost three times as large as the population of Kenya, there are more Germans than Kenyans in each of the age groups in Table 1.13. But it is not the *number* of Germans of a given age we want to compare with the *number* of Kenyans of the same age. Rather, we want to compare the *percentage* of the total number of Germans in each age group with the *percentage* of the total number of Kenyans in the same age group. To make this comparison, we shall construct a density histogram for each population. In such histograms, the areas of the blocks represent percentages. We proceed in three steps.

Step 1. Compute the percentage of Germans in each age group by dividing the number of Germans in that age group by the total number of Germans and then multiplying by 100%. For example, the percentage of Germans under 15 years of age is the relative frequency times 100%:

$$\frac{13.01}{82.80} \times 100\% = 15.7\%$$

Step 2. Find the height of the histogram bar over each class interval. Using the principle that Area = Percent, we use the formula

$$\text{Percent} = \text{Area of rectangle} = \text{Width} \times \text{Height}$$

In other words

$$\text{Height} = \text{Percent/Width}$$

For example, for the age group from 0 to 15 years, we get for the German population

$$\text{Height} = 15.7\%/15 = 1.05$$

Combining the two steps for the German population we get Table 1.14.

We note that for a density histogram the class intervals do *not* have to be of the same size.

Step 3. Draw the histogram. We mark age on the horizontal axis. The horizontal scale runs from 0 to 100 years. The unit on the vertical axis is "percent per year."

We note that the height of the appropriate rectangle of the histogram tells us the approximate percentage of Germans of any specified age in the rectangle: The percentage of 10-year-olds is

Table 1.14 Density Table Heights for Age Distribution of German Population (Year 2000)

Age group	Percent	Height
0 to 15⁻	15.7%	1.05
15 to 30⁻	17.4%	1.16
30 to 50⁻	31.8%	1.59
50 to 65⁻	18.9%	1.26
65 to 80⁻	12.7%	0.85
80 to 95⁻	3.5%	0.23

roughly 1.05%, and the percentage of 40-year-olds is roughly 1.59%. The percentage of 60-year-olds is approximately 1.26%. And the percentage of all Germans who are *between* their 5th and their 10th birthdays is roughly $5 \times 1.05\% = 5.25\%$, namely the area of the shaded block:

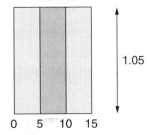

Looking at Figure 1.12, we notice that the two blocks covering the ages from 0 to 30 years are not as tall as the two blocks covering the ages from 30 to 65 years. Thus the percentage of Germans of any particular age between 30 years and 60 years is greater than the percentage of Germans of any particular age under 30. For example, the percentage of 40-year-olds is greater than the percentage of 10-year-olds. Essentially, this is because Germans are having fewer children today than they had in the past. At the same time, people are living longer. As a result, in twenty years there will be fewer workers and more retired people than there are today. This becomes a serious social and political problem. It will be difficult for the working people to pay for the health care and the pensions of the retired people. To maintain their economic growth and standard of living, the Germans will have to recruit foreign workers—a prospect they do not yet seem to have come to terms with. Many European countries face the same problem. Unlike the United States, many of the European countries do not have a tradition of recruiting skilled foreign labor.

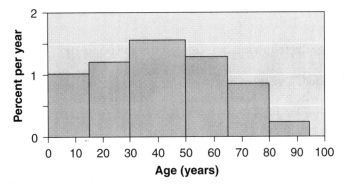

Figure 1.12 Distribution of German population by Age: 2000, as shown by a density histogram.

Table 1.15 Density Table Heights for Age Distribution of Kenyan Population (Year 2000)

Age group	Percent	Height
0 to 15-	42.8%	2.85
15 to 30-	31.6%	2.11
30 to 50-	17.2%	0.86
50 to 65-	5.7%	0.38
65 to 80-	2.4%	0.16
80 to 95-	0.4%	0.03

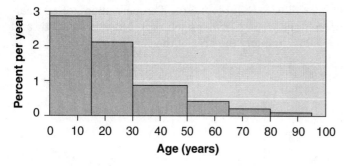

Figure 1.13 Distribution of Kenyan population by Age: Year 2000, as shown by a density histogram.

The age distribution of Kenya tells a very different story. The numbers in Table 1.15 are derived from Table 1.13. Again, the height of the histogram tells us the approximate percentage of Kenyans of any age. The percentage of 5-year-olds is roughly 2.85%; the percentage of 20-year-olds is roughly 2.11%; and so on. Figure 1.13 shows that the population of Kenya is very young. As can be computed from the graph, more than half of all Kenyans are under the age of 20 years. The population of Germany is much older: Around half of all Germans are 40 years old or older. While the population of Germany is shrinking, the population of Kenya is growing rapidly, mostly because the birth rate is high. If the current rate of growth persists, the population of Kenya will double in size in 30 years.

The statistical lesson here was that the density histogram is an ideal tool for comparing the distributions of a variable in two settings, such as the age distributions of Germany and Kenya.

Section 1.5 Exercises

1. Use Table 1.14 and Figure 1.12 to answer the following questions about the population of Germany.
   a. What percentage of Germans are under 30 years of age?
   b. Estimate the percentage of Germans who are at least 30 years old, but not yet 40 years old.
   c. Estimate the percentage of Germans under 40.
   d. Are there more 10-year-olds than 40-year-olds?
2. Use Table 1.15 and Figure 1.13 to answer the following questions about the population of Kenya.
   a. Is the percentage of 10-year-olds in Kenya much smaller than, about the same as, or much greater than the percentage of 40-year-olds?
   b. What percentage of Kenyans are under 30 years of age?

c. Estimate the percentage of 16-year-olds.
   d. Estimate the percentage who are at least 15 years old, but not yet 18 years old.
   e. Estimate the percentage who are under 18.
3. Below is a density histogram for the age distribution of the U.S. population in 2000.

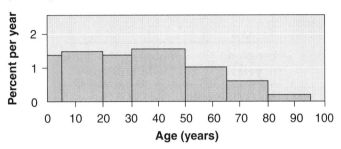

Source: U.S. Census Bureau, Year 2000.

   a. The percentage of 40-year-olds is around
      1%   1.5%   15%   30%   45%
   (Choose one option and explain.)
   b. The percentage who are at least 30, but not yet 50, is around
      1%   1.5%   15%   30%   45%
   c. The percentage of 55-year-olds is around
      1%   1.5%   15%   30%   45%
   d. The percentage who are at least 50, but not yet 65, is around
      1%   1.5%   15%   30%   45%
   e. The percentage who are at least 30, but not yet 65, is around
      1%   1.5%   15%   30%   45%
   f. Does the United States have rapid population growth like Kenya, slow growth, negative growth like Germany, or something else?
4. In a medical study the systolic blood pressure (that is, the pressure while the heart is contracting and pumping blood) of a couple of thousand young women was measured. The unit for such measurements is "millimeter mercury" (mmHg). The figure below is the density histogram for the data.

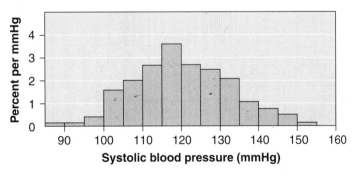

   a. How would you describe the shape of the histogram?
   b. The percentage of women with blood pressure between 90 mmHg and 150 mmHg was around
      1%   20%   50%   80%   99%
   (Choose one option.)
   c. The percentage of women with blood pressure under 120 mmHg was around
      1%   20%   50%   80%   99%

d. The percentage of women with blood pressure over 130 mmHg was around
   1%   20%   50%   80%   99%

e. The percentage of women with blood pressure between 105 mmHg and 110 mmHg was around
   1%   2%   5%   10%   20%

5. The projected age distribution of U.S. residents in 2075 by the Census Bureau is as shown below.

Age group (in years)	Number of people (in millions)
Under 10 years	34.9
10–19	35.7
20–29	36.8
30–39	38.1
40–49	37.8
50–59	37.5
60–69	34.5
70–79	27.2
80–89	18.8
90–99	7.7
100–109	1.7
Total	310.6

a. Construct a density histogram.

b. Compare the histogram with the density histogram in Exercise 3. Is the population getting older or younger?

6. A major attraction at the Yellowstone National Park is the eruption of the geyser Old Faithful. The table on the following page summarizes a sample of 400 times (in minutes) between eruptions.

Time (minutes)	Frequency
30–39	4
40–49	15
50–59	85
60–69	49
70–79	15
80–89	211
90–99	18
100–109	3

a. Construct a density histogram.

b. What percentage of the time was the waiting time less than an hour?

7. Major active volcanoes in Alaska have the following heights above the sea level.

4265, 3490, 4450, 3995, 150, 5315, 2560, 7985, 5710, 3545, 5370, 5775, 10, 140, 8185, 2945, 4450, 7540, 885, 4025, 4885, 3945, 6050, 7295, 6720, 2759, 3540, 8960, 7050, 5030, 7545, 10265, 2015, 3465, 9430, 11070, 7015, 11413, 6830, 8450, 6965, 5055, 1980

Construct a density histogram using the class intervals 0 to 1000⁻, 1000 to 2000⁻, and so forth.

8. The following are the IQ scores of 50 randomly sampled undergraduates:.

124, 106, 108, 87, 114, 102, 132, 117, 99, 102, 128, 119, 105, 84, 93, 135, 110,
97, 103, 130, 92, 103, 102, 116, 93, 111, 88, 109, 108, 142, 90, 130, 107, 109,
114, 98, 103, 136, 110, 107, 140, 109, 115, 147, 113, 119, 115, 125, 128, 98

Construct a density histogram using the class intervals 80 to 90⁻, 90 to 100⁻, and so forth.

9. A company had 105 sales last year, grouped by price as follows:

Price range (in thousands of dollars)	Number of products sold
0–50⁻	21
50–100⁻	47
100–150⁻	15
150–200⁻	12
200–250⁻	10

a. Construct a density histogram. Describe the shape.
b. Estimate the total annual sales. *Hint:* Make the rough approximation that the cost of every item is the midpoint of its interval.

10. The distribution of a company's stock to shareholders according to the number of shares held is as shown in the table below.

Number of shares held	Percentage of stockholders
0–25⁻	39
25–50⁻	27
50–100⁻	20
100–500⁻	12
500–700⁻	2

a. Construct a density histogram.
b. How would you describe the shape of the distribution?

1.6 MISUSING STATISTICS

Descriptive statistics can easily be used to give misleading or wrong impressions about the world. This happens all too frequently in books, reports, and the media. In this section we present examples of how statistical techniques have been misused, so you will not be misled as easily when you encounter statistics in newspapers or magazines or on radio or television, and you may be better able to interpret them correctly.

Many of the misuses of statistics are achieved by graphical distortions, but there are other ways as well.

The Area Fallacy

In each of the bar graphs and histograms we have constructed in this chapter, the widths of the bars were equal. Only the heights differed from bar to bar. Thus the area of each bar represented the relative contribution of that bar accurately, just as the height did. One of the most common

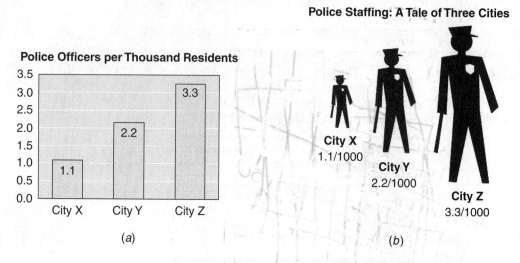

Figure 1.14 Two ways to display statistics about numbers of people: (a) truthful but boring bar chart; (b) lively but misleading pictorial.

forms of distorted graphical presentation is a figure in which the *length* (or height) of each symbol indicates the quantity the figure is supposed to represent, but the *area* does not. This type of distortion often occurs in media stories in which the statistics of interest are numbers of people. For example, suppose that the story is about the different numbers of police officers per thousand residents in three cities:

City X	City Y	City Z
1.1	2.2	3.3

A fair bar chart of these data is shown in Figure 1.14(a), but many newspapers would instead use a "livelier" display of drawings of people, such as using stylized figures of police officers with heights proportional to the numbers, as in Figure 1.14(b). The officer for City Z is correctly three times as tall as the officer for City X but (because the figures are similarly shaped) three times as wide as well and therefore the officer of City Z looks *nine* times as large. Worse yet, the figures are not placed on the same baseline, obscuring the proper (height) relationship among them even further.

This kind of distortion is often not the result of a conscious attempt to deceive the reader, but rather of a thoughtless attempt to make the graphic display eye-catching by substituting pictures for text and abstract bar shapes.

The Missing Baseline

A reader glancing at a bar graph is likely to base a first impression on the assumption that the vertical scale in the graph extends all the way to zero, because that is the way we think of graphs. Often, however, much of the bottom of the vertical scale is cut off. Such a graph makes a difference look far more dramatic than it really is. We illustrate with an example.

Example 1.9 The Most Expensive Big Cities for Drivers

Figure 1.15 appeared in *USA Today*, May 3, 1995. Once you have *found* the bar graph among all the cartoon cars, buildings, and seagulls (called "chart clutter" by Edward R. Tufte, one of the world's experts on the efficient and appropriate display of information), you might think that

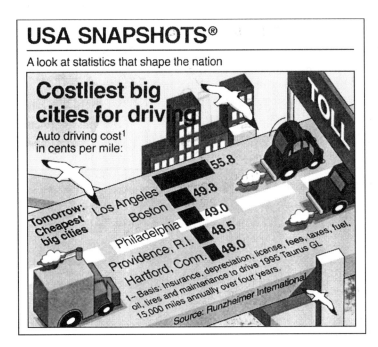

Figure 1.15 Bar graph of driving expenses (cents per mile) in various U.S. cities. (*Source:* USA Today, May 3, 1995.)

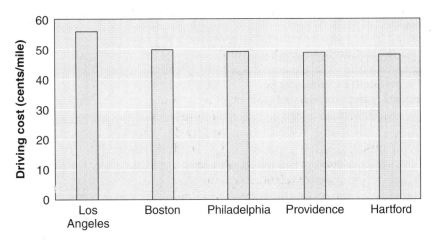

Figure 1.16 Correctly drawn bar graph of the driving expense data in Figure 1.15.

you could cut your driving expenses by almost two-thirds by living and working in Hartford rather than Los Angeles. In fact, you would save only about 14% (see Figure 1.16 for an ordinary bar chart that does not cause a distorted impression).

The intention behind these graphs, as with the graphs that use pictures instead of bars, is not always malicious. Graphic designers are trying first of all to make their displays attract the reader's attention. Parts of a graph that show a change catch the eye. Parts that do not show a change are often regarded as boring and a waste of display space, even though they are needed if they are to make an impression that agrees with reality.

The Combination Graph

Often the changes in two different measurements over time are shown on a single graph. Such a graph can be very useful for giving a view of how one quantity may be affecting another (later in this book we will discuss statistical techniques for determining the amount of influence of one variable on another). Each of the two graphs, however, is susceptible to the zero point fallacy just described. By carefully choosing the scales and vertical placement for the two plots, a presenter can make an impression, or its opposite, that has nothing to do with the data.

Example 1.10   SAT Scores vs. Public School Funding

Suppose one wishes to assess the influence of public education spending on student preparedness for college. Table 1.16 shows the amount of money spent per pupil on public education in the United States for school years ending between 1990 and 1995. Table 1.17 shows the national average composite SAT (Scholastic Aptitude Test then, now called the Scholastic Assessment Test) scores for college-bound students taking the tests over the same range of years.

Figure 1.17 shows three combination graphs constructed from the data in Tables 1.16 and 1.17. Figure 1.17(a) might have been prepared to support an article opposing increases in public school funding by suggesting that SAT scores have changed little. Figure 1.17(b), on the other hand, might have been prepared to support an article in praise of increased per-pupil spending by suggesting that SAT scores have greatly increased as a consequence. In both graphs, a wide range is used for the measurement the presenter wants to look unchanged, while a narrow range is used for the measurement that the presenter wants to make dramatic. Figure 1.17(c) is an impartial plot of the same data, achieved by starting both scales at 0. Neither measurement changed dramatically in this graph.

Table 1.16   Per-Pupil Expenditures on Public Education

School year ending	Per-pupil expenditures on elementary and secondary education, 1997–1998 dollars
1990	6591
1991	6626
1992	6587
1993	6587
1994	6633

Source: *New York Times Almanac 2000*, p. 368.

Table 1.17   National Average Composite SAT Scores

Year	Average composite SAT score, all students
1990	900
1991	896
1992	899
1993	902
1994	902
1995	910

Source: U.S. Department of Education, *The Condition of Education 1996*, Indicator 22.

Section 1.6 Misusing Statistics 41

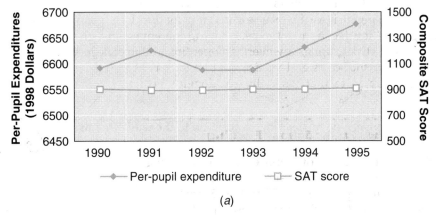

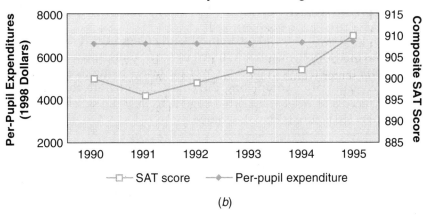

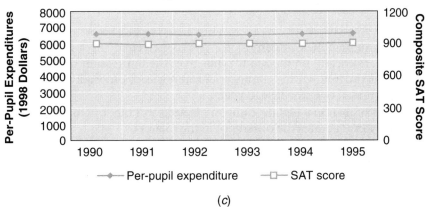

Figure 1.17 Combination graphs of per-pupil public education expenditures in constant dollars versus composite SAT scores.

One might wonder whether these two indices are the best indices of whether public school funding directly influences how well students are prepared for college. Rather than the per-pupil expenditure for only the year in which the test is taken, one might construct an index based on averaging the secondary school expenditures (in constant dollars) for the four years before the student took the test, for example. Further, as an index of college preparedness in a given year, one might use an index based on college performance rather than on a test used in college admissions.

Other Tricks

There are a few other ways a graph can be misleading in addition to the ways we have discussed. For example, a pie chart is often shown at an angle so as to look more like a pie. The wedges on the front facing the reader then look wider than the wedges on the sides.

Sometimes the distortion does not depend on the graph. Numerical indices can be computed in a variety of ways. It is not difficult to construct an index that makes the point one wants to make statistically no matter what the facts of the situation are.

Example 1.11 How Fast Can You Read?

A speed reading training school once used in its advertisements a reading index that measured reading level by the score on a test of comprehension multiplied by the reading speed.

Measured by that index, "readers" who learned to skim over the text extremely rapidly could show much higher reading level scores after completing the course than before they took the course, even if they understood far less of what was on the page than they did when reading at their ordinary pace and comprehension before taking the course.

Of course, the specific ways of distorting the underlying truth of the data presented here are just a sampling of the many ways that casually or dishonestly presented and casually examined statistical information can mislead us. One goal of this textbook is to produce **statistical literacy**. One aspect of statistical literacy is not to be easily misled by improper conclusions drawn from data, often supported by false ways of interpreting and displaying data.

Section 1.6 Exercises

1. A commercial released by an automobile company says, "Ninety percent of all our cars sold in this country in the last 10 years are still on the road." This gives the impression that the company's cars are built to last. Explain how this commercial can mislead, especially if the company's sales have rapidly increased over the last 10 years.
2. The average weekly earnings of men in a certain company is $800 and that of women is $710. Draw a bar chart that visually exaggerates the difference in the salaries.
3. A manager who has achieved a small growth in sales over the past two years wants to show his performance in a better light.
   a. What trick might he adopt when he is showing his performance graphically?
   b. If he has had a steep decrease in his sales, what trick might he use when showing his performance graphically?
4. The U.S. map at the top of the following page (distributed recently by First National Bank of Boston) purports to show what portion of income earned by U.S. citizens was being taken and spent by the federal government. The shading is to indicate that federal spending has become equal to the total incomes of all people residing in the states west of the Mississippi River, excepting Louisiana,

Arkansas, and a part of Missouri. This map certainly gives a very distorted picture. Can you figure out why? Is the federal government in fact spending an amount that is over half the entire income of its citizens? *Hint:* What do you know about the population densities of various states?

5. The market share of a toothpaste company for the past few years has been as shown in the accompanying figure. This is an honest bar graph with no intention to mislead. But how can the performance be represented so that it seems better than it actually was? *Hint*: Will you begin the vertical axis from zero?

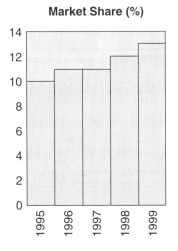

6. The profit or loss made by a company in the last five years is as follows:

Year	Profit or loss ($)
1995	−97,563
1996	28,576
1997	30,379
1998	41,274
1999	49,273

Note that the company had a loss in the year 1995. The chart the company publishes to show this data is shown in the accompanying figure.

44 CHAPTER 1 EXPLORING DATA: TABLES AND GRAPHS

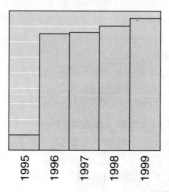

Profit/Loss

a. What is misleading about this bar graph?
b. Draw an appropriate graph. *(Hint:* For a negative amount, let the bar lie below the axis.)

7. A government agency reports population trends and value of agency services on the same graph, similar to those shown for SAT scores and school funding in Figure 1.17. The data are as shown below in the following table.
    a. Graph these data in a way that makes the government agency look better than it should.

	1960	1970	1980	1990	2000
Government spending (in billions of $)	6.4	6.45	6.51	6.54	6.57
Population (in millions)	92.1	96.4	100.1	112.5	120.1

   b. Graph it honestly.
   c. Propose an appropriate index (a statistic) to measure value of agency services adjusted for population growth.
8. Find examples from newspapers, magazines, and similar materials of presentations of statistics that are subject to the kinds of misinterpretation presented in this section.
9. Imagine that the marketing department of a corporation wants to mislead people as to how the increasing market share of the firm have changed over the decades. The graph they produced is shown here.

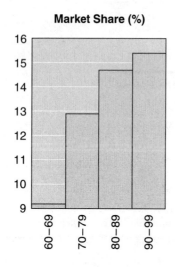
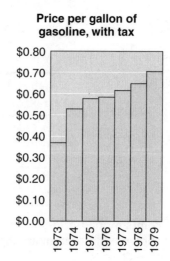

a. Why is this graph misleading?
b. Redraw the graph so that a fair visual image is presented.
c. Think of another example where such a trick can be used.

10. Discuss whether the accompanying graph (a similar graph appeared in a major national magazine) fairly represents the price increase of gasoline at the pump over several years.

Chapter 1 Summary

A **variable** is a characteristic that changes from subject to subject in a study.

Quantitative variables take numerical values. (Examples: height, income, IQ.)

Categorical variables place each object (person perhaps) in one of several groups or categories. (Examples: gender, religion, state of residence.) If the categories are **ranked**, then the variable is said to be **ordered** or **ranked categorical**.

Categorical data are best displayed in **bar graphs** (sometimes **Pareto charts**) or **pie charts**.

Quantitative data are displayed in **dotplots, stem-and-leaf plots,** or **histograms**. We look for **symmetry** or **skewness**. We also look for **gaps** in the distribution and exceptional values (**outliers**).

Histograms for counting variables center their rectangles width one at the integers 0,1,2,3,...

A histogram is **symmetric** if its left and right sides are approximately mirror images of each other. (Examples: bell-shaped, uniform, U-shaped.)

The histogram is **right-skewed** (skewed to the right, positively skewed) if it has a long tail to the right.

The histogram is **left-skewed** (skewed to the left, negatively skewed) if it has a long tail to the left.

Density histograms are very useful and are essential when comparing how data is distributed in different settings.

Displaying a data set is not an end in itself. The purpose of such displays is to help us recognize the overall pattern of the data so that we can understand the information contained in the data.

Beware of picture graphs and distorted scales!

CHAPTER REVIEW EXERCISES

1. Here are the speeds of cars traveling down a street with a restricted speed zone when school children are present:

    28, 26, 29, 27, 22, 24, 23, 29, 34, 24, 23, 22, 27, 26, 22,
    36, 54, 29, 28, 16, 41, 34, 27, 29, 20, 21, 34, 29, 21, 28

   a. Construct a stem-and-leaf plot.
   b. Is the shape what you would expect?

2. The following data were collected to determine whether the chance that a horse will win a race is influenced by its starting position. Data represent 134 races each with exactly eight horses in the starting line-up. Starting position 1 is closest to the rail on the inside of the track.

	Starting Position							
	1	2	3	4	5	6	7	8
Number of wins	29	19	18	25	17	10	15	11

Source: New York Post, August 30, 1955, 42.

a. Draw a pie chart for the data.
b. Draw a bar graph.
c. Can you see any apparent relationship between the chances of winning and the starting position?

3. Identify if the following statements are true or false, and give reasons why they are true or false.
    a. In a frequency histogram, the height of a bar is always equal to the frequency of the class the bar represents.
    b. In a density histogram, the area of each bar equals the percentage of observations in the corresponding class interval.

4. A sequence of 300 pseudo-random digits (the 10 digits are to appear random and hence each should have the same likelihood of occurring) was generated on a Casio calculator.

Digit	0	1	2	3	4	5	6	7	8	9
Frequency	25	28	29	35	35	31	27	33	32	25

Source: I.R. Dunsmore, F. Daly, and the M345 Course Team (1987), M345 Statistical Methods, Unit 9: Categorical data, In Milton Keynes: *The Open University*, Table 1.5.

a. Construct a pie chart and a bar graph.
b. What can you infer, if anything?

5. People have noticed that sports teams tend to do better at home in their own stadiums than when they play in their opponents' stadiums. Part of the reason may be in the fans' effect on the referees. To study this possibility, videotape of a particular soccer match was shown to 11 experts, 5 of whom were shown the videotape with the crowd noise (the "noise" group), and 6 of whom watched without any sound (the "no noise" group). Each person judged 52 incidents in which there may have been a foul, and all were asked to decide whether a foul had occurred. (A. Neville, N. Balmer, and M. Williams, 1999, Crowd influence on decisions in Association football, *The Lancet*, Vol. 353.)

In half of the incidents, all the experts agreed. For the other half of the incidents, the following were the results:

- Potential fouls against the home team: 61% were deemed fouls by the "no noise" group, and 47% were deemed fouls by the "noise" group.
- Potential fouls against the visiting team: 36% were deemed fouls by the "no noise" group, and 58% were deemed fouls by the "noise" group.

    a. Look at the data on the potential fouls against the home team. What does the fact that the "no noise" group decided there were more fouls than the "noise" group suggest about the effect of crowd noise on the experts' judgments?
    b. Look at the data on the potential fouls against the visiting team. What does the fact that the "no noise" group decided there were fewer fouls than the "noise" group suggest about the effect of crowd noise on the experts' judgments?
    c. Does it appear that crowd noise biases the experts' judgment in favor of the home team, in favor of the visiting team, or neither?
    d. For the potential fouls against the home team, the referees in the actual game decided that 53% were fouls. Were the referees' decisions closer to the "noise" group or "no noise" group?
    e. For the potential fouls against the visiting team, the referees in the actual game decided that 60% were fouls. Were the referees' decisions closer to the "noise" group or "no noise" group?
    f. Overall, does it appear that the crowd noise tends to lead the referees to favor the home team?

6. In a histogram, explain how the selected width of the class interval affects
    a. the preservation of detail contained in the original data.
    b. accidental irregularities in the heights of the bars.

7. The frequency table below shows the number of children per household for 50 households interviewed in a survey.

Number of children	f	Percent
0	3	
1	8	
2	26	
3	10	
4	2	
5	0	
6	1	
Total	50	

   a. Complete the third column to show the percentage of families with each number of children.
   b. Construct a density histogram of the data. (Center each rectangle at an integer: 0, 1, ..., 6.)
   c. From your graph in part b, find the proportion of families having two or fewer children.

8. Here are data on average fuel economy and average vehicle purchase price for all cars made by a certain automobile manufacturer in selected years.

	1970	1975	1980	1985	1990	1995
Fuel economy (MPG)	16.8	19.2	23.1	25.1	25.9	25.7
Price (thousands of 1995 dollars)	15	16.5	18.1	19.2	20.5	15

   a. Using combination graphing (see Figure 1.17), draw a graph that stresses gains in fuel economy while de-emphasizing rising prices.
   b. Draw a combination graph in a way that a political group wanting to stress poor advances in fuel economy and the high price of cars might draw the graph.
   c. Draw it honestly.

9. Identify the true and false statements among the following statements, and give reasons why they are true or false.
   a. In a bar graph, the width of the bars is unimportant, but the choice of width is important in the case of a histogram.
   b. A histogram is used to display the distribution of quantitative data like measurements or counts.
   c. In a histogram, all the bars must touch each other, but in a bar graph for categorical data they need not do so. (See Figure 1.16.)

10. The following data concern 40 heavy men, each weighing at least 225 pounds. Their cholesterol levels (measured in milligrams per 100 milliliters) were recorded and were categorized as belonging to either the Type A behavior category or the Type B behavior category. (Type A behavior is characterized by urgency, aggression, and ambition, while Type B behavior is characterized as relaxed, noncompetitive, and less hurried.)

Type A behavior:	233, 291, 312, 250, 246, 197, 268, 224, 239, 239, 254, 276, 234, 181, 248, 252, 202, 218, 212, 325
Type B behavior:	344, 185, 263, 246, 224, 212, 188, 250, 148, 169, 226, 175, 242, 252, 153, 183, 137, 202, 194, 213

Source: S. Selvin (1991) *Statistical Analysis of Epidemiological Data,* New York: Oxford University Press, Table 2.1.

a. Construct a back-to-back stem-and-leaf plot for Type A behavior and Type B behavior. *Hint:* Drop the ones digits. For example, 188 becomes 180; then 180 is assigned a hundreds stem of 1 and a tens leaf of 8.

b. What do you infer?

11. In a national survey, men were asked why they exercise. Here are the reasons they gave:

Reason	Percentage responding
Health	51
Stress relief	25
Weight loss	20
Other	4

Draw a pie chart of these data.

12. You have learned that usually the heights of a picture-based "histogram" are accurate and the distortion arises because our perception keys in on area.

a. Is this the case with the following picture?

Comparative Annual Cost per Capita for Care of Insane in Pittsburgh City Homes and Pennsylvania State Hospitals

$147 South Mountain $172 Pittsburgh $198 Harrisburg $213 Norristown $214 Warren

Pittsburgh Civic Commission, *Report on Expenditures of the Department of Charities* (Pittsburgh, 1911), p. 7.

b. Draw an appropriate bar graph for the given institutions. (Note that this was drawn in 1911. Would such a graph contain the word "insane" today?)

13. A six-sided die is rolled 120 times, and the following outcomes are obtained:

Outcome	f	Proportion
1	15	
2	21	
3	23	
4	19	
5	17	
6	25	
Total	120	

a. Draw a density histogram of the data.

b. From the graph, find (i) the percentage of rolls that give 3 or less, (ii) the percentage of rolls that give 5 or more, and (iii) the percentage of rolls that give an even number.

c. How would you roughly describe the shape of the density histogram?

14. The following table gives the percentage breakdown of the cigarette smoking habits of young adults (ages 18–24). They are broken into three groups: (i) those who currently smoke, (ii) those who are former smokers, and (iii) those who have never smoked. Two years (1965 and 1991) and gender are considered (from the 1998 paper "Trends in Cigarette Smoking" by the American Lung Association).

Smoking status	1965, Males	1965, Females	1991, Males	1991, Females
Current	54.1	38.1	23.5	22.4
Former	7.6	6.2	8.0	7.5
Never	38.3	55.7	68.5	70.1

a. Create pie charts for the 1965 males and the 1965 females. On which categories do males and females differ the most?
b. Create a pie chart for the 1991 males. What changes do you see from 1965 to 1991 for the males?
c. Create a pie chart for the 1991 females. What changes do you see from 1965 to 1991 for the females?
d. Now compare the males and females in 1991. What differences do you see? Are the differences larger than in 1965 (see part a) or smaller?
e. Briefly summarize what these data indicate.

PROFESSIONAL PROFILE

Dr. Andrew Harris
Volcanologist
Institute of Geophysics and Planetology
University of Hawaii, Honolulu

Dr. Harris uses an infrared temperature gun to measure the temperature of active lava on Kilauea Volcano, Hawaii

Dr. Andrew Harris

"We want to know how volcanoes work," explains Dr. Andrew Harris. Dr. Harris collects thermal data to calculate the size and temperature of hot features on volcanoes, including lava flows and vents. He collects some of the data personally, using infrared temperature guns and thermocouple probes, but he also works with a network of satellites that can measure the temperature of the volcano surface from space. He has collected this data from volcanoes in Italy (Etna, Stromboli, and Vulcano), Guatemala (Pacaya, Fuego, and Santiaguito), Iceland, Africa, Central and South America, and Antarctica.

With thousands of pieces of temperature information coming in every month, statistics is an important tool for Dr. Harris: "Each year, I collect around 1000 temperature measurements across the main crater of the Sicilian volcano, Vulcano. Statistical analyses allow me to determine whether changes are taking place. If the mean is rising and the frequency distribution is becoming skewed towards higher temperatures, this may show that the volcano is heating up." These, and other statistical tests, are used by Dr. Harris to determine how likely it is that a volcano will erupt. The science of statistics enables Dr. Harris and his colleagues to examine large data sets, extract useful information, examine spatial and temporal activity trends, test hypotheses, and assess the confidence with which they can answer the question, "Will it erupt?"

2

Summarizing Data by Numerical Measures: Centers and Spreads

Society suffers from the curse of the arithmetic mean. Such compression necessarily ignores the fact that most measurements involve a spread of values around the mean—arising from real differences and random fluctuations.

R. E. Beard

Objectives

After studying this chapter, you will understand the following:

- The importance of assessing the center and the spread of a data set
- The two most important measures of center: the mean and the median
- A resistant measure of spread: the interquartile range
- The most widely used measure of spread: the standard deviation
- The 68-95-99.7% rule for the standard deviation
- How to read and construct boxplots
- How to apply the tools of Chapters 1 and 2 to describe and compare data sets

2.1 CENTERS OF DATA 53

2.2 MEAN VERSUS MEDIAN AS A MEASURE OF CENTER 60

2.3 MEASURING SPREAD: THE STANDARD DEVIATION 69

2.4 NORMAL DISTRIBUTIONS 82

2.5 BOXPLOTS: THE FIVE-NUMBER SUMMARY 89

2.6 DESCRIPTIVE DATA ANALYSIS 98

CHAPTER REVIEW EXERCISES 107

Key Problem

The following data consist of the ages of actors and actresses who have won Academy Awards since 1969.

Actors:

46	60	45	31	38	37	52	54	42	32	51	43
61	35	45	52	49	76	37	42	40	30	60	38
56	48	48	41	43	63						

Actresses:

25	34	39	49	45	35	33	29	42	80	26	41
21	61	38	49	33	74	31	33	41	31	35	41
42	37	26	34	34	35						

Do winning actresses tend to be younger than winning actors? Is there less spread in the age distribution for actresses than for actors? How do we summarize the data so we can answer such comparative questions?

The data are presented graphically in Figure 2.1 in graphs called **boxplots** or **box-and-whiskers plots**. In the boxplot, the line inside the box locates the **median** (a measure of center that will be explained in Section 2.1), and the ends of the box locate the first and third **quartiles** (two descriptive statistics that give the range covered by the middle half of the data, as will be explained in Section 2.4). Outside the boxes, four data points are marked with special symbols (° or *), indicating values so far from the bulk of the data that they are viewed as not really belonging with the rest of the data and hence are called extreme values or outliers (see Section 1.2). We extend lines (the whiskers) from the ends of the box to the minimum and maximum values of the data, excluding the outliers.

Do you think these two boxplots help answer the two questions posed above about differences between male and female Academy Award winners? Do they help us compare the distribution of the two data sets?

The goal of descriptive statistics is to summarize data so that we can recognize the information the data contain. In Chapter 1 we learned how to **organize** raw data into frequency tables and **display** them graphically. Exploratory data analysis always begins with visual displays. For

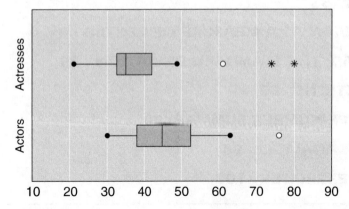

Figure 2.1 Boxplot comparison of ages of male vs. female Academy Award winners.

quantitative data we try to characterize the shape of the distribution and we look for gaps and exceptional observations (outliers).

In this chapter we shall add numbers to our description of data. Such numbers can indicate where the **center** of the data set lies, indicate the **spread** of the data, or measure the **skewness** of the data, to list a few possibilities.

2.1 CENTERS OF DATA

In this section we shall discuss three commonly used measures of center: the median, the mean, and the mode.

Example 2.1

In 1850 the population of the United States was much younger than it is today. Typically, a family had many children, and people only lived half as long as they do today. The age distribution of the United States looked in 1850 the way the age distribution of Kenya looks today (see Figure 1.13). The median age was 19 years: Half of all Americans were under the age of 19, and half were 19 years old or older. The life expectancy at birth was 39 years. (The life expectancy for a particular year is the average number of years that everyone born in that year would be expected to live if death rates remained constant.)

The twentieth century brought dramatic improvements in public health, and life expectancy doubled between 1850 and 2000. This, combined with a decline in the number of children born, caused a doubling of the median age in the United States (see Table 2.1). The line graphs in Figure 2.2 display the change over time in median age and life expectancy.

Table 2.1

Year	Median Age (years)	Life Expectancy (years)
1850	19	39
1900	23	47
1950	30	68
2000	35	76

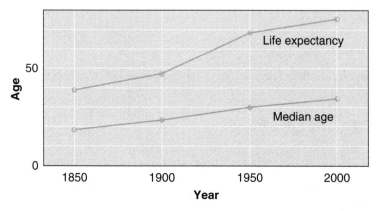

Figure 2.2 Median age and life expectancy at birth from 1850 to 2000.

Median

The **median** M of a data set is the middle value when the data are arranged according to size. Half of the observations are smaller than the median, and half are larger. This idea must be made into a precise rule. To find the median of a list of numbers, do the following:

Step 1. Arrange all observations according to size, from smallest to largest.

Step 2. If the number of observations is odd, then the median M is the observation located in the exact middle of the ordered list. If the number of observations is even, then the median M is the number halfway between the two middle numbers on the list.

The median is often interpreted as the "typical" value of a data set.

Example 2.2

The number of home runs hit by Babe Ruth during his 15 years with the New York Yankees, from 1920 to 1934, are as follows:

1920	1921	1922	1923	1924	1925	1926	1927	1928	1929	1930	1931	1932	1933	1934
54	59	35	41	46	25	47	60	54	46	49	46	41	34	22

Arranged in increasing order, the data are:

$$22 \quad 25 \quad 34 \quad 35 \quad 41 \quad 41 \quad 46 \quad \underset{M}{\overset{\uparrow}{46}} \quad 46 \quad 47 \quad 49 \quad 54 \quad 54 \quad 59 \quad 60$$

Here we have 15 observations, an odd number. The median is the exact middle number: $M = 46$. The median is a good indicator of Ruth's "typical" annual home run production because he often hit between 40 and 50 home runs in a season.

Ruth's 1927 record of 60 home runs in a single season remained unbroken until 1961, when Roger Maris, another Yankee, hit 61 home runs. Maris's home run totals for his 10 years in the American League are, in increasing order:

$$8 \quad 13 \quad 14 \quad 16 \quad 23 \quad \underset{M}{\overset{\uparrow}{}} \quad 26 \quad 28 \quad 33 \quad 39 \quad 61$$

Here we have 10 observations, an even number. The median is halfway between the fifth smallest and the fifth largest observations:

$$M = \frac{23 + 26}{2} = \frac{49}{2} = 24.5$$

The median is a fairly good indicator of Maris's typical annual home run production. His record of 61 home runs in 1961 is an exceptional achievement—an outlier. Overall, Ruth was a much better home run hitter than Maris, as we can see by comparing their medians.

Mean

The **arithmetic mean** is the most commonly used measure of the "center" of a data set. We derive it by adding up all the values and dividing the sum by the number of values. The arithmetic mean is usually referred to as simply the "mean" or the "average." Because the word "average" can sometimes be misunderstood, we will usually prefer the term "mean."

Example 2.3

During the summer of 1998, the attention of sports fans was caught by a race between the Chicago Cubs' Sammy Sosa and the St. Louis Cardinals' Mark McGwire to beat Roger Maris's single season home run record. Both succeeded. Sosa got 66 home runs and McGwire hit 70, a record that stood until 2001 when Barry Bonds hit 73 home runs. To compare the overall home run productions of Sosa and McGwire, we shall compute the mean annual number of home runs for each.

The numbers of home runs hit by Sosa from 1993 to 2001 are as follows:

1993	1994	1995	1996	1997	1998	1999	2000	2001
33	25	36	40	36	66	63	50	64

The mean number of home runs hit by Sosa is

$$\bar{x} = \frac{33 + 25 + 36 + 40 + 36 + 66 + 63 + 50 + 64}{9} = \frac{413}{9} = 45.9$$

The numbers of home runs hit by McGwire from 1987 to 2001 are as follows:

1987	1988	1989	1990	1991	1992	1993	1994	1995	1996	1997	1998	1999	2000	2001
49	32	33	39	22	42	9	9	39	52	58	70	65	32	29

McGwire's mean number of home runs is

$$\bar{x} = \frac{49 + 32 + 33 + 39 + 22 + 42 + 9 + 9 + 39 + 52 + 58 + 70 + 65 + 32 + 29}{15} = \frac{580}{15} = 38.7$$

We note that the two means are close. The two players' median number of home runs are even closer: 40 for Sosa and 39 for McGwire.

Notation

The mean is usually denoted $\bar{x}$ (read "x-bar") if the data represent a repeated trials experiment (as in Example 2.3, where the baseball season is repeated year after year and each season, or trial, the player's number of home runs x is obtained) or if the data set is a **sample** from a larger population (such as all American households). If we have all the values from an entire population, then the mean is denoted μ (the Greek letter "mu," corresponding to the Roman letter "m," that is, "m" for "mean"). For example, the average (mean) size of all the 100 million households in the United States would be denoted μ, whereas the average (mean) size of the 60,000 households interviewed by the Current Population Survey during a given month would be denoted $\bar{x}$.

We shall introduce some more notation: Let x denote the variable of interest. Then $\sum x$ denotes the sum of all the values of the variable x. (The symbol $\sum$ is the capital Greek letter "sigma," corresponding to the Roman letter "S," that is, "S" for "Sum"). If we have n date values, the formula to compute the mean is

$$\bar{x} = \frac{\sum x}{n}.$$

The mean of all the values in a population is

$$\mu = \frac{\sum x}{N},$$

where N denotes the population size. Roughly speaking, the two formulas are the same, except for whether we have some (as with data) or all possible numbers (as with a population).

Mode

The **mode** is another measure of the "typical value." It is the value (or values) that occurs most frequently. There may be more than one mode, and the mode(s) may be far from the center of the distribution of the values. For that reason, the mode is not often used as a measure of center for quantitative data. Note, however, that the mode is the one measure that we can compute for categorical data. (See Section 1.3.)

Example 2.4

Figure 2.3 displays a density histogram that shows the educational level (number of years of schooling completed) for people 25 years old and older in 1998. The histogram's spike at 12 years of schooling represents all those who quit school after high school. The mode is 12: More people have 12 years of schooling than any other number, as one might expect (why?).

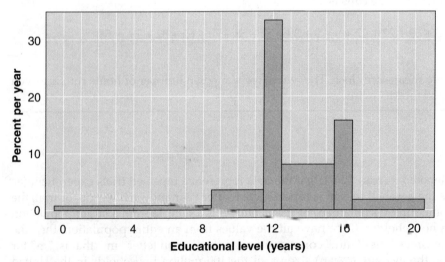

Figure 2.3 Distribution of people age 25 and older in the United States by educational level in 1998.

Section 2.1 Exercises

1. Below is a stem-and-leaf plot of the annual earnings of a sample of 50 recent MBAs (some working only part-time)
   a. Find the mode, the mean, and the median earnings. Compare the three values.
   b. Which measure (mode, mean, or median) is a better indicator of the typical earnings?

Stem	Leaf
1	2,6,7,8,9,9
2	2,3,5,7,7,7,7,8,9,9
3	1,2,3,4,5,6,6,7,7,7,9
4	0,3,3,3,4,5,6,6,7
5	3,3,7,7,8
6	0,2,6
7	4,7
8	7
9	
10	4
HI	156, 248

   Key: "1 6" stands for $16,000.

2. The histogram in Figure 2.3 has a second spike at 16 years of schooling.
   a. Which group of people does this spike represent?
   b. Is the percentage of people age 25 and over who are not high school graduates closest to (choose one) 17%, 24%, or 34%?
   c. Is the percentage of people age 25 and over with a bachelor's degree (earned by completing 4 years of college) or higher closest to (choose one) 17%, 24%, or 34%?

3. Here are the salaries of the Chicago Cubs for the 1996 season. The figures are in millions of dollars.

Player	Salary (millions $)	Player	Salary (millions $)
Ryne Sandberg	5.975	Derrick May	0.300
Mark Grace	4.400	Mark Parent	0.250
Randy Myers	3.583	Anthony Young	0.230
Jose Guzman	3.500	Rey Sanchez	0.225
Mike Morgan	3.375	Frank Castillo	0.225
Sammy Sosa	2.950	Willie Banks	0.190
Steve Buechele	2.550	Karl Rhodes	0.145
Shawon Dunston	2.375	Jim Bullinger	0.145
Dan Plesac	1.700	Steve Trachsel	0.112
Glenallen Hill	1.000	Eddie Zambrano	0.112
Willie Wilson	0.700	Jose Hernandez	0.112
Jose Bautista	0.695	Jessie Hollins	0.109
Rick Wilkins	0.350	Blaise Ilsley	0.109
Shawn Boskie	0.300		

   a. Construct a frequency histogram using the intervals $0 - 0.5^-, 0.5 - 1.0^-, 1.0 - 1.5^-, \ldots$
   b. Find the mean and the median salaries.
   c. Do you think the mean or the median is the "better" statistic to use to report the "typical" salary for the Cubs? What about the mode?

4. The number of campsites in each of five campgrounds in Yoho National Park is:

Campground	Number of campsites
Kicking Horse	92
Hoodoo Creek	106
Chancellor Park	64
Takakkaw Falls	35
Lake O'Hara	30

   a. What is the mean number of campsites per campground?
   b. What is the median number of campsites per campground?

5. The number of calories in a serving of various kinds of cheese is given by the table below:
   a. What is the mean number of calories per serving?
   b. What is the median number of calories per serving?
   c. Compare the mean and the median.

Kind of cheese	Number of calories
American	106
Cream	99
Feta	75
Monterey	106
Ricotta (whole milk)	216
Swiss	107

6. a. For the Key Problem find the mean and median ages of actors and actresses upon winning an Academy Award.
   b. What do you learn from these measures of central tendency?

7. Several years ago the number of millionaires per 1000 persons in the 10 states with the highest density of millionaires were reported as shown in the accompanying table below.
   a. Without calculating, guess which is larger: the mean or the median of the 10 numbers.
   b. Find the mean and the median.
   c. How many of these numbers are less than the mean?
   d. Does the mean or the median give a better idea of the center of these data?

State	Number of millionaires per 1000 persons
Idaho	27
Maine	8
North Dakota	7
Nebraska	7
Minnesota	6
Indiana	5
Wisconsin	4
Iowa	4
New Jersey	4
Connecticut	3

8. Here are the numbers of earthquakes that were recorded as 7 or greater on the Richter scale during the years 1900–1909 and 1980–1989:

Year	Number of earthquakes	Year	Number of earthquakes
1900	13	1980	18
1901	14	1981	14
1902	8	1982	10
1903	10	1983	15
1904	16	1984	8
1905	26	1985	15
1906	32	1986	6
1907	27	1987	11
1908	18	1988	8
1909	32	1989	7

   a. Make a back-to-back stem-and-leaf plot for the data.
   b. Compare the years 1900–1909 to the years 1980–1989 using either means or medians. Which decade tended to have more earthquakes?

9. Here is the frequency histogram for the durations, in minutes, of 230 eruptions of the geyser Old Faithful. The intervals are 0–0.5⁻, 0.5–1⁻, and so forth. The mean of these durations is 3.26 minutes. Is that a good measure of the typical duration?

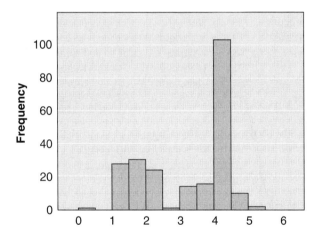

10. a. Use the data in Example 1.8 to estimate the mean U.S. household size in 1999. (Pretend that no household had over 9 people.)
    b. What was the median household size?
    c. What was the mode? *Hint:* Our best estimate (guess) of the mean size of all 100 million U.S. households in 1999 is the mean size of the households in the survey, which equals

$$\frac{\text{number of persons in the sample}}{\text{number of households in the sample}}.$$

   The number of households is 50,785. To find the total number of people in the sample, first count the number of people in all the sample households of size one (12,709). Then count the number of people in all the sample households of size two (2 × 16,517 = 33,034) and add that number to the first. Likewise, add the number of people in all the sample households of size three (3 × 8,435 = 25,305) to the total, and so forth.

2.2 MEAN VERSUS MEDIAN AS A MEASURE OF CENTER

The mode is of interest when we look at categorical data. For example, in Example 1.3 we learned that heart disease is the most common cause of death in the United States. The mode is also of interest when we look at "counting variables" such as "household size" (see Example 1.7) or "number of years of schooling completed" (see Example 2.4). But the mode is not really a measure of the center of a data set—even though the mode sometimes happens to be near the center.

The most common measures of the center of a distribution (of data) are the median and the mean. If the distribution is roughly symmetric and without outliers, then the mean and the median will be close together.

Example 2.5

The ages of American presidents when they died are as follows:

Washington	67	Fillmore	74	Roosevelt	60
Adams	90	Pierce	64	Taft	72
Jefferson	83	Buchanan	77	Wilson	67
Madison	85	Lincoln	56	Harding	57
Monroe	73	Johnson	66	Coolidge	60
Adams	80	Grant	63	Hoover	90
Jackson	78	Hayes	70	Roosevelt	63
Van Buren	79	Garfield	49	Truman	88
Harrison	68	Arthur	57	Eisenhower	78
Tyler	71	Cleveland	71	Kennedy	46
Polk	53	Harrison	67	Johnson	64
Taylor	65	McKinley	58	Nixon	81

We can summarize the data in the following stem-and-leaf plot (here we use such a plot, rather than an ordinary histogram, so we can still see the actual ages):

Stem	Leaf
4	6,9
5	3,6,7,7,8
6	0,0,3,3,4,4,5,6,7,7,7,8
7	0,1,1,2,3,4,7,8,8,9
8	0,1,3,5,8
9	0,0

The mean age at death of the 36 presidents is $2490/36 = 69.17$ years. The median age at death is the average of the 18th and the 19th youngest ages, that is $(67 + 68)/2 = 67.5$.

Since three presidents died at age 67 and no more than two died at any other age, the mode is 67. In this example, the mode happens to be a "central value." Note, however, that if another president dies at age 78, then there will be two modes because 67 and 78 will each occur three times. And if another president dies at age 90, then 90 will also be a mode.

Notice that it is easy to find the median from a stem-and-leaf plot because the data are arranged in increasing order. If a distribution is skewed or if there are outliers, then the mean and the median need not be close together. The reason is that *the mean is very sensitive to the "pull" of observa-*

tions far from the center. To illustrate this, consider the home run data for Roger Maris in Example 2.2. His total of 61 home runs in 1961 is an exceptional performance, an outlier. If we exclude this score as atypical, then the remaining 9 scores have a roughly symmetric distribution with median 23 and mean 22.2. If we include his record-setting score of 61 home runs, then the median increases a little bit from 23 to 24.5, but the mean jumps from 22.2 to 26.1. A single outlier increases Roger Maris's mean annual home run production by almost 4 home runs!

An observation whose presence greatly changes the value of a statistic (in this case, the mean) is said to be **influential.** Roger Maris's total of 61 homers in 1961 is an influential observation. Because the mean cannot resist the influence of **extreme** (very large or very small) **values,** we say that the mean is not a **resistant** measure of center. In a skewed distribution, the extreme values pull the mean away from the median toward the long tail.

The following examples show that income distributions are usually skewed to the right and therefore have higher mean incomes than median incomes.

Example 2.6

Here are the salaries of the Chicago Cubs for the 2000 season:

Player	Salary	Player	Salary
Sammy Sosa	$11,000,000	Matt Karchner	$776,000
Kevin Tapani	$ 6,000,000	Kerry Wood	$690,000
Ismael Valdes	$ 5,737,500	Felix Heredia	$625,000
Mark Grace	$ 5,300,000	Brian Williams	$600,000
Henry Rodriguez	$ 4,600,000	Willie Greene	$395,000
Eric Young	$ 4,500,000	Jeff Huson	$300,000
Jon Lieber	$ 3,833,333	Jeremi Gonzalez	$265,000
Rick Aguilera	$ 3,500,000	Andrew Lorraine	$245,000
Ricky Gutierrez	$ 2,500,000	Kyle Farnsworth	$235,000
Joe Giradi	$ 2,000,000	Jose Nieves	$217,500
Mark Guthrie	$ 1,600,000	Roosevelt Brown	$210,000
Glenallen Hill	$ 1,500,000	Cole Liniak	$205,000
Damon Buford	$ 1,100,000	Jose Molina	$205,000
Shane Andrews	$ 1,000,000	Tarrik Brock	$200,000
Jeff Reed	$ 1,000,000	Danny Young	$200,000

We can display the salaries in a frequency histogram (Figure 2.4) with class intervals 0-1⁻, 1-2⁻, etc. The median salary is $888,000. The mean salary is $2.0 million. The reason that the mean is so much higher than the median is clear: Most players have salaries in the low end, so only a few players have very high salaries. Thus the distribution is skewed to the right. Two-thirds of the Cubs each earn $1.6 million or less. Sammy Sosa alone earns about as much as these 20 players combined! Both the skewness and the outlying Sosa salary "pull" the mean to the right.

During the 1994–1995 Major League Baseball players' strike, players and owners used different statistics to represent the "typical" player's salary. In the public relations war, the players wanted to demonstrate that they were underpaid, so they used the median salary. The owners wanted to show that the players were well paid, so the owners used the higher number, the mean salary. Because even the "low" median salary was ten times higher than the typical family income in the United States, the players lost the public relations battle.

Besides public relations, the players and owners also had legitimate reasons for using different measures of the typical salary. The median salary is a good indicator of what the vast majority of players will earn during their first few years in Major League Baseball. For the owners who pay

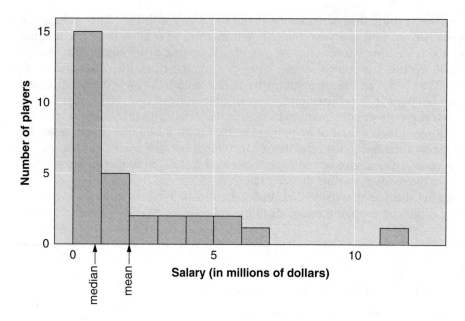

Figure 2.4 Distribution of salaries of the Chicago Cubs for the 2000 season.

the salaries, however, it is the mean salary that counts, not the median. For example, the total payroll for the Chicago Cubs is

$$\text{total payroll} = (\text{number of players}) \times (\text{mean salary})$$
$$= 30 \times 2.0 \text{ million}$$
$$= \$60 \text{ million}.$$

When we compute the total payroll, the median salary is irrelevant.

Example 2.7

Each year, the Current Population Survey estimates the previous year's distribution of household incomes. (See Example 1.8.) The income distribution for 1998 is displayed in a density histogram in Figure 2.5 (noting that the interval 0–10 has been split in two to better display the shape of the data, and that the interval 100–160 is split into 100–125⁻ and 125–160 because of limited data in the high income range).

The income distribution is clearly skewed to the right. In 1998, the median household income was around $39,000. The mean household income was around $52,000.

The distribution of house prices in a community is also usually skewed to the right. There will be many moderately priced houses and a smaller number of expensive houses. The few expensive houses pull the mean up, but have little effect on the median. So the mean house price will be higher than the median house price.

Which number is more relevant, the median price or the mean price? That depends on your point of view. A young couple moving into a community will probably be more interested in the

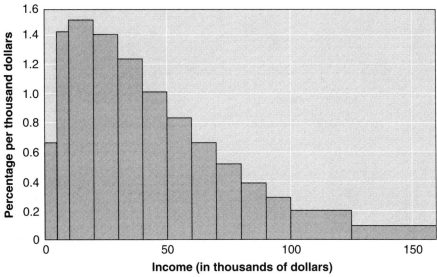

Figure 2.5 Distribution of households in the United States by income in 1998.

median home price. The median tells them how much it will typically cost to buy a home. But if you are a member of the school board, then you will be more interested in the mean home price. The amount of property taxes that can be raised to support local schools is determined by the total market value of all the homes, which can be computed from the mean:

$$\text{Total market value of all homes} = (\text{Number of homes}) \times (\text{Mean home price}).$$

In symbols,

$$\sum x = n \times \bar{x}.$$

This property of recovering the total from the mean is often useful.

Figure 2.6 shows the relative positions of the median and the mean for right-skewed, symmetric, and left-skewed distributions. Notice that in a skewed distribution, the extreme values in the tail pull the mean away from the median toward the long tail. Note that the mean and the median are the same for a symmetric distribution *regardless* of whether it is high in the middle, flat, or low in the middle.

Outliers

Roger Maris's record-setting 61 home runs in 1961 (see Example 2.2) raised the question: What do we do with outliers? Do we include them, or do we leave them out because they are atypical and distort the mean as an indicator of the typical value? In Roger Maris's case we have no excuse for leaving out his exceptional achievement in 1961. It did not come completely out of the blue: The year before, he had scored 39 home runs, his second best score overall and the following year he scored 33 home runs, his third best score overall. Besides, the 61 home runs is certainly the most interesting number in the data set! The line graph in Figure 2.7 illustrates that over just a couple of seasons, Maris peaked and then faded as a home run hitter.

The best way to characterize Roger Maris as a home run hitter is to note his record-setting 61 home runs in 1961, but to use the median as the indicator of his typical annual home run production. What is unusual about him is that he peaked and then faded so rapidly, like a supernova in the sky.

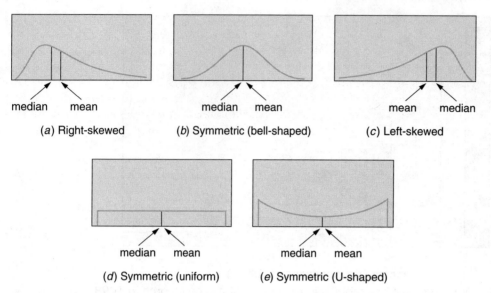

Figure 2.6 Relative positions of the median and the mean for right-skewed, symmetric, and left-skewed distributions.

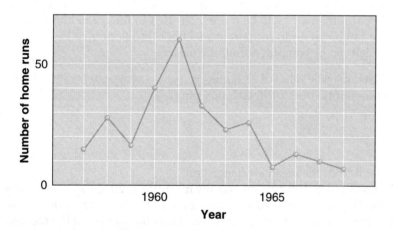

Figure 2.7 Roger Maris's annual home run production from 1957 to 1968.

The next example looks at measurement errors. As anyone who has stepped on a bathroom scale knows, repeated measurements of the same thing differ slightly from each other. One of the first scientists to deal with this problem was the Danish astronomer Tycho Brahé. His careful observations were used by Johannes Kepler to derive his three fundamental planetary laws that later enabled Isaac Newton to formulate his theory of gravitational force.

Example 2.8

It has been known since the late seventeenth century (the 1600s) that light travels at a finite speed and is not transmitted instantaneously. It takes about eight minutes for light from the sun to reach Earth and about one second for light to travel from the moon to Earth. Around 1880, Simon Newcomb and A. A. Michelson conducted experiments that gave the first accurate determination of the speed of light.

Newcomb's experiment consisted of bouncing a light beam off a rapidly rotating mirror at a right angle to a distant fixed mirror, and back to the rotating mirror (see Figure 2.8). Because the rotating mirror would have turned a little bit by the time the light beam is bounced back to the rotating mirror, the returning light would be reflected in a slightly different direction than its original source. Newcomb measured the angle between the returning light beam and the source. This angle is proportional to the passage time (the time the light beam took to travel from the rotating mirror to the fixed mirror and back again). That is, if the light takes three times as long, the angle is three times as large. This relation allows measurement of the passage time. The speed of light can then be computed.

$$\text{Speed of light} = (\text{Distance traveled})/(\text{Passage time})$$

Newcomb made 66 determinations of the time a light beam takes to travel to a fixed mirror 3721 meters away and back. His passage times in nanoseconds (billionths of seconds) are 24,800 plus the values in Table 2.2. Thus, the number 28 in Table 2.2 stands for 24,828 nanoseconds, and the number –44 stands for 24,756 nanoseconds.

Why didn't Newcomb get the same number each time? Part of the reason is that it is impossible to measure the rotational speed of the mirror or the angular displacement of the returning light beam with complete accuracy. Furthermore, the speed of light traveling through air varies a tiny bit with the atmospheric conditions. Newcomb's data are displayed in the histogram of Figure 2.9 where the interval widths are: –45 to –42, –42 to –39, and so forth. Newcomb's data seem to follow a bell-shaped distribution except for two outliers: –2 and –44. He was unable to identify any equipment failure that might have caused these two outliers.

So how do we estimate "the true passage time"? The mean of the 66 observations is 26.21. Newcomb discarded the most extreme outlier (–44) and computed the mean of the remaining 65 observations. He got 27.29. Another possibility is to use the median of all 66 observations, which equals 27.

Influential extreme values pose a problem in statistics. The problem is deciding what to do with them. Some outliers are simply mistakes. For example, the infant mortality rate of Monaco in Table 1.5 (copied from the *New York Times Almanac 2000*) is almost surely an error. Such outliers should be discarded. Other outliers, like Roger Maris's 61 home runs (see Example 2.2 and Figure 2.7),

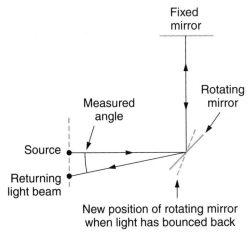

Figure 2.8 Schematic of Newcomb's experiment.

Table 2.2 Newcomb's measurements of the passage time of light

28	22	36	26	28	28
26	24	32	30	27	24
33	21	36	32	31	25
24	25	28	36	27	32
34	30	25	26	26	25
–44	23	21	30	33	29
27	29	28	22	26	27
16	31	29	36	32	28
40	19	37	23	32	29
–2	24	25	27	24	16
29	20	28	27	39	23

Source: Stigler, S. M. (1977). Do robust estimators work with real data? *Ann. Stat.* 5, No. 6, 1055–1098.

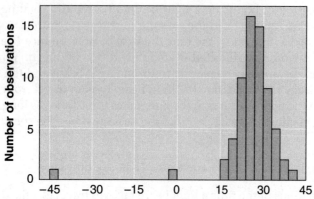
Figure 2.9 Newcomb's passage times (in nanoseconds) minus 24,800.

represent unique events of interest. Such outliers should not be discarded but should be studied separately. Finally, there are unexplained outliers (which could be either error or important to consider) like Simon Newcomb's two lowest passage times in Example 2.8. We might be tempted to think that Newcomb's outliers were caused by carelessness. It turns out, however, that outliers are an unavoidable part of measurement work. Even at the National Institute of Standards and Technology (NIST), where things are measured as carefully as possible, extreme values still occur. NIST discards outliers only "for cause, such as door-slam or equipment malfunction." That leaves a small number of outliers that cannot be explained away and therefore cannot be discarded.

Section 2.2 Exercises

Technically, "average" is another word for the mean of a set of data, but in Exercises 1 and 2 the word "average" is used in its everyday nontechnical way and may refer to any measure of center, and your answer can be either "mean" or "median."

1. Which measure of center (mean, median, or mode) would be the most appropriate measure of center to use in each of the following cases? Explain your answer. In some cases, both the mean and the median can be appropriate, depending on the purpose, or because they are about the same value.
   a. The average number of students in statistics classes at the College of Lake County, Illinois, is 27.4.
   b. The average home value in Lake County, Illinois, is $165,000.
   c. The most common shoe size sold in Marshall Field's department store is 8.
   d. The average yearly wage in a company is $30,000.
   e. The average age of new faculty members at the University of Illinois is 28.3.
   f. The average weight of adult men is 169 pounds.
2. Describe which measure (mean, median, or mode) was used in each situation (or state that it could have been either of two measures).
   a. Half of the workers in the company make more than $30,000, and half make less.
   b. The "average" number of children per family in Champaign, Illinois, is 1.9. *Hint:* The only possible values are integers. If so, how could 1.9 occur?
   c. More people prefer red cars than any other color.
   d. The "average" age of a college student is 20.3 years.
   e. More people vote Republican in Lake County, Illinois, than Democratic or Independent.

3. Three students, Bruce, Ross, and Carmen, are taking a statistics course. Their scores on five exams are given in the following table.

	Exam I	Exam II	Exam III	Exam IV	Exam V
Bruce	84	88	68	68	95
Ross	81	79	81	79	81
Carmen	79	80	78	90	80

Each student wants to claim the best performance in the class. Compute the mean, median, and mode for each student. Which measure would each use to compare the three students to substantiate each student's claim of being the best? Which student's claim is clearly without any validity?

4. The mean salary for 920 major league baseball players in 1999 was $1,567,873. What was the total salary for major league baseball players?

5. A physician's office reported that the mean cost of a physical exam was $150. If 90 patients were examined last month, how much money did the physician collect on these exams?

6. According to *Monitoring the Future*, a study published by the University of Michigan Institute for Social Research and National Institute on Drug Abuse, the following data represent the percentages of high school seniors in the United States who had ever used marijuana or hashish.

1975	1980	1990	1991	1992	1993	1994	1995	1996	1997	1998
47.3	60.3	40.7	36.7	32.6	35.3	38.2	41.7	44.9	49.6	49.1

a. Compute the mean and the median percentage of high school seniors using marijuana or hashish.
b. Which is larger and why?

7. The following represent actual values from a sample of white blood cell counts (thousands of cells/microliter) for 25 different patients in a northern Illinois hospital.

4.1	19.5	7.0	7.0	1.5	13.9	4.5	4.2	10.2	6.0	12.9	25.3	6.2
5.6	10.0	6.7	54.1	12.9	3.2	5.5	11.0	9.8	9.9	72.6	50.3	

a. Construct a histogram using the intervals 0 to 5.0⁻, 5.0 to 10.0⁻, 10.0 to 15.0⁻,...
b. Compute the mean and median for these counts. Is the distribution symmetric, skewed to the left, or skewed to the right? Is the location of the mean relative to the median as described in Figure 2.6?
c. Compute the mean and median if the values of 19.5, 25.3, 54.1, 72.6, and 50.3 are removed. How do the values of the mean and median compare with part **b**?

8. Suppose a doctor measures a patient's diastolic blood pressure on 5 different occasions and gets the following results:

$$85 \quad 79 \quad 93 \quad 82 \quad 86$$

a. Compute the mean and median.
b. Suppose the doctor inadvertently recorded 97 instead of 79. What would the mean and median be now?
c. If the doctor was having a really bad day, the blood pressure might have been recorded as 790 instead of 79. In that case, what would the values be for the mean and median, assuming

that the calculation is automatically done by computer and that nobody has noticed this absurd error?

d. What can you say about the influence of extreme values on the values of the mean and median?

9. According to *Consumer Reports*, April 1995, the following represents the manufacturer's ratings of electrical power used (in amps) for each of 22 upright vacuum cleaners.

7.8	7	12	10	12	7	7.2	7.3	11	7	12
11	9.5	9	9	10	7.2	9.5	11	10	7.5	10

a. Construct a stem-and-leaf plot for the data.
b. Compute the mean and median. Are the two values close? Explain your answer in terms of the amount of skewness that seems to exist in the data.
c. Suppose the four values of 10 had been inadvertently recorded as 19 instead of 10. Now what would the mean and median be? Which measure would be most affected by the change?

10. A study in Switzerland examined the number of cesarean sections in deliveries of babies performed in a year by 15 doctors.

27	50	33	25	86	25	85	31
37	44	20	36	59	34	28	

a. Draw a stem-and-leaf plot of the data. Do you see any potential outliers?
b. Compute the mean and median.
c. Remove the two largest values from the data set and compute the mean and median again. What can you say about the resistance of each measure to extreme values?

11. The distribution of individual incomes in the United States is skewed toward the low end. In 1997 the mean and median values of the top 1% of all incomes in the United States were $330,000 and $675,000. Which of the values represents the mean and which represents the median?

State	Number of millionaires per 1000 persons
Idaho	27
Maine	8
North Dakota	7
Nebraska	7
Minnesota	6
Indiana	5
Wisconsin	4
Iowa	4
New Jersey	4
Connecticut	3

12. The states having the highest densities of millionaires were reported in the table above.
    a. Without doing any calculations, can you decide which is larger: the mean or the median?
    b. Calculate the mean and median and see whether your conclusion was correct.

13. Below is sketched the age distribution for the people who died of heart disease in 1996. Which is greater, the mean or the median age?

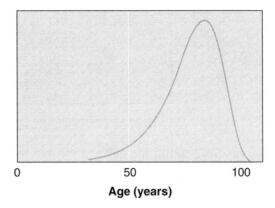

Age (years)

14. Use Newcomb's data in Example 2.8 to estimate the speed of light. *Hint:* There is more than one reasonable way to proceed (mean versus median, one versus two outliers removed).

2.3 MEASURING SPREAD: THE STANDARD DEVIATION

One mystery that has fascinated baseball fans for half a century is the disappearance of the so-called 0.400 hitter. A baseball player's batting average for a season is his total number of hits divided by his number of times at bat. Between 1901 and 1930, seven players had seasonal batting averages exceeding 0.400. In 1941 Ted Williams hit 0.406. Since 1941, though, no player has managed to hit above 0.400. Why is that? Are there no great hitters anymore? Williams himself concluded that today's players have plenty of power but lack finesse. In other words, today's players are not smart enough to hit above 0.400.

Harvard's Stephen Jay Gould has a different explanation for the disappearance of the 0.400 hitter. In the book *Full House* he looks at the distribution of batting averages for all Major League Baseball players in a given year.

Figure 2.10 displays the distribution of the batting averages of 263 Major League Baseball players in 1985. The distribution is roughly bell-shaped and is without outliers. A bell-shaped distribution can be described in terms of two numbers (two statistics): its **center**, indicated by the **mean** $\bar{x}$,

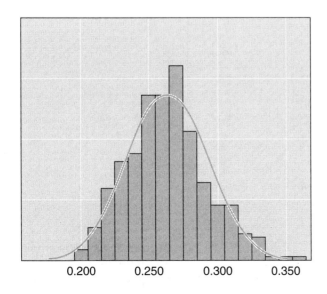

Figure 2.10 Batting averages of 263 Major League Baseball players in 1985.

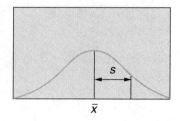

Figure 2.11 Two bell-shaped distributions showing the mean $\bar{x}$ and the standard deviation s.

and the **spread** of the data around the mean, measured by the **standard deviation** s. Figure 2.11 shows that the more spread out a bell-shaped data set is, the larger its standard deviation is.

If the histogram of a data set follows a bell curve without outliers, then a majority of the observations will be at most one standard deviation away from the mean. Very few observations will be more than two standard deviations away from the mean. The **empirical rule for bell-shaped distributions** (or 68-95-99.7% rule) makes this precise.

Empirical Rule for Bell-Shaped Distributions (or 68-95-99.7% Rule)

1. About 68% of the data will lie within 1 standard deviation (in either direction) of the mean.
2. About 95% of the data will lie within 2 standard deviations of the mean.
3. About 99.7% of the data (for practical purposes, all of the data!) will lie within 3 standard deviations of the mean.

Figure 2.12 illustrates the empirical rule. For example, for the batting averages in Figure 2.10, $\bar{x} = 0.263$ and $s = 0.029$, so

$$\bar{x} - s = 0.263 - 0.029 = 0.234$$
$$\bar{x} + s = 0.263 + 0.029 = 0.292.$$

It turns out that 178 of the 263 players had batting averages between 0.234 and 0.292, which is 67.7%—very close to the 68% the empirical rule predicts.

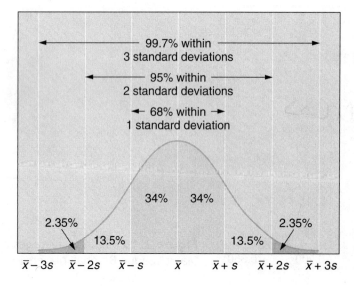

Figure 2.12 The empirical rule (or 68-95-99.7% rule) for bell-shaped distributions without outliers.

$$\bar{x} - 2s = 0.263 - 0.058 = 0.205$$
$$\bar{x} + 2s = 0.263 + 0.058 = 0.321$$

Of the 263 players, 251 had batting averages between 0.205 and 0.321, which is 95.4%—very close to the 95% the empirical rule predicts.

Many sportswriters have speculated that the reason for the disappearance of the 0.400 hitter in baseball is that tougher competition and more distractions have tilted the balance between hitting and pitching against the hitters. It is true that pitching has improved dramatically over the last 50 years. But Gould argues that the balance has not tilted. If it had, then the mean batting average should have declined, and that has not happened. Gould notes that the mean batting average has been fairly steady at 0.260 for over 100 years. So what is going on? Gould argues that as baseball has come of age, **both** pitching and hitting have improved to near the optimal level, maintaining a balance between hitting and pitching.

As overall play has improved, the difference between the highest and the lowest seasonal batting average has shrunk: The best hitters today face much better trained pitchers than their counterparts did 75 years ago. At the same time, the worst hitters today are much better than their counterparts 75 years ago. The result is that the bell curve for batting averages in a year has shrunk around its mean: As an examination of batting average data over the years shows, the standard deviation has decreased from around 0.05 in the 1870s to around 0.03 today, a 40% drop. Figure 2.13 compares the distribution of batting averages today (solid curve) with the distribution of batting averages at the beginning of the twentieth century (dashed curve).

We can now explain the disappearance of the 0.400 hitter in terms of the 68-95-99.7% rule: A hundred years ago a hitter had to score 3 standard deviations above the mean to hit above 0.400. According to the 68-95-99.7% rule, roughly 100%–99.7% = 0.3%, or 3 out of 1000 batting averages, will be more than 3 standard deviations away from the mean. Half of these, or roughly 1 out of 700 batting averages, will be more than 3 standard deviations above the mean. So at the beginning of the twentieth century it was not surprising that a couple of times per decade, some player hit over 0.400. Today, however, a hitter has to score more than 4 standard deviations above the mean to hit above 0.400, and that is extremely rare. This example illustrates that in

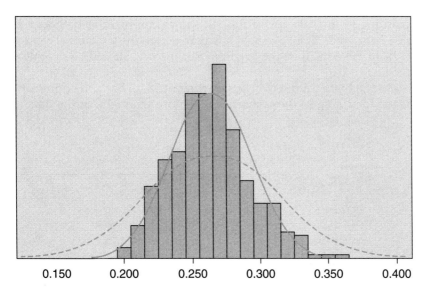

Figure 2.13 Distribution of batting averages in 1985 (solid curve) and batting averages at the beginning of the 20th century (dashed curve). For the past 50 years there has been less variation (spread) than there was 100 years ago.

order to understand a bell-shaped distribution, we need to know both the center and the variation (spread) of the data.

Variation is also the issue in the next example. Many banks used to have their customers wait in separate lines at each teller's window. It frustrated the customers to have to guess which line would move faster. (And the fastest lines by chance often tend to move much faster than the slow lines!) Therefore the banks switched to a single waiting line for all the tellers. The change neither sped up nor slowed down the tellers, so the mean waiting time for the customers did not change. Because there was no longer a fast line or a slow line, however, everyone waited in line for about the same amount of time. In other words, the variation in waiting time was reduced, and the customers were spared a lot of frustration. The new standard deviation for the customers' waiting time will be a lot less, but the new mean waiting time will be about the same.

Consider next the distribution of percent total return for all common stock on the New York Stock Exchange over a recent year (see Figure 2.14). The total return on a stock over a year is the change in its market price (up or down) plus any dividend payments made. It is expressed as a percentage of the beginning price. A return less than zero means a loss. The variation in the returns of all stocks over a year reflects the "firm-specific risk" or "unique risk" of holding one individual stock. It can be reduced by diversification, that is, by splitting one's investment into a portfolio of 20 to 30 stocks. The risk that remains even after extensive diversification is called "market risk." It basically represents the potential variation of the stock market as a whole from one year to the next and is also called "systematic risk." It comes from major economic factors such as the business cycle (expansion, depression, recession, etc.), the inflation rate, and interest rates.

Figure 2.15 illustrates that fully diversified portfolios (dashed curve) have about half the risk (variation) of one-stock portfolios (solid curve). Can you imagine why some people would prefer a one-stock investment strategy while most would prefer a portfolio strategy?

For data sets with roughly bell-shaped histograms and no outliers we can use the 68-95-99.7% rule to estimate $\bar{x}$ and s from the histogram. A usual inspection of the histogram won't tell us the exact values of $\bar{x}$ and s, but we can guess approximate values. To illustrate how we do this, consider the histogram in Figure 2.16, which represents the heights of a couple of thousand college women.

Since $\bar{x}$ is near the center we estimate that $\bar{x} \approx 64.5$ inches. We will use the 95% rule to guess an approximate value of s. According to the rule, as Figure 2.12 shows, approximately 95% of the data lie in the interval from $\bar{x} - 2s$ to $\bar{x} + 2s$. It follows that if we can determine an interval centered at $\bar{x}$ that contains roughly 95% of the data, then the length of this interval should be roughly $4s$. We shall apply this method to Figure 2.16.

Clearly the combined area of the histogram rectangles starting at 58 inches and ending at 71 inches is more than 95% of the total area of the histogram. So more than 95% of the data are

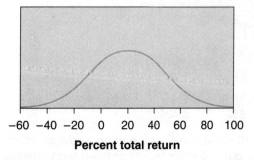

Figure 2.14 Distribution of percent total return.

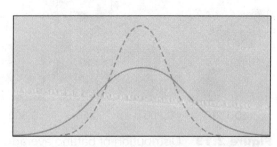

Figure 2.15 Distribution of percent returns on one-stock portfolios (solid curve) and diversified portfolios (dashed curve).

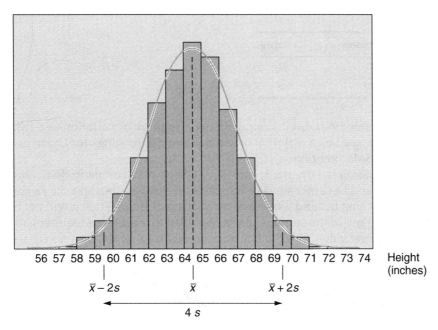

Figure 2.16 Estimating $\bar{x}$ and s from the histogram.

between 58 inches and 71 inches. Therefore the computation

$$4s = 71 - 58 = 13,$$

which yields $s = 3.25$, gives us too large a value for s.

On the other hand, the combined area of the histogram rectangles starting at 61 inches and ending at 68 inches is less than 95% of the total area of the histogram: The combined area of the rectangles below 61 inches and those above 68 inches is greater than 5%. (It looks closer to 10%.) So fewer than 95% of the data are between 61 inches and 68 inches. Therefore the computation

$$4s = 68 - 61 = 7,$$

which yields $s = 1.75$, gives us too small a value for s.

We can get a rough estimate of s by "splitting the difference" between 1.75 inches and 3.25 inches. We get

$$s \approx 2.5 \text{ inches.}$$

This is only a guess. All we can say with confidence is that s is between 2 and 3 inches.

We shall now consider some numerical examples.

Example 2.9

Two patients, Dylan and Laverne, are hospitalized in Mercy Community Hospital. In order to monitor their conditions, the doctor has the nurse monitor their pulse rates three times a day. The results of one day's readings are shown in Table 2.3.

If we order the data, we see that the median pulse rate of each patient is 72. In fact, if we compute the mean for each, we also see that each mean is also 72. Obviously there is more that needs to be described than the center.

The basic difference in the two patients is in the variation of the two data sets. Laverne's readings are clustered together much more closely than are Dylan's.

Table 2.3 Pulse Readings

	Morning	Afternoon	Evening
Dylan	72	58	86
Laverne	70	74	72

Why should we be concerned about variation? We can sometimes think of variation as a measure of consistency. In Example 2.9, the large variation or inconsistency in readings for Dylan may indicate a medical problem that needs attention.

As another example where variation is important, consider an aerospace company that manufactures a part for the space shuttle. The dimensions of the parts must fall within specific ranges, or **tolerances,** to ensure that everything fits and works properly. If a part is too large, it will not fit; if it is too small, it will not work. Many of the supplied parts could fail to meet these requirements, even if they average out to their target value. Their actual measurements must also be very close to this mean value (thus making the mean value truly typical of almost all of the values).

How do we obtain a measure like s that tells us the spread? The simplest measure is the **range,** which is the difference between the highest value and the lowest value.

$$\text{Range} = (\text{highest value}) - (\text{lowest value})$$

The range is often used in weather reports, which give us the low and high temperatures of the day.

For the two data sets in Example 2.9, the ranges are easily computed.

$$\text{Dylan:} \quad \text{Range} = 86 - 58 = 28$$
$$\text{Laverne:} \quad \text{Range} = 74 - 70 = 4$$

Although the range is easy to obtain, it is generally not used outside weather reporting. The reason is that since the range is computed from the two most extreme values, the range often exaggerates the spread of large data sets. For example, the vast majority of the women in Figure 2.16 were between 60 inches and 69 inches tall, a spread of 9 inches. But the tallest woman was 73 inches and the shortest 56 inches. So the range was 17 inches. The range gives us the misleading impression that college women vary more in height than they typically do. We are (usually) interested in measuring the **typical variation**, rather than the *most extreme variation*. The range does not tell us the typical variation. The range is very sensitive to extreme values, because it is computed by using *only* the most extreme values. Thus, any extreme, and hence, influential, value that is in error could totally change our measure of the size of the spread. We will not use the range as a measure of spread.

Another approach that we could take, which leads us to a formula for s, is to see how far away each data value is from the mean. Let us try this approach with the data from Example 2.9. The mean, $\bar{x}$, was computed to be 72 in each case. Then we find $x - \bar{x}$, called the **deviation from the mean** for each data point x, shown in Table 2.4.

Table 2.4 Deviations in Pulse Readings

Dylan		Laverne	
x	$x - \bar{x}$	x	$x - \bar{x}$
72	0	70	-2
58	-14	74	2
86	14	72	0
Sum	0	Sum	0

We could attempt to get a measure of spread by adding up the deviations. But in each case the sum of the deviations is 0. In fact, it can be shown that this will always be the case. Therefore we do not obtain a good measure of spread by adding or averaging the deviations—they always add to 0. The sum of the positive deviations offsets the sum of the negative deviations.

To overcome this problem, we square each deviation. The **variance** of the data is the average of the squared deviations and the **standard deviation** is the square root of the variance. The precise formulas we use to compute the variance and the standard deviation vary a little bit, depending on whether the data we are looking at is from the complete population, from a sample of a population, or from a repeated-trials "experiment." For simplicity we will call data from either a sample from a population or from repeated trials a sample.

Population Variance and Population Standard Deviation

Imagine that we are looking at complete data from a population of size N. For example, consider the age distribution of the population of the United States. (See Exercise 1.5.3) Let N denote the total number of Americans. (N is around 280 million.) The average age of all Americans is

$$\mu = \frac{\sum_1^N x}{N}$$

and the variance of the complete population age distribution is

$$\sigma^2 = \frac{\sum_1^N (x - \mu)^2}{N}.$$

The symbol σ is the lowercase Greek letter "sigma." The **population standard deviation** is

$$\sigma = \sqrt{\sigma^2} = \sqrt{\frac{\sum_1^N (x - \mu)^2}{N}}.$$

Here are the steps for finding the **population** variance and standard deviation:

In Words	In Symbols
Step 1. Find the mean of the population data set.	$\mu = \frac{\sum_1^N x}{N}$
Step 2. Find the deviation of each entry.	$x - \mu$
Step 3. Square each deviation.	$(x - \mu)^2$
Step 4. Add over all population members to get the **sum of squares.**	$\sum_1^N (x - \mu)^2$
Step 5. Divide by N to get the **population variance.**	$\sigma^2 = \frac{\sum_1^N (x - \mu)^2}{N}$
Step 6. Find the square root of the variance to get the **population standard deviation.**	$\sigma = \sqrt{\frac{\sum_1^N (x - \mu)^2}{N}}$

Sample Variance and Sample Standard Deviation

It is unusual to have complete population data. Consider now the more common situation where we only have data from a sample of size n from a much larger population of size N or from n repeated trials like n die throws. To estimate the value of the **population mean** μ we use the **sample mean**

$$\bar{x} = \frac{\sum_1^n x}{n}$$

and to estimate the **population variance** σ^2 we use the **sample variance**

$$s^2 = \frac{\sum_1^n (x - \bar{x})^2}{n - 1}.$$

Why do we divide by $n - 1$ and not n? The answer is that, for technical reasons, most statisticians think that we get a better estimate of σ^2 using s^2 defined by dividing by $n - 1$. We note that if the sample size n is at least 30, then it makes little difference whether we divide by n or by $n - 1$ in defining s^2. If we have to calculate s^2 using a calculator, then the following "computational" formula is more convenient:

$$s^2 = \frac{n\sum_1^n x^2 - (\sum_1^n x)^2}{n(n - 1)}.$$

Here are the steps for finding the **sample variance** and standard deviation:

In Words	In Symbols
Step 1. Find the mean of the sample data set.	$\bar{x} = \frac{\sum_1^n x}{n}$
Step 2. Find the deviation of each entry.	$x - \bar{x}$
Step 3. Square each deviation.	$(x - \bar{x})^2$
Step 4. Add over all sample members to get the **sum of squares**.	$\sum_1^n (x - \bar{x})^2$
Step 5. Divide by $n - 1$ to get the **sample variance**.	$s^2 = \frac{\sum_1^n (x - \bar{x})^2}{n - 1}$
Step 6. Find the square root of the variance to get the **sample standard deviation**.	$s = \sqrt{\frac{\sum_1^n (x - \bar{x})^2}{n - 1}}$

The formulas for sample variance and sample standard deviation are not only used for samples from a population but also for repeated measurement data, such as Newcomb's data in Example 2.8 and the pulse measurements in Example 2.9.

Let us return to the Dylan and Laverne data in Example 2.9. The squared deviations $(x - \bar{x})^2$ and their sums $\sum(x - \bar{x})^2$ are shown in Table 2.5. The variances and standard deviations for these repeated trials are, noting that $n - 1 = 3 - 1 = 2$,

Dylan: $s^2 = \frac{392}{2} = 196,$ Laverne: $s^2 = \frac{8}{2} = 4$

$s = \sqrt{196} = 14,$ $s = \sqrt{4} = 2.$

Section 2.3 Measuring Spread: The Standard Deviation

Table 2.5 Squared Deviations in Pulse Readings

Dylan			Laverne		
x	$x - \bar{x}$	$(x - \bar{x})^2$	x	$x - \bar{x}$	$(x - \bar{x})^2$
72	0	0	70	−2	4
58	−14	196	74	2	4
86	14	196	72	0	0
	Sum	392		Sum	8

The standard deviation is not a resistant measure of spread. It is easily distorted by outliers, as the next example illustrates.

Example 2.10

Consider the sample of Newcomb's 66 repeated measurements (trials) in Example 2.8. If we exclude the two outliers −44 and −2, then the remaining 64 observations have a roughly bell-shaped distribution. (See Figure 2.9.) We will use the computational formula for s^2 to save time. For these 64 observations we get

$$\sum x^2 = 50{,}912, \qquad \sum x = 1776, \qquad \bar{x} = \frac{1776}{64} = 27.75,$$

$$s^2 = \frac{64 \cdot 50{,}912 - (1776)^2}{64 \times 63} = 25.84, \qquad s = 5.08.$$

The 64 observations obey the 68-95-99.7% rule reasonably well: In the interval $\bar{x} \pm s$ we find 46 out of 64 observations, or 72%—which is not too far from 68%. In the interval $\bar{x} \pm 2s$ we find 60 out of 64 observations, or 94%—which is close to 95%.

If we include the two outliers, then we get

$$\sum x^2 = 52{,}852, \qquad \sum x = 1730, \qquad \bar{x} = 26.21,$$

$$s^2 = \frac{66 \cdot 52{,}852 - (1730)^2}{66 \times 65} = 115.46, \qquad s = 10.7.$$

Note that the inclusion of the two outliers doubled the standard deviation. Indeed, the standard deviation is even more vulnerable to the influence of outliers than the mean! Also note that if we include the two outliers, then the data set does not obey the 68-95-99.7% rule: In the interval 26.21 ± 10.7, we find 61 out of 66 observations, or 92%—which is far from 68%.

Linear Transformations and Change of Scale

Strictly speaking, in Examples 2.8 and 2.10 we found the mean, the median, and the standard deviation for the values in Table 2.2—not for Newcomb's passage times. We can easily determine those now. Because

passage time = 24,800 + (value in Table 2.2) (in nanoseconds)

we get

mean passage time = 24,800 + (mean value for Table 2.2)

median passage time = 24,800 + (median value for Table 2.2)

standard deviation of passage times = standard deviation of values in Table 2.2.

Adding 24,800 to each value in Table 2.2 adds 24,800 to the mean and the median, but leaves the standard deviation unchanged. Adding 24,800 to each value shifts the center of the distribution (mean or median) by that amount, but has no effect on the spread.

What would happen to the mean, the median, and the standard deviation if the passage times were recorded in seconds instead of nanoseconds? Because

$$\text{passage time in seconds} = (\text{passage time in nanoseconds}) \times 10^{-9}$$

we get the new mean, median, and standard deviation simply by multiplying the old ones by 10^{-9}.

To summarize,

1. Adding or subtracting the same number from each value in a data set shifts the center (mean or median) by the same amount while the spread (standard deviation) remains the same.
2. Multiplying each value by a positive constant C also multiplies the mean, median, and standard deviation by the same factor C.

Example 2.11

In 1987 the daily maximum temperature in Mayville, North Dakota, was recorded each day. For the year, the mean of the daily maximum temperatures was 58.21 °F. The median was 60 °F, and the standard deviation was 24.39 °F. The formula for converting temperatures from Fahrenheit to Celsius is

$$°C = \frac{5}{9}(°F - 32).$$

Imagine that each of the 365 daily maximum temperatures is converted from Fahrenheit to Celsius. For the new data set we get

$$\text{new mean} = \frac{5}{9}(\text{old mean} - 32) \ °C$$

$$= \frac{5}{9}(58.21 - 32) \ °C = 14.56 \ °C$$

$$\text{new median} = \frac{5}{9}(\text{old median} - 32) \ °C$$

$$= \frac{5}{9}(60 - 32) \ °C = 15.56 \ °C$$

$$\text{new standard deviation} = \frac{5}{9}(\text{old standard deviation})$$

$$= \frac{5}{9} \times 24.39 \ °C = 13.55 \ °C.$$

Section 2.3 Exercises

Note: Use of a calculator is needed for many of the exercises.

1. The National Association of Colleges and Employees has reported the following average salary offers for new graduates with bachelor degrees for 1997 and 1998 (rounded to the nearest $1000).

Section 2.3 Measuring Spread: The Standard Deviation

Discipline	1997	1998
Accounting	30,000	32,000
Advertising	25,000	23,000
Biology	26,000	24,000
Business administration	29,000	29,000
Chemistry	34,000	31,000
Communications	26,000	25,000
Computer science	37,000	40,000
Computer programming	35,000	40,000
Criminal justice	25,000	26,000
Economics	31,000	33,000
Bioengineering	37,000	38,000
Chemical engineering	43,000	45,000
Computer engineering	40,000	41,000
Electrical engineering	40,000	41,000
Mechanical engineering	38,000	40,000
English	24,000	35,000
Health technology	25,000	19,000
History	25,000	24,000
Human resources	27,000	26,000
Journalism	24,000	23,000
Mathematics	32,000	36,000
Nursing	31,000	37,000
Psychology	23,000	26,000
Social work	21,000	25,000
Sociology	24,000	27,000

    a. Compute the mean, median, and standard deviation for 1997.
    b. Compute the mean, median, and standard deviation for 1998.
    Do there appear to be any significant changes?

2. A math instructor gave an exam. A sample of 10 exam grades is given here.

    62 54 76 69 43 77 61 73 81 56

    a. Compute the mean, median, and standard deviation.
    b. Because the test was hard, the instructor decided to adjust the grades by adding 10 points to each score. Without recomputing, what will be the values of the mean, median, and standard deviation for the new scores?

3. An instructor gave an exam with a possible perfect score of 33, with the following results:

    25 30 22 14 17 29 33 11 20 22

    a. Compute the mean, median, and standard deviation.
    b. The instructor rescales the scores by tripling each of the scores. Without recomputing, what are the values of the mean, median, and standard deviation now?

4. The United States Environmental Protection Agency has reported the number of days metropolitan areas have failed to meet adequate air quality standards.
    Compute the mean, median, and standard deviation for each year. What do the results suggest?

SUMMARIZING DATA BY NUMERICAL MEASURES: CENTERS AND SPREADS

Area	1988	1997	Area	1988	1997
Atlanta	44	26	New Haven	26	19
Bakersfield	126	55	New York	57	23
Baltimore	60	30	Orange County, CA	56	3
Boston	28	30	Philadelphia	53	32
Chicago	40	9	Phoenix	29	15
Dallas	37	15	Pittsburgh	43	20
Denver	35	0	Riverside, CA	185	106
Detroit	35	12	Sacramento	88	2
El Paso	15	3	St. Louis	44	15
Fresno	110	50	Salt Lake City	16	1
Hartford	39	16	San Diego	123	14
Houston	72	47	San Francisco	1	0
Las Vegas	30	0	Seattle	20	1
Los Angeles	239	63	Ventura	108	44
Miami	8	3	Washington, DC	56	28
Minneapolis	11	0			

5. Produce five data points with a mean of 12 and a standard deviation of 0. Is there any other set of five points that achieve this?

6. *MacWorld* in September 1996 reported the average cost per megabyte of storage on hard drives available for home computers:

1992	1993	1994	1995	1996
$5.07	$2.40	$1.14	$0.53	$0.36

   a. Compute the mean, median, and standard deviation of costs given above.
   b. Do you think that these are appropriate measures to reflect the real story in these data?

7. The earned run averages (ERAs) of various Chicago Cubs pitchers (number of runs per game that a pitcher allows the opposing team to score) in the 2000 season through April 29 were:

   3.86   6.45   8.71   0.00   2.77   2.84   8.53   3.48   5.33   5.75   14.46

   a. Compute the mean, median, and standard deviation.
   b. Suppose you eliminate 14.46 as an outlier. Recompute the mean, median, and standard deviation. Discuss how each has changed and why.

8. An article in *Farming* (September/October 1994) listed the following in-state tuition and fees per school for 14 land-grant universities:

1554	2291	2084	4443	2884	2478	3087
3708	2510	2055	3000	2052	2013	2550

   a. Compute the mean, median, and standard deviation.
   b. Suppose each university decided to increase the tuition and fees by $500 in 1996. What would the new values of the mean, median, and standard deviation be?
   c. We know that tuition and fees have steadily risen. Suppose each university had increased its tuition and fees for the year 2000 by 10% (multiply the 1994 tuition by 1.1). Now what would the values be for the mean, median, and standard deviation?

9. Is it possible for the standard deviation to be negative? Explain.

Section 2.3 Measuring Spread: The Standard Deviation

10. Consider a sample of 5 exam grades in each of two classes:

Class I:	90	92	87	88	95
Class II:	50	70	82	55	80

    Without doing any calculations, which set of scores has the higher mean? Which set has the larger standard deviation? Why?

11. Briefly describe the difference between what the mean and the standard deviation measure. Explain why reporting both is important.

12. a. Compute the standard deviation for the presidents' data in Example 2.5.
    b. What percentage of the data lies within one standard deviation of the mean?
    c. What percentage of the data lies within two standard deviations of the mean?

13. a. Compute the standard deviation for Roger Maris' home runs in Example 2.2.
    b. Leave out Maris' exceptional 61 home runs and compute the standard deviation for the remaining 9 scores.
    c. Compare the two answers. Is the standard deviation a resistant measure of spread?

14. Scores on IQ tests have a symetric distribution that is roughly bell shaped. Reported IQs are adjusted so that the distribution has a mean of 100 and a standard deviation of 15. Suppose that a standard IQ test is given to 2000 students. Use the 68–95–99.7% rule to answer the following questions:
    a. About how many students in the group will have an IQ between 70 and 130?
    b. About how many students in the group will have an IQ between 100 and 130?
    c. About how many students in the group will have an IQ greater than 145? *Hint:* Find the approximate number between 55 and 145 first.

15. The following observations were made of the weights (in kilograms) of 12-year-old boys.

40.96	33.54	37.64	26.82	42.99	44.39	32.04	31.02	38.53	34.20
31.54	34.66	44.81	39.69	40.73	38.73	37.06	38.15	34.85	33.01
44.57	42.96	49.89	50.05	42.11	49.99	54.18	35.13	41.06	37.14
39.58	41.42	48.64	32.82	41.62	38.07	44.61	40.61	38.93	44.10
36.13	52.28	29.71	35.03	50.87	45.82	34.87	49.15	31.00	41.86

    a. Construct a histogram using the intervals 25.0–27.5⁻, 27.5–30.0⁻, 30.0–32.5⁻,...
    b. Find the mean and the standard deviation.
    c. What percentage of the data lies within 1 standard deviation of the mean? What percentage of the data lies within 2 standard deviations of the mean? How does this compare with what we would expect of a normal (bell-shaped) distribution?

16. The multiple choice questions in this exercise refer to the histograms that follow, which show the distribution of the daily high temperatures in 1987 for Olga, Washington (top), Minneapolis, Minnesota (middle), and Belle Glade, Florida (bottom). All three histograms use the same temperature scale shown at the bottom of the Belle Glade histogram.
    a. The average high temperature for Olga is closest to (choose one)

    $$45°F \quad 60°F \quad 75°F \quad 83°F \quad 95°F$$

    Hint: Figure 2.16 indicates how $\bar{x}$ and s can be estimated from the histogram.
    b. The standard deviation of the high temperatures for Olga is closest to (choose one).

    $$1°F \quad 3°F \quad 5°F \quad 10°F \quad 20°F$$

c. Is the average high temperature for Minneapolis at least 20 degrees less than, about the same as, or at least 20 degrees higher than the average high temperature for Olga? (Choose one.)
d. Is the standard deviation of the high temperatures for Minneapolis about half of, about the same as, or about twice the standard deviation for Olga?
e. Is the average high temperature for Belle Glade at least 20 degrees less than, about the same as, or at least 20 degrees higher than the average high temperature for Olga?
f. Is the standard deviation for Belle Glade about half of, about the same as, or about twice the standard deviation for Olga?

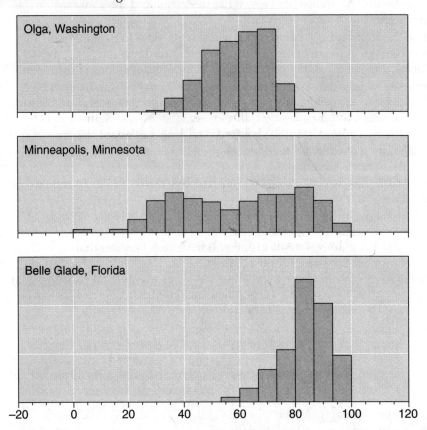

17. a. The average height of the women in Figure 1.9 (at the end of Section 1.4) is closest to (choose one):

$$60" \quad 62.5" \quad 65"$$

(Figure 2.16 indicates how $\bar{x}$ and s can be estimated from the histogram.)
b. The standard deviation of the heights is closest to:

$$0.5" \quad 1" \quad 2.5" \quad 4"$$

2.4 NORMAL DISTRIBUTIONS

If the histogram of a data set follows a bell curve without outliers, then the mean $\bar{x}$ and the standard deviation s summarize the data: All we need to be able to apply the 68-95-99.7% rule from Section 2.3 is the two numbers $\bar{x}$ and s. We illustrated the 68-95-99.7% rule using the baseball batting averages in Figure 2.10. We shall give two more examples.

Section 2.4 Normal Distributions

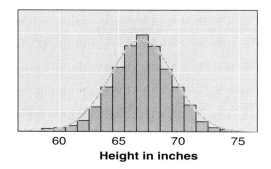

Figure 2.17 Density histogram of the heights of 8585 19th-century British men.

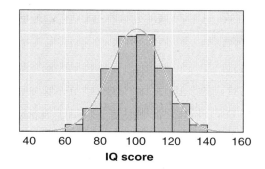

Figure 2.18 Density histogram of 2200 children's Wechsler IQ scores.

Figure 2.17 shows the density histogram of the heights of 8585 nineteenth century British men, collected by Sir Francis Galton (1822–1911), a pioneer of statistics. Figure 2.18 is the density histogram of 2200 children's IQs on a test called the Weschler intelligence test. The superimposed bell curves follow both histograms extremely well.

The Belgian scientist A. Quetelet (1796–1874) first noted that if the heights of a large group of people are measured, the density histogram of these heights will resemble a bell-shaped curve. He also studied the distributions of other physiological measurements, such as weight (in recent years, usually skewed just a bit right: do you know why?), chest girth, and arm length, and found that each approximately follow the same type of bell-shaped distribution.

Because so many physiological data sets have bell-shaped distributions, Galton coined the term "normal distributions" for bell-shaped distributions. They are also called Gaussian distributions, after the German mathematician and astronomer Carl Friedrich Gauss (1777–1855), who encountered such distributions in his study of measurement errors in science.

The fact that most physiological characteristics are normally distributed has many practical applications. For example, the designer of an airplane cockpit must arrange it so that most pilots are comfortable and can reach all of the controls. Clearly, this requires knowledge of average heights, average arm lengths, and so on, as well as knowledge of the variability around these averages so that *most* pilots will be accommodated. Since the physiological variables in question are normally distributed, all that the cockpit designers need to know is the mean and the standard deviation of each variable.

Table 2.6 uses the heights in Figure 2.17 and the IQ scores in Figure 2.18 to illustrate the 68% rule. We are also including the baseball data we already looked at in Section 2.3. For each of the 3 data sets we have counted the number of observations in the interval from $\bar{x} - s$ to $\bar{x} + s$.

As Table 2.6 shows, 5778 out of the 8585 men in Galton's study, or 67.3%, had heights between $\bar{x} - s$ and $\bar{x} + s$. This is very close to the 68% predicted by the empirical rule. The two other data sets also conform to the 68% rule.

Table 2.6 The $(\bar{x} = s)$:68% Rule Empirically Demonstrated

Data set	$\bar{x}$	s	$\bar{x} - s$	$\bar{x} + s$	Between $\bar{x} - s$ and $(\bar{x} + s)$		
					Count	Total	Proportion
Batting averages	0.263	0.029	0.234	0.292	178	263	0.677
Heights	67.02	2.564	64.46	69.58	5778	8585	0.673
IQs	100.47	15.189	85.27	115.66	1481	2200	0.673

84 SUMMARIZING DATA BY NUMERICAL MEASURES: CENTERS AND SPREADS

Table 2.7 The $(\bar{x} \pm 2s)$:95% Rule Empirically Demonstrated

Data set	$\bar{x}$	s	$\bar{x} - 2s$	$\bar{x} + 2s$	Between $\bar{x} - 2s$ and $(\bar{x} + 2s)$		
					Count	Total	Proportion
Batting averages	0.263	0.029	0.205	0.321	251	263	0.954
Heights	67.02	2.564	61.89	72.15	8087	8585	0.942
IQs	100.47	15.189	70.09	130.85	2105	2200	0.957

Table 2.7 illustrates the 95% rule using the same three data sets.

In all three cases, very close to 95% of the observations are between $\bar{x} - 2s$ and $\bar{x} + 2s$. But we can do much more. Using only $\bar{x}$ and s, we can for each of these three data sets (batting averages, male heights, and IQ scores) approximately tell the percent of observations in *any interval*, not just in the intervals $(\bar{x} - s, \bar{x} + s)$ and $(\bar{x} - 2s, \bar{x} + 2s)$. To do this we first have to introduce *standard scores* or *z-scores*.

z-Scores

In high school Yoko scored 29 on the mathematics section of the ACT college entrance exam. Her friend Soledad took the SAT and scored 650 on the mathematics section. Nationwide the ACT scores followed a normal curve with mean 21 and standard deviation 6. The SAT scores were normally distributed with mean 500 and standard deviation 100. If we assume that the ACT and the SAT measured the same kind of ability, who did better?

To compare the two scores we have to convert them into z-scores. The z-score tells us how many standard deviations the original (raw) score is above or below the mean:

$$z\text{-score} = \frac{\text{raw score} - \text{mean}}{\text{standard deviation}}.$$

For Yoko we get

$$z\text{-score} = \frac{29 - 21}{6} = \frac{8}{6} = 1.33.$$

For Soledad we get

$$z\text{-score} = \frac{650 - 500}{100} = \frac{150}{100} = 1.50.$$

Since Soledad's score is 1.50 standard deviations above the mean and Yoko's score is 1.33 standard deviations above the mean, Soledad did better.

Using the Standard Normal Table

Let us return to the Major League Baseball batting averages in 1985 that we discussed in the beginning of Section 2.3 and again in this section. We shall show how we can estimate the proportion of players whose batting average was between 0.225 and 0.265, for example, using only the overall mean $\bar{x} = 0.263$ and the standard deviation $s = 0.029$, plus Table E (in Appendix E) for the standard normal curve.

We have chosen a vertical scale in Figure 2.19 that gives the histogram a total area equal to 1. The area under the superimposed normal curve is also equal to 1. The area of each histogram bar equals the proportion of players with batting averages in the corresponding interval on the horizontal axis. The proportion of players with batting averages between 0.225 and 0.265 equals the combined area of the four shaded histogram bars. Since the superimposed normal curve follows

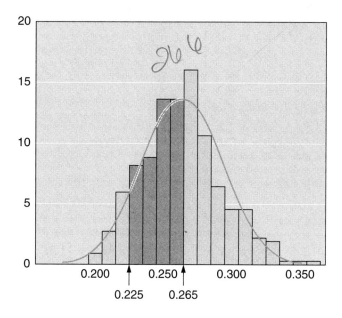

Figure 2.19 Batting averages between 0.225 and 0.265 of 263 Major League Baseball players in 1985.

the histogram closely, the area of the four shaded bars is close to the area under the normal curve from 0.225 to 0.265. So we can find the proportion of players with batting averages in an interval approximately by finding the area of an interval under an appropriate normal curve.

Important Fact: We can determine the area of an interval under any normal curve in two steps: First we convert the interval endpoints into a z-score interval under the normal curve that has mean 0 and standard deviation 1. The latter is called the **standard normal curve**.

The standard normal curve is displayed in Figure 2.20. It has the following basic properties:

- The curve is bell-shaped and symmetric around 0; the part of the curve to the right of the dashed line in Figure 2.20 is the mirror image of the part to the left of the dashed line.
- The curve extends indefinitely in both directions, always above the horizontal axis, approaching, but never touching the axis.
- The total area under the curve is 1.
- Almost all of the area (in fact 99.7% of the area) under the curve is between –3 and 3.

The z-scores corresponding to 0.225 and 0.265 are

$$\frac{0.225 - 0.263}{0.029} = -1.31 \text{ and } \frac{0.265 - 0.263}{0.029} = 0.07.$$

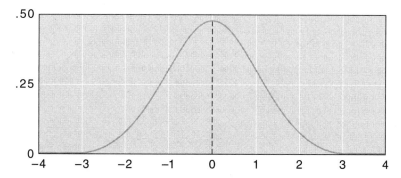

Figure 2.20 The standard normal curve.

86 SUMMARIZING DATA BY NUMERICAL MEASURES: CENTERS AND SPREADS

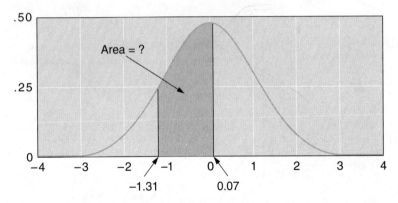

Figure 2.21 The area from −1.31 to 0.07.

The batting averages between 0.225 and 0.265 correspond to the z-scores between −1.31 and 0.07. Therefore the proportion of batting averages between 0.225 and 0.265 equals the proportion of z-scores between −1.31 and 0.07, which in turn approximately equals the *area under the standard normal curve* from −1.31 to 0.07, indicated in Figure 2.21.

Table E in the back of the book gives us the area under the standard normal curve that lies to the left of a specified z-score. A portion of Table E is reproduced as Table 2.8 below. To find the area to the left of $z = 0.07$ as shown in Figure 2.22, locate 0.0 in the table column under z, enter the table horizontally and locate the number under 7. We get 0.5279. The area to the left of 0.07 is 0.5279.

The area to the left of 0.07 is the sum of the area to the left of −1.31 and the area between −1.31 and 0.07, as shown in Figure 2.23. Therefore

(Area between −1.31 and 0.07) = (Area to the left of 0.07) − (Area to the left of −1.31).

Table 2.8 Standard Normal Table

z	0	1	2	3	4	5	6	7	8	9
0.0	.5000	.5040	.5080	.5120	.5160	.5199	.5239	.5279	.5319	.5359
0.1	.5398	.5438	.5478	.5517	.5557	.5596	.5636	.5675	.5714	.5733
0.2	.5793	.5382	.5871	.5910	.5948	.5987	.6026	.6064	.6103	.6141
0.3	.6179	.6217	.6255	.6293	.6331	.6368	.6406	.6443	.6480	.6517
0.4	.6554	.6591	.6628	.6664	.6700	.6736	.6772	.6808	.6844	.6879
0.5	.6915	.6950	.6985	.7019	.7054	.7088	.7123	.7157	.7190	.7224
0.6	.7257	.7291	.7324	.7357	.7389	.7422	.7454	.7486	.7517	.7549
0.7	.7580	.7611	.7642	.7673	.7704	.7734	.7764	.7794	.7823	.7852
0.8	.7881	.7910	.7939	.7967	.7995	.8023	.8051	.8078	.8106	.8133
0.9	.8159	.8186	.8212	.8238	.8264	.8289	.8315	.8340	.8365	.8389
1.0	.8413	.8438	.8461	.8485	.8508	.8531	.8554	.8577	.8599	.8621
1.1	.8643	.8665	.8686	.8708	.8729	.8749	.8770	.8790	.8810	.8830
1.2	.8849	.8869	.8888	.8907	.8925	.8944	.8962	.8980	.8997	.9015
1.3	.9032	.9049	.9066	.9082	.9099	.9115	.9131	.9147	.9162	.9177
1.4	.9192	.9207	.9222	.9236	.9251	.9265	.9279	.9292	.9306	.9319
1.5	.9332	.9345	.9357	.9370	.9382	.9394	.9406	.9418	.9429	.9441
1.6	.9452	.9463	.9474	.9484	.9495	.9505	.9515	.9525	.9535	.9545
1.7	.9554	.9564	.9573	.9582	.9591	.9599	.9608	.9616	.9625	.9633
1.8	.9641	.9649	.9656	.9664	.9671	.9678	.9686	.9693	.9699	.9706
1.9	.9713	.9719	.9726	.9732	.9738	.9744	.9750	.9756	.9761	.9767
2.0	.9772	.9778	.9783	.9788	.9793	.9798	.9803	.9808	.9812	.9817

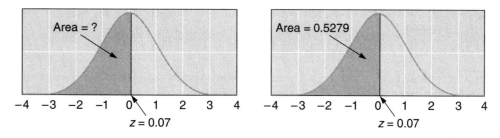

Figure 2.22 Finding the area under the standard normal curve to the left of 0.07.

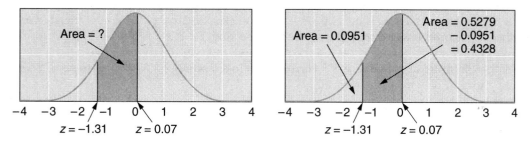

Figure 2.23 Finding the area under the standard normal curve that lies between −1.31 and 0.07.

According to the normal Table E in Appendix E, the area under the standard normal curve to the left of −1.31 is 0.0951. Therefore the area between −1.31 and 0.07 is

$$0.5279 - 0.0951 = 0.4328.$$

Based on the z-score of −1.31 and the normal table value of 0.0951 given above we estimate that approximately 9.51% of the batting averages were below 0.225. This happens to be exactly right: 25 out of 263 were, which is 0.0951 to three significant figures! We estimate that 43.3% of the batting averages were between 0.225 and 0.265. In fact, 115 out of 263 players, or 43.7%, had batting averages in this range.

Finally let us estimate the percentage of the players whose batting average was over 0.295. We first convert 0.295 into z-score

$$z = \frac{0.295 - 0.263}{0.029} = \frac{32}{29} = 1.10.$$

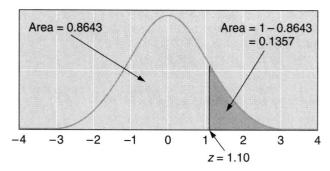

Figure 2.24 Finding the area under the standard normal curve to the right of 1.10.

Since the total area under the standard normal curve equals 1, the area under the standard normal curve to the right of 1.10 is 1 minus the area to the left of 1.10, as indicated in Figure 2.24. We estimate that 13.6% of the players had batting averages over 0.295. In fact, 38 out of 263 (or 14.4%) had.

Let us take another look at Yoko and Soledad's math scores. What was Soledad's *percentile rank* on the math SAT? In other words, what percentage of all the students who took the SAT scored lower than Soledad on the mathematics part? We can find the answer to this question in the standard normal Table E in the back of the book. Since Soledad's z-score is 1.50, locate 1.5 in the table column under z, enter the table horizontally and locate the number in the column under 0. We get .9332. That is 93.32% or about 93%. So around 93% of all students who took the SAT scored lower than Soledad on the math part.

Yoko's z-score was 1.33. We locate 1.3 in the column under z, enter the table horizontally and locate .9082 in the column under 3. That is 90.82% or about 91%. So around 91% of all the students who took the ACT scored lower than Yoko on the math section.

How high did a student have to score on the math SAT to be in the top 5%? We need a score above the 95th percentile. In the table, .9505 corresponds to a z-score of 1.65. To get the SAT score corresponding to $z = 1.65$ we use the formula

$$\text{raw score} = \text{mean} + (\text{z-score}) \times (\text{standard deviation}).$$

We get

$$\text{SAT score} = 500 + 1.65 \times 100$$
$$= 500 + 165$$
$$= 665.$$

Table 2.9 Standard Normal Table

z	0	1	2	3	4	5	6	7	8	9
0.0	.5000	.5040	.5080	.5120	.5160	.5199	.5239	.5279	.5319	.5359
0.1	.5398	.5438	.5478	.5517	.5557	.5596	.5636	.5675	.5714	.5733
0.2	.5793	.5382	.5871	.5910	.5948	.5987	.6026	.6064	.6103	.6141
0.3	.6179	.6217	.6255	.6293	.6331	.6368	.6406	.6443	.6480	.6517
0.4	.6554	.6591	.6628	.6664	.6700	.6736	.6772	.6808	.6844	.6879
0.5	.6915	.6950	.6985	.7019	.7054	.7088	.7123	.7157	.7190	.7224
0.6	.7257	.7291	.7324	.7357	.7389	.7422	.7454	.7486	.7517	.7549
0.7	.7580	.7611	.7642	.7673	.7704	.7734	.7764	.7794	.7823	.7852
0.8	.7881	.7910	.7939	.7967	.7995	.8023	.8051	.8078	.8106	.8133
0.9	.8159	.8186	.8212	.8238	.8264	.8289	.8315	.8340	.8365	.8389
1.0	.8413	.8438	.8461	.8485	.8508	.8531	.8554	.8577	.8599	.8621
1.1	.8643	.8665	.8686	.8708	.8729	.8749	.8770	.8790	.8810	.8830
1.2	.8849	.8869	.8888	.8907	.8925	.8944	.8962	.8980	.8997	.9015
1.3	.9032	.9049	.9066	.9082	.9099	.9115	.9131	.9147	.9162	.9177
1.4	.9192	.9207	.9222	.9236	.9251	.9265	.9279	.9292	.9306	.9319
1.5	.9332	.9345	.9357	.9370	.9382	.9394	.9406	.9418	.9429	.9441
1.6	.9452	.9463	.9474	.9484	.9495	.9505	.9515	.9525	.9535	.9545
1.7	.9554	.9564	.9573	.9582	.9591	.9599	.9608	.9616	.9625	.9633
1.8	.9641	.9649	.9656	.9664	.9671	.9678	.9686	.9693	.9699	.9706
1.9	.9713	.9719	.9726	.9732	.9738	.9744	.9750	.9756	.9761	.9767
2.0	.9772	.9778	.9783	.9788	.9793	.9798	.9803	.9808	.9812	.9817

Section 2.4 Exercises

1. On a midterm exam the mean score was 50 and the standard deviation was 15.
   a. Convert each of the following exam scores to z-scores: 50, 35, 80.
   b. Find the exam scores corresponding to the z-scores 0, 0.6, 1, –2.
2. Convert each entry on the following list into z-scores: 5, 2, 6, 8, 5, 4. (Use the mean and the standard deviation of the list.)
3. IQ scores are normally distributed with mean 100 and standard deviation 15. Find the percentage of IQ scores that are
   a. below 120.
   b. below 90.
   c. between 90 and 120.
4. How high must a person's IQ score be to fall in the top 5% of all IQ scores? How high to fall in the top 2½ %? (See the previous exercise.)
5. The heights of young women are approximately normally distributed with mean 65 inches and standard deviation 2.5 inches. Find the percentage of women who are
   a. under 69 inches tall.
   b. under 62 inches tall.
   c. between 62 and 69 inches tall.
6. The Public Health Service examined a representative cross section of several thousand American men aged 18 to 74. The systolic blood pressure of these men followed a normal curve with mean 130 and standard deviation 18. Any blood pressure of 160 or above is considered "high." Any blood pressure from 140 to 159 is considered "borderline."
   a. Estimate the percentage of men with high systolic blood pressure.
   b. Estimate the percentage with borderline systolic blood pressure.
7. The length of human pregnancies from conception to birth follow a normal curve with mean 266 days and standard deviation 16 days.
   a. Estimate the percentage of pregnancies that last less than 240 days. (That is roughly 8 months.)
   b. Estimate the percentage of pregnancies that last between 240 days and 270 days. (That is about 8 to 9 months.)
8. The systolic blood pressure of men aged 18 to 24 follows a normal curve with mean 124 and standard deviation 14, according to the Public Health Service and the National Center for Health Statistics. (See Exercise 6.)
   a. Estimate the percentage of men aged 18 to 24 with high systolic blood pressure, that is, 160 or above.
   b. Estimate the percentage of men aged 18 to 24 with borderline systolic blood pressure, that is, from 140 to 159.

2.5 BOXPLOTS: THE FIVE-NUMBER SUMMARY

We saw in Section 2.3 and 2.4 that if a data set has a bell-shaped distribution without outliers, the mean and the standard deviation *summarize* the main features of the distribution. For skewed distributions and distributions with outliers, however, the mean and standard deviation will not give us a meaningful summary. The problem is that it is difficult to interpret the mean and the standard deviation when there are outliers or the distribution is skewed. Instead we shall use the *quartiles* to summarize the data. The quartiles Q_1, Q_2, and Q_3 divide the data into four equal parts (Figure 2.25). Q_1 separates the bottom 25% of the data from the top 75%. Q_2 is the median, which separates the bottom 50% from the top 50% of the data. Finally, Q_3 separates the bottom 75% from the top 25% of the data.

SUMMARIZING DATA BY NUMERICAL MEASURES: CENTERS AND SPREADS

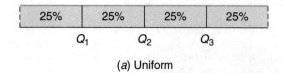
(a) Uniform

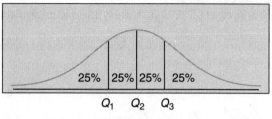
(b) Bell-shaped

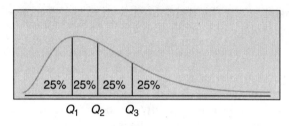
(c) Right-skewed

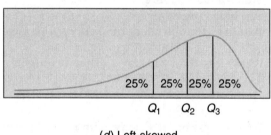
(d) Left-skewed

Figure 2.25 Quartiles for (a) uniform, (b) bell-shaped, (c) right-skewed, and (d) left-skewed distributions.

We will give a very simple method for determining the quartiles. Some computer programs and calculators might compute the values in a slightly different manner, but the results will be close.

1. Sort the data from smallest to largest and compute the median. This is Q_2.
2. The first quartile, Q_1, is defined to be the median of all the data to the left of Q_2.
3. The third quartile, Q_3, is defined to be the median of all the data to the right of Q_2.

Example 2.12 illustrates how the quartiles are determined.

Example 2.12

The number of home runs hit by Babe Ruth during his 15 years with the New York Yankees were, in increasing order:

$$22 \quad 25 \quad 34 \quad \underset{\underset{Q_1}{\uparrow}}{35} \quad 41 \quad 41 \quad 46 \quad \underset{\underset{Q_2}{\uparrow}}{46} \quad 46 \quad 47 \quad 49 \quad \underset{\underset{Q_3}{\uparrow}}{54} \quad 54 \quad 59 \quad 60$$

Section 2.5 Boxplots: The Five-Number Summary

With 15 observations, the median is the eighth smallest observation. The first quartile (Q_1) is the median of the first seven observations and the third quartile (Q_3) is the median of the last seven observations. Note that even though the seventh, eighth, and ninth smallest observations have the same value, namely 46, the seventh observation is considered to be to the left of Q_2, and the ninth observation is considered to be to the right of Q_2.

From 1987 to 2000, the number of home runs hit by Mark McGwire were, in increasing order:

$$9 \quad 9 \quad 22 \quad 32 \quad 32 \quad 33 \quad 39 \quad\quad 39 \quad 42 \quad 49 \quad 52 \quad 58 \quad 65 \quad 70$$
$$\uparrow \quad\quad\quad\quad\quad\quad\quad \uparrow \quad\quad\quad\quad\quad\quad\quad \uparrow$$
$$Q_1 \quad\quad\quad\quad\quad\quad\quad Q_2 \quad\quad\quad\quad\quad\quad\quad Q_3$$

With 14 observations, the median is the average of the seventh and the eighth smallest observations. The first quartile (Q_1) is the median of the first seven observations and the third quartile (Q_3) is the median of the last seven observations. We get $Q_1 = 32$, median $= 39$, and $Q_3 = 52$. Here the seventh observation is considered to be to the left of Q_2, and the eighth observation is considered to be to the right of Q_2.

To compare the performance of Babe Ruth and Mark McGwire, we could draw back-to-back stem-and-leaf plots or we could draw side-by-side **boxplots**.

We need five values to provide a boxplot:

1. The minimum value
2. The first quartile
3. The median
4. The third quartile
5. The maximum value

The boxplot based on these five values is often referred to as the **five-number summary** or **five-number boxplot**.

To draw the five-number boxplot, draw a box around the first and third quartiles. Draw a vertical line through the box at the median value, and draw horizontal lines (the whiskers) to connect the edges of the box (at the first and third quartiles) to the minimum and maximum values. Boxplots give a rough idea of the shape of the data set. In Figure 2.26, boxplots are sketched for five common data shapes.

In addition to using the five numbers, we sometimes want to specially identify outliers as was done in Figure 2.1. Figure 2.1 displays **outlier boxplots**. **Outliers** are indicated by asterisks (∗) and **potential outliers** by circles (∘). The whiskers extend from the box to the largest and the smallest *non-outlier* values. (Outliers and potential outliers will be defined below.)

Example 2.13

Table 2.10 gives the number of farms (in thousands) for individual states in 1998. For example, Alabama has 49,000 farms.

$$\text{Minimum} = 0.6, \text{ maximum} = 226, \text{ first quartile} = 9.6,$$
$$\text{median} = 38.75, \text{ third quartile} = 65.$$

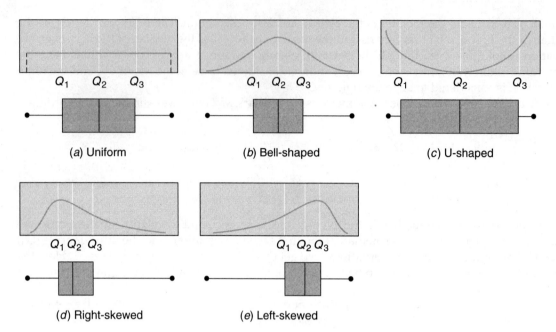

Figure 2.26 Distribution shapes and typical boxplots for (a) uniform, (b) bell-shaped, (c) U-shaped, (d) right-skewed, and (e) left-skewed distributions.

Table 2.10 Number of Farms per State, 1998 (in thousands)

Number of Farms per State, 1998 (in thousands)			
Alabama	49.0	Nebraska	55.0
Alaska	0.6	Nevada	3.0
Arizona	7.9	New Hampshire	3.1
Arkansas	49.5	New Jersey	9.6
California	89.0	New Mexico	16.0
Colorado	29.5	New York	38.0
Connecticut	4.1	N. Carolina	58.0
Delaware	2.7	N. Dakota	31.0
Florida	45.0	Ohio	80.0
Georgia	50.0	Oklahoma	83.0
Hawaii	5.5	Oregon	39.5
Idaho	24.5	Pennsylvania	60.0
Illinois	79.0	Rhode Island	0.8
Indiana	66.0	S. Carolina	25.0
Iowa	97.0	S. Dakota	32.5
Kansas	65.0	Tennessee	91.0
Kentucky	90.0	Texas	226.0
Louisiana	30.0	Utah	15.0
Maine	6.9	Vermont	6.7
Maryland	12.5	Virginia	49.0
Massachusetts	6.0	Washington	40.0
Michigan	52.0	W. Virginia	21.0
Minnesota	80.0	Wisconsin	78.0
Mississippi	42.0	Wyoming	9.2
Missouri	110.0		
Montana	27.5	United States	2192

Source: National Agricultural Statistics Service, U.S. Department of Agriculture.

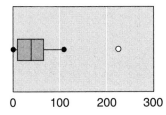

Figure 2.27 Boxplot of number of farms per state.

The outlier boxplot is drawn in Figure 2.27. We see that there is one extreme value, or outlier, with value 226.0 (Texas). The precise rule for deciding when a point is labeled an outlier is given below. If we separately label the outlier (as was done in the Key Problem at the beginning of this chapter), the maximum of the remaining data is 110.0 (Missouri). Note that the whisker is drawn to this remaining state's maximum and **not** to the outlying value (see also Figure 2.1). Note also that Q_1, Q_2, and Q_3 were computed using all 50 states, including the outlying Texas value.

When we look at the data, we notice that this outlying value corresponds to the state of Texas. It is useful to call attention to outliers, like Texas in this example, because we may want to understand why they are so different from the rest of the data.

The boxplot gives us a graphical method for looking in broad strokes at variation within the data (see Figure 2.26). The box itself represents the middle 50% of the data. We define the **interquartile range** (IQR) as the difference between the third and first quartiles:

$$\text{IQR} = Q_3 - Q_1.$$

In Example 2.13, IQR = 65 − 9.6 = 55.4.

The interquartile range is a much more resistant measure of variation than the standard deviation is, because, unlike the standard deviation, it is not influenced by the largest and smallest values in the data set. We will see later, though, that the standard deviation is a more important measure of variation for statistical inference when we study estimation and hypothesis testing.

Example 2.14

The following data consist of the percentages of eighth-grade students in various states (all states that reported data) who scored at or above proficient levels in national tests in 1998 (source: *National Assessment of Educational Progress*, National Center for Education Statistics, U.S. Department of Education).

Reading:

```
21 28 23 22 30 42 25 12 23 25 19 35 29
18 42 31 36 37 19 29 24 24 34 31 29
33 30 22 26 28 31 33 32 27 33 29 31
```

Science:

```
18 31 23 22 20 32 36 21  5 21 21 15 30 36 23
13 41 25 37 32 37 12 28 41 35 19 27 24 41
32 26 17 22 23 32 34 27 27 21 39 34 27
```

We can compare the two sets of data by using boxplots, drawn in Figure 2.28.

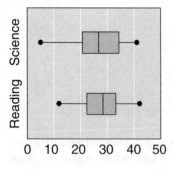

Figure 2.28 State percentages of eighth-grade students scoring at or above proficient levels on national tests.

There is not a great deal of difference between the medians of the two groups of scores, but because the science box is wider, there appears to be more variation in the science scores. When we compare the medians, we are comparing the centers of the two groups. However, when we look at the size of each box (the IQR), we are considering the variation (or spread) within each group of scores.

In this case the difference in variation might be of considerable importance to educators, suggesting the need for more remedial and gifted programs in science, perhaps.

In the Key Problem at the start of the chapter (in particular, see Figure 2.1), it appears that there is a significant difference between the median ages of award-winning actors and actresses and possibly a difference between the variations within the groups. Even though a visual inspection of the difference between the centers seems too large to have occurred just by chance, we cannot make a definitive conclusion at this time. To judge whether there is a difference too large to be explainable by chance, we must use statistical hypothesis testing methods developed later in this text.

We have considered the possible influences and importance of extreme values (outliers) in this chapter. What we need is a method when constructing an outlier boxplot to determine whether a value is indeed so extreme that it needs to be specially labeled to call the user's attention to it. The interquartile range gives us a starting point to do this. Here is the procedure that many use.

- Label a value to be an **outlier** if it is more than 3 times the IQR in distance from the nearer of the first and third quartiles. Denote it by an asterick (∗).
- If a value is more than 1.5 and less than 3 times the IQR in distance from the nearer of the first and third quartiles, label it to be a **potential outlier** and denote it by a circle (∘).

Figure 2.29 clearly shows the procedure for a typical boxplot. This somewhat arbitrary choice of 1.5 and 3 has become standard practice in defining outliers when doing a boxplot. The TI-83 calcu-

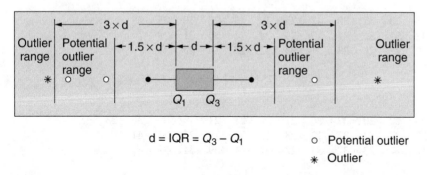

Figure 2.29 How outliers and potential outliers are identified in an outlier boxplot.

Section 2.5 Boxplots: The Five-Number Summary

lator and many statistical computer programs will also show outliers on a modified boxplot, as was done in the Key Problem. In this case, the whiskers are drawn to the smallest and largest values that are not outliers. Q_1, Q_2, and Q_3 are computed using all the data, including potential and actual outlier values. Keep firmly in mind that the labeling of points as outliers is tricky and subjective; the 1.5, 3 rule above is just one possible approach that statisticians use.

Caution: A **boxplot** displays the **center** and the **spread** of a set of numbers. An **outlier boxplot** also indicates **outliers**. But a boxplot does not tell us whether the data set has **clusters** or **gaps**. As a tool for determining the **shape** (or **distribution**) of a data set, boxplots can be deceptive. For example, consider the following back-to-back stem-and-leaf plots:

Leaf (First Data Set)	Stem	Leaf (Second Data Set)
5,6,9	0	5
	1	6
	2	2,7
7,7,8,9	3	2,5,7
0	4	1,3,5,9
	5	0,2,4,6,8
8,9	6	1,4,6,8
1,3,3	7	3,6,9
1,3	8	4,8
	9	5
1	10	1

The first data set has clusters and gaps, whereas the second data set is bell-shaped with no gaps. But the two data sets have the same five-number summary and therefore the same boxplot! Obviously, the boxplot does not tell the whole story.

Example 2.15

Consider the ages of the male Academy Award winners in the Key Problem at the start of the chapter. The data has been summarized in Table 2.11.

$$\text{Actors: } \text{IQR} = 52 - 38 = 14$$
$$1.5 \times \text{IQR} = 1.5 \times 14 = 21$$
$$Q_1 - (1.5 \times \text{IQR}) = 38 - 21 = 17$$
$$Q_3 + (1.5 \times \text{IQR}) = 52 + 21 = 73$$
$$3 \times \text{IQR} = 3 \times 14 = 42$$
$$Q_1 - (3 \times \text{IQR}) = 38 - 42 = -4$$
$$Q_3 + (3 \times \text{IQR}) = 52 + 42 = 94$$

Table 2.11 Key Problems Statistics

	Minimum	First Quartile	Median	Third Quartile	Maximum
Actors	30	38	45	52	76
Actresses	21	33	35	42	80

Thus any award winner older than 94 or younger than −4 is an outlier, an impossibility for −4 and extremely unlikely for 94. Any age between −4 and 17 or between 73 and 94 is a potential outlier. Looking at the data, we see that the age 76 case is a potential outlier for the actors. Since there have been famous child stars, we could have had a potential low outlier, of course, but in this case we did not. A similar analysis is possible for the actress data.

Section 2.5 Exercises

1. Construct a boxplot for the white blood cell count data of Exercise 7 of Section 2.2. Be sure to indicate outliers and potential outliers. What do the outliers indicate in this example?
2. Repeat Problem 1 using the data on drug use in Exercise 6 of Section 2.2 (high school marijuana use).
3. The U.S. Bureau of Labor Statistics gave these unemployment rates for the years in the table below:

Year	Unemployment Rate (Percent)
1975	8.5
1980	7.1
1981	7.6
1982	9.7
1983	9.6
1984	7.5
1985	7.2
1986	7.0
1987	6.2
1988	5.5
1989	5.3
1990	5.6
1991	6.8
1992	7.5
1993	6.9
1994	6.1
1995	5.6

   a. Draw a boxplot and label the corresponding values.
   b. Determine the interquartile range.
   c. Determine whether there are any outliers or potential outliers.
4. Draw boxplots using the same scale (see Figure 2.1) for the data in Exercise 1 of Section 2.3 for both 1997 and 1998. Compare the centers and variations of each (medians and IQR) to determine whether there was any difference. Are there any outliers or possible outliers for either of these years?
5. Repeat the previous exercise for the data in Exercise 4 of Section 2.3.
6. The percentages of the popular vote won by the winning candidate in presidential elections from 1948 to 1996 are given below.

1948	1952	1956	1960	1964	1968	1972	1976	1980	1984	1988	1992	1996
49.6	55.1	57.4	49.7	61.1	43.4	60.7	50.1	50.7	58.8	53.9	43.2	49.2

a. What are the mean, median, and standard deviation of the winning percentages?
b. Compute the first and third quartiles. Which elections fall below the first quartile (close elections)? Which fall above the third quartile (landslides)?

7. The U.S. Department of Energy released fuel consumptions (in miles per gallon) for midsize cars and for four-wheel drive sport utility vehicles (SUVs) in 1998.

Midsize:

25	26	29	24	26	29	30	28	28	29	27	28
23	23	25	26	33	29	26	24	28	26	16	25
30	25										

SUV:

19	20	19	17	18	19	20	19	19	18	19	16
19	21	20	19	26	26	22					

Discuss any significant differences you can find between fuel consumption of midsize cars and sport utility vehicles, using indices and techniques you have learned to use in this chapter.

8. *USA Today* (February 17, 1995) reported the sizes of the police forces in the 10 largest cities in the United States in 1993 (in hundreds).

29.3   7.6   12.1   4.7   6.2   1.9   3.9   2.8   2.0   1.7

a. Construct a boxplot for the data.
b. Find any outliers and potential outliers.
c. Find the mean and median of the data. What can you tell about the shape of the distribution? Is it skewed to the right, skewed to the left, or symmetric?

9. When should we use the interquartile range as a measure of variability instead of the standard deviation?
10. Draw a boxplot for the salaries of the Chicago Cubs for the 2000 season. (See Example 2.6.) Explain how the shape of the boxplot reflects the shape of the histogram. (See Figure 2.26.)
11. Draw a boxplot for Newcomb's measurements in Table 2.2 (Example 2.8). Determine the interquartile range and identify any outliers.
12. Draw a boxplot for Roger Maris's home run data. (See Example 2.2.) Determine the interquartile range. Is Maris's record-setting total of 61 homers an outlier according to our technical definition?
13. Sketch a boxplot for the age distribution of deaths due to heart disease in 1996. (See Exercise 13 of Section 2.) Make sure that the shape of the boxplot reflects the shape of the histogram. (See Figure 2.26.)
14. The boxplots at the top of the next page represent four different data sets. One distribution is uniform, one is bell shaped, one is right skewed, and one is left skewed. Which is which? *Hint:* See Figure 2.26.

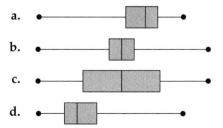

15. Do the analysis of Example 2.15 for the Academy Award actress age data.

2.6 DESCRIPTIVE DATA ANALYSIS

Describing data and then interpreting these descriptions can be an art as well as a science. By this we mean that it often depends on sound individual judgment, not a cut-and-dried set of rules. We have seen several ways of summarizing data graphically (for example, stem-and-leaf plots, histograms, and boxplots) and several numerical measures (mean, median, IQR, standard deviation). Which methods are best? Unfortunately, there is no easy answer. The statistician must judge how best to represent the given data so that they reveal, or at least suggest, the underlying story they have to tell. Histograms and stem-and-leaf plots are helpful if we are interested in the shape of the distribution of the data or if we want to look at where there are concentrations of data. Boxplots, on the other hand, are useful if we want to focus mainly on centering and variation without being too concerned about details concerning the shape of the distribution.

Just as there can be several good ways to proceed, there can be useless approaches as well. If the statistician does not interpret the results correctly, having the data does us no good. To modify a well-known saying: "The numbers do *not* speak for themselves." They need the attention of somebody good at statistics if they are to reveal their secrets.

Statistics is a powerful tool to analyze data, but we need to be very careful how we interpret our results. Often we see examples of data analyses (especially medical research) in newspaper articles and in news stories on television. You should be especially cautious about believing everything you read, especially if you do not know how the data were collected and how they were interpreted.

We often use graphical methods and numerical summaries to compare the characteristics of two (sometimes more) different sets of data effectively. We can look at our graphs and see whether there appears to be a difference between their means, medians, standard deviations, or interquartile ranges.

Even competent statisticians can disagree on the conclusions that can be validly drawn. For years it was thought that statistical studies had shown that a high-fiber diet helps prevent colon cancer and that taking estrogen helps women prevent heart disease. Both of these claims may in fact be true, but recent statistical studies have questioned these long-standing beliefs. It is very possible, for example, that those on high-fiber diets may have other characteristics that could also help explain the results. It is very difficult in looking at data to "prove" that there is a cause-and-effect relationship—that is, that one variable causes another to occur. Also, be very cautious in accepting the data-driven conclusions of statisticians with vested interests in the companies and institutions whose products and policies they are evaluating, because their job is to defend and validate these products and policies.

Examples of Data Analyses Using the Descriptive Methods Learned

Let us look at some data sets and see how we can analyze them.

Example 2.16

A group of 10 students in a statistics course decided to work on their own before an exam and had the scores below:

80 72 95 40 45 100 90 50 60 70

A second group of 10 students decided to form a study group before an exam. They received the following scores:

74 68 70 65 60 75 80 71 62 77

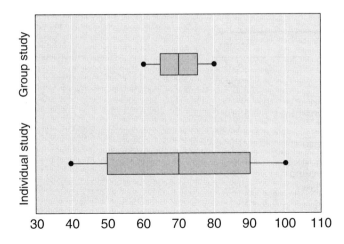

Figure 2.30 Comparison of grade distributions for self-study versus group study.

We first draw boxplots of both sets of exam scores to get a rough comparison of the two data sets (Figure 2.30).

From looking at the boxplots, it appears that the center of each set of scores is almost the same. However, the IQR (or the width of the box) shows that the variation of the first set is considerably larger than that of the second group. What does this mean? Is this evidence that study groups produce more similar exam scores? Without knowing more about how the students ended up in each group, it is very difficult to draw any inferences at all. But perhaps students working together tend to learn the same things and to think about the material in the same way and thus respond more similarly on an exam. The negative aspect is that we might wonder whether the good students in such a study group tend to be drawn downward to the average because everyone was doing things the same way. Nonetheless, the boxplots seem to indicate clearly a great difference in the variation in exam scores between the isolated-study and the group-study sets of students.

Example 2.17

A mathematics teacher looking toward retirement wants to know where older people are living. Table 2.12 represents the percentage of the population at least 65 years old by state in 1996. A boxplot is drawn in Figure 2.31.

It appears in this case that the spread is quite narrow; that is, most states have similar rates of the percentage of the population over 65. This in itself seems an interesting and perhaps surprising finding. However, in contrast to previous cases, we have potential outliers on both the lower and the upper ends of the data. It is probably not too surprising that Florida provides the potential outlier at the upper end; it is common knowledge that many people retire to this area. Can you guess the state in which people do not want to live as they get older (the lower potential outlier)? From a look at the table, the lowest value is easy to spot, corresponding to Alaska. (Can you think of reasons why people don't usually retire in Alaska?)

Example 2.18

The data in Table 2.13 consist of the numbers of hysterectomies performed in a year by a sample of male and female doctors in Switzerland.

Table 2.12

State	Percent	State	Percent
Alabama	13.0	Montana	13.2
Alaska	5.2	Nebraska	13.8
Arizona	13.2	Nevada	11.4
Arkansas	14.4	New Hampshire	12.0
California	10.5	New Jersey	13.8
Colorado	11.0	New Mexico	11.0
Connecticut	14.3	New York	13.4
Delaware	12.8	North Carolina	12.5
Florida	18.5	North Dakota	14.5
Georgia	9.9	Ohio	13.4
Hawaii	12.9	Oklahoma	13.5
Idaho	11.4	Oregon	13.4
Illinois	12.5	Pennsylvania	15.9
Indiana	12.6	Rhode Island	15.8
Iowa	15.2	South Carolina	12.1
Kansas	13.7	South Dakota	14.4
Kentucky	12.6	Tennessee	12.5
Louisiana	11.4	Texas	10.2
Maine	13.9	Utah	8.8
Maryland	11.4	Vermont	12.1
Massachusetts	14.1	Virginia	11.2
Michigan	12.4	Washington	11.6
Minnesota	12.4	West Virginia	15.2
Mississippi	12.3	Wyoming	11.2

Source: Statistical Abstract of the United States, 1997.

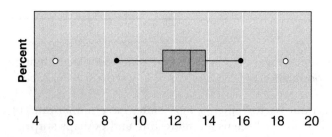

Figure 2.31 Distribution of retirement populations by state.

Table 2.13 Hysterectomies in Switzerland

Male doctors:									
	27	50	33	25	86	25	85	31	37
	44	20	36	59	34	28			
Female doctors:									
	5	7	10	14	18				
	19	25	29	31	33				

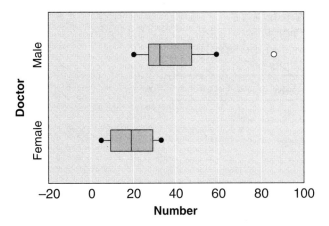

Figure 2.32 Hysterectomies performed by female and male doctors in Switzerland.

How do these values compare? Boxplots are shown in Figure 2.32. The boxplots show a striking difference between the female and male doctors. The IQRs (indicating spread) do not seem very different in this case, but the male physicians as a group certainly appear to perform considerably more hysterectomies than the female doctors. This certainly should be a cause of concern for women patients and would probably cause some concern in general that female and male doctors could have different attitudes toward their women patients and the need for the medical procedure. Also, the question arises whether the results generalize beyond Switzerland. Further, before we can really draw any conclusions, we need to know how the male and female doctors were selected for the study. We must be satisfied that they are representative of male and female doctors who perform hysterectomies in Switzerland. Further, what the data do is display a striking difference between male and female doctors; the data in and of themselves do not provide a criticism of either group of physicians. Indeed, the data do not even show that male doctors perform hysterectomies at a higher rate—male doctors simply may have larger numbers of patients.

In the following examples, we are interested in the shape of a distribution beyond what a boxplot reveals.

Example 2.19

π (pi) is an irrational number that occurs frequently in mathematical applications. An irrational number cannot be represented by a fraction; its decimal expansion has no repeating pattern of digits. Before calculators became widely used, the fraction 22/7, which differs from π by about one part in 2500, was often used as an approximation for π to make computations easier.

Is there a difference between the distributions of the digits in the decimal expansions of the two numbers? Table 2.14 gives the frequencies with which the digits occur within the first 1000 digits of $\pi >$. The density histogram, given in Figure 2.33, is rather flat or uniform, suggesting that all the digits tend to occur about equally often, each about 1/10 of the time. Such a distribution is called a **uniform distribution.** To tell whether the first 1000 digits really are typical for a uniform distribution, we must use statistical methods developed later in the course (the chi-square test, discussed in Chapter 10). In fact, the frequencies do "fit" the uniform distribution model, easily. Indeed, high-powered mathematics can be used to "prove" that the digits 0–9 do occur uniformly as the number of digits gets larger and larger.

The distribution for the fraction 22/7, where we only look at the first 100 digits, looks quite different. The frequencies are listed in Table 2.15, and the density histogram is shown in Figure 2.34.

Table 2.14
Digit Frequencies within the First 1000 Digits of π

Digit	Frequency
0	93
1	116
2	103
3	102
4	93
5	97
6	94
7	95
8	101
9	106

Table 2.15
Digit Frequencies within the First 100 Digits of 22/7

Digit	Frequency
0	0
1	17
2	17
3	1
4	17
5	16
6	0
7	16
8	16
9	0

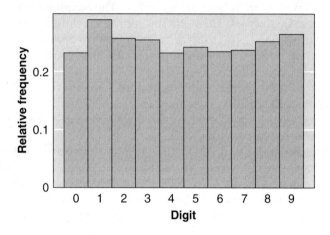

Figure 2.33 Density histogram of the first 1000 digits of π.

There is no tendency at all toward uniformity. There are gaps in the distribution for 0, 6, and 9, which do not appear in the histogram for π, and 3 only appears once (the 3 in 3.14...). These gaps and this extreme nonuniformity are not a chance accident, because the decimal expansion of 22/7 has a sequence of integers, 142857, that is repeated endlessly in the decimal expansion.

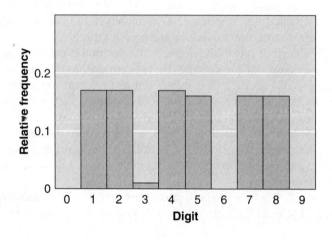

Figure 2.34 Density histogram of the first 100 digits of 22/7.

Section 2.6 Descriptive Data Analysis

Example 2.20

The following data represent actual scores given on a statistics exam at the College of Lake County. The points scored were out of a total of 120.

117	115	114	109	105	104	104	103	102	100	97	96	96	94
81	79	78	78	75	74	71	70	67	65	61	50	41	35

We could look at an outlier boxplot (Figure 2.35) to see what the data distribution looks like. We see that the distribution is clustered toward the upper end of the scale and that the values of 12 and 16 are potential outliers on the lower end. This does not tell the whole story of the distribution, however. Figure 2.36 is a histogram of the data, which can reveal more details about the general shape of the data distribution.

We now see that there are two peaks in the data. There is a concentration around 73 as well as one around 108. If a set of data has two areas of high concentration, it is called **bimodal**. Bimodal data distributions are relatively uncommon and are usually caused by the blending of two different populations, such as heights of men and women. In this case the concentrations seem to represent two different levels of understanding of the material by two groups of students on the exam.

The purpose of this section was to show how we can begin to compel the data to tell their story, using the techniques of the first two chapters. It must be emphasized that without studying

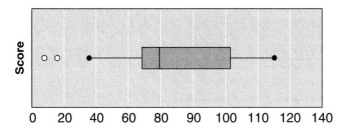

Figure 2.35 Outlier boxplot of statistics examination scores for Example 2.20.

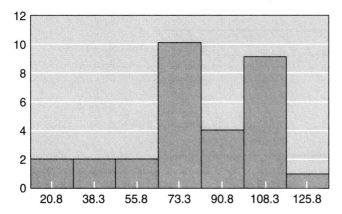

Figure 2.36 Frequency histogram of statistics examination scores for Example 2.20.

statistical inference, the main topic of this textbook, we cannot completely trust the story our pictures and indices (statistics) seem to be telling us, because we simply do not have any objective way of assessing the strength of the evidence. What seems obvious may not in fact have strong evidence to support it. Moreover, our descriptive approaches may entirely miss important characteristics for which there is strong statistical evidence, because the descriptive approaches depend on our informal visual judgment. Nonetheless, the descriptive techniques of Chapters 1 and 2 are essential in helping us organize data and see what seems likely to be true about the data.

Exercises for Section 2.6

The exercises for Section 2.6 do not necessarily all have clear answers. They are meant more to stimulate thinking and provide a sample of the challenging issues that arise when we try to draw conclusions via descriptive statistics alone. They emphasize the need for statistical inference techniques, because the methods of descriptive statistics alone often fail to provide a clear answer.

1. The concept of risk is important in finance. "Risk" is defined as the variability of returns (that is, the variablility of the profit or loss expressed as percentage of investment). Risk is typically measured by the standard deviation. People prefer greater returns to less, of course, and they generally prefer less risk to more risk, because the greater the risk, the greater the uncertainty regarding what the return would be and the greater possibility of loss.

    The following are the returns (in percentage of original investment) on two stocks in the last 6 years.

Stock A:	4.6	4.0	10.3	10.7	7.5	10.7
Stock B:	9.2	7.5	6.6	6.2	7.8	10.3

    If the past pattern of returns is expected to continue in the future, which seems the more preferred investment (other factors being the same)?

2. Two sales representatives have the same sales target. The percentage of the targets achieved by these representatives in the last eight quarters is as follows:

Representative A	Representative B
60	126
85	93
93	92
110	98
109	103
156	100

    a. Compare their average performances.
    b. Using some method of assessing variability, which representative is a more consistent performer?
    c. Is it clear which representative is the better performer?

3. The following data set of heights (in inches) of brothers and sisters comes from a historic study by the famous statistician Karl Pearson.

Family	Height of brother	Height of sister
1	71	69
2	68	64
3	66	65
4	67	63
5	70	65
6	71	62
7	70	65
8	73	64
9	72	66
10	65	59
11	66	62

Source: K. Pearson, and A. Lee, A "On the Laws of Inheritance in Man," *Biometrika*, 2, 357 (1902–1903).

What can you infer from the data? *Hint:* Graph the pairs of heights and look for a pattern.

4. **Mystery.** The following tables contain the casualty data from a certain disaster. People are classified by gender, economic status (I = high, II = middle, III = low, other = ?), and age (adult or child). Look over the tables, and then answer the questions below.

By economic status and sex:

Economic status	Population exposed to risk			Number of deaths			Deaths per 100 exposed to risk		
	Male	Female	Both	Male	Female	Both	Male	Female	Both
I (high)	180	145	325	118	4	122	65	3	37
II	179	106	285	154	13	167	87	12	59
III	510	196	706	422	106	528	83	54	73
Other	862	23	885	670	3	673	78	13	76
Total	1731	470	2201	1364	126	1490	80	27	67

By economic status and age:

Economic status	Population exposed to risk			Number of deaths			Deaths per 100 exposed to risk		
	Adult	Child	Both	Adult	Child	Both	Adult	Child	Both
I (high)	319	6	325	122	0	122	38	0	37
II	261	24	285	167	0	167	64	0	59
III	627	79	706	476	52	528	76	66	73
Other	885	0	885	673	0	673	76	—	76
Total	2092	109	2201	1438	52	1490	69	48	67

Source: Data from Robert J. Mac G. Dawson, *Journal of Statistics Education* 3, no. 3 (1995).

a. Look at the "Population exposed to risk" parts of the tables.
   i. What percentage of the total was from the "other" economic status?
   ii. What is the percentage of females among those in economic status group I? Group II? Group III? The others? What trend do you see?
   iii. What is the percentage of children among those in economic status group I? Group II? Group III? The others? What trend do you see?
b. Now look at the "Deaths per 100 exposed to risk" parts of the tables.
   i. Which status group had the lowest overall death rate? Which had the highest?
   ii. Is the overall death rate higher among males or among females?
   iii. Compare the death rates of males and females for economic group I; for economic group II; for economic group III. What difference do you see among the groups when comparing these male/female rates?
   iv. Look at the death rates for children in economic groups I, II, and III. (There are no children among the others.) What do you notice?
c. Guess what caused all these deaths. Explain what evidence you used to make your guess.
d. Who do you think the "others" were? Why?

CHAPTER 2 SUMMARY

Descriptive data analysis begins with graphic displays. We then add numbers to the description. The most common descriptions of center and spread are the **mean and standard deviation** and the **five-number summary**. The five-number summary consists of the median, the first and third quartiles, and the smallest and largest observations.

Both the mean and the median are measures of the center. The **median** separates the bottom 50% of the data from the top 50% of the data. The **mean** of the data is the usual arithmetic average. For approximately symmetric distributions without outliers, both the mean and the median will be close to the center of the distribution. In skewed distributions and distributions with outliers, the mean is affected by the "pull" of extreme (very large or very small) observations. In skewed distributions, the mean is pulled away from the median toward the long tail. For such distributions, the median is a better measure of the center than the mean.

Unlike most distributions, approximately bell-shaped distributions can be summarized by just two numbers, the mean and the standard deviation.

For approximately bell-shaped distributions without outliers, the **standard deviation** describes the spread of the data around the mean. The 68-95-99.7% Rule applies to such distributions.

z-scores tell how many standard deviations each observation is above or below the mean.

$$z\text{-score} = \frac{\text{raw score} - \text{mean}}{\text{standard deviation}}$$

The **percentile rank** of an observation is the percentage of all the data that are less than the observation.

For approximately bell-shaped distributions without outliers we can estimate the percent of the data that are in a given interval, using only the mean and the standard deviation (and standard normal tables.) For such distributions, the mean and the standard deviation contain (almost) all the information in the data set.

The **first quartile** Q_1 separates the bottom 25% of a data set from the top 75%. The **third quartile** Q_3 separates the bottom 75% of a data set from the top 25%.

For skewed distributions and distributions with outliers, the quartiles and the **interquartile range** IQR = $Q_3 - Q_1$ give a better description of the spread of the data than the standard deviation.

Boxplots, based on the five-number summary, are useful for comparing skewed distributions.

The mean and standard deviation can be affected a lot by a few outliers. We express this by saying that the mean and standard deviation are not **resistant measures**. (They cannot resist the influence of a few outliers.) In contrast, the median and quartiles are resistant.

Measures of Center

Measure	Sensitivity to Extreme Values
Mean	Sensitive
Median	Resistant

Measures of Spread

Measure	Sensitivity to Extreme Values
Range	Very Sensitive
Standard Deviation	Sensitive
Interquartile Range	Resistant

CHAPTER REVIEW EXERCISES

1. On April 16, 2000, the winning teams in the National Basketball Association scored the following number of points.

   95   102   101   114   100   105   85   104   112

   a. Compute the mean, median, standard deviation, and interquartile range.
   b. Suppose the team with 114 points had shot better and scored 133 points. Then what would the above measures be? What does this say about the resistance of these measures to extreme values?

2. A mathematics teacher is considering retirement. He has been told that San Diego and Las Vegas are both good retirement locations, so he decides to do some research on climate. He dislikes big temperature changes from summer to winter. He obtains the following average high temperatures for each month.

	Jan	Feb	Mar	Apr	May	June	July	Aug	Sept	Oct	Nov	Dec
L.V.	55	60	69	79	88	98	105	102	95	81	66	57
S.D.	65	65	67	68	70	72	76	77	77	73	71	67

   Draw boxplots for each set of data and compare the results.

3. Combine the data for both the actors and actresses in the Key Problem into one data set.
   a. Compute the mean, median, interquartile range, and standard deviation for the combined data set.
   b. Draw a box plot for the data.

4. Draw a density histogram for the combined data in Exercise 3. What can you determine about the distribution?

5. Which measure of the center of a distribution (mean, median, or mode) is most likely the best to use in the following cases?
   a. When we have very large or very small values in relation to the rest of the data
   b. When we are concerned with which value occurs most frequently

c. When the data distribution is sharply skewed to the right
d. When the data distribution is sharply skewed to the left
e. When the data distribution is approximately symmetric

6. Is the mode a good measure for the center of a distribution? Explain your answer.
7. *Monitoring the Future*, from the University of Michigan Institute for Social Research and National Institute on Drug Abuse, contains the following data on the percentages of European American (E) and African American (A) 12th-graders heavily using alcohol.

	1980	1987	1988	1989	1990	1991	1992	1993	1994	1995	1996	1997	1998
E	44.6	41.2	38.8	36.9	36.2	32.9	31.3	31.3	31.7	32.9	34.0	36.1	36.6
A	17.0	15.5	14.9	16.6	11.6	11.8	10.8	14.6	14.2	15.5	15.1	12.0	12.7

a. Find the minimum, maximum, median, first and third quartiles, and interquartile range for European American students and for African American students.
b. Construct a boxplot for both sets of data on the same set of axes.
c. Discuss any differences you see between the two groups. *Note:* One could ask about possible trends over time as well. That is the subject of Chapter 3.

8. Discuss why we are interested in resistant measures when doing statistics.
9. The U.S. Federal Trade Commission has released the following data on the amount of tar and nicotine of various brands of cigarettes. All measurements are in milligrams.

Tar	Nicotine	Tar	Nicotine
16	1.2	11	0.9
16	1.2	2	0.2
9	0.8	18	1.4
1	0.1	15	1.2
8	0.8	13	1.1
10	0.8	15	1.0
16	1.0	17	1.3
14	1.0	9	0.8
13	1.1	12	1.0
15	1.2	14	1.0
16	1.2	5	0.5
9	0.7	6	0.6
8	0.7	18	1.4

a. Compute the mean, median, and standard deviation for both tar and nicotine content.
b. Are there any outliers or potential outliers in either group?

10. The *Universal Almanac*, 1993, gave the winning times for the Boston Marathon. The following figures represent the number of minutes over 2 hours for the winning male runner.

Years 1953–1972:

| 18 | 20 | 18 | 14 | 20 | 25 | 22 | 20 | 23 | 23 |
| 18 | 19 | 16 | 17 | 15 | 22 | 13 | 10 | 18 | 15 |

Years 1973–1992:

| 16 | 13 | 9 | 20 | 14 | 10 | 9 | 12 | 9 | 8 |
| 9 | 10 | 14 | 7 | 11 | 8 | 9 | 8 | 11 | 8 |

a. Draw boxplots for both sets of data.
b. Can you determine any appreciable differences between the two different groups?

11. The following data consist of the ages of a random sample of 50 drivers arrested while driving under the influence of alcohol (DUI). These data were given in the *Statistical Abstract of the United States*, 112th edition.

46	16	41	26	22	33	30	22	36	34	63	21	26	18	27	24	31
38	26	55	31	47	27	43	35	22	64	40	58	20	49	37	53	25
29	32	23	49	39	40	24	56	30	51	21	45	27	34	47	35	

a. Construct a boxplot of the data. Are there any outliers or potential outliers?
b. Construct a stem-and-leaf plot of the same data. What can you say about the shape of the distribution?

12. The body weights of fawns between 1 and 5 months old in Mesa Verde National Park have an approximately bell-shaped distribution with a mean of 27.2 kilograms and a standard deviation of 4.3 kilograms (based on information from *The Mule Deer of Mesa Verde National Park*, 1981). Based on that information, compute
a. The approximate percentage of mule deer in Mesa Verde that weigh more than 35.8 kilograms or less than 18.6 kilograms.
b. The approximate percentage of mule deer fawns that weigh between 22.9 kilograms and 31.5 kilograms.

13. Consider the data of the percentage of residents age 65 and older given in Example 2.16. Draw a stem-and-leaf plot for the data. What can you say about the general shape of the distribution?

14. The peak wind gusts (in miles per hour) at Chicago's O'Hare International Airport and Denver's Stapleton Airport for February 1995 are given in the following table.

Day	Chicago	Denver	Day	Chicago	Denver
1	15	32	15	25	14
2	15	39	16	12	17
3	30	15	17	17	22
4	30	27	18	26	36
5	26	17	19	22	17
6	18	16	20	33	20
7	23	14	21	33	13
8	21	14	22	18	18
9	27	32	23	34	12
10	34	26	24	24	14
11	32	20	25	24	17
12	21	12	26	25	25
13	26	11	27	15	26
14	24	38	28	17	13

a. Find the mean and median of the peak wind gusts for Chicago and Denver for February 1995.
b. Find the interquartile range of the peak wind gusts for each city.
c. Draw two box plots, one for each city. What can you conclude about differences between the peak wind gusts for Chicago and Denver for this month?

15. To investigate the work done by income tax preparation specialists, *Money* magazine sent one individual's financial data to 50 tax preparers. Below are the values of the calculated income tax due (in thousands of dollars) by the 50 specialists.

```
12.5  14.7  16.0  16.6  16.7  16.8  16.9  17.3  17.3  17.5  18.4  18.9  19.1
19.1  19.2  19.7  21.1  21.8  21.9  22.6  22.7  22.7  22.7  22.7  22.8  22.9
22.9  22.9  22.9  23.1  23.3  23.4  23.5  23.5  23.5  23.5  23.6  23.7  24.0
24.0  24.0  24.0  24.2  24.4  24.5  25.0  25.3  25.7  25.9  35.8
```

    a. Find the mean and median amount of tax due.
    b. Find the interquartile range of the tax due.
    c. Draw a boxplot of the given data.
    d. Suppose the person with this income tax report was audited by the IRS. How could he or she use this study to his or her advantage?

16. The following table gives the ages of presidents of the United States at their inaugurations.
    a. Construct a stem-and-leaf plot of the ages. What can you conclude about the distribution? From the shape of the distribution, do you expect the median and mean to be about the same, the median to be much greater than the mean, or the mean to be much greater than the median?
    b. Find the mean, median, standard deviation, and interquartile range.
    c. Determine any outliers and potential outliers. Was President Clinton especially young when he was inaugurated compared to the others? Was President Reagan especially old?

Washington	57	Buchanan	65	Harding	55
J. Adams	61	Lincoln	52	Coolidge	51
Jefferson	57	A. Johnson	56	Hoover	54
Madison	57	Grant	46	F. Roosevelt	51
Monroe	58	Hayes	54	Truman	60
J.Q. Adams	57	Garfield	49	Eisenhower	61
Jackson	61	Arthur	51	Kennedy	43
Van Buren	54	Cleveland	47	L. Johnson	55
W. Harrison	68	B. Harrison	55	Nixon	56
Tyler	51	Cleveland	55	Ford	61
Polk	49	McKinley	54	Carter	52
Taylor	64	T. Roosevelt	42	Reagan	69
Fillmore	50	Taft	51	Bush	64
Pierce	48	Wilson	56	Clinton	46

17. A large statistics class took an exam. The scores followed a normal curve with mean 78 and standard deviation 10. Students who scored 90 or above got an A on the exam. Those who scored between 80 and 90 got a B, between 70 and 80 a C, between 60 and 70 a D, and below 60 an F. Estimate the percentage of students who got each of the grades A, B, C, D, and F.

18. In 1994, the Math SAT scores of male high school students nationwide followed a normal distribution with mean 500 and standard deviation 120. The Math SAT scores of female high school students followed a normal distribution with mean 460 and standard deviation 110.
    a. Estimate the percentage of the men who scored 800 or above. (Any score above 800 is reported as 800.)
    b. Estimate the percentage of the women who scored 800 or above.

19. In 2001, the Math ACT scores followed a normal curve with mean 21 and standard deviation 5. What score would place you in the top 1%?

PROFESSIONAL PROFILE

Alan F. Hoskin
Manager
Research and Statistics
Department
National Safety Council
Itasca, Illinois

Statistical methods were used to estimate that child safety seats have saved more than 4000 lives since 1975. Adult safety belts have saved an additional 112,000 lives.

Alan F. Hoskin

In 1999, over 95,000 people died from unintentional injuries in the United States. That's the equivalent of the entire city of Davenport, Iowa. For all the tragedy that represents, it is an improvement over 1969, when more than 116,000 people died unintentionally. The death rate from unintentional injuries has gone down 39% in this nation during the past 30 years. This advance can be credited in large part to the work of the National Safety Council research and statistics department headed by Alan Hoskin.

The National Safety Council looks at the risks we face at home and at work, while driving and flying, while on vacation or enjoying a break from our workaday lives. Under Alan Hoskin's leadership, the Council analyzes the information it receives on accidents, injuries, and deaths in order to make informed suggestions to save lives and prevent injuries.

National Safety Council investigations helped provide proof that car seatbelts save lives. National Safety Council research showed all 50 states that they could save lives by raising the drinking age to 21. Today, seatbelts are a part of our driving experiences, and all states have raised the drinking age to 21.

Alan Hoskin notes that many statistical techniques are critical to making sense of the data they receive. They use descriptive statistics to outline the nature of safety problems. On the research side, the National Safety Council uses such techniques as time series analysis to produce the widely reported estimates of holiday fatalities. These reports, while they may seem gruesome, are a key element in the National Safety Council's countermeasures and safety programs. For these reports, Alan Hoskin *wants* his numbers to be wrong and to have far fewer lives lost. "We've had a major impact," says Hoskin. "People are much less at risk than when I started 27 years ago. There has been an improvement in the quality of life for people in this country." Applying statistics can save lives.

3

Linear Relationships: Regression and Correlation

One has to draw the line somewhere.

Objectives

After studying this chapter, you will understand the following:

- The scatterplot as a graph of two-variable data
- Correlation as a measure of how well a straight line fits two-variable data
- Finding the best-fitting regression line using the least squares method
- Uses and importance of regression
- Coefficient of determination
- Standard deviation of residuals
- Correlation does not imply causation

3.1 SCATTERPLOTS 115

3.2 THE CORRELATION COEFFICIENT 120

3.3 REGRESSION 132

3.4 THE QUESTION OF CAUSATION 148

CHAPTER REVIEW EXERCISES 151

Key Problem

Building a Linear Model of Data

Do heavier cars get poorer gas mileage? We might think so, but how can we be more certain? Table 3.1 gives the weight and city gas mileage of 16 car models. Going through this table of data as it stands, it is hard to tell how the two variables, weight and mileage, are related. In particular, can we estimate the average fuel economy of all vehicles of a particular weight? How much can we trust our estimate? Would the Environmental Protection Agency (EPA) be willing to use our results? As a first step, we graph the data to see whether there is an underlying pattern in the data.

Figure 3.1 is a **scatterplot** of the data. Each point in the plot represents one of the cars from Table 3.1. For example, the point on the far left of the graph (and with the highest fuel economy) is the Geo Metro. From the table we see that the Geo Metro weighs 1695 pounds and gets 46 miles per gallon in city driving. As a review of graphing points, note that one can approximately read these coordinates of the Geo Metro from the graph: about (1700,46).

We can look at the scatterplot of the data and ask whether there appears to be an approximately linear relationship (straight line) between car weight and gas mileage. We see in the plot that, *generally*, the heavier the car, the lower the gas mileage, and we see that this trend seems rather linear in nature. Our Key Problem is to find out how to summarize such a linear relationship between two variables by graphing a line in the scatterplot and writing an equation for this line. This so-called **regression** problem has two interrelated components in its solution: (1) the relationship itself, as described by the equation and graph of the regression line, and (2) a measure of how well the regression line fits the data and, hence, how useful it is. For example, the EPA might want to estimate the average gas mileage for all cars weighing 3500 pounds and then assess the accuracy of this estimate.

Table 3.1 Auto Weights and Mileages

Car	Weight (pounds)	City MPG
Acura Integra	2705	25
Buick LeSabre	3470	19
Cadillac Seville	3935	16
Chevrolet Lumina	3195	21
Chevrolet Astro	4025	15
Dodge Colt	2270	29
Ford Festiva	1845	31
Ford Aerostar	3735	15
Geo Metro	1695	46
Hyundai Scoupe	2285	26
Lincoln Continental	3695	17
Mazda 626	2970	26
Mitsubishi Mirage	2295	29
Pontiac Firebird	3240	19
Subaru Justy	2045	33
Volvo 240	2985	21
Mean	2899	24.25

Source: Journal of Statistical Education (1993) online data sets.

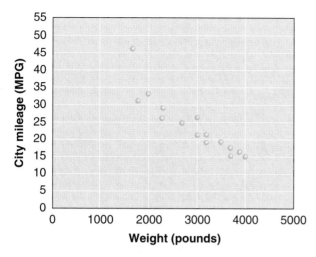

Figure 3.1 Weight versus miles per gallon for city driving for cars in Table 3.1.

3.1 SCATTERPLOTS

So far in this book, we have been dealing with one variable at a time. However, in many situations we want to know how two (or more) variables are related. For example, a pediatrician needs to know how the height and weight of little girls typically increase with age so that she can make sure that her young patients are developing normally.

The focus of this chapter is to study sample data that come in **pairs** like the data in Table 3.1. In the Key Problem we think of weight as the **explanatory variable** and city mileage as the **response variable** since the weight of a car helps explain its gas mileage. In Figure 3.1 we plot the explanatory variable on the horizontal axis (the x axis) and the response variable on the vertical axis (the y axis).

Figure 3.1 shows a **negative correlation** between weight and mileage. Two variables are said to be **negatively correlated** if above-average values of one variable tend to accompany below-average values of the other. The scatterplot slopes downward from left to right.

Two variables are said to be **positively correlated** if above-average values of one variable tend to accompany above-average values of the other and below-average values of one variable tend to accompany below-average values of the other. The scatterplot slopes upward from left to right.

Example 3.1

Cancer is one of the leading causes of death in the United States (See Example 1.3) and worldwide. Although overall cancer rates do not vary much around the world, the types of cancer do. In most affluent countries, cancers of the lung, colon, breast, and prostate are more common. In poorer countries and in the Far East, cancers of the stomach, liver, mouth, esophagus, and uterine cervix are most common. Heredity no doubt plays a role, but migrant studies show that a lot of variation is attributable to environmental factors. For example, women of Puerto Rican descent living in New York City have a much higher breast cancer rate than women living in Puerto Rico. As people around the world adopt Western eating habits and lifestyles they develop the same types of cancer as people in the West. Researchers have tried to identify environmental risk factors for cancer. They suspect that a diet rich in animal fat increases a person's risk for colon, prostate, breast, or ovarian cancer. The scatterplot in Figure 3.2 relates the age-adjusted annual breast cancer mortality rates (MR) for 30 countries and the average daily consumption (in grams) of animal fat per capita.

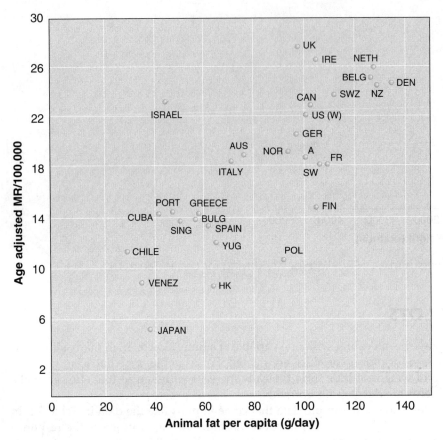

Figure 3.2 Scatterplot of age-adjusted breast cancer mortality rates (per 100,000 population) for 30 countries against average daily consumption of animal fat per capita. (Source: Rose et al. Comparisons of Mortality Rates. *Cancer* 1986; 58:2363–2371.)

Figure 3.2 shows a **positive correlation** between animal fat consumption and breast cancer rates: Countries with low per capita intake of animal fat have lower breast cancer rates than countries with high per capita intake of animal fat. (Israel is an outlier with a high rate of breast cancer in spite of low average animal fat consumption.)

Figure 3.3 shows that there appears to be no relationship between breast cancer risk and consumption of vegetable fat. Greek women, in particular, seem to suffer no ill effects from the large amounts of olive oil they consume.

Epidemiologic studies such as the one summarized in Figure 3.2 make animal fat a suspect in breast cancer deaths. But there are other suspects such as early onset of puberty, lack of regular physical activity, and excess weight, all of which are common in the West.

Section 3.1 Exercises

1. What two variables are of interest when each of these questions is asked? Identify the explanatory variable and the response variable in each case.
   a. Do children who view a lot of television do poorly in schoolwork?
   b. Does a copper bar expand when heated?
   c. Do people who drink more milk become sick less often?
   d. Do tires with higher inflation (more air in them) give better mileage than those with lower inflation?

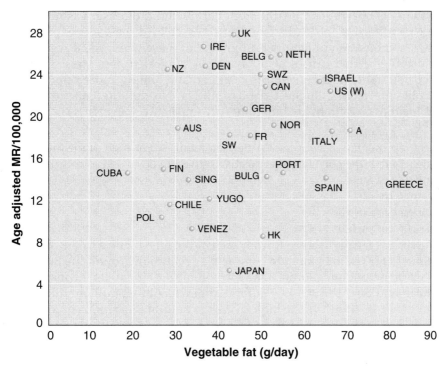

Figure 3.3 Scatterplot of age-adjusted breast cancer mortality rates (per 100,000 population) for 30 countries against average daily consumption of vegetable fat per capita.

2. Professor Eron of the University of Illinois did a study of the television viewing habits of 875 third-grade children. The following excerpt reports his conclusions.

   There is a strong positive relationship between the violence rating of favorite programs watched, whether reported by mothers or fathers, and aggression of boys as rated by their peers in the classroom. There was no significant relationship when TV habits of girls were reported either by mothers or fathers. (*Journal of Abnormal and Social Psychology*, vol. 67, 1963, p. 195.)

   What two pairs of variables are of interest in this study?

3. The following extract reports on a relationship between variables. What are the variables, and what is the relationship? (*Hint:* Pick whatever pair of variables you like.)

   As people get older there is a marked decline in sports involvement—except for the viewing of sports on television. The average time devoted to active sports per week is 5.1 hours for those aged 18 to 24 compared with 1.7 hours for those 25 and over. But the time spent watching TV sports increases somewhat for all those over 25.

4. Identify three sets of variables whose relationships are discussed in the following extract.

   During the months immediately following the 1974 gasoline shortage it was noted that deaths from vehicular accidents were on a significant decline for the first time in decades. The phenomenon was hailed at the time as one of the few benefits from the fuel crunch.

   More recently notice has been paid to other statistics from that period which indicate a drop in deaths from causes other than highway accidents, but which some researchers believe could have a tie-in to reduced driving habits. San Francisco County, for one, noted a 13 percent decline in deaths from all causes during a three-month period at the height of the gasoline shortage.

A decrease of almost 33 percent in deaths from chronic lung disease was noted, along with a 16 percent decline in deaths from cardiovascular diseases. It is possible, of course, that a combination of factors brought about these unexpected results. But a number of driving-related causes also have been suggested: less stress from driving less, fewer pollutants in the air, more walking, and simply more opportunity to relax at home are among suggested possibilities. No definite conclusions have been reached, but the correlations are intriguing. (Champaign-Urbana *News-Gazette*, October 21, 1975.)

5. Imagine that we have a representative sample of American families. For each of the following pairs of variables, state whether the correlation is positive or negative.
   a. Age of husband and age of wife.
   b. Age and price of family car.
   c. Age and systolic blood pressure of wife.
   d. Height of father and (adult) height of son.

6. The following table gives the mean distance R from each of the nine planets to the Sun and the mean length of time T it takes the planet to complete a revolution around the Sun. The distance R is measured in astronomical units. One astronomical unit equals the mean distance between the Earth and Sun, about 149.6 million kilometers or 93 million miles. The time T is measured in years. In addition, the table gives the natural logarithms of R and T, $\ln(R)$ and $\ln(T)$. Draw a scatterplot for the paired values of $\ln(R)$ and $\ln(T)$—nine points in all, one for each planet. Does there appear to be a linear relationship between $\ln(R)$ and $\ln(T)$?

Planet	Distance R (astronomical units)	Time T Years	$\ln(R)$	$\ln(T)$
Mercury	0.387	0.241	−0.95	−1.42
Venus	0.723	0.616	−0.32	−0.48
Earth	1.000	1.000	0.00	0.00
Mars	1.523	1.88	0.42	0.63
Jupiter	5.20	11.86	1.65	2.47
Saturn	9.55	29.46	2.26	3.38
Uranus	19.22	84	2.96	4.43
Neptune	30.11	165	3.40	5.11
Pluto	39.4	248	3.67	5.51

Use the scatterplot to estimate the slope m of the equation

$$\ln(T) = m \cdot \ln(R).$$

(You are deriving Kepler's third law of planetary motion, a famous milestone in astronomy. Kepler deduced his three planetary laws 400 years ago using Tycho Brahe's careful astronomical observations—made without the aid of a telescope! Kepler's laws, in turn, enabled Newton to formulate his theory of gravitational force.)

7. Ask at least five of your friends or relatives their weights and heights. Draw a scatterplot for weight versus height. Is there a linear relationship between the two variables?

8. Find an article in a newspaper or a magazine that deals with a relationship between two variables. What are those variables? What relationship between those two variables is presented or discussed?

9. The following figure shows the amount of nicotine (in milligrams) and tar (in milligrams) for 25 brands of cigarettes.

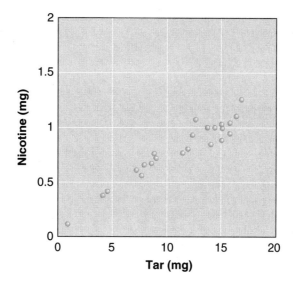

Is the correlation between amounts of tar and nicotine positive or negative?

10. The following figure shows the city and highway mileages (miles per gallon) for 92 cars (1993 models).

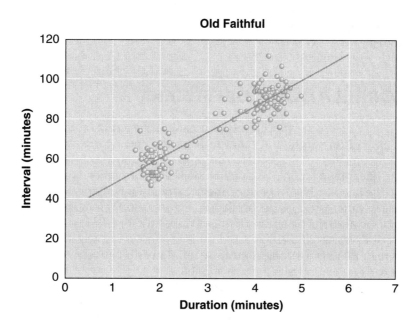

Is the correlation between city mileage and highway mileage positive or negative?

11. For each of the three response variables, blood pressure, height, and weight, pick at least three possible explanatory variables from the following list: age, height, father's height, mother's height, level of stress, number of calories consumed per day, number of hours of exercise per week, weight. For each pair of variables, indicate whether the correlation is positive or negative.

12. The following figure shows the infant mortality rate (deaths per 1000 of population for babies less than one year old) and the female life expectancy in years for 97 countries.

120 CHAPTER 3 LINEAR RELATIONSHIPS: REGRESSION AND CORRELATION

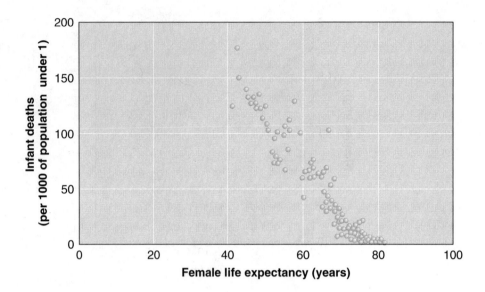

a. What is the range in infant mortality for these countries?
b. Estimate the median infant mortality rate for these countries.
c. What is the range in female life expectancy for these countries?
d. Estimate the median female life expectancy.
e. What is the general relationship between infant mortality and female life expectancy? Is the correlation positive or negative?
f. Does this relationship make sense? Explain.

3.2 THE CORRELATION COEFFICIENT

Suppose that the points of a scatterplot can be described as clustering around a sloped straight line as in Figures 3.1 and 3.2. In addition to characterizing the linear correlation as either positive or negative, depending on whether the scatterplot slopes up or down, we would like to characterize the strength of the linear correlation. We say that the linear correlation is **strong** if the points cluster tightly around the line. We have a **perfect linear correlation** if all the points lie on the line. Figure 3.4 illustrates the different types of correlation. Figure 3.4(*h*) reminds us that the relationship between two variables need not be linear. (Kepler's third law: $T = R^{3/2}$, is an example of a nonlinear relationship. See Exercise 6 of Section 3.1.) In this text, however, we shall only study linear relationships. Although there is a **strong nonlinear relationship** between x and y in Figure 3.4(*h*), there is no **linear correlation** between the two variables.

To obtain an objective measure of the linear correlation between paired x- and y-values, the statistician Karl Pearson introduced the **linear correlation coefficient** r. Later in this section we shall give the formula for computing r. The value of r is always between –1 and +1. If there is a positive correlation between x and y (the scatterplot slopes upward), then r is positive. If there is a negative correlation between x and y (the scatterplot slopes downward), then r is negative. A value of +1 corresponds to a perfect positive linear correlation. The tighter the points of a scatterplot cluster around a straight line with a positive slope, the closer r will be to +1. A value of –1 corresponds to a perfect negative linear correlation. The tighter the points of a scatterplot cluster around a straight line with negative slope, the closer r will be to –1. A correlation coefficient close to 0 indicates that the points are not clustered around any sloped straight line. In short, r tells us how tightly the points of a scatterplot cluster around a sloped line. That is, r measures the strength of the **linear relationship** between the paired x- and y-values. We can classify the linear

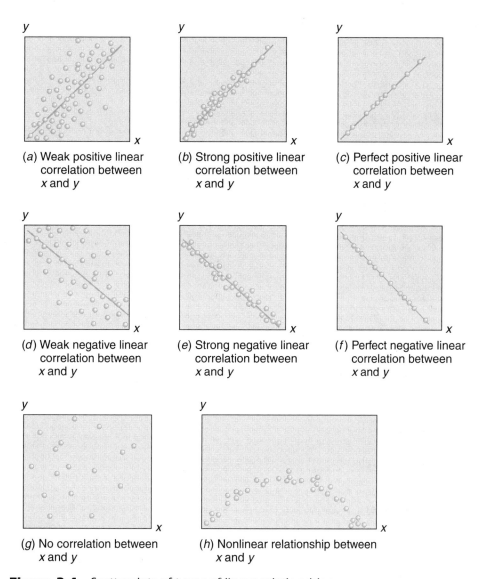

Figure 3.4 Scatterplots of types of linear relationships.

Figure 3.5 Range of values for the correlation coefficient.

correlation as strong, weak to moderate, or negligible, according to the magnitude of r, as indicated in Figure 3.5.

As shown in Figure 3.5, there are five natural categories: strong positive, weak to moderate positive, negligible, weak to moderate negative, and strong negative correlation. (Well-informed statisticians might differ on where to put the cutoff point from one category to the next). As an

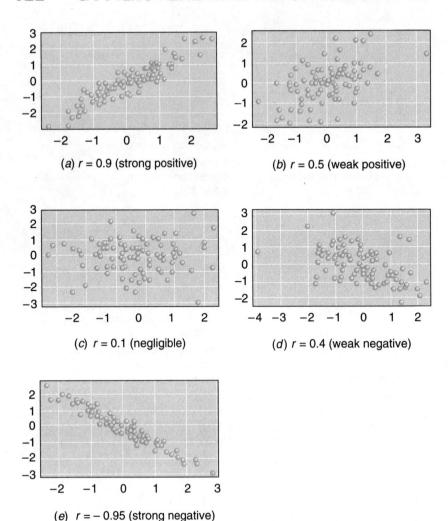

Figure 3.6 Scatterplots illustrating five levels of correlation.

illustration, Figure 3.6 contains five scatterplots, each of which contains 100 simulated data points and represents one of the five categories, with $r = 0.9$ in (a), 0.5 in (b), 0.1 in (c), -0.4 in (d), and -0.95 in (e).

For the Key Problem, the data in Figure 3.1 seem to lie rather close to a negatively sloped straight line; the correlation $r = -0.92$ is close to -1. This is an example of a strong negative correlation.

After we consider more real data examples, you will learn how to compute the correlation.

Example 3.2

Figure 3.7 shows measurements of the length (x) and width (y) of the palate (the roof of the mouth) of each of 44 male kangaroos. (We would have to ask a zoologist exactly how one defines these two quantities.) As expected, they are positively related. Both reflect the overall size of the kangaroo. So

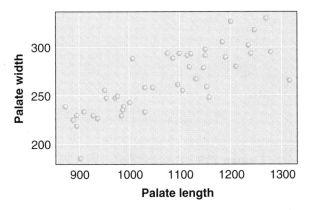

Figure 3.7 Palate width versus palate length of male kangaroos.

a kangaroo with a larger palate length is likely a larger kangaroo and hence tends to have a larger palate width as well. The correlation coefficient between x and y, when calculated, is 0.81, indicating a strong positive correlation.

Example 3.3

For male high school students there is a strong positive correlation between performance in high jump and performance in long jump since both reflect overall ability and training in track-and-field: Students with ability and training do well in both and students without do poorly in both. In the men's decathlon at the Olympics, however, all competitors are both highly trained and very talented. Unusual skill relative to other decathlon contestants in one of the 10 events of the decathlon does not necessarily translate into similar superior skill in the other events at this high level of performance.

Figure 3.8 shows the scatterplot of high jump and long jump results of 34 participants in the men's decathlon in the 1988 Olympics. The variables are measured in meters. There is clearly a positive linear relationship between long jump and high jump performance, but it is not as strong as in Example 3.2. The correlation coefficient between the two variables is only 0.47. This is a weak to moderate positive correlation.

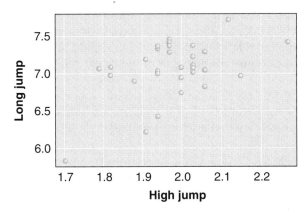

Figure 3.8 Long jump versus high jump results in 1988 men's Olympic decathlon.

124 CHAPTER 3 LINEAR RELATIONSHIPS: REGRESSION AND CORRELATION

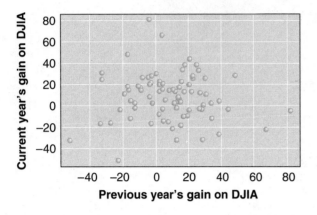

Figure 3.9 Current versus previous year annual returns on Dow Jones Industrial Average.

Example 3.4

People are always searching for systems to make money investing in the stock market. The Dow Jones Industrial Average (DJIA), reported continuously, is a widely quoted stock market index based on the combined worth of 30 of America's giant "blue chip" companies. One might think that if the DJIA increases by a large percentage in one year, the companies it comprises are profitable, so it is likely to increase again in the next year. Similarly, a decrease last year should predict a poorer than typical performance this year.

Figure 3.9 shows 84 points, one for each year from 1916 through 1999. Each point has the previous year's percentage increase of the DJIA as its x coordinate and the current year's percentage increase of the DJIA as its y coordinate. So, for example, the uppermost point of approximately (−4, 80) represents a pair of years, in the first of which the DJIA lost a bit (−4%) but gained 80% in the second. Somewhat surprisingly, perhaps, there appears to be no positively sloped linear relationship in this scatterplot. The correlation coefficient, when calculated, is close to zero ($r = -0.03$), so it represents a negligible correlation. A big gain in the stock market last year does not tell us anything about the likely overall gain this year!

Example 3.5

Figure 3.10 shows a scatterplot for the variables x and y (no physical interpretation given). There is a strong bowl-shaped (parabolic) relationship between x and y. But the correlation coefficient, when calculated, is close to 0. Why does r fail to be positive when in fact there is a strong relationship? Is there a flaw in the correlation coefficient? No. This is because the correlation coefficient measures the strength of a **linear** relationship, not other types of relationships. Thus, you should always be careful to refer to the correlation coefficient r as describing the strength of the **linear** relationship between two variables.

Note also that if the scatterplot is shown, one is unlikely to conclude mistakenly that there is no relationship in the case of a small r when in fact a significant nonlinear relationship holds. Even a zero correlation coefficient does not rule out the possibility that a strong nonlinear relationship exists between x and y, as shown in Example 3.5.

Often a strong nonlinear relationship will also produce a large correlation coefficient, thus suggesting a linear relationship. But we do not want to report incorrectly that there is a linear relationship if the true relationship is a strong nonlinear one. Therefore, the scatterplot is essential in analyzing a relationship.

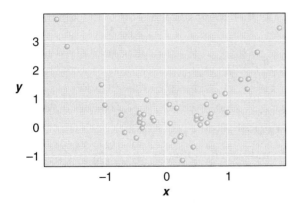

Figure 3.10 A scatterplot with a strong nonlinear relationship but zero correlation.

How to Calculate the Correlation Coefficient r

Recall that for each variable x and y, we can calculate their means $\bar{x}$ and $\bar{y}$ and their standard deviations s_x and s_y using the formulas for mean and standard deviation discussed in Section 2.3. Note that we subscript s with either x or y here because there are two standard deviations, one for the x's and one for the y's. The subscript helps us distinguish them from each other.

The correlation coefficient is defined

$$r = \frac{1}{n-1}\sum\left(\frac{x-\bar{x}}{s_x}\right)\left(\frac{y-\bar{y}}{s_y}\right) = \frac{1}{(n-1)}\frac{1}{s_x s_y}\sum(x-\bar{x})(y-\bar{y}).$$

where n is the number of (x,y) pairs in the data and $\sum$ indicates summation over all the data points. For example, if the set of x's is $\{1, 4, 7, 9\}$ then $\sum x = 1 + 4 + 7 + 9 = 21$. If you are used to seeing a $\times$ sign or a $\cdot$ sign for multiplication, as in 2×3 or $x \cdot y$, note that in complicated formulas the multiplication sign is often omitted, so the product of x and y is indicated simply by xy.

Example 3.6

We shall compute r for the following hypothetical set of 6 data points (x, y):

x	5	2	4	5	6	8
y	12	4	10	10	8	16

It is easily seen that $\bar{x} = 5$, $s_x = 2$, $\bar{y} = 10$, and $s_y = 4$. We standardize each of the x's by first subtracting $\bar{x}$ and then dividing by s_x. We standardize each of the y's by first subtracting $\bar{y}$ and then dividing by s_y. Next we compute the products of the corresponding standardized scores, add up the products, and divide by $n - 1$.

x	y	$\dfrac{x-5}{2}$	$\dfrac{y-10}{4}$	$\left(\dfrac{x-5}{2}\right)\left(\dfrac{y-10}{4}\right)$
5	12	0.0	0.5	0.00
2	4	−1.5	−1.5	2.25
4	10	−0.5	0.0	0.00
5	10	0.0	0.0	0.00
6	8	0.5	−0.5	−0.25
8	16	1.5	1.5	2.25
				4.25

$$r = 4.25/5 = 0.85$$

Look at the two expressions that are multiplied together to make each term in the summation that defines r:

$$\frac{x - \bar{x}}{s_x} \quad \text{and} \quad \frac{y - \bar{y}}{s_y}$$

Each of these expressions is called a **standardized score**.

Computing r using this definition is a four-step spreadsheet-like procedure:

Step 1: Standardize the x's by first subtracting $\bar{x}$ and then dividing by s_x.
Step 2: Standardize the y's by first subtracting $\bar{y}$ and then dividing by s_y.
Step 3: Multiply each standardized x by its associated standardized y.
Step 4: Divide the sum of the products from Step 3 by $n - 1$ (rather than by n, just as we did when we defined the variance and the standard deviation in Section 2.3).

Note that r is almost the average of the products of the standardized scores; if we divided by n rather than by $n - 1$, it would be exactly the average.

Generally you will use computer programs, such as the easy-to-use instructional software available with this textbook, to find statistics such as r. However, it can be dangerous just to hit a computer key and get an instant "answer" unless you truly understand what the computer is computing from the data. When you learn a new statistical procedure, it *always* helps in the learning process to do a couple of problems "by hand" using a calculator or, perhaps, a spreadsheet.

To compute r using the four steps above takes a great deal of computational work. There is an easier way to compute these quantities, however. Take the definition of the standard deviation of x from Section 2.3:

$$s_x = \sqrt{\frac{\sum(x - \bar{x})^2}{n - 1}}.$$

Applying some algebra converts this formula to

$$s_x = \sqrt{\frac{n\sum x^2 - (\sum x)^2}{n(n - 1)}}.$$

This version may look more complicated, but it really requires less computing and saves time. You no longer have to subtract the mean from each entry in the list of numbers to compute the standard deviation. You can set up a spreadsheet-like table with columns for x and x^2, step through the data once to fill out those columns, and use the sums to find the mean and the standard deviation all at once at the end.

Further algebra work along the same lines leads to the computational expression for the correlation coefficient r:

$$r = \frac{n\sum xy - (\sum x)(\sum y)}{\sqrt{[n\sum x^2 - (\sum x)^2][n\sum y^2 - (\sum y)^2]}}$$

where, of course, $\sum xy$ simply means we multiply each point's coordinates together and then add these products over all the points. Although this looks complicated (and certainly need not be memorized!), it is determined by five easy-to-compute sums that can be viewed as summary statistics for the paired variables x and y. This expression is faster to compute if only a hand calculator is available.

The following example shows the use of this formula to compute r.

Example 3.7

Find the correlation between the number of weeks spent in a weight-loss program (x) and the weight lost (in pounds) by participants (y) as given in Table 3.2 and shown in Figure 3.11. As practice, find the standard deviation of x, using the computationally friendly formula.

Solution: We first use the following spreadsheet-like table to compute the five sums with $n = 9$.

Computation of r for Weight Loss Data

Person number	x	y	x^2	y^2	xy
1	12	19	144	361	228
2	4	18	16	324	72
3	8	8	64	64	64
4	4	16	16	256	64
5	12	26	144	676	312
6	7	12	49	144	84
7	5	12	25	144	60
8	9	29	81	841	261
9	14	31	196	961	434
Sum	75	171	735	3771	1579
	$= \sum x$	$= \sum y$	$= \sum x^2$	$= \sum y^2$	$= \sum xy$

Now, using the computational formula, we find the correlation:

$$r = \frac{(9 \times 1579) - (75 \times 171)}{\sqrt{[(9 \times 735) - (75 \times 75)][(9 \times 3771) - (171 \times 171)]}}$$

$$= \frac{1386}{\sqrt{990 \times 4698}} \approx 0.64.$$

Table 3.2 Weight Loss versus Number of Weeks in Program

Number of weeks in program (x)	Total weight loss (y)
12	19
4	18
8	8
4	16
12	26
7	12
5	12
9	29
14	31
Total 75	171
Mean 8.33	19.0

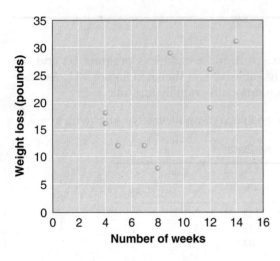

Figure 3.11 Weight loss versus number of weeks in program for nine people in a weight-loss program.

Correlation Is Scale-Free

The correlation coefficient is a pure number, without units, like inches or pounds. It is not affected by:

- adding the same number to all the values of one variable
- multiplying all the values of one variable by the same positive number
- switching the two variables

Example 3.8

We can calculate the correlation between car weight and mileage given in Table 3.1 to get $r = -0.92$. (Try it for practice, using our computational formula.) Here, the x variable is weight in pounds and y is miles per gallon. Suppose that a Japanese scientist is studying the same data set but with the weight reported in kilograms and gas mileage reported in kilometers per liter of gas. The numbers will all be different, and in fact there is a scale change in both variables x and y. For example, a weight of 2500 pounds becomes approximately 1136 kilograms, because 1 kilogram ≈ 2.2 pounds and each mile per gallon equals 0.425 kilometers per liter. Would the Japanese scientist obtain a different correlation for the same variables measured in different scales? We certainly hope not! Imagine the chaos that would result if the computed strength of the linear relationship changed based on the units of measurement used. Fortunately, the correlation stays the same regardless of the scales used for x and for y. The reason is that, in the definition of r, the variables are divided by their standard deviations, which are in the same units as the variables themselves, so the quotients are not influenced by scale changes. Thus, the resulting correlation coefficient is not affected by any scale change in either variable.

Rule: r is the same for all choices of scales for x and for y.

Section 3.2 Exercises

1. The following figure shows graphs of different collections of data in two variables. For each graph (a) through (d), tell by inspection whether the correlation between the two variables is strong positive; weak to moderate positive; weak to moderate negative; or strong negative.

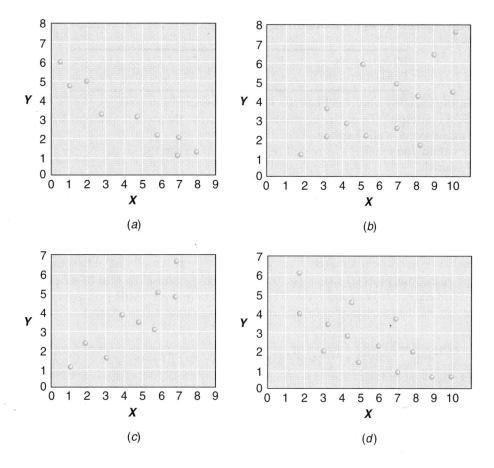

(a) (b) (c) (d)

2. Tell whether you would expect the correlation between the two variables in each of these sets of data to be strong positive, weak to moderate positive, weak to moderate negative, or strong negative. *Hint*: There may be no clear answer for some figures, in that either of two categories of correlation may be reasonable. The point is to get you to think hard about the degree of association.
   a. Height and weight of children from ages 4 to 12 years
   b. Time spent studying for an exam and score on that exam
   c. Attitude toward mathematics (whether or not you like mathematics) and final grade in mathematics
   d. Popularity in school or college and academic achievement
   e. Price of a watch and how many minutes it gains or loses per week
   f. Age of a car and cost per month of repairs
   g. Air pressure kept in automobile tires and gasoline mileage associated with the tires

3. The following table gives the daily high and low temperatures in degrees Fahrenheit (°F) in Champaign, Illinois, for five consecutive days in April 2000.

   | Low | 35 | 38 | 38 | 40 | 42 |
   | High | 63 | 66 | 68 | 69 | 72 |

   a. Calculate the correlation coefficient between the daily high and low temperatures in two ways. Verify that the answers are the same.
   b. Convert the temperatures to degrees Celsius (°C = °F − 32). Recompute the correlation between the two variables now measured in Celsius degrees. Is this what you expected? Why?

4. The following table gives the midterm and final exam scores of eight students in a mathematics class.

Midterm score X	55	60	69	74	79	88	89	95
Final exam score Y	62	74	66	77	82	91	93	60

   a. Compute the correlation between the midterm and final exam scores.
   b. Draw a scatterplot. Does the correlation summarize the overall relationship between the two exam scores well? Explain.
   c. How would you explain the overall relationship between the students' performance in the two exams?
   d. Suppose the teacher decides to assign the following simple scoring system to the exam scores: Any score up to 59 is converted into 1; any score between 60 and 69 is converted into 2; any score between 70 and 79 is converted into 3; any score between 80 and 89 is converted into 4; and any score above 90 is converted into 5. What is the correlation of the new exam scores? *Hint:* The new correlation should be close, but not identical, to the old, because this is only approximately a rescaling. Hence you actually have to compute it!

5. For each of the four plots (a)–(d) in the following figure, choose the best description: (i) strong positive correlation; (ii) moderate positive correlation; (iii) very little positive or negative correlation; (iv) moderate negative correlation; (v) strong negative correlation.

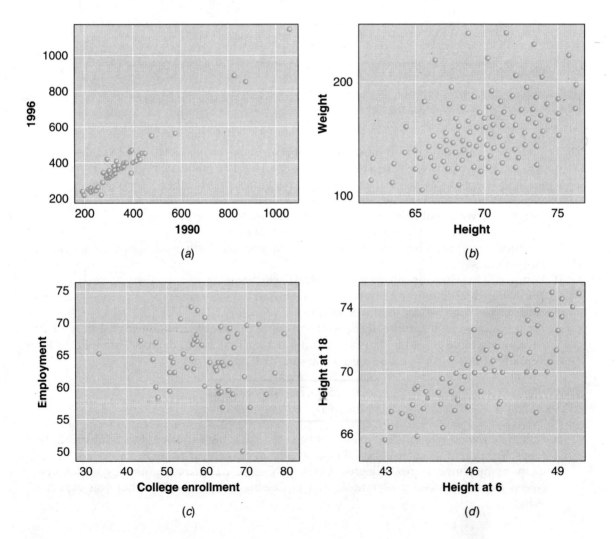

a. Energy usage in 1996 versus energy usage in 1990 for the 50 states
b. Weight versus height for a representative sample of male college students
c. Percentage of adults employed versus percentage of college-age people enrolled in college, for the 50 states (from U.S. Bureau of the Census)
d. Height at age 18 versus height at age 6 for a group of boys participating in a study on human growth

6. Consider the following scatterplots:

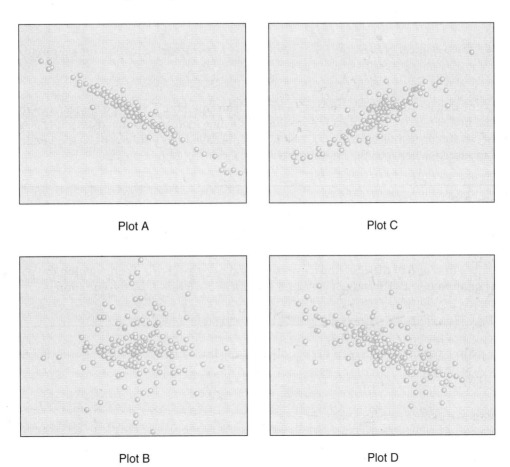

Plot A

Plot C

Plot B

Plot D

The correlation coefficients for the graphs are 0.92, –0.05, –0.83, and –0.98 in some order. Match the scatterplots and the correlation coefficients.

7. Imagine that we have a large representative sample of American families and that we have computed the following correlation coefficients:
   a. between age of husband and age of wife
   b. between height of husband and height of wife
   c. between height of husband today and his height at age 6
   d. between height and weight of husband

The correlation coefficients are 0.25, 0.4, 0.8, and 0.95 in some order. Which is which? *Hint:* Look at the scatterplots in Exercise 5.

8. An article in the August 1986 issue of the *Journal of Marriage and the Family* studied whether there is a link between the loneliness of college females and that of their parents. The participants in the study were 130 female undergraduates and their parents. Students and parents completed questionnaires measuring loneliness, depression, self-esteem, and social skills.

a. The correlation between mother's loneliness score and daughter's loneliness score was found to be 0.26. Interpret this value in layman's terms. (How strong is the correlation?)
b. The correlation between daughter's loneliness score and mother's self-esteem score was found to be –0.14. Interpret this value in layman's terms.
c. The correlation between daughter's loneliness score and depression score was found to be 0.69. Interpret this value.

3.3 REGRESSION

Before we discuss linear regression, let us review the *slope-intercept* equation for a straight line,

$$y = mx + b.$$

Every line that is not parallel to the y-axis has such an equation. The **slope** m measures the steepness of the line; it tells us how much the y-value on the line increases (or decreases, if m is negative) when the x-value increases by one unit. The number b, called the **y-intercept,** is the y-value of the point of intersection between the line and the y-axis. Figure 3.12 illustrates this.

Example 3.9

Find the equation for the line with slope $m = 2$ that goes through the point (3,5).

Solution

First we plug in $m = 2$ in the equation $y = mx + b$ and get $y = 2x + b$. Then we plug the point (3,5) into the equation $y = 2x + b$ and get $5 = 2 \cdot 3 + b$ or $b = 5 - 6 = -1$. So the equation is

$$y = 2x - 1.$$

Some people prefer to memorize the equation of the line with slope m going through the point (x_0, y_0) in the form

$$y - y_0 = m(x - x_0).$$

This equation can of course be rewritten

$$y = mx + (y_0 - mx_0).$$

In other words, the y-intercept is

$$b = y_0 - mx_0.$$

Each student must decide which formula is easier to remember.

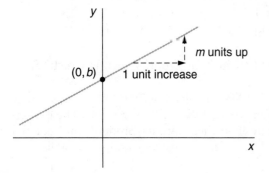

Figure 3.12 The line $y = mx + b$.

Example 3.10

One of the main attractions in Yellowstone National Park is the geyser Old Faithful, which erupts roughly once every 75 minutes. In fact, the intervals between eruptions last between 45 minutes and 2 hours and the eruptions themselves last between 1.5 and 5.5 minutes. Since a long eruption depletes more water from the geyser's underground reservoir than a short eruption, it takes longer to refill the reservoir after a long eruption than after a short eruption. So after a long eruption we should expect a longer wait before the next eruption. As a service to park visitors the Park Service continually posts the estimated time of the next eruption so that visitors know when to gather around the geyser.

Until the Borah Peak earthquake in 1983, the Park Service predicted eruptions using the equation

$$\hat{y} = 10x + 0,$$

where $\hat{y}$ denotes the estimated time interval (in minutes) from the end of the last eruption to the beginning of the next eruption and x denotes the duration (in minutes) of the previous eruption. About two-thirds of the predictions were correct within 5 minutes.

The 1983 Borah Peak earthquake shifted the underground circulation of hot water away from the geyser. As a result, the intervals between eruptions have become slightly longer and less predictable. The Park Service now makes its predictions with a leeway of 10 minutes. In an exercise, we shall return to Old Faithful and look at data from recent eruptions to derive a new formula for predicting the length of the waiting time between eruptions.

Fitting a Line to Data

In Example 3.10 the Park Service had a linear equation that summarized the relationship between the waiting time until next eruption and the duration of last eruption. It allowed the park rangers to predict the waiting time to the next eruption with considerable accuracy. Finding such a linear equation that summarizes the relationship between two variables can be accomplished by drawing a line through a scatterplot and determining the equation of that line. Unless we have a perfect linear correlation, no line will pass through all the points. **Fitting a line** to the data means drawing a line that comes as close as possible to the points. We shall illustrate this basic idea.

The following data were obtained from an experiment that you could easily do yourself. Students in a high school statistics class collected cylindrical objects, such as coffee cans, juice cans, and a rolled oats container. They then measured the circumference (distance around) and the diameter (distance across) of each object and prepared a table like Table 3.3. They used string to measure the circumferences. One purpose of this experiment was to approximately verify the basic relationship between diameter and circumference, namely that circumference = $\pi \times$ (diameter). That is, they were using statistical methods with real data to confirm a basic geometric fact. Another purpose was for the students to discover how unreliable actual measurement can be, even for such a clear and simple geometric concept as the relationship between circumference and diameter.

The class prepared a graph to show how the data are related, as shown in Figure 3.13. As you can see in the graph, the data roughly lie along a straight line.

Table 3.3 Diameters and Circumferences of Selected Cylindrical Objects

Object	Diameter (centimeters)	Circumference (centimeters)
Pill bottle	3.0	10.0
Coffee can (small)	5.0	16.0
Tomato juice can	10.8	32.5
Coffee can (large)	13.0	41.0
Rolled oats container	10.0	32.3
Soup can	6.8	21.0
Candle	4.5	18.0

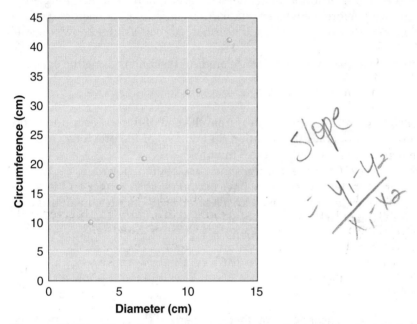

Figure 3.13 Diameters and circumferences of selected cylindrical objects.

Next they looked for a rule (a straight-line equation) that would fit the data. They tried drawing a line through the scatterplot in such a way that it would pass as close to the points as possible. One student took a piece of thread and held it taut over the data points in such a way that it was possible to see where the "best" line should be drawn. When the thread appeared to pass closest to the points, she held the thread as a guide to draw that "best" straight line. Note that she made this judgment informally and subjectively, not guided by any rule or criterion. Imagine, for example, that the student picked the line that goes through the points (3,10) and (13,41). We want to determine the *slope-intercept* equation for this line using the general equation,

$$y = mx + b.$$

We plug the two points into the equation. This gives us

$$10 = 3m + b \quad \text{and} \quad 41 = 13m + b.$$

These two equations in the two unknowns m and b have the solutions $m = 3.1$ and $b = 0.7$. So the equation for the line through the points (3,10) and (13,41) is

$$y = 3.1x + 0.7.$$

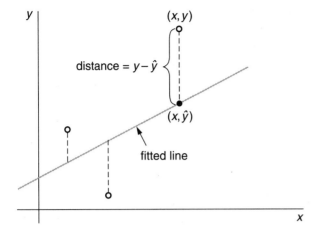

Figure 3.14 The residuals of three points.

We note that this is obviously not the correct formula for computing the circumference y from the diameter x of a circle. But as anyone who has worked in a lab knows, we can not expect a formula derived from real data to be 100% accurate.

Given a scatterplot, a **regression line** is a line that "fits the data" by passing close to the points. A regression line summarizes the relationship between the two variables. The standard statistical convention is to denote the points of the scatterplot by (x,y) and to write the equation of the regression line

$$\hat{y} = mx + b.$$

Thus (x,y) is an actual observed data point, whereas $(x,\hat{y})$ is not an observed data point, but instead gives the estimated y value at x. It is important to remember that, for each observed x, the corresponding value of $\hat{y}$ (called the **predicted value, estimated value,** or **fitted value**) may not be the same as the observed value of y. For example, when $x = 4.5$ centimeters, we obtain from the student's constructed line that $\hat{y} = (3.1 \cdot 4.5) + 0.7 = 14.65$ centimeters. However, the candle with a measured diameter of 4.5 centimeters had a measured circumference of 18.0 centimeters. The **error of estimation** between the actual measured value and the estimated value by the line in this case is

$$y - \hat{y} = 18.0 - 14.65 = 3.35 \text{ centimeters}.$$

This estimation error, the deviation from the fitted line, is often called a **residual.** The residual is the amount that remains unexplained by the estimated rule used to "fit" the data. Note that if we use the correct equation, $\hat{y} = \pi \cdot x$, then $\hat{y} = 14.1$ when $x = 4.5$, and there is still a large estimation error (18.0 cm − 14.1 cm = 3.9 cm). This demonstrates that the students' measuring process was quite inaccurate. The next section provides a statistical method to find the best-fitting line for data when measurement error is present.

In Figure 3.14, we see for each data point that the residual $y - \hat{y}$ is the vertical distance from the data point (x,y) to the fitted line. For points above the line, the residual is positive because $y > \hat{y}$. For points below the line, the residual is negative because $y < \hat{y}$.

The Least Squares Regression Line

Our method of fitting a line to the data in the previous example was subjective. Different students might draw different lines by eye through a scatterplot. This is especially true for scatterplots whose points are more widely scattered than those in Figure 3.13. The almost universally used least squares method gives us an objective way to fit a regression line to the data. We shall explain the idea.

How can we tell whether a straight line fits the data well? Clearly, the "overall average residual size," measured in some way, should be small. Finding a measure for the "overall" residual size is a problem similar to the problem of finding an "overall" deviation from a mean value for a single variable, as discussed in Section 2.3. The most used way of describing the overall size of the residuals is to *square* the estimation errors $y - \hat{y}$, as we did with deviations, and find the sum of all the squared estimation errors. This resulting measure of overall deviation of the observed points from the line is called the **residual sum of squares**, also called the **error sum of squares**, or **SSE**:

$$SSE = \sum(y - \hat{y})^2.$$

Geometrically, this is the sum of the squared vertical distances between the data points (x,y) and the line. (See Figure 3.14.)

For each candidate straight line $\hat{y} = mx + b$, we can calculate the corresponding residual sum of squares. Our best-fitting line is then the line with the smallest possible residual sum of squares. This line is called the **least squares regression line.** The name is appropriate because we are choosing the line that makes the error sum of squares as small as possible. If we imagine several lines, we can see that each line will have its own error sum of squares based on its own set of vertical distances to the data points. Consider the following example for data (0,0), (1,0), (2,3):

For the line $\hat{y} = x$,

$$\sum(y - \hat{y})^2 = (0 - 0)^2 + (0 - 1)^2 + (3 - 2)^2 = 2.$$

For the line $\hat{y} = 1.5x - 0.5$,

$$\sum(y - \hat{y})^2 = (0 + 0.5)^2 + (0 - 1)^2 + (3 - 2.5)^2 = 1.5.$$

The second line (which happens to be the least squares regression line) has a much smaller error sum of squares and, hence, is a better fitting line than the first line. The residuals for both lines are indicated in Figure 3.15.

The following method of finding the least squares regression equation is not the most convenient if you have to use a calculator to determine it from paired data. However, it is easier to understand than the computational version of the equation given later.

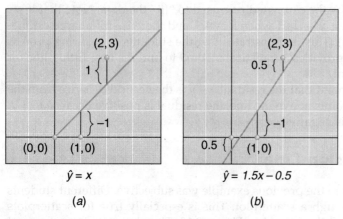

Figure 3.15 Comparison of residuals for two lines.

> **Equation of the Least-Squares Regression Line**
>
> Consider a data set of paired x- and y-values. Let the mean and standard deviation be $\bar{x}$ and s_x for x and $\bar{y}$ and s_y for y. Let r denote the correlation coefficient. The least squares regression line for estimating y from x goes through the point $(\bar{x}, \bar{y})$ and has slope $m = r(s_y/s_x)$. Its equation is $\hat{y} = mx + b$, where $b = \bar{y} - m\bar{x}$, as can be see by plugging $(\bar{x}, \bar{y})$ into the above equation, and then solving for b.

When we use the least squares regression line to estimate or predict one variable from another, our first step is to denote the variable whose value we know by x and the variable whose unknown value we are trying to estimate by y.

$$x = \text{known variable}$$
$$y = \text{unknown variable}$$

We can then compute the slope

$$m = r \frac{\text{SD of unknown variable } y}{\text{SD of known variable } x}$$
$$= r \frac{s_y}{s_x}$$

and the y-intercept $b = \bar{y} - m\bar{x}$. Finally we can estimate y from x:

$$\hat{y} = mx + b = \text{estimated (predicted) } y\text{-value}.$$

In the next example the known variable x is the age of the husband and the unknown variable y is the age of the wife.

Example 3.11

For a representative cross section of several hundred married couples from Illinois, the average age of the husbands was 46 years with a standard deviation of 13 years. The average age of the wives was 43 years with a standard deviation of 13 years as well. The correlation between age of husband and age of wife was 0.95. The least squares regression line for estimating the age y of the wife from the age x of the husband has the equation

$$\hat{y} = 0.95x - 0.70.$$

Here $m = 0.95$ and $b = -0.70$.

Figure 3.16 displays the scatterplot and the regression line. Note how strong the linear relationship is. If a randomly chosen husband from Illinois happens to be 26 years old, then we would predict the age of his wife to be around $(0.95)(26) - 0.70 = 24$ years. If he happens to be 66 years old, then we would predict the age of his wife to be around $(0.95)(66) - 0.70$ years $= 62$ years.

Finally, imagine that you have to guess the age of a randomly chosen wife from Illinois without knowing the age of her husband. What should you guess? The age distribution of married women in Illinois looks something like Figure 3.17. If we have no information about the married woman whose age we are trying to guess, then we might as well use the mean age $\bar{y}$ of all married women as our guess. So we guess that the woman is around 43 years old. This guess could very well be off by 10 to 20 years. In contrast, if we knew the age of her husband and used the regression equation to guess the age of the wife, then our guess would probably only be off by around 3 to 6 years, a dramatic improvement in accuracy.

138 CHAPTER 3 LINEAR RELATIONSHIPS: REGRESSION AND CORRELATION

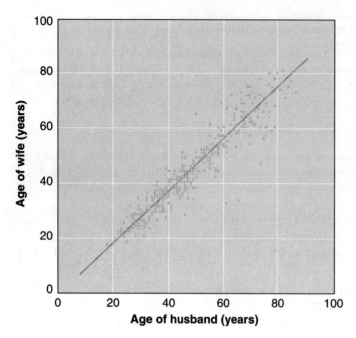

Figure 3.16 The regression line for estimating the age of the wife from the age of the husband.

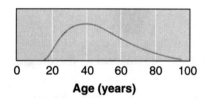

Figure 3.17 Age distribution of married women in Illinois.

Example 3.11 illustrates the fact that if there is a *strong* linear correlation between two variables, then we can use the value of the known variable x to compute a fairly accurate prediction $\bar{y}$ of the unknown variable y.

For large data sets of paired x- and y-values, the regression line can also be used to estimate the *average* value for y corresponding to each value of x, as illustrated by the following example where x = height of father and y = height of son.

Example 3.12

In their 1903 paper, "On the Laws of Inheritance in Man," Pearson and Lee studied the correlation between the heights of 1078 fathers and sons. The average height of the fathers was 67.68 inches with a standard deviation of 2.70 inches. The average height of the adult sons was 68.65 inches with a standard deviation of 2.71 inches. The correlation between height of father and son was 0.514. The regression line for estimating the height of a son from the height of his father has slope

$$m = r\frac{s_y}{s_x} = 0.514\frac{2.71}{2.70} = 0.516,$$

and y-intercept

$$b = \bar{y} - m\bar{x} = 68.65 - (0.516)(67.68) = 33.73.$$

The regression equation is

$$\hat{y} = 0.516x + 33.73.$$

We note that r indicates only a weak to moderate correlation. Therefore the prediction of the son's height y for a *particular father* of height x will not be very accurate. (This is not surprising since the height of the son depends just as much on the height of his mother as on the height of his father.) But the regression equation will estimate with great accuracy the *average height* of the sons of the more than 100 fathers of height $x = 69$ inches (say) in Pearson and Lee's study. Similarly, the regression equation will estimate with great accuracy the average height of the sons whose fathers are 65 inches tall. Figure 3.18 illustrates this.

The points in Figure 3.18 show the *average* height of the sons, for each value of father's height (rounded to the nearest inch.) The points closely follow the regression line $\hat{y} = 0.516x + 33.73$. Associated with each increase of one inch in height of the father there is, on average, an increase of m inches in the height of the son.

Figure 3.19 is a rough sketch of the scatterplot for Pearson and Lee's data on the heights of fathers and sons with the ellipse showing where most of the points lie. The regression line goes through the midpoint of each vertical strip of the elliptical scatterplot, as highlighted for one strip with its midpoint shown.

We thus have two closely related and important uses for the regression line:

i. predicting a new or future y when its x is known (as in Examples 3.10 and 3.11)
ii. estimating the average y for all, or at least a large number of y's, paired with the same specified x (as in Example 3.12)

In both cases, the solution is $\hat{y}$, computed from the regression formula

$$\hat{y} = mx + b.$$

If we know the correlation coefficient r and the two standard deviations s_x and s_y, then we can compute the slope m of the regression line using the formula $m = r(s_y/s_x)$. A more direct and computationally faster formula for computing the slope m from paired x- and y-values is

$$m = \frac{n\sum xy - (\sum x)(\sum y)}{n\sum x^2 - (\sum x)^2}$$

where n, of course, is the number of observations.

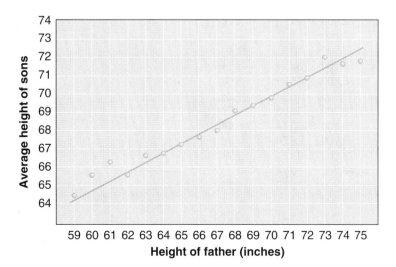

Figure 3.18 Average height of the sons for each value of the father's height.

140 CHAPTER 3 LINEAR RELATIONSHIPS: REGRESSION AND CORRELATION

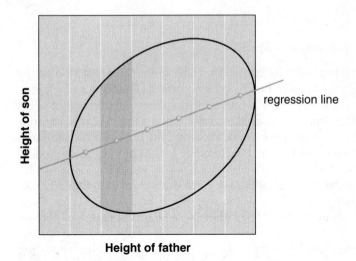

Figure 3.19 The regression line estimates the average value for *y* corresponding to each value of *x*.

We now return to our Key Problem as an example. For the sake of simplicity, we round off the car weight variable to the nearest 100 pounds. We use Table 3.4 to calculate the five sums for the two variables *x* and *y*. The line is now easy to find via the following steps, based on the computationally friendly formula for *m*.

Step 1: Find the means of *x* and *y*. From the table, we have

$$\bar{x} = 46{,}300/16 = 2890$$
$$\bar{y} = 388/16 = 24.30.$$

Table 3.4 Finding the Regression Line for the Key Problem

x	y	x^2	y^2	xy	$\hat{y}$	$y - \hat{y}$
2700	25	7,290,000	625	67,500	26.18	−1.18
3500	19	12,250,000	361	66,500	18.26	.74
3900	16	15,210,000	256	62,400	14.30	1.70
3200	21	10,240,000	441	67,200	21.23	−.23
4000	15	16,000,000	225	60,000	13.31	1.69
2300	29	5,290,000	841	66,700	30.14	−1.14
1800	31	3,240,000	961	55,800	35.09	−4.09
3700	15	13,690,000	225	55,500	16.28	−1.28
1700	46	2,890,000	2116	78,200	36.08	9.92
2300	26	5,290,000	676	59,800	30.14	−4.14
3700	17	13,690,000	289	62,900	16.28	.72
3000	26	9,000,000	676	78,000	23.21	2.79
2300	29	5,290,000	841	66,700	30.14	−1.14
3200	19	10,240,000	361	60,800	21.23	−2.23
2000	33	4,000,000	1089	66,000	33.11	−0.11
3000	21	9,000,000	441	63,000	23.21	−2.21
46,300	388	142,610,000	10,424	1,037,000		
Σx	Σy	Σx^2	Σy^2	Σxy		

Step 2: Find the slope using the computational formula from the preceding page.

$$m = \frac{(16 \times 1{,}037{,}000) - (46{,}300 \times 388)}{(16 \times 142{,}610{,}000) - (46{,}300 \times 46{,}300)}$$
$$= -0.0099$$

Step 3: Find the intercept by plugging $(\bar{x}, \bar{y})$ into $\hat{y} = mx + b$ and solving for b.

$$b = \bar{y} - m\bar{x} = 24.30 - (-0.0099) \cdot 2890 = 52.91$$

Step 4: Plugging in m and b, the least squares line is

$$\hat{y} = 52.91 - 0.0099x.$$

Suppose the Environmental Protection Agency would like to know the average fuel economy for *all* cars weighing $x = 3500$ pounds. Substituting into the regression equation, we estimate that the average fuel economy will be 18.26 miles per gallon (compute this yourself). This 18.26 is also the prediction of fuel economy for a specific car weighing 3500 pounds. The correlation is strong and (x,y) points tend to be close to the line. Thus, this prediction will be fairly good. That is, the 3500-pound cars will not vary much in fuel economy and will all be rather close to the predicted 18.26 miles per gallon.

In this example, we can also compute the standard deviations, $s_x = 758.5$ and $s_y = 8.226$, and the correlation, $r = -0.917$. The slope can then also be computed from its conceptual formula, namely

$$m = r\frac{s_y}{s_x} = -0.917 \, \frac{8.226}{758.5} = -0.0099$$

In Table 3.4 we also give the predicted values $\hat{y}$ and the residuals $y - \hat{y}$.

*Explained and Unexplained Variation

Tall men tend to weigh more than short men. But a man's weight also depends on other factors besides height such as whether he leads a physically active life or not. Let us try to find out to what extent a man's weight can be explained by his height.

A study looked at the relationship between height x and weight y for a large representative cross section of college-age males. The regression formula for computing the estimated weight $\hat{y}$ (in pounds) for a young man of height x (in inches) was

$$\hat{y} = 4.5x - 150.$$

The average height was $\bar{x} = 70$ inches and the average weight was $\bar{y} = 165$ pounds. The correlation was only $r = 0.45$.

The slope of the regression line is $m = 4.5$, which tells us that corresponding to each increase of one inch in height there is, on average, an increase of 4.5 pounds in weight. We expect a young man of average height (70 inches) to be of average weight (165 lbs). We expect a 71-inch tall man to weigh 165 lbs + 4.5 lbs, a 72-inch tall man to weigh 165 lbs + 9 lbs, a 73-inch tall man to weigh 165 lbs + 13.5 lbs, and so forth. (To check this, just plug into the regression formula.)

If a 72-inch tall man happens to weigh 165 lbs + 9 lbs = 174 lbs, then we say that his weight is completely explained by his height. If, on the other hand, the 72-inch tall man happens to weigh 185 lbs, for example, which is 20 lbs above average, then we say that 9 lbs out of the 20 lbs above average are explained by his height and the remaining 11 lbs above average are unexplained. Thus

142 CHAPTER 3 LINEAR RELATIONSHIPS: REGRESSION AND CORRELATION

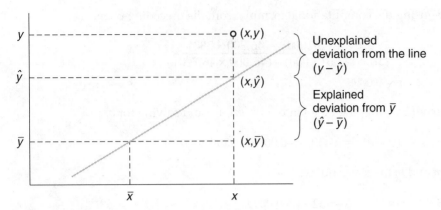

Figure 3.20 Deviations for the regression line.

we split the **total deviation** of the weight y from the mean weight $\bar{y}$, $y - \bar{y}$ ($= 20$ lbs), into the explained deviation from $\bar{y}$, $\hat{y} - \bar{y}$ ($= 9$ lbs), and the unexplained deviation from $\bar{y}$, $y - \hat{y}$ ($= 11$ lbs), as illustrated by Figure 3.20. Thus

$$y - \bar{y} = (y - \hat{y}) + (\hat{y} - \bar{y}),$$

namely,

total deviation = unexplained deviation + explained deviation.

The **total variation** (about the mean $\bar{y}$) of the y-values, $\sum(y - \bar{y})^2$, can also be split into two parts

$$\sum(y - \bar{y})^2, = \sum(y - \hat{y})^2 + \sum(\hat{y} - \bar{y})^2.$$

The last sum, $\sum(\hat{y} - \bar{y})^2$, is called the **explained variation.** It is the part of the variation of the y-values that can be explained by the regression formula from the variation of the corresponding x-values. The sum $\sum(y - \hat{y})^2$ is called the **unexplained variation.** It is the part of the variation of y-values that cannot be explained by the variation of the corresponding x-values.

The unexplained deviation $y - \hat{y}$ shown in Figure 3.20 is of course just another name for what was previously referred to as the residual.

It can be shown that the proportion of the total variation that can be explained by the regression line equals r^2:

$$r^2 = \frac{\text{explained variation}}{\text{total variation}} = \frac{\sum(\hat{y} - \bar{y})^2}{\sum(y - \bar{y})^2}.$$

This quantity is often called the **coefficient of determination.** The closer r^2 is to 1, the tighter the points of the scatterplot cluster around the line, and the better the fit of the line. In other words, the better the regression line determines y. The quantity $100r^2$ is referred to as **"the percentage of the total variation in y that can be explained by the variation in x."**

In the Key Problem, $r = -0.92$ and $r^2 = 0.85$. So 85% of the total variation in gas mileage can be explained by the variation in weight. In Example 3.12, $r = 0.51$ and $r^2 = 0.26$. So only 26% of the total variation in height of the sons can be explained by the variation in height of the fathers. Fully 74% of the total variation remains unexplained.

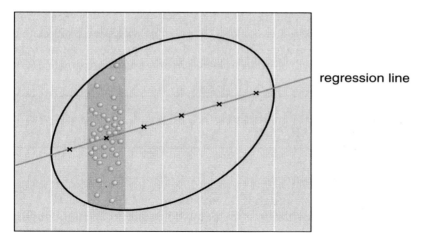

Figure 3.21 For large data sets with elliptical scatterplot, the residuals in each vertical strip have mean close to 0 and a standard deviation close to $s_y\sqrt{1-r^2}$.

In our example on the heights and weights of college-age men, $r = 0.45$ and $r^2 = 0.20$. So only 20% of the total variation in weight can be explained by the variation in height. Fully 80% of the variation in weight is caused by factors other than height.

*The Residuals

Consider again a sample n of paired x- and y-values with least squares regression equation

$$\hat{y} = mx + b$$

for estimating the response variable y from the explanatory variable x.

Fact 1. The residuals $y - \hat{y}$ have mean 0.

Fact 2. The standard deviation of the residuals $\left(\sqrt{\Sigma((y-\hat{y})^2/(n-1))}\right)$ can be computed

$$(\text{SD of residuals}) = s_y\sqrt{1-r^2}.$$

For **large data sets** we can apply the 68-95-99.7% rule to the residuals:

Around 68% of the residuals are less than 1 SD in size and around 95% of the residuals are less than 2 SDs in size. If the scatterplot is *elliptical*, as shown in Figure 3.21, and we use the regression equation to estimate future values of the response variable y from the explanatory variable x, then the SD of past residuals tells us the likely size of future estimation errors.

Example 3.13

Consider again the data on the ages of husbands and wives in Example 3.11. If we have to guess the age of a randomly chosen wife without knowing the age of her husband, then our best guess is $\bar{y} = 43$ years. The likely size of our estimation error is $s_y = 13$ years. If on the other hand we know the age of her husband and use regression to estimate her age, then the likely size of our estimation error is

$$(\text{SD of residuals}) = s_y\sqrt{1-r^2}$$
$$= (13 \text{ years})\sqrt{1-(0.95)^2} = 4 \text{ years}.$$

If the husband is 26 years old, then the regression prediction of his wife's age is 24 years. (See Example 3.11.) If the husband in 66 years old, then the regression prediction of his wife's age is 62 years. If we predict the age of each Illinois wife from the age of her husband using the regression formula, then approximately 68% of all our predictions will be off by at most 4 years. And approximately 95% of all our predictions will be off by at most 8 years.

Section 3.3 Exercises

1. **a.** Find the equation of the line that passes through (1,3) with the slope 4/3.
   b. Find the equation of the line that passes through the points (5,3) and (6,2). *Hint:* Find its slope using the two points.
   c. Which of the following points are on the line $y = 2 - 3x$: (1,1), (1,–1), (0,2)?

2. The following table gives the length (from nose to tail) and weights of eight laboratory mice.

Length (centimeters), x	Weight (grams), y
16	32
15	26
20	40
13	27
15	30
17	38
16	34
21	43

   a. A regression equation $\hat{y} = 1.5x + 8.81$ was proposed to predict the weight of a mouse given its length. The following table was constructed using this regression line.

x	y	$\hat{y}$	$y - \hat{y}$	$(y - \hat{y})^2$
16	32	32.81	–0.81	0.6561
15	26	31.31	–5.31	28.1961
20	40	38.81	1.19	1.4161
13	27	28.31	–1.31	1.7161
15	30	31.31	–1.31	1.7161
17	38	34.31	3.69	13.6161
16	34	32.81	1.19	1.4161
21	43	40.31	2.69	7.2361

   b. What is the error sum of squares for estimating weight using this equation? (x = length in centimeters, y = weight in grams.)
   c. The regression equation

   $$\hat{y} = 1.9x + 2.16$$

   is used in the following table to estimate the weight of a mouse given its length. What is the error sum of squares for this equation?

Section 3.3 Regression   145

x	ȳ	ŷ	y − ŷ	(y − ŷ)²
16	32	32.56	−0.56	0.3136
15	26	30.66	−4.66	21.7156
20	40	40.16	−0.16	0.0256
13	27	26.86	0.14	0.0196
15	30	30.66	−0.66	0.4356
17	38	34.46	3.54	12.5316
16	34	32.56	1.44	2.0736
21	43	42.06	0.94	0.8836

d. Which of the two equations

$$\hat{y} = 1.5x + 8.81$$

$$\hat{y} = 1.9x + 2.16$$

fits the mice data better, in terms of error sum of squares?

3. Refer back to Exercise 2.
   a. Find the least squares regression line for predicting the weights of the laboratory mice from their lengths.
   b. What percentage of variability in the weight of those mice can be explained by their length? (*Hint:* Find r^2.)
   c. Find the least squares regression line for predicting the lengths of the laboratory mice from their weights.
   d. What percentage of variability in the lengths of those mice can be explained by their weights?

4. The data in the table below are the numbers of push-ups y that could be done by a sample of 12 male instructors at Howard Community College. (The 38-year-old teacher was the track coach.)
   a. Draw the scatterplot. Identify the outlier.
   b. Find the correlation of these data.
   c. Find the slope of the least squares best-fitting line for predicting number of push-ups from age.
   d. Find the equation of the regression line with this slope. Then use the regression line to estimate the number of push-ups that can be done on average by teachers of each of the following ages:
       i. 26 years
       ii. 39 years
       iii. 50 years

Age	Number of push-ups
21	10
25	8
22	11
28	6
30	7
38	15
22	9
27	6
44	4
48	3
35	8
48	5

146 CHAPTER 3 LINEAR RELATIONSHIPS: REGRESSION AND CORRELATION

5. For a representative cross section of several thousand American men, the average height was found to be 68.8 inches with a standard deviation of 2.8 inches. The average weight was 171 pounds with a standard deviation of 30 pounds. The correlation between height and weight was 0.4.
   a. Find the equation of the regression line for estimating weight (y) from height (x).
   b. Estimate the average weight of all the men who are 66 inches tall.
   c. Predict (guess) the weight of a randomly chosen man who happens to be 66 inches tall.
   d. What percentage of the variation in weight can be explained by the variation in height?
   e. Predict (guess) the weight of a man who is 72 inches tall.
6. In a study of the stability of IQ scores, a large group of people was tested at age 18 and again at age 35. At both age 18 and age 35, the average IQ score was around 100 with a standard deviation around 15. The scatterplot was elliptical with a correlation around 0.8.
   a. Find the regression equation for estimating the IQ score at age 35 (y) from the IQ score at age 18 (x).
   b. Estimate the average score at age 35 for all the people who scored 120 at age 18.
   c. Predict the IQ score at age 35 for someone who scored 120 at age 18.
7. The following six statements refer to scatterplots and their regression lines in general. For each statement, determine whether it is true or false.
   a. If $r = 0$, then all the points are on the regression line.
   b. If all of the residuals are 0, then all the points are on the regression line.
   c. If all of the residuals are 0, then r is either -1 or $+1$.
   d. If the SD of the residuals is 0, then $r = 0$.
   e. If the slope of the regression line is 0, then the SD of the residuals is 0.
   f. If $r = 0$, then the slope of the regression line is 0.
8. Here is a scatterplot showing the relationship between body temperature (degrees Fahrenheit) and heart rate (beats per minute) for a group of people. Body temperature is the x variable, and heart rate is the y variable. The average body temperature is 98.25 degrees Fahrenheit, and the SD of body temperature is 0.73 degrees Fahrenheit. The average heart rate is 73.8 beats per minute, and the SD of the heart rates is 7.03 beats per minute. The correlation coefficient is 0.25. The regression line for predicting heart rate from temperature is shown.

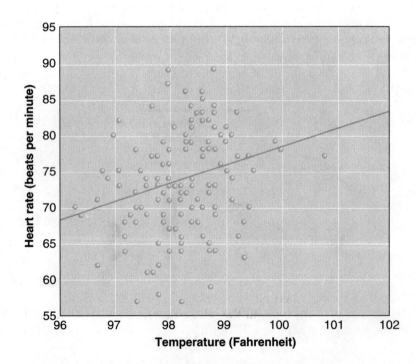

a. What percentage of the variation in heart rate can be explained by the variation in temperature?
b. Find the equation of the regression line for estimating heart rate from temperature.
c. Estimate the average heart rate (in beats per minute) for all people with a body temperature of 98.75°F.
d. Estimate the average heart rate (in beats per minute) for all people with a body temperature of 97.5°F.
e. Predict the heart rate (in beats per minute) for a person with a body temperature of 98.75°F.

9. In one large study of identical male twins, the average height was found to be about 175 cm, with an SD of about 7.5 cm. The correlation between heights of twins was about 0.95. The scatterplot was elliptical. (In twin studies, the convention is to plot each twin pair twice: once as (x,y), and once as (y,x). That is, the John/Jack twin pair with John being 175 cm tall and Jim 172 cm tall produces **two** data points in the plot: (175,172) and (172,175). Thus, $\bar{x} = \bar{y} = 175$ and $s_x = s_y = 7.5$.)
   a. Using the regression line, the average height of those men whose identical twin was 175 cm tall was around:
       165 cm   170 cm 160 cm   180 cm   175 cm
   b. Using the regression line, the average height of those men whose identical twin was 165 cm tall was around:
       164 cm   164.5 cm   165 cm   165.5 cm   166 cm
   c. Using the regression line, the SD of the heights of those men whose identical twin was 175 cm tall was around:
       2.5 cm   5 cm   7.5 cm   10 cm   12.5 cm
   d. If you had to guess the height of a randomly chosen subject from this study, your best guess would be:
       165 cm   170 cm   160 cm   180 cm 175 cm
   e. The likely size of the prediction error in **d** is:
       2.5 cm   5 cm   7.5 cm   10 cm 12.5 cm
   f. A randomly chosen subject stands before you. He is 165 cm tall. You have to guess the height of his twin brother. What should you guess?
       164 cm   164.5 cm   165 cm   165.5 cm   166 cm
   g. The likely size of the prediction error in **f** is:
       2.5 cm   5 cm   7.5 cm   10 cm   12.5 cm

10. The following scatterplot shows the relationship between the duration of an eruption of Old Faithful (in minutes) and the length of the time interval (in minutes) till the next eruption for 150 eruptions in July 1995. The average duration was 3.35 minutes with a standard deviation of 1.157 minutes. The average time interval till the next eruption was 77.94 minutes with a standard deviation of 16.58 minutes. The correlation was 0.9186. The regression line for estimating the time interval till the next eruption from the duration of the last eruption is shown.
   a. Look at the scatterplot. Is the mean duration of 3.35 minutes the "typical" length of an eruption? (Do a lot of eruptions last between 2.5 minutes and 4 minutes?)
   b. What percentage of the variation of the length of the time interval till the next eruption can be explained by the variation in duration of the previous eruption?
   c. Find the equation of the regression line for estimating the time interval till the next eruption from the duration of the last eruption.
   d. Predict (guess) the time interval till the next eruption if the duration of the last eruption was 2 minutes.

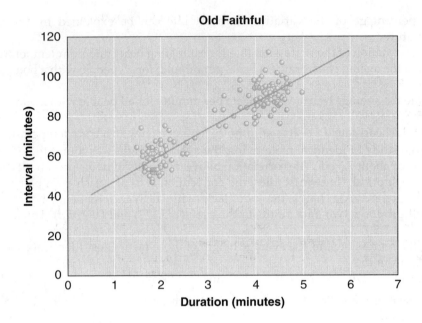

e. Predict (guess) the time interval till the next eruption if the duration of the last eruption was 4.5 minutes.
f. Compute the SD of the residuals.
g. If we use the regression equation to predict the time interval till the next eruption, what is the likely size of our prediction error? Apply the 68-95-99.7% rule to make this precise.

3.4 THE QUESTION OF CAUSATION

In the Key Problem we found a strong correlation between the weight of a car and its gas mileage, reflecting a **cause-and-effect** relationship between weight and mileage: The heavier the car is, the more energy it takes to drive it, and hence the fewer miles per gallon the car will get. There is also a cause-and-effect relationship between the duration of an eruption of Old Faithful (Example 3.10) and the length of the time interval until the next eruption. In Example 3.1 we wondered whether the strong correlation between animal fat consumption and breast cancer rates meant that the consumption of animal fat **causes** breast cancer.

Epidemiologists use statistical methods to search for causes of diseases. Several years before the discovery of the role of bacteria in disease transmission, the nineteenth-century English physician John Snow was able to locate the source of the 1854 outbreak of cholera in London.

A week into the outbreak, Snow obtained the addresses of all the people who had died from cholera. He found that nearly all the deaths had taken place in the neighborhood of Broad Street, Golden Square, and the adjoining streets. By interviewing survivors Snow found that the few cholera deaths outside this area were in households that usually sent for water from the much-frequented street-pump in Broad Street, since they preferred the water from the Broad Street pump to that of nearer pumps. (In those days most people got their water from public wells.) Since the only circumstance all the cholera victims seemed to share was that they drank water from the Broad Street pump, Snow concluded that the Broad Street pump was somehow the source of the cholera outbreak. He persuaded authorities to remove the handle of the pump to prevent people from using the pump. This brought the cholera outbreak to an end.

John Snow's detective work was classical epidemiology. Even though no one at the time had identified the water-borne bacteria that cause cholera, Snow was able to link people's risk of becoming cholera victims during the 1854 outbreak to whether or not they drank water from the Broad Street pump. Unfortunately, such positive correlational statistical links between risk factors (drinking water form the Broad Street pump) and diseases (cholera) can be misleading and may not represent cause-and-effect relationships.

Some medical studies in the 1970s found a link between coffee drinking and heart disease: The rate of heart disease was higher for coffee drinkers than for nondrinkers. And among coffee drinkers, the rate was highest for those who drank a lot of coffee. In other words, the researchers found a positive correlation between the amount of coffee people drank and their risk of heart disease. Did this mean that coffee drinking **causes** heart disease? Further studies revealed that the coffee drinkers tended to smoke more than the nondrinkers. It turned out that it was their smoking that gave the coffee drinkers heart disease, not the coffee. The number of cigarettes smoked per day is the **lurking variable** (or **hidden variable**) that explains the correlation between the number of cups of coffee per day and the risk of heart disease.

A **lurking variable** in a study is one that is not included in the study initially (the number of cigarettes smoked per day), but that affects the response variable (the risk of heart disease), and is strongly correlated with the suspected explanatory variable (the number of cups of coffee per day).

If we cannot separate the effects of two or more variables, then we say that they are **confounded.** For example, studies have found that women who eat a high-fiber diet have a reduced risk of heart disease. Does that mean that dietary fiber protects a woman against heart disease? Or could the explanation be that women who choose their diet carefully also exercise more and avoid smoking, factors that do reduce the rate of heart disease? In this example, the effects of a high-fiber diet are confounded with the effects of exercise and an otherwise healthy lifestyle.

Strong correlation suggests but does not prove causation. To establish causation statistically, one must perform randomized controlled experiments (discussed in Chapter 7), in which the suggested causal variable is varied and the effect is observed. If it is not possible to conduct such experiments, then we must have other scientific proof. One of the best-known examples of the confusion about correlation and cause regards smoking and cancer. For the past 50 years or more, research projects have produced data showing a strong positive correlation between the amount a person smokes and that person's risk of getting cancer. On one side of the debate, medical associations and consumer groups have insisted that smoking causes cancer. On the other side—notably the tobacco industry—it has, until recently, been argued that although smoking and cancer are statistically linked, smoking had not been established scientifically as a cause of cancer because other explanations for the observed correlations between smoking and lung cancer were possible. The problem was that the best way to prove that smoking causes cancer is to conduct an experiment in which scientists tell the participants how much to smoke and then record who gets cancer and who does not. Obviously, nobody would participate in such an experiment. Instead scientists experiment on animals. Carefully designed studies on other mammals have shown strong evidence of causality in these species, which likely applies to humans too. The study of smoking and lung cancer from the statistical perspective is discussed at length in Chapter 7.

Section 3.4 Exercises

1. Several medical studies have found that snoring is a risk factor for heart disease. The more a person snores, the higher the risk of heart disease. (See Exercise 2 in Section 1.1.) Does this mean that snoring causes heart disease? If not, identify a possible lurking variable.
2. For American cities, there is a strong positive correlation between the number of churches in the city and the number of violent crimes reported to the police. Does this mean that if we

build more churches, then there will be an increase in violent crime? Identify the lurking variable that explains the correlation.

3. For school children there is a positive correlation between shoe size and vocabulary size (the number of different words the child uses). Identify the lurking variable that explains the correlation.

4. There is a positive correlation between the weight of a Thanksgiving turkey and the amount of time it takes to cook it. Is this a cause-and-effect relationship, or can you think of a lurking variable?

5. There is a negative correlation between the number of influenza cases reported in Chicago each month of the year and the amount of ice cream sold in Chicago that month. Does this mean that ice cream protects you against the flu? If not, identify the lurking variable.

6. For the world's nations there is a negative correlation between the infant mortality rate (number of infant deaths per 1000 live births) and the number of TV sets per capita. Does this mean that we can lower the infant mortality rate in Ethiopia by giving the Ethiopians more TV sets? If not, identify the lurking variable.

7. People who use artificial sweeteners tend to weigh more than people who use sugar. Does this mean that if you switch from sugar to artificial sweeteners, then you will put on weight? Explain. (The explanation may be slightly tricky—be creative!)

8. For the last 30 years the number of infant deaths in the United States has declined from year to year, whereas the number of deaths due to diabetes has increased from year to year. Is this negative correlation between number of infant deaths and number of deaths due to diabetes a cause-and-effect relationship?

CHAPTER 3 SUMMARY

Paired observations of two quantitative variables can be displayed in a **scatterplot.** We use the scatterplot to study the relationship between the two variables. If we suspect a **cause-and-effect** relationship between the two variables, then we plot the suspected **explanatory variable** on the horizontal axis (the x axis) and the **response variable** on the vertical axis (the y axis). If there is no explanatory-response distinction, then either variable can go on the horizontal axis.

If the scatterplot slopes upward from left to right, then we say that the variables are **positively correlated.** If the scatterplot slopes downward from left to right, then we say that the variables are **negatively correlated.** If the points of the scatterplot are tightly clustered around a sloped line, then we say that there is a **strong linear correlation** between the variables.

The **linear correlation coefficient** r measures the strength and direction (up or down) of a straight-line relationship. The value of r ranges from -1 (when all the points lie on a line that slopes down) to $+1$ (when all the points lie on a line that slopes up). The closer r is to -1, the tighter the points cluster around a line that slopes down. The closer r is to $+1$, the tighter the points cluster around a line that slopes up. If r is close to 0, then the points of the scatterplot are not clustered around any sloped line.

The **least squares regression line** for estimating y from x goes through the point $(\bar{x}, \bar{y})$ and has slope $m = r(s_y/s_x)$. Its equation is

$$\hat{y} = mx + b,$$

where $b = \bar{y} - m\bar{x}$. We call $\hat{y}$ the **fitted** y-value, the **estimated** y-value, or the **predicted** y-value corresponding to x. The regression line (and its equation) can be used to predict **individual** y-values from the corresponding x-values.

For **large** data sets of paired x- and y-values, the regression line estimates the **average** value of y corresponding to each value of x.

We call $r^2 = $ (explained variation)/(total variation) $= \sum(\hat{y}-\bar{y})^2 / \sum(y-\bar{y})^2$ the **coefficient of determination**. The quantity $r^2 \cdot 100\%$ is **the percentage of the total variation of y that can be explained by the variation of x**. For each data point (x,y) the **error of prediction**, $y - \hat{y}$, between the actual y-value and the estimated value is called the **residual**. The residuals have mean 0 and standard deviation

$$(\text{SD of residuals}) = s_y \sqrt{1-r^2}.$$

For **large** data sets with **elliptical** scatterplot, the SD of past residuals tells us the **likely size of the future prediction errors**, either for a specified x or for all x's.

CHAPTER REVIEW EXERCISES

1. For each of the following questions, identify the two variables of interest, and state whether you would expect the correlation coefficient to be (i) close to 1 (strong positive), (ii) close to 0 (negligible), (iii) close to –1 (strong negative), (iv) positive but close to neither 1 nor 0 (weak to moderate), (v) negative but close to neither –1 nor 0 (weak to moderate).
   a. Do taller basketball players block more shots?
   b. Will the number of times a student skips a class affect his or her score on the final exam?
   c. Do students from wealthier families perform better on standardized tests?
   d. Do people who have more years of education earn more money?
   e. Is there a relationship between the population of a county and the number of farms in that county?

2. Respond *true* or *false* to each of the following statements. If the statement is false, explain why.
   a. The correlation coefficient can be useful in determining how well a regression line fits the data.
   b. It can happen that the correlation coefficient is negative and the slope of the least squares best-fitting line is positive.
   c. The method of least squares is usually effective and is the most widely used method for finding regression lines.

3. The following table contains verbal (x) and quantitative (y) scores on the SAT from five students. A researcher wants to answer this question: "Is there a relationship between verbal and quantitative scores on the SAT?" For statistical evidence, the researcher wants to calculate the correlation coefficient (r) for these data. This exercise will take you through the four steps of calculating r.

Student	x	y	$x - \bar{x}$	$y - \bar{y}$	$(x - \bar{x})(y - \bar{y})$	$(x - \bar{x})^2$	$(y - \bar{y})^2$
1	670	710					
2	550	500					
3	720	620					
4	410	490					
5	520	560					
Sum	2870	2880					

   a. Complete the table.
   b. Using the information from the table you just formed, find the variance s_y^2 of y, the variance s_x^2 of x, the standard deviation s_x of x, and the standard deviation s_y of y.
   c. Find the correlation between x and y.

d. What kind of relationship between mathematical and verbal scores on the SAT does this value of r imply?
   e. Find the least squares regression line.
   f. Bob gets a verbal score of 700. Predict his quantitative score.
4. In their study of over 1000 pairs of husbands and wives, Pearson and Lee found the average height of the husbands to be 67.7 inches with a standard deviation of 2.7 inches. The average height of the wives was 62.5 inches with a standard deviation of 2.4 inches. The correlation between height of husband and height of wife was 0.28.
   a. Is this a strong or a weak correlation?
   b. Determine the regression equation for estimating the height of a wife from the height of her husband.
   c. What percentage of the total variation in height of the wives can be explained by the variation in height of their husbands?
   d. Estimate the average height of the wives whose husbands were 73 inches tall (to the nearest inch).
   e. Estimate the average height of the wives whose husbands were 65 inches tall (to the nearest inch).
5. The following table is a list of 10 baseball teams and their total number of runs scored and home runs hit from the 1996 season.
   a. The regression line obtained by the method of least squares is $\hat{y} = 2x + 455$. Which of the 10 observations has the largest squared error? Which has the smallest?
   b. A team not on the list, the Atlanta Braves, hit 197 home runs. Predict how many runs they scored.

Team	Number of home runs (x)	Total number of runs (y)	$\hat{y}$	$(y - \hat{y})^2$
Colorado Rockies	221	961	897	4096
Pittsburgh Pirates	138	776	731	2025
Los Angeles Dodgers	150	703	755	2704
St. Louis Cardinals	142	759	739	400
Cincinnati Reds	191	778	837	3481
Chicago Cubs	175	772	805	1089
Florida Marlins	150	688	755	4489
San Diego Padres	147	771	749	484
New York Mets	147	746	749	9
Houston Astros	129	753	713	1600

6. Find a newspaper article that cites a study in which two variables are related to each other.
   a. What are the two variables that are being related?
   b. Even if there is a relationship between these two variables, does that mean that one is causing the other? Do you believe that the authors of this article answered this question of causation adequately?
7. Consider the following section from a published health study:

   Among the current smokers who had high fish intake, the number of cigarettes smoked per day had no effect on the incidence of heart disease and the death rates from heart problems. As seen with lung diseases, the consumption of fish seems to protect smokers from the harmful effects of smoking on the heart and blood vessels. It has been suggested that fish oils alter some metabolic processes that result in beneficial effects including more dilatation (or relaxation) of blood vessels, less platelet adhesiveness, reduced inflammatory response to the injury caused by smoking, lower triglyceride and fibrinogen levels, and lower blood pressure (especially in patients who have mildly elevated blood pressure). (Source: www.healer-inc.com/a039705a.htm.)

a. What are the two variables being compared in this study?
b. Could you conclude from this study that if you were a smoker and you ate a lot of fish, you are definitely at a reduced chance of suffering from heart disease? Why or why not?

8. The following table contains the sizes (in carats) of nine diamonds and their prices.

Size (x)	Price (y)
0.17	$353
0.16	$328
0.17	$350
0.18	$325
0.25	$642
0.16	$342
0.15	$322
0.19	$485
0.21	$483

Source: www2.ncsu.edu/ncsu/pams/stat/info/jse/datasets.index.html.

a. Draw a scatterplot showing the relationship between these two variables.
b. Find the least squares regression line. What is the equation for the line?
c. You find a diamond of size 0.23 carat in a store, and the salesperson is going to sell it to you for $500. Is this a good deal, according to the regression line found above?
d. Which of the nine diamonds listed above was the worst deal for the person who bought it, according to your regression line?

9. What is the correlation between the size of a diamond and its price, using the data from Exercise 8?

10. The chapter discusses two methods for finding the slope of the least squares best-fitting line. Explain both of the methods briefly, stressing advantages when they exist.

11. The following table gives the weights (x) of eight adult men and the amounts of weight they are able to lift (y).
a. What is the correlation between x and y?
b. What is the slope of the least squares best-fitting regression line?
c. Using the fact that the line must pass through $(\bar{x},\bar{y}) = (171,184)$, determine the equation for the least squares best-fitting regression line.
d. If another male 10th-grade student weighs 156 pounds, what would be the best prediction of how much he will be able to lift?

Man	x	y	$x - \bar{x}$	$y - \bar{y}$	$(x - \bar{x})(y - \bar{y})$	$(y - \bar{y})^2$
1	150	172	−21	−12	252	144
2	174	210	3	26	78	676
3	163	159	−8	−25	200	625
4	167	175	−4	−9	36	81
5	210	200	39	16	624	256
6	189	220	18	36	648	1296
7	140	150	−31	−34	1054	1156
8	175	185	4	1	4	1
Sum	1368	1472	0	0	2896	4235

12. The following table shows the return and risk for selected investments during the period 1960–1990. (The risk variable is a specially defined index intended to summarize in a single number the amount of risk for that type of investment. Higher values indicate higher risk.)

	Return (x)	Risk (y)
Common stock	9.8	15.8
Long-term corporate bonds	6.9	11.1
Long-term Treasury bonds	6.4	10.9
Short-term Treasury bonds	6.4	2.9
Residential real estate	10.6	10.7
Farm real estate	9.9	8.5
Business real estate	8.7	4.9

    a. What sort of relationship would you expect between return and risk for investments? Why?
    b. What is the correlation between return and risk?
    c. Does your answer agree with what you guessed in part **a**?

13. The following table shows the expenditures (in billions of dollars) for advertising and promoting smoking in the United States for the years 1975 to 1996, and the percentages of high school seniors who smoke daily in the subsequent year.
    a. The means for x and y are 2.80 and 21.3, respectively; the standard deviations for x and y are 1.697 and 3.428, respectively; and the correlation between x and y is -0.478. What does it suggest?
    b. Find the intercept and slope of the least squares line for estimating the percentage of seniors who smoke from expenditures for advertising.
    c. Plot the points, and draw the least squares line on the plot.
    d. Find the predicted percentage of seniors who smoke for the years 1975, 1983, and 1996 using the least squares line. How far off are the estimates from the actual values?
    e. Is the least squares line appropriate for these data? Why or why not?

	x Expenditures	y % Smoking
1975	0.49	28.8
1976	0.64	28.8
1977	0.78	27.5
1978	0.88	25.4
1979	1.08	21.3
1980	1.24	20.3
1981	1.55	21.1
1982	1.80	21.2
1983	1.90	18.7
1984	2.10	19.5
1985	2.48	18.7
1986	2.38	18.7
1987	2.58	18.1
1988	3.27	18.9
1989	3.62	19.1
1990	3.99	18.5
1991	4.65	17.2
1992	5.23	19.0
1993	6.01	19.4
1994	4.83	21.6
1995	4.90	22.2
1996	5.11	24.6

Do these results suggest that higher spending on cigarette advertising causes fewer high school seniors to smoke? Explain.

II

Probability Modeling and Obtaining Data

CHAPTER 4 PROBABILITIES AND SIMULATION   156

CHAPTER 5 EXPECTED VALUES AND SIMULATION   218

CHAPTER 6 PROBABILITY DISTRIBUTIONS: THE ESSENTIALS   248

CHAPTER 7 OBTAINING DATA: RANDOM SAMPLING AND RANDOMIZED CONTROLLED EXPERIMENTS   292

PROFESSIONAL PROFILE

Dr. Paul W. Chodas
Solar System
Dynamics Group
Jet Propulsion Laboratory
Pasadena, California

A close-up view over the horizon of the asteroid Eros, taken by the NEAR spacecraft. This asteroid, first discovered in 1898 and roughly the size of Manhattan Island, is being studied closely by Dr. Chodas and other scientists. Their work centers on information sent back to Earth by the NEAR spacecraft, which orbited Eros in 2000.

Dr. Paul W. Chodas

It's a classic science fiction story. Our heroes are on a world that will be hit by a huge asteroid. Their entire world is threatened with oblivion. It makes a good science fiction story. But what about reality? Could a large asteroid *really* hit the Earth and destroy it? That's a question that Dr. Paul W. Chodas studies for the Solar System Dynamics Group at Jet Propulsion Laboratory (JPL). His primary responsibility is to calculate orbits for asteroids and comets and to compute the probability that one of these might impact the Earth.

In 1998, Dr. Chodas helped dispel an alarming report that a newly discovered kilometer-sized asteroid might impact the Earth in 30 years. His statistical analysis showed that although a near miss was possible, an impact in 2028 could be ruled out. Another sizable asteroid was discovered in 1999. For this new asteroid, Dr. Chodas calculated that there was a two-in-a-million chance that it would hit the Earth in 2044. Dr. Chodas used basic statistical techniques to calculate probabilities. "I ran some 100,000 Monte Carlo orbits consistent with the data, selected those which brought the asteroid to the vicinity of the Earth in 2044, and calculated the density of cases which cut across the Earth," he explains. Fortunately, when additional tracking data were obtained several months later, the possibility of collision was ruled out.

With advance warnings from Dr. Chodas and others, NASA and JPL might be able to divert an asteroid that posed a threat to Earth. Without the information, there might be too little time to respond. Although much of Dr. Chodas's work involves applying orbital mechanics, statistical methods are essential in his analyses. "When I took statistics at college," he recalls, "I had no idea how useful and important it would be in my professional career"—or how it could change the fate of the Earth in the years ahead.

4

Probabilities and Simulation

Objectives

After studying this chapter, you will understand the following:

- The experimental probability $\hat{P}(A)$ as an estimator of the theoretical probability $P(A)$
- The five-step simulation method for computing experimental probabilities
- Estimating probability distributions approximately via five-step simulation
- How to simulate probability problems using coins, dice, random digits, and especially using box models

- Rules for doing elementary theoretical probability computations
- The theoretical probability model of equally likely outcomes
- Independence of events
- Sampling from a population with and without replacement
- Introduction to conditional probability

4.1 EXPERIMENTAL PROBABILITY   159

4.2 PROBABILITY MODELS   166

4.3 SIMULATION: A POWERFUL TOOL FOR LEARNING AND DOING STATISTICS   182

4.4 SIMULATING RANDOM SAMPLING VIA A BOX MODEL   200

4.5 RANDOM SAMPLING WITH OR WITHOUT REPLACEMENT   206

CHAPTER REVIEW EXERCISES   212

Key Problem

The Tired MDs

Program directors at teaching hospitals insist that the mental alertness of young doctors does not suffer when the doctors are on call and work for over 36 hours.

A group of 25 doctors was given a battery of mental tests before they started their 36-hour call and a similar battery of tests of the same difficulty as the first at the end of their call period. Twenty of the doctors did better on the tests they took before their call than on the tests they took after their call, and five did better on the second set of tests.

If sleep deprivation does not affect a doctor's mental abilities, as the program directors insist, then each doctor would be just as likely to do better on the second set of tests as on the first set. If the claim of no loss in mental abilities were true, how likely would it be that only 5 or fewer out of the 25 doctors would do better on the second set of tests? If it is highly unlikely, then we will find it difficult to accept the directors' claim.

We live in a world of uncertainty. For example, no one knows how many hurricanes will threaten the Atlantic coast next hurricane season. But uncertainty is not the same as total unpredictability. A recent news item predicted an increase in the annual number of hurricanes over the next decade. Supposedly, the coming hurricane seasons will resemble those between 1920 and 1970 more than those of the last thirty years.

Meteorologists (weather forecasters) make such forecasts using probability models and data on current weather patterns. Each year the National Oceanographic and Atmospheric Administration (NOAA) spends $100 million trying to forecast the weather as far in advance as possible. They have models for predicting next year's weather and models for predicting tomorrow's weather. When a hurricane approaches, they use their probability models and current weather data to assess where it might hit land so that the coastal residents can be warned of the risk. Unfortunately for weather forecasters, the weather is fundamentally chaotic: No amount of data collection and computing power will ever give 100% accurate forecasts. Weather forecasters can tell when tornadoes are *likely* in the Midwest, but they cannot tell for sure when and where a tornado will strike.

Phenomena such as the weather, the stock market, or even automobile traffic are affected by a large number of causes that are not all known and that vary in time in a complicated manner. The behavior of such phenomena cannot be predicted in detail in advance. They are called **random phenomena** and are said to be "affected by chance." Statisticians look for patterns in the chaos of raw data obtained from measuring such phenomena, and they make **statistical inferences** (draw conclusions) about causes and effects. To make statistical inferences on a data set, the statistician first needs to choose a probability model for how the data would be generated by chance. Then information from the probability model is combined with information from the data to produce a statistical inference, as Figure 4.1 shows schematically.

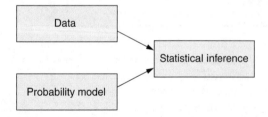

Figure 4.1 Model-based statistical inference.

In this chapter we introduce a five-step simulation method that has the helpful feature of forcing us to think carefully and explicitly about our choice of the probability model. Hence this chapter is an important prelude to doing effective, model-based statistical inference.

4.1 EXPERIMENTAL PROBABILITY

Every day, people make decisions and act in the face of an uncertain future on the basis of perceived risks and chances. We invest in startup stocks on the chance that we will make a lot of money, or in U.S. Treasury bonds to reduce the risk of losing the money we have. We get flu shots to reduce our risk of getting the flu. We send out numerous résumés with the hope of increasing our chances of getting a good job. We buy life insurance to protect our dependents in case we should die. To be able to quantify and compare their concerns about uncertainty, people talk about the "chance," "risk," "likelihood," "probability," or "odds" of an event taking place. The terms are almost but not quite interchangeable. "Probability" is almost always expressed as a number, whereas "risk" often is used in the more qualitative sense, for example, of being "high" or "low." "Chance" can be used in a qualitative sense as well ("a slim chance"), but it can also mean the same numeric concept as "probability" ("a 30% chance of rain"), and that is how it will be used in this book.

Consider questions such as the following:

- If the federal government introduces random drug testing for all its employees, what is the chance that someone who tests positive actually is drug free?
- If a man and a woman are both carriers of Tay-Sachs disease, what is the risk that the child will be born with the deadly disease?
- If I show up at the airport with only carry-on luggage 30 minutes before departure, what is the chance that, because of security or overbooking, I will not be able to get a seat on the airplane?
- If a quality control inspector randomly chooses 10 of the 500 VCRs for inspection, what is the probability that none of the inspected ones will be defective if there are 50 defectives?
- If I have a kidney transplant, what are my chances of surviving 5 years?

As these questions illustrate, probability is important in a wide variety of fields, including medicine, manufacturing, and public policy, as well as in daily life. Probability assesses the likelihood of an event happening: The smaller the probability of something, the less likely it is to happen.

In this book we use an experimental, or relative frequency, interpretation of probability. For example, out of the hundreds of tornadoes that touch down every year, data show that roughly 6 out of 10 strike between 2 P.M. and 8 P.M. We therefore say that the probability that a tornado will touch down between 2 P.M. and 8 P.M. is 6/10.

We often think of a random phenomenon as a **random experiment.** An experiment can (at least in principle) be repeated many times, and probabilities are used to describe the results we expect to see over many repetitions of the experiment. Each repetition is often called a **trial**.

Some of the simplest such experiments are found in games of chance. We shall often use examples from gambling to teach important probability concepts because in games of chance the nature of "chance variation" can be clearly understood.

Gambling is an ancient human practice. Archaeologists have found artifacts used in games of chance in Egypt dating back to 3500 B.C. Egyptian tomb paintings show gods and people tossing *astragali*, small irregular six-sided sheep ankle bones. Cubical dice of wood or pottery have been found in Iraq and India, dating back to 3000 B.C. The Aryans who invaded India around 1500 B.C. loved to gamble. The Rgveda, a collection of Aryan hymns, includes a poem, "The Gamester's Lament," in which the sun god Savitr scolds the gambler, telling him to stop playing with dice and go home to his wife and kids!

For thousands of years astragali, dice, and "throwing-sticks" of wood or ivory were used both in gambling and in divining the future. They were also used in the casting of lots. The Bible has numerous references to the casting of lots as a way to settle disputes or divide property.

During the Middle Ages a popular game of dice in Western Europe was called *hazard*. The game was similar to modern-day *craps* and its name was derived from the Arabic *al-zahr*, which means "the die." The game is believed to have been brought to Europe by soldiers returning from the Crusades.

It was in the seventeenth century that gamblers (and mathematicians and philosophers) began to think about probability in a more systematic way—with the obvious hope of improving their winnings! The serious study of probability apparently began in the seventeenth century when Galileo in Italy and Pascal and Fermat in France were asked by gamblers to help them figure the odds in various games of chance involving dice. In the process of helping the gamblers, Galileo, Pascal, and Fermat laid the groundwork for the formal rules of probability theory.

The **experimental probability** of any particular outcome of an experiment is the fraction of times (in the media, it is often given as the percentage of the time) the outcome occurs in a series of independent repetitions. This fraction is also called the **relative frequency** of the outcome.

The **theoretical probability of an outcome** is the number that the experimental probability of the outcome approaches as the number of independent repetitions gets large. The larger the number of repetitions, the closer the experimental probability is likely to be to the theoretical probability. Thus, for practical purposes we can treat the experimental probability we obtain as the theoretical probability when the number of repetitions is very large. Often we refer to a theoretical probability simply as a **probability**. We shall illustrate with some examples.

Example 4.1

What is the probability that a pregnant woman will have a boy?

Solution: It is impossible to predict the sex of a child at the time of conception. But if we look at all the children born in a given year, we see that the relative frequency of male births is almost the same from year to year. This striking relative frequency is illustrated by Table 4.1. (The relative frequency of male births is of course the number of male births divided by the total number of births.)

Based on these experimental probabilities, we estimate that the theoretical probability that a newborn baby will be a boy is between 0.512 and 0.513, or, for those who prefer their probabilities in percents, between 51.2% and 51.3%.

Note that this does not tell us how many boys the woman will have if she has two children. Probabilities do not tell us what will happen in the short run. Probabilities refer to what will hap-

Table 4.1 Male Births in the United States, 1940–1990

Year	Live births in thousands	Male births in thousands	Relative frequency (experimental probability)
1940	2559	1313	0.513
1950	3632	1863	0.513
1955	4097	2099	0.512
1960	4258	2180	0.512
1965	3760	1927	0.513
1970	3731	1915	0.513
1975	3144	1613	0.513
1980	3612	1853	0.513
1985	3761	1928	0.513
1990	4158	2129	0.512

Source: Statistical Abstract of the United States, 1995.

pen in the long run, when we look at thousands of babies. For example, it is a high probability that approximately 51% of all the babies born next year in New York City will be boys. This is an example of the **law of statistical regularity**, also known as the **law of large numbers.** This law tells us that if a chance experiment (here the birth of a child) is repeated over and over, then in the long run the proportion of time the experiment has a particular outcome (a boy) will stabilize around a value we call the probability (here 0.513).

Example 4.2

When we toss a coin, it can land either heads or tails. The outcome is unpredictable. What is the probability that it lands heads?

Solution: Two hundred and fifty years ago, the French naturalist Count Buffon tossed a coin 4040 times, getting 2048 heads. His experimental probability was 2048/4040 = 0.5069. A hundred years ago, the English statistician Karl Pearson tossed a coin 24,000 times, getting 12,012 heads. His experimental probability was 12,012/24,000 = 0.5005.

During World War II, the English mathematician John Kerrich tossed a coin 10,000 times while being held a prisoner of war. He got 5067 heads, corresponding to an experimental probability of 0.5067. Let us take a closer look at John Kerrich's data.

To see what can happen as the number of coin tosses increases, refer to Table 4.2. The table tracks Kerrich's results of tossing a coin 10,000 times. The first row gives the number of heads that

Table 4.2 10,000 Tosses of a Coin

Number of tosses	Number of heads	Relative frequency (experimental probability)
10	4	0.400
20	10	0.500
30	17	0.567
40	21	0.525
50	25	0.500
100	44	0.440
200	98	0.490
300	146	0.487
400	199	0.498
500	255	0.510
600	312	0.520
700	368	0.526
800	413	0.516
900	458	0.509
1000	502	0.502
2000	1013	0.507
3000	1510	0.503
4000	2029	0.507
5000	2533	0.507
6000	3009	0.502
7000	3516	0.502
8000	4034	0.504
9000	4538	0.504
10,000	5067	0.507

Source: J.E. Kerrich, *An Experimental Introduction to the Theory of Probability,* J. Jorgenson and Co., Copenhagen, 1946, p. 14.

162 PROBABILITIES AND SIMULATION

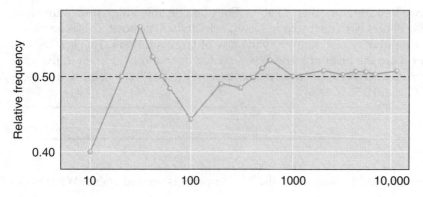

Figure 4.2 The relative frequency of heads as a function of the number of coin tosses in Kerrich's data.

appeared in the first 10 tosses, the second row gives the number appearing in the first 20 tosses, and so on. Notice the sequence of relative frequencies (each an experimental probability of heads). At first they show considerable variability as they dance up and down. However, after hundreds of coin tosses, the experimental probabilities, although continuing to vary a little, appear to approach a number near 0.500, possibly 1/2 itself. This statistical regularity is a characteristic of random phenomena: As the number of independent repetitions is increased (see Table 4.2), the continuously updated experimental probabilities approach a fixed number, the probability.

The process of the experimental probability of heads approaching the probability, a number near 0.500, is shown graphically in Figure 4.2. Note that the horizontal scale is compressed (a so-called logarithormic scale) as the number of tosses gets large, for easy viewing.

We are not surprised to find that, when we toss a coin, the probability we get heads seems to be 1/2. Our intuition tells us that heads and tails are equally likely outcomes of the coin toss. Most people would assume this to be true. Unfortunately, we cannot always rely on our intuition. Only by looking at data (such as Kerrich's coin tosses and the birth statistics of the United States) are we able to tell that heads and tails are indeed equally likely outcomes of a coin toss, whereas boys are slightly more likely than girls among newborn babies. Probabilities must be determined experimentally. This basic experimental principle is the foundation of statistical inference, a major topic of this textbook.

The next example is a game of chance that is so complicated that our uninformed intuition does not even tell us whether the chances (the theoretical probability) of winning are favorable (better than 1/2) or unfavorable (less than 1/2). Observing the experimental probability of winning for a large number of repetitions can, however, approximate the unknown chance of winning!

Example 4.3

In the game of *craps* the player repeatedly throws a pair of dice and adds up the scores (numbers of spots on the top faces) on the two dice. Figure 4.3 shows the 36 possible outcomes for a pair of dice.

If the sum (score) is 7 or 11, the player wins immediately. If the sum is 2, 3, or 12, the player loses immediately. If the sum is 4, 5, 6, 8, 9, or 10, then that number becomes the player's *point*. Once the point is established, the player has to keep throwing the pair of dice until he or she either wins by repeating the point, or loses by throwing a 7. In a casino setting, the (very rich) "house" collects the bets and doles out winnings to players.

For example, based on the 10 repetitions listed in Table 4.3, we estimate that the probability of winning in craps is 0.40 (the experimental probability). To improve the accuracy of our estimate (guess)

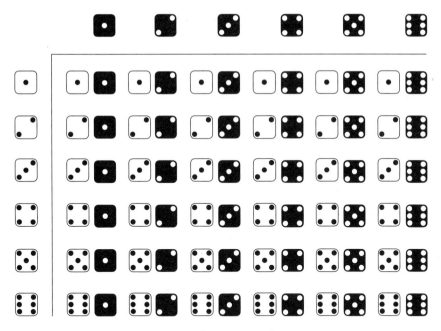

Figure 4.3 Throwing a pair of dice: List of outcomes.

Table 4.3 Estimating the Probability of a Player Winning in Craps

Repetition	Scores	Outcome
1	11	Win
2	6, 4, 3, 7	Loss
3	4, 8, 5, 10, 12, 4	Win
4	10, 6, 4, 7	Loss
5	10, 4, 5, 9, 7	Loss
6	9, 3, 7	Loss
7	6, 7	Loss
8	7	Win
9	8, 5, 5, 5, 3, 8	Win
10	6, 8, 9, 9, 7	Loss

we would need more repetitions. As our experience with Kerrich's coin-tossing data in Table 4.2 suggests, we need hundreds of repetitions before we can be confident that the experimental probability will be likely to be within a desired accuracy, for example 0.05, of the theoretical probability.

Our chance of winning in a real game of craps also depends on using fair dice, that is, in the long run the six faces will all turn up with the same relative frequency, 1/6.

Summary

The outcome of a random phenomenon (such as having a heart attack or being in a car accident) is unpredictable. However, in a very long series of independent repetitions a stable pattern emerges that can be described in terms of observed relative frequencies of the various possible outcomes.

Consider a particular outcome. We are interested in the theoretical probability of the outcome:

$$P(\text{outcome}) = \text{fraction of times the outcome would occur in an infinitely long series of independent repetitions.}$$

To estimate (guess) the probability we consider a number of independent repetitions and compute the **experimental probability**,

$$\hat{P}(\text{outcome}) = \frac{\text{number of successful repetitions}}{\text{total number of repetitions}}$$

$$= \text{relative frequency of the outcome.}$$

Here, a "successful repetition" means a repetition in which the outcome of interest occurs. The more repetitions we observe, the closer the experimental probability will get to the theoretical probability. The symbol $\hat{P}$ is read **"P hat"**; the "hat" merely reminds the statistician that this is not the theoretical probability P but an estimate of P using experimental data. The phrase **"independent repetitions"** means that the outcome of one repetition has no influence on which outcome occurs in subsequent repetitions.

Section 4.1 Exercises

1. Look up the *Statistical Abstract of the United States* (in a library, or online at http://www.census.gov/statab/www/) to determine the relative frequency of male births for each of the last 10 years.

2. Roll a die 60 times and observe how often you obtain each of the six numbers 1, 2, 3, 4, 5, and 6. Use your data to compute the following experimental probabilities:

    a. $\hat{P}$(rolling a 1); $\hat{P}$(rolling a 2); $\hat{P}$(rolling a 3); $\hat{P}$(rolling a 4); $\hat{P}$(rolling a 5); $\hat{P}$(rolling a 6)

    b. $\hat{P}$(number on die is even)

    Hint for part b: The number on the die being even is an "event" that occurs for any of several outcomes, but the principle is the same. We use the $\hat{P}$(outcome) formula with "event" replacing "outcome."

3. Shuffle a standard deck of 52 cards and draw one card. Repeat at least 50 times. Estimate the probability of drawing an ace. Is your answer consistent with what you think is the theoretical probability of drawing an ace?

4. Toss three coins 50 times and keep a tally of the number of heads on each toss of the three. Then use your data to compute the following experimental probabilities of events:

    a. $\hat{P}$(all three are heads)   **c.** $\hat{P}$(exactly one is heads)

    b. $\hat{P}$(exactly two are heads)   **d.** $\hat{P}$(none of the three is heads)

    The probability from parts **a** and **d** should be fairly equal, and the probabilities from parts **b** and **c** should be close to equal. Is this true for your 50 repetitions? (*Hint*: Getting exactly one head means you got exactly two tails.)

5. A building supplies manager wants to estimate the probability that a certain type of carpet tack will fall point-up when dropped on a smooth wood floor. A worker drops one of the tacks 500 times and finds that it falls point-up (its head flat on the floor) 35 times. What is the experimental probability that the tack will fall point-up? Do you think that this estimate will be fairly accurate?

6. Over his career, a basketball player has made 1210 free throws and missed 214. What is his experimental probability of making a free throw? Do you think this experimental probability is a good estimate?

7. Toss two dice 50 times and record the sum of the dots appearing each time. Then compute the following experimental probabilities:

    a. $\hat{P}$(the sum of the two dice equals 2)   **c.** $\hat{P}$(the sum of the two dice equals 10)

    b. $\hat{P}$(the sum of the two dice equals 7)   **d.** $\hat{P}$(the sum of the two dice equals 6)

8. Define or explain the concepts of relative frequency, experimental probability, statistical regularity, and (theoretical) probability. Relate the concepts to each other.

9. Play craps 100 times and estimate the theoretical probability of winning.

10. Toss a new coin. If it lands heads up, then stop. Otherwise keep tossing the coin until you get heads. Count the number of tosses. Repeat this experiment 50 times and keep a record of the number of tosses in each of the 50 trials. Use your data to compute the following experimental probabilities:

    a. $\hat{P}$ (we need only 1 toss to get a head)   d. $\hat{P}$ (we need 4 tosses)
    b. $\hat{P}$ (we need 2 tosses)   e. $\hat{P}$ (we need 5 tosses)
    c. $\hat{P}$ (we need 3 tosses)   f. $\hat{P}$ (we need 6 tosses)

    Is $P(1000 \text{ tosses needed}) > 0$? Explain. *Note:* **a–f** combine to produce an example of estimating a probability distribution, that is, estimating $P(\text{outcome})$ for all possible outcomes.

11. Hold a penny on its edge on a tabletop with your left index finger and flick it with your right index finger so that the penny spins upright for several seconds before it falls. Note whether the penny falls heads up or tails up. Repeat this experiment 100 times and estimate the probability that the penny falls heads up. Do heads and tails seem equally likely (that is, does each have theoretical probability $1/2$) in this experiment? *Hint*: If you get many more or many less than 50 heads, the answer is "no."

12. An article in a medical journal reported on 492 patients who underwent a specific heart valve procedure. Of these patients, 33 died during or soon after the procedure and 84 suffered nonfatal complications. Based on these data, estimate the probability that a patient who undergoes this procedure

    a. will suffer fatal or nonfatal complications.
    b. will die.

13. In April 1992, *USA Today* presented the data on over 17,000 tornadoes touching down over a 20-year period. Note that these experimental probabilities should be very good estimates of the theoretical probabilities, because 17,000 is a huge number of repetitions. Thus, you can assume that these are the true probabilities.

    a. What is the most likely two-hour period for a tornado to touch down? What is this probability?
    b. If a tornado touches down on a given day, what is the chance that it happens between 2 P.M. and 8 P.M.? What is the chance that it happens between noon and 10 P.M.?

Time of day tornadoes touch down	Experimental probability
Midnight to 2 A.M.	0.029
2 to 4 A.M.	0.021
4 to 6 A.M.	0.022
6 to 8 A.M.	0.022
8 to 10 A.M.	0.031
10 A.M. to noon	0.045
Noon to 2 P.M.	0.097
2 to 4 P.M.	0.179
4 to 6 P.M.	0.224
6 to 8 P.M.	0.191
8 to 10 P.M.	0.098
10 P.M. to midnight	0.041

4.2 PROBABILITY MODELS

We describe, or "model," a random phenomenon by listing the possible outcomes and assigning probabilities to them. This is called a **probability model**. It is the result of certain assumptions made about the nature of the phenomenon being modeled. For example, certain outcomes may be assumed equally likely. We call the set, or collection, of all possible outcomes the **sample space**. An **event** is merely a collection of some of the outcomes. We want to be able to compute the probability of an event occurring. By an **event occurring**, we mean that the random experiment results in one of the event's listed outcomes.

Equally Likely Outcomes

If we expect two outcomes of a chance experiment to occur with the same relative frequency in the long run, then we say that the two outcomes are equally likely. Many random phenomena are modeled as having a finite number of possible outcomes, all equally likely. The flip of a coin, the toss of a casino die, and the spin of a roulette wheel come to mind. As we have seen, among newborn babies, boys and girls are almost equally likely. If all outcomes are equally likely, then the probability that any particular event A will take place can be computed by the fundamental **rule for equally likely outcomes**:

$$P(A) = \frac{\text{number of different outcomes for which } A \text{ occurs}}{\text{total number of possible outcomes}}.$$

In particular,

$$P(\text{outcome}) = \frac{1}{\text{total number of possible outcomes}}$$

for each outcome.

Example 4.4

The roulette wheel in an American casino has 38 pockets numbered 1 through 36 plus 0 and 00. The croupier sets the wheel in motion and tosses a ball inside in the direction opposite the motion of the wheel. Bets are made on the number of the pocket where the ball comes to rest. The sample space of this chance experiment is

$$S = \{0, 00, 1, 2, 3, \ldots, 36\}.$$

Since the casino makes every effort to ensure that the wheel is perfectly balanced, we assign the probability $1/38$ to each of the 38 outcomes in S. One of the many ways you can place a bet is to bet on the block of numbers from 31 to 36. Your chance of winning this bet is

$$P(\{31, 32, \ldots, 36\}) = 6/38$$

according to the rule for equally likely outcomes.

Example 4.5

Consider the chance experiment of having two children. If we focus on the children's sex, then the sample space S for this experiment has four outcomes,

$$S = \{GG, GB, BG, BB\}.$$

Here, GB indicates that the first child born is a girl and the second is a boy. The event A that the two children are of the same sex is a collection of two outcomes:

$$A = \{GG, BB\}.$$

The event A will occur if we either have two girls or two boys.

We usually assign the probability $1/4$ to each of the four outcomes in S because we pretend that the four outcomes are equally likely, even though in reality they are only approximately so. As we saw in Example 4.1, this is not quite accurate, because boys are slightly more probable than girls. To get better estimates of the four probabilities we would need data from a large number of two-child families. Note also in this context that the occasional occurrence of identical twins means that the sex of the firstborn and the sex of the second child are not totally independent, as they would need to be if we could legitimately assume that all four outcomes are equally likely. These two rough approximations (boys and girls equally likely, independent of the sex of the firstborn and second child) allow us to model the four outcomes as being equally likely. In fact, for most purposes this approximate model will do and is the one people use. Note that $P(A) = 2/4$ by the rule for equally likely outcomes.

Incidentally, our intuitive notion that the sexes of siblings are independent is completely wrong in the case of crocodiles. The sex of a crocodile hatchling is determined by the temperature at which it hatches. Since the temperature varies within the crocodile nest, the nest usually produces both male and female baby crocodiles. But if the weather is unusually cold or unusually warm, all the hatchlings will have the same sex. This is another illustration that reminds us that it is risky to casually make assumptions without thorough knowledge of the real-world setting.

Example 4.6

If we toss a die, then the sample space is

$$S = \{1, 2, 3, 4, 5, 6\}.$$

If the die is a casino die, then we assign the probability $1/6$ to each of the six outcomes because we assume that the outcomes are equally likely. The list of all possible outcomes and the corresponding probabilities is called the **probability distribution**:

Score	1	2	3	4	5	6
Probability	1/6	1/6	1/6	1/6	1/6	1/6

Example 4.7

On a Las Vegas roulette wheel, the numbers 0 and 00 are colored green, 18 of the other numbers are colored red, and 18 are black. If you bet on "red," you win if the ball comes to rest in a red slot. Assuming that the roulette wheel is perfectly balanced and all outcomes are equally likely, your chance of wining a bet on "red" is $18/38$, since there are 18 outcomes for which the color is red out of the 38 possible outcomes. That is, $P(\text{red}) = 18/38$ by the rule for equally likely outcomes.

Example 4.8

A family has three children. What can we say about the children's sexes? We can use a so-called **tree diagram** to represent the eight possible outcomes (Figure 4.4).

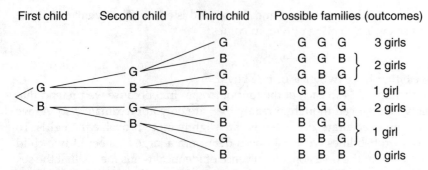

Figure 4.4 Tree diagram for possible three-child families.

Let us assume, as an approximation to reality, that the eight outcomes are equally likely. In that case, each must have probability 1/8. It follows that

$$P(\text{three girls}) = P(GGG) = 1/8$$
$$P(\text{two girls}) = (\text{Number of outcomes with 2 girls, 1 boy})/8$$
$$= 3/8$$
$$P(\text{one girl}) = (\text{Number of outcomes with 1 girl, 2 boys})/8$$
$$= 3/8$$
$$P(\text{no girls}) = P(BBB) = 1/8.$$

We thus have found the probability distribution for the number of girls in a three-child family:

Number of girls	0	1	2	3
P(number of girls)	1/8	3/8	3/8	1/8

Notice that the probabilities add up to 1.

Tree diagrams are often used to represent the outcomes of an experiment that is carried out in stages. In particular, as illustrated in Example 4.8, they help us count the number of outcomes on an event A and in the sample space S, as needed when outcomes are equally likely.

Not Equally Likely Outcomes

To guide us in assigning probabilities to the outcomes of a random phenomenon when the outcomes are not equally likely, we note two fundamental rules: The probabilities we assign to outcomes must all be between 0 and 1, and they must add up to 1. That is,

$$0 \leq P(\text{outcome}) \leq 1,$$

$$\sum_{\substack{\text{all outcomes} \\ \text{in } A}} P(\text{outcome}) = 1.$$

Here, $\sum$ is the summation symbol we introduced in Chapter 2, which merely tells us to add up what is to its right. A subscript, such as "all outcomes in S" in this case, tells us what values to include in the total.

The reason the two fundamental rules must hold is that relative frequencies obey these rules and that probabilities, being long-run relative frequencies, inherit all the properties of relative frequencies.

A fundamental principle for all random phenomena is that the probability that any particular event A will take place equals the sum of the probabilities of the outcomes in the event. That is, for any event A,

$$P(A) = \sum_{\substack{\text{all outcomes} \\ \text{in } A}} P(\text{outcome}).$$

This formula is a consequence of the fact that the corresponding formula holds for experimental probabilities. We shall use the formula in the next example.

Example 4.9

A dollar coin is flipped successively until the first head is observed. If we count the number of flips needed to get heads for the first time, then the sample space consists of all the counting numbers,

$$S = \{1, 2, 3, 4, 5, 6, 7, \ldots\}.$$

Since there are infinitely many outcomes, they certainly cannot be equally likely. In Exercise 10 of Section 4.1, we try to estimate the probability of some of the outcomes of this chance experiment. The theoretical probability distribution for the number of tosses it takes to get a head is actually given by

Number of Tosses	1	2	3	4	5	6	...
Probability	1/2	1/4	1/8	1/16	1/32	1/64	...

Again, these probabilities add up to 1: $1/2 + 1/4 + 1/8 + 1/16 + 1/32 + \ldots = 1$. Using these probabilities we can compute the probability of any event A. For example,

$$P(\text{we need at most 3 tosses}) = P(1) + P(2) + P(3) = 1/2 + 1/4 + 1/8 = 7/8.$$

Complementary Event

Imagine that we play 1000 games of craps (see Example 4.3) and we win 474 times and lose 526 times. That gives us the experimental probabilities

$$\hat{P}(\text{we win}) = 0.474 \quad \text{and} \quad \hat{P}(\text{we lose}) = 0.526.$$

Consider any event A (like winning), associated with any random phenomenon. Clearly

$$\hat{P}(A \text{ does not take place}) = 1 - \hat{P}(A \text{ takes place}).$$

Because theoretical probabilities are (very) long-run relative frequencies,

$$P(A \text{ does not take place}) = 1 - P(A \text{ takes place}).$$

The event "A does not take place" is called the **complement** of A. This is another one of the basic rules useful for computing probabilities of events. This **rule of complementary events** is written briefly as

$$P(\text{not } A) = 1 - P(A).$$

The Addition Rule (The "or" Rule) for Probabilities

In probability theory there is both an **addition rule** and a **multiplication rule**. You need these rules to find complicated probabilities such as the theoretical probability of winning a game of craps with a pair of symmetric casino dice. We shall first learn the addition rule, which we introduce by an example.

Example 4.10

Before a person can safely receive a blood transfusion it is necessary to determine his or her ABO blood group. There are four ABO blood groups: A, B, O, and AB. A person of blood type O can receive blood only from a donor of type O. The possible donor groups for each blood group are given in Table 4.4.

The distribution of the four ABO blood types varies with ethnic groups. For example, Peruvian Incas are all type O. The blood group distributions for African Americans and for European Americans are given in Table 4.5. What is the probability that a randomly chosen European American is of type O? What is the probability that he or she is not of type O?

Solution: The "European American" row of proportions in Table 4.5 provides the probability distribution for the blood type of a randomly sampled European American.

We have just applied a fundamental principle that is a consequence of the rule for equally likely outcomes: **When randomly sampling from a population (and all members of the population are equally likely to be selected), the population proportions become the probability distribution for one randomly sampled member of the population.** (Remember this key principle!)

Note that the four outcomes (A, B, O, AB) are *not* equally likely. The probability that the randomly chosen European American is type O is 0.45. The probability of not getting a person of type O is, by the rule of complementary events,

$$P(\text{not type O}) = 1 - P(\text{type O}) = 1 - 0.45 = 0.55.$$

There is another way to do it:

$$P(\text{not type O}) = P(\text{type A}) + P(\text{type B}) + P(\text{type AB})$$
$$= 0.42 + 0.10 + 0.03 = 0.55.$$

The last expression follows from the fact that every person who is not type O is either type A, type B, or type AB. That is, the event

$$(\text{not type O}) = (\text{Type A}) \text{ or } (\text{Type B}) \text{ or } (\text{Type AB}).$$

No one has more than one blood type, so these three events on the right have no outcomes in common. Thus we can add their probabilities to obtain $P(\text{not type O})$. This is the **addition rule for probabilities**.

Table 4.4 The ABO Groups in Transfusion

Recipient	Possible donors	Forbidden donors
O	O	A, B, and AB
A	O or A	B and AB
B	O or B	A and AB
AB	All groups	None

Table 4.5 ABO Blood Type Population Proportions for African Americans and European Americans

Ethnicity	Blood type probability			
	O	A	B	AB
African American	0.49	0.27	0.20	0.0
European American	0.45	0.42	0.10	0.0

Let's carefully state the **addition rule for probabilities of mutually exclusive events** in the simplest case of two events with no outcomes in common.

Suppose two events have no outcomes in common. Then the probability that one or the other occurs is the sum of the probabilities of each event. That is, for events labeled A and B,

$$P(A \text{ or } B) = P(A) + P(B), \text{ when } A \text{ and } B \text{ are mutually exclusive.}$$

Here **"mutually exclusive"** means "having no outcomes in common." Note that the addition rule applies for more than two mutually exclusive events, as we made use of in the above example. For example, for mutually exclusive events A, B, and C, we have

$$P(A \text{ or } B \text{ or } C) = P(A) + P(B) + P(C).$$

If the events A and B are not mutually exclusive, then the **addition rule for probabilities** gets a little more complicated. We have

$$P(A \text{ or } B \text{ or both}) = P(A) + P(B) - P(\text{both } A \text{ and } B).$$

To illustrate, imagine that we toss a pair of symmetric casino dice, one white and one black. Consider the events

$$A = (\text{the white die shows 5 dots})$$
$$B = (\text{the black die shows 5 dots}).$$

Since the 36 possible outcomes are equally likely,

$$P(A) = P(B) = 6/36 = 1/6 \text{ and } P(\text{both } A \text{ and } B) = 1/36.$$

Further,

$$P(\text{we get at least one 5}) = P(A \text{ or } B \text{ or both})$$
$$= P(A) + P(B) - P(\text{both } A \text{ and } B) = 1/6 + 1/6 - 1/36 = 11/36.$$

Notice that $P(A) + P(B) = 6/36 + 6/36 = 12/36$ counts the outcome of both dice being 5 *twice*.

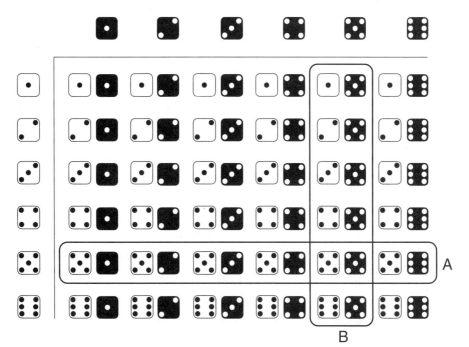

Example 4.11

Probability models play an important role in the genetic theory of heredity, originated by Gregor Mendel (1822–1884). Many futurists predict that the twenty-first century will be the century of genetics, with breakthroughs in genetic engineering, gene therapy, cloning, and the like.

Let us review some elementary genetics. Consider an individual (a person, an animal, or a plant) that is the result of sexual reproduction (as opposed to cloning). The individual's genetic makeup is inherited from its parents. For every genetic trait, each parent has a gene pair. Each parent contributes half of his or her gene pair, with equal probability for the two halves, to the offspring, who therefore gets a gene pair with one gene from each parent. The two parents make their contributions independently of each other.

What trait appears in the offspring depends on which of two or more forms (called *alleles*) its two genes have. For many traits, the genes have two alleles, one *dominant* (often denoted A) and the other *recessive* (often denoted a). A recessive trait becomes apparent in the offspring only when *both* halves of the offspring's inherited gene pair have the recessive form. Otherwise, the dominant trait appears.

An important example of a recessive genetic disease caused by a single gene pair is sickle-cell anemia, a severe and potentially deadly disorder that, in the United States, affects roughly one in 500 African Americans. The gene responsible for sickle-cell anemia comes in two alleles: the dominant healthy allele A, and the recessive defective allele a. A baby born with genotype aa will develop sickle-cell anemia and probably die young. A person of genotype Aa (A from mother, a from father) or aA (a from mother, A from father) will not develop the symptoms of sickle-cell anemia, but he or she may pass the disease on to the next generation. He or she is thus a *carrier* of the disease. (There is no difference between genotypes Aa and aA; it does not matter which parent contributed which form.) A person with genotype AA will neither develop sickle-cell anemia nor be able to pass the disease on to the next generation.

Data suggest that 1 in 10 African Americans is a carrier of sickle-cell anemia. Imagine that two healthy sickle-cell anemia carriers are expecting a child. What is the chance that the child will be healthy?

Solution: Both parents have the genetic makeup either Aa or aA. As far as the sickle-cell anemia gene pair is concerned, we can view having a child as a random experiment carried out in two independent stages:

First stage:	Inherit a gene from the mother
Second stage:	Independently inherit a gene from the father

We can use a tree diagram to represent all possible genotypes for the child. The dominant healthy gene is denoted by A. The recessive sickle-cell anemia gene is denoted by a.

```
First stage         Second stage        Outcome for
(gene from mother)  (gene from father)  child

                    ─── A               AA    Healthy non-cancer
         ── A ──                        Aa  ⎫
                    ─── a                   ⎬ Healthy carrier
         ── a ──    ─── A               aA  ⎭
                    ─── a               aa    SCA
```

According to Mendel's theory, the four outcomes are equally likely. Therefore

$$P(\text{child is healthy and not a carrier}) = P(AA) = 1/4$$
$$P(\text{child has sickle-cell anemia}) = P(aa) = 1/4$$
$$P(\text{child is a healthy carrier}) = P(aA \text{ or } Aa) = 2/4 = 1/2.$$

By the addition rule,

$$P(\text{child is healthy}) = P(\text{healthy carrier or healthy noncarrier})$$
$$= P(\text{healthy carrier}) + P(\text{healthy noncarrier})$$
$$= 1/2 + 1/4 = 3/4.$$

The a allele is not an unmitigated curse. It turns out that healthy carriers (with genotype Aa or aA) have higher resistance to malaria than noncarriers (with genotype AA). If a person lives in a region with high incidence of malaria, such as Central Africa, then that person is better off being Aa than AA.

Is there a cure for sickle-cell anemia? The following medical case offers hope. An eight-year-old girl in Boston suffering from both leukemia and sickle-cell anemia was treated by destroying her cancerous bone marrow and replacing it with healthy marrow from a donor. This cured both her leukemia *and* her sickle-cell anemia! For most people it is too risky to treat sickle-cell anemia with bone marrow transplants. In the future, gene therapy may offer a cure.

Cystic fibrosis and Tay-Sachs disease are other examples of important recessive single-gene diseases. Cystic fibrosis affects mainly Caucasians, and Tay-Sachs mainly Jews of Eastern European ancestry. Huntington disease is an example of a *dominant* disease. If A denotes the lethal Huntington gene and a denotes the corresponding healthy gene, then everyone with genetic makeup AA, Aa, or aA will develop this degenerative neurological disorder when they reach middle age. That is, the presence of even one A allele suffices to cause the disease.

Genetics is complex but fascinating, and our developing understanding of it is vital for finding medical cures for inherited diseases.

*Probability Trees

It is often helpful to draw a tree diagram to represent the possible outcomes of an experiment that is carried out in stages, as we have seen in Figure 4.4. A **probability tree** for a multistage experiment displays on each branch of the tree the probability of the next stage outcome, assuming that the present stage outcome has occurred.

Consider, for example, a diagnostic urine test for bladder cancer. Ideally, every person with bladder cancer should test positive if given this test, and every person without bladder cancer should test negative. But no diagnostic test is that good. The **sensitivity** of the test is the probability that a person with cancer will test positive. The **specificity** of the urine test is the probability that a person who does not have cancer will test negative. Imagine that the sensitivity of the test is 0.95 and the specificity of the test is 0.90. Finally, let us assume that 1 out of 5000 adult Americans has bladder cancer, that is, $P(\text{cancer}) = 1/5000 = 0.0002$ for a randomly chosen adult. Suppose that we give a randomly chosen adult this screening test for bladder cancer. The probability tree is easy to construct and is shown in Figure 4.5. The terms commonly used for the four possible multistage outcomes, such as "true positive" and "false negative", are also given in Figure 4.5.

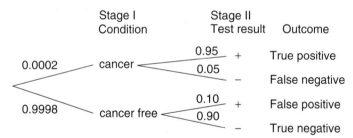

Figure 4.5 Testing a randomly chosen adult for bladder cancer.

Note that 0.9998, 0.05, and 0.1 in Figure 4.5 were all obtained from the above three given probabilities of 0.0002, 0.95, and 0.9 using the rule

$$P(\text{not } A) = 1 - P(A).$$

If we randomly test a very large number of adults, then about 95% of those who have bladder cancer will test positive. Since 1 in 5000 adults has bladder cancer, it follows that the proportion of adults in the population of tested adults who *both* have cancer and test positive will be around

$$95\% \text{ of } (1 \text{ in } 5000) = 95\% \text{ of } 0.0002 = 0.95 \times 0.0002 = 0.00019.$$

Moreover, because population proportions yield probabilities for a randomly sampled person we conclude that the probability that a random screening gives us a true positive is 0.00019. That is, $P(\text{true positive}) = 0.00019$. Similarly, the probability of getting false negative test result is

$$P(\text{false negative}) = 0.05 \times 0.0002 = 0.00001.$$

The probability of getting a false positive test result is

$$P(\text{false positive}) = 0.10 \times 0.9998 = 0.09998.$$

We have discovered the general **multiplication rule for probabilities:**

> We can compute the probability of any multistage outcome of the tree by multiplying the probabilities of the branches along the route that leads to the outcome. That is,
>
> $$P(\text{outcome}) = \text{product of probabilities along the branches producing the outcome.}$$

Recall that we can compute the probability for any event (in particular, for an event made up of outcomes at the far right in the probability tree) by adding the probabilities of the outcomes (the addition rule) that produce the event of interest. That is,

$$P(\text{event}) = \sum_{\substack{\text{outcomes} \\ \text{in event}}} P(\text{outcome}).$$

So for example, looking at Figure 4.5 we see that the probability a randomly chosen adult will test positive is

$$P(\text{positive}) = P(\text{true positive or false positive})$$
$$= P(\text{true positive}) + P(\text{false positive})$$
$$= 0.00019 + 0.09998 \approx 0.1$$

where 0.00019 and 0.09998 were obtained from the multiplication rule, as shown previously. That is, if we randomly screen a large number of individuals for bladder cancer using this test, around 10% of those screened will test positive in spite of the fact that only 1 out of 5000 has bladder cancer. Most of the positive test results will be false alarms!

Let us return to genetics. The gene for albinism in humans is recessive, like the gene for sickle-cell anemia (see Example 4.11). Imagine that a husband and a wife both are carriers of the albinism gene and that they have two children. Assume that the children are not identical twins so that they inherit their genotypes independently of each other. We therefore get the probability tree shown in Figure 4.6.

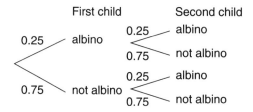

Figure 4.6 Albinism in a two-child family.

The probability that both children are albino is, by the multiplication rule for trees,

$$0.25 \times 0.25 = 0.0625.$$

The probability that the first child is not an albino but the second is equals

$$0.75 \times 0.25 = 0.1875.$$

We have discovered the **product rule for independent events:**

> If two events C and D are independent, then
> $$P(\text{both } C \text{ and } D \text{ occur}) = P(C) \times P(D).$$

By "independent events," we mean that the occurrence of one event has no influence on whether the other occurs or not. A coin turning up heads and a die showing at least five dots are clearly independent events, but rain in San Francisco and rain in Oakland, across the San Francisco Bay, will not be independent. In the above example whether or not the second child will be an albino is not influenced by the first child's condition, since the children inherit their genes independently of each other. Whenever we know events C and D are independent, we are allowed to use the product rule to solve for $P(\text{both } C \text{ and } D)$. We will learn more about the important concept of independence later.

As another illustration of independent events, ABO blood type (see Example 4.10) is independent of the Rh factor. That is, the distribution of the Rh factor in the European American population (0.85 Rh+ and 0.15 Rh–) is the same in each of the ABO blood groups (see Table 4.5). Therefore, knowing a person's ABO blood type does not influence the probability distribution of the Rh factor: $P(\text{Rh+}) = 0.85$ for *each* of the ABO blood types. Imagine that we select a European American at random. We use the product rule for independent events to compute the probabilities for the various combination of Rh factor and ABO blood type.

$$P(\text{type O and Rh+}) = P(\text{type O}) \times P(\text{Rh+})$$
$$= 0.45 \times 0.85 = 0.3825.$$

Similarly,

$$P(\text{type AB and Rh–}) = P(\text{type AB}) \times P(\text{Rh–})$$
$$= 0.03 \times 0.15 = 0.0045.$$

Genetic theory tells us that ABO blood type, Rh factor, and gender are all independent. For a randomly chosen European American, thus

$$P(\text{type A and Rh+ and male}) = P(\text{type A}) \times P(\text{Rh+}) \times P(\text{male})$$
$$= 0.42 \times 0.85 \times 0.50 = 0.1785.$$

*Conditional Probability

Let us return to the bladder cancer screening example (Figure 4.5). If a randomly screened person tests positive, how likely is it that the test result is correct? That is, what is the (conditional) probability that the person has bladder cancer, given that he or she tested positive?

To answer this question, imagine that we randomly screen a million adults. We will get around

$$P(\text{positive tests}) \times 1{,}000{,}000 = 0.100 \times 1{,}000{,}000 = 100{,}000$$

positive tests results. Out of these, the number who actually have cancer, that is, the number of true positive test results will be around

$$P(\text{true positive}) \times 1{,}000{,}000 = 0.00019 \times 1{,}000{,}000 = 190.$$

So only around 190 out of 100,000 persons who test positive will actually have cancer! The rest will be false alarms. Thus, the **conditional probability** that a randomly chosen adult has bladder cancer, given that he or she tested positive is denoted by $P(\text{cancer} \mid \text{positive test})$, and

$$P(\text{cancer} \mid \text{positive test}) = \frac{190}{100{,}000}$$

Because this fraction is so small, it should be noted that doctors would not use this test for random screening of the general population. Rather they would focus on people at greater risk of bladder cancer because of clinical indications, or family history, ethnicity, and so forth.

Let us generalize the argument we just used to determine the conditional probability. Consider a random phenomenon or chance experiment (such as screening a randomly chosen adult for bladder cancer). Consider two events, A (the person tests positive) and B (the person has cancer) in the chance experiment. If we repeat the chance experiment a large number of times, the **experimental conditional probability of B given A** is

$$\hat{P}(B|A) = \frac{\text{number of times both } A \text{ and } B \text{ occur}}{\text{number of times } A \text{ occurs}}$$

$$= \frac{\hat{P}(\text{both } A \text{ and } B) \times (\text{number of repetitions})}{\hat{P}(A) \times (\text{number of repetitions})} = \frac{\hat{P}(\text{both } A \text{ and } B)}{\hat{P}(A)}.$$

Since theoretical probabilities are long-run experimental probabilities, we get the following formula for the theoretical **conditional probability of B, given A**:

$$P(B|A) = \frac{P(\text{both } A \text{ and } B)}{P(A)}.$$

We shall consider an elementary illustration. Imagine that you deal a card to a friend from a well-shuffled deck of 52 cards. The deck of cards has four suits: spades, hearts, diamonds, and clubs. In each suit there are 13 cards: 2 through 10, jack, queen, king, and ace. Jacks, queens, and kings are "face cards." Your friend picks up the card you dealt, but she does not show or tell you which card it is. Let B denote the event that she got a king. Since there are 4 kings in the deck and all 52 cards are equally likely,

$$P(B) = 4/52 = 1/13.$$

Imagine now that after picking up the card your friend exclaims "I got a face card!" How likely is it now that she got a king? (Your friend obviously knows whether she got a king or not. But you don't. The question is: How likely is it *from your perspective* that she got a king?)

Let A denote the event that she got a face card. Since the 12 face cards are equally likely, the conditional probability that she got a king, given that she got a face card, is

$$P(B|A) = 4/12 = 1/3.$$

This example illustrates that if all outcomes of a chance experiment are equally likely, then we can compute conditional probabilities:

$$P(B|A) = \frac{\text{number of outcomes that are both in } A \text{ and } B}{\text{number of outcomes in } A}.$$

Of course, we can also use the formula

$$P(B|A) = P(\text{Both } A \text{ and } B)/P(A) = \left(\frac{4}{52}\right)\bigg/\left(\frac{12}{52}\right).$$

*The Multiplication Rule (The "and" Rule) for Probabilities of Two Events

Earlier we learned the multiplication rule for independent events. There is a similar rule that works in general, that is, when two events, say A and B, are not independent.

> If we multiply $P(B|A)$ by $P(A)$, then we get the general case of the **multiplication rule for probabilities**:
>
> $$P(\text{Both } A \text{ and } B) = P(A) \times P(B|A).$$

Similarly, for any 3 events A, B, and C,

$$P(A \text{ and } B \text{ and } C) = P(A) \times P(B|A) \times P(C|A \text{ and } B).$$

For example, suppose that the cards are in a deck of 52 playing cards shuffled well and that three cards are dealt one at a time. Let A denote the event that the first card dealt is a heart, B the event that the second card dealt is a heart, and C the event that the third card dealt is a heart.

$$\begin{aligned} P(\text{all three cards are hearts}) &= P(A \text{ and } B \text{ and } C) \\ &= P(A) \times P(B|A) \times P(C|A \text{ and } B) \\ &= (13/52) \times (12/51) \times (11/50) \end{aligned}$$

Similarly,

$$P(\text{all three cards are aces}) = (4/52) \times (3/51) \times (2/50).$$

Also,

$$\begin{aligned} &P(\text{first card is a spade, second card a heart, and third card a diamond}) \\ &= (13/52) \times (13/51) \times (13/50). \end{aligned}$$

Summary

The probabilities assigned to the outcomes of a random phenomenon must be between 0 and 1 and must add up to 1:

$$0 \leq P(\text{outcome}) \leq 1,$$

$$\sum_{\text{all outcomes}} P(\text{outcome}) = 1.$$

PROBABILITIES AND SIMULATION

Equally Likely Outcomes If all outcomes are equally likely, then we can compute the probability of any event A:

$$P(A) = \frac{\text{number of different outcomes in } A}{\text{total number of outcomes}}.$$

The probability of any event equals the sum of the probabilities of the outcomes in the event:

$$P(\text{event}) = \sum_{\substack{\text{all outcomes} \\ \text{in event}}} P(\text{outcome}).$$

Complement Rule The probability that a given event A does not occur is 1 minus the probability that A does occur:

$$P(\text{not } A) = 1 - P(A).$$

Addition Rule (The "or" Rule) If two events A and B have no outcomes in common (are mutually exclusive), they cannot both occur at the same time. The probability that one or the other occurs is the sum of the probabilities of two events:

$$P(A \text{ or } B) = P(A) + P(B).$$

If the events A and B are not mutually exclusive, they can both occur at the same time.

$$P(A \text{ or } B \text{ or both}) = P(A) + P(B) - P(\text{both } A \text{ and } B)$$

The Multiplication Rule (The "and" Rule) If the events A and B are independent, then

$$P(\text{both } A \text{ and } B) = P(A) \times P(B).$$

In general,

$$P(\text{both } A \text{ and } B) = P(A) \times P(A | B).$$

A caution: The addition rule applies in one situation and the product rule in another. To apply either rule correctly, one thus must make sure that the appropriate situation holds.

Rules for Probability Trees
- Multiply along the branches to find $P(\text{outcome})$.
- Add across the appropriate top branches to find $P(\text{event})$ as a sum of $P(\text{outcomes})$.

Section 4.2 Exercises

The probabilities required in these exercises are theoretical, not experimental.

1. Identify three probability experiments having multiple stages (such as tossing two dice) and equally likely outcomes. Draw a tree diagram for each to justify your choices. Compute a couple of probabilities of interest to you.
2. A four-sided die (What must it be shaped like?) is called a tetrahedral die. Assume that the possible outcomes of tossing a four-sided die are 1, 2, 3, and 4. Under the assumption of equally likely outcomes, find the following:
   a. $P(1)$
   b. $P(2)$
   c. $P(3)$
   d. $P(4)$
   e. $P(\text{even number})$
   f. $P(\text{number less than 4})$
3. A ten-spinner is shown below. The arrow is free to spin randomly until it stops in one of the 10 sectors. Assume that the arrow is equally likely to stop in each of the 10 sectors.

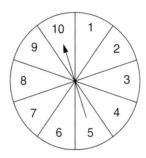

    a. What is the theoretical probability of obtaining each of the digits 0 through 9?
    b. What is the theoretical probability of obtaining an odd digit?
4. Suppose you have 19 classmates and the instructor is about to choose two members of your class to present problems on the board.
    a. What is the theoretical probability that you and your friend Tasha will both be chosen, assuming that each class member has an equal chance of being selected? (*Hint:* How many ways can two people be chosen by choosing one person and then a second from the remainder?)
    b. The class has four seniors. One person is randomly chosen to critique the solutions. What is the probability a senior will be chosen to critique the solutions?
5. A regular deck of playing cards has 52 cards. These cards consist of four suits, called spades, hearts, diamonds, and clubs. Each suit has 13 cards with values 2, 3,... , 10, jack, queen, king, and ace. Hearts and diamonds are red, whereas spades and clubs are black.

    Suppose you deal the top card from a well-shuffled deck that is face down. Find the following theoretical probabilities:
    a. *P*(black card)   **c.** *P*(ace)
    b. *P*(heart)   **d.** *P*(king of diamonds)
6. Suppose you throw a pair of casino dice (see Example 4.3). Compute:
    a. *P*(the sum of the two dice equals 2)   **d.** *P*(the sum of the two dice equals 6)
    b. *P*(the sum of the two dice equals 7)   **e.** *P*(the sum is at most 5)
    c. *P*(the sum of the two dice equals 10)   **f.** *P*(the sum is at least 10)
Hint: Use the following table to see all of the possible outcomes of throwing a pair of dice and the corresponding sum in each case.

	Die 2					
Die 1	1	2	3	4	5	6
1	2	3	4	5	6	7
2	3	4	5	6	7	8
3	4	5	6	7	8	9
4	5	6	7	8	9	10
5	6	7	8	9	10	11
6	7	8	9	10	11	12

7. In roulette, half of the numbers between 1 and 36 are odd, half are even. The numbers 0 and 00 are considered neither odd nor even. If you bet on "odd," what is your chance of winning? If you bet on the four numbers {17, 18, 20, 21}, what is your chance of winning? (You win if the ball comes to rest in any of the those four pockets.)
8. A coin is tossed four times, and the sequence of heads and tails is recorded.
    a. List each of the 16 sequences in the sample space *S*.
    b. Find the probability of getting two heads and two tails in any order.
    c. Find the probability of getting heads on the second toss.

d. Find the probability of getting at most one tail.
e. Find the probability of getting at least one tail.
Hint: It may help to draw a tree diagram.

9. Consider the coin toss experiment in Example 4.7 and in Exercise 10 of Section 4.1. Someone suggests that we assign the following probabilities to the outcomes in the sample space $S = \{1, 2, 3, 4, 5, \ldots\}$:

Outcome	1	2	3	4	5	...
Probability	1/2	1/3	1/4	1/5	1/6	...

What is wrong with this assignment? Would $1/2, 1/4, 1/8, \ldots$ be a logically correct assignment? (Do these probabilities add up to 1?)

10. The gene responsible for eye color comes in two forms (alleles): dominant brown (B) or recessive blue (b). A person with genotype BB, Bb, or bB has brown eyes, whereas a person with genotype bb has blue eyes. (This model is obviously an oversimplification. Eyes come in many shades of color. But the model is a good first approximation.) Suppose that a blue-eyed man and a brown-eyed woman have a blue-eyed child.
    a. What is the genotype of the woman: BB or Bb? (By Bb here we really mean either Bb or bB.)
    b. What is the chance that the couple's second child will also have blue eyes?
    c. Suppose that two brown-eyed parents have a blue-eyed child. What is the chance that the couple's second child will have brown eyes?

11. If two healthy cystic fibrosis carriers have a baby, there is a 3/4 chance that the child will be healthy and a 1/4 chance that the child will develop cystic fibrosis. Does this mean that if the couple has four children, then three will be healthy and one will get cystic fibrosis? Explain.

12. (Complex but interesting genetic problem) In humans there is a special chromosome pair that determines sex. The female pair is XX and the male pair is XY. A child inherits one of the mother's X chromosomes and, with equal probability, the father's X chromosome to become a female or the father's Y chromosome to become a male. (The different birth rates for boys and girls is explained by different prenatal survival rates.) Some genes are carried only by the X chromosome. These genes are said to be **sex-linked.** Females have a pair of such genes; males have only one. Females can have genotype AA, Aa, or aa, but males have only two genotypes: A and a. Many sex-linked genes, such as hemophilia (a condition in which blood cannot coagulate, so one easily suffers life-threatening bleeding) and color blindness, are recessive and cause harm.

    Let a be such a gene. Then all a males and all aa females show the defect. A males and AA, Aa, or aA females will be healthy.
    a. Can a boy inherit hemophilia from his father? Explain.
    b. Can a boy inherit hemophilia from a healthy mother?
    c. Imagine that the mother and the maternal grandparents of a newborn baby boy are all healthy, but that one of the boy's maternal uncles has hemophilia. What is the chance that the boy has hemophilia?

13. Explain why breast cancer screening for younger women will detect few cancers but will produce many false positive test results.

14. Recall there are four ABO blood types: A, B, O, AB. The gene for ABO blood type has three forms (alleles): A, B, and O. O is recessive. A person with genotype AA or AO (in either order) will have blood type A. A person with genotype BB or BO (in either order) will have blood type B. A person with genotype AB (one parent contributed A, one contributed B) will have blood type AB. A person with genotype OO will have blood type O. Two parents have blood type AB. Find the probability that their first child has
    a. blood type A.   b. blood type B.   c. blood type AB.

15. A symmetric casino die is thrown four times. Compute the probability of getting
    a. four sixes.   b. no sixes.   c. at least one six.

16. Assume that both male and female African Americans have the ABO blood type distribution shown on the "African American" line of Table 4.5. Also assume that the blood types of husbands and wives are independent. Find the probability that in a randomly chosen African American married couple,
    a. the wife is type AB.
    b. both husband and wife are type O.
    c. the wife is type A and the husband is either type O or type A.
    d. the wife is type B and the husband is either type O or type B.
    e. the wife can receive a blood transfusion from her husband (see Table 4.4).
17. A wildcatter (oil prospector not employed by a major oil company) is drilling an exploratory oil well in a region not known to contain oil or gas. His chance of drilling through shale is 0.6. If the well hits shale, the chance of striking oil is 0.3. If the well does not hit shale, the chance of striking oil is only 0.05. Draw a probability tree in which the first pair of branches represents hitting shale or not hitting shale and the second set of branches represents striking oil or not striking oil. What is the chance that the wildcatter strikes oil?
18. About 1 in 10 African Americans is a carrier of sickle-cell anemia (see Example 4.11).
    a. What is the chance that in a randomly chosen African American couple both husband and wife are carriers of sickle-cell anemia? Recall that if a couple of healthy carriers have a child, the risk that the child will have sickle-cell anemia is 0.25.
    b. Draw a probability tree in which the first branches represent the sickle-cell anemia status of a randomly chosen African American couple. Let the second set of branches represent the sickle-cell anemia status of the couple's firstborn. What is the chance that the child of a randomly chosen African American couple has sickle-cell anemia? *Hint*: Are both carriers or not?
19. Here are the number (in thousands) of college degrees awarded in 1996, classified by level and sex of the degree recipient.

	Bachelor's	Master's	Professional	Ph.Ds	Total
Male	522	179	45	27	773
Female	642	227	32	18	919
Total	1164	406	77	45	1692

   a. If you select a degree recipient at random, what is the chance that you pick a woman?
   b. What is the conditional probability that you picked a woman, given that you selected a Ph.D.?
   c. What is the conditional probability that you picked a bachelor's degree recipient, given that you selected a woman?
20. What is the probability that a family with three children has all boys if you know that at least one of the children is a boy? (See Figure 4.4)
21. Compute the probability of winning in craps. (See Example 4.3) Draw a probability tree in which the first set of branches represents your score on the first toss of the dice and the second set of branches represent winning and losing.
22. Sometimes a conditional probability can be found theoretically by simple logical reasoning. Out of four bottles of milk, two are spoiled. First John buys one of the bottles, and then Jack buys one of the remaining bottles. Find
    a. P(Jack gets spoiled milk | John got spoiled milk)
    b. P(Jack gets spoiled milk | John got fresh milk)

23. A somewhat absent-minded hiker forgets to bring his insect repellent on 30% of his hikes. The probability of being bitten is 0.9 if he forgets the repellent, and 0.2 if he uses the repellent. Perform at least 50 trials simulating whether the hiker is bitten or not.
    a. What is the experimental probability that he will be bitten?
    b. Now look only at the times when the hiker was bitten. What is the total number of times he was bitten? Of these times, how many times did he forget to bring his repellent?
    c. Using the information from part b, calculate the experimental probability that he forgot his repellent given that he was bitten.
24. Consider the hiker in the previous exercise. Draw a probability tree and compute the theoretical probability that
    a. he will forget the insect repellent and will get bitten.
    b. he will get bitten.
    c. given that he gets bitten, he forgot the repellent.
25. ELISA (enzyme-linked immunosorbent assay) tests are used to screen donated blood for the AIDS virus. The probability that an infected person tests positive is 0.997. The probability that a person who is not infected tests negative is 0.985. *Hint:* These last two probabilities really are conditional probabilities.
    a. What is the probability that a person who is not infected will test positive?

    Suppose that 1% of the U.S. population is infected and that we test a randomly chosen individual. Draw a probability tree (see Figure 4.5) and compute the probability that we get
    b. a true positive.
    c. a false positive.
    d. a positive test result.
    e. If the person tests positive, how likely is it that he or she is infected?
26. Consider the AIDS test in the previous exercise. Imagine that we randomly test a person from a country where 10% of the population is infected. Find the probability that we get
    a. a true positive.
    b. a false positive.
    c. a positive test result.
    d. If the person tests positive, what is the conditional probability that he or she is infected?
27. Two cards are dealt from a deck of 52 cards. Find the probability that
    a. both are aces.
    b. neither is an ace.
    c. at least one of them is an ace.
    Hint: Use $P(\text{both } A \text{ and } B) = P(B|A)P(A)$.

4.3 SIMULATION: A POWERFUL TOOL FOR LEARNING AND DOING STATISTICS

Imagine that we want to find the theoretical probability of winning a game of craps (see Example 4.3). To do so, we could play a lot of craps games and use the experimental probability we get as our estimate of the theoretical probability. Unfortunately, we would need at least 400 craps games to be reasonably confident that our experimental probability is within 0.05 of the theoretical probability. The five-step simulation method presented shortly will allow us to carry out such simulation-based estimations of unknown probabilities. If we want to be reasonably sure that we are within 0.01 of the theoretical probability, we need at least 10,000 craps games. This approach does not look very promising.

Fortunately, we can make a computer do it for us. In a few seconds a fast computer can "play" 100,000 craps games and report back to us the fraction of times it won. This is called **simulation,** and it is useful for studying complex chance experiments that are built from simpler underlying chance experiments. (The game of craps is a complex chance experiment based on the simple, well-understood chance experiment of tossing a pair of dice.) This simulation technique is often called **Monte Carlo simulation**. It was first used under that name by physicists studying nuclear chain reactions.

A computer uses a random number generator program to produce digits to simulate the toss of a pair of dice or any other chance experiment. To show how the computer does this, we will use a table of random digits to do the simulations. The numbers in Table B.2 in Appendix B look just like numbers produced by 1800 independent tosses of a symmetric casino die. The random number generator in a computer can produce many more such numbers.

We can simulate the toss of a pair of casino dice by picking two digits from Table B.2. Here are the first two rows from this table.

Row									
1	66533	45332	24614	22231	26431	35541	12165	62116	16111
2	61261	22613	26252	14622	32262	33244	34614	13316	41136

To use a random number table, you may start with any row and at any place in that row. Once you pick a starting point, however, you should keep moving in a fixed way, such as from left to right, from that place through the rest of the table. You can move right or left or diagonally, or in whatever direction you wish; the important thing is to move systematically so that you do not allow for any human selective behavior that could invalidate independent selections.

We use these digits to simulate 10 games of craps (see Example 4.3). Let us pick the digits from left to right in the first row and then continue from left to right in the second row, and so forth. The results are shown in Table 4.6. Based on these 10 simulated craps games, we estimate that the probability of winning in craps is around 0.30. To get a more accurate estimate, we should make the computer simulate at least 1000 craps games. (The theoretical probability of winning with casino dice is close to 0.49. If we use 1000 simulations, we can be reasonably confident that the experimental probability will likely be within 0.03 of the true probability.)

Example 4.12

Safety of Blood Transfusions

Suppose you are injured and lose so much blood that you need several blood transfusions, and you are in a poor underdeveloped country where blood supplies are suspect. Let us assume that

Table 4.6 Simulating 10 Craps Games

Repetition	Scores	Outcome
1	(6,6)	Loss
2	(5,3) (3,4)	Loss
3	(5,3) (3,2) (2,4) (6,1)	Loss
4	(4,2) (2,2) (3,1) (2,6) (4,3)	Loss
5	(1,3) (5,5) (4,1) (1,2) (1,6)	Loss
6	(5,6)	Win
7	(2,1)	Loss
8	(1,6)	Win
9	(1,6)	Win
10	(1,1)	Loss

the local blood supply is such that 1/6 (a made-up probability) of the available transfusions contain an infectious organism such as hepatitis B virus or HIV. Assuming that you need five transfusions, what is the chance that all five transfusions are free from the infectious agent?

Solution: Having one transfusion is like rolling a casino die once. If the die shows 1 through 5, you are not exposed to the infectious agent; if the die shows a 6, you are exposed. We can therefore simulate one transfusion by selecting a digit from Table B.2 of Appendix B. Select a 6 and you are exposed; select 1 through 5 and you are not exposed. We can simulate receiving five transfusions by selecting five digits from Table B.2. If none of the digits is a 6, then you are not exposed at all. We shall use the first of 5 digits of each row of Table B.2 to repeat the simulation. For the first simulation we use the digits 66533 in the first row. For the second simulation we use the digits 61261 in the second row, and so forth. The first 7 simulations are shown in Table 4.7.

If we use all 40 rows in the first 5-digit column of Table B.2 to repeat the simulation 40 times, you escape exposure 15 times and are exposed 25 times. We therefore estimate the probability of receiving safe blood to be around $15/40 = 0.375$. (The theoretical probability is around 0.40.)

The Five-Step Simulation Method

Example 4.12 illustrates our five-step method for doing a simulation study:

Step 1. Choice of a Probability Model: First we identify receiving a blood transfusion with selecting one of the numbers 1, 2, 3, 4, 5, 6 at random with equal probability. We identify getting a 6 with receiving contaminated blood and getting 1 through 5 with receiving safe blood.

Step 2. Definition of One Simulation: Five transfusions corresponds to selecting five random digits as described in Step 1.

Step 3. Definition of the Event of Interest: We get five safe transfusions (are not exposed) if none of the five digits chosen in Step 2 is a 6.

Step 4. Repetitions of the Simulation: In Table 4.7 we listed seven simulations. The greater the number of simulations, the more accurate our estimate of the theoretical probability will be. If we use a computer, we should do at least 1000 simulations.

Step 5. Find the Experimental Probability of the Event of Interest: In Table 4.7, $\hat{P}$(5 safe transfusions) = 2/7, a number we cannot even begin to trust as having any accuracy! With 40 simulations, $\hat{P}$(5 safe transfusions) comes to 0.375, which begins to approach the theoretical probability.

In the next example we will use simulation to estimate a probability distribution (that is, the probabilities for the various possible values of a statistic of interest).

Table 4.7 Simulating 5 Blood Transfusions

Repetitions	Digits	Outcome
1	66533	Exposed
2	61261	Exposed
3	61144	Exposed
4	21341	Not Exposed
5	15365	Exposed
6	51612	Exposed
7	31244	Not Exposed

Example 4.13

Multiple-Choice Quiz

A multiple-choice quiz has five questions. To each question there are six options to choose from, one of which is correct. Imagine that you have no clue whatsoever what the correct answers are and blindly (randomly) guess the answer to each of the five questions. How likely are you to get 0, 1, 2, 3, 4, or 5 correct answers? That is, find the probability distribution of the number of correct answers.

Solution: We will use the five-step method.

Step 1. Choice of Model: Since there are six answers to choose from, the probability that we guess the correct answer to any particular question is $1/6$. We can therefore identify trying to guess an answer with selecting one of the numbers 1, 2, 3, 4, 5, or 6 at random with equal probability. We identify getting a 6 with guessing the correct answer. (You might, of course, have selected one of the other digits, such as 1.)

Step 2. Definition of One Simulation: Trying to guess the answers to five questions corresponds to selecting five random numbers as described in Step 1. Each 6 corresponds to a correct answer.

Step 3. Definition of the Statistic of Interest: The statistic of interest is the number of correct answers. It corresponds to the number of sixes among the 5 digits chosen in Step 2.

Note: Because we are looking at the entire **distribution of a statistic of interest** rather than a particular event, the title for Step 3 is changed from "Event of Interest" to "Statistic of Interest."

Step 4. Repetitions of the Simulation: We shall use the second 5-digit column in Table B.2 of Appendix B to simulate 40 values of the statistic of interest, the number of correct answers. We list the first five simulations.

Repetitions	Digits	Number of correct answers
1	45331	0
2	22613	1
3	46631	2
4	66222	2
5	46135	1

Step 5. Find the Experimental Probability Distribution of the Statistic of Interest: We put the results of the 40 simulations in Table 4.8.

The title of Step 5 has changed because we now want the experimental probabilities for all possible values of the Step 3 "Statistic of Interest." We have also put the corresponding theoretical probabilities in Table 4.8 for comparison. Note that the theoretical probabilities are rounded to a

Table 4.8 Experimental and Theoretical Probabilities for the Number of Correct Answers to the Quiz

Number of Correct Answers	Experimental frequency	Experimental probability	Theoretical probability
0	15	0.375	0.402
1	14	0.350	0.402
2	7	0.175	0.161
3	3	0.075	0.032
4	1	0.025	0.003
5	0	0.000	0.000

certain number of significant digits. For example,

$$P(5 \text{ correct}) = (1/6)^5 \approx 0.000.$$

Because we only simulated 40 observations of the statistic of interest the experimental probability distribution is only moderately close to the theoretical probability distribution. (The theoretical distribution is a binomial distribution; we shall discuss binomial distributions in Chapter 6.)

We can also compare the two distributions by looking at the corresponding histograms. We can draw a density histogram (see Figure 4.7) to show the shape of the experimental probability distribution of the statistic interest, and we can draw a **probability histogram** to show the shape of the theoretical probability distribution of the statistic of interest (see Figure 4.8). When the possible values of the statistic of interest are all integers, as is often the case, then the rectangles provide a convenient visual impression of the probability distribution's center, spread, and shape, just as a density histogram does with data.

In a probability histogram the area of each rectangle equals the theoretical probability of obtaining its value. Because the sum of all outcome probabilities is 1, it is easy to show that the sum of the areas of the rectangles of a probability histogram equals 1. We note that the experimental probability histogram of Figure 4.7 is quite similar to the theoretical probability histogram of Figure 4.8. In fact, if the number of simulations had been large, (say, 400), they would have been very close.

Occasionally the outcome values of a probability distribution are not all integers. Then we can simply use the heights of lines at the values to graph the distribution. To illustrate, consider the distribution that is graphed in Figure 4.9.

The number of correct guesses on the quiz in Example 4.13 is an example of a **random variable.** That is, a random variable is a commonly used name for a statistic of interest. As such, a random variable is a numerical outcome affected by chance, such as the number of tornadoes to touch down in Kansas next spring or Sammy Sosa's batting average next season.

There is a mathematical formula for computing the theoretical probabilities in Table 4.8, but we will not learn it until Chapter 9. For now we can use the experimental probabilities obtained by simulation to estimate the theoretical probabilities. To get accurate estimates we need a large number of simulations, but that is no problem if we have a computer program do the simulations for us (as the optional instructional software supplied as ancillary to the book does).

We now give an example that illustrates how the experimental probability distribution gets closer and closer to the theoretical probability distribution as we have the computer increase the number of simulations from 100 to 100,000.

We shall focus on a random variable that is useful for studying the game of craps. In the game of craps (see Example 4.3) we repeatedly throw a pair of dice and add up the spots on the two dice. This sum of the scores on the two dice is a random variable. Let us assume that the dice are fair dice (sym-

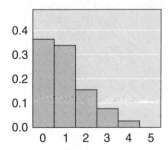

Figure 4.7 Experimental probability histogram derived from Table 4.8.

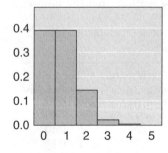

Figure 4.8 Theoretical probability histogram derived from Table 4.8.

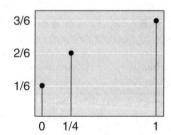

Figure 4.9 Graph of probability distribution whose outcome values are not integers.

metric casino dice are very nearly fair) and that the 36 outcomes in Figure 4.3 are equally likely. In that case we can find the theoretical probability distribution for the sum of two dice. This is an example of a theoretical probability distribution of a random variable. In the standard notation for displaying such a probability distribution, the possible values of the random variable are denoted by x and the associated probabilities by $p(x)$. (Note that we used the notation $P(A)$ with a capital P for the probability **of an event** A, whereas we use the notation $p(x)$ with a lowercase p for the probability **at a value** x for a probability distribution of a random variable.) This variable has a limited number of possible values x, so we can list all of them with their associated probabilities $p(x)$ (see Exercise 6, Section 4.2):

Sum of dice (x)	P(sum = x) (denoted $p(x)$)
2	1/36
3	2/36
4	3/36
5	4/36
6	5/36
7	6/36
8	5/36
9	4/36
10	3/36
11	2/36
12	1/36

Note that $\sum_{\text{all } x} p(x) = 1$.

We can simulate one observation of the sum by picking a pair of digits from the random number table for a fair six-sided die Table B.2 in Appendix B and adding the two digits. We can simulate 100 observations of the sum by picking 100 such pairs of digits and computing the sum of each pair.

Our instructional computer software simulated 100,000 such sums of two tossed fair dice. Table 4.9 shows the relative frequency distributions obtained for the first 100, 1000, 10,000, and the entire 100,000 simulations. The next-to-last column provides the theoretical probability, rounded to five significant figures for easy comparison. The last column, which computes the error of estimation $\hat{p}(x) - p(x)$ when estimating the true probability, shows the very high accuracy achieved with 100,000 simulations (the largest error being 0.00137 in size.) As this example convincingly demonstrates, simulation is a powerful tool of great practical use for finding probability distributions accurately. Professional statisticians make heavy use of such simulation analyses in settings in which they are unable to derive the true probability distribution mathematically.

Table 4.9 Relative Frequency Distributions for the Sum of the Scores on Two Casino Dice

Sum	100 simulations	1000 simulations	10,000 simulations	100,000 simulations	Theoretical probability	Error (Exp.–theor.)
2	0.02	0.025	0.0298	0.02789	0.02778	0.00011
3	0.07	0.053	0.0549	0.05610	0.05556	0.00054
4	0.05	0.093	0.0853	0.08318	0.08333	–0.00015
5	0.12	0.104	0.1112	0.10974	0.11111	–0.00137
6	0.15	0.127	0.1311	0.13834	0.13889	–0.00055
7	0.19	0.169	0.1626	0.16590	0.16667	–0.00077
8	0.11	0.128	0.1369	0.13901	0.13889	0.00012
9	0.14	0.129	0.1134	0.11118	0.11111	0.00007
10	0.07	0.081	0.0848	0.08319	0.08333	–0.00014
11	0.07	0.062	0.0581	0.05665	0.05556	0.00109
12	0.01	0.029	0.0319	0.02882	0.02778	0.00104

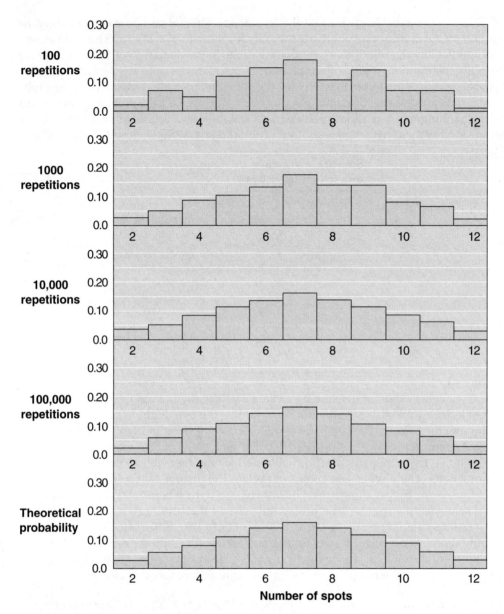

Figure 4.10 Experimental density histogram (top four) and theoretical probability histogram (bottom) derived from Table 4.9.

To the eye, the 100,000-simulation-based density histogram and the probability histogram are indistinguishable (see Figure 4.10). The last column of Table 4.9 makes this closeness numerically clear.

Experiments with Two Equally Likely Outcomes

Many experiments consist of repeated trials in which there are only two equally likely outcomes per trial. The numbers in Table B.1, the two-digit (0 or 1) random number in Appendix B, look just like numbers produced by 1800 independent tosses of a fair coin, if we identify heads with the digit 1 and tails with the digit 0. We can use the digits in Table B.1 to simulate the outcomes of independent repetitions of any random phenomenon with two equally likely outcomes.

Example 4.14

True-False Quiz

Imagine that you blindly guess the answer to each of the five questions on a true-false quiz. How many correct answers will you likely get?

Solution: We use the five-step method to estimate the distribution of the random variable defined as the number of correct answers.

Step 1. Choice of Model: The probability that we guess the correct answer to any particular question is 1/2. We can therefore identify guessing an answer with selecting one of the digits 0 or 1, with equal probability. Let us identify the digit 0 with a wrong answer and the digit 1 with a correct answer.

Step 2. Definition of One Simulation: Guessing the answers to five questions corresponds to selecting five digits according to the probability model described in Step 1. Each 1 occurring corresponds to a correct answer.

Step 3. Definition of Statistic of Interest: The number of correct answers corresponds to the number of 1s among the five digits chosen in Step 2.

Step 4. Repetitions of the Simulation: We shall use the 40 rows of the second 5-digit column in Table B.1 to simulate 40 repetitions of the quiz.

Repetition	Random Digit	Number of Correct Answers
1	10000	1
2	00111	3
3	11000	2
4	00101	2
5	11011	4
.	.	.
.	.	.
.	.	.
40	10100	2

Step 5. Find the Experimental Probability Distribution of the Statistic of Interest: We put the results of the 40 simulations in Table 4.10 (rounding the theoretical probabilities to three significant digits).

The theoretical distribution is again a binomial distribution, to be studied in Chapter 6. Note that, with the relatively small number of repetitions of simulations, some of the errors are fairly large, for example −0.088.

Table 4.10 Experimental and Theoretical Probabilities for the Number of Correct Answers to the Quiz

Number of correct answers	Experimental frequency	Experimental probability	Theoretical probability	Error (Exp. − theor.)
0	1	0.025	0.031	−0.006
1	8	0.200	0.156	0.044
2	14	0.350	0.313	0.037
3	9	0.225	0.313	−0.088
4	6	0.150	0.156	−0.006
5	2	0.050	0.031	0.019

190   PROBABILITIES AND SIMULATION

We consider a second example with two likely outcomes per trial, this time with the number of trials random.

Example 4.15

Family Size

A young couple wants to have a daughter but does not want a large family. If their firstborn is a girl, they will not attempt to have any more children. If their firstborn is a boy, they will try again until they have a girl. How many children will they likely have?

Solution: We use the five-step method.

Step 1. Choice of Model: We assume that each child is equally likely to be male or female. If the couple has more than one child, we assume that the children's sexes are independent (ignoring the slight possibility of identical twins). We identify having a child with selecting one of the digits 0 or 1, with equal probability. We identify the digit 0 with a boy and the digit 1 with a girl.

Step 2. Definition of One Simulation: If the first digit we select is a 1, then we stop. Otherwise we continue selecting digits until we first get a 1. Note that the number of trials per simulation is random here.

Step 3. Definition of Statistic of Interest: The statistic of interest is the number of children the couple has, which corresponds to the number of digits in the simulation sample.

Step 4. Repetitions of the Simulation: We pick digits from left to right in the first row of Table B.1 (in Appendix B) to simulate 10 families. Table 4.11 shows the simulation of 10 families. Computer software simulated 100,000 families. Table 4.12 shows the density histogram for the first 100, 1000, 10,000, and 100,000 simulated families. In Table 4.12, we have combined 9, 10,... into one category called "Over 8." We have to do this somewhere for distributions where the number of values for the random variable is high or even infinite. One should stop at a point when the experimental probabilities have begun to be very small; here we have stopped at 8.

Step 5. Find the Experimental Probability Distribution of the Statistic of Interest: The results of the full 100,000 simulations provide accurate estimation of the theoretical probabilities, as one can see from the last three columns of Table 4.12.

Some of the columns in Table 4.12 do not add up to 1, because of rounding (but are close to 1). With 1000 simulations, all of the experimental probabilities differ from the corresponding theoretical probabilities by less than 0.01. With 100,000 simulations, the average size of difference is

Table 4.11   *Simulating 10 families*

Repetitions	Digits	Number of Children
1	01	2
2	1	1
3	1	1
4	01	2
5	00000001	8
6	01	2
7	01	2
8	1	1
9	1	1
10	0001	4

Table 4.12 Relative Frequency Distributions ($\hat{p}(x)$'s) and Theoretical Distribution $p(x)$'s of the Number of Children

No. of children	100 simulations	1000 simulations	10,000 simulations	100,000 simulations	Theoretical probability $p(x)$	Error* ($\hat{p}(x) - p(x)$)
1	0.44	0.502	0.503	0.501	0.500	0.001
2	0.25	0.249	0.246	0.249	0.250	−0.001
3	0.13	0.123	0.127	0.124	0.125	−0.001
4	0.13	0.055	0.062	0.062	0.063	−0.001
5	0.03	0.034	0.029	0.032	0.031	0.001
6	0.02	0.019	0.017	0.015	0.016	−0.001
7	0.00	0.004	0.008	0.008	0.008	0.000
8	0.00	0.006	0.004	0.004	0.004	0.000
Over 8	0.00	0.008	0.003	0.004	0.004	0.000

*$\hat{p}(x)$ in the right-hand column is for 100,000 simulations.

slightly less than 0.001, certainly impressive accuracy. Figure 4.11 presents a visual comparison of true and estimated probability distributions. Again, the density histogram of the statistic of interest for a large number of simulations is indistinguishable from the probability histogram of the statistic of interest. By contrast, note that with only 100 simulations there are a couple of rather large errors.

The point of the last two examples is that for all practical purposes, we can discover the entire probability distribution of a statistic of interest via simulation.

Example 4.16

What Looks Random?

If we toss a fair coin three times, there are $2 \times 2 \times 2 = 8$ possible outcomes, shown in Figure 4.12. All of the outcomes are equally likely. The probability of getting HHH is the same as getting HTH, namely 1/8.

If we toss the coin 10 times, there are $2 \times 2 \times 2 \times 2 \times 2 \times 2 \times 2 \times 2 \times 2 \times 2 = 1024$ equally likely possible outcomes. Each of the three outcomes

THTHHTHTHT   HHHHHHHHHH   HHHTTTTTTT

has the same probability, namely $1/1024 \approx 0.001$. Most people would think that the first outcome must be more likely because it looks "more random" than the other two outcomes. But this impression is wrong, because each specific outcome (result of 10 tosses) has the same very small probability. Our intuition tells us that heads and tails should tend to alternate. This is "dead wrong," because the coin has no memory of its previous tosses! A head is just as likely to be followed by another head as by a tail. If the first five tosses happen to result in heads, then it is still just as likely that the sixth toss will result in another head as a tail. In real life, a string of bad luck can in fact be followed by another bad luck event!

The third outcome is a run of three heads followed by a run of seven tails. Is having at least one such run of at least three heads or at least three tails unusual? More precisely, what is the probability that in 10 tosses of a fair coin we get at least one run of 3 or more consecutive heads or consecutive tails? Note here that there are many outcomes causing this event to happen in addition to the

192 PROBABILITIES AND SIMULATION

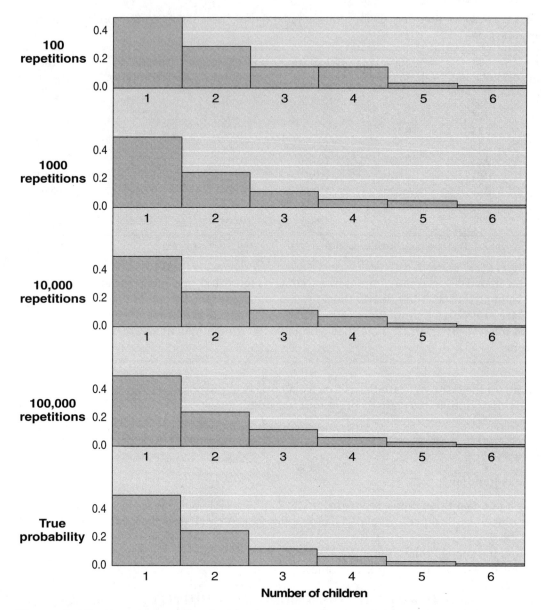

Figure 4.11 Experimental density histograms (top four) and theoretical probability histogram (bottom) derived from Table 4.12.

Figure 4.12 Possible outcomes of three coin tosses.

second and the third outcome above. This is actually a fairly challenging problem to solve theoretically.

Solution: We use the five-step method.

Step 1. Choice of Model: We identify tossing the coin once with selecting one of the digits 0 or 1, with equal probability. We shall identify the digit 0 with tails and the digit 1 with heads.

Step 2. Definition of One Simulation: Tossing the coin 10 times corresponds to selecting 10 digits as described in Step 1. Each 0 corresponds to a tail, and each 1 corresponds to a head.

Step 3. Definition of the Event of Interest: We are looking for runs of three or more consecutive 0s or 1s.

Step 4. Repetitions of the Simulation: We use the 40 rows of third and fourth 5-digit columns in Table B.1 to do 40 simulations.

Repetition	Random Digit	Run of 3 or more?
1	00010 10111	Yes
2	10111 10011	Yes
3	01101 01101	No
4	00001 01100	Yes
5	10100 00110	Yes
.	.	.
.	.	.
.	.	.
40	01101 00110	No

Step 5. Find the Experimental Probability of the Event of Interest: In 35 out of the 40 repetitions we get a run of three or more. We therefore estimate the probability of getting a run of three or more consecutive heads or tails in 10 tosses to be around $35/40 = 0.875$. Of course, doing 1000 simulations would produce a much more accurate probability estimate. (Careful counting reveals that out of the 1024 possible equally likely outcomes, 846 outcomes have a run of 3 or more, so the theoretical probability is $846/1024 = 0.826$.) That is, most of the time (over $4/5$) we will see at least one run of three. Does this surprise you? Even longer runs are not that unusual. For example,

$$P(\text{a run of at least 6 heads or at least 6 tails}) = 96/1024 \approx 0.094,$$

about 10% of the time.

Psychologists have done studies that show that most people do not realize how likely such runs are. People usually do not realize how often true randomness produces streaks, like a "streak of bad luck." For example, if people are asked to write down "random" sequences of 10 heads or tails, they invent fewer long runs (most people try to avoid runs of 4 or more) than will naturally occur with 10 independent tosses of a fair coin. Would you create anywhere near 9 sequences with a run of at least six heads or six tails in a row if you made up 100 sets of 10 coin tosses? Probably not! You probably wouldn't make any! It just seems too "nonrandom."

More Simulations

Tables B.1 and B.2 in Appendix B are very convenient for simulating random phenomena that have either two (like the toss of a fair coin) or six (like the roll of a fair die) equally likely outcomes. We will now learn how to simulate random phenomena with different numbers of outcomes and random phenomena whose outcomes are not equally likely. To simulate those experiments, we use the random numbers in Table B.3 of Appendix B.

First we discuss how we might generate random numbers such as those in Table B.3. Seeing how a random number table can be constructed from a physical process helps us understand its basic character better.

Many state lotteries have a daily "pick three" game in which a new winning three-digit number is selected every day. The lottery uses a set of 10 ping-pong balls that are identical except for being numbered 0, 1, 2, 3, 4, 5, 6, 7, 8, 9. The balls are placed in a large glass bowl, where they are tossed about and mixed by an air jet until one is forced out at random; the number on that ball is the first digit of the winning number. The ball is then returned to the glass bowl, and the process is repeated, generating the second digit independently of the first. The second ball is returned, and the process is repeated, generating the third digit independently of the first two. This is called **random sampling with replacement** and will be discussed in greater detail in Section 4.5.

For developing our intuition about random numbers, we may reason as if the digits in Table B.3 were generated as in the state lottery and as if the random number generator in a computer produces digits that way as well. In other words, the random digits are independent of each other and equally likely to be any of the numbers from 0 to 9.

Example 4.17

Nine of ten Americans are right-handed, and one of ten is left-handed. If we randomly select 15 Americans, how many will be left-handed?

Solution: We use the five-step method.

Step 1. Choice of Probability Model: We represent selecting one person at random by selecting one of the numbers 0, 1, 2, 3, 4, 5, 6, 7, 8, 9 at random with equal probability. We interpret getting 1 through 9 as selecting a right-handed person, and we interpret getting 0 as selecting a left-handed person. Therefore our chance of picking a right-handed person is $9/10$, and our chance of picking a left-handed person is $1/10$.

Step 2. Definition of One Simulation: Selecting 15 persons corresponds to selecting 15 digits as described in the probability model of Step 1. Each 0 corresponds to a left-handed person.

Step 3. Definition of Statistic of Interest: The number of left-handed persons corresponds to the number of zeros among the 15 digits chosen in Step 2.

Step 4. Repetitions of the Simulation: We shall use the 40 rows of fourth, fifth, and sixth 5-digit columns in Table B.3 to simulate 40 samples of 15 persons each.

Repetition	Random Digits	Number of left-handed persons
1	91807 57883 65394	1
2	78007 58644 73823	2
3	47004 48304 77410	4
4	78232 57097 01430	3
5	56211 85446 13656	0
.	.	.
.	.	.
.	.	.
40	03265 12748 77513	1

Step 5. Find the Experimental Probability Distribution of the Statistic of Interest: The results of the 40 simulations are shown in Table 4.13.

Note: As remarked earlier, when the number of values of a random variable is large, we sometimes lump its values beyond a certain point in a single category. Thus, we create the "Over 5" category here.

Table 4.13 Experimental and Theoretical Probabilities for the Number of Left-handed Persons in a Random Sample of 15 Persons

Number of left-handed persons	Experimental frequency	Experimental probability	Theoretical probability	Error $(\hat{p}(x) - p(x))$
0	9	0.225	0.206	0.019
1	14	0.350	0.343	0.007
2	11	0.275	0.267	0.008
3	5	0.125	0.129	−0.004
4	1	0.025	0.043	−0.018
5	0	0.000	0.010	−0.010
Over 5	0	0.000	0.002	−0.002

We often have to select digits in pairs or triples, and so forth, to simulate correctly the probabilities of the various outcomes of a random phenomenon, as the next example shows.

Example 4.18

Imagine that two healthy cystic fibrosis (CF) carriers marry and have four children. CF is a recessive disease like sickle-cell anemia (see Example 4.11). How many of their children will get cystic fibrosis?

Solution: We use the five-step method.

Step 1. Choice of Model: The children inherit their genotypes independently of each other (we ignore the possibility of identical twins). For each of the four children the probability of getting cystic fibrosis is $1/4$. We represent determining a child's CF status by selecting a pair of random digits from Table B.3 (in Appendix B). There are 100 equally likely pairs: 00, 01, 02, ..., 98, 99. If we pick 01 through 25, the child has cystic fibrosis. If we pick 00 or 26 through 99, the child does not have cystic fibrosis. Convince yourself that this yields $P(CF) = 0.25$ exactly. How would you model a probability of 0.38? of 0.615?

Step 2. Definition of One Simulation: Checking the CF status of four children corresponds to selecting four pairs of random digits as described in Step 1.

Step 3. Definition of Statistic of Interest: The number of children with CF corresponds to the number of two-digit numbers chosen in Step 2 that are between 01 and 25.

Step 4. Repetitions of the Simulation: We shall use the first eight digits in each row of Table B.3 to form four pairs of digits. We shall simulate 40 families of four children each.

Repetition	Digits	Number of children with cystic fibrosis
1	32, 23, 61, 26	1
2	40, 33, 65, 06	1
3	88, 79, 59, 37	0
4	12, 80, 76, 51	1
5	65, 92, 99, 67	0
.	.	.
.	.	.
.	.	.
40	04, 86, 16, 38	2

Table 4.14 Experimental and Theoretical Probabilities for the Number of Children with Cystic Fibrosis

Number of children with cystic fibrosis	Experimental frequency	Experimental probability	Theoretical probability	Error ($\hat{p}(x) - p(x)$)
0	11	0.275	0.316	−0.041
1	19	0.475	0.422	0.053
2	8	0.200	0.211	−0.011
3	1	0.025	0.047	−0.022
4	1	0.025	0.004	0.021

Step 5. Find the Experimental Probability Distribution of the Statistic of Interest: The results of the 40 simulations are shown in Table 4.14.

The experimental probability that at least one of the four children has cystic fibrosis is

$$1 - 0.275 = 0.725$$

or

$$0.475 + 0.200 + 0.025 + 0.025 = 0.725.$$

According to Table 4.14, the theoretical probability is

$$1 - 0.316 = 0.684$$

or

$$0.422 + 0.211 + 0.047 + 0.004 = 0.684.$$

Note that we found the probability of the event of interest in two ways: (1) using the complement rule and (2) breaking the event into mutually exclusive events of 1, 2, 3, or 4 children having CF and then applying the addition rule.

If we want to simulate the random event that determines whether a child gets cystic fibrosis or not, we have several other possibilities. We could take single random digits from Table B.3, interpret 1 as the child having CF, interpret 2 through 4 as the child not having CF, and ignore 0 and 5 through 9. We could also take pairs of digits from Table B.1, use 11 to represent the child having CF, and use the other three combinations 00, 01, and 10 to represent the child not having CF.

Now we consider an example of disease transmission. Although it is a simplistic example, it will serve to introduce the use of statistics in the field of epidemiology. Probability models and statistical inference are vital to epidemiology. We shall only spell out Steps 1 through 3 of the five-step method.

Example 4.19

The chance of contracting strep throat when coming into contact with an infected person is estimated as 0.15. Suppose the four children of a family come into contact with an infected person. What is the chance of at least one of the four children getting the disease?

Solution:

Step 1. Choice of a Model: Use pairs of random digits, sampling with replacement, with the following rule:

    01 through 15:   Child gets strep throat.
    16 through 99, plus 00:   Child does not get strep throat.

Step 2. Definition of one Simulation: Read four pairs of random digits, one for each child in the family.

Step 3. Definition of Event of Interest: The event of interest occurs (some of the children in the family get sick with strep throat) if at least one of the four pairs of digits is in the range 01 through 15.

Step 4. Repetitions of the Simulation: The more the better.

Step 5. Find the Experimental Probability of the Event of Interest: Compute

$$\frac{\text{number of times event of interest occurs}}{\text{total number of repetitions}}.$$

Here is another model for the spread of an infectious disease.

Example 4.20

Consider one person infected by human papilloma virus (HPV). As our probability model, based on past experience, we assume that a person infected with HPV will infect one other person with probability of 0.20, will infect two other persons with probability 0.45, will infect three persons with probability 0.30, and will not infect anyone with probability 0.05. Using a pair of random digits, we can simulate the number of persons infected by this person:

    00 through 04:   no one is infected.
    05 through 24:   one person is infected.
    25 through 69:   two persons are infected.
    70 through 99:   three persons are infected.

If the first person infects anyone, we can simulate the number of persons infected by each of these, and so on. Will we have an epidemic (many become infected), or will the disease die out quickly without infecting many people?

    The disease could die out entirely by chance, especially early on. For example, P(nobody infected) $= 0.05$; that is, there is a 0.05 chance that the disease dies out immediately! If we simulate the spread of the disease many times, however, we will find that on average the first person will infect two persons (this a topic for Chapter 5); each of these will on average infect two persons; and so on. The disease will tend to spread at an exponential rate, (tending to double the number infected at every stage). As a sophisticated theoretical probability analysis can also show us, this simplistic model for disease spread has the potential to create an unlimited epidemic. One could use the five-step simulation approach here to develop this into a nice group project.

Example 4.21

Data collected show that a secretary received telephone calls at a rate described by the following table:

Calls in an hour	Percent of the time
0	15%
1	35%
2	30%
3	15%
4	5%

Estimate the theoretical probability that he will receive four or more calls during a two-hour period.

Solution:

Step 1. Choice of a Model: Using a random number table, assign the digits in the following way:

Calls in an hour	Percentage of the time	Random digits
0	15%	00–14
1	35%	15–49
2	30%	50–79
3	15%	80–94
4	5%	95–99

Step 2. Definition of One Simulation: Read two pairs of random digits, one for each of the two hours.

Step 3. Definition of Event of Interest: Using the table in Step 1, convert the random two-digit numbers into a number of calls in each of the two hours. Add these together, and if it is at least 4, then the simulation is a "success" in the sense that the event of interest occurs. Otherwise it is called a "failure." (The "success/failure" terminology is convenient and we will use it from time to time.)

Step 4. Repetition of the Simulation: We shall use the first four digits in each row of Table B.3 (in Appendix B) to form two pairs of digits. We simulate five repetitions.

Repetition	Random digits	Calls in Hour 1	Calls in Hour 2	Total number of calls	Success?
1	32 33	1	1	2	No
2	40 33	1	1	2	No
3	88 79	3	2	5	Yes
4	12 80	0	3	3	No
5	65 92	2	3	5	Yes

Step 5. Find the Experimental Probability of the Event of Interest: Based on only five repetitions, we estimate the probability of getting four or more calls to be around 2/5 = 0.40. By sheer coincidence, this happens to be almost the theoretical answer. The theoretical probability is 0.405.

Summary

We use simulation to estimate theoretical probabilities. We can simulate independent repetitions of a random phenomenon by using a random number table. Several random number tables are given in Appendix B.

Section 4.3 Simulation **199**

The Five-Step Method

Step 1. Choice of Probability Model: When using a random number table, we identify each possible outcome of the random phenomenon with a block of digits in the random number table. This is done in such a way that the known probability of the outcome equals the probability of the corresponding block of digits.

Step 2. Definition of One Simulation: Take one block of digits at a time from the random number table, progressing from block to block in a consistent manner that rules out the influence of human judgment in each case.

Step 3. Definition of Event or **Definition of Statistic of Interest:** The event or statistic of interest is the main focus of the problem being solved.

Step 4. Repetitions of the Simulation: One thousand or more simulations are desirable if using computer software.

Step 5. Find the Experimental Probability of the Event of Interest or **Find the Experimental Probability Distribution of the Statistic of Interest**: These experimental probabilities can be used as estimates of the theoretical probabilities.

Section 4.3 Exercises

1. Estimate the probability of receiving 10 safe blood transfusions under the conditions described in Example 4.12. Make 40 simulations by using the last two five-digit columns in Table B.2 of Appendix B. For the first simulation use 62116 16111 in the first row. Identify 10 safe transfusions with not finding any sixes among the 10 digits. For the second simulation use 13316 41136 in the second row, and so on.

2. Estimate the probability distribution of the number of sixes in 10 tosses of a fair die (see Example 4.13). Make 40 simulations by using the third and the fourth five-digit columns in Table B.2. For the first simulation use 24614 22231 in the first row. Count the number of sixes. For the second simulation use 26252 14622 in the second row, and so on.

3. Estimate the probability distribution of the number of heads in 10 tosses of a fair coin. Make 40 simulations using the last two five-digit columns in Table B.1 of Appendix B. For the first simulation use 10001 10110 in the first row. Identify heads with 1 and tails with 0. For the second simulation use 00000 01111 in the second row, and so on.

4. Estimate the probability distribution of the number of girls in a family with four children. Make 40 simulations using the first four digits in each row in Table B.1. For the first simulation use 0111 in the first row. Identify 1 with a girl and 0 with a boy. For the second simulation use 1110 in the second row, and so on.

5. Estimate the probability that in 10 tosses of a fair coin we get at least one run of four or more consecutive heads or tails (see Example 4.16). Make 40 simulations using the fifth and sixth five-digit columns in Table B.1. For the first simulation use 00010 11001 in the first row. Identify 0 with tails and 1 with heads. For the second simulation use 11001 11011 in the second row, and so on. (The first simulation has a run of three zeros (tails) and the second simulation has a run of three ones (heads), but neither has a run of four or more consecutive zeros or ones.)

6. Estimating the probability distribution of the number of left-handed persons in a random sample of five persons (see Example 4.17). Make 40 simulations using the first five-digit column in Table B.3 of Appendix B. For the first simulation use 32236 in the first row. Identify each 0 with a left-handed person. Count the number of left-handed persons (zeros) in the sample. For the second simulation use 40336 in the second row, and so on.

For each of the following exercises, spell out Steps 1 through 3 of our five-step method. You do not need to carry out simulations. Indeed, imagine the instructional software doing them for you, with the number of repetitions being 1000 or more.

7. A manufacturer of 60-second film advertises that 95 out of 100 prints will develop successfully. (By this, they mean that the probability of a print developing successfully is 0.95.) Suppose you buy a pack of 12 and find that two prints do not develop. If the manufacturer's claim is true, what is the probability that two or even more prints do not develop in a pack of 12? Based on the experimental probability, do you have reason to complain about the company's advertising?

8. A pharmaceutical company knows from previous testing that a certain antibiotic capsule falls below prescribed strength 6% of the time, making the dosage ineffective. What is the probability that a prescription of 20 such capsules will contain two or more with ineffective dosages?

9. The following table gives the proportion of each type of hit Babe Ruth achieved during the 1928 season:

Type of hit	Percentage of total hits
Single	50
Double	15
Triple	4
Home run	31

Assume that these sample-based percentages represent the probability hitting distribution for Babe Ruth. If in a given game Ruth had three hits, estimate the probability that he had at least one home run.

10. Using the data from Exercise 9, estimate the probability that Ruth had no singles in a game in which he had exactly two hits.

4.4 SIMULATING RANDOM SAMPLING VIA A BOX MODEL

The box models presented in this section represent probability models for random phenomena and, as such, they help us understand what happens when a random phenomenon is repeatedly observed. But the box models also have another very important purpose: If we use the instructional software accompanying this book to simulate a random phenomenon on a computer, the first step in the simulation is to design a box model for the random phenomenon. This box model informs the computer what the possible outcomes of the random phenomenon are and what the corresponding probabilities are. With this information, the computer can then simulate the random phenomenon in question.

Recall the state lottery machine described in the preceding section: the glass bowl with the 10 hollow plastic balls (ping-pong balls) numbered from 0 to 9. The device is used to generate the digits of the winning "Pick 3" number, one digit at a time. Three times, one of the 10 balls is ejected from the bowl at random by an air jet, the number on the ball is noted, and the ball is returned to the bowl. This process generates three-place numbers in which each of the 10 digits has an equal probability of appearing in each place, independently of which digits appear in the other two places.

Imagine a simpler and cheaper version of the machine. Instead of a glass bowl, the balls are contained in a sturdy cardboard box. Rather than having an air jet shake up the balls until one is ejected, we shake the box, open it, reach in without looking, and randomly select a ball. We record the number on the ball and put the ball back. We could use this kind of arrangement to represent probability models of a wide variety of experiments, not just those involving dice, coins, or digits. Such an imagined device is called a **box model**. A box model is really a probability model made concrete.

Tossing a fair casino die, for example, can be modeled by drawing a ball at random from a box containing six balls numbered 1 through 6.

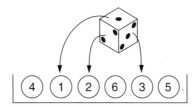

Many board games, such as Monopoly®, and many gambling games, such as craps (see Example 4.3), involve tossing two dice. Tossing two dice and recording both scores can be simulated by first drawing one ball at random, noting the number on it, and returning the ball to the box. After the ball has been returned, we again have six balls in the box. Then a second drawing is made at random from the box and the number on that ball is recorded.

Tossing a die 10 times (or equivalently tossing 10 different dice) and recording each die score can be modeled by 10 times drawing one of the six balls at random, noting the number on the ball, and replacing the ball in the box before the next drawing. This is called **random sampling with replacement**.

Uninformed guessing of the answers to 15 true-false questions on a quiz can be simulated by performing 15 random draws with replacement from the box:

Each guess corresponds to the random draw of one ball. If ball number 1 is chosen, the guessed answer is correct. If ball number 0 is chosen, the guessed answer is wrong. We record the number that is on the ball and return the ball to the box before the next drawing. Then the sum of the 15 numbers drawn will equal the number of correct answers on the 15-question quiz. It is vital to realize that there is no difference in the data produced if a real student guesses on 15 true-false questions or if we do 15 random draws with replacement from the box model. The simulation is a perfect imitation of the randomly guessing student.

We can use the same box model for simulating tossing of a coin if we identify heads with ball number 1 and tails with ball number 0. To simulate the number of heads in 50 tosses of a fair coin, we can look at the sum of the numbers on 50 balls drawn at random with replacement from the box. Similar to the above example of the guessing student, a box-model-based simulation is a perfect stand-in for actual coin tossing.

We can also use the same simple 0,1 box model to solve the Key Problem.

Example 4.22

The Tired MDs

Assume for the moment that the program director at the teaching hospital is correct when he insists that the young doctors are just as alert at the end of their 36-hour call as at the beginning. Then, if we gave 25 doctors the same type of mental tests at the beginning of their call and again at the end, each doctor would be just as likely to do better on the second battery of tests as on the first set of tests.

With these two equally likely outcomes assumed, we can use the 0,1 box model shown above to simulate the situation. We draw randomly with replacement from the box 25 times, once for each of the 25 doctors. If we draw ball number 0, the doctor does better on the first set of tests. If we draw ball number 1, the doctor does better on the second set of tests. The number of doctors who do better on the second set of tests will equal the number of ones in these 25 draws. Even more simply, the sum of the numbers on the 25 balls drawn equals the number of doctors who do better on the second set of tests, because summing these 25 zeros and ones is identical to just counting the ones.

Recall that only 5 of the 25 doctors did better on the second set of tests in the actual data collected by the program director. That is surprisingly few, if we believe the program director's claim that the young doctors are just as mentally alert at the end of their 36-hour call as at the beginning of the call. Could this have happened just by chance? This leads us as statisticians to ask how likely it is to observe behavior by the 25 doctors this extreme or even more extreme if the program director's claim is true. That is, if we assume that the program director is correct, what is the probability that only five or fewer than five of the 25 doctors do better on the second set of tests?

We can investigate the likelihood of the program director's claim being true by simulation, using our box model just described. We could use the 0,1 random numbers in Table B.1 of Appendix B to simulate 25 draws from the box and compute the sum of the 25 digits drawn. The sum would be the number of doctors doing better on the second set of tests. This would be one simulation of comparing the 25 doctors' performances on the two batteries of tests. But there are only enough digits in Table B.1 to repeat the simulation 72 times! We need a computer to do a large number of simulations to obtain good accuracy. Based on 50,000 computer simulations, the experimental probability estimate is

$$\hat{P}(5 \text{ or fewer do better on the second battery of tests}) = 0.002.$$

This result actually agrees with the exact theoretical probability to three significant digits.

Since this probability is so small we conclude that more than 5 of the 25 doctors would have done better on the second test if there had been no decline in mental alertness. Thus, the observed data of only five doctors doing better on the second battery of tests contradicts the program director's claim! As we will see in Chapter 9 and beyond, this line of reasoning is called **hypothesis testing** and is one of the central ideas of statistical inference.

Keno is a popular game in casinos. A player bets on one or more of the numbers from 1 to 80. The casino (the "house") randomly picks 20 of the numbers from 1 to 80. If the player bets on a single number, say 33, the player wins if 33 is among the 20 numbers chosen by the house. The probability of winning is $20/80 = 1/4$. The house pays $3 to a winner for every $1 bet. Thus, if the player wins on a $1 bet, he gains $2: his dollar back plus two dollars more. If he loses, the house keeps his dollar. We can simulate a player's gain or loss on a $1 bet by drawing from the box:

The amount he gains on the bet is the number on the chosen ball. (A gain of –1 dollar is of course a loss of one dollar.) Notice that the chance of gaining $2 is the same as whether he draws from the box or plays Keno, namely $p(2) = 1/4$. And the chance of gaining –1 dollar (losing a dollar) is the same whether he draws from the box or plays Keno, namely $p(-1) = 3/4$. Thus, if you bet $1 on number 33 ten times, then we can simulate your net gain or loss by randomly drawing from the box ten times with replacement and adding the resulting ten numbers.

A Keno player who bets on a pair of numbers wins if both her numbers are chosen by the house. Her theoretical probability of winning can be computed using advanced mathematical reasoning; in fact, it is very close to 0.06. The bet pays $12 if she wins, producing a gain of $11. We can simulate her gain or loss on a $1 bet on a pair of numbers by drawing one ball at random from the box:

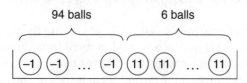

With this box, her probability of gaining $11 is 0.06, and her probability of losing $1 is 0.94. If she bets $1 on a pair of numbers 30 times, we can simulate her gain or loss by randomly drawing from the box 30 times with replacement and adding up the resulting 30 numbers. It may help to emphasize that what is being simulated here is not the Keno game itself, but rather the probability distribution of one's chances of winning and losing, the essence of the situation actually. That is, we are simulating 30 draws from the distribution

x	-1	11
$p(x)$	0.94	0.06

Building a Box Model and Doing Five-Step Simulation with It

When we make a box model to simulate a random phenomenon, we have to decide:

What distinct numbers should be on the balls?
How many balls of each kind should there be?

Answering these two questions allows us to specify the box model completely, thus completing Step 1 of the five-step method. To carry out one simulation we have to decide:

How many draws from the box?
Do we draw with or without replacement between draws?

This completes Step 2.

In the second Keno example, the numbers on the balls were the different amounts of money that could be gained or lost on each play, namely –1 and 11. To figure out how many balls we should have of each kind, we note that the probability of drawing a ball with any particular number, say 11, should be the same as the probability of winning that amount on one play: 6/100 in the Keno example. Thus, six of the 100 balls are 11s, and 94 of them are –1s. If we place a $1 bet on the same pair of numbers in several **independent** Keno games, then we can simulate our net gain by drawing repeatedly **with replacement** from the box and adding up the resulting numbers. Finally, the number of draws from the box must equal the number of Keno plays.

Example 4.23

Boy or Girl?

The probability that a newborn baby is a girl is 0.487 (see Example 4.1). What is the probability that out of 100 newborn babies, at least half are girls? What is the probability that out of 2500 newborn babies, at least half are girls? How about 10,000? Note that we are here interested in a subtle effect: As more and more children are born it becomes more and more likely that the proportion of girls will be close to 0.487. It thus becomes less and less likely that at least half of the children will be girls.

Solution: In the long run, more boys than girls will be born, but in the short run we may beat the odds and have at least as many girls as boys. In order to model $P(\text{girl}) = 0.487$, we construct a box with 487 ones and 513 zeros. To simulate counting the number of girls among 100 newborns, we randomly draw with replacement 100 times from the box and add up the 100 numbers we get. (Remember that this adding just counts the ones.) Here, each 0 corresponds to a boy and each 1 to a girl. Note that the proportion of 1s in the box was set equal to the probability of a girl, and similarly for a boy. We draw **with replacement** to guarantee that the simulated births are **independent** of each other. The event of interest is getting at least 50 girls. Simulating 50,000 times reveals that the probability that at least half of the 100 newborns are girls is very close to 0.44.

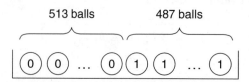

To simulate counting the number of girls among 2500 newborns, we draw 2500 times with replacement from the box. The event of interest is getting at least 1250 girls. The probability is very close to 0.10, as 50,000 simulations reveals to us.

To simulate counting the number of girls among 10,000 newborns, we draw 10,000 times with replacement from the box. The event of interest is getting at least 5000 girls. The probability is very close to 0.005.

Clearly, "beating the odds" and having at least as many girls as boys gets progressively harder as the number of draws increases from 100 to 2500 to 10,000.

Although the balls in the box models we have seen so far have been marked with numbers, they do not have to be, even when we do computer simulation (such as by using the textbook's accompanying instructional software).

In the following situation, numbering the balls is artificial and distracting. Four-o'clocks are ornamental plants. If we cross red-flowered ones (genotype RR) with white-flowered ones (genotype WW), all the offspring will be pink-flowered (genotype RW or WR). If we cross two pink-flowered four-o'clocks (both parents of genotype RW or WR), a tree diagram analysis like the one in Example 4.11 tells us that the offspring will have white flowers with probability 0.25, pink flowers with probability 0.5, and red flowers with probability 0.25. (Do you see why?) To simulate growing 100 offspring having two pink-flowered four-o'clock parents, we can sample 100 times at random with replacement from the box.

We then count the number of times we get a white ball, a pink ball, and a red ball. The point is that these color values are most naturally represented not as numbers but rather as nonnumerical categories (three different colors).

Summary

A sampling **box model** is a boxfull of balls (for concreteness, they can be viewed as ping-pong balls) that are all equally likely to be chosen. The number of balls can be large. The balls are marked to represent the different possible outcomes (usually, but not necessarily, with numbers), and the numbers of balls with the different marks are in proportion to the probabilities of the different outcomes.

If we know the possible outcomes of a random phenomenon and the corresponding probabilities, then we can make a box model for simulating independent repetitions of the random phenomenon. Each repetition (one simulation) corresponds to randomly drawing a certain number of times from the box. Specifying the box really amounts to a concrete representation of a probability distribution. It allows us to use the accompanying simulation software.

Section 4.4 Simulating Random Sampling via a Box Model

When we make a Step 1 box model for a random phenomenon, we have to decide:

What numbers should be on the balls?
How many balls should there be of each kind?

When we specify one simulation in Step 2, we need to decide:

How many draws will be made from the box?
Do we draw with or without replacement between draws?

The number of draws in Step 2 always equals the number of trials or draws that constitute one simulation of the random phenomenon, such as the number of children in a family or the number of persons interviewed in a population survey.

For gambling problems in which the player makes the same bet in several independent plays,

- The numbers on the balls are the amounts that can be won or lost on a single play.
- The fraction of balls with any particular number equals the probability of winning that amount on a single play.
- The number of draws equals the number of plays.
- We draw with replacement.
- The gain or loss is the sum of the numbers on the balls drawn.

Gambling problems are very helpful in our learning about probability modeling and simulation.

Section 4.4 Exercises

Note: In the gambling problems below, losing an amount, say losing $2, is the same as "winning" –2 dollars; hence the ball to be used has the number –2 on it.

1. In the casino game High-Low, a pair of fair six-sided dice is rolled and the sum of the two scores computed (see Figure 4.3). If you bet $1 on Low, you win $1 (get your dollar and one more back, with a net gain of $1) if the sum is 2, 3, 4, 5, or 6. If you bet $1 on High, you win $1 if the sum is 8, 9, 10, 11, or 12. If you bet $1 on 7, you win $4 if the sum is 7. For each of the three bets, "Low", "High", or "7", design a box model (that is, specify which balls are present and how many there are of each) for your gain or loss.
2. A roulette wheel has 38 slots, of which 18 are red, 18 are black, and 2 are green. (See Examples 4.6 and 4.9.) A bet on "red" pays *even money:* If you bet $1 on red and a red number comes up, you win $1. If a black or a green number comes up, you lose $1. Design a box model for a $1 bet on red.
3. A bet on a single number at roulette (38 slots, recall) pays "35 to 1": If you bet $1 on 3, say, and the ball ends up in slot number 3, you get your $1 back plus $35 in winnings. If the ball ends up in any other slot, you lose the dollar. Design a box model for a $1 bet on a single number.
4. A "split" is two adjacent numbers on a roulette (38 slots, recall) table, such as 2 and 3. If you bet on a split and either number comes up, you get your $1 back plus $17 in winnings. Design a box model for a $1 bet on a split.
5. A quiz has 10 multiple-choice questions. To each question there are five suggested answers, one of which is correct. Imagine that you blindly guess the answer to each of the 10 questions. Design a box model for simulating the number of correct answers you get on the quiz.
6. A husband and a wife are both carriers of the albinism gene. Any child of theirs has a 0.25 chance of being an albino. Imagine that the couple has five children. Design a box model for simulating the number of albinos among the five children.

7. A traffic light at an intersection is green 58% of the time for people traveling on the main street. Imagine that you pass this light twice a day: once in the morning and once in the afternoon. How would you set up a box model and simulate how many times in the 5-day work week the light is green?
8. Consider the following probability distribution:

x	1	6	7	12
$p(x)$	0.33	0.17	0.49	0.01

How would you construct a box model and simulate drawing from the corresponding population until the first 7 is obtained?
9. Consider the following box model.

Find the probability distribution it is simulating by finding the probability $p(x)$ for each x occurring in the box.
10. Consider the following box model:

Find the probability distribution it is simulating by finding the probability $p(x)$ for each x occurring in the box.

4.5 RANDOM SAMPLING WITH OR WITHOUT REPLACEMENT

In Section 4.4 we saw how we could simulate independent repetitions of a random phenomenon by drawing at random with replacement from a box of numbered balls. Box models can also be used to study population surveys in which a representative cross section of the population is interviewed (see Example 1.7). In such surveys the people who will be interviewed are selected at random but **without** replacement between draws, since we do not want to contact the same person twice. We shall now consider **random sampling without replacement.**

To see the difference, consider a box containing three balls: one red, one black, and one white.

Drawing two balls at random **with** replacement, as we have been doing, means first drawing one ball at random and noting its color before replacing the ball in the box. Then a second drawing is made. Because the first ball has been put back in the box, the result of the first draw in no way influences the result of the second draw: The two draws are independent! Figure 4.13 shows a tree diagram that represents the nine possible outcomes, each with probability 1/9.

Drawing two balls at random **without** replacement means that the first ball is **not** replaced in the box before the second drawing is made. There are six possible outcomes, all equally likely, with probability 1/6 (see Figure 4.14). When we draw **without** replacement the draws are *not* independent!

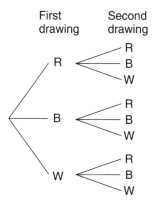

Figure 4.13 Drawing two balls at random with replacement.

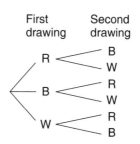

Figure 4.14 Drawing two balls at random without replacement.

Sample Surveys

Each month the *Current Population Survey* interviews a random sample of over 100,000 Americans to help policy makers in the government judge the economic health of the nation. The size of the labor force, the unemployment rate, the distribution of income, and educational level are all estimated from these survey results.

Market surveys use random samples to discover consumer preferences and usage of products. Public opinion polls ask people's opinions on a variety of issues such as candidate preferences; satisfaction with prominent elected officials; and attitudes on controversial social issues such as gun control, abortion, and the death penalty.

All sample surveys use random samples drawn without replacement. By sampling without replacement, the survey organizations avoid, as they certainly should, selecting any person more than once. Unfortunately, the needed probability computations for doing statistical inference are much easier if the sampling is with replacement.

Simulation is also simpler if we sample with replacement. Fortunately, whenever the sample size is less than 5% of the population size it makes little difference whether we sample with or without replacement: Even if we draw with replacement, not too many persons will be drawn more than once. So when we analyze sample survey data from samples of less than 5% of the population we can presume that the sample was drawn with replacement.

Example 4.24

Flag Signals

How many different signals can be made using four flags of different colors lined up on a vertical flagpole if exactly three flags are used for each signal?

Solution: Let the four colors be R (red), O (orange), Y (yellow), and B (blue). To make a flag signal, we first pick the flag to go on top; then we pick the flag to go in the middle; finally, we pick the flag to go at the bottom.

In other words, we sample three flags, one at a time without replacement. A total of $4 \times 3 \times 2 = 24$ different signals are possible; they are shown in Figure 4.15.

Whether we sample with or without replacement, there are usually too many possible outcomes to draw a tree diagram. Instead we can use the **multiplication rule** to compute the number of possible outcomes.

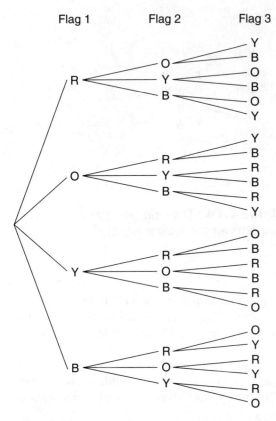

Figure 4.15 Possible three-flag signals without replacement.

Multiplication Rule for Multiple-Stage Experiments

Suppose an experiment is conducted in k stages. Suppose that stage 1 can be performed in N_1 ways; for each of these, stage 2 can be performed in N_2 ways; for each of the ways of performing stages 1 and 2, stage 3 can be performed in N_3 ways; and so forth. Then there are

$$N_1 \times N_2 \times N_3 \times \cdots \times N_k$$

ways of performing stages 1 through k.

To illustrate, if we toss a die five times, then there are six possible outcomes for the first toss, six possible outcomes for the second toss, and so on. If we record the five scores in order, then there are

$$6 \times 6 \times 6 \times 6 \times 6 = 7{,}776$$

possible outcomes. Nobody wants to draw the tree diagram for this situation!

A regular deck of playing cards has 52 cards. Each card has a suit and a face value. The four suits are spades, hearts, diamonds, and clubs. There are 13 cards in each suit with face values 2, 3,..., 10, jack, queen, king, and ace. Then the number of ways of dealing five cards, one at a time without replacement, from a deck of 52 cards is

$$52 \times 51 \times 50 \times 49 \times 48 = 311{,}875{,}200$$

(the total number of outcomes for the experiment of dealing five-card hands) if we keep track of the order in which the cards are dealt. The first card can be chosen in 52 ways, the second card in

51 ways, the third card in 50 ways, and so on. Thirteen of the 52 cards are hearts. The number of ways of dealing five hearts is

$$13 \times 12 \times 11 \times 10 \times 9 = 154{,}440.$$

Using $P(A) = $ (Number in A)/(Number in S), permitted because all five-card hands are equally likely,

$$P(5 \text{ hearts}) = \frac{\text{number of 5-card hands with 5 hearts}}{\text{number of 5-card hands}}$$
$$= \frac{154{,}400}{311{,}875{,}200} \approx 0.0005 = 1/2000.$$

Here is another probability problem with cards involving simple counting: Four of the 52 cards are aces. It follows that 48 cards are not aces. The number of ways that we can pick five cards and not get any aces is

$$48 \times 47 \times 46 \times 45 \times 44 = 205{,}476{,}480$$

if we keep track of the order in which we pick the cards. (The first card can be chosen in 48 ways, the second card in 47 ways, and so on.) Thus

$$P(\text{no aces among 5 cards}) = 205{,}476{,}480 / 311{,}875{,}200 \approx 0.659.$$

Thus, about 2/3 of the time, we will be dealt a hand with no aces in it.

Traditional statistics courses often go much further in developing counting principles useful for solving such equally likely outcome problems. But we shall not do so, because it is a distraction. Moreover, we can always use simulation to estimate probabilities relating to sampling without replacement.

Example 4.25

What is the probability of getting at least one ace in a hand of five cards dealt from an ordinary deck of 52 playing cards?

Solution: The drawing involved here is without replacement, because each card can occur only once in a hand of five cards. We could estimate the probability of getting at least one ace in a hand of five cards by dealing out five-card hands to find out the number of times one or more aces occur, but this would take a huge amount of time. Another approach would be to use a table of random numbers and follow the five-step procedure.

Step 1. Choice of a Model: Use two-digit random numbers from 01 to 52, inclusive. Ignore all others.

01–04:   ace
05–52:   remaining cards in deck

If the first six digits are 09 75 48, we treat this as 09 48, because 75 is greater than 52 and hence is ignored. Thus, we now have 52 equally likely outcomes by this trick of ignoring 53–99 and 00. This is a powerful tool for obtaining equally likely probabilities when the number of outcomes is not 2, 6, 10, 100, or 1000, say.

Table 4.15 Estimating the Probability of at Least One Ace in Five Cards

Repetition	Random number	Success?
1	49 29 25 02 52	Yes
2	42 45 40 49 07	No
3	37 20 30 38 21	No
4	48 32 07 30 22	No
5	43 49 04 26 09	Yes
6	09 38 44 22 36	No
7	39 16 51 19 06	No
8	10 09 49 50 24	No
9	01 21 03 26 02	Yes
10	10 02 16 47 13	Yes

Step 2. Definition of a Simulation: A simulation consists of reading off five random numbers between 01 and 52, ignoring duplicates. That is, if the first six digits in the table are 03 03 27, then we treat this as 03 27 because the second 03 is a duplicate.

Step 3. Definition of Event of Interest: A successful simulation occurs when at least one of the five two-digit random numbers is between 01 and 04, inclusive (that is, when at least one of the numbers obtained represents an ace).

Step 4. Repetitions of the Simulation: Do at least 100 simulations. Suppose 10 repetitions produced the results listed in Table 4.15. Here we have removed all pairs larger than 52. In four of the 10 repetitions, at least one of the random numbers is less than or equal to 4 (repetitions 1, 5, 9, and 10). Therefore, in four of the repetitions we have drawn at least one ace.

Step 5. Find the Experimental Probability of the Event of Interest: Based on only 10 simulations, we estimate that the probability of getting at least one ace is around 0.4. Since we computed

$$P(\text{no aces}) \approx 0.659$$

it follows from $P(\text{not } A) = 1 - P(A)$ that

$$P(\text{at least one ace}) = 1 - P(\text{no aces}) \approx 0.341.$$

Note: We could take a box model approach in Step 1 using the box

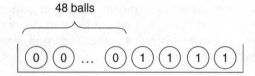

where 1 represents an ace and 0 represents a non-ace. Then Step 2 amounts to randomly sampling five times *without* replacement. The rest of the steps are the same. Moreover, by using box-model-based simulation software, we can now do 1000 or more repetitions of the basic Step 2 simulation in Step 4.

Summary

The multiplication rule is used to count the number of different outcomes in a multistage experiment. The rule applies to both sampling with and without replacement.

Section 4.5 Exercises

You may find some of these probability problems difficult to solve theoretically, especially when the sampling is without replacement.

Try to solve them theoretically first, but you will often need to resort to the five-step method. If you need to use the five-step method without simulation software, you may want to state just the first three steps carefully; if you do the entire five-step method, use common sense in deciding how many simulations to do in Step 4.

1. A bag contains five black marbles, four red marbles, and three white marbles. Three marbles are drawn in succession, producing a sample space of $12 \times 12 \times 12$ outcomes if we replace, and $12 \times 11 \times 10$ outcomes if we do not replace.
   a. If each marble is replaced before the next one is drawn, what is the probability that at least one of three is white? *Hint:* What is the complement of at least one being white?
   b. If each marble is not replaced before the next one is drawn, what is the probability that at least one of the three is white?
   c. If each marble is not replaced before the next one is drawn, what is the probability that all three are of the same color? *Hint:* The event that all three are the same color is the event (all are black) or (all are red) or (all are white).

2. A student has to match three terms that she has never seen before with their definitions. If she guesses, find her probability of
   a. getting all three correct.
   b. getting none correct.
   c. getting exactly one correct.
   d. getting exactly two correct.
   Hint: Label the terms 1, 2, 3 for convenience. Write the student's answer as a sequence such as 2, 3, 1, meaning the student assigns definition 2 to term 1, definition 3 to term 2, and so on. List all the different possible orders of 1, 2, 3 with which the student can respond. For example, the response 3, 2, 1 produces one match: that for term 2 and its correct definition.

3. A man has 10 keys, and exactly one of them fits the lock on his office door. He tries the keys one at a time, never trying any key more than once.
   a. Is this sampling with or without replacement? Why?
   b. What is the probability that he will obtain the correct key within his first three choices?
   Hint: Find the probability of the complement first.

4. In a small town with only 100 registered voters, two candidates are running for mayor: Mr. Jones and Mrs. Smith. You know that out of the 100 registered voters, 40 would vote for Mr. Jones and 60 for Mrs. Smith, but not everyone will actually vote. In fact, because of a terrible snowstorm, only 10 of the voters are able to vote. For each of the following questions, assume that the people who show up to vote are a random sample from the population of 100.

   Simulate the election using at least 50 trials, recalling that this is sampling without replacement (since each person can vote only once). Use a random number table, and assign numbers 00 through 39 to Mr. Jones and the rest to Mrs. Smith. Sampling without replacement means that if you see a number you have already seen (within the same trial), you have to throw out that number and go to the next.
   a. What is the probability that Mr. Jones will win the election (that is, receive six or more votes)?
   b. What is the probability that Mrs. Smith will win?
   c. What is the probability that they each receive five votes?
   d. Do you think that the low voter turnout helped Mr. Jones's chances or hurt them?
   (*Hint:* What would his chances have been if all 100 had voted?)

212 PROBABILITIES AND SIMULATION

5. Assume the setting of Exercise 4.
   a. Could you legitimately have done the simulation by sampling with replacement, assuming that your goal is to obtain answers that are approximately correct?
   b. If your answer in part **a** is "yes," build as simple a box model as possible to simulate the 10 voters. In particular, use as few balls as possible (let 0 denote a vote for Jones and 1 a vote for Smith).
   c. Find the theoretical probability that Mrs. Smith gets all 10 votes. *Hint:* Can you assume sampling with replacement to do your computation? If so, recall that the 10 trials are independent, allowing use of the multiplication rule for independence of the different voters.
6. A group of 10 men are choosing teams to play a game of basketball, and they decide to assign people randomly to either team, with five players on each team. Two of the players are taller than the rest, both of them being 6 feet, 5 inches tall. What is the probability that they will end up on the same team? *Hint:* This is hard to do theoretically. Simulate it with a 0,1 box of size 10, where 1 indicates being of height 6 feet 5 inches!
7. A "population" of 10 consists of five in favor and five opposed to raising the drinking age.
   a. Two are sampled randomly with replacement. What is the probability the sample is evenly split?
   b. Answer part **a** for sampling without replacement. *Hint:* Sample the two people one at a time.
8. Four cards are dealt one at a time. Find *P*(all are aces).
9. John has three pairs of shoes, five shirts, and four pairs of pants. How many ways can John dress?

CHAPTER REVIEW EXERCISES

If the problem calls for simulation to estimate a probability, design a box model (specify the values of the balls and how many balls there are of each value in the box), and solve the problem by doing 1000 computer simulations.

1. Assume you know that 90% of the entire student population at a large university is right-handed and 10% is left-handed. Your class has 10 students, and the room the class is assigned to has 10 desks, all designed for right-handed people.
   a. What is the probability that all the students in your class will be right-handed?
   b. What is the probability that in your class of 10 students there will be at least one left-handed person?
   c. Assume now that there is one left-handed desk in the room. Estimate the probability that there will be a perfect match between the students and their preferred types of desk. Do 1000 computer simulations.
   d. The probabilities in parts **b** and **c** should be different. Why?
2. State each of the following probability rules:
   a. Complement rule
   b. Addition rule
   c. Product rule—in general and for independent events
   d. *P*(event) when outcomes are equally likely
3. You are interested in estimating the percentage of married couples in which both husband and wife are Democrats, and the percentage in which both husband and wife are Republicans. You know that in the population in general, 50% of people are Democrats.
   a. You should not assume that the political party of the wife is independent of the party of the husband. Why not?
   b. Since the husband's and wife's political parties are not independent, this situation cannot be simulated by simply flipping two coins, one for the husband and one for the wife. Assume that you know that 75% of the time, the wife will have the same political party as the

husband. So, to estimate the desired percentages, begin as follows: Flip a coin 25 times, letting heads indicate that the husband is a Democrat and tails indicate that the husband is a Republican. Now, for each of the husbands, simulate the political party of the wife. (*Hint:* This is equivalent to performing 25 trials, with each trial having a 75% chance of the outcome "same party" and a 25% chance of "different party.")

    c. Using your results, estimate the probability of couples in which both husband and wife are Democrats; both Republicans. What is the proportion in which both are of the same party?

    d. Solve part **c** theoretically using a probability tree argument.

4. A friend brings you a coin and claims it is not fair (not equally likely to give heads or tails). He claims that he flipped it five times and saw only one head.

    a. Use the five-step method with 25 trials to estimate the probability of seeing at most one head in five flips of a fair coin.

    b. Does your friend have a legitimate claim? Why or why not?

    c. Solve part **a** using a theoretical probability argument.

5. Explain the difference between sampling with replacement and sampling without replacement. Give an example of where each method is used.

6. The following table presents the results of a survey asking people whether they have volunteered or not during the past 12 months:

By gender	
Men	45% volunteered
Women	52% volunteered

By employment status	
Full-time	50% volunteered
Part-time	58% volunteered
Not employed	46% volunteered
Retired	40% volunteered

Source: Champaign-Urbana News-Gazette, March 6, 1997, B1.

    a. Assume that these percentages represent true probabilities. If you ask a woman at random, what is the probability that she has not volunteered during the past 12 months?

    b. Estimate the probability that if you ask five retirees (men and women) if they have volunteered in the past year, at least one of them will tell you that he or she has. Perform at least 50 trials using a random number table, or design a box model.

7. Now we want to calculate the probability in part **b** of Exercise 6 theoretically.

    a. Write out the complement of the event "at least one of the five retirees has volunteered during the past 12 months."

    b. Assume that the retirees are independent of each other with respect to whether or not they have volunteered. Using the product law of independent outcomes, write the probability of the event you found in part **a** as the product of five separate probabilities.

    c. Finally, use the property of complementary events for theoretical probability to calculate P(at least one of the five retirees has volunteered).

8. Once again, look at the data in Exercise 6. In a certain neighborhood, the breakdown of employment status (for both men and women) is as follows:

Full-time	65%
Part-time	10%
Not employed	5%
Retired	20%

a. Using a random number table, estimate the probability that if you draw an individual at random from the neighborhood, that person has volunteered during the past 12 months. (*Hint:* Use four consecutive digits from the random number table. The first two tell you what employment status the person has, and the second two tell you whether or not the person has volunteered, noting that how one interprets the second two digits depends on the first two.)

b. Using the results from part **a**, estimate the probability that, given that a person volunteers, he or she has a full-time job.

c. Suppose you walk around the neighborhood asking people whether they have volunteered or not. Are you sampling with or without replacement?

d. Use a probability tree to solve part **a** theoretically.

9. Five friends go to a restaurant where there are 10 choices on the menu. Estimate the probability that none of the five chooses the same item, assuming each chooses one item and that they all make their choices independently of each other.

10. You and some friends begin to draw straws to see who has to go pick up the pizza you ordered. There are a total of 10 people in the group, and there are 10 straws. The loser is whoever draws the one long straw.

    a. Before any straws are drawn, what is the theoretical probability that you will be the one who has to go pick up the pizza? Assume that each straw is equally likely to be drawn.

    b. Three of your friends draw first, and none of them draws the long straw. What is your probability of being the loser now? In other words, what is P(drawing long straw | first three drew short straws)?

11. A bus is scheduled to arrive at a bus stop at exactly 8:24 A.M., but it actually arrives according to the probability distribution at the left. You arrive at the stop to catch the bus according to the probability distribution at the right.

x	$p(x)$	x	$p(x)$
8:21	0.05	8:21	0.05
8:22	0.10	8:22	0.15
8:23	0.15	8:23	0.30
8:24	0.35	8:24	0.20
8:25	0.20	8:25	0.20
8:26	0.10	8:26	0.05
8:27	0.05	8:27	0.05

a. Do at least 50 trials to estimate your probability of catching the bus. (If you arrive at the same time the bus does, assume that you catch the bus.)

b. Estimate the probability that you will have to wait 3 minutes or more for the bus. (If you arrive at 8:21 and the bus comes at 8:24, you have waited exactly 3 minutes.)

c. Solve parts **a** and **b** theoretically using a probability tree argument.

12. Return to the bus data in Exercise 11. Estimate the following conditional probabilities:

    a. P(you catch the bus | the bus arrives at 8:24)

    b. P(you miss the bus | you arrive at the bus stop at 8:25)

    c. P(you are forced to wait 3 minutes or more | the bus arrives at 8:26)

    d. Solve parts **a** through **c** theoretically using the definition of conditional probability.

13. Four football teams are in a league together. Based on the strengths and weaknesses of each team, you deduce the following probabilities:

    a. What is P(Team C beats Team A)?

P(Team A beats Team B) = 0.75
P(Team A beats Team C) = 0.40
P(Team A beats Team D) = 0.50
P(Team B beats Team C) = 0.60
P(Team B beats Team D) = 0.30
P(Team C beats Team D) = 0.40

    b. Assume that each game is independent of the others and that each team plays each of the other teams twice during the season. Estimate the probability that Team A will finish the season with two or fewer losses (that is, a 4-2 record or better). Do at least 50 trials, with each trial being the six-game season that Team A has to play.

14. Return to the data in Exercise 13. The four teams are to play a single-elimination tournament with the following form. The teams are to be randomly assigned to the brackets.

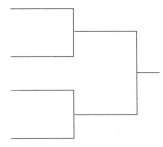

    a. Draw out 25 bracket diagrams such as the one shown above. In each case, assign the teams randomly to the four starting positions. Is this assignment of four teams sampling with replacement or without replacement?
    b. Simulate each of the 25 tournaments of part **a** using the probabilities given in Exercise 13. What is your estimate for the probability that Team C wins the tournament?
    c. Using the 25 simulations you just performed, what is your estimate for the conditional probability P(Team C wins the tournament | Team C plays Team A in the first round)?

15. Jan plays a game in which she spins a wheel with 10 equally likely outcomes, numbered 0 through 9. Whatever number she spins is the amount of money she wins. After her first spin, she gets to decide whether she wants to spin again and try to improve on the amount she has won. If she spins a second time, she wins *only* the amount she receives on the second spin, not what she would have received from the first spin.
    a. Assume that Jan always decides to spin the second time (clearly a poor strategy if she got a 9 on the first spin). Estimate the probability that she will reduce the amount of money she wins on her second spin. Do at least 50 trials. This can be done very easily by means of a random number table.
    b. Find the following theoretical conditional probabilities:

P(on the second spin she increases her winnings | on first spin she won $3)

P(on the second spin she increases her winnings | on first spin she won $4)

P(on the second spin she increases her winnings | on first spin she won $5)

    c. Jan decides to continue if the probability of increasing her winnings is $> 1/2$. On the basis of the results of part **b**, what should Jan's rule be for deciding whether to spin the wheel a second time?

16. Design box models and, if appropriate, describe the sampling scheme for
    a. Exercise 8, part **a** (*Hint:* This requires two boxes, with the second box dependent on the result of the first box draw.).
    b. Exercise 9.

c. Exercise 10, part **a**.
d. Exercise 11 (*Hint:* This requires two independent boxes.).
17. A box contains six balls labeled 1 through 6. What is the probability that if you draw out two of the balls, the sum of the numbers on the two is equal to or greater than 6 if you
    a. draw with replacement?
    b. draw without replacement?
    Perform at least 50 trials for each part to estimate the probabilities, or solve the problem theoretically.
18. A study of 100 married couples is conducted as to whether they exercise at least once a week. The results are as follows:

$$P(\text{wife exercised}) = 0.56$$
$$P(\text{husband exercised}) = 0.49.$$

   a. Is it reasonable to assume that the behaviors of the husband and wife in one couple are independent?
   b. Suppose that, within each couple, the husband's and wife's exercise behaviors are independent.
      i. What is $P(\text{both exercise})$?
      ii. What is $P(\text{neither exercise})$?
19. Write out the complement of each of the following events, but do not use the word *not*.
    a. "There was at least one game in which he scored a point."
    b. "At least five of the houses on the block are painted white."
    c. "The first card dealt from the deck of cards is a diamond."
    d. "Exactly 7 of the 10 members of the class passed the final exam."
    e. "There is at least one boy among the three children."
20. The ABO blood type distribution for African Americans is shown in Table 4.5. What is the probability that a randomly chosen African American is
    a. either type O or type A?
    b. not type O?
    If we randomly choose two unrelated African Americans, what is the probability that
    c. both are type O?
    d. neither is type O?
21. Deal one card from a well-shuffled deck of 52 cards. What is the probability that you get
    a. the queen of hearts?   d. a club?
    b. an ace?   e. a red card?
    c. a face card?
    (Spades and clubs are black; hearts and diamonds are red; jacks, queens, and kings are face cards.)
22. Deal two cards, one at a time without replacement. What is the probability that
    a. the first card is the queen of hearts and second card is the ace of spades?
    b. both cards are aces?
    c. both cards are hearts?
    d. neither card is a heart?
23. One out of ten Americans is left-handed, and nine out of ten are right-handed. Draw a probability tree for selecting two unrelated persons at random and determining whether they are left-handed or right-handed. What is the probability that
    a. both are left-handed?
    b. both are right-handed?
    c. one of them is left-handed and the other right-handed?
24. Draw a probability tree for selecting three unrelated persons at random and determining whether they are left-handed or right-handed. What is the probability that

a. all three are left-handed?
  b. all three are right-handed?
  c. one of three is left-handed and the other two right-handed?
  d. two are left-handed and one is right-handed?
25. On a slot machine there are three reels with digits 0, 1, 2, 3, 4, and 5 and a flower on each reel. When a coin is inserted and the lever pulled the three reels spin independently and each comes to rest at one of the seven positions mentioned with equal probability. What is the probability that
  a. a flower shows on all three reels?
  b. no flower shows?
  c. a flower shows on one of the three reels (either the first, the second, or the third reel) but not on the other two reels?
  d. a flower shows on two of the three reels but not on the third? *Hint*: To answer the last two questions you will need to draw a probability tree.
26. The probability that a newborn baby is a boy is 0.51. If a family has five children, what is the chance that
  a. the firstborn is a girl?
  b. the oldest and the youngest are girls?
  c. all of the children are boys?
  d. all of the children are girls?
  e. all five children have the same sex?
27. One third of all adults in Kentucky smoke. A telephone interview is conducted with 450 randomly chosen adults from Kentucky. Design a box model for simulating on a computer the selection of the sample and the determination of the percentage of the people in the sample who smoke. Use as few balls as possible. Define one simulation. Define the statistic of interest. Do 1000 computer simulations to estimate the probability that the sample percentage will be between 30% and 36% inclusive.
28. Repeat the previous exercise with a random sample of 300 adults from Utah. In Utah only one sixth of all adults smoke. Do 1000 computer simulations to estimate the probability that over 25% of the people in the sample smoke.
29. Hospital patients face a 1 in 20 risk of acquiring an infection in the hospital. Each year, over 2 million infections are acquired in hospitals in the United States, leading directly or indirectly to over 80,000 deaths. Design a box model for randomly selecting the records of 50 former hospital patients and determining how many of the 50 patients got infected in the hospital. Define one simulation. Define the statistic of interest. Do 1000 computer simulations to estimate the probability that at most 5 of the 50 patients got infected.
30. Repeat the previous exercise with a random sample of 30 intensive care unit patients, who face a 1 in 10 risk of getting infected in the hospital. Estimate the probability that at least 5 of the 30 intensive care patients got infected.
31. It is estimated that 40% of the 13,000 women in the United States who were diagnosed with AIDS in 1997 got infected through heterosexual contact. Imagine that we randomly select the medical records of 80 women who were diagnosed with AIDS in 1997 and check how many of these 80 women got infected through heterosexual contact. Design a box model. Define one simulation. Define the statistic of interest. Do 1000 computer simulations to estimate the probability that at most half of the 80 women got infected through heterosexual contact.
32. In the 1980s the percentage of preschool children in the United States who were immunized for measles decreased to 70%, apparently because the public became less fearful of the disease. As a result, the number of reported cases of measles increased and reached 30,000 in 1990. Imagine that we randomly select the 1990 medical records of 180 preschoolers and check how many of the children were immunized for measles. Design a box model. Define one simulation. Define the statistic of interest. Do 1000 computer simulations to estimate the probability that at least 120 of the preschoolers were immunized.

PROFESSIONAL PROFILE

Dr. J. Steven Landefeld
Director
Bureau of Economic Analysis
Washington, D.C.

Dr. J. Steven Landefeld

How are we doing as a nation? What does the economic report card on our economy look like?

This is a question that J. Steven Landefeld can answer. He's the Director of the Bureau of Economic Analysis (BEA) in the Department of Commerce in Washington, D.C. The BEA has the job of collecting and analyzing economic data for the nation. One of their best-known products, the monthly Gross Domestic Product (GDP) report, is the most comprehensive measure of the economy. The GDP is widely quoted and frequently cited as the reason that interest rates and stocks go up or down, or that other changes occur in our economy. Another widely reported number from the BEA is the quarterly Balance of Payments report, which summarizes how much the United States has exported and imported, and how that trade has been financed.

Dr. Landefeld notes that the Bureau of Economic Analysis was founded in 1913. "We are," he says, "the nation's accountant." For its first 30 years, the Bureau provided a variety of reports to help national leaders make sound economic decisions. However, the early reports could not give a full picture of the national economic position. Under the pressure of World War II, a new comprehensive measure of the economy—the gross national product—was introduced to facilitate wartime planning. Today, under Dr. Landefeld's leadership, the BEA gives a monthly look at U.S. economic health in the form of the GDP report.

"What is most important about our numbers," says Dr. Landefeld, "is the trend. Many people put too much emphasis on any one point. Any point estimate can and will be revised. But, in any quarter, we can tell if the growth in our economy is speeding up or slowing down and what components are responsible for the change."

This is a challenge Dr. Landefeld enjoys. "I've held all sorts of research and policy positions, ranging from those in academia to the White House," he explains, "but this is the most interesting and exciting job I've ever had."

5

Expected Values and Simulation

*Measure, measure, measure.
Measure again and again.*
Galileo

Objectives

After studying this chapter, you will understand the following:

- The law of large numbers for independent repetitions of a random variable
- The expected value (theoretical mean) of a random variable as a measure of its center and the limiting value of $\bar{X}$
- The five-step simulation method for estimating the mean of a random variable
- The theoretical formula for computing the mean of a random variable
- The standard deviation (SD) of a random variable as a measure of its spread
- The five-step method for estimating the SD of a random variable
- The theoretical formula for computing the SD of a random variable
- The mean and SD of a random variable computed from its box model

---

5.1 THE EXPECTED VALUE (THEORETICAL MEAN) OF A RANDOM VARIABLE 221

5.2 USING FIVE-STEP SIMULATION TO ESTIMATE MEAN VALUES 232

5.3 THE STANDARD DEVIATION OF A RANDOM VARIABLE 239

CHAPTER REVIEW EXERCISES 244

Key Problem

The Hermits' Epidemic

Six hermits live on an otherwise deserted island. An infectious disease strikes the island. None of the hermits has ever contracted the disease before. No one is immune. Once a person is exposed and contracts the disease, he becomes infectious and also becomes immune to the disease in the future. Suppose one of the hermits gets the disease. While infectious, he randomly visits one of the other five hermits. The visited hermit contracts the disease and, while infectious, visits one of the other five hermits at random. If the visited hermit has not yet had the disease, he gets it now and is infectious. The disease is transmitted in this manner until an infectious hermit visits an immune hermit (who has already had the disease), at which time the disease dies out.

Assuming this pattern of visits, how many hermits, on average, can be expected to get the disease?

Probability modeling is very important in epidemiology, the study of the spread of diseases. Our interesting but unrealistic model of an epidemic is such an example. Try imagining the kinds of probability modeling assumptions you might make to turn the Hermit problem into a realistic model for the spread of a contagious disease.

Scientists like to count and measure things. Gregor Mendel, the founder of the science of genetics, discovered the laws of heredity by counting pea plants. Counts and measurements are easy to analyze and easy to communicate to other people. Johannes Kepler deduced his three planetary laws from Tycho Brahe's careful astronomical measurements.

Alternatively, subjective information, such as a burn patient's pain level, is not as easy to interpret. Is it low, moderate, or severe? How will a doctor interpret the patient's description of his or her pain level? One person's "moderate" pain might be another person's "severe" pain. We can assign an integer indicating the pain's level or rank. For example, a burn victim's pain in a study on pain reduction might be scaled

None	0
Low	1
Moderate	2
Severe	3

But the use of such a scale does not eliminate the subjectivity of the pain assessment. In the social sciences such a rank-based scale is often called a **Likert scale.** If we use such a scale each patient is assigned an integer to characterize his or her level of pain. Similarly, a university course evaluation form might use rankings from 1 (lowest) to 5 (highest). In a study on urbanization, families might be classified as living in a rural, suburban, or urban setting. We can assign these categories numbers such as 1, 2, and 3 for rural, suburban, and urban. Such data are called **ranked** or **ordinal** (meaning ordered) data. Just as we can compare measurements and counts we can compare ranked data. Pain level 2 is more severe than pain level 1; residential category 2 is less urban than residential category 3. We can easily handle such data statistically as well; we would use the median and interquartile range as measures of center and spread.

In this manner, any random phenomenon being observed can be viewed as producing numerical data, whether it is measurements, counts, or rankings. This leads us to the importance of the concept of a random variable.

5.1 The Expected Value (Theoretical Mean) of a Random Variable

As stated in Chapter 4, a **random variable** is a numerical description of the outcome of a random experiment. As such, a "random variable" is just a name for a statistic of interest. We usually denote random variables by capital letters such as X, Y, and Z. As we learned in Chapter 4, the **probability distribution** of a random variable X lists the possible values X can take on (the x's) and the corresponding probabilities (the $p(x)$'s).

In Chapter 4 we considered several random variables and their probability distributions. The examples included:

- The sum of the scores on two dice
- The number of correct guesses on a five-question true-false quiz
- The number of left-handed persons in a random sample of 15 persons
- The number of children with cystic fibrosis in a 4-child family

In this chapter we take a closer look at random variables. In particular, we study two quantities that describe the center and spread of a random variable, or more accurately of the probability distribution (law) of the random variable. We shall define the mean and standard deviation of a random variable X and, hence, of its probability distribution. They are usually denoted by μ_X ("mu sub X") and σ_X ("sigma sub X"), with the subscript often dropped to produce μ and σ if it is obvious which random variable X we are referring to.

In Chapter 4 we either used real data or simulated data to estimate event probabilities and probability distributions of random variables: We used the experimental probability $\hat{P}(A)$ of an event A of interest to estimate its theoretical probability $P(A)$.

In this chapter we shall either use real data or simulated data, obtained by the five-step method, to estimate the theoretical mean μ_X and the theoretical standard deviation σ_X of a random variable X of interest. We shall use the sample mean $\overline{X}$ and the sample standard deviation S_X of repeated observations of X as our estimates. In short, we use the experimental values to estimate the theoretical values:

$$\hat{P}(A) \approx P(A), \quad \overline{X} \approx \mu_X, \quad S_X \approx \sigma_X.$$

5.1 THE EXPECTED VALUE (THEORETICAL MEAN) OF A RANDOM VARIABLE

On a pack of GE 40-watt Soft White light bulbs you can find the statement "Avg. life 1000 hrs." General Electric is not promising that every one of its 40-watt Soft White bulbs will give you 1000 hours of lighting. Some may last for fewer than 1000 hours, and some will last for more than 1000 hours. The exact number of hours may depend on many factors—slight variations in the manufacturing process, how gently the light bulbs were handled in transport, how oily your fingers were when you inserted the bulb in the socket, or how often you turn the light on and off, for example. What General Electric is promising is that *in the long run* people will on average get (about) 1000 hours of lighting per light bulb.

Statisticians imagine finding the average (mean) life of all 40-watt GE light bulbs, past and future. They call this mean the **expected value** of the life of the light bulbs. The intuitive notion is that the number of hours of life we will get from an individual light bulb will be somewhere near the "expected life." In this sense, the expected value indicates the typical values of the random variable.

Consider now any random variable X determined by the outcome of some random phenomenon. We define the **theoretical mean** of X (also called the **expected value** of X) as the average score of X in a very, very long series of independent repetitions of the random phenomenon. We usually use the notation μ_X or $E(X)$ to denote the theoretical mean or expected value of X.

The following example illustrates how we sometimes can compute μ_X. The example also illustrates the interpretation of μ_X as the "typical" value of X, when X is observed repeatedly.

Example 5.1

In the card game of bridge, a well-shuffled deck of 52 cards is dealt, one card at a time, to four players, who receive 13 cards each. A bridge hand is thus a random sample of 13 cards drawn without replacement. The number of hearts X varies from bridge hand to bridge hand. We are interested in two related questions. First, if we play a lot of bridge hands, how many hearts on average will we get per hand? That is, what is $E(X)$? Secondly, what is the distribution of the numbers of hearts in a typical bridge hand as it relates to the central value $E(X)$?

The first question is easy to answer. Because $P(\text{heart}) = 1/4$, in the long run, as we are dealt hundreds of bridge hands, we know from Chapter 4 that very close to $1/4$ of the cards we are dealt will be hearts, just as very close to $1/4$ of the cards will be spades, diamonds, and clubs. Since there are 13 cards in a bridge hand, in the long run, the average number of hearts per bridge hand (i.e., the expected number) is

$$E(X) = \frac{1}{4} \cdot 13 = 3.25.$$

Now, the second question is trickier. You cannot be dealt 3.25 hearts. How likely is it that the number of hearts dealt will be close to 3.25? For example, is being dealt 0 hearts or 8 hearts unusual? To help answer the second question, we will simulate a large number of bridge hands, focusing on the experimental probability distribution of the number of hearts obtained. To simulate drawing a bridge hand and counting the number of hearts, we can use a box model, where a 1 indicates a heart and a 0 indicates a nonheart.

Note that we do not need to have the balls in the box indicate irrelevant aspects, such as the face value of the card and whether a nonheart is a spade, a diamond, or a club. The box models only those aspects of the physical reality that are important to us!

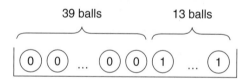

We simulate a bridge hand by drawing 13 balls at random without replacement. The number of hearts corresponds to the sum of the numbers on the 13 balls drawn. If we do 10,000 simulations, we get the probability distribution of X (given in Table 5.1 and Figure 5.1).

Table 5.1 Number of Hearts X in a Bridge Hand

Number of hearts, x	Probability, $p(x)$
0	0.01
1	0.08
2	0.21
3	0.29
4	0.24
5	0.12
6	0.04
7	0.01
Over 7	0.00
Total	1.00

5.1 The Expected Value (Theoretical Mean) of a Random Variable   223

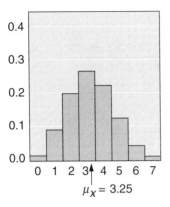

Figure 5.1 Probability distribution and mean of the number of hearts in a bridge hand.

We see that $p(2) + p(3) + p(4) + p(5) = 0.86$; that is, in the long run, we are dealt between 2 and 5 hearts 86% of the time. Thus the number of hearts is usually close to the mean (or expected value) 3.25.

The fact that, in the long run, the average number of hearts per bridge hand will be close to $\mu_X = E(X) = 3.25$ is an example of what statisticians call the law of large numbers. This law holds for any sequence of independent repetitions of a numerically valued random phenomenon. For example, we simulated 1000 bridge hands, observing the number of hearts X 1000 times. We obtained $\overline{X} = 3.17 \approx 3.25 = \mu$. A more detailed example of the law of large numbers is given at the end of this section.

The Law of Large Numbers for Independent Repetitions

Consider a random variable X (such as the number of hearts dealt in a bridge hand) determined by the outcomes of some random phenomenon. In a very long sequence of independent repetitions of the random phenomenon, the average $\overline{X}$ of the observed values of X will be very close to the theoretical mean μ_X.

Important Expected Value Formulas

The argument we used in Example 5.1 to compute the expected number of hearts in a bridge hand can be used in many situations. Consider the following examples.

Example 5.2

A multiple choice exam has 40 questions. To each question there are 5 answers to choose from, one of which is correct. Imagine that you have no clue whatsoever what the correct answers are and blindly (randomly) guess at the answer to each of the 40 questions. How many correct answers should you expect?

Solution:
$$40 \times 1/5 = 8$$

Reason: If each one of thousands of people guesses at the answers to all 40 questions, then very close to 1/5 of all the tens of thousands of guesses will be correct; that is,

$$\frac{\text{number of correct guesses}}{\text{number of guesses}} = 1/5.$$

It follows that the average number of correct answers per person is

$$\frac{\text{number of correct guesses}}{\text{number of persons}} = \frac{\text{number of guesses}}{\text{number of persons}} \times \frac{\text{number of correct guesses}}{\text{number of guesses}}$$

$$= 40 \times \frac{\text{number of correct guesses}}{\text{number of guesses}} \approx 40 \times 1/5 = 8.$$

This shows that the expected number of correct answers on the exam is $40 \times 1/5 = 8$.

Example 5.3

A fair coin is tossed 25 times. What is the expected number of heads?

Solution:

$$25 \times 1/2 = 12.5$$

Reason: If thousands of persons each tossed a fair coin 25 times, then very close to 1/2 of the tens of thousands of coin tosses would result in heads. It follows that the average number of heads per person would be very close to $25 \times 1/2$.

Example 5.4

A fair die is tossed 45 times. What is the expected number of sixes?

Solution:

$$45 \times 1/6 = 7.5$$

Reason: In the long run, very close to 1/6 of the tosses of a fair die will result in a six. If thousands of persons each tossed a fair die 45 times, their average number of sixes would be very close to $45 \times 1/6$.

Example 5.5

Ten percent of all Americans are left-handed. What is the expected number of left-handed persons in a random sample of 85 Americans?

Solution:

$$85 \times 0.1 = 8.5$$

Example 5.6

Thirty-five percent of all American women aged 25 to 29 are single. What is the expected number of single women in a random sample of 60 American women aged 25 to 29?

Solution:

$$60 \times 0.35 = 21$$

The solutions to Examples 5.1 through 5.6 all have a common form.

> **The n Times p Formula for the Expected Number of Occurrences of an Event in Multiple Trials**
>
> Consider an event that has probability p of occurring in each of n trials. Let X denote the number of times that the event of interest occurs in the n trials. Then the expected value of X (or the theoretical mean of X) is given by
>
> $$\mu_X = np.$$

In Example 5.1 we counted the number of times we got a heart in $n = 13$ drawings from a well-shuffled deck of cards. In each drawing the chance of getting a heart was $p = 1/4$.

In Example 5.5 we counted the number of times we selected a left-handed person in $n = 85$ drawings from the American population. In each drawing the chance of getting a left-handed person is $p = 0.1$.

The n times p formula for μ_X only applies when we are counting the number of times that something happens out of n trials. We shall need two more formulas for computing the theoretical mean in general.

> **The Distribution Formula for the Expected Value of a Random Variable**
>
> If the possible values of X are $x_1, x_2, \ldots, x_k$ with corresponding probabilities $p(x_1), p(x_2), \ldots, p(x_k)$, (that is, we are given the distribution of X) we can compute the expected value of X by multiplying each possible value by its probability and adding up all the products. That is,
>
> $$\mu_X = x_1 p(x_1) + x_2 p(x_2) + \cdots + x_k p(x_k)$$
> $$= \sum x p(x).$$

Recall the box models in Section 4.4 and in Example 5.1. We can use box models to simulate the number of hearts in a bridge hand, the number of correct guesses on a 15-question true-false quiz, our net gain on 30 $1 bets on a pair of numbers in Keno, and the number of girls among 100 newborns. In each case we compute the sum of the numbers drawn from the box. This leads us to the next expected value formula.

> **The Box Model Formula for the Expected Value of a Random Variable**
>
> If X is the sum of repeated random draws from a box (made with or without replacement between draws), then
>
> $$\mu_X = \text{(number of draws)} \cdot \text{(box mean)}.$$
>
> Here the box mean is just the average (mean) of all the numbers in the box. In particular if we have just one draw ($n = 1$), then
>
> $$\mu_X = \text{box mean}.$$

Population Surveys and Random Variables

Many national surveys are carried out via random sampling from various populations. These surveys have great influence on public policy.

Example 5.7

A "household" is either a single person living alone or a group of people living together, regardless of their relationship to each other. There are over 100 million households in the United States. What is the average size of American households?

Solution: Here is the probability distribution of American households by size according to the U.S. Census Bureau:

Household size, x	1	2	3	4	5	6	7
Fraction of households, $p(x)$	0.26	0.32	0.17	0.15	0.07	0.02	0.01

Suppose we select a household at random. That selection is equivalent to drawing a ball from a box containing over 100 million balls, one for each American household, each ball numbered with the size of the corresponding household. The size of the household drawn is a random variable X with probability distribution given by the preceding table. The mean μ_X equals the average size of all American households according to the box model formula. By our distribution formula, we get

$$\begin{aligned}\mu_X &= \sum x p(x) \\ &= 1(0.26) + 2(0.32) + 3(0.17) + 4(0.15) + 5(0.07) + 6(0.02) + 7(0.01) \\ &= 2.55.\end{aligned}$$

We have ignored households with eight or more members, because less than half a percent of all households are that large. Figure 5.2 locates the mean on the probability histogram.

How are incomes distributed in the United States? Are household incomes keeping up with inflation? How do the incomes of men and women compare? How do the incomes of various minority groups compare with those of the rest of the population? What is the unemployment rate for various population groups? The monthly Current Population Survey attempts to answer such questions using a nationwide random sample of about 57,000 households. A large amount of economic and social data is published in the annual *Statistical Abstract of the United States* (see the U.S. Census Bureau Web site, www.census.gov).

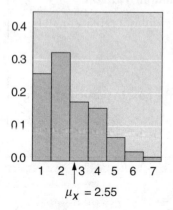

Figure 5.2 Probability distribution of U.S. households by size.

5.1 The Expected Value (Theoretical Mean) of a Random Variable

Suppose we select one of the over 100 million U.S. households at random. The income of the chosen household is a random variable X, whose mean μ_X equals the average income of all American households. As in Example 5.7, selecting one household corresponds to drawing from a box model with one ball for each of the over 100 million households. In this case the number on each ball is the previous year's income for the corresponding household. If we sample many households at random without replacement (as the Current Population Survey does), then the average income for the households in the sample, denoted $\overline{X}$, will be close to the population mean μ_X. Thus, this is an application of the law of large numbers that was stated above.

Law of Large Numbers for Population Surveys

Draw observations $X_1, X_2, \ldots, X_n$ at random without replacement from a very large population with mean μ. If the sample size n is large, then the average $\overline{X} = \sum_i^n X/n$ of the observed values will be close to the μ.

Box Models and Expected Values

Consider again the game of Keno, described in Section 4.4. We can simulate a player's gain on a $1 bet on a single number by drawing one ball at random from the following box:

$$\boxed{(-1)\ (-1)\ (-1)\ (2)}$$

The corresponding probability distribution for the player's gain X is

x	−1	2
p(x)	3/4	1/4

We can simulate a player's average gain (or loss) on 100 independent $1 bets on a single number by drawing 100 times at random with replacement from the box and computing the average of the resulting 100 numbers. What can we say about this average? It is obviously a random variable. Its value depends on chance, but it will obey the law of large numbers. That is, the average gain $\overline{X}$ will be close to

$$\mu_X = \text{box mean} = \frac{(-1-1-1+2)}{4} = -\frac{1}{4}.$$

Here we could just as easily have computed μ_X using the distribution formula:

$$\mu_X = (-1)p(-1) + 2p(2) = \left(-1 \cdot \frac{3}{4}\right) + \left(2 \cdot \frac{1}{4}\right) = -\frac{1}{4}$$

We conclude that in a very long series of independent $1 bets on a single number in Keno, your average winning per bet will approach −1/4 dollar. In other words, in the long run you will on average lose a quarter per $1 bet. If you make 100 $1 bets on a single number in Keno, then by the box model formula your expected loss is 100 × (−1/4) = $25. If you make 1000 $1 bets, your expected loss is $250.

Example 5.8

If you repeatedly bet $1 on a pair of numbers in Keno, how much will you lose on average per bet?

Solution: Recall the discussion of Keno in Section 4.4. We learned that we can simulate a gambler's gain on a $1 bet on a pair of numbers by drawing one ball at random from the following box:

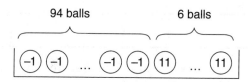

The gambler's expected gain on a single $1 bet therefore equals the box mean:

$$\mu_X = \frac{(94 \cdot (-1)) + (6 \cdot 11)}{100} = \frac{-94 + 66}{100} = -\frac{28}{100} = -0.28.$$

In the long run the gambler will on average lose $.28 per $1 bet.

In the short run it is possible (if not likely) that the gambler wins more money than he loses and comes out ahead. But the more bets the gambler makes, the less likely it is that he comes out ahead. For example, as we shall see in Chapter 6 (or we can show using simulation) the gambler has a 46% chance (that is, probability of 0.46) of making money on 10 independent $1 bets on a pair of numbers, a 15% chance of making money on 100 bets, but only 1 chance in 1000 of making money on 1000 bets. If the gambler bets on a single number, then his chances of coming out ahead are even less: The gambler has a 22% chance of making money on 10 independent $1 bets on a single number, a 2.5% chance of making money on 100 bets, and almost no chance at all of making money on 1000 bets. You cannot beat the odds in the long run.

This explains why gambling is a can't-lose business for the casinos; it is not gambling for them at all. Imagine, for example, a thousand gamblers each playing Keno 10 times, making $1 bets on either a single number or a pair of numbers independently of each other. Each of the gamblers has a reasonable chance of making a little money, but most will not. On 10,000 independent $1 bets (1000 gamblers betting 10 times), it can be shown that the house is guaranteed to make over $2000. (If you do not believe this, simulate 5000 one-number plays and 5000 two-number plays!)

As thousands of gamblers independently make bets, the **law of large numbers** guarantees the casino a gain of around a quarter per dollar bet on a single number in Keno and around 28 cents per dollar bet on a pair of numbers. In contrast, the law of large numbers does not apply to the individual gambler who only makes a few bets. A few gamblers do make money. That is why gamblers flock to casinos. Most don't make money, of course.

Insurance companies work much the same way as casinos. The company is betting that not too many of its customers will collect on their policies. Using life tables, which list the probabilities of dying within given numbers of years for persons of various ages and conditions in life, the life insurance company computes the expected pay-out on each policy and sets the insurance rates high enough so that in the long run the company will make a profit. With thousands of policyholders, the insurance company "plays" often enough to rely on the law of large numbers, where X is the amount of money the insurance company makes (could be negative) on a policy. The probabilities

5.1 The Expected Value (Theoretical Mean) of a Random Variable

in the life tables are experimental probabilities obtained from the careful analysis of large numbers of lifetimes. They are developed by professionals known as *actuaries*, who are trained in statistics.

Finally we shall illustrate the **law of large numbers** by simulating 100,000 tosses of a symmetric casino die. The box for simulating the score X on a fair die is

$$\boxed{①\ ②\ ③\ ④\ ⑤\ ⑥}$$

But then because μ_X = box average, this yields

$$\mu_X = \frac{1 + 2 + \cdots + 6}{6} = \frac{21}{6} = 3.5.$$

Below we have tabulated the sample mean $\bar{x}$ and the difference between the sample mean and the theoretical mean $\mu_X = 3.5$, namely ($\bar{x} - 3.5$), for 10, 50, 100 simulations, up to 100,000 simulations. We see that $\bar{x}$ dances around the theoretical mean 3.5, getting closer as the number of trials increases.

Number of throws	$\bar{x}$	$\bar{x} - 3.5$
10	2.800	−0.700
50	3.460	−0.040
100	3.230	−0.270
500	3.450	−0.050
1000	3.493	−0.007
5000	3.540	0.040
10,000	3.496	−0.004
50,000	3.498	−0.002
100,000	3.503	0.003

Figure 5.3 displays the average score $\bar{x}$ as a function of the number of tosses.

Summary

Consider a random variable X determined by the outcome of some random phenomenon. The mean of X denoted by μ_X, or the expected value of X, denoted E(X), is the average score of X in a very long series of independent repetitions of the random phenomenon.

The n Times p Formula for the Expected Number of Occurrences of an Event in Multiple Trials Consider an event that has probability p of occurring in each of n trials. If x = number of times that the event of interest occurs in the n trials, then

$$\mu_X = np.$$

The Distribution Formula for the Expected Value of a Random Variable If the possible values of X are $x_1, x_2, \ldots, x_k$ with corresponding probabilities $p(x_1), p(x_2), \ldots, p(x_k)$,

$$\mu_X = x_1 p(x_1) + x_2 p(x_2) + \cdots + x_k p(x_k)$$
$$= \sum x p(x).$$

The set of $(x, p(x))$ values is called the **probability distribution** of X.

230   CHAPTER 5 EXPECTED VALUES AND SIMULATION

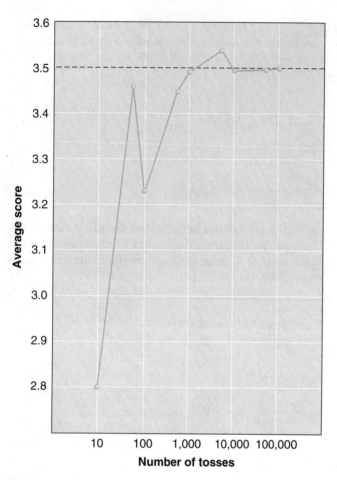

Figure 5.3 Simulated $\bar{x}$'s as a function of the number of tosses.

The Box Model Formula for the Expected Value of a Random Variable If X is the sum of repeated random draws from a box (made with or without replacement between draws), then

$$\mu_X = (\text{number of draws}) \cdot (\text{box mean}).$$

Section 5.1 Exercises

Note: In the exercises in which a computer is required, if you do not have access to simulation software, just describe the box model and the sampling from the box needed to carry out the simulation. Your explanation should be complete enough that a fellow student with a computer and five-step software could use your answer to solve the problem numerically.

1. What is the expected number of aces in a bridge hand?
2. A fair die is rolled once. What is the expected value?
3. Compute the expected gain for each of the three $1 bets in the game of High-Low described in Exercise 1 of Section 4.4.
4. Compute the expected gain for a $1 bet on red in roulette. (See Exercise 2 of Section 4.4.)

5.1 The Expected Value (Theoretical Mean) of a Random Variable

5. Compute the expected gain for a $1 bet on a single number in roulette. (See Exercise 3 in Section 4.4.)
6. Compute the expected gain for a $1 bet on a split in roulette. (See Exercise 4 in Section 4.4.)
7. A true-false quiz has 15 questions. Imagine that you blindly guess the answer to each of the 15 questions. How many correct answers can you expect to get?
8. Simulate your net gain on 10 independent $1 bets on a pair of numbers in Keno. (See Example 5.8.) Use a computer to repeat the simulation 10,000 times. Estimate the probability distribution of your net gain. What are your chances of coming out ahead? What is your expected gain?
9. Using a computer, simulate the casino's net gain on 10,000 independent $1 bets on a pair of numbers in Keno (this is one simulation). Repeat the simulation 10 times. What is the casino's expected gain?
10. Simulate your net gain on 10 independent $1 bets on a single number in Keno. Use a computer to repeat this 10-play simulation 10,000 times. Estimate the probability distribution of your net gain. What is your probability of coming out ahead? What is your expected gain?
11. Simulate the casino's net gain on 10,000 independent $1 bets on a single number in Keno. Use a computer to repeat the simulation 10 times if you have access to simulation software. What is the casino's expected gain?
12. Five symmetric casino dice are tossed and the five scores added up. What is the expected value of the sum?
13. A quiz has 15 multiple choice questions. Each question has five answers to choose from, one of which is correct. A correct answer is worth four points, but a point is taken off for each wrong answer. What is your expected score if you blindly (randomly) guess the answer to each of the 15 questions?
14. On the day before her 50th birthday a business woman buys a life insurance policy that pays $1 million if she dies before her 55th birthday. The policy costs $8000 per year for up to 5 years, payable on her 50th, 51st, 52nd, 53rd, and 54th birthday. The amount X that the insurance company earns on this policy is $8000 per year for up to 5 years as long as the woman is alive minus the $1 million the company has to pay if she dies before her 55th birthday. Here is the distribution of X = amount of insurance company claims, computed using data from the U.S. National Center for Health Statistics. For example, 0.00686 is the probability that she will die at age 51. Fill in the missing probability that she lives to be 55 or older and compute the insurance company's expected earnings (μ_X) on this policy.

Age at death	Earnings	Probability
50	−992,000	0.00638
51	−984,000	0.00686
52	−976,000	0.00732
53	−968,000	0.00777
54	−960,000	0.00822
55 or older	40,000	

15. In a state lottery's Pick 3 game a three-digit integer is selected at random. All three-digit numbers from 000 to 999 are equally likely. If a player bets $1 on a particular number and if that number is selected, the payoff is $500 minus the $1 paid for the lottery ticket. Let X equal the payoff to the bettor, namely −$1 or $499. Compute the expected payoff μ_X.

5.2 USING FIVE-STEP SIMULATION TO ESTIMATE MEAN VALUES

Consider a random variable X determined by the outcome of some random phenomenon. We can estimate the mean μ_X by computing the average score of X in a long series of independent repetitions of the random phenomenon.

Example 5.9

The Cereal Box Problem

Suppose "Tripl Crisp Cereal" is running a promotion on its cereal by including one of six different colored pens in each box.

Fire Orange	Brilliant Yellow	Baby Blue
Shell Pink	Rosie Red	Grassy Green

Assume that when you buy a box of cereal, your chances of getting any of the six colored pens are equal. About how many boxes of cereal would you expect to have to buy to get a complete set of all six colored pens? We let the random variable X = number of cereal boxes needed.

The first thing to notice is that this problem involves chance. What is the *smallest* number of boxes of cereal that you would have to buy? Clearly, six. It is possible (but not very likely!) that each box purchased would have a different colored pen. In that case six boxes would do it. On the other hand, what can we say about the largest number of boxes that you might have to buy? It is possible (but again, not likely) that you could buy 1000 or even 10,000 boxes and not get that last color of pen! So for any one person the answer is at least six boxes, but it could be an arbitrarily large number of boxes.

How can we find the expected number of boxes needed in our cereal box problem? There is a theoretical answer, but it requires sophisticated mathematical reasoning. We prefer an experimental data-collecting approach, because it is actually far more helpful in learning statistics.

We could go on a shopping trip and start repeatedly buying cereal boxes until we get a pen of each color. We could record our purchases as in Table 5.2. Here is how Table 5.2 works. For each cereal box, a tally mark (/) shows what color of pen was obtained. Table 5.2 shows that so far, in shopping trip 1, three boxes have been purchased, with one red and two green pens

A shopping trip ends when we have a complete set of all six colors. Table 5.3 shows the results of shopping trip 1 completed. We see that we bought 21 boxes of cereal before we had a complete set of six colors. We also see, for example, that we have three orange pens and two blue pens.

We now have one estimate of how many boxes of cereal are needed to get a complete set of six pens: 21 boxes. But we could have been lucky, or unlucky. With this single shopping trip we have no way of knowing whether 21 is unusually large, unusually small, or neither. In order to get a better overall estimate of the long-run average (the expected value) of the number of boxes needed, we can go on several more shopping trips and then find the average number of boxes of

Table 5.2 *Partial Results of One Cereal-Buying Shopping Trip*

Shopping trip	Outcomes (pens obtained)						Number of boxes of cereal
	Orange	Yellow	Blue	Pink	Red	Green	
1					/	//	3

5.2 Using Five-Step Simulation to Estimate Mean Values

Table 5.3 Results of One Cereal-Buying Shopping Trip

Shopping trip	Outcomes (pens obtained)						Number of boxes of cereal
	Orange	Yellow	Blue	Pink	Red	Green	
1	///	/	//	////	//// //	////	21

cereal that were purchased. If we carry out enough trips, the average will get as close to the theoretical expected value as we wish. Collecting a large amount of data can therefore substitute for a mathematical derivation of the unknown expected value! That is, we substitute an average of data for a theoretically derived mathematical average.

It is not practical to solve this problem in this way (actually going out and buying cereal). It is too time-consuming, too expensive, and not sensible (unless, of course, we *really* like Tripl Crisp Cereal).

Another approach to solving this problem is to do a simulation as a substitute for the real experiment of actually going shopping for cereal. In particular, we can use a physical simulation model—a fair, six-sided die—to substitute for the process of purchasing boxes of cereal.

We use the five-step method.

Step 1. Choice of Box Model: There are six different pens to collect. Assume that when we buy a box of Tripl Crisp, our chances of getting any of the six colors of pen are equal. We can imagine a box model with six balls, each marked with a color, but we can more easily use a familiar six-sided die as a physical model for buying cereal boxes with their pens. Let each side of the die correspond to one of the six pen colors:

1 = orange	2 = yellow	3 = blue
4 = pink	5 = red	6 = green

One toss of the die will correspond to the purchase of one box of cereal. An equivalent way of simulating the purchase of boxes of cereal is to use the random digits (1 – 6) in Table B.2 in Appendix B. Each digit will correspond to a color.

Step 2. Definition of One Simulation: A completed shopping trip corresponds to rolling the die until all six sides are obtained.

Step 3. Definition of the Statistic of Interest: Record the number of rolls of the die (number of cereal boxes purchased) in Step 2. This is the statistic of interest in the cereal box problem.

Step 4. Repetitions of the Simulation: It takes 16 rolls to obtain all six sides in our first simulated shopping trip. We let this count as shopping trip 2, shown in Table 5.4.

Table 5.4 Outcomes of Rolls of a Die to Simulate a Complete Shopping Trip

Shopping trip	Outcomes of rolls of a die						Number of boxes of cereal
	Orange	Yellow	Blue	Pink	Red	Green	
2	//	///	///	/	////	//	16

CHAPTER 5 EXPECTED VALUES AND SIMULATION

The next three shopping trips we simulate by using the random digits in the first two rows of Table B.2:

Row									
1	66533	45332	24614	22231	26431	35541	12165	62116	16111
2	61261	22613	26252	14622	32262	33244	34614	13316	41136

We regard the first five numbers in the first row, 66533, as indicating the first five pen colors obtained. Since we are using the rules 1 = orange, 2 = yellow, 3 = blue, 4 = pink, 5 = red, and 6 = green, the pens corresponding to (6, 6, 5, 3, 3) are (green, green, red, blue, blue). This can be the beginning of shopping trip 3, which is shown completed in Table 5.5.

We now go to the next group of five numbers, 4 5 3 3 2, and record the outcomes in Table 5.5. Then we go to the next group, 2 4 6 1 4. In this group we need only the first four numbers, since the 1 gives us the color needed to complete the set. We can draw a loop around these 14 numbers in the table to show that this is a completed shopping trip.

We now begin shopping trip 4 with the number 4 and continue in the same way to complete trips 4 and 5. The results are recorded in Table 5.5. The first set of all six colors was obtained in 14 rolls of the die, the second set was obtained in 13 rolls, and the third set was obtained in 28 rolls.

Step 5. Finding the Mean (Average) of the Statistic of Interest: In the cereal box problem the statistic of interest is the number of boxes (as modeled by random number selection, say) that have to be obtained to get a pen of every color. Divide the total number of boxes obtained in all the simulations (shopping trips) by the number of simulations (shopping trips), thereby obtaining the average number of cereal boxes per shopping trip.

In Table 5.5 three shopping trips are listed, in which 14, 13, and 28 boxes were purchased. This gives an average of $\hat{E}(X) = (14 + 13 + 28)/3 = 18.3$ boxes. If we include the first two shopping trips in Table 5.3 and Table 5.4, we get an average of $\hat{E}(X) = (21 + 16 + 14 + 13 + 28)/5 = 18.4$ boxes. Here $\hat{E}(X)$ denotes the experimental expected value, just as $\hat{P}(A)$ denotes the experimental probability of A.

The theoretical expected value can be shown by advanced probability techniques to be equal to 14.7. Our simulated average of 18.4 is not that close. A run of 400 simulations, which are not shown here, produced an average of

$$\hat{E}(X) = 14.73,$$

which is much closer to the theoretical 14.7. Thus, by the law of large numbers for the simulated average $\hat{E}(X)$ we get

$$\hat{E}(X) \approx E(X)$$

by doing a large number of simulations.

Table 5.5 Outcomes from Using a Random Number Table to Simulate a Complete Shopping Trip

Shopping trip	Outcomes of rolls of a die						Number of boxes of cereal
	Orange	Yellow	Blue	Pink	Red	Green	
3	/	//	////	//	//	///	14
4	//	////	///	//	/	/	13
5	//// //// //	////	/	/	//	//// //	28

5.2 Using Five-Step Simulation to Estimate Mean Values

It is interesting to note the variations that occur from simulation to simulation. Only five lucky shoppers out of 400 got all their pens in only six boxes. Thus,

$$\hat{P}(6 \text{ different pens in 6 boxes}) = \frac{5}{400} = 0.0125,$$

which is close to the theoretical probability (an equally likely outcome situation $6^6 = 6,656$ for all the ways of choosing 6 pens).

$$P(\text{different pens in 6 boxes}) = \frac{6 \times 5 \times 4 \times 3 \times 2 \times 1}{46656}$$

In our (not shown) simulations, one unlucky shopper needed to buy 56 boxes!

Example 5.10

Tickets, Please!

Airline flights are sometimes overbooked. That is, some airlines expect that a certain percentage (the no-show rate) of ticketed passengers will not show up to claim their seats. So, to ensure that as many seats as possible are filled with paying passengers, the airline issues more tickets for the flight than there are seats on the airplane. Thus, there is a slight chance that a seat will not be available on a flight for which a person has a confirmed reservation. Often, a person denied boarding on a flight is entitled to a compensatory payment.

The no-show rate for each regularly scheduled flight is an experimental probability $\hat{P}(\text{no show})$ that the airline computes from its daily records for every day on which that flight was made. Suppose the airplane has 40 seats, and the airline's records indicate a no-show rate for a certain flight of 10%. The airline accepts 43 reservations for this flight. We will find for this flight (a) the expected number of empty seats and (b) the expected number of ticketed passengers who do not get a seat. Both (a) and (b) are of interest, because the airline expects to lose the amount in (a) times the price of a ticket if there are empty seats and the amount in (b) times the penalty paid to each passenger denied a seat if more passengers show up than there are seats.

Solution: If more passengers show up than there are seats, the procedure we use for solving this problem is, as usual, the five-step method:

Step 1. Choice of Model: Use a table of random digits. One of several appropriate assignments is

$$1 = \text{no show}$$
$$0, 2\text{--}9 = \text{person appears for reservation.}$$

Letting 1 denote a no show and 0 a person appearing, an appropriate box model is

9 balls

⟨ 0 0 ... 0 1 ⟩

Note that $p(0 = 0.9, p(1) = 0.1$ as required.

Step 2. Definition of One Simulation: If using an ordinary random number table (Table B.3 of Appendix B), a simulation consists of reading 43 digits, one for each reservation made. In the case of the box model, 43 balls are sampled with replacement from the box above.

Step 3. Definition of Statistics of Interest: We are interested in the number of no-shows, that is, the number of zeros among the 43 digits or balls. (With a no-show rate of 10%, the expected number of no-shows is $43 \times 0.1 = 4.3$ by our basic $n \times p$ formula.) If there are more than 3 no-shows, then there will be empty seats on the flight, the first statistic of interest. If there are fewer than 3 no-shows, then there will be people with reservations who will not get a seat, this the second statistic of interest.

Step 4. Repetitions of the Simulation: We'll use the random number table approach. For the first simulation we use the first 43 digits in the first row in Table B.3 in Appendix B. There is only one zero in the first row, corresponding to only one no-show. This means that 42 persons show up. Out of these, 40 will be seated and 2 will not be seated. There will be no empty seats.

For the second simulation we use the first 43 digits in the second row. There are 5 zeros, corresponding to 5 no-shows. So 38 persons show up, everyone gets seated, and there are 2 empty seats.

For the third simulation we use the first 43 digits in the third row, and so forth. Table 5.6 shows the results of 10 simulations using the first 10 rows.

Step 5. Finding the Mean of the Statistic(s) of Interest: The average number of no-shows in the 10 simulations is 4.1, which is close to the expected number of no-shows, 4.3. The average number of empty seats per flight is 1.6 (the theoretical expected value is 1.56), this being the experimental expected value for the first statistic of interest. The average number of persons not seated per flight is 0.5. (the theoretical expected value is found to be 0.26 by advanced probability techniques.) It is instructive to note how much the number of no-shows in Column 2 varies. That is why it is called a random variable!

To gain somewhat more accuracy, 100 simulations were done via computer. Let X denote the number of empty seats and Y the number of persons not seated. The result of the 100 simulations was

$$\hat{E}(X) = 1.5 \text{ (theoretical } E(X) = 1.56)$$
$$\hat{E}(Y) = 0.31 \text{ (theoretical } E(X) = 0.26).$$

There was one case of 10 no-shows, producing 7 empty seats.

Table 5.6 Airline Seating with Overbooking Strategy

Repetition	Number of zeros (no-shows)	Persons seated	Empty seats	Persons not seated
1	1	40	0	2
2	5	38	2	0
3	4	39	1	0
4	8	35	5	0
5	0	40	0	3
6	5	38	2	0
7	4	39	1	0
8	6	37	3	0
9	4	39	1	0
10	4	39	1	0
Sum	41	384	16	5
Average	4.1	38.4	1.6	0.5

Example 5.11

The Hermits' Epidemic (revisited)

This problem is somewhat similar to the cereal box problem. We can use a fair die or Table B.2 in Appendix B to simulate the spread of the disease. Number the hermits 1, 2, ..., 6. For the first simulation we can use the numbers in the first row of Table B.2:

$$66533 \quad 45332\ldots$$

The first hermit to get the disease is number 6. He then visits hermit number 5. (We ignore the second 6, because hermit number 6 cannot visit himself.) Hermit number 5 visits hermit number 3, who visits hermit number 4. (We ignore the second 3.) Finally, hermit number 4 visits hermit number 5, who already had the disease and by now is immune, and the disease dies out. In this simulation, four hermits got infected and two did not. That is, $X = 4$, if we let the random variable X denote the number of infected hermits.

For the second simulation we use the numbers in the second row of Table B.2:

$$61261 \quad 22613$$

In this simulation, hermits 6, 1, and 2 get infected and hermits 3, 4, and 5 do not, so three hermits get infected. If we do 50,000 simulations (or use a theoretical approach for finding the expected value), we will find that the expected number of infected hermits $E(X) = 3.51$.

The Key Problem could be changed in a variety of ways to make it more realistic as a model for the spread of disease. For example, the number of hermits on the island could be increased. Suppose there were 100 hermits. We could use the digits of a random number table in pairs (00, 01, 02, ..., 99) to designate which hermit is visited each time. With 100 hermits, how many do you think will get infected? Thirty? Forty? Or fifty? Actually, since each infected hermit only gets one chance to pass the disease on, the disease will die out quickly. There is a probability of 0.50 that at most 12 hermits will be infected, and the expected number of infected hermits is around 13.

If we increase the number of hermits to 1000, then only 5% to 7% of the hermits will get infected before the disease dies out. If we consider one million hermits, then only around 0.2% (this is $0.2 \times 0.01 = 0.002 =$ one out of every 500) of the hermits will get infected before the disease dies out.

This model for the spread of a disease is realistic only for hermits! To make our model more realistic, we have to allow for the possibility that an infected person passes the disease on to more than one person.

Other possible modifications include: making the number of days a hermit is infectious be random; assuming that every member of the hermit society makes a random number of visits each day (that is, comes into contact with a random number of other hermits); assuming that the disease is initiated with several infectious hermits; and assuming that the disease is transmitted by a visit with probability less than 1. Now the model begins to seem fairly realistic in its attempt to simulate the way a disease actually spreads.

Summary

Consider a random variable X determined by the outcome of some random phenomenon. We can estimate the theoretical mean $E(X)$ by the average score of X, namely $\bar{X}$, denoted as $\hat{E}(X)$ in a long series of independent repetitions (for example independent simulations using the five-step method) of the random phenomenon.

Section 5.2 Exercises

For all exercises, use the five-step method. In most cases the simulation software will be needed.

1. The following table gives the results of five simulated shopping trips in the cereal box problem.

Trip	1. Orange	2. Yellow	3. Blue	4. Pink	5. Red	6. Green	Number of boxes
1	///	/	//	////	//// //	////	21
2	//	///	///	/	////	//	16
3	/	//	////	//	//	///	14
4	//	////	///	//	/	/	13
5	//// //// //	////	/	/	//	//// //	28

   a. Estimate the expected number of boxes of cereal needed to get the complete set of six pens.
   b. Using a six-sided die, do another 20 trials. Extend the table to provide the information for the additional trials.
   c. Find the (experimental) expected number of boxes of cereal needed to be purchased, based on 25 trials.
   d. It can be shown by advanced mathematics (probability theory) that the experimental expected number of boxes required to obtain all six colors becomes closer and closer to 14.7, the theoretical expected value, as more and more trials (shopping trips) are carried out. Compare your answer in part **c** to the theoretical result of 14.7.

2. Suppose that instead of six different colors of pens, Tripl Crisp Cereal decides to use *four* different colors of pens in its boxes. What is the average number of boxes of cereal one would expect to have to buy in order to get the complete set of four pens?
   Make 40 simulations by using the digits 1 through 4 and ignoring 5 and 6 in Table B.2 of Appendix B. Use the fourth, fifth, and sixth five-digit columns. For the first simulation use 22231 26431 35541 in the first row. For the second simulation use 14622 32262 33244 in the second row, and so forth.

3. For the airline reservation problem in Example 5.10, make the following changes, but still assume that the aircraft has 40 seats. For each part below, set up the first three steps to find the expected number of empty seats and the expected number of persons not getting a seat. Do 10 simulations.
   a. Suppose the airline accepts 45 (instead of 43) reservations for a flight. Assume a no-show rate of 10%. For the first simulation use the 45 digits in the eleventh row of Table B.3 (in Appendix B). Identify each zero with a no-show. For the second simulation use the 45 digits in the twelfth row in Table B.3.
   b. Suppose the airline accepts 45 reservations, but that the no-show rate is 1 in 6 (instead of 10%). For the first simulation use the 45 digits in the eleventh row of Table B.2. Identify each 6 with a no-show. For the second simulation use the 45 digits in the twelfth row in Table B.2, and so forth.

4. In Example 5.10, assume that the airline accepts 43 reservations for the 40 seats, but that the no-show rate is only 5% (instead of 10%). Make a box model for the number of no-shows. Use a computer to simulate the number of no-shows. Repeat 20 times. For each simulation, find the number of empty seats and the number of persons not getting a seat.
   a. Estimate the expected number of empty seats.
   b. Estimate the expected number of persons not getting a seat.

5. A newly married couple agree that they want to have children. The husband wants to have at least one son, and the wife wants to have at least one daughter. They are interested in finding how many children they can expect to have before they will have at least one boy and at least one girl. Assume that the probability of having a boy or a girl is 0.5. For the first simulation use the digits in the first row of Table B.1 (in Appendix B) from left to right. Identify 0 with a boy

and 1 with a girl. For the second simulation use the digits in the second row of Table B.1. Do 20 simulations. Estimate the expected number of children.

6. Imagine a place where the chance of rain is 1 in 6 every day of the year, independently of the weather the previous days. A newcomer to the place fails to bring an umbrella. How many days of dry weather can he or she expect before the first rainy day? For the first simulation use the digits in the first row of Table B.2, going from left to right. Identify 3 with "rain" and 1, 2, 4, 5, and 6 with "no rain." For the second simulation use the digits in the second row. Do 20 simulations. Each time count the number of dry days before the first rainy day.

7. Simulate the Hermits' Epidemic (Example 5.11) with 10 hermits labeled 0, 1, ..., 9. For the first simulation use the digits in the twenty-first row of Table B.3, going from left to right. For the second simulation use the digits in the twenty-second row of Table B.3. Do 10 simulations. Each time count the number of hermits who get infected. Estimate the expected number.

*5.3 THE STANDARD DEVIATION OF A RANDOM VARIABLE

Consider a random variable X determined by the outcome of some random phenomenon. We define the theoretical standard deviation of X, denoted σ_X or SD of X, as the sample standard deviation of a very, very long series of independent observations of X. We shall need a couple of formulas to compute σ_X.

The Distribution Formula for the Standard Deviation of a Random Variable If the possible values of X are $x_1, x_2, ..., x_k$ with corresponding probabilities $p(x_1), p(x_2), ..., p(x_k)$, then we can compute the mean of X by

$$\mu_x = x_1 p(x_1) + x_2 p(x_2) + \cdots + x_k p(x_k) = \sum x p(x).$$

The variance of X is computed by

$$\sigma_X^2 = (x_1 - \mu_X)^2 p(x_1) + (x_2 - \mu_X) p(x_2) + \cdots$$
$$+ (x_k - \mu_X)^2 p(x_k) = \sum (x - \mu_x)^2 p(x)$$
$$= \sum (x - \mu_x)^2 p(x).$$

The standard deviation σ_X of X is the square root of the variance:

$$\sigma_X = \sqrt{\sum (x - \mu_X)^2 p(x)}.$$

As an illustration, let X denote the value on a symmetric casino die. In that case, $\mu_X = 3.5$, as we showed at the end of Section 5.1. The possible values of X are 1, 2, 3, 4, 5, and 6, each with probability 1/6. The variance of X is

$$\sigma_X^2 = [(1 - 3.5)^2 \times 1/6] + [(2 - 3.5)^2 \times 1/6] + [(3 - 3.5)^2 \times 1/6]$$
$$+ [(4 - 3.5)^2 \times 1/6] + [(5 - 3.5)^2 \times 1/6] + [(6 - 3.5)^2 \times 1/6]$$
$$= [(-2.5)^2 \times 1/6] + [(-1.5)^2 \times 1/6] + [(-0.5)^2 \times 1/6]$$
$$+ [(0.5)^2 \times 1/6] + [(1.5)^2 \times 1/6] + [(2.5)^2 \times 1/6]$$
$$= 35/12.$$

The standard deviation of X is $\sigma_X = \sqrt{35/12}$.

At the end of Section 5.1 we showed that in a series of tosses of a fair die the sample mean $\overline{X}$ gets closer and closer to the theoretical mean μ_X as the number of tosses increases. In the same way, the sample standard deviation s_X gets closer and closer to the theoretical standard deviation σ_X as the number of tosses increases. This fact is illustrated by the following data obtained by simulation.

	s_X	$s_X - \sqrt{35/12}$ (error)
10	1.989	0.281
50	1.695	−0.013
100	1.799	0.091
500	1.761	0.053
1,000	1.733	0.025
5,000	1.721	0.013
10,000	1.708	0.000
50,000	1.705	−0.003
100,000	1.704	−0.004

Recall the Empirical Rule from Section 2.3, that allows us to interpret the average $\bar{x}$ and the standard deviation s of a data set when the histogram of the data is not too far from being "bell shaped":

• About (68%) of the data are in the range $\bar{x} \pm 1s$.

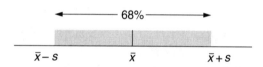

• About 95% of the data are in the range $\bar{x} \pm 2s$.

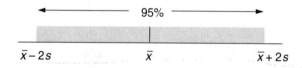

In summary, the average of the data locates the center of the data, and the standard deviation of the data measures the spread around the average.

Now consider any random variable. We saw in Chapter 4 that if we have tens of thousands of independent observations of X, then the experimental probability histogram for the data will be very close to the theoretical probability histogram of X. So we can think of the theoretical distribution of X as the experimental distribution of a very large number of independent observations of X. Think of μ_X as the mean of these observations and σ_X as the standard deviation of the data. It follows that if the probability histogram of X is not too far from being "bell-shaped," then there is

• about a 68% chance that X will fall in the range $\mu_X \pm \sigma_X$.
• about a 95% chance that X will fall in the range $\mu_X \pm 2\sigma_X$.

In other words, μ_X locates the "center" of the distribution of X, and σ_X measures the "spread" of the distribution of X around μ_X. We express this loosely by saying that "X will be around μ_X," with the typical size of the distance between X and μ_X being σ_X. We sometimes write "give or take σ_X or so."

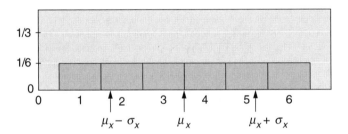

Figure 5.4 The probability histogram of the score on a fair die.

For the number X rolled on a fair die, the theoretical mean μ_X and standard deviation σ_X are

$$\mu_X = 3.5 \text{ and } \sigma_X = \sqrt{35/12} = 1.71, \text{ respectively.}$$

We note that even though the probability histogram is not bell-shaped, the chance that X falls in the range $3.5 - 1.7$ to $3.5 + 1.7$ is $4/6$, which is very close to 0.68. And the chance that X falls in the range $\mu_X - 2\sigma_X$ to $\mu_X + 2\sigma_X$, or $3.5 - 3.4$ to $3.5 + 3.4$, is 1, which is not too far from 0.95. This probability histogram is illustrated in Figure 5.4.

We shall often consider random variables that can be represented as the sum of numbered balls drawn from a box. It will therefore be useful to have formulas for the standard deviation of such variables.

Consider a box with N numbered balls. The **box mean** is

$$\mu_{\text{box}} = \frac{\sum x}{N},$$

where we sum over all the balls in the box. The **box variance** is

$$\sigma_{\text{box}}^2 = \frac{\sum (x - \mu_{\text{box}})^2}{N}.$$

Important: Notice that we use the formula in Section 2.3 for a population variance by dividing by N, not $N - 1$. The box standard deviation is the square root of the box variance.

$$\sigma_{\text{box}} = \sqrt{\sigma_{\text{box}}^2}$$

Consider again the box we use to simulate the toss of a fair die.

$$\boxed{1\ 2\ 3\ 4\ 5\ 6}$$

With $N = 6$ balls in the box we get

$$\mu_{\text{box}} = (1 + 2 + 3 + 4 + 5 + 6)/6 = 21/6 = 3.5$$
$$\sigma_{\text{box}}^2 = [(1 - 3.5)^2 + (2 - 3.5)^2 + \cdots + (6 - 3.5)^2]/6$$
$$= 35/12.$$

Thus,
$$\sigma_{box} = \sqrt{35/(12)}.$$

We already know that if X is the sum of the numbers on n randomly selected balls from a box (drawn with or without replacement between draws), then
$$\mu_x = n \cdot \mu_{box}.$$

The formula for σ_X depends on whether we replace or not as we see next.

The Box Model Formulas for σ_X If X is the sum of the numbers on n randomly selected balls from a box, drawn *with replacement* between draws, then
$$\sigma_{box} = \sqrt{n} \cdot \sigma_{box}.$$

If X is the sum of the numbers on n randomly selected balls from a box, drawn *without replacement* between draws, then
$$\sigma_{box} = \sqrt{n} \cdot \sigma_{box} \cdot \sqrt{\frac{N-n}{N-1}},$$

where N = number of balls in the box. If we draw without replacement, then the sample size n cannot be greater than the total number of balls N, of course.

To simulate the sum of the scores on a pair of symmetric casino dice we can draw at random twice with replacement from the box we just looked at and compute the sum X of the two numbers drawn from the box. We get

$$\mu_x = 2 \cdot \mu_{box} = 2 \cdot 3.5 = 7$$

$$\sigma_X = \sqrt{2} \cdot \sigma_{box} = \sqrt{2} \cdot \sqrt{35/12} \approx \sqrt{35/6} \approx 2.4.$$

The probability histogram for X was given in Figure 4.10, shown here as Figure 5.5. Clearly this distribution is roughly bell-shaped. We shall see how well the 68-95-99.7% rule applies.

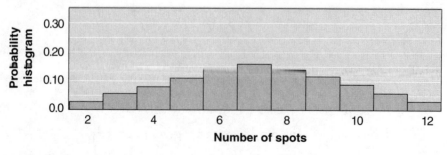

Figure 5.5 Probability distribution of the sum of two fair dice.

According to the rule, there should be around a 68% chance that X falls in the interval $M_X \pm \sigma_X$, or between 7 − 2.4 = 4.6 and 7 + 2.4 = 9.4. Similarly, there should around a 95% chance that X falls in the interval $\mu_x \pm 2\sigma_x$, or between 7 − 4.8 = 2.2 and 7 + 4.8 = 11.8.

This turns out to be almost exactly right. In fact, if we look at Figure 4.3 at the beginning of Chapter 4, we see that the probability that X takes one of the values 5, 6, 7, 8, or 9 is 24/36 = 2/3 or 66.7%. And the chance that X takes one of the values from 3 to 11 is 34/36 or 94.4%.

Example 5.12

A fair die is tossed 100 times. The sum of the scores most likely will be around _____, give or take _____ or so.

Solution: The sum X of the 100 scores will be around its expected value μ_X, give or take one standard deviation μ_X or so (the typical size of the variation of X from μ_X). We can simulate X by drawing 100 times at random with replacement from the box

$$\boxed{\;①\;②\;③\;④\;⑤\;⑥\;⑦\;}$$

and adding up to 100 numbers we get

$$\mu_X = 100 \times \mu_{\text{box}} = 100 \times 3.5 = 350,$$
$$\sigma_x = \sqrt{100} \times \sigma_{\text{box}} = \sqrt{100} \times \sqrt{35/12} \approx 17.$$

The sum of the 100 scores most likely will be around 350, give or take 17 or so. There is about a 95% chance that the sum will be between $\mu_X + 2\sigma_X \approx 350 - 34 = 316$ and $\mu_X + 2\sigma_X \approx 350 + 34 = 384$.

Summary

The theoretical variance of a random variable X is conveniently given by

$$\sigma_X^2 = \text{Var}(X) = \sum(x - \mu_X)^2 p(x).$$

The theoretical standard deviation σ_X of X is then equal to $\sqrt{\text{Var}(X)}$.

Just as we can estimate the theoretical mean $E(X)$ of a random variable X by the simulated average score of X, namely $\overline{X}$, in a long series of independent repetitions of the random phenomenon (using the five-step method), we can also estimate the standard deviation σ_X by the simulated s_X. This estimate of σ_X, is accurate if the number of independent repetitions is large.

If X is the sum of the numbers on n randomly selected balls from a box, drawn **with** replacement between draws, then

$$\sigma_X = \sqrt{n} \cdot \sigma_{\text{box}}.$$

If X is the sum of the numbers on n randomly selected balls from a box, drawn **without** replacement between draws, then

$$\sigma_X = \sqrt{n} \cdot \sigma_{\text{box}} \cdot \sqrt{\frac{N-n}{N-1}}$$

where N = number of balls in the box.

In either case, X is likely to be around μ_X, give or take σ_X or so.

If the probability histogram of a random variable X is roughly bell shaped then the 68–95–99.7% rule applies using μ_X and σ_X.

Section 5.3 Exercises

1. A professor has a key ring with 10 keys, one of which opens his office door. Without his reading glasses he cannot tell the keys apart, so he tries to use them, one at a time, to open the door. He is equally likely to try the correct key first, second, third, or last. How many wrong keys will he try before he finds the correct one? We can simulate the number of wrong keys by picking one random number from Table B.3 in Appendix B. Do 45 simulations using the digits in the last row of Table B.3, going from left to right. Estimate the expected number of wrong keys (The theoretical expected value is 4.5). Estimate the standard deviation for the number of wrong keys.

2. A student blindly guesses all the answers to a 15-question true-false quiz. Use Table B.1 (in Appendix B) to simulate the number of correct answers he or she gets. Estimate the expected number, the variance, and the standard deviation. Do 25 simulations. For the first simulation use the first 15 digits in the first row of Table B.1. For the second simulation use the first 15 digits in the second row, and so forth. Identify each 0 with a wrong answer and each 1 with a correct answer.

3. Let X be the sum of the numbers on 225 balls drawn at random with replacement from the box.

$$\boxed{①\ ②\ ③\ ④}$$

   a. What is the smallest X can be?
   b. What is the largest X can be?
   c. Most likely, X will be around _____, give or take _____ or so.

4. A true-false quiz has 36 questions. Imagine that you blindly guess the answer to each question. The number of correct answers you get most likely will be around _____, give or take _____ or so.

5. A fair die is tossed 250 times. The sum of the 250 scores is likely to be around _____, give or take _____ or so.

6. A fair die is tossed 180 times. The number of sixes is likely to be around _____, give or take _____ or so.

CHAPTER REVIEW EXERCISES

1. A fair coin is tossed 45 times. What is the expected number of heads?
2. A fair die is tossed 60 times. What is the expected number of sixes?
3. Recall from Table 4.5 that the ABO blood type distribution for African Americans is

Blood type	O	A	B	AB
Probability	0.49	0.27	0.20	0.04

For a random sample of 150 African Americans, compute the expected number of persons of each of the four ABO blood types.

4. Assume that both male and female African Americans have the blood type distribution in Exercise 3.
   a. If we randomly choose an African American couple, what is the chance that both are type O?
   b. In a random sample of 50 African American couples, what is the expected number of couples where both are type O?

5. A study on heart attacks examined the records of 2254 patients admitted to a coronary care unit over a 10-year period. Assume that heart attacks are equally likely to occur on any day of the week. How many of the 2254 heart attacks would we expect to have occurred
   a. on Mondays?
   b. on weekends (Saturdays or Sundays)?

6. According to the U.S. Census Bureau, the distribution of American families by size is

Family size, x	2	3	4	5	6	7
Fraction of families, $p(x)$	0.43	0.23	0.21	0.09	0.03	0.02

(A family is defined as a group of two or more related persons living together. We have ignored the few families with 8 or more members.) Compute the average size of all American families.

7. Each kernel on a corncob is the result of an independent event. A certain genetic experiment with corn produces four types of kernels when we consider both color and shape: purple/normal, purple/shrunken, yellow/normal, and yellow/shrunken, with probabilities 9/16, 3/16, 3/16, and 1/16, respectively. If there are 624 kernels on a cob, compute the expected number of each of the four types of kernels.

8. Estimate the expected number of times you have to roll a fair die before one of the six sides appears for the second time. (That is, how many times do you have to roll before a side that you have already seen comes up again?)
   a. What is the smallest number possible?
   b. What is the largest number possible?
   c. Do 10 simulations. For the first simulation use the digits in the first row in Table B.2 (in Appendix B). For the second simulation use the digits in the second row. Estimate the mean, the variance, and the standard deviation.

9. John is taking a test consisting of 10 multiple-choice questions with four choices for each question. For each question he gets right, he receives four points, but for every one he answers incorrectly, he loses one point.
   a. If John does not know the answer to a question and blindly guesses the answer, what is his expected score for that question?
   b. If he does not answer a question, then he neither loses nor gains any points for that question. Which is the better strategy if he does not know the answer to a question, guessing the answer or not answering the question?

10. If you bet one dollar on a three-digit number in the Illinois state lottery game "Pick 3" and the lottery matches your 3 digits in exact order, you win $500. You do not get your $1 purchase price back.
    a. What is your chance of winning a single bet?
    b. Imagine that each day you make five $1 bets on five different three-digit numbers. What is your chance of winning on a given day?
    c. Make a box model for your gain on a given day. (You gain either $495 or –$5.)
    d. What is your expected gain on a given day?
       Imagine that you bet on five numbers on 300 different days over a year.
    e. If you have simulation software, do 10,000 simulations on a computer to estimate the distribution and the expected value of the number of times you win over 300 days.

11. One in five automobile accidents involve a rear-end collision. Use the third and the fourth five-digit columns in Table B.3 (in Appendix B) to simulate 10 automobile accidents. Identify a zero and a one with a rear-end collision. Do 20 simulations. For the first simulation use the digits in the first row (of the third and fourth five-digit columns.) Count the number of rear-end collisions. For the second simulation, use the digits in the second row. Estimate the distribution and the expected value of the number of rear end collisions.

12. Are fatal automobile accidents involving drunk drivers caused by weekend binge drinkers or by everyday drunks (equally likely to be drunk on any day of the week)? A study looked at a random sample of 133 fatal-accident reports involving drunk drivers. Assuming that all the accidents were caused by everyday drunks (not true of course), how many of the accidents would we expect to have occurred
    a. on Wednesdays?
    b. on weekends (Fridays, Saturdays, or Sundays)?

PROFESSIONAL PROFILE

Jennifer A. Harris Trosper
Mars 2001 Operations Development Manager
Jet Propulsion Laboratory
Pasadena, California

Pathfinder on Mars (image courtesy of NASA/JPL/Caltech)

Jennifer A. Harris Trosper

"It's fun!" So says Jennifer A. Harris Trosper of her work on a series of Mars exploration projects. She was the flight director for operations on the *Mars Pathfinder* Project in charge of deploying the rover.

Jennifer Harris Trosper's special area of interest is solving problems in technical areas. She uses statistics to help manage risks and to weigh the probabilities of various events happening. Based on a statistical analysis of data, she and other members of exploration team can weigh the importance of various possible events and develop strategies to prepare for them. "You learn where to focus your attention," she explains.

This is critical in managing operations on Mars. It can take up to 20 minutes for a radio signal to reach Mars, depending on where Mars is in relationship to Earth. It takes an equal amount of time for a signal to return. With time lags running between 14 and 40 minutes, you must anticipate events. It's like trying to drive a car when all you can see is what was going on fifteen minutes ago.

Jennifer Harris Trosper is enthusiastic about her work and its importance for the future. Many of the technologies developed for the Mars projects can be used by others right here on Earth. "And," she adds, "I believe that understanding our solar systems helps us understand our Earth better. I also believe our work inspires people to do great things." She encourages everyone to have challenging goals. "If you have the desire and the passion, don't let anybody tell you you can't do something. You can accomplish whatever you really want."

6

Probability Distributions: The Essentials

You have a very serious disease. Of ten persons who get the disease only one survives. But do not worry, it is lucky you came to me for I've recently had nine patients with the disease and they all died of it.

G. Polya

Objectives

After studying this chapter, you will understand the following:

- Binomial distribution of the number of successes in independent trials
- Sampling and the binomial distribution
- Geometric distribution of the time until a first success
- Poisson distribution of the number of random occurrences in an interval
- The Poisson approximation to binomial probabilities
- The normal distribution (bell-shaped curve)
- The computation of normal probabilities using a standard normal table
- Practical applications of the normal distribution
- The normal approximation to binomial probabilities

6.1 THE BINOMIAL DISTRIBUTION 251

6.2 THE GEOMETRIC DISTRIBUTION 259

6.3 THE POISSON DISTRIBUTION 263

6.4 THE NORMAL PROBABILITY DISTRIBUTION 273

  CHAPTER REVIEW EXERCISES 288

Key Problem

In the 1898 book *Das Gesetz der Kleinen Zahlen* (*The Law of Small Numbers*), Dr. L. von Bortkiewicz summarized some data on 14 Prussian cavalry units over the 20 years from 1875 to 1894, producing 280 observations. For each unit and year, he presented the number of unfortunate soldiers who died from blows by horses, presumably kicks. In most of the units, no one was kicked to death by horses. The worst were two units that had four such fatalities (unit 11 in 1880 and unit 14 in 1882). The table shows the number of units that had 0, 1, 2, 3, or 4 of these deaths. A unit is counted each year. Thus, if a unit had 0 fatalities in 17 years, it contributes 17 to the 144 units with 0 fatalities.

Number of soldiers kicked to death	Number of units
0	144
1	91
2	32
3	11
4	2
5 or more	0
	280

According to the table, in 144 cases there were no deaths, 91 cases had one, and so on.

Could Dr. Bortkiewicz have used this data to estimate the probability of a unit in the next year having at least one soldier kicked to death? Do these frequencies show a predictable pattern? Can we fit a probability model to these data?

Certain distributions of random variables are used so often that their probability distributions have been given names. In this chapter we introduce the binomial distribution, the hypergeometric distribution, and the Poisson distribution. These distributions are used to model the number of times an outcome of interest occurs in a particular chance experiment. Later in the chapter, we resume our discussion of the normal distribution that we used (Chapter 2) to model physiological variables and repeated measurements of some entity.

The **factorial** notation is convenient for representing certain discrete probability distributions. For $k = 1, 2, 3, \ldots$, the product of the first k positive integers is called "k factorial" and is written $k!$ For example,

$$1! = 1$$
$$2! = 1 \times 2 = 2$$
$$3! = 1 \times 2 \times 3 = 6$$
$$4! = 1 \times 2 \times 3 \times 4 = 24.$$

Finally, "zero-factorial" is defined as 1:

$$0! = 1.$$

For two positive integers $x \leq n$, the **binomial coefficient**

$$\binom{n}{x} = \frac{n!}{x!(n-x)!}$$

gives the number of different ways of selecting x objects from a collection of n distinct objects when the order in which the x objects are chosen does not matter. (This can be proved using the multiplication rule for a multi-stage experiment of Section 4.5.) The binomial coefficient is read as "n choose x." For example, there are "3 choose 2" = $3!/(2! \times 1!) = 3$ ways of selecting 2 objects from (a, b, c), namely (a, b), (a, c), (b, c). As an example of computing a binomial coefficient,

$$\binom{5}{2} = \frac{5!}{2! \times 3!} = \frac{5 \times 4 \times 3 \times 2 \times 1}{2 \times 1 \times 3 \times 2 \times 1} = \frac{5 \times 4}{2} = 10.$$

6.1 THE BINOMIAL DISTRIBUTION

In statistics we often consider repeated independent performances of a chance experiment in which an event of interest may or may not occur. Each repetition of the experiment is usually called a **trial**. We are often interested in the number of times the event of interest occurs in n independent trials.

Imagine, for example, that you make three independent one-dollar bets on "red" in roulette. Each time you bet, your chance of winning a dollar is $p = 18/38$ and your chance of losing a dollar is $1 - p = 20/38$. You have to win at least two out of the three times to make money. If we list the outcomes of the three bets, then the three-bet chance experiment has eight possible outcomes:

{LLL, WLL, LWL, LLW, WWL, WLW, LWW, WWW},

where L stands for loss and W stands for win.

Let X denote the number of times you win. Since the three bets are independent, we can use the product rule for independent events and the addition rule for mutually exclusive events (Section 4.2). For example, the event $(X = 2)$ occurs if any of the three mutually exclusive events WWL, WLW, and WWW occurs. That is $(X = 2) = ((WWL) \text{ or } (WLW) \text{ or } (LWW))$.

Also (WWL) = ((win on first bet) and (win on second bet) and (win on third bet)), each of these three events being independent. Therefore,

$$p(2) = P(X = 2) = P(WWL) + P(WLW) + P(LWW)$$
$$= p^2(1-p) + p(1-p)p + (1-p)p^2$$
$$= 3p^2(1-p) = 3 \cdot \left(\frac{18}{38}\right)^2 \frac{20}{38} = 0.3543.$$

Since

$$p(3) = P(X = 3) = P(WWW) = p \cdot p \cdot p = \left(\frac{18}{38}\right)^3 = 0.1063,$$

your chance of making money is

$$P(X \geq 2) = P(X = 2 \text{ or } X = 3) = p(2) + p(3) = 0.3543 + 0.1063 = 0.4606;$$

that is, you have about a 46% chance of making money. We could also have obtained this result by simulating our net gain on three one-dollar bets on a computer 10,000 times using a box model of eighteen 1s and twenty –1s.

The Binomial Probability Distribution

Consider n **independent** trials, each resulting in one of two possible outcomes, which for convenience we call "**success**" and "**failure.**" Assume that the probability of success, p, is the same for all n trials. Let the random variable X denote the total number of successes in n trials. Then X has the probability density function (pdf):

$$p(x) = P(X = x)$$
$$= P(x \text{ successes and } n\text{-}x \text{ failures})$$
$$= \binom{n}{x} p^x ((1-p))^{n-x}, \text{ for } x = 0, (1), \ldots, n.)$$

We say that X is **binomially distributed** with parameters n and p. The expected value of X and the standard deviation of X are

$$\mu_X = np \quad \text{and} \quad \sigma_X = \sqrt{np(1-p)}, \text{ respectively.}$$

> A **binomial experiment** has the following assumptions:
> - Each trial has two possible outcomes, generically denoted "success" and "failure."
> - The number of trials n is fixed in advance.
> - The success-probability p is the same from trial to trial.
> - The trials are independent.
> - The random variable X of interest is the total number of successes in the n trials.
>
> If these 5 assumptions hold for our real world, then we compute $P(X = x)$ using the binomial probability density function.

In our roulette example there are three independent trials (bets). A win is a "success" and a loss is a "failure." The probability of a success is $18/38$ for each trial. It follows that X, the number of times you win, is binomial with $n = 3$ and $p = 18/38$.

Now imagine that you make 35 one-dollar bets on a pair of numbers in Keno (Section 4.4). The number of wins X is binomial with $n = 35$ and $p = 0.06$. The probability of losing all 35 Keno bets is

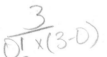

$$p(0) = \binom{n}{0} p^0 (1-p)^n = (0.94)^{35} = 0.1147.$$

The chance of winning once and losing 34 times is

$$p(1) = \binom{n}{1} p (1-p)^{n-1} = 35(0.06)(0.94)^{34} = 0.2562.$$

The chance of winning twice and losing 33 times is

$$p(2) = \binom{n}{2} p^2 (1-p)^{n-2} = \frac{35 \cdot 34}{2} (0.06)^2 (0.94)^{33} = 0.2780.$$

Thus, the probability that you lose money on the 35 bets occurs if $X \leq 2$ and is given by,

$$P(X \leq 2) = p(0) + p(1) + p(2)$$
$$= 0.1147 + 0.2562 + 0.2780 = 0.6489.$$

Therefore, your chance of making money on the 35 bets, noting that $(X \geq 3)$ is the complement of $(X \leq 2)$, is

$$P(X \geq 3) = 1 - P(X \leq 2) = 1 - 0.6489 = 0.3511.$$

You have only about a 35% chance of coming out ahead! Again, we could have obtained this result by simulating our net gain on the 35 one-dollar bets on a computer 10,000 times, using the appropriate box model.

Example 6.1

Four-o'clocks are ornamental plants. If we cross two pink-flowered four-o'clocks, the chance that an offspring will have red flowers is $1/4$, pink flowers $1/2$, and white flowers $1/4$. (See the discussion at the end of Section 4.4.) Different offspring inherit their genetic make-up independently. Imagine that 150 offspring are produced. Then the number X of red-flowered offspring is binomial with $n = 150$ and $p = 1/4$. (Denote each red-flowered offspring a "success" and each pink-flowered or white-flowered offspring a "failure.") The number Y of white-flowered offspring is also binomial with $n = 150$ and $p = 1/4$. Finally, the number Z of pink-flowered offspring is binomial with $n = 150$ and $p = 1/2$.

Example 6.2

A symmetric casino die is tossed 15 times and we count the number X of sixes. If we assume that the outcomes of successive tosses are independent, and we for each toss identify "success" with getting a six and "failure" with not getting a six, then it follows that X is binomial with $n = 15$ and $p = 1/6$.

Example 6.3

A true-false quiz has 10 questions. Imagine that you do not know any of the answers and blindly guess each of the 10 answers. Then your number X of correct answers will be binomial with $n = 10$ and $p = 1/2$.

Binomial variables are "counting variables." We count the number of times an event of interest occurs in repeated trials. But not all counting variables are binomial. It is important to be able to tell whether a counting variable is binomial or not. A counting variable X is binomial only when the five bulleted assumptions of a binomial experiment hold.

Example 6.4

You have $5 and you make repeated one-dollar bets on "red" in roulette until you either double your money (gain $5) or you go broke. Here the success-failure trials (the individual bets) are independent with the same success-probability $p = 18/38$ in each trial. But *the number of trials has not been fixed in advance* and X, the number of bets you win, is **not** binomial.

Example 6.5

A bridge hand of 13 cards is dealt at random, one card at a time, from a well-shuffled deck of 52 cards. Let X denote the number of hearts among the 13 cards. Each card dealt is either a heart or not a heart. If we identify getting a heart with a "success" and not getting a heart with a "failure," then we have 13 success-failure trials. But since we do not replace cards between draws, successive draws from the deck are *not independent* and X is *not* binomial. (In fact, X has a hypergeometric distribution that was tabulated in Table 5.1 in Section 5.1.)

Binomial Distribution Histograms

Figures 6.1, 6.2, and 6.3 display the probability histograms of three different binomial distributions. The heights of the various rectangles equal the corresponding binomial probabilities. A missing rectangle means that the corresponding probability is less than 0.005. Notice that the histogram in Figure 6.1 is symmetric, almost bell-shaped, whereas the shapes of the histograms in Figures 6.2 and 6.3 are skewed.

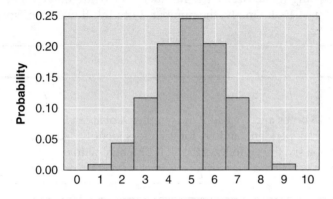

Figure 6.1 Binomial distribution histogram for $p = 0.5$, $n = 10$.

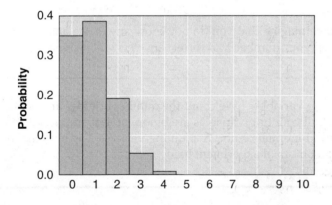

Figure 6.2 Binomial distribution histogram for $p = 0.1$, $n = 10$.

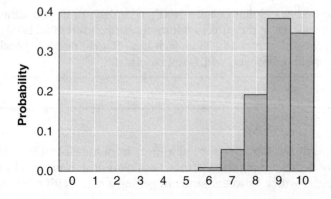

Figure 6.3 Binomial distribution histogram for $p = 0.9$, $n = 10$.

Using the Tables for Cumulative Binomial Probabilities

If the binomial parameter n is not too large, then it is easy to compute binomial probabilities using a calculator. But it may be possible to save time by using the binomial tables in Appendix G.

Example 6.6

Consider 25 married couples where both husband and wife are carriers of cystic fibrosis (see Example 4.11). If each of the couples has a child, what is the probability that at most eight of the 25 children develop the deadly disease? What is the chance that at least four of the children get cystic fibrosis?

Solution: Each child has a 25% risk of getting cystic fibrosis. If X denotes the number of the 25 children who get cystic fibrosis, then X is binomial with $n = 25$ and $p = 0.25$. The pdf of X is

$$p(x) = P(X = x) = \binom{25}{x}\left(\frac{1}{4}\right)^x \left(\left(\frac{3}{4}\right)\right)^{25-x}, \text{ for } x = 0, (1), \ldots, 25.$$

Suppose we are interested in $P(X \leq 8)$ and $P(X \geq 4)$. These probabilities can easily be obtained from the binomial tables. Table 6.1 provides the relevant binomial table from Appendix G.

Table 6.1 Cumulative Binomial Probabilities. The Table Gives the Probability of Obtaining x or Fewer Successes in n Independent Trials, namely $P(X \leq x)$, Where $p =$ Probability of Success in a Single Trial.

						p				
n	x	0.05	0.10	0.15	0.20	0.25	0.30	0.35	0.40	0.45	0.50
25	0	0.2774	0.0718	0.0172	0.0038	0.0008	0.0001	0.0000	0.0000	0.0000	0.0000
	1	0.6424	0.2712	0.0931	0.0274	0.0070	0.0016	0.0003	0.0001	0.0000	0.0000
	2	0.8729	0.5371	0.2537	0.0982	0.0321	0.0090	0.0021	0.0004	0.0001	0.0000
	3	0.9659	0.7636	0.4711	0.2340	(0.0962)	0.0332	0.0097	0.0024	0.0005	0.0001
	4	0.9928	0.9020	0.6821	0.4207	0.2137	0.0905	0.0320	0.0095	0.0023	0.0005
	5	0.9988	0.9666	0.8385	0.6167	0.3783	0.1935	0.0826	0.0294	0.0086	0.0020
	6	0.9988	0.9905	0.9305	0.7800	0.5611	0.3407	0.1734	0.0736	0.0258	0.0073
	7	1.0000	0.9977	0.9745	0.8909	0.7265	0.5118	0.3061	0.1536	0.0639	0.0216
	8	1.0000	0.9995	0.9920	0.9532	(0.8506)	0.6769	0.4668	0.2735	0.1340	0.0539
	9	1.0000	0.9999	0.9979	0.9827	0.9287	0.8106	0.6303	0.4246	0.2424	0.1148
	10	1.0000	1.0000	0.9995	0.9944	0.9703	0.9022	0.7712	0.5858	0.3843	0.2122
	11	1.0000	1.0000	0.9999	0.9985	0.9893	0.9558	0.8746	0.7323	0.5426	0.3450
	12	1.0000	1.0000	1.0000	0.9996	0.9966	0.9825	0.9396	0.8462	0.6937	0.5000
	13	1.0000	1.0000	1.0000	0.9999	0.9991	0.9940	0.9745	0.9222	0.8173	0.6550
	14	1.0000	1.0000	1.0000	1.0000	0.9998	0.9982	0.9907	0.9656	0.9040	0.7878
	15	1.0000	1.0000	1.0000	1.0000	1.0000	0.9995	0.9971	0.9868	0.9560	0.8852
	16	1.0000	1.0000	1.0000	1.0000	1.0000	0.9999	0.9992	0.9957	0.9826	0.9461
	17	1.0000	1.0000	1.0000	1.0000	1.0000	1.0000	0.9998	0.9988	0.9942	0.9784
	18	1.0000	1.0000	1.0000	1.0000	1.0000	1.0000	1.0000	0.9997	0.9984	0.9927
	19	1.0000	1.0000	1.0000	1.0000	1.0000	1.0000	1.0000	0.9999	0.9996	0.9980
	20	1.0000	1.0000	1.0000	1.0000	1.0000	1.0000	1.0000	1.0000	0.9999	0.9995
	21	1.0000	1.0000	1.0000	1.0000	1.0000	1.0000	1.0000	1.0000	1.0000	0.9999
	22	1.0000	1.0000	1.0000	1.0000	1.0000	1.0000	1.0000	1.0000	1.0000	1.0000
	23	1.0000	1.0000	1.0000	1.0000	1.0000	1.0000	1.0000	1.0000	1.0000	1.0000
	24	1.0000	1.0000	1.0000	1.0000	1.0000	1.0000	1.0000	1.0000	1.0000	1.0000
	25	1.0000	1.0000	1.0000	1.0000	1.0000	1.0000	1.0000	1.0000	1.0000	1.0000

To determine $P(X \leq 8)$, locate the binomial table with $n = 25$. From the number in the $x = 8$ row and in the column under $p = 0.25$, we get

$$P(X \leq 8) = 0.8506.$$

To determine $P(X \geq 4)$, find the number in the $x = 3$ row and in the column under $p = 0.25$ to get

$$P(X \geq 4) = 1 - P(X \leq 3) = 1 - 0.0962 = 0.9038.$$

Note that $P(4 \leq X \leq 8) = P(X \leq 8) - P(X \leq 3) = 0.8506 - 0.0462 = 0.8044$. This shows how to use the binomial tables to solve for a binomial random variable that falls within an interval.

The binomial distribution can also be used to model random sampling from a finite population if the population is much larger than the sample (population size $\geq 20n$ is the rule of thumb) and if each member either has or lacks an attribute of interest.

Sampling and the Binomial Distribution

Consider a population of size N of which N_1 members of the population have a certain attribute and the remaining $N_2 = N - N_1$ members do not. For example, we might consider a population of $N = N_1 + N_2$ persons of which N_1 smoke and N_2 do not smoke. Or we might consider a shipment of $N = N_1 + N_2$ identical electronic devices of which N_1 devices are defective and the remaining N_2 devices are not defective.

Suppose we randomly draw a sample of n members from the population, one at a time *without replacement* between draws. This method of sampling is called **simple random sampling** or just **random sampling.** When we sample in this way, then all possible samples of size n are equally likely to be obtained.

Let X denote the number of population members in the simple random sample who have the attribute in question. That is, the number of smokers X in the sample or the number of defective electronic devices in the sample X is said to have a **hypergeometric distribution** with pdf

$$p(x) = P(X = x) = \binom{N_1}{x}\binom{N_2}{n-x} \Big/ \binom{N_1 + N_2}{n},$$

for $x = 0, 1, \ldots, n$. Note that if we randomly sample from the population, one at a time *with replacement* between draws, then the draws will be *independent* and, hence, we have a binomial experiment. This X, the number of times we select a population member with the attribute, will be binomial with parameters n and $p = N_1 / (N_1 + N_2)$.

If the sample size n is less than 5% of the population size N, it makes very little difference in computing $P(X = x)$ whether we sample with or without replacement. So even if we sample without replacement, X is approximately binomial with parameters n and $p = N_1 / (N_1 + N_2)$.

We state this basic principle allowing finite population sampling to be modeled by the binomial distribution:

> Suppose that a random sample (without replacement) of size n is taken from a finite population in which the proportion of members that have a specified attribute is p. If the sample size n is less than 5% of the population size, then the number X of members in the sample that have the attribute is approximately binomial with parameters n and p.

Example 6.7

As mentioned at the end of Section 4.2, 85% of all Americans have Rh+ blood type. If we take a simple random sample of 25 Americans. What is the chance that all 25 are Rh+? What is the chance that at least 20 of the persons in the sample are Rh+? What is the chance that at most 20 of the persons in the sample are Rh+?

Solution: If X denotes the number of persons in the sample who are Rh+, then, noting that the population size N is huge, X is (almost) binomial with $n = 25$ and $p = 0.85$. Since the binomial tables only cover binomial distributions with $p \leq 0.50$, we shall consider instead the number Y of persons in the sample who are not Rh+ (and hence are Rh–). Clearly the probability of being Rh– is $1 - 0.85 = 0.15$. Thus Y is (almost) binomial with $n = 25$ and $p = 0.15$. For example, since $X + Y = 25$, the binomial tables give us

$$P(X = 25) = P(Y = 0) \approx 0.0172,$$
$$P(X \geq 20) = P(Y \leq 5) \approx 0.8385, \text{ and}$$
$$P(X \leq 20) = P(Y \geq 5) = 1 - P(Y \leq 4)$$
$$\approx 1 - 0.6821 = 0.3179.$$

Example 6.8

A bridge hand of 13 cards is dealt at random, one card at a time, from a well-shuffled deck of 52 cards. Let X denote the number of hearts among the 13 cards. X has a hypergeometric distribution with $N_1 = 13$, $N_2 = 39$, and $n = 13$.

$$p(x) = P(X = x) = P(\text{we get } x \text{ hearts and } n - x \text{ nonhearts})$$
$$= \binom{13}{x}\binom{39}{13-x} \bigg/ \binom{52}{13}$$

for $x = 0, 1, 2, \ldots, 13$. This distribution is tabulated in Table 5.1 in Section 5.1. Here the sample size n is 25% of the population size and we cannot use the binomial approximation.

Summary

Let X denote the number of times an event of interest occurs in a series of trials. Assume that

- the number of trials n is fixed in advance;
- the trials are independent;
- for each trial, the probability that the event of interest will occur equals the same value p.

Then X has the **binomial probability density function**

$$p(x) = P(X = x) = P(\text{the event occurs exactly } x \text{ times})$$
$$= \binom{n}{p} p^x ((1-p))^{n-x}, \quad \text{for } x = 0, (1), \ldots, n.)$$

258 PROBABILITY DISTRIBUTIONS: THE ESSENTIALS

We say that X is **binomially distributed** with parameters n and p. The mean and the standard deviation of X are

$$\mu_X = np \quad \text{and} \quad \sigma_X = \sqrt{np(1-p)}.$$

Imagine that we randomly draw a sample of n members from a population, one at a time without replacement between draws. This method of sampling is called **random sampling without replacement**. Let p denote the proportion of population members that have a certain attribute. If the sample size n is less than 5% of the population size, then the number X of members in the sample that have the attribute is approximately binomial with parameters n and p.

Section 6.1 Exercises

1. Compute the binomial coefficients $\binom{4}{0}, \binom{4}{1}, \binom{4}{2}, \binom{4}{3}$, and $\binom{4}{4}$.

2. A fair coin is tossed five times. Let X denote the number of times the coin lands heads up. For $x = 0, 1, 2, 3, 4, 5$ compute

$$p(x) = P(X = x) = P(\text{we get } x \text{ heads and } 5 - x \text{ tails in any order}).$$

3. A fair coin is tossed 20 times. Let X denote the number of times the coin lands heads up.
   a. What is the expected number of heads μ_X?
   b. Use the binomial tables to determine

$$P(X \leq 9), P(X \leq 10), \text{ and } P(X = 10).$$

   Hint: $P(X = 10) = P(X \leq 10) - P(X \leq 9)$.

4. For families with four children, what are the separate probabilities that a randomly selected family will have 0, 1, 2, 3, or 4 girls, assuming that boys and girls are equally likely at each birth and ignoring the possibility of identical twins? Given 2000 families, each with four children, how many families would you expect to have no girls? One girl? Two girls? Three girls? Four girls?

5. Your chance of winning a bet on a single number in Keno is $p = 1/4$. (See Section 4.4.) The bet pays 2 to 1: If you bet one dollar and you win, then you gain two dollars. (You get your dollar back plus two additional dollars.) Of course if you lose, your gain is -1. Imagine that you make 25 independent one-dollar bets on a single number in Keno, one dollar at a time.

   a. What is your net gain if you win 8 times and lose 17 times?
   b. What is you net gain if you win 9 times and lose 16 times?
   c. What is your net gain if you win 10 times and lose 15 times?
   d. How many times do you have to win to make money?
   e. What is your chance of making money on the 25 bets? (Use the tables.)

6. Ninety percent of all Americans are right-handed. Imagine that we take a random sample of 25 Americans. Use the tables to determine the probability that
   a. all 25 are right-handed.
   b. at least 20 of the 25 are right-handed.
   c. at most 20 of the 25 are right-handed.
   d. exactly 20 of the 25 are right-handed. *Hint:* Consider the number of persons in the sample who are **not** right-handed.
   e. Explain why you were allowed to use the binomial probability distribution.

7. A multiple-choice test has 25 questions. To each question there are five answers to choose from, one of which is correct. Imagine that you do not have a clue what the answers are and you blindly guess the answer to each of the 25 questions.
   a. What is the chance that you guess the answer to the first question correctly?
   b. What is your expected number of correct answers on the test? (Compute the expected value.)
   c. What is the probability that you get at most eight correct answers on the test?
8. There is about an 80% chance that a randomly chosen person aged 20 will still be alive at age 65, according to the U.S. National Center for Health Statistics. Suppose that we randomly select fifteen 20-year-olds.
   a. How many do you expect to still be alive at age 65? (Compute the expected value.)
   b. What is the chance that all 15 are alive at age 65?
   c. What is the chance that at least 12 of the 15 are alive at age 65?
   d. What is the chance that at most 12 of the 15 are alive at age 65?
   e. Why is it legitimate to use the binomial probability distribution?
9. It is estimated that 40% of all U.S. females get pregnant at least once before reaching the age of 20. Suppose that we randomly select 16 U.S. females who are at least 20 years old. Find the probability that the number of women in the sample who got pregnant at least once before reaching the age of 20 will be
   a. at most eight.
   b. at most four.
   c. between five and eight, inclusive.
10. In the casino game chuck-a-luck, three fair, six-sided dice are rolled. If you bet a dollar on fives, then you lose your dollar if no fives are rolled. If at least one five is rolled, you get your dollar back plus one dollar for each five shown by the three dice. If X denotes your net gain on a one-dollar bet on fives, then X can equal -1, 1, 2, or 3.
    a. Compute the corresponding probabilities, $p(-1)$, $p(1)$, $p(2)$, and $p(3)$. *Hint:* What is the probability that you get no fives? One five? Two fives? Three fives?
    b. Compute your expected net gain μ_X using the distribution-based formula ($\mu_X = \sum xp(x)$) in Chapter 5.
11. A labor dispute has arisen concerning the allegedly discriminatory way that 20 laborers at a construction site were given their job assignments. Six of the 20 job assignments were considered highly undesirable, whereas the remaining 14 jobs were considered desirable. The dispute was triggered by the fact that all four African American laborers working on the site were given undesirable job assignments. The grievance states that "if the jobs were assigned without regard to race or color, then it is highly unlikely that all four African American workers would have received undesirable assignments."

    Imagine that the 20 jobs were randomly assigned to the 20 workers. Let X denote the number of African-American workers who were assigned undesirable jobs.
    a. What is the distribution of X?
    b. Compute $P(X = 4)$.
    c. Do we have reason to doubt that the jobs were randomly assigned? Explain.

*6.2 THE GEOMETRIC DISTRIBUTION

If you buy a Pick 3 lottery ticket and bet on one of the numbers from 000 to 999, your chance of winning is $p = 1/1000$. Imagine that you buy a lottery ticket five days a week, every week of the year, year in and year out, until you win a bet. How many Pick 3 lottery tickets do you have to buy to win once? We can list the possible outcomes of this chance experiment:

$$S = \{W, LW, LLW, LLLW, LLLLW,\ldots\}.$$

Here LW means that you lose your first bet, but win the second bet; LLLW means that you lose your first three bets, but win the fourth bet, and so forth. Let X denote the number of bets you have to make to win once. Since the daily lottery drawings are **independent**, we can use the product rule for independent events to get

$$P(X = 1) = P(W) = P(\text{win first bet}) = p.$$

$$P(X = 2) = P(LW) = P(\text{lose first bet}) \times P(\text{win second bet}) = (1-p)p.$$

$$P(X = 3) = P(LLW)$$
$$= P(\text{lose first bet}) \times P(\text{lose second bet}) \times P(\text{win third bet})$$
$$= (1-p)(1-p)p.$$

Generalizing this, we see that X has the probability density function (pdf)

$$p(x) = P(X = x) = P(\text{you lose first } x - 1 \text{ bets and win bet number } x) = (1-p)^{x-1}p,$$

for $x = 1, 2, 3, 4,\ldots$.

We say that X is **geometrically distributed** with success-probability $p = 1/1000$. Our analysis did not depend on the specific value of the success-probability p in the lottery. We can introduce the geometric distribution with parameter p as follows:

> Consider a sequence of independent success-failure trials with the same success-probability p in each trial. Let X denote the number of trials it takes to observe the first success. Then X has a **geometric distribution** with pdf:
>
> $$p(x) = P(X = x) = (1-p)^{x-1}p \quad \text{for } x = 1, 2, 3,\ldots.$$
>
> Furthermore, the mean and standard deviation of X are
>
> $$\mu_X = 1/p \quad \text{and} \quad \sigma_X = \sqrt{(1-p)/p^2}.$$

Note using the product rule for independent events that for $x = 1, 2, 3,\ldots$

$$P(X > x) = P(\text{first } x \text{ trials result in failure})$$
$$= P[(\text{first a failure}) \text{ and } (\text{second a failure}) \text{ and } \cdots (x\text{th trial a failure})]$$
$$= (1-p)^x$$

for $x = 1, 2,\ldots$.

As Figures 6.4 and 6.5 show, geometric distributions are very skewed (and not bell-shaped). For that reason we cannot use the 68-95-99.7% rule to interpret the value of the mean and the standard deviation.

Example 6.9

Toss a pair of casino dice and add up the scores on the two dice. How many times do we have to toss the pair to get a sum of 7?

Section *6.2 The Geometric Distribution   261

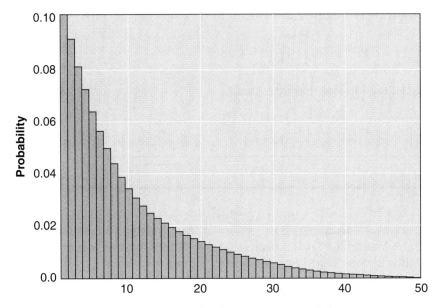

Figure 6.4 Geometric probability histogram, $p = 0.1$.

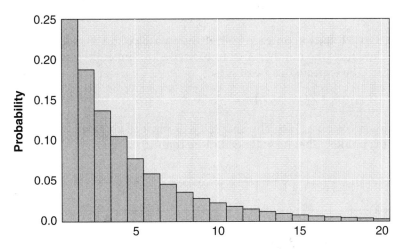

Figure 6.5 Geometric probability histogram, $p = 0.25$.

Solution: As Figure 4.3 in Section 4.1 shows, when we toss a pair of fair dice, the chance of getting a sum of 7 is $p = 6/36 = 1/6$. The number X of tosses needed to get a sum of 7 is geometric with parameter $p = 1/6$. The pdf of X is

$$p(x) = \frac{1}{6}\left(\frac{5}{6}\right)^{x-1}, \quad \text{for } x = 1, 2, 3, \ldots$$

The mean and standard deviation of X are $\mu_X = 6$ and $\sigma_X = \sqrt{30} \approx 5.5$. To get a sum of 7 we will most likely need around 6 tosses, give or take 5.5 tosses or so. It is unlikely that we will need exactly 6 tosses. In fact, the chance that we need exactly 6 tosses is only

$$p(6) = \frac{1}{6}\left(\frac{5}{6}\right)^5 = 0.0670.$$

It is quite possible that we will need more than 10 tosses to get a sum of 7:

$$P(X > 10) = \left(\frac{5}{6}\right)^{10} = 0.1615.$$

But it is unlikely that we will need more than 20 tosses:

$$P(X > 20) = \left(\frac{5}{6}\right)^{20} = 0.0261.$$

Example 6.10

Recall the game of craps, discussed in Example 4.3 in Section 4.1. A player (called the shooter) rolls a pair of fair, six-sided dice. A first roll of 7 or 11 means that the shooter wins, whereas 2, 3, or 12 on the first roll means that the shooter loses. Anything else becomes the shooter's point, and in that case the shooter continues to roll the dice until either a 7 turns up, in which case the shooter loses, or the point turns up again and the shooter wins (this is called making the point). For example, if the first roll is a 5, then the shooter continues to roll until he or she either makes another 5 and wins or makes a 7 and loses. Assume that 5 has just been established as the shooter's point. How many more rolls will the shooter have to make to finish the game?

Solution: As Figure 4.3 in Section 4.1 shows, on each roll of the two dice the chance of either making a 5 or 7 is $p = 10/36$. Therefore, the number X of rolls needed to finish the game is geometric with parameter $p = 10/36$. The expected value of X is

$$\mu_X = 1/p = 36/10 = 3.6$$

and the standard deviation is $\sigma_X = \sqrt{9.36} = 3.06$. The shooter will most likely need around 3.6 rolls, give or take 3 rolls or so. It is unlikely that he will need more than 10 rolls:

$$P(X > 10) = \left(1 - \frac{10}{36}\right)^{10} = \left(\frac{26}{36}\right)^{10} = 0.0386.$$

Summary

Let X denote the number of trials it takes for an event of interest to occur in a series of trials. Assume the following:

- The trials are independent.
- In each trial the chance that the event of interest occurs equals the same probability p.

Then X has the **geometric probability density function**

$$p(x) = P(X = x) = p(1-p)^{x-1}, \quad \text{for } x = 1, 2, 3, \ldots$$

Section 6.2 Exercises

1. How many days until the next day of measurable rain? Should this be modeled by a geometric probability distribution? Explain.

2. A deck of cards is dealt one card at a time. Each card is either an ace or not an ace. We are interested in the number of cards until the first ace. Should this situation be modeled by a geometric probability distribution? Explain.

3. A couple is determined to have children until they have a girl. We are interested in the number of children they will have. Should this situation be modeled by a geometric probability distribution? Explain.

4. To reduce population growth, China implemented a "one-child" per family policy that has been extremely unpopular in parts of the country. Some rural communities have suggested changing the policy to a "one-son" policy. This would mean that families could have children until they had a son. Suppose that the probability of having a boy is 0.51 (boys are slightly more likely than girls) and that the sex of a child is independent from birth to birth.
   a. Describe a five-step simulation that would give an experimental expected value for the number of children in families that follow the one-son policy. Assume that families have children until they have a boy, and there are no twins, triplets, and so forth.
   b. Describe a five-step simulation that would give a table for the experimental probabilities for the possible values (see Example 4.15). (From Clifford Korda, *The Mathematics Teacher* 87, 1994, pp. 232–235.)

5. a. Write the mathematical function for the theoretical pdf for the number of children in a family that follows the one-son policy mentioned in Exercise 4. Assume that families have children until they have a boy, and there are no twins, triplets, and so forth.
   b. What is the theoretical expected number of children in such a family? What is the theoretical standard deviation?
   c. Make a pdf table of the probabilities for 1, 2, 3, 4, and "5 or more" children.
   d. Make a graph of the histogram of the theoretical pdf (the probability histogram).

6. In a game of craps, bystanders can also place bets at almost any point of the game. They can bet on the shooter or against the shooter. Many people prefer to wait to see the result of the shooter's first shot before placing a bet, because some points (first rolls of 4, 5, 6, 8, 9, or 10) are easier to make than others.
   a. Suppose the shooter's first roll is a 4. What is the probability of shooting a 4 on any given roll? (*Hint:* There are 36 possible outcomes. How many of these are equal to 4?)
   b. Theoretically, how many rolls should we expect before the shooter rolls another 4? (Do not count the first roll.) Remember that earlier we figured out that it typically takes six rolls before the first 7.
   c. Suppose the shooter's first roll is a 6. Theoretically, how many rolls should we expect before the shooter rolls another 6? Which is the better situation to be in: first roll a 4 or first roll a 6?
   d. Graph the theoretical pdf's for the number of rolls until the first 4, the number of rolls until the first 6, and the number of rolls until the first 7, side by side.

7. Describe a five-step simulation to determine experimentally the expected number of rolls of two dice before the first 7.

8. Suppose you are tossing a fair coin until the first heads. You have tossed it twice and got tails both times. How many more times should you expect to toss the coin until a heads appears? Explain.

6.3 THE POISSON DISTRIBUTION

The Poisson distribution provides easy-to-compute approximations for hard-to-compute binomial probabilities if n is large and p is small. Binomial probabilities are easy to compute or look up in a binomial table if n is small. But when n is large, it is time-consuming to evaluate binomial probabilities, even with a calculator. Imagine the task facing scientists 200 years ago when they had to evaluate binomial probabilities!

For example, consider the problem of computing the probability that out of 2500 randomly chosen newborn babies, at least half are girls. The number X of girls among the 2500 newborns is binomial with $n = 2500$ and $p = 0.487$. (See Example 4.1, Section 4.1.) Then

$$p(x) = \binom{2500}{x}(0.487)^x((0.513)^{2500-x}, \quad x = 0, (1), \ldots, 2500.)$$

The probability that at least half of the children are girls is

$$P(X \geq 1250) = p(1250) + p(1251) + p(1252) + p(1253) + \cdots + p(2500).$$

It would be very difficult to compute this probability on a hand-held calculator. Today's computers can estimate this probability using five-step simulation. (It is close to 0.1.) A theoretical way of estimating binomial probabilities for large n was worked out early in the eighteenth century by the French Protestant Abraham de Moivre, who had fled to England to avoid religious persecution. De Moivre's approximation, which works for large n and p around $1/2$, will be presented in Section 6.4. In this section we shall use an approximation found in the early nineteenth century by the French mathematician and physicist Siméon Poisson. Poisson's approximation only works for large n and small p.

The Poisson Approximation to the Binomial

Let X be binomial with parameters $n \geq 25$ and $p \leq 0.1$. If $np \leq 10$ (i.e., the larger n is, the smaller p must be!), we can use the Poisson method to approximate binomial probabilities; that is, in terms of pdf's,

$$p(x) = P(X = x) = \binom{n}{x}p^x(1-p)^{n-x} \approx \frac{\lambda^x}{x!}e^{-\lambda}, \text{ for } x = 0, (1), \ldots, n$$

where $\lambda = \mu_X = np$. (Here "$\approx$" means "is approximately equal to.") The mean of the approximating Poisson pdf is $\mu_X = \lambda$, and the standard deviation is $\sigma_X = \sqrt{\lambda}$.

Example 6.11

Hospital patients face a 1 in 20 risk of acquiring an infection in the hospital. Suppose we randomly select the medical records of 30 former hospital patients and determine the number X of these patients who got infected during their hospital stay. X is approximately binomial with $n = 30$ and $p = 1/20$. The criteria ($p \leq 0.1$, $n \geq 25$, $np \leq 10$) are met for the Poisson approximation to be accurate, with $\lambda = np = 1.5$. In Table 6.2 we present the corresponding binomial and Poisson probabilities. In this case it is not difficult to compute the binomial probabilities. Notice that even though n is only 30, the approximation is quite good (< 0.01 error in each row).

Example 6.12

According to the Canadian Immunization Guide, about 1 in 24,000 people develop transient thrombocytopenia from a measles shot. This medical condition involves a temporary suppression of the production of clot-forming blood platelets, leading to increased bruising and bleeding. If 36,000 children are given measles shots, what is the chance that at most four of the vaccinated

Table 6.2 Binomial Probabilities with $n = 30$ and $p = 1/20$ and Corresponding Poisson Probabilities with $\lambda = np = 1.5$

x	$\binom{n}{x}p^x(1-p)^{n-x}$	$\dfrac{\lambda^x}{x!}e^{-\lambda}$
0	0.215	0.223
1	0.339	0.335
2	0.259	0.251
3	0.127	0.126
4	0.045	0.047
5	0.012	0.014
6	0.003	0.004
7	0.000	0.001

children will develop transient thrombocytopenia? What is the probability that at least two of the children develop this condition?

Solution: If X denotes the number of vaccinated children who develop this side effect, then X is binomial with $n = 36{,}000$ and $p = 1/24{,}000$. The corresponding Poisson distribution has $\lambda = np = 1.5$. In Table 6.3 we present the corresponding binomial and Poisson probabilities. Notice that with such a large value of n and small p, the difference between the binomial probabilities and the corresponding Poisson probabilities is very small (< 0.0001).

The approximating probability that at most four of the vaccinated children suffer side effects is

$$P(X \leq 4) = p(0) + p(1) + p(2) + p(3) + p(4) \approx 0.981424,$$

whereas the probability that at least two children suffer side effects is

$$P(X \geq 2) = 1 - P(X \leq 1) = 1 - (p(0) + p(1)) \approx 0.442175.$$

The approximating probabilities $p(0), p(1), p(2), \ldots$, are listed in the right-hand column of Table 6.3. But the quickest way to determine $P(X \leq 4)$ and $P(X \geq 2)$ is to use the Poisson tables (see Appendix J), a part of which is displayed in Table 6.4.

Find the number in the row with $x = 4$ in the column under $\lambda = 1.5$ to get $P(X \leq 4) = 0.981$. To determine $P(X \geq 2) = 1 - P(X \leq 1)$, find the number in the row with $x = 1$ in the column under

Table 6.3 Binomial Probabilities with $n = 36{,}000$ and $p = 1/24{,}000$ and Corresponding Poisson Probabilities with $\lambda = np = 1.5$

x	$\binom{n}{x}p^x(1-p)^{n-x}$	$\dfrac{\lambda^x}{x!}e^{-\lambda}$
0	0.223123	0.223130
1	0.334699	0.334695
2	0.251028	0.251021
3	0.125512	0.125511
4	0.047065	0.047067
5	0.014119	0.014120
6	0.003529	0.003530
7	0.000756	0.000756
8	0.000142	0.000142

Table 6.4 Cumulative Poisson Probabilities. The Table Gives the Probability of x or Fewer Events When the Expected Number of Such Events is λ

	\multicolumn{10}{c}{λ = E(X)}
x	1.1	1.2	1.3	1.4	1.5	1.6	1.7	1.8	1.9	2.0
0	0.333	0.301	0.273	0.247	0.223	0.202	0.183	0.165	0.150	0.135
1	0.699	0.663	0.627	0.592	0.558	0.525	0.493	0.463	0.434	0.406
2	0.900	0.879	0.857	0.833	0.809	0.783	0.757	0.731	0.704	0.677
3	0.974	0.966	0.957	0.946	0.934	0.921	0.907	0.891	0.875	0.857
4	0.995	0.992	0.989	0.986	0.981	0.976	0.970	0.964	0.956	0.947
5	0.999	0.998	0.998	0.997	0.996	0.994	0.992	0.990	0.987	0.983
6	1.000	1.000	1.000	0.999	0.999	0.999	0.998	0.997	0.997	0.995
7	1.000	1.000	1.000	1.000	1.000	1.000	1.000	0.999	0.999	0.999
8	1.000	1.000	1.000	1.000	1.000	1.000	1.000	1.000	1.000	1.000

$\lambda = 1.5$ to get 0.558 and, hence, $P(X \geq 2) = 1 - 0.588 = 0.442$. As you can see, both approximations will be extremely accurate.

Example 6.13

In the United States the annual death rate due to leukemia is 7.5 per 100,000 population. What is the expected number of leukemia deaths in a simple random sample of 10,000 Americans over a 4-year period? Determine the probability that a simple random sample of 10,000 Americans over a 4-year period has at most five leukemia deaths.

Solution: Over a 4-year period the death rate due to leukemia is $4(7.5) = 30$ per 100,000 population. Let X denote the number of leukemia deaths in a simple random sample of 10,000 Americans over a 4-year period. X is binomial with $n = 10,000$ and $p = 30/100,000$—which is approximately Poisson with $\lambda = np = 3$. The expected number of deaths due to leukemia is $\mu_X = \lambda = np = 3$. According to the statistical tables,

$$P(X \leq 5) = 0.916.$$

Poisson distributions are also used to model the number of times an event of interest occurs during a predetermined amount of time or area or volume. Typical examples include

- the number of α-particles emitted by a tiny radio-active source during a specified time interval;
- the number of traffic accidents at a busy intersection over a year;
- the number of an insurance company's policy holders who die on a given day;
- the number of patients who arrive at a hospital emergency room between 7 P.M. and 8 P.M. on a Tuesday evening;
- the number of telephone calls received at a switchboard between 9 A.M. and 10:30 A.M;
- the number of flaws in a square meter of cloth;
- the number of nematodes in a cubic centimeter of earth.

That these phenomena follow a Poisson model is based on **Siméon Poisson's Limit Theorem:** For fixed $\lambda > 0$ and $x = 0, 1, 2, 3, \ldots$

$$\binom{n}{x}\left(\frac{\lambda}{n}\right)^x\left(1 - \frac{\lambda}{n}\right)^{n-x} \longrightarrow \frac{\lambda^x}{x!}e^{-\lambda}$$

as n gets larger. (Note that the left-hand side is a binomial probability with $p = \lambda/n$ and the right-hand side is a Poisson probability. Because λ is fixed, p gets small as n gets large.

Example 6.14

One milligram of Uranium 238, containing around 2.5×10^{18} atoms, undergoes spontaneous radioactive decay at an average rate of 13 disintegrations per second. Each disintegration results in the emission of an α-particle which can be detected by a Geiger counter. Let X denote the number of α-particles emitted during a specified one-second time period. Thus $E(X) = 13$. What is the disintegration rate of X?

Solution: Break the one-second time period into one million consecutive microsecond periods (a very large n). If an emission is recorded during a microsecond period, say that a success has occurred. The average number of emissions per microsecond is obviously $13/1,000,000$. Make the following assumptions (justified by the physical reality):

- The number of emissions during nonoverlapping time intervals are independent.
- During any specified microsecond time period the probability of a single emission is the same; call it p. (That is, the probability of a single emission is stationary over time.)
- It is highly unlikely that there will be more than one emission during any one-microsecond time period.

Under these assumptions X is essentially the number of successes in $n = 1,000,000$ independent success-failure trials with a very small success-probability p in each trial. Thus X is binomial with $n = 1,000,000$; and $13 = E(X) = np$. Hence, $p = 13/1,000,000$. But $p = \lambda/n$ and hence $\lambda = 13$. It follows that X is (almost) Poisson with parameter $\lambda = n \cdot p = 13$.

We have learned the following conditions for when a **Poisson experiment** occurs.

> **Conditions of a Poisson Experiment.** The number of occurrences of an event during a preset amount of time will have a Poisson distribution if the following real-world conditions are judged to be satisfied:
>
> - The probability of an occurrence within a small time interval of fixed length is the same regardless of when the time interval begins. That is, the rate of occurrence is constant over time.
> - An occurrence within a given interval is independent of whether there is an occurrence in another non overlapping interval (even if the intervals are close to each other). That is, past occurrences do not influence the likelihood of future occurrences.
> - It is much less likely that there will be two or more occurrences in a small interval than that there will be one occurrence (the real-world ideal is that simultaneous occurrences are impossible).

Example 6.15

Imagine standing on a bridge over a rural interstate and counting the number of cars, X, that pass under it in an hour. Compared with a one-hour interval, a one-second interval can be considered small. Condition 1 means that the chance that a car passes under the bridge during any one-second period is the same for every one-second period in the hour. This condition would be violated if, for example, half-way through the hour an athletic event at a nearby high school ended.

Condition 2 means that a car passing during a particular second has no effect on the chance that one passes during the next (or any other) second. That would not be the case if you are counting regularly scheduled buses stopping at a bus stop. If a bus has just passed, then for the next several minutes it would be very unlikely that another one would come.

Condition 3 is valid for a small time interval because the cars are not attached to each other and hence multiple occurrences are impossible. If instead of counting cars, you counted the number of people in cars, condition 3 would not hold because there could be several people in one car. Since we are confident that the assumptions of a Poisson Experiment hold, we can use the Poisson probability distribution.

Let us now consider the Key Problem involving the Prussian calvary officer.

Example 6.16

Assume that the 14 Prussian calvary corps had approximately the same number of men and horses, and that the risk of someone being kicked to death was the same each day of the 20 years for each of the 14 calvary corps. In other words, assume that condition 1 is satisfied with the small time interval being a single day. Conditions 2 and 3 also follow if we assume that one soldier being kicked to death is independent of another soldier being kicked to death. The number X of fatalities due to horse kicks in a given calvary corps in a given year is therefore approximately Poisson. We can consider the data in the Key Problem as 280 independent observations of X. The theoretical mean of X is $\lambda = \mu_X$, which is estimated by the sample mean $\bar{x}$. There were a total of $(1 \times 91) + (2 \times 32) + (3 \times 11) + (4 \times 2) = 196$ fatalities in 14 calvary corps over 20 years ($14 \times 20 = 280$ units, recalling that each unit is counted in each of the 20 years). Therefore, $\bar{x} = 196/280 = 0.7$ per unit per year.

In Table 6.5, we have listed the Poisson probabilities corresponding to $\lambda = 0.7$. We have also computed the expected number of units with x fatalities by multiplying each Poisson probability by the number of units, 280.

The Poisson model fit the data reasonably well, as illustrated by Figure 6.6.

It is tempting to use the Poisson distribution any time the description of the problem sounds superficially like the horse kick example. But consider the number of runs the Oakland Athletics baseball team scored in each of the 144 games they played in 1995. "How many runs were scored

Table 6.5 Prussian Soldier Data

Number of soldiers x kicked to death	Number of units	Experimental probability $P(X = x)$	$P(X = x) = \dfrac{(0.7)^x}{x!} e^{-0.7}$	Expected number of units
0	144	0.514	0.497	139.0
1	91	0.325	0.348	97.3
2	32	0.114	0.122	34.1
3	11	0.039	0.028	7.9
4	2	0.007	0.005	1.4
5 or more	0	0.000	0.001	0.2
Total	280	0.999	1.001	279.9

Note: $E(X) = nP(X = x)$.

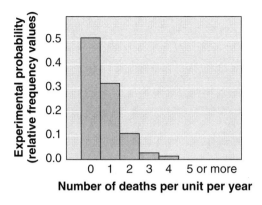 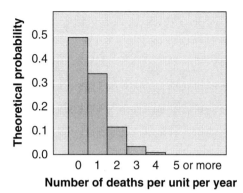

Figure 6.6 Prussian soldier data density histogram (top) and theoretical Poisson probability histogram for $\lambda = 0.7$ (bottom).

per game?" sounds like "How many soldiers were kicked to death per unit per year?" Because the Poisson distribution worked so well in the latter case, one might be inclined to assume that it applies to the former. The question is whether the number of runs per game is reasonably modeled as Poisson.

The time interval here is one game. The average number of runs per game scored by the Athletics was 5.07; let this be our estimate of λ. In Table 6.6 we have, for each possible number of runs, the number of games in which the Athletics scored that number of runs as well as the observed proportion of games (that is, the experimental probability of that number of runs per game) and the theoretical Poisson probability of that number of runs assuming a theoretical mean of 5.07. Figure 6.7 has the corresponding density histogram constructed from these data and the theoretical Poisson probability histogram for $\lambda = 5.07$. They really do not look close! Ignoring the random upward and downward fluctuations of the experimental probability (as one should), the two shapes are still quite different.

Table 6.6 Number of Runs Scored per Game by Oakland Athletics in 1995

Number of runs x	Number of games	Proportion of games	Theoretical probability $P(X = x)$ when $\lambda = 5.07$
0	8	0.0556	0.0063
1	14	0.0972	0.0319
2	14	0.0972	0.0808
3	14	0.0972	0.1365
4	18	0.1250	0.1730
5	11	0.0764	0.1754
6	20	0.1389	0.1482
7	13	0.0903	0.1073
8	14	0.0972	0.0680
9	5	0.0347	0.0383
10	2	0.0139	0.0194
11	6	0.0417	0.0089
12	1	0.0069	0.0038
13	3	0.0208	0.0015
14	1	0.0069	0.0005
15 and over	0	0.0000	0.0003

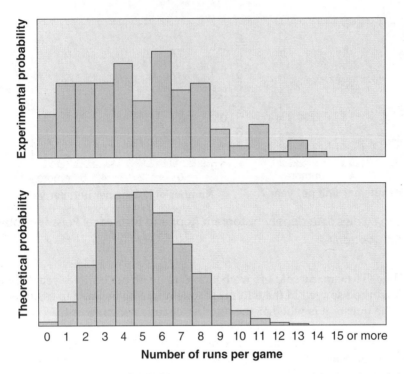

Figure 6.7 Oakland Athletics data density histogram (top) and theoretical Poisson probability histogram for $\lambda = 5.07$ (bottom).

The last two columns in Table 6.6 show some fairly large discrepancies. For example, the Athletics were shut out (had 0 runs) eight times, or about 5.6% (relative frequency = 0.056) of the time. The Poisson distribution predicts shutouts only about 0.6% (0.006) of the time. The Athletics scored four runs 12.5% of the time and five runs 7.6% of the time, while the Poisson distribution yields probabilities of over 0.17 for four runs and also over 0.17 for five runs. What other discrepancies are there? You should try to explain how this situation might violate some of the three conditions for the Poisson distribution, and explain why the Poisson therefore does not model these data well.

Summary

The pdf for a Poisson random variable X with mean $\lambda > 0$ is

$$p(x) = P(X = x) = \frac{\lambda^x}{x!} e^{-\lambda}, \quad \text{for } x = 0, 1, 2,\ldots$$

The mean of X is $\mu_X = \lambda$, and the standard deviation of X is $\sigma_X = \sqrt{\lambda}$.

Poisson distributions are often used to model the number of times an event of interest occurs during a predetermined time interval. The number of occurrences will have a Poisson distribution if the following real-world conditions are judged to be satisfied: the rate of occurrences is constant over time; past occurrences do not influence the likelihood of future occurrences; and simultaneous occurrences are impossible.

Sometimes these three assumptions are loosely characterized as implying that the times of the occurrences are locally isolated and totally random. Poisson distributions are also used to model the number of misprints in a book, the number of bacterial colonies on a petri dish, the number of diseased trees on an acre of woodland, and the number of parts per million of some toxin found in the water or air emissions from a factory, and so forth.

Finally, the Poisson distributions provide easy-to-compute approximations for hard-to-compute binomial probabilities if n is large and p is small: If X is binomial with $n > 25$, $p \leq 0.1$, and $np \leq 10$, then

$$P(X = x) = \binom{n}{p} p^x (1-p)^{n-x} \approx \frac{\lambda^x}{x!} e^{-\lambda},$$

where $\lambda = \mu_X = np$.

Section 6.3 Exercises

1. According to the Statistical Abstract of the United States, the infant mortality rate of Australia is 5.0 deaths of children under 1 year of age per 1000 live births. Use the Poisson approximation to determine the probability that among 2000 live-born infants selected at random there will be
   a. at most 15 infant deaths.
   b. at least five infant deaths.
   c. between 5 and 15 infant deaths, inclusive.

2. You play the Pick 3 lottery game by betting on one of the numbers from 000 to 999. Each number has the probability $p = 1/1000$ of being the winning number. Each bet costs you $1. If you win you receive $500. (You do not get the purchase price of your winning lottery ticket back.) A man bets on 3 different numbers, 4 days a week, 50 weeks a year, for 2 years. On a given day his gain is either $497 or –$3.
   a. What is the chance that he wins on a given day?
   b. What is his expected net gain over the 2 years?
   c. Estimate the probability that he does not win a single bet over the 2 years.
   d. Estimate the probability that he makes money over the 2 years. (*Hint:* How many times does he have to win to make money?)

3. In America smallpox vaccination was discontinued 25 years ago because smallpox no longer posed a threat. If bioterrorists now unleash the smallpox virus on an American city, hundreds of thousands of people might die. We could restart the vaccination program, but vaccination is not completely risk free. According to the *CDC Morbidity and Mortality Weekly Report,* June 22, 2001, the risk of dying from a smallpox vaccination is one in a million. Imagine that 10 million individuals are vaccinated against smallpox. Determine the probability that among the vaccinated individuals
   a. at most 12 die from the vaccination.
   b. at least 10 die from the vaccination.
   c. between 4 and 17, inclusive, die from the vaccination.

4. In the United States the annual death rate due to diabetes is 24 per 100,000 population.
   a. What is the expected number of deaths due to diabetes over a 2-year period in a simple random sample of 10,000 Americans?

   Determine the probability that over a 2-year period a simple random sample of 10,000 Americans has
   b. no deaths due to diabetes.
   c. at most eight deaths due to diabetes.
   d. at least five deaths due to diabetes.

5. During the twentieth century an average of 1.6 hurricanes per year struck the U.S. mainland. Estimate the probability that in a given year
   a. no hurricane will strike the U.S. mainland.
   b. at most one hurricane will strike the U.S. mainland.
   c. at least two hurricanes will strike the U.S. mainland.
6. The world's most active volcano above sea level is Kilauea on Hawaii that, since the mid-eighties, has erupted an average of 3.5 times per year. (An eruption denotes vigorous activity either from a new event or from an established event after a slowdown.)
   a. What is the expected number of eruptions during the next 24 months?
   b. Estimate the probability that over the next 24 months Kilauea will have between 4 and 10 eruptions, inclusive.
7. A Nevada roulette wheel has 38 pockets; one is numbered 0, another is numbered 00, and the rest are numbered from 1 through 36. The croupier spins the wheel, and drops a ball inside. The ball is equally likely to land in any one of the 38 pockets. Before it lands, bets can be placed. If you bet a dollar on a single number at roulette, and the number comes up, you get the $1 back together with winnings of $35. If any other number comes up, you lose the dollar. Gamblers say that a single number pays 35 to 1. Suppose you play roulette 152 times, betting a dollar on the number 17 each time.
   a. How many times can you expect to win? (Compute the expected value.)
   b. What is the chance that you don't win any bets?
   c. How many times do you have to win to come out ahead (to win more money than you lose)?
   d. What is your chance of coming out ahead?
   e. Compute your expected gain (positive or negative).
8. Thirty-five out of every 10,000 births result in a pair of identical twins, on average. Determine the probability that out of 3000 births selected at random there will be
   a. at most five pairs of identical twins.
   b. between 5 and 16 pairs of identical twins, inclusive.
9. Go through the three conditions for a Poisson random variable and give possible reasons why in the example of the Oakland Athletics the data do not seem to follow the Poisson distribution. *Hint:* Are the number of runs scored by the Oakland A's during the last three innings of a game independent of the number of runs they scored during the first three innings? Explain.
10. Rebecca works as a greeter at a grocery store. She theorizes that during the 2:00 P.M. to 4:00 P.M. time period, customers arrive at a rate of two per minute. She is willing to make the three Poisson assumptions.
    She keeps track of the number of customers who arrive each morning during a 15-minute period. Her results are shown in the table below.

Minute	Number of customers	Minute	Number of customers
1	2	9	1
2	0	10	2
3	4	11	3
4	5	12	4
5	1	13	2
6	0	14	1
7	2	15	0
8	3		

   a. In how many 1-minute periods did no customer arrive? In how many 1-minute periods did one customer arrive? Two customers? Three customers? Four customers? Five customers? What is the average number of customers per minute?

b. Using the answers from part **a**, find the experimental probability that no customer arrives in a 1-minute period. Calculate such a probability for every number of customers up to five.

c. According to the Poisson distribution with a theoretical mean of 2 (our estimate of λ in part **a**, recall), the theoretical probabilities of the six outcomes are as follows:

Number of customers in a minute	Theoretical probability
0	0.135
1	0.271
2	0.271
3	0.180
4	0.090
5	0.036

Create two histograms, one showing the experimental probabilities, and the other showing the theoretical probabilities.

d. Compare the two histograms you created. Does this evidence seem to support Rebecca's hypothesis concerning the arrival of customers?

11. The following table shows experimental probabilities based on 1000 observations. The table also has the corresponding theoretical probabilities under the Poisson distribution with a mean of 3.4.

Number of occurrences x	Experimental probability $\hat{P}(X = x)$	Theoretical probability $P(X = x)$
0	0.053	0.033
1	0.153	0.113
2	0.138	0.193
3	0.103	0.219
4	0.225	0.186
5	0.214	0.126
6	0.068	0.072
7	0.047	0.035

a. Make a density histogram of the experimental probabilities, and a histogram of the theoretical probabilities.

b. Compare the two histograms. Do the obtained data seem to follow the Poisson distribution with mean $\lambda = 3.4$?

12. Since seismologists in 1935 started using the Richter scale to measure the magnitude of earthquakes, no earthquake has ever registered higher than 9. On average there is an earthquake of magnitude 7.5 or greater somewhere in the world four times every 5 years. Estimate the probability that in a given year there are

a. no earthquakes of magnitude 7.5 or greater.

b. at least two earthquakes of magnitude 7.5 or greater.

c. at most two earthquakes of magnitude 7.5 or greater.

6.4 THE NORMAL PROBABILITY DISTRIBUTION

In Sections 2.3 and 2.4 we looked at different data sets whose histogram followed a normal (bell-shaped) curve. Examples included the batting averages of Major League Baseball players, IQ scores, and the heights of men and women, considered separately. Many other variables are also approximately normally distributed. If we consider men and women separately, cholesterol level, systolic blood pressure, reaction time, and lung capacity are all approximately normally distributed. Weight distributions are skewed to the right.

Suppose we have the heights of a random sample of 100 women. If we group the data into class intervals 2 inches wide, we would get the density histogram in Figure 6.8(a). If we had a sample of 1,000 women, and if we group their heights into class intervals of 1 inch, we would get the density histogram in Figure 6.8(b). (See, for example, Pearson and Lee's histogram in Figure 1.9 for the heights of 1,052 women, recorded to the nearest quarter of an inch.) If we had a sample of 50,000 women and we grouped their heights into class intervals of 1/2 inch, then we would get the density histogram in Figure 6.8(c). Notice that as n gets large and the class width gets small, the density histogram increasingly approximates a normal curve. Finally, if it were possible to measure the exact height (with no round-off error) of the over 100 million adult women in the United States, then we would get closer and closer to the limiting normal density histogram in Figure 6.8(d) as the class width approaches 0.

Let μ denote the mean height of the over 100 million women in the United States and σ the standard deviation of the heights. The normal curve in Figure 6.8(d) is scaled so that the total area under the curve equals 1 is given by the equation

$$f(x) = \frac{1}{\sigma\sqrt{2\pi}} e^{-(x-\mu)^2/(2\sigma^2)},$$

where $e \approx 2.71828$ and $\pi \approx 3.14159$. This **normal probability density curve (or normal curve)**, with mean μ and standard deviation σ, is displayed in Figure 6.9.

The curve has the following properties:

- It is bell-shaped and symmetric around the mean μ.
- The curve extends indefinitely in both directions, always above the horizontal axis, approaching, but never touching the axis.

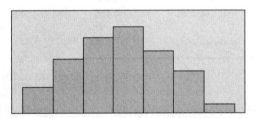

(a) Random sample of 100 women

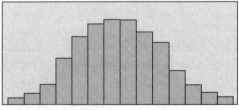

(b) Sample size increased and class width decreased

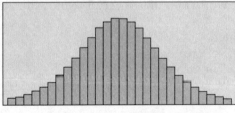

(c) Sample size increased and class width decreased further

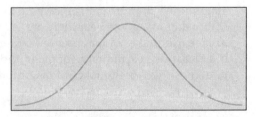

(d) Normal distribution for the population

Figure 6.8 Histograms for the heights of women.

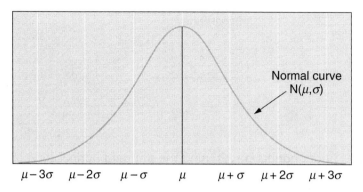

Figure 6.9 The normal probability density curve with mean μ and standard deviation σ.

- The total area under the curve is 1.
- Almost all of the area (in fact, 99.7% of the area) under the curve is between $\mu - 3\sigma$ and $\mu + 3\sigma$.

The standard normal curve (discussed in Section 2.4) is a special curve with mean $\mu = 0$ and standard deviation $\sigma = 1$.

Recall that for density histograms, which have total area 1, that the **area of the bars** over a given interval equals the **proportion of observations** in that interval. If we apply this to the probability density curve in Figure 6.8(d), which we view as the histogram for the exact heights of the more than 100 million U.S. women, we conclude that the proportion of women who have heights between two given limits, a and b, equals the area from a to b under the probability density curve, as shown in Figure 6.10.

As we explained in Section 2.4, we can determine areas under any normal curve in two steps First, we convert the interval endpoints into z-scores. Then, we find the area over the z-score interval under the standard normal curve. This method is illustrated by Figure 6.11.

We learned how to find areas under the standard normal curve in Section 2.4. Figure 6.12 displays how the 68-95-99.7% rule for the normal curve with mean μ and standard deviation σ follows from the 68-95-99.7% rule for the standard normal curve.

Now let X denote the exact height (measured without round-off error) of a randomly selected woman. X is said to be **normally distributed with mean μ and standard deviation σ**. The probability that the woman's height is between a and b is

$P(a < X < b)$ = proportion of women in the population with height between a and b

= area from a to b under the normal curve with mean μ and standard deviation σ

= area from $(a - \mu)/\sigma$ to $(b - \mu)/\sigma$ under the standard normal curve.

We learned to find areas under the standard normal curve in Section 2.4.

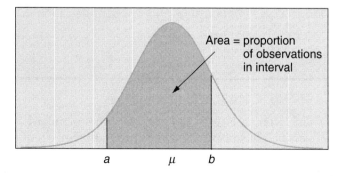

Figure 6.10 The proportion of observations in an interval.

276 PROBABILITY DISTRIBUTIONS: THE ESSENTIALS

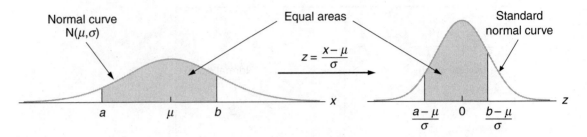

Figure 6.11 Finding areas under normal curves using the standard normal curve.

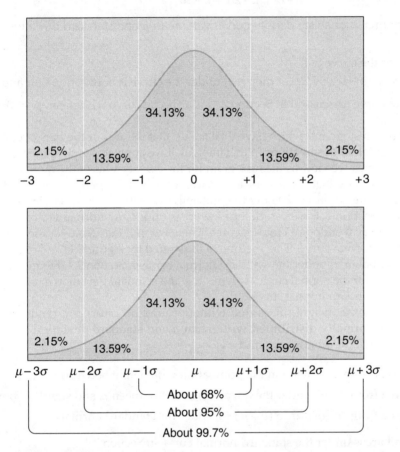

Figure 6.12 Areas under the standard normal curve derived from the normal table and the corresponding areas under the normal curve with mean μ and standard deviation σ.

Example 6.17

Assume that the mean height of U.S. women is $\mu = 63.7$ inches with a standard deviation of $\sigma = 2.6$ inches. Let X denote the exact height (measured without round-off error) of a randomly selected woman. Find

$$P(60 < X < 66) = \text{proportion of women with height between 60 inches and 66 inches.}$$

Solution: We shall proceed in three steps. (Learn these steps for the many similar problems to be solved.)

Step 1. Draw a number line and shade the interval of interest.

```
        60                              66
```

Step 2. Convert the interval endpoints into z-scores.

$$(60 - \mu)/\sigma = (60 - 63.7)/2.6 = -1.42$$
$$(66 - \mu)/\sigma = (66 - 63.7)/2.6 = 0.88$$

```
       -1.42                            0.88
```

Step 3. Find the area under the standard normal curve over the z-score interval.

$$P(60 < X < 66) = 0.7328$$

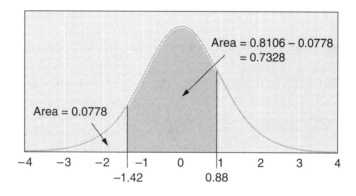

Normally Distributed Random Variables

Consider a **continuous random variable** X determined by the outcome of some random phenomenon. Recall that a continuous random variable is one for which all its probabilities $P(a \leq X \leq b)$ are found by finding the area under its probability density function from a to b. If the density histogram of tens of thousands of independent observations of X follows the normal probability density curve depicted in Figure 6.9, then we say that X is **normally distributed with mean** μ **and standard deviation** σ. We shall use the shorthand notation $N(\mu, \sigma)$ to denote the normal distribution with mean μ and standard deviation σ. We shall write: X is $N(\mu, \sigma)$.

Important Fact: If a random variable X is $N(\mu, \sigma)$, then the standardized variable

$$Z = \frac{X - \mu}{\sigma}$$

has the standard normal distribution $N(0,1)$. It is this fact that allows us to determine any normal probability from areas under the standard normal density curve.

Example 6.18

According to the National Center for Health Statistics, the systolic blood pressure of adult women is (approximately) normally distributed with mean $\mu = 123$ mmHg and standard deviation $\sigma = 21$ mmHg. If we randomly select an adult woman and measure her systolic blood pressure X, find

$$P(X > 150) = \text{proportion of adult women with systolic blood pressure over 150.}$$

Solution: The random variable X is (approximately) $N(123,21)$. We shall proceed in three steps.

Step 1. Draw a number line and shade the interval of interest.

150

Step 2. Convert the interval endpoint into z-score.

$$(150 - \mu)/\sigma = (150 - 123)/21 = 1.29$$

1.29

Step 3. Find the area under the standard normal curve over the z-score interval.

$$P(X > 150) = 1 - P(X \leq 150) = 1 - P(Z \leq 1.29) = 1 - 0.9015 = 0.0985$$

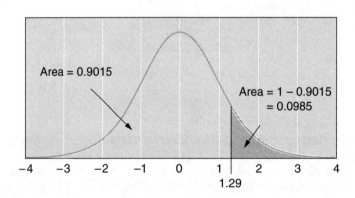

The Normal Approximation to Binomial Distributions

We introduced the binomial distribution in Section 6.1. A binomial random variable X (see Section 6.1) with parameters n and p can be thought of as the number of successes in n independent success-failure trials with success-probability p in each trial. The pdf of X is

$$p(x) = P(X = x) = \binom{n}{x} p^x (1-p)^{n-x},$$

for $x = 0, 1, \ldots, n$. As we discussed at the beginning of Section 6.3, this formula becomes awkward to use as the number of trials n gets very large. In Section 6.3 we presented Poisson's approximation

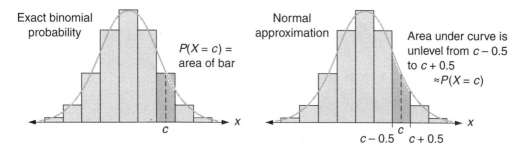

Figure 6.13 The normal approximation to a binomial probability.

to binomial probabilities, an approximation that works well if p gets small as n gets large. Here we shall present the normal approximation to binomial probabilities, discovered 300 years ago by Isaac Newton's contemporary, Abraham de Moivre. His approximation works for any fixed value p as n gets large.

Recall that the mean and standard deviation of a binomial variable are

$$\mu = np \text{ and } \sigma = \sqrt{np(1-p)}, \text{ respectively.}$$

As indicated in Figure 6.13, the basic idea of the normal approximation is that we can replace the probability histogram of a binomial variable X with the normal density curve that has the same mean and standard deviation as X.

The bars of a binomial probability histogram are centered at the values $0, 1, \ldots, n$ and each bar has a width of one unit. The bar centered at the integer c covers the interval from $c - 0.5$ to $c + 0.5$ and has an area equal to $P(X = c)$. If the corresponding continuous normal curve follows the histogram closely, then the area of the bar centered at the integer c must be close to the area under the normal curve from $c - 0.5$ to $c + 0.5$. We can therefore estimate $P(X = c)$ by the area under the corresponding normal curve from $c - 0.5$ to $c + 0.5$. The same continuity connection applies to binomial probability of the form $P(a \leq X \leq b)$ where a and b are integers: $P(a \leq X \leq b)$ = Area under normal curve with an interval from $a - 1/2$ to $b + 1/2$.

Example 6.19

Let X denote the number of heads in 100 tosses of a fair coin. Estimate $P(45 \leq X \leq 55)$.

Solution: X is binomial with $n = 100$ and $p = 1/2$. X has mean $\mu = np = 50$ and standard deviation $\sigma = \sqrt{np(1-p)} = \sqrt{25} = 5$. As we can see in Figure 6.14, the normal curve $N(50,5)$ follows the probability histogram of X very closely.

The probability $P(45 \leq X \leq 55)$ equals the combined area of the 11 histogram bars centered over the integer values 45 through 55 in Figure 6.14 and Figure 6.15. These 11 bars cover the interval from 44.5 to 55.5. We can therefore estimate $P(45 \leq X \leq 55)$ by the area under the normal curve $N(50,5)$ from 44.5 to 55.5.

To determine the area under the normal curve $N(50,5)$ from 44.5 to 55.5 we proceed in three steps.

Step 1. Draw a number line and shade the interval of interest.

```
    44.5                              55.5
```

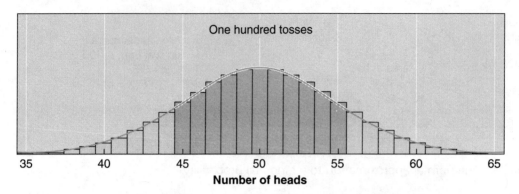

Figure 6.14 The probability histogram for the number of heads in 100 tosses, with the normal curve $N(50,5)$ superimposed.

Step 2. Convert the original endpoints into z-scores.

$$(44.5 - \mu)/\sigma = (44.5 - 50)/5 = -1.1$$
$$(55.5 - \mu)/\sigma = (55.5 - 50)/5 = 1.1$$

Step 3. Find the area under the standard normal curve over the z-score interval.

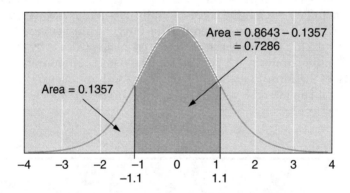

The normal approximation gives us $P(45 \leq X \leq 55) = 0.7286$ and the binomial probability is 0.7287, to four decimals. The normal approximation is extremely accurate in this case.

The normal approximation does not work equally well for all binomial distributions. If the binomial probability histogram is skewed because of a small or large p, then the normal approximation will be poor. Figure 6.16 displays three binomial probability histograms with superimposed normal curves for $p = 0.1$. They are all skewed to the right, but less so for the larger n's. The binomial probability histogram for $p = 0.1$ and $n = 10$ can be found in Figure 6.2 in Section 6.1. In Figure 6.15, in both the cases when $n = 10$ and $n = 25$, the normal approximation is poor. The normal approximation is not too bad when $n = 100$, but the Poisson approximation, which works well for large n and small p, is actually better. In the case when $n = 400$, the normal approximation is quite good.

For fixed p, the normal approximation gets better as n gets larger. How can we tell whether n is large enough to guarantee that the normal approximation will be reasonably accurate? How

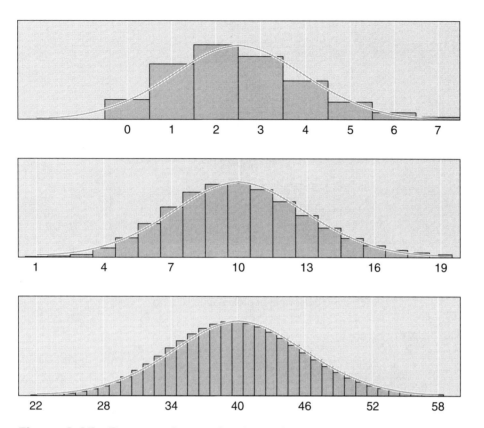

Figure 6.15 The normal approximation to binomial distributions with $p = 0.1$ for $n = 25$, 100, and 400.

large n should be depends on how small p is (or how close p is to 1). The following is a useful rule of thumb.

Recall that a normal variable is 99.7% certain to fall in the range $\mu \pm 3\sigma$ (See Figure 6.12.) A binomial variable has values in the range 0 to n. If the normal range $\mu \pm 3\sigma$ is contained in the binomial range,

$$0 \leq \mu - 3\sigma \leq \mu + 3\sigma \leq n$$

that is, if

$$0 \leq np - 3\sqrt{np(1-p)} \leq np + 3\sqrt{np(1-p)} \leq n,$$

then the normal approximation will be reasonably accurate. But if the normal extends below $0 - 1/2 = -1/2$ or above $n + 1/2$, clearly the approximation can be poor.

Remark. Most textbooks state the recommended condition in terms of np and $n(1-p)$. The condition $0 \leq np - 3\sqrt{np(1-p)} \leq np + 3\sqrt{np(1-p)} \leq n$ can be restated as

$$np \geq 9(1-p) \text{ and } n(1-p) \geq 9p.$$

These two inequalities are almost the same as an easier to memorize rule, namely

$$np \geq 9 \text{ and } n(1-p) \geq 9.$$

These will be our rule of thumb used in solving problems.

282 PROBABILITY DISTRIBUTIONS: THE ESSENTIALS

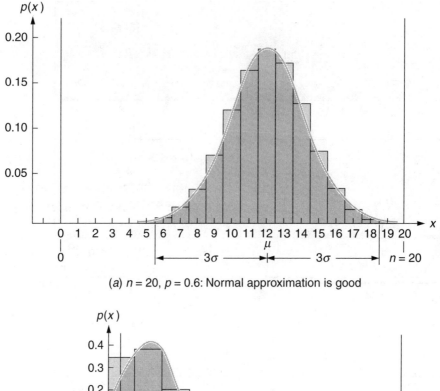

(a) $n = 20$, $p = 0.6$: Normal approximation is good

(b) $n = 10$, $p = 0.1$: Normal approximation is poor

Figure 6.16 Rule of thumb for normal approximation to binomial probabilities.

Figure 6.16 illustrates our rule of thumb. In Figure 6.16(a), we consider the binomial distribution with $n = 20$ and $p = 0.6$. The interval $\mu \pm 3\sigma = 12 \pm 3\sqrt{4.8}$ or $(5.43, 18.57)$ lies within the range 0 to normal approximation to the binomial distribution with $n = 10$ and $p = 0.1$. Here, $\mu \pm 3\sigma = 1 \pm 3\sqrt{0.9}$ or $(-1.85, 3.85)$ goes below the binomial range 0 to 10. Thus, the normal approximation is not good, as we see. For example, $P(X = 0) = 0.349$ while the normal approximation gives 0.2421, a serious error.

Summary

Consider a continuous random variable X determined by the outcome of some random phenomenon. If the density histogram of tens of thousands of independent observations of X very closely follows the bell-curve $f(t)$ with mean μ and standard deviation σ,

$$f(t) = \frac{1}{\sigma\sqrt{2\pi}} e^{-(x-\mu)^2/(2\sigma)} \quad \text{for } -\infty < x < \infty,$$

then we say that X is normally distributed with mean μ and standard deviation σ. We use the shorthand notation: **X is N(μ,σ)**.

If the random variable X is N(μ,σ), then the standardized variable

$$Z = \frac{X-\mu}{\sigma}$$

has the standard normal distribution N(0,1). If X is N(μ,σ), then

$P(a < X < b)$ = area from a to b under the normal curve with mean μ and standard deviation σ
= area from $(a-\mu)/\sigma$ to $(b-\mu)/\sigma$ under the standard normal curve.

In particular, if X is N(μ,σ), then there is about a 68% chance that X will be in the range $\mu - \sigma$ to $\mu + \sigma$, a 95% chance that X will be in the range $\mu - 2\sigma$ to $\mu + 2\sigma$, and a 99.7% chance that X will be in the range $\mu - 3\sigma$ to $\mu + 3\sigma$. This is the 68-95-99.7% rule for normal random variables. It is the theoretical companion to the 68-95-99.7% rule for approximately bell-shaped data sets.

The Normal Approximation to Binomial Distributions. Let X be binomial with parameters n and p. Then X has mean and standard deviation

$$\mu = np \text{ and } \sigma = \sqrt{np(1-p)}.$$

If $0 \leq \mu \pm 3\sigma \leq n$, which is essentially $np \geq 9$ and $n(1-p) \geq 9$, then we can accurately estimate $P(a < X < b)$ by areas under the normal density curve with the same mean and standard deviation as X.

Remark. Most textbooks instead give the following easy-to-use requirement for when to use the normal approximation to binomial probabilities:

$$np \geq 5 \text{ and } n(1-p) \geq 5.$$

However, if $p \leq 0.1$ and $5 \leq np \leq 10$, then the Poisson approximation is slightly more accurate than the normal approximation, suggesting that the rule should be modified, as has been done in this chapter.

Section 6.4 Exercises

1. A random variable is assumed to have the normal distribution with mean 5.0 and standard deviation 1.5.
   a. State a range in which 68% of the observations generated by this random variable will lie.
   b. State a range in which 95% of the observations generated by this random variable will lie.
2. Another random variable is assumed to have the normal distribution, this time with mean –3.5 and standard deviation 0.5.
   a. State a range in which 68% of the observations of this random variable will lie.
   b. State a range in which 95% of the observations of this random variable will lie.

3. A random variable was observed 100 times. Before gathering the data, the researcher hypothesized that the data would follow the normal distribution with mean 45 and standard deviation 3. The observations are given in the following table.

43.50	45.44	45.82	45.75	44.01
42.17	41.50	41.13	44.42	41.98
47.52	48.27	42.18	44.54	47.94
44.47	47.51	51.13	41.00	42.79
49.31	47.30	41.50	39.93	48.67
44.51	42.23	45.59	48.92	44.49
41.59	46.89	44.59	40.75	40.48
44.31	48.07	45.38	47.00	43.51
53.50	44.97	42.97	45.50	41.09
41.18	43.53	49.56	49.58	40.68
43.73	45.10	39.82	50.41	48.66
48.56	47.53	41.79	47.78	44.95
40.82	41.22	42.99	43.04	39.42
46.58	42.12	50.90	47.80	42.93
46.52	44.73	43.30	42.31	45.72
41.16	43.92	43.77	47.39	41.55
45.63	40.97	45.72	43.56	45.28
44.88	52.84	41.55	43.80	43.87
41.70	42.43	44.60	46.89	44.25
48.43	49.38	44.26	40.80	43.61

a. If the hypothesis is correct, what percentage of observations should lie between 42 and 48?
b. What percentage of the values really lie between 42 and 48?
c. If the hypothesis is correct, what percentage of the observations should lie between 39 and 51?
d. What percentage of the values really lie between 39 and 51?
e. Does the researcher's hypothesis seem reasonable?

4. A random variable X has a normal distribution with mean 10 and standard deviation 2. Calculate the following probabilities:
   a. $P(X < 10)$ c. $P(8 < X < 10)$
   b. $P(X > 10)$ d. $P(6 < X < 12)$

5. The 50 observations in the following table were made of the weights (in kilograms) of 12-year-old boys. The sample mean of this set of data is 40, and the standard deviation is 6.43.

40.96	33.54	37.64	26.82
42.99	44.39	32.04	31.02
38.53	34.20	31.54	34.66
44.81	39.69	40.73	38.73
37.06	38.15	34.85	33.01
44.57	42.96	49.89	50.05
42.11	49.99	54.18	35.13
41.06	37.14	39.58	41.42
48.04	32.82	41.62	38.07
44.61	40.61	38.93	44.10
36.13	52.28	29.71	35.03
50.87	45.82	34.87	49.15
31.00	41.86		

a. What percentage of the observations fall within one standard deviation of the mean?
b. What percentage of the observations fall within two standard deviations of the mean?
c. Do these data seem to follow the normal distribution in the sense of obeying the 68% and 95% rules?

6. A random variable Z has the standard normal distribution.
   a. What is the mean of Z?
   b. What is the standard deviation of Z?
   c. What is the probability of Z being between –1 and 1?

7. A random variable has the normal distribution with mean 3 and standard deviation 2. Convert the following observations of this random variable into standard units (z-scores):
   a. 3.234
   b. 5.193
   c. 1.401
   d. –0.0184

8. Repeat Exercise 7, this time for a normal random variable having mean 6 and standard deviation 3.
   a. 5.290
   b. 2.816
   c. 8.791
   d. 10.271

9. Repeat Exercise 7, this time for a normal random variable having mean –5 and standard deviation 4.
   a. 4.823
   b. –5.972
   c. –11.732
   d. 1.672

Assume that Z is standard normal. In every case, first draw a simple sketch of the given information and the area (probability) or z value you are able to find.

10. Use the normal table to find each probability.
    a. $P(Z < 1.96)$
    b. $P(Z < -1.96)$
    c. $P(Z < 1.0)$
    d. $P(Z < -1.0)$
    e. $P(Z < 0.5)$
    f. $P(Z < -0.5)$
    g. $P(Z < 0)$

11. Use the normal table to find each probability.
    a. $P(Z > 1.96)$
    b. $P(Z > -1.96)$
    c. $P(Z > 1.0)$
    d. $P(Z > -1.0)$
    e. $P(Z > 0.5)$
    f. $P(Z > 0)$

12. Use the normal table to find each probability.
    a. $P(Z < 0.68)$
    b. $P(Z > 0.68)$
    c. $P(Z < -0.68)$
    d. $P(Z > -0.68)$

13. Use the normal table to find z.
    a. $P(Z < z) = 0.95$
    b. $P(Z < z) = 0.90$
    c. $P(Z < z) = 0.98$
    d. $P(Z < z) = 0.66$
    e. $P(Z < z) = 0.50$

14. Use the normal table to find z.
    a. $P(Z < z) = 0.05$
    b. $P(Z < z) = 0.25$
    c. $P(Z < z) = 0.01$
    d. $P(Z < z) = 0.10$
    e. $P(Z < z) = 0.5$

15. Use the normal table to find z.
    a. $P(Z > z) = 0.95$
    b. $P(Z > z) = 0.90$
    c. $P(Z > z) = 0.99$
    d. $P(Z > z) = 0.05$
    e. $P(Z > z) = 0.01$
    f. $P(Z > z) = 0.10$

16. Use the normal table to solve for z. But first decide whether z is positive, negative, or zero.
    a. $P(Z < z) = 0.68$
    b. $P(Z < z) = 0.16$

For each of the following problems, if the exact value is not available in the table, you may use the closest one (that is, you do not need to interpolate).

17. The heights of a group of male students follow the normal distribution with mean 70 inches and a standard deviation 3.1 inches.
    a. What percentage of the students would you expect to be shorter than 68 inches?
    b. What percentage of the students would you expect to be taller than 73.5 inches?
    c. What height are 31% of the students shorter than, and what height are 69% of them taller than?
18. To help ensure that boxes of bananas weigh at least 40 pounds upon arrival at their destination, a packing plant might adopt this rule: Pack boxes to have a weight of 41.5 pounds of bananas with a maximum permissible range of 3 ounces above or below 41 pounds, 8 ounces. (That is, pack boxes to have at least 41 pounds, 5 ounces, but no more than 41 pounds, 11 ounces of bananas.) With this rule and the shrinkage in travel (mostly due to evaporation of water), the distribution of box weights upon arrival may be assumed to be approximately normal with a mean of 41 pounds and a standard deviation of 4 ounces.

    Suppose 30 million boxes of bananas are packed each year. Given this packing plant rule and the assumption about the distribution of box weights upon arrival, how many boxes would be expected to weight less than 40 pounds upon arrival? (*Note:* A more complete table tells us that for a standard normal Z, $P(Z < -4.00) = 0.000032$. Also, recall that the expected number of objects with a certain property is given by np, where n = total number of objects and p is a probability to be solved for.)
19. A car manufacturer is producing a piston for its engines. The piston is supposed to have a diameter of 5.300 inches, but because of variability, the diameter of a piston actually follows a normal distribution with a mean of 5.300 and a standard deviation of 0.010. If a piston is more than 0.025 inch away from the needed size of 5.300 inches, the piston is rejected.
    a. What percentage of the pistons would you expect to be too large?
    b. What percentage of the pistons would you expect to be too small? Hint: Recall that a probability is converted to a percentage by moving the decimal point two places right.
20. Recall the situation presented in Exercise 19. Suppose instead, because of a mechanical problem with the machine producing the pistons, the pistons have diameters that follow a normal distribution with a mean of 5.290 inches. The standard deviation is still 0.010. The specifications still require the diameter of the piston to be 5.300 plus or minus 0.025.
    a. What percentage of the pistons would you expect to be too large in this situation?
    b. What percentage of the pistons would you expect to be too small?
21. A set of final exam scores has a sample mean of 52 and a sample standard deviation of 6. The scores are normally distributed. If a teacher wants to assign a grade of A to the top 15% of the scores, what score should be the lowest A? If the bottom 15% are to be F's, what test score should be the highest F?
22. Pipettes are used in chemical and biological laboratories to dispense controlled amounts of liquid. Each pipette is carefully calibrated, but, because it is a physical measuring instrument, it will not dispense the exact amount set by the user. The error in amount is well described by the normal distribution.

    Suppose a scientist is using a pipette known to dispense amounts with a standard deviation of 0.05 microliter (a microliter is a millionth of a liter). The scientist is able to set the pipette to dispense varying amounts of liquid.
    a. If the pipette is set to dispense 200 microliters, what percentage of the time will it dispense greater than 200.075 microliters?
    b. If the pipette is set to dispense 175 microliters, what percentage of the time will it dispense less than 174.90 microliters of liquid?
    c. The scientist wanted to set the pipette to dispense 200 microliters, but she accidentally set it to 199. Is there any chance that the pipette will dispense as much as 200 microliters?

23. Suppose diastolic blood pressures (the lowest pressures between heartbeats) for women age 18 to 74 are normally distributed with mean 76 and standard deviation 12.
   a. Suppose high blood pressure is defined to be 90 or above. Find the probability that a randomly selected woman has high blood pressure.
   b. Suppose low blood pressure is defined to be at or below 60. What proportion of the population has low blood pressure?
   c. Suppose "typical" blood pressure is defined to be between 60 and 90. What proportion of the population has typical blood pressure?

24. Suppose a random variable X is normally distributed with mean 50 and standard deviation 8. Find the following:
   a. $P(X < 58)$
   b. $P(X < 46)$
   c. $P(46 < X < 56)$
   d. $P(40 < X < 46)$
   e. $P(X < 54)$
   f. $P(X > 40)$

25. Let X be binomial with parameters n and p. Use the correction for continuity and match each of the four binomial probabilities below with the corresponding normal probability with $\mu = np$ and $\sigma = \sqrt{np(1-p)}$.

Binomial Probability	Normal Probability
$P(X > 78)$	$P(X > 77.5)$
$P(X \geq 78)$	$P(X < 77.5)$
$P(X \leq 78)$	$P(X > 78.5)$
$P(X < 78)$	$P(X < 78.5)$

26. Let X be binomial with parameters n and p, as specified in parts **a** through **j**. For which cases would it be appropriate to use the normal approximation?
   a. $n = 20, p = 0.6$
   b. $n = 10, p = 0.1$
   c. $n = 10, p = 0.5$
   d. $n = 25, p = 0.1$
   e. $n = 100, p = 0.5$
   f. $n = 45, p = 0.1$
   g. $n = 100, p = 0.1$
   h. $n = 10, p = 0.9$
   i. $n = 100, p = 0.01$
   j. $n = 400, p = 0.1$

27. A fair coin is tossed 25 times. Compute the chance that you get at least 15 heads
   a. using the binomial tables.
   b. using the normal tables.

28. A multiple choice test has 18 questions. To each question there are three possible answers to choose from, one of which is correct. Imagine that you do not know any of the answers and blindly guess the answer to each of the 18 questions. Compute the probability that you get at most eight correct answers
   a. using the binomial tables.
   b. using the normal tables.

29. Ten percent of all Americans are left-handed. Imagine that we pick a simple random sample of 100 Americans. Let X denote the number of left-handed persons in the sample. Compute $P(12 \leq X \leq 14)$ using
   a. binomial probabilities.
   b. the Poisson approximation.
   c. the normal approximation.

30. The probability that a newborn baby will be a girl is $p = 0.487$. If 2,500 babies are born in a city, compute the chance that at least half of them will be girls.

31. Repeat Exercise 30, this time for a city in which 10,000 babies are born. What is the chance that at least half of them will be girls? Repeat for 40,000 babies.

32. A manufacturer receives a shipment of 50,000 parts from an outside subcontractor. When the shipment arrives, the manufacturer wants to have assurance that the parts meet agreed-to specifications. A simple random sample of $n = 400$ parts is inspected. Let X denote the number of defective parts among the 400 inspected. The whole shipment is accepted if $X \leq 5$. If $X \geq 6$, then the whole shipment is rejected and returned to the outside subcontractor. (This is an example of an important area of statistics known as acceptance sampling.) Let p denote the (unknown) proportion of defective parts among the 50,000 in the shipment. Compute the probability that the shipment is accepted if

   a. $p = 0.005$.
   b. $p = 0.010$.
   c. $p = 0.015$.
   d. $p = 0.020$.
   e. $p = 0.025$.
   f. $p = 0.030$.
   g. $p = 0.035$.
   h. $p = 0.040$.

   Hint: for the smallest p use the Poisson approximation. For larger p use the normal approximation.

33. The risk of dying from a smallpox vaccination is one in a million. Imagine that 225 million individuals are vaccinated against smallpox. Determine the probability that among the vaccinated individuals
   a. at least 180 die from the vaccination.
   b. between 200 and 250, inclusive, die from the vaccination.

34. In the United States the annual death rate due to leukemia is 7.5 per 100,000 population.
   a. What is the expected number of leukemia deaths in a simple random sample of 100,000 Americans over a 4-year period?

   Determine the probability that a simple random sample of 100,000 Americans over a 4-year period has

   b. fewer than 20 leukemia deaths.
   c. between 20 and 40 leukemia deaths, inclusive.

CHAPTER REVIEW EXERCISES

1. For each of the following distributions give the value of the theoretical mean and standard deviation:
   a. Binomial distribution with $n = 15$ and $p = 0.35$.
   b. Poisson distribution with parameter $\lambda = 4.3$.
   c. The standard normal distribution.

2. Convert the following sampled values from a normal distribution with mean 5 and standard deviation 2 into standardized units (z-scores):
   a. 4.392
   b. 6.921
   c. 8.936
   d. 0.0638

3. Calculate the following probabilities, where Z has the standard normal distribution:
   a. $P(Z < -1.43)$
   b. $P(Z > 0.97)$
   c. $P(Z < -2.10)$
   d. $P(Z > 1.75)$

4. Solve for z:
   a. $P(Z > z) = 0.1131$
   b. $P(Z > z) = 0.3557$
   c. $P(Z > z) = 0.0250$
   d. $P(Z > z) = 0.0104$

5. Consider the following situation: An engineer keeps track of how often an unacceptable part comes off the production line. He hypothesizes that there are 2.5 bad parts per hour.

a. What distribution might the number of bad parts in an hour be assumed to have? Explain?
b. What is the theoretical mean and standard deviation, based on the engineer's hypothesis?
c. Suppose, instead, we addressed the problem this way: The engineer knows that 1000 parts come through an hour, and he suspects that 0.0025 of them are defective. He is interested in finding out the probability of no defectives being produced in a given hour. What distribution would he use in this case? Explain.

6. When Jill leaves home in the morning, there is a 40% ($p = 0.4$) chance that the traffic light she comes to first will be green. Based on a 5-day work week, calculate the following information:
   a. What is the mean number of times she will come to the light when it is green? What is the standard deviation?
   b. What is the probability that on all 5 days she will reach the light when it is green?
   c. What is the probability that she will reach the light when it is green on exactly 4 out of the 5 days?
   d. What is the probability that she will reach the light when it is green on exactly 3 out of the 5 days?

7. A random variable is hypothesized to have the Poisson distribution with mean 1.3. The variable is observed 100 times, with the following distribution of results, along with the theoretical proportions:

Value	Observed proportion	Theoretical proportion
0	0.23	0.27
1	0.41	0.35
2	0.21	0.23
3	0.08	0.10
4	0.05	0.03
≥ 5	0.02	0.01

   a. Draw a density histogram for both the observed and the theoretical proportions. Do they seem to be similar?
   b. Calculate the value of the chi-square statistic based on these data.

8. For each of the following random variables, tell whether it is continuous or discrete:
   a. The amount of time you have to wait in line.
   b. The number of people who are in front of you when you get into line.
   c. The total number of cars that pass by in an hour.
   d. The average speed of the cars that pass by in an hour.

9. A random variable is suspected to follow the normal distribution with mean 6 and standard deviation 0.5. Once the random variable is observed many times, it is calculated that 53% of the observations lie between 5.5 and 6.5. Furthermore, 83% of the observations lie between 5 and 7. Do these observations of the random variable seem to give evidence for or against the case that the random variable has the normal distribution? Explain.

10. Convert the following observed values sampled from the normal distribution with mean -6.5 and standard deviation 2.4 into standard units (z-scores):
    a. -2.381
    b. -12.905
    c. -0.0964
    d. Consider the value from part b. What is the probability of seeing a random variable from this distribution smaller than that?

11. Calculate the following probabilities from the normal distribution specified:
    a. $P(X < 4.0)$, where X has mean 5.0 and standard deviation 1.0
    b. $P(X > 6.5)$, where X has mean 5.5 and standard deviation 0.75

c. $P(X > -5.44)$, where X has mean -7 and standard deviation 0.6
d. $P(4 < X < 7)$ where X has mean 6 and standard deviation 1.5

12. An engineer is designing a pedestrian walkway over an expressway. He knows that the heights of semis passing underneath follow the normal distribution with a mean of 14.5 feet and a standard deviation of 0.5 feet.
    a. If he builds the walkway with a clearance of 15.5 feet, what percentage of the trucks will not fit?
    b. How much clearance should the engineer build under the walkway if he wants to be sure that 99.9% of the trucks will fit?

13. Recall the situation in Exercise 2. The engineer determined how high the walkway should be, but now he wants to figure out how much weight the walkway should be able to hold. The walkway must be large enough that 50 people can be on it at once. The engineer figures that the weight of 50 people follows the normal distribution with a mean of 7500 pounds and a standard deviation of 175 pounds.
    a. If the walkway is built to carry a weight of 7815 pounds, what percentage of the time will 50 people be able to overload it?
    b. Suppose the engineer makes some mistakes in his calculations and has the bridge built to support only a weight of 7400 pounds. What percentage of the time will 50 people on the walkway overload it now?

14. A gambler is playing a game in which he has a 20% ($p = 0.2$) chance of winning. Perhaps he is willing to play because when he does win, he wins big. He decides to play the game seven times.
    a. What is the probability that he will lose all seven games?
    b. What is the probability that he will win exactly two games?
    c. What is the probability that he will win exactly one game?
    d. How many games will he win on average?

15. Calculate the following probabilities. The variable X has the specified normal distribution:
    a. $P(X > 9.5)$, where X has mean 8.4 and standard deviation 0.5
    b. $P(X < -6.7)$, where X has mean -6.0 and standard deviation 0.4
    c. $P(X > 0)$, where X has mean -0.4 and standard deviation 2

16. Suppose a drug is tested on 10 people with probability of 0.75 of a cure.
    a. Find the probability that exactly eight people are cured.
    b. Find the probability that at least seven people are cured.
    c. Find the probability that six, seven, or eight people are cured.

17. How would you simulate the drug trials of Exercise 16?

18. Consider a Poisson random variable with mean $\lambda = 5$. Find
    a. P(at least three occurrences).
    b. P(two, three, or four occurrences).
    c. P(one occurrence).
    d. P(less than three occurrences).

19. Let Z be standard normal. Find
    a. $P(0.5 < Z < 1.5)$.
    b. $P(-0.5 < Z < 1)$.
    c. $P(Z < -1.5 \text{ or } > 0.5)$. *Hint:* What is the complement?

20. Let Z be standard normal. Let $P(-a < Z < a) = 0.8$. Find a.

21. Let X be normal with mean 10 and standard deviation 75. In each case find a.
    a. $P(X < a) = 0.9$
    b. $P(X > a) = 0.05$
    c. $P(-a < X - 10 < a) = 0.95$

22. On average, six major hurricanes strike the U.S. mainland every 10 years. Compute the probability that in a given year
   a. no major hurricane will strike the U.S. mainland.
   b. at most two major hurricanes will strike the U.S. mainland.
   c. at least one major hurricane will strike the U.S. mainland.
23. A leading genetic cause of mental retardation is Fragile X Syndrome, which affects one in 1500 males worldwide. Consider a simple random sample of 9000 males. Compute the probability that at least eight of the men in the sample suffer from Fragile X Syndrome
   a. by computing binomial probabilities (if you can).
   b. by using the Poisson approximation.
   c. by using the normal approximation.
   d. Which approximation is slightly more accurate in this case, the Poisson approximation or the normal approximation?
24. Over the past century and a half, the Hawaiian volcano Mauna Loa has erupted, on average, once every 5 years. Compute the probability that in a given year Mauna Loa will
   a. not erupt.
   b. will erupt exactly once.

PROFESSIONAL PROFILE

Mr. Bruce W. Hoynoski
Vice-President and
Chief Statistician
Nielsen Media Research
Schaumburg, Illinois

Bruce Hoynoski sets a "People Box."

Mr. Bruce W. Hoynoski

Just how many people *really* watched a particular television program? That number is very important to television executives, to advertisers, and, ultimately, to the public.

The man in charge of determining that number for Nielsen Media Research is Bruce W. Hoynoski. His job is to collect accurate data about the television viewing habits of Americans. He is also responsible for the proper statistical processing of that data.

Under Bruce Hoynoski's direction, Nielsen Media research collects viewing information from households in 210 regions covering all 50 of the United States of America. The information comes from paper logs kept by participants and from data collection boxes that record which shows are on a particular TV set. There are also "People Boxes," which ask individuals to record who is watching each program. People keeping paper viewing diaries do this for a week. Those participating in electronic collection programs do so for two to five years.

Bruce Hoynoski notes that there have been changes in the average number of people watching individual television programs. Years ago, highly rated programs could receive a 30+ rating (over 30-million households watching). Today, a "hit" show will receive a third of that. "It's not," he explains, "that there are fewer people watching television these days. We actually have slightly more people regularly watching prime time television. The ratings are smaller because there are more choices and people are taking advantage of those choices."

As a statistician who reports on the habits and choices of Americans, Bruce Hoynoski observes that statistics and their interpretation play a major role in our lives. "Not only do they influence what we have available in our media, but influence our choices in politics, the products we can buy, and more."

7

Obtaining Data: Random Sampling and Randomized Controlled Experiments

The hypothesis is that 60 is divisible by all positive integers below it. I tried 1, 2, 3, 4, 5, and 6. Encouraged, I tried 10, 20, and 30. Based on this systematic sample from the population 1, 2,..., 60, I infer that the hypothesis is true.

Imaginary quote

Objectives

After studying this chapter, you will understand the following:

- Randomization: the key to validity, believability, and statistical accuracy assessment
- The dangers in drawing inferences from observational studies
- Various nonrandom (and inappropriate) methods for sampling from a population
- How obtaining data by nonrandom methods can lead to confounding and bias
- Simpson's paradox and its relationship to the dangers of a third variable in an observational study
- Randomized controlled experiments and their advantages
- Probability sampling: sampling from a population using randomization

7.1 INTRODUCTION 294

7.2 EXPERIMENTAL DESIGN: OBSERVATIONAL STUDIES VERSUS RANDOMIZED CONTROLLED EXPERIMENTS 298

7.3 SAMPLING FROM A REAL POPULATION: PROBABILITY, SAMPLING VERSUS NON-PROBABILITY SAMPLING 312

7.4 FROM RESEARCH QUESTION TO DATA COLLECTION: SOME EXAMPLES 320

CHAPTER REVIEW EXERCISES 326

> ## Key Problem
>
> ### Is Smoking Good for You?
>
> A study on thyroid and heart disease among women in Whickham, an area in England, was conducted from 1972 to 1974. Twenty years later, in 1994, a follow-up study of the same people highlighted an interesting relationship between smoking and death. Of the 1314 women in the original study, 28% (369) were dead 20 years later. However, of those women who were smokers originally (582), only 24% (139) had died, while 31% (230) of the nonsmokers (732) had died.[1]
>
> Does smoking help keep people alive? A naïve interpretation of the data seems to suggest this obviously incorrect conclusion. The focus of this chapter is on obtaining valid data that is capable of producing valid statistical inferences. Is the obviously incorrect inference that smoking increased life expectancy caused by invalid data, or by an invalid statistical inference using data that is capable of producing a valid statistical inference, or maybe both?

7.1 INTRODUCTION

Our experience with obtaining randomly generated data up to now has been limited to the simulated data of Chapters 4 and 5. Future chapters explain how to make statistical inferences from real data (that is, data collected by sampling from a population or by conducting an experiment) under the assumption that the data have been generated by some random mechanism. The validity of statistical inferences greatly depends on the quality of the data, which in turn depends on how the data have been collected. If the data are bad (as they can be in a variety of ways, some of which are not obvious to the non–statistically trained), conclusions from them are not to be trusted!

This chapter provides tools for judging the quality of data. This chapter also provides introductory material on the data collection aspects of planning and conducting your own statistical studies. The emphasis is on obtaining high-quality data capable of producing statistical inferences that can be trusted.

Section 7.4 is a modest introductory attempt to teach you how to collect your own data in ways that lead to valid and believable inferences. Besides producing valid and believable inferences, you should be capable of providing an honest statement of the accuracy of your statistical inferences. It should be noted that entire courses on statistical experimental design and sampling from real-world populations (like voters) are devoted to the topic of obtaining quality data.

The key to statistically useful data collection is to use an objective probabilistic **randomization** mechanism, rather than depending on subjective human judgment, even expert judgment. Three main benefits follow from having a carefully planned statistical study that collects its needed data using randomization:

1. *Validity:* Valid statistical inferences are only possible from high-quality data. That is, through the use of appropriate statistical methods, information that is accurate and trustworthy can be extracted. Alternatively, if data have been improperly collected, little or no valid information can be extracted, regardless of the statistical methods used on the data.
2. *Believability:* Outside observers need to be convinced that the data have provided statistically valid conclusions. That is, scientific journals, government agencies, businesses, or any other

---

[1] D. R. Appleton, J. M. French, and P. J. Vanderpump, "Ignoring a Covariate: An Example of Simpson's Paradox," *The American Statistician*, vol. 50, 1996, pp. 340-341.

consumer of the statistical study will find its statistical inferences valid and, thus, convincing. Note that believability flows from the validity of the statistical inference, which flows from the data being of good quality, which flows from the data having been properly collected using a random mechanism.

3. *Prediction of Accuracy:* Researchers can determine how likely it is that the statistical inferences made from randomly obtained data are accurate or correct. Put simply, the use of a randomization mechanism in collecting data allows us to obtain probability statements about the accuracy or correctness of our inference.

There are two major kinds of statistical studies. One kind considered in this chapter is called a **randomized controlled experiment**; the other is called **probability sampling**. Both kinds of statistical studies are widely and frequently employed by statisticians and both require the use of randomization mechanisms to ensure believability and accuracy.

A *randomized controlled experiment* is used to estimate the **effect** of a **treatment** when it is applied to a **unit**. The effect of a treatment is typically measured by comparing the **outcomes** produced by the treatment to the outcomes produced when either no treatment is used or a standard treatment is used. The units that receive the treatment of interest are called the **treatment group**; and the units that receive either no treatment or the standard treatment are called the **control group**. Randomization works here to assign units to the treatment and control groups.

There are many different kinds of treatments, units, and outcomes that can be studied in a randomized controlled experiment. For example, the units could be voting adults, sick children, cars, houses, or plots of farmland. The units are usually not randomly chosen, for example 200 sick children who volunteered from five Chicago hospitals. The corresponding treatments of interest that might be applied to these units could be, respectively, the way a question is asked, a new drug, a new gasoline additive, a new type of exterior paint, or a new strain of corn seed. The respective outcomes could be the answer to the question, whether the disease is cured, the miles per gallon obtained, the number of years elapsed before a new coating of paint is needed, or the number of bushels of corn crop that are produced.

The second type of statistical study, *probability sampling*, is used to estimate some unknown numerical population characteristic by *randomly* sampling units from the population of all units and measuring this characteristic on the sampled units. The unknown value of the characteristic for the population is referred to as the **population parameter**, whereas the value of the characteristic for the sample is referred to as a **sample statistic**. We say *the* population parameter because there can be only one value for the characteristic when it is measured by using all the units in a population. We say *a* sample statistic because there can be many possible values for any sample statistic used since there are many possible samples that can be drawn from a population.

There are many different kinds of populations and characteristics that can be studied by probability sampling. A distinction is made between a **real population** and a **conceptual population**. A real population has a finite number of units that actually exist as identifiable entities. Examples of real populations could be all the dairy cattle in Pennsylvania, all the registered voters of Chicago, all women of childbearing age in Ontario, or all the computers of a certain type that are manufactured by a particular company.

Their corresponding characteristics could be, respectively, the amount of milk produced by a cow in one week, the percentage of the sampled units favoring registration of hand guns, the age of the woman at the time of birth of her first child, or the percentage of sampled units with a particular manufacturing flaw.

A conceptual population, however, exists only as a concept and it is imagined to have an infinite number of units. An example of a conceptual population would be the potentially infinite number of times a coin could be tossed. Each time the coin is tossed, the outcome is either a head or a tail. A sample from this imaginary population is drawn by tossing the coin a finite number of times. A characteristic (or parameter) of interest for this population might be the proportion of the units that are heads (0.5 for a fair coin). The sample statistic would be the proportion of heads in the finite sample.

In sampling data from either a real or conceptual population, the only statistically acceptable sampling methods must use an objective (completely specified, leaving nothing to human judgment) random mechanism. A random sample is often more difficult to obtain from a real population than from a conceptual population. This is because the units to be sampled from a conceptual population are usually generated by a simple random experiment being carried out, which makes it easy to ensure that the proper randomness is employed. For sampling from a real population, however, the units to be sampled are usually not so easily accessible to the person carrying out a simple random experiment, so that much more care is needed to ensure that proper randomization occurs. Thus, the part of this chapter that deals with random sampling will focus much more on random sampling from real populations than from conceptual populations.

These two types of statistical studies, randomized controlled experiments and probability sampling, are best illustrated by examples.

Designing a Randomized Controlled Experiment: Testing a New Vaccine

Suppose a new vaccine is being developed to reduce the prevalence of a dangerous disease; for example, an effective AIDS vaccine. How will we know whether such a vaccine was truly effective? The only sure way to know is to give some people the vaccine (the treatment group) and the rest a placebo (the control group) and compare their disease rates. That is, we must do a controlled experiment (in the case of a clinical trial) where we **control** who gets the vaccine and who does not. For this method to work, however, it is essential to ensure that the only systematic difference between the treatment and control groups is due to the treatment. The two groups must be as alike as possible in all other ways. The best way to ensure that this happens is to **randomly assign** the people enrolled in the experiment to the treatment and control groups.

Common sense might lead us to believe that we could do better than randomization if we used the opinion of medical public health experts to carefully and systematically balance the two groups with respect to all characteristics judged likely to influence the rate of occurrence of the disease, such as age, socioeconomic status, and overall level of health. However, randomization automatically does the required balancing, even of important variables that the experts may have failed to think about! For example, because each person is randomly assigned either to the treatment or control group, randomization forces the average age to be about the same in both groups.

The AIDS study is an example of a randomized controlled experiment: the treatment is the vaccine; the units are the people enrolled in the experiment; the outcome measured on each unit is whether or not the disease was contracted; and, most importantly, *the effect of the treatment is the change in the rate at which the disease is contracted in the treatment group as compared to in the control group.*

The results of a randomized controlled experiment are used to answer the question: "Is the treatment effect large enough so that we can be confident that the treatment is effective?" As we will see in Chapter 9 on hypothesis testing, the results of a randomized controlled experiment will include not only an estimate of the size of the treatment effect (called the **effect size**), but also an estimate of the **statistical significance** (strength of evidence) of the effect.

If the effect size is large enough to be of practical medical importance and the effect is statistically significant, then the treatment is declared to be effective. The decision to recommend widespread vaccination against the disease is based on the statistically valid and believable inference that the vaccine truly lowers one's chances of contracting the disease; that is, we require *strong statistical evidence* the vaccine does indeed lower the rate at which the disease is contracted.

Sampling from a Real Population: A Preelection Poll

Consider two candidates running for the presidency. Just before the election, a poll of 1500 potential voters is taken to assess the relative popularity of the candidates. To be sure that the people sampled are as similar as possible to the entire voting population, the best approach is to choose the 1500 people completely at random from the population of all potential voters. Once again, systematically choosing the members of the sample to have characteristics similar to the population is inferior to randomly choosing the sample from the population!

Probability sampling tries to answer, "Is one candidate ahead of the other? If so, by how much?" Clearly, this study will be useful only if we can be reasonably confident that the sample is *representative* of the population. For example, if 40% of the sample favor Candidate A, then we want to be able to conclude with confidence that something close to 40% of the population also favor Candidate A.

Randomization Is the Key

One major aim of this chapter is to enable you to conduct your own statistical studies by collecting data, either by designing a randomized controlled experiment or by probability sampling from a population of interest. Then, when you complete your statistical analysis of the data, you will be able to make valid inferences about the population.

Without careful planning of how the data will be collected, inferences from the resulting data will often be untrustworthy. Without randomization, the treatment effect may be confounded with other factors or variables. In the case of a nonrandomized controlled experiment (an **observational study**), a non-treatment variable (e.g., age) could differ between the treatment group and the control group. Then, if the treatment group has a significantly lower disease rate, is it because of age, the treatment, or both? Probabilistically balancing out all non-treatment variables is the beauty of using randomization.

In the case of sampling from a real population, it is easy to see how nonrandom (i.e., non-probability) sampling can result in samples that are seriously nonrepresentative of the population. For example, if 30% of 100 students questioned outside the college's Fine Arts building state that they use marijuana recreationally at least once a week, can we conclude that approximately 30% of the students at the college do likewise? Probably not, because randomization was not used in selecting the sample.

In the case of sampling from a conceptual population, it is also important that the sample be randomly drawn by having the data generating mechanism employ randomization. For example, it is usually not difficult to generate random data by tossing a coin, spinning a roulette wheel, rolling two dice, or randomly choosing balls from a box model.

Bias, that is, systematic error in the quantity being estimated, can occur when non-probability sampling is used or when observational studies (nonrandomized experiments) are used. As in the non-probability sampling example, the 30% of the students using marijuana in the sample is very likely a biased estimate that overestimates the entire student body's marijuana usage rate. In an observational study, the observed reduction of disease occurrence in the nonrandomly selected vaccine treatment group could in fact be caused by the difference in the ages of the individuals in the treatment and control groups. (The disease occurs more often in older people.) Thus, this reduction is likely a **biased estimate** of the vaccine effect.

This chapter explains and illustrates principles of good data collection, which

- allows you to evaluate the validity and believability of a statistical study;
- allows probability-based estimation of the strength of statistical evidence for the stated statistical inference;

- allows you to plan your own statistical study so that it will withstand the scrutiny of others by being valid and believable; and
- enables your statistical study to provide its own measure of how strongly its inferences and conclusions can be trusted.

Randomization is the key to success.

Section 7.1 Exercises

1. Give an example of a controlled experiment (look for examples in newspapers, magazines, the Internet, television, etc.). Clearly state the goal of the experiment and the treatment to be given. Note whether randomization seems to have been used or not.
2. Give an example of probability sampling in the form of a survey. Clearly state the question that the survey tries to answer. Note whether randomization seems to have been used or not.
3. Find an example where confounding seems a definite possibility (look for examples in newspapers, magazines, the Internet, television, etc.). Identify the variable that is suggested to be the cause of what is observed. State a second variable that could also be a partial confounding cause.
4. Find examples in which the data, although clearly not produced by randomization, are being used by the writer to suggest the validity of a statistical inference about a larger population (look for examples in newspapers, magazines, the Internet, television, etc.). *Hint:* Magazine and Internet surveys are notorious for letting their readers or viewers believe their surveys apply to broader populations.
5. Explain confounding, and how confounding can lead to a biased assessment regarding the magnitude of influence for a variable.

7.2 EXPERIMENTAL DESIGN: OBSERVATIONAL STUDIES VERSUS RANDOMIZED CONTROLLED EXPERIMENTS

Experimental design refers, in general, to the use of an experiment to determine the effectiveness of a treatment. As noted above, when a randomized mechanism is properly used in an experimental design, the experiment is called a **randomized controlled experiment.** If no randomization mechanism is used, the experiment is called an **observational study.** The purpose of this section is to provide a closer look at observational studies and randomized controlled experiments by presenting a detailed analysis of a real-world example for each type of experiment. These two examples show the difficulties of using an observational study in contrast to the advantage of using a randomized controlled experiment to answer an experimental question.

Smoking and Lung Cancer

A major ongoing public health controversy of the twentieth century has been whether smoking causes lung cancer. In 1957 the *British Medical Journal* editorialized that smoking does indeed cause lung cancer, citing "the painstaking investigations of statisticians that seem to have closed every loophole of escape for tobacco."

The basis of this claim was a study conducted by British statisticians R. Doll and A. B. Hill from 1948 to 1949. The basic observational data consisted of a selected set of 709 people with lung cancer and a selected matching set of 709 people without lung cancer. In each set there were 649 men and 60 women. The treatment variable was whether an individual smoked or not. Among the 1230 men from both groups who smoked (most of the men in both groups smoked), 647 had lung cancer, this being almost 52% (a very high rate, but note that half of the people in the study were selected because they had lung cancer). However, among the 68 nonsmoking men, only two had lung cancer, a percentage of just under 3%. Thus, the experimental probabilities of lung cancer differed substantially from smoking to nonsmoking men. The principles of good science require us not to presume a causal explanation here, no matter how plausible it seems. But we can certainly say that for the men, smoking and lung cancer seem heavily associated.

For the women, the rate of lung cancer among the smokers was about 59%, compared to a 37% lung cancer rate among the nonsmokers. Again the association is clear, but with the intriguing observation that a much higher percentage of the nonsmoking women had lung cancer than their nonsmoking male counterparts. (A possible reason is that the women may have had more exposure to secondhand smoke.) These data, as well as data from many other studies, establish an association between smoking and lung cancer.

Sir Ronald Fisher (1890–1962) was perhaps the most famous and influential statistician of all time (and a famous geneticist as well). In his role as careful scientist and eminent statistician, Fisher was appalled by the strong causal conclusions drawn by the *British Medical Journal.* He forcefully pointed out (and continued pointing out for the rest of his life) that just because smoking and lung cancer are associated, one cannot necessarily conclude that smoking causes lung cancer.[2]

In the absence of further scientific evidence, it could just as well be that lung cancer causes smoking, or that a third factor causes both to tend to occur together. For example, people in the early stages of a long-developing disease such as lung cancer could experience severe physical discomfort, and smoking could help alleviate the discomfort. If this explanation were true, lung cancer would indeed cause cigarette smoking. Moreover, this rather unusual causal mechanism would produce a positive association between cigarette smoking and lung cancer. To quote Fisher, "And to take the poor chap's cigarettes away from him would be rather like taking away his stick from a blind man. It would make an already unhappy person a little more unhappy than he need be."

As for a third factor causing both smoking and lung cancer, Fisher went to great lengths to argue that one's genetic makeup could predispose one toward or away from smoking, and this same genetic makeup could also affect susceptibility to lung cancer. To understand this idea, imagine a gene whose possession causes a high rate of lung cancer and facilitates a high rate of cigarette smoking. Interestingly, some geneticists currently claim that a tendency toward addictive behaviors is genetically influenced, and it is well known that certain cancers have a genetic predisposition. Any positive linkage between genetic characteristics contributing to addictive behaviors and those contributing to cancer would make Fisher's claim at least somewhat true.

Other possible third factors include type of employment (people working in coal mines, foundries, and other settings with polluted air may be more likely to smoke than people who work in offices or other professional settings), or whether one lives in an urban or rural environment (people living in an urban area, who live amid more pollution, may also be more likely to smoke).

Science is conservative in its progress, as our approach to statistical decision making in the next chapter will emphasize. It accepts causal explanations in data-driven studies only when either the statistical or the scientific evidence, or a combination of both, is very strong. Sir Ronald Fisher was in this sense acting as the good, cautious statistical scientist, even though today the evidence is overwhelming that cigarette smoking does indeed cause lung cancer, as discussed later. From the perspective of wanting valid statistical inferences, the big problem with the 1948–1949 study is that it was an observational study rather than an experimental study using randomization.

---

[2]See R. D. Cook, "Smoking and Lung Cancer," in *R. A. Fisher: An Appreciation*, ed. S. E. Fienberg and D. V. Hinkley (New York: Springer-Verlag, 1980), pp. 182–191; P. D. Stolley, "When Genius Errs: R. A. Fisher and the Lung Cancer Controversy," *American Journal of Epidemiology*, vol. 133, pp. 416–425; and B. W. Brown Jr., "Statistics, Scientific Method, and Smoking," in *Statistics: A Guide to the Unknown*, ed. J. M. Tanur, F. Mosteller, W. H. Kruskal, R. F. Link, R. S. Pieters, and G. R. Rising (San Francisco: Holden-Day, 1972), pp. 40–51. These references discuss Fisher's role in the smoking/lung cancer controversy, and provide many additional references.

Observational Studies

Fisher's concerns are relevant for any observational study. An observational study has the same basic two-group structure as a randomized controlled experiment. In both types of study, some units are assigned to receive the treatment, thus forming the **treatment group**, and other units are not assigned to receive the treatment, thus forming the **control group**. Members of the control group are sometimes called **controls**. The key distinction is that each unit in a randomized controlled experiment is assigned randomly by the investigator to the treatment group or to the control group, whereas in an observational study each unit falls into the treatment group or into the control group as a result of what happened in the natural course of events. For example, some people in an observational study on smoking have chosen to smoke, thus becoming members of the treatment group.

Some observational studies (often reported in the media) fail to have a control group. For example, Freudian psychoanalysis was often defended as effective by citing the "statistic" that 75% of the people undergoing psychoanalysis got better. But note that there is no reference to a control group in this statement. It was pointed out by skeptics that around 75% of people with mental illnesses tend to get better in one year without any psychotherapy. If true, this claim would certainly put the claimed 75% cure rate for psychoanalysis in a different light.

Without randomization, valid statistical inference is difficult or impossible. Observational studies are often statistically analyzed anyway, because investigators of some phenomena are forced to use previously collected data. Historic observational data are usually plentiful and cheap, whereas the cost of careful statistical experimentation is often high. However, with observational data the researcher does not randomly assign available subjects to the treatment and control groups. For example, researchers for the 1948–1949 lung cancer study did not decide who would and would not smoke (the groups self-selected). Of course, randomly assigning people to smoke or not would be both unethical and impossible in practice. How were the people chosen for this observational study? The researchers found 709 people with lung cancer in various London hospitals. They matched them to 709 people without lung cancer. For example, there were 60 women with lung cancer, so the researchers deliberately balanced for gender (a possible risk factor) by finding 60 women without lung cancer. Although the researchers did decide who was to be in their matched study, the people either did or did not smoke (which determines the assignment to the treatment and the control groups).

By contrast, randomized assignment of people to smoking or not, if it had been possible, would have had the advantage of simultaneously controlling for all important variables or factors that could be influencing the rate of occurrence of lung cancer. For example, if age were a major determining factor, then because of randomization, the average age in the treatment group would be about the same as in the control group, by the law of large numbers.

As Fisher correctly claimed, there is nothing in the *observational data themselves* that rules out lung cancer causing smoking or a third factor causing both smoking and lung cancer. And good science requires that these possibilities be ruled out before we can begin to claim that smoking causes lung cancer.

Analyzing the Key Problem

The data described in the Key Problem are also from an observational study. Clearly, smoking is positively associated with being alive 20 years later (about 31% of smokers versus about 24% of nonsmokers alive). But is it reasonable to conclude from this fact that smoking prolongs life? Note that the researchers in the study in no way controlled which of the women smoked. In particular, this is clearly not a controlled experiment where the random assignment of units (women) to treatment versus control (smoking or not) occurs. Hence, other variables could easily differ between the two groups. In fact, we will see shortly that the smokers tended to be younger on average and that this third variable (age) obviously has a powerful influence on the 20-years death rate because as one grows older, one is less likely to live another 20 years, that is, age is a confounding variable. Table 7.1 provides an expanded version of the data, which helps us disentangle the influence of age and smoking on dying within 20 years.

Look first at the women who were young (18–34) in 1974. Almost all were still alive 20 years later, whether they smoked or not. Look next at the women over 65. Most of them (about 85–86%) were dead 20 years later, again whether they smoked or not. However, there is a sizable difference in the death rate between the smoking and nonsmoking women who were middle-aged (35–64) in 1974, that is, 26% of the smokers, compared to 18.4% of the nonsmokers were dead.

This new perspective of looking at death rates within the different age categories (age being a likely confounding variable) produces very believable smoking versus nonsmoking rates within category. Statisticians describe this as *controlling for age*. This new perspective suggests that we can look at death rates within each age group and claim there is no evidence that smoking is harmful for the young and old, but that there is evidence it is harmful for the middle-aged. Contrast this with our initial examination of the data in the Key Problem, which suggests that smoking reduces death rates for women.

To help understand this seeming contradiction, we use the theoretical probability technique of probability trees to solve for probabilities of events that are the result of a multistage process. Here we will build two trees using the observed proportions in Table 7.1, one for the smokers and one for the nonsmokers, to explain the paradox.

The probability tree for smokers is shown in Figure 7.1. The event of interest is

$$E_S = \text{(dead after 20 years)},$$

the S reminding us the analysis is for smokers. Stage 1 is the classification of the person by age group as young, middle-aged, or elderly. Stage 2 gives, within each age group, the probability of

Table 7.1 20-year Death Rates of Smoking and Nonsmoking Women

	Smokers			
Age category in 1974	Number in category	Proportion in category	Number dead in 1994	Proportion dead in 1994
18–34	179	0.31	5	0.028
35–64	354	0.61	92	0.260
65+	49	0.08	42	0.857
Totals	582	1.00	139	0.239

	Nonsmokers			
Age category in 1974	Number in category	Proportion in category	Number dead in 1994	Proportion dead in 1994
18–34	219	0.30	6	0.027
35–64	320	0.44	59	0.184
65+	193	0.26	165	0.855
Totals	732	1.00	230	0.314

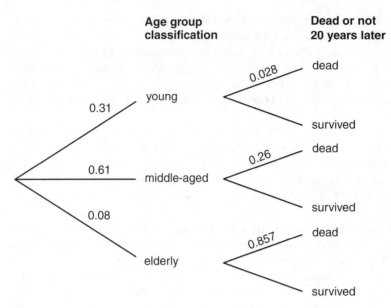

Figure 7.1 Probability tree for death among smokers.

survival. We solve for $P(E_S)$ by combining the multiplication and addition rules for probability trees given in Section 4.2.

Remember that you first identify the outcomes at the far right of the tree that make up the event E_S. Next, for each of these outcomes, identify the branches that form the route that leads to the outcome. Then, the probability for each outcome is obtained by multiplying the probabilities of the branches along the route that leads to the outcome (this is the multiplication rule). Finally, $P(E_S)$ is obtained by adding up the probabilities of the outcomes that make up the event (the addition rule).

We already know that $P(E_S) = 0.24$ from the information given in the Key Problem. This probability is computed by adding the contributions of the young, $(0.31)(0.028) = 0.0087$; the middle-aged, $(0.61)(0.26) = 0.159$; and the elderly, $(0.08)(0.857) = 0.069$ (see Figure 7.1). By the addition rule,

$$P(E_S) = (0.31)(0.028) + (0.61)(0.26) + (0.08)(0.57) \approx 0.24$$

except for round-off error.

Figure 7.2 gives the analogous probability tree for the nonsmokers, for which we already know that $P(E_{NS}) = 0.31$. Our calculations here show that the young contribute $(0.30)(0.027) = 0.008$ to the total, the middle-aged $(0.44)(0.184) = 0.081$, and the elderly $(0.26)(0.855) = 0.223$. Thus

$$P(E_{NS}) = (0.30)(0.027) + (0.44)(0.184) + (0.26)(0.855) \approx 0.31,$$

except for round-off error, as shown in Figure 7.2.

This paradox resolves itself by noting that the differing probabilities of death 20 years later (0.24 versus 0.31) are each determined by *both* the proportions of people in each age category *and* the death rates within each such age category. As we would suspect, all of the within-age category rates of death are higher for the smokers group. The fact that a much higher proportion of the nonsmokers were elderly (26% versus 8% of the smokers being elderly) gave the nonsmokers a higher death rate overall. Thus, the confounding variable age, which in this observational study was not balanced across the two groups, affected the total group probabilities.

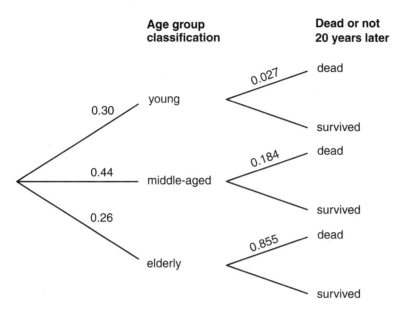

Figure 7.2 Probability tree for death among nonsmokers.

There is a powerful lesson here in the misuse of statistics. The erroneous and superficial suggested conclusion in the Key Problem statement, that smoking seems to prolong life, resulted *in part* from using bad data (being observational and hence unable to balance out other variables—age here). It also resulted in part from using faulty statistical reasoning that ignored the possibility of confounding smoking with age as a cause of death 20 years later.

Statisticians and philosophers refer to this setting as an example of **Simpson's paradox**. In general, Simpson's paradox occurs whenever the difference in the rate of occurrence of something between two groups is nullified or reversed when the influence of a third previously overlooked variable or factor is removed (controlled for). Simpson's paradox often leads to naïve and incorrect analyses of causal factors.

Our probability tree analysis shows one way to handle such a third factor when we have identified it—namely, to study the association between the two variables of interest (such as smoking and lung cancer) within separate restricted ranges or categories (as for age) of the third factor. However, this approach is effective only if we can be clever enough to figure out *all* such potentially trouble-making third factors, a challenge at which we are likely to fail (and at which even the experts can fail). It is important to realize, though, that this strategy of studying the association of interest within narrow ranges of a measured variable or in the differing categories of a possibly confounding third factor is often used effectively in statistical studies that are forced to depend on observational data.

Fortunately, as we have been stressing, there is a better way: the randomized controlled experiment. It prevents the possibility of a damaging and confounding third factor by forcing the two groups being compared to be approximately alike in every way except the treatment aspect under study. Here is an illustration of such a randomized controlled experiment.

Randomized Controlled Experiments: The Salk Polio Vaccine

Polio is a scary disease that primarily strikes children, often leaving its victims paralyzed. Throughout the first half of the twentieth century, it claimed many victims, including Franklin Delano

Roosevelt. By the late 1940s and early 1950s, approximately 30,000 to 60,000 people per year were contracting polio. Research revealed that polio was caused by a virus, leading to the search for a vaccine.[3] In the early 1950s Dr. Jonas Salk developed a killed-virus vaccine against polio.

In 1954 it was decided that a large-scale study should be conducted to assess the effectiveness of the Salk polio vaccine. Some observational studies were contemplated but rejected. One such proposal was to use a given year, say 1953, to provide controls (children not vaccinated) and use the next year, 1954, as the treatment year (children are vaccinated). In the treatment year, the vaccine would be widely dispersed, and whoever wished to use it could do so. One could then compare the polio rates in the two years to see if the rate decreased. The problem here was that the groups of children for the two years could be very different in ways other than just having or not having the vaccine. For example, look at Figure 7.3. If the control year had been 1946, and the treatment year 1947, then it would look as if the vaccine were quite effective even if it were worthless. If the control year were 1947, and the treatment year 1948, then even if the vaccine had cut the polio rate in half, it would have appeared ineffective. This problem exists because polio, like many other diseases, has an epidemic character and thus can vary greatly in incidence from year to year.

Another possibility was to give the children in some regions of the country the vaccine, leaving the rest of the country as the controls. The problem with this experimental design is that regions are different in more ways than just whether they have the vaccine. In particular, polio is prone to regional epidemics.

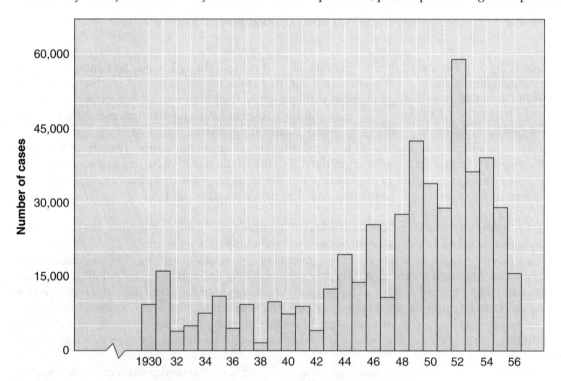

Figure 7.3 Annual occurrences of polio in the United States.

---

[3] A vaccine is a substance that causes the body to believe it has contracted the virus. The body reacts by developing antibodies in the blood that are specifically targeted to kill that virus. These antibodies remain in the body, so that if later the person does contract the virus, the antibodies will kill it before it can cause harm. Vaccines can be made from small amounts of the live virus, or of killed virus. The danger in live virus is that the vaccine may cause the person to contract the disease before enough antibodies can be aroused. The danger in killed virus is that it may not be able to fool the body into creating the antibodies. See P. Meier, "The Biggest Public Health Experiment Ever: The 1954 Field Trial of the Salk Poliomyelitis Vaccine," in *Statistics: A Guide to the Unknown*, ed. Tanur, Mosteller, Kruskal, Link, Pieters, and Rising (San Francisco: Holden-Day, 1978), pp. 2–13, for an account of the Salk experiment.

If the vaccine worked, it was important to have strong evidence that it worked, so that it could be widely used immediately to stop polio. Ambiguous results could have delayed the adoption of a vaccine for years. Since valid and believable statistical evidence was required, these observational approaches would not do.

NFIP Study In 1938 President Roosevelt established the National Foundation for Infantile Paralysis (NFIP), now known as the March of Dimes Birth Defects Foundation. (In honor of President Roosevelt and his association with the March of Dimes, the United States placed Roosevelt's image on the dime in 1946.) The NFIP planned a massive study that would deal with the possible effects of year and region. At schools willing to participate, the second graders would be the group getting the vaccine, and the first and third graders would be the controls. There could be some differences between the treatment and control groups, such as age and possibly more contagion within certain grades than in others, but because the control ages bracketed the treatment ages and everyone was within the same schools, those factors were expected to be minimal. However, an ethical problem with scientific implications arose: People could not be forced to take the vaccine. More precisely, children could not be given the vaccine without their parents' permission. For this reason the NFIP asked parents of second-graders to volunteer their kids for the study. About 64% did volunteer. Thus the NFIP plan:

- *Treatment group:* Second graders, with their parents' permission, are given the vaccine.
- *Control group:* The first and third graders at the same schools are not given the vaccine.

Although not randomized, the design for this statistical experiment is fairly good, but there are still some important ways in which the treatment and control groups may differ, three of which are next.

The Volunteer Effect The people in the treatment group receiving the vaccine were volunteers. The control group, by contrast, was made up of some people who would have volunteered and some who would not. Volunteers are somewhat different from nonvolunteers in ways that matter medically. In this instance, volunteers tended to be more well-to-do, but surprisingly, polio has been shown to be more likely to strike the more affluent.[4] Thus this design is somewhat biased against the vaccine because the treatment group, being more affluent on average than the control group, has a greater likelihood of polio for its members.

Change in Children's Behavior In the NFIP study the parents knew whether their child was vaccinated or not. This knowledge may have changed their behavior. For example, if the parents knew their child was vaccinated, they may have been more likely to let the child engage in more risky behavior, such as attending summer camp, where exposure to the virus may have been more likely. This effect could also work against the vaccine.

Effect on Diagnosis The doctors evaluating the children at the end of the study to determine who contracted polio usually knew who got the vaccine. Depending on the attitude of the doctor toward the effectiveness of the vaccine, this knowledge could sway the doctor's diagnosis in borderline cases, perhaps with the doctor being totally unaware of being so influenced.

The Need for a Randomized Controlled Experiment Some health departments that would be involved in the study objected to the NFIP plan for the previous reasons. It was too important to obtain

---

[4] One usually associates affluence with better health, but in the case of polio it works the other way around. The explanation is that in less affluent areas the polio virus is more likely to be present (because of poorer hygiene). Hence, children are likely to contract polio while they are still young and protected by antibodies transmitted at birth from their mothers, when it is relatively harmless, at which time they will develop their own polio antibodies to protect them from contracting the disease when they are older, when it is more harmful.

a clear conclusion to allow these effects to muddy the waters. Mindful of the ethical problems but aware that randomization was essential, a second design was proposed with the following features:

1. Randomization
2. Placebo
3. Double-blind protocol

These features are aimed at preventing confounding variables from arising between the treatment and control groups. What are these features? How are they used? The proposal was to take a large group of children and randomly assign about half to the vaccine and half to the control. *Randomization* has a specific meaning: An objective probabilistic mechanism assigns people to the groups. Conceptually, one places the names of all the children in the study in a box (perhaps using balls) and randomly draws half of the names without replacement. The children whose names are drawn receive the treatment (vaccine), and the ones whose names remain in the box do not receive the treatment and thus become the control group.

In practice, people use tables of random numbers (recall Chapters 4 and 5) or let a computer do the randomization. The important notion is that the subjects themselves do not decide (nor do their families or doctors) whether to take the vaccine. Moreover, the subjective judgments of the researchers have *no* effect on who receives the vaccine.

In order to achieve this design for this study, people were asked to volunteer their children, and only those volunteered children were randomly assigned to the two groups. The parents were explicitly told that their volunteered child may or may not receive the vaccine. Notice that this approach balances out the volunteer effect: Both groups contained all volunteers. Randomization will also tend to even out other important variables between the treatment and control groups, such as overall health, age, sex, and level of affluence. Thus we simply do not have to be concerned about the confounding influence of other health-related variables.

To eliminate differences in behavior based on knowing whether one received the vaccine, it was important that neither the children nor their parents knew who was in which group. There is no way a child would not notice being given an injection, so the plan was to give everyone in the study a seemingly identical injection. The treatment group had the vaccine in their injections, while the control group had plain saltwater. The saltwater injection is an example of a **placebo.** A placebo is an inert treatment (that is, having no active medicinal ingredients) that, to the recipient, looks and feels the same as the real treatment (see the later section, "Those Amazing Placebos"). The placebo produces a **single-blind randomized controlled experiment**; that is, the subjects are blind as to whether they received the treatment or control. Thus, for example, the risk-taking behavior and the psychological state of the two groups should be the same.

Finally, the experimenters who carry out the study (those giving the injections and the doctors making the diagnoses of the children) are not told which children received the vaccine, and which received the placebo. Hence the single-blind experiment becomes a **double-blind experiment,** the second "blindness" being that of the people carrying out the experiment.

Table 7.2   Results of NFIP and Randomized Control Polio Studies

	Number of subjects	Number with polio	Polio rate per 100,000
NFIP study			
Vaccinated (2nd-grade volunteers)	221,998	56	25
Control (1st- and 3rd-graders)	725,173	391	54
Randomized control study			
Vaccinated (volunteers)	200,745	57	28
Control (volunteers)	201,229	142	71

Thus we have the plan for the randomized controlled experiment:

- *Treatment group:* Vaccine is given to half of the volunteers, randomly selected.
- *Control group:* Placebo is given to the other half of the volunteers.

The two plans (nonrandom NFIP and randomized) were implemented, each in about the same number of schools. The results of the two studies were announced on April 12, 1955, the 10th anniversary of President Roosevelt's death. Some of the data are in Table 7.2. The experiment was a success! The randomized control study showed that the polio rate among the nonvaccinated children was about $71/28 \approx 2.54$ times greater than for the vaccinated ones. The NFIP study had also showed that the vaccine group did better, but not by quite as wide a margin: $54/25 \approx 2.16$. It appears that the volunteer effect, and possibly the other effects, did lessen the vaccine's apparent effectiveness. More important, the randomized control study was not subject to the criticisms aimed at the NFIP study, and thus its validity and believability were much greater.

Other indications point to the volunteer effect. The polio rates in the vaccinated groups of the two studies were very close (25 versus 28 per 100,000), but the polio rates in the control groups were different: 54 in the NFIP study compared to 71 in the randomized control study (Table 7.2). Since neither control group received the vaccine, the difference is likely due to the volunteer effect: The controls in the NFIP design consisted of everyone, whereas those in the randomized control study were all volunteers. Recall that volunteers were on average more affluent and hence more likely to contract polio.

The goal of finding convincing evidence that the vaccine worked was achieved, so that in the subsequent year the vaccine was widely disseminated throughout the nation. Unfortunately, dissemination was abruptly cut short when a bad batch of the vaccine caused 79 children to contract polio. Later the vaccine was given out again, and new and improved vaccines were developed a few years later. Subsequent widespread use of the vaccine eventually led to the virtual eradication of polio from the United States. As a student of statistics, note the vital role that a randomized controlled experiment played in this nationally important public health issue.

Back to Smoking

The studies showing association between smoking and lung cancer are not as convincing as the randomized control Salk vaccine study. Why not execute a similar study for smoking? Imagine such a study. A group of people is identified (volunteers?) to be in the study, say, all about 18 years old. Half are randomly assigned to smoke for the rest of their lives; half are randomly assigned to not smoke. Since they should not know their group, we need a placebo. We create placebo cigarettes that look (and taste) just like regular cigarettes and have the same detectable effects on people. (If not, then it would not be long before the subjects figured out their group.) But these placebo cigarettes cannot have the cancer-causing ingredients of regular cigarettes. (Do we really know what those are?) These people would have to be followed and supplied with the correct type of cigarettes until they die or, say, for 40 years. Then the researchers would analyze who died of lung cancer and who did not. Because of the randomization and the use of a placebo, it would finally be scientifically safe to conclude the effect of smoking. Thus, if carried out, this study would provide powerful evidence one way or the other. Unfortunately, for both practical and ethical reasons, such a study is impossible to carry out—studies involving cigarette smoking must be observational!

The collective force of many observational studies can, however, be convincing. The U.S. surgeon general appointed a blue-ribbon committee in 1962 to review the scientific (experiments on rats, tissue studies on deceased lung cancer victims, etc.), medical, and statistical evidence of the health effects of smoking and to arrive at a summary conclusion. The committee surveyed a large number of various kinds of studies, and in 1964 it issued a comprehensive report flatly stating that "cigarette smoking is causally related to lung cancer in men; the magnitude of the effect of smoking far outweighs all other factors."

Taking the studies as a whole, the idea that the relationship between smoking and lung cancer is causative is convincing because of a number of considerations:

1. The association appears in many types of studies using many types of subjects.
2. The association is so strong (the mortality due to lung cancer among smokers is 10 to 20 times that among nonsmokers) that it is unlikely that the effect could be totally explained by other factors.
3. The suspected cause (smoking) occurs before the effect (lung cancer).
4. Tobacco smoke contains substances that are known to cause cancer in animals (established by doing randomized controlled experiments).
5. People are more likely to contract lung cancer the more they smoke per day, the earlier they start smoking, and the more they inhale (shown observationally).

Thus the overwhelming and multifaceted evidence allows us to bypass the need for the usually required randomized controlled experiment with treatment and control groups and yet still conclude causality. Successful bypassing of randomized experimentation is unusual, and this instance required an enormous investment of scientific and medical resources.

As an interesting aside, recall that R. A. Fisher made the case that according to some of the Doll and Hill 1948–1949 data, noninhalers seemed more likely to contract lung cancer than inhalers. This is an example of an association that did not hold up under more careful examination. In fact, even Doll and Hill had additional data showing the difference was minimal. By contrast, the surgeon general's report cited a large body of evidence that showed that the more people inhaled, the higher the rate of lung cancer. It seems likely that some subtle third-variable influence was present in the earlier data or, as happens occasionally in statistical studies, the statistical "gods of chance" conspired to fool us—that is, the natural randomness that produced the data yielded an unlikely result (like nine heads in 10 tosses of a fair coin).

The case against smoking was strong, but there were still doubters, in particular the tobacco companies (e.g., elevated disease rates due to other factors no one had considered). In 1979 the surgeon general produced another (heftier) report, even more comprehensive and even more damning for smoking.[5] Still, it was not until January 1998 that even the tobacco companies had to admit, "We recognize that there is a substantial body of evidence which supports the judgment that cigarette smoking plays a causal role in the development of lung cancer and other diseases in smokers."[6] After so many years of statistical research, at least no one can argue with the pundit who said, "It is now proved beyond doubt that smoking is one of the leading causes of statistics."

Observational studies are valuable, but it takes many more of them, and a wide variety of types, to collect evidence equal to good randomized controlled experiments.

Those Amazing Placebos

The notion that an inert substance can heal because the patient believes it can is an old one. Such inert substances were dubbed placebos, from the Latin "I shall please," because a healer, lacking a truly curative substance, would prescribe a placebo to make the patient happy. Over the years placebos have included "usnea (moss from the skull of victims of violent death), Gascoyne's powder (bezoar, amber, pearls, crabs' eyes, coral, and black tops of crabs' claws), triangular Wormian bone from the juncture of the sagittal and lambdoid sutures of the skull of an executed criminal, wood

---

[5] *Smoking and Health: Report of the Advisory Committee to the Surgeon General of the Public Health Service*, U.S. Department of Health, Education, and Welfare, Public Health Service publication no. 1103, 1964, p. 106; *Smoking and Health: A Report of the Surgeon General*, U.S. Department of Health, Education, and Welfare publication no. (PHS) 79–50066, 1979.

[6] Statement before the U.S. House of Representatives Commerce Committee, January 19, 1998, by Geoffrey C. Bible, chairman and chief executive officer of Philip Morris Companies, Inc.

lice, human placenta and perspiration,"[7] and many other strange and unpleasant substances. The placebos worked in the sense that people would often feel better after having taken them, even though now we know these "medicines" were basically useless. The placebo effect is real; that is, true healing can occur simply by believing the placebo will work. Psychologists and medical researchers have studied the influence of mental functioning on the immune system and on other physiological characteristics related to disease.

When a new drug or medical procedure is introduced, people often find a high cure rate. For example, one treatment for angina (suffocating chest pains associated with heart disease) was to tie off the mammary artery. Two studies *without any controls* reported 68% and 91% of the patients improving from the surgery. These rates seem impressively high, and this procedure subsequently became quite popular in the years 1955 to 1960. Popularity plummeted after two more experiments *with controls* showed 67% improvement in the treatment group, but 71% in the control group. The controls received a placebo in the form of a skin incision that did not affect the artery.[8] It appears as if the improvements people felt were based on the placebo effect. The actual surgery did not provide the relief. Without carefully controlled experiments comparing the treatment to the placebo, surgeons would likely have continued performing a dangerous but useless procedure.

Chemonucleolysis is a treatment for alleviating the pain of a slipped disc in the spine. It involves injecting an enzyme directly into the disc.[9] Between 1963 and 1975, almost 17,000 people had this treatment, with the studies reporting success rates from 50% to 80%. These studies did not use controls. Before the procedure could be approved for use in the United States, it needed to be evaluated with a randomized controlled experiment. In 1975 a double-blind randomized controlled experiment using 106 patients (53 in each group) was performed in which the placebo was an injection without the active enzyme. The treatment group had 60% success, and the placebo control group had 50% success. Thus the treatment appears to be a little better than the placebo. The difference of 10% was not statistically significant (i.e., the statistical evidence was weak; see the next chapter), and the treatment failed to be approved.

A later study, again without controls, was reported in 1977 to have a 70% success rate with the procedure. Still, no approval for its use was forthcoming. Two more double-blind randomized controlled experiments (with placebos) were conducted. Combining the three placebo-control studies (this combining called a **meta-analysis**), the success rate for the chemonucleolysis treatment was 70%, and that for the placebo was 47%, enough to produce strong statistical evidence in favor of the treatment. Finally, the treatment was approved. This appears to be a case in which the treatment really is effective, but again the message is clear: We found this out only by doing careful randomized double-blind controlled experiments.

Notice that the uncontrolled studies for both the angina surgery and the chemonucleolysis injection showed success rates of 50% to 90%. The angina surgery actually did slightly worse than the placebo, suggesting that the surgery was useless at best and possibly even harmful. For the disc treatment, the placebo still did well, but not quite as well as chemonucleolysis. Chemonucleolysis was only about 23% better than a placebo. Both examples suggest a strong placebo effect. Thus, by not comparing the treatment of interest to a placebo, one often obtains overly optimistic, or even wrong, impressions of the effectiveness of a drug or medical treatment.

Lessons

In the introduction to this chapter, we mentioned three benefits to running a well-designed randomized controlled experiment. The examples in this section illustrate the first two: validity and believability.

---

[7]A. K. Shapiro and E. Shapiro, "The Placebo: Is It Much Ado about Nothing?" in *The Placebo Effect*, ed. A. Harrington (Cambridge, Mass.: Harvard University Press, 1997), pp. 13–14.

[8]Ibid.

[9]R. L. Sanford, "The Wonders of Placebo," in *Statistics in the Pharmaceutical Industry*, ed. C. R. Buncher and J.-Y. Tsay (New York: Marcel Dekker, 1994).

- **Salk polio vaccine.** *Validity:* Both studies gave evidence that the vaccine was effective, but the randomized controlled experimental study was more scientifically valid than the NFIP study. *Believability:* The medical profession, and indeed the entire country, immediately accepted the effectiveness of the vaccine.
- **Smoking studies.** *Validity:* The Key Problem illustrates the dangers of taking observational studies at face value; and any single observational study has little validity in arguing that smoking causes lung cancer. *Believability:* Because randomized controlled experiments were not feasible, many varied observational studies over decades were necessary to achieve common acceptance (believability) of the harmful effects of smoking.
- **Surgery for angina.** *Validity:* Uncontrolled experiments suggested that this surgery was effective, but the randomized controlled experiments showed that the effectiveness could be accounted for by the placebo effect. Hence, there is no statistical validity argument for making the surgery the procedure of choice! *Believability:* None.
- **Chemonucleolysis.** *Validity:* Uncontrolled experiments showed roughly a 70% success rate for this procedure, and randomized controlled experiments eventually showed that the placebo success rate was about 47%, suggesting a 23% increase due to the treatment. Thus a validity argument supporting the procedure was produced. *Believability:* If more controlled experiments instead of just observational studies had been run early on (1964), then approval for use of the procedure probably would not have taken until 1982. Poor planning delayed approval for 18 years, denying unknown numbers of people an effective treatment.

Section 7.2 Exercises

1. The following is a quote about a recent study on breast cancer. Use this paragraph to answer the following questions.

   > The cancer institute's study involved 13,338 women in the United States and Canada, making it one of the largest cancer prevention studies ever. Some women were given tamoxifen, others placebos. For those given tamoxifen over a five-year period, one in 236 developed breast cancer. The placebo breast cancer rate was one in 130 women. There were significant reductions in the occurrence of both invasive and non-invasive breast cancers in every age group, from 35–45 to the over-60 group.

   a. Identify the treatment group and the control group in this study.
   b. How were placebos used in this study? Do you think it was necessary to use them?
   c. Name three additional factors that could influence the outcomes of this study.
   d. Write down one question you would like to ask the researchers about their study.

2. The following is a quote about a recent study on the effect of the vitamin folic acid on heart disease and homocysteine levels in the blood. Use this paragraph to answer the questions below.

   > In the study, researchers fed breakfast cereal daily to 75 men and women with heart disease at the Providence St. Vincent Medical Center in Portland, Oregon. They found that the more folic acid there was in their cereals, the more their blood homocysteine levels declined. They also found that while cereals with the standard level of fortification had little effect, adding nearly five times as much, or a total of 665 micrograms of folic acid, cut homocysteine levels by 14 percent. (Source: http://www.cnn.com/HEALTH/9804/08/cereal.heart/index.html)

   a. Identify the treatment group and the control group in this study.
   b. From the information in the paragraph, do you think that placebos were used in the study? Explain your answer.
   c. Name three additional factors that could influence the outcome of this study.

d. There was a large amount of press coverage on this study, and several doctors on television were recommending increasing one's folic acid intake to at least 400 micrograms. Do you think these recommendations were reasonable given the information from the study?

3. This following is a quote about a recent study on the effect of smoking on the heart attack rate of men and women. Use this paragraph to answer the questions that follow.

> Women who smoke have a 50 percent higher risk of having a heart attack than male smokers, according to a report in the *British Medical Journal*. Dr. Eva Prescott and colleagues at the Institute of Preventive Medicine in Copenhagen concluded that women may be more sensitive to the harmful effects of cigarettes because of an interaction between components of tobacco and hormones. "There is growing epidemiological evidence that women who smoke are relatively deficient in estrogen," Prescott said. Doctors have known that estrogen deficiency is associated with cardiovascular disease, and women's risk of having a heart attack increases after the menopause when estrogen levels fall. Studies of hormone replacement therapy have shown it lowers the chance of suffering a heart attack.
>
> The researchers studied 25,000 men and women over a 12-year period and compared the risk of heart attack among smokers and non-smokers. The women's 50 percent increased risk did not depend on age and was not influenced by high blood pressure, cholesterol, height, weight, exercise or alcohol consumption.

a. Identify the treatment group and the control group in this study.
b. Is this study an observational study or a randomized controlled experiment?
c. Could one use placebos in this study? Why or why not?
d. Name three additional factors that could influence the outcomes of this study.
e. Write down one question you would like to ask the researchers about their study.

4. How would you conduct an observational study to determine the effects of alcohol consumption on heart disease? Consider the following questions: How do you choose the control group and the treatment group? What information should you get from the subjects?

5. Design a randomized controlled experiment to determine the effects of alcohol consumption on heart disease. In the design of your experiment, consider the following questions: How do you choose the control group and the treatment group? How do you choose these groups to minimize the volunteer effect? Would a placebo be necessary for this kind of experiment? If so, what would you use as a placebo for this experiment? What limitations or obstacles would you face with this experiment?

6. A study to determine whether eating fish is associated with lower blood pressure compared the inhabitants' diets in two villages in Tanzania. One village was on the shore of a lake and had a diet consisting primarily of fish, and the other village was inland and had a diet consisting primarily of vegetables. Of the people in the fish-eating village, only 3% had high blood pressure, whereas in the vegetable-eating village, 16% of the people had high blood pressure.
a. Is there a control group in this experiment? If so, describe it?
b. Is this an observational study, or a randomized controlled experiment?
c. Is there a placebo? If so, what is it?
d. How strong is the evidence in this study concerning the benefits of eating fish on high blood pressure?
e. What other factors could effect blood pressure?

7. A study was conducted to see whether a new drug, Copolymer 1, was effective in preventing relapses among people with multiple sclerosis. One group of patients was given the drug and another was given a placebo. Patients were randomly assigned to the two groups. After two years, 33.6% of the people taking the drug were relapse free, and 27% of the people taking the placebo were relapse free.
a. Is this an observational study, or a randomized controlled experiment?
b. Is this a study in which people may have been randomly assigned to treatment and controls?

c. What other factors could be affecting these results? Is there any reason to think these other factors have occurred?

8. An article in the October 7, 1997, *Chicago Tribune* reported a study trying to determine whether children behave differently when they are older than their classmates. The article said that a study in the October issue of the journal *Pediatrics* found that 12% of the children who started school when they were a year or more older than their classmates displayed extreme behavior problems, compared with 7% of children whose ages were normal for their grade. According to the study, the problems become more apparent as the children grow older.
   a. Was there a control group in this study?
   b. Is this study a randomized controlled experiment, or an observational study?
   c. Does the study prove that being older than one's classmates tends to cause bad behavior? What other explanation could there be?

9. Let us consider a University of California graduate admissions study.

	Men		Women	
	Number of applicants	Percent admitted	Number of applicants	Percent admitted
Major A	325	37	593	34
Major B	560	63	25	68
Major C	825	62	108	82

   a. Compare the overall female and male admission rates for these three majors. Is this strong evidence of bias against female admission to graduate school?
   b. Do a probability tree analysis for males and another for females to show how a breakdown by major produces the overall female and male admission rates. Try to explain how the rates within a major can favor women but yet the overall admission rate can still favor men.
   c. From the viewpoint of the original question of possible admission bias against women, there is an interesting anomaly in these data—do you see it? Can we even claim this anomaly provides evidence of admission rate bias? Why or why not?

7.3 SAMPLING FROM A REAL POPULATION: PROBABILITY SAMPLING VERSUS NON-PROBABILITY SAMPLING

Sampling is used to estimate some unknown numerical characteristic of a real population. When a randomization mechanism is properly used in collecting the sample, the sampling is called probability sampling. If a randomization mechanism is not used, serious error, called bias, can occur.

In this section we take a closer look at sampling, describing in detail through actual real-world examples the problems that can occur, how proper probability sampling techniques can be used to minimize these problems, and the dangers and difficulties to look out for when you choose to use (or are forced by circumstances to use) non-probability sampling techniques.

Preelection Polls

Let us go back to the 1936 presidential election between Franklin Roosevelt (the Democrat) and Alf Landon (the Republican). The *Literary Digest* was a popular magazine that, for years, had accurately predicted the outcome of presidential elections based on their polls. In 1936 it sent out approximately 10 million sample ballots on postcards asking people for their presidential choice, and received fully 2,376,523 responses. Meanwhile, George Gallup, who had been experimenting with new sampling techniques, sent out only 3000 such cards before the *Literary Digest* poll began. Based on this much smaller sample, Gallup boldly declared that the *Literary Digest* poll would be wrong in its prediction about the election. Moreover, Gallup declared that the *Literary Digest* poll would show about 44% for Roosevelt and 56% for Landon. The editor of the *Literary Digest* was understandably peeved at this young upstart, saying, "Never before has anyone foretold what our poll was going to show even before it started.... Our fine statistical friend should be advised that the *Digest* would carry on with those old fashioned methods that have produced correct forecasts exactly one hundred percent of the time." Gallup had the last laugh.[10] The *Digest's* poll showed an overwhelming win for Landon, with numbers remarkably close to what Gallup predicted:

Roosevelt 42.9%
Landon 57.1%

Gallup was correct that the *Digest* would be wrong. Based on another small but carefully selected sample, Gallup's prediction of the actual result was 55.7% for Roosevelt to 44.3% for Landon. The actual election had Roosevelt winning in a landslide:

Roosevelt 62.5%
Landon 37.5%

The *Literary Digest* was very far off, missing Roosevelt's percentage by 19.6 percentage points. Gallup was much closer. He predicted the correct winner, and was only 6.8 percentage points off (large by today's polling standards, but tolerable in this election).

A preelection poll can fail to predict an election accurately since people may change their minds between the polling and the elections; the people polled may not be representative of the people who actually vote; people may not tell the truth; or too few people may be polled to achieve much accuracy. The last reason is not the case for the *Literary Digest* poll. Two and a half million is huge compared to typical preelection polling sample sizes of 8000 to 9000 used for polls that predict with high accuracy. Gallup accurately estimated the results of the poll discussed above using only 3000 people. Since the *Literary Digest's* sample size was much larger, could the other reasons stated add up to an error of over 19%? Or are there other important problems with the sampling?

Selection Bias A poll will not be an accurate reflection of the election if those polled (the sample) are not representative of the voting population, the *Literary Digest* sent cards to people using lists of telephone and automobile owners. That approach may be reasonable today, but in the depths of the Depression affluent people were much more likely to have telephones and cars. The 10 million people polled were thus more affluent on average than the electorate at large. This poll is said to suffer from **selection bias**. If there is selection bias, the sample is not representative of the population in important ways. Note that the *Literary Digest* poll certainly did not sample randomly from the voting population.

Affluence was an important factor in the 1936 election. Roosevelt was much more popular than Landon among the less affluent, and Landon was much more popular among the affluent. A high percentage of wealthy people hated the New Deal, which was Roosevelt's attempt to help mitigate the effects of the Depression. Thus the selection bias in this poll worked against Roosevelt.

Nonresponse Bias Notice that 10 million cards were sent out, but only about 2.5 million were returned. Could those responding to the polls be different in important ways from those who did not respond? For example, people who volunteered to be in the polio experiment tended to be more af-

---
[10]George Gallup, *The Sophisticated Poll Watchers Guide* (Ephrata, Pa.: Princeton Opinion Press, 1972), p. 66.

fluent and, hence, more susceptible to polio than people who did not volunteer. Here, people who returned their cards (a kind of volunteering) tended to be more affluent than those who did not return their cards. Thus the *Literary Digest* poll was likely to be still further biased against Roosevelt. This source of bias is termed **nonresponse bias**: The people who did not respond to the poll were not representative of those actually polled.

In his effort to predict the results of the *Literary Digest* poll, Gallup sent cards to people on the same list, but to only 3000 of them (presumably randomly selected). One of his purposes was to show the power of a small, appropriately selected sample in accurately predicting a large population. Gallup was being especially clever, for the population he wanted to predict was not the 10 million recipients, but rather those who would actually respond to the *Digest* poll. His sample was also subject to nonresponse and selection bias and, hence, yielded percentages for Landon and Roosevelt similar to those of the huge *Digest* poll. The important statistical lesson for us is that no matter how large the sample size is, if the sampling plan is flawed, predictions will be invalid with respect to their goal of predicting the desired population. Even if all 10 million had responded, the *Literary Digest* poll would have still been in trouble because this huge sample would have not been representative of the actual population of interest, namely those voting in the election.

Probability Sampling

A good sample is one that reflects (that is representative of) the population of interest. To the extent possible, the subjects that are chosen for the sample should be like the subjects that are not chosen. This idea is similar to that of a good experimental study (Section 6.2); the treatment subjects should be like the control subjects except that they receive the treatment and the control subjects do not. In the case of choosing a sample from a population, the sample should be like the population in all possible ways. The procedural problem with the *Literary Digest* poll was that the people in the sample were generally wealthier than the people in the general population, and wealthier people were less likely to support Roosevelt. By today's standards, the *Literary Digest* poll was incredibly naïve in its approach.

Simple Random Sample In this section we discuss how probability sampling techniques use randomization mechanisms to ensure that the sample is representative of the population. As we shall see, the randomization mechanism may be either simple or complex, depending on the difficulties involved in obtaining the sample.

The simplest procedure for obtaining a good sample is the **simple random sample**. We will describe it as a box model (Section 4.3). The members of the real-world population are represented by the balls in the box. On each ball a number represents the characteristic we wish to estimate or predict (voting preference, height, income, age, and so on). To obtain a sample of size n, we draw n balls randomly from the box without replacement (see Section 4.5 discussion on sampling with and without replacement). The drawing is done without replacement because we do not wish the same person to appear more than once.

This plan guarantees that each individual in the population has the same probability of being chosen, or equally that each sample of size n is equally likely. Thus the sample is chosen without prejudice or selection bias and even a relatively small sample will represent the population well.

In practice, one first constructs a list of everyone in the population. For example, if the population consists of the students at a particular university, then a roster of the enrolled students could be used. If the population consists of households in an area of town, then a list of all addresses in that area could be used. One easy way to proceed is to assign a distinct random number to each individual and to choose the individuals assigned to the n lowest random numbers to be the sample. Computers can automate this task.

Example 7.1

A class has 25 students, and an outside evaluator wishes to interview a representative 10 to obtain in-depth opinions about the class. The evaluator uses a table of random numbers to assign one number to each student in the class. Using the random numbers from the third column in Table B.3, we obtain the following assignment:

A: 41919	F: 19650*	K: 24052	P: 22852	U: 04016*
B: 32089	G: 90804	L: 05112*	Q: 84507	V: 09508*
C: 22189	H: 10731*	M: 63664	R: 90819	W: 12940*
D: 58586	I: 55223	N: 09799*	S: 55182	X: 07237*
E: 94452	J: 02692*	O: 70899	T: 09024*	Y: 19982

The 10 students assigned the lowest random numbers are then chosen for the sample, that is, students F, H, J, L, N, T, U, V, W, and X. (Can you think of other ways to use a random number table to obtain a simple random sample of size 10?)

A properly drawn simple random sample will never have selection bias. However, it may have nonresponse bias if the individuals selected must, then, volunteer to provide information. Thus it is important that after choosing the sample, a strong effort is made to obtain responses. Generally, the best response rates are (in decreasing order) from: in-person interviews, telephone interviews, and mailings. Professional samplers incorporate follow-ups into their plans; for example, they may try numerous times to contact the people chosen for the sample. Only when the nonresponse percentage is fairly low (e.g., 5% to 10% seems good) can we safely presume that nonresponse bias is not an issue.

Not-So-Simple Random Samples The simple random sample is easy to understand, but it may be difficult to implement. For example, virtually every year the National Opinion Research Center (NORC) conducts the General Social Survey, which consists of extensive in-person interviews with people all over the country; for example, in 1990, 1372 people were interviewed. Taking a simple random sample from the population of everyone in the United States is not practical, primarily because interviewers would have to travel many miles just to interview selected individuals (to Alaska or Hawaii, for example). Further, we would need a list of the names of everybody in the population (totally impractical) in order to have the computer randomly select the sample. The sampling procedure NORC uses is quite complicated, but it is still a valid probability sample because the choice of who is sampled and who is not is entirely governed by an objective probabilistic mechanism. The sampling in 1990 proceeded in stages.

First, the United States was broken into 2489 "primary sampling units" (PSUs), each one being essentially a single county or a collection of counties. These PSUs were grouped into "strata," a stratum being a group of similar PSUs. The strata were defined based on variables such as the region of the country, metropolitan (or not), percentage of minority residents, and income. That is, counties were bundled into strata on the basis of the purpose of the social survey. A certain number of PSUs were randomly chosen from each stratum.

Next, each PSU was broken into "segments," each segment consisting of one or more blocks (small geographic units like ordinary city blocks). A certain number of segments were then randomly chosen from each PSU.

Finally, a certain number of households were randomly chosen from each segment. Interviewers went directly to those households to interview people, and one person was interviewed in each household. Subjectivity was not permitted even at the household level; for example, whoever answers the door must be the one interviewed.

The key to obtaining an objective probabilistic sample is that the sampling units (PSUs, strata, segments, households) were defined before any sampling took place, and there was a prespecified random mechanism that would choose the units at each stage. No subjective discretion was allowed on the part of the interviewers regarding whom to interview or where to go to find the people to be interviewed.

In this book, we will concentrate on simple random sampling. However, the principles learned about simple random sampling apply to more complicated probability sampling approaches, such as the stratified and multilevel sampling approach used by NORC. Moreover, the sample size required for high accuracy is usually very similar for the various sampling methods.

Non-Probability Sampling

Probability samples (those using randomization to select units from a well-defined population) have the virtue of yielding results that are valid and believable. Moreover, a probability sample can be analyzed to determine its accuracy. (Randomized controlled experiments have these same virtues.) However, non-probability samples abound, and they should be looked at with the same critical and skeptical eye that one uses for observational studies. Beware the results of the magazine reader survey, for example. We next describe some basic types of non-probability samples.

Judgment and Quota Sampling One of the main problems with the *Literary Digest* poll was that the people who responded (the sample) were not representative of the people who voted (the population). A natural alternative would be to choose people nonrandomly such that they reflect the overall population. Thus, if the population is half male, one would choose a sample with half males; if the population is 17% African American, then the sample should be approximately 17% African American. Similarly, the sample should reflect the population in terms of income, education, residence (urban, suburban, rural), and a number of other factors. Such a sample is called a *judgment sample*.

In **judgment sampling,** the samplers depend on their expert wisdom and, thus, use their own discretion to decide whom to choose for the sample. A particular form of judgment sampling is **quota sampling,** wherein the interviewers doing the sample have to interview demographically specific types of individuals (for example, three white men with high school educations who live in the city, two African American women with college educations who live in the suburbs, and so on) but are otherwise free to choose anyone who fits the descriptions. Judgment sampling appears to be a good method and sometimes produces reasonable results. However, it often creates serious bias, and, hence, randomization is better.

The preelection polls by Gallup and others from 1936 to 1948 used quota sampling and sample sizes varying from 3600 to 50,000. The errors in Gallup's prediction for the Republican candidate—Gallup's prediction minus the actual results—were 6.8%, 3%, 2.3%, and 5.3% for 1936, 1940, 1944, and 1948, respectively. These numbers are reasonably small (and much smaller than the 19.6% for the *Literary Digest* poll), but they are all positive—that is, the samples are slightly biased in favor of the Republicans. For the years 1936, 1940, and 1944, the errors were small enough that Gallup predicted the correct winner (Roosevelt each time). In 1948 the Democratic nominee was Harry S. Truman, who had taken over as president when Roosevelt died in 1945, and the Republican was Thomas E. Dewey. The Gallup poll (and similarly, other major polls) predicted Dewey would get 49.5% of the vote and Truman would get 44.5%. It was a close race and indeed led to the *Chicago Tribune's* infamous headline on November 3, 1948, "Dewey Defeats Truman."

Yes: The polls were embarrassed, as was the *Tribune*, and the polls came under quite a bit of criticism. They were only off by about 5%, closer than in 1936 when Gallup was so lauded, but they did get the wrong winner. According to *New York Times* columnist William Safire, "Sampling is no science; ask President Dewey and Prime Minister Peres."[11] (The authors, however, strongly

---

[11]William Safire, "Sampling Is Not Enumerating," *The New York Times*, Dec. 7, 1997.

disagree with Mr. Safire on this point.) As he points out, there never was a President Dewey. In fact, the actual vote reversed the prediction: Truman had 49.8% and Dewey 44.5%.

Let's critique Safire's statement. What went wrong? Were the pollsters simply unlucky? Every poll will be off by a certain amount, and in a close election that amount may be enough to make the wrong prediction. Another possibility is the polling technique. Quota sampling is *not* probability sampling; it allows individual discretion on whom to interview, which can lead to bias. Even though demographically the sample looks representative, there is still a slight bias towards Republicans. It could be that Republicans are more likely to be at home, or may live in "nicer" houses, or be "better" dressed and, hence, preferred by the interviewer, perhaps unconsciously.

After 1948, Gallup and the other major polling organizations moved to probability sampling. In the years 1952 to 1996, the average size of the error of Gallup's prediction of the winning candidate was 2.23%, compared to 4.35% from 1936 to 1948. In addition, in five of the elections the error was in favor of the Republican, and in seven of them the error was in favor of the Democrat. Thus the bias toward the Republicans found in the quota sampling has disappeared. Another interesting fact is that the Gallup poll of 1948 was based on a sample of size 50,000, whereas the probability polls were based on many fewer, around 3000 to 4000 in the past 20 years. Thus, the probability-based samples are less biased, more accurate, and more efficient than the quota samples. Probability samples of three to four thousand come within about 2% of the correct percentage of a population of close to 100 million voters, which seems miraculous.

Safire's criticism of the polls in the 1948 election is on target in that there will be error if one does anything other than count the voting preferences of everyone (an impossible task), but sampling *is* a science that provides accurate estimates, and it has improved a great deal since 1948. The Israeli elections for prime minister in 1996 pitted Shimon Peres against Benjamin Netanyahu. Peres was favored in the polls, with the last polls before the election predicting Peres with 51.5% and Netanyahu with 48.5%. These polls clearly declared that their results were accurate only to within a margin of error of 3%. The final results showed Peres with 49.5% of the vote and Netanyahu with 50.5%. Thus, the polls predicted the wrong winner, but were off by only 2%, within the declared margin of error. The science of sampling did not fail; the election was merely so close that the polls had too large a margin of error to make a reliable prediction of who would win.

Haphazard Samples and Samples of Convenience A sample can be easily formed from people who are convenient to sample. For example, one might stand on a corner and interview anyone who will stop, or just sample people in one's class or workplace or neighborhood. Or, a magazine may ask its readers to fill out a survey and send it in to the magazine: very convenient for the magazine, but hardly representative of any meaningful population! These samples are non-probability samples because there is no objective random mechanism choosing the people for the sample. The standing-on-the-corner poll may seem random, but it is not random in the probabilistic sense; it is just haphazard. Such polls are subject to bias (often very large bias), and there is no statistical way to analyze the data. In particular, we cannot report a margin of error, as Gallup polling does.

Self-selected Samples It is fairly common to be watching television and be invited to make a telephone call to register your opinion on some issue. In these polls, it is not the pollsters who decide who will be in the sample, but the people themselves. Such self-selected sampling polls are subject to all kinds of biases: What type of person watches the show? What type of person is most likely to call? What prevents a group of people from orchestrating a campaign to make hundreds of calls on a particular side of the issue being raised? The polling of magazine readers also suffers from serious self-selection bias.

A similar phenomenon is the Internet-based poll, in which a Web page is put up, inviting people to respond. Only people who happen on the page and decide to vote are in the sample. An example is the Scotty J. survey. One question asks, "Do you believe that marijuana should be legalized?" The following results are based on 5152 responses:

Yes, I strongly support this:	70%
Yes, I think this is a good idea:	19%
Don't know / not sure:	4%
No, I think this is a bad idea:	3%
No, I strongly oppose this:	4%

Source: http://www.legalize.com/

Thus 89% believe marijuana should be legalized, and 7% think it should not. Is this representative of the entire population of the United States? Other questions on the poll lend insight into the demographics of those responding: 70% are male and 30% are female; 89% are white, 3% Hispanic, 2% black, and 2% Asian; 51% are under 20 years old, 32% are between 20 and 29, and 10% are between 30 and 39. There is an overrepresentation of young white males relative to the nation as a whole—that is, statistically speaking, the people most likely to be surfing the Web. Looking around further at the Scotty J. Web site makes it clear that it is set up to advocate legalizing marijuana. Scotty J. is upfront about the fact that this is not a scientific poll but one designed to persuade people. It is likely to attract people who are already leaning toward a particular viewpoint. Hence, here lies its most serious bias of all: A disproportionate number of surfers at this site favor the legalization of marijuana!

Complete Enumeration: Census

Because the purpose of a sample is to learn about a specified population, the best approach would appear to be to sample the entire population. At times it is easy and hence desirable to sample the entire population; for example, when a teacher surveys all students at the end of class to find out their opinions. However, a sample typically is taken when the population is so large that it is too costly or not physically possible to interview everyone. For these same reasons, it is important to plan the sampling procedure carefully.

The U.S. Constitution (written long before the statistical science of survey sampling was developed and refined!) requires an "actual enumeration" of all people in the United States to be undertaken every 10 years in order to apportion the correct number of members of Congress to each state. This enumeration is the census, which is carried out every decade. Over the years, the census has collected more information than just a raw count of people, including questions on education, income, family, ethnicity, and other matters.

One might expect the census to be more accurate than sampling. Surprisingly, that is not always the case. The problem is that in spite of good intentions and an enormously costly effort, not everyone is counted in the census, and those not counted tend to be disproportionately children and minorities. It is estimated that as many as 4.7 million people (roughly 2% of the population!) were not counted in the 1990 census. The National Academy of Sciences produced a report for the Bureau of the Census that recommended incorporating probability sampling to augment the enumeration in order to produce more accurate results. Although this recommendation was not adopted because of political considerations, well-designed probability sampling would undoubtedly increase the accuracy of the census. Safire is correct when he says "Sampling is not enumeration." In some cases, it is better!

Lessons

In the introduction to this chapter, we mentioned three benefits to using randomization to obtain data. The sampling examples in this section illustrate the first two: validity and believability.

- **The *Literary Digest* poll.** *Validity:* The poll did not use a probability sample, and it had a huge prediction error of 19.6%. It was not valid. *Believability:* None.

- **The Gallup (and other) polls.** *Validity:* Preelection polls for 1936 through 1948 were not probability polls, and they systematically overestimated the Republican candidates' strength. The polls since 1948 have been probability polls, and they have been more accurate and have shown no bias toward or against Republicans. *Believability:* Politicians and the media pay a lot of money for good probability samples, an indication that they believe them.
- **The Scotty J. survey.** *Validity:* This survey is a non-probability self-selected poll and is not reflective of the entire country. It does not validly predict for any well-defined population of interest, for example, for males under 25. *Believability:* Such polls do not convince policy makers or many others.
- **The Peres-Netanyahu election.** *Validity, believability, and prediction of accuracy of inference:* When an election is too close to call, probability polls are able to quantify the inability to resolve the closeness and predict a winner by the standard error, a concept extensively studied later in the book. This reflects the third important property of randomization, namely its ability to predict its own likely error size in whatever inference is made. Thus, the poll was valid and believable in its ability to likely have come within 3% of the true voting percentages. Unfortunately, this was not accurate enough to predict this election; a probability sample of 50,000 would have been!

Section 7.3 Exercises

1. Suppose you are conducting a survey on the spending habits of college freshmen. After taking a simple random sample of 200 students, you have the choice of conducting the interviews in person or sending questionnaires to your sample. Give two advantages of each method. Based on your answer, choose a method and explain your choice.
2. Design a survey using quota sampling to determine whether students on your campus approve or disapprove of the president's job performance. In order to use quota sampling, you will first need to determine the percentages of the different sex, race, and ethnic categories on your campus that seem relevant to the poll and choose the characteristics of your sample based on these percentages. How will you choose the sample? How many respondents do you need for the survey? How will you phrase the question? Will you ask the question verbally, or will you have the survey respondents fill out a questionnaire? What kinds of biases will this survey contain?
3. Design a survey using convenience sampling to determine whether people approve of the use of Native American names and likenesses as symbols or mascots for college or professional sports teams. In the design of your survey, consider the following questions: Where will you conduct the survey? How many respondents do you need for the survey? How will you phrase the survey question? Will you ask the question verbally, or will you have the survey respondents fill out a questionnaire? Can your survey be used to make valid and believable inferences?
4. It has been shown by different studies that the characteristics or behavior of the survey taker can influence the outcome of the survey. In Exercises 2 and 3 above, what characteristics of the survey taker could influence the outcome? For each characteristic, would the effect be greater if the questions are asked verbally or if the respondent fills out a questionnaire? How could you minimize the effect of the survey taker on the outcome of the survey?
5. Design a simple random sample survey to determine whether students on your campus feel that marijuana should be legalized. How do you choose the sample? What questions do you want to ask? How can you reduce the nonresponse bias of the sample? Complete the survey and report your results. Make sure to include the nonresponse rate for your sample.

6. Design a simple random sample survey to determine whether faculty on your campus approve or disapprove of making final exams optional for students. How do you choose the sample? What questions do you want to ask? How can you reduce the nonresponse bias of the sample? Complete the survey and report your results. Make sure to include the nonresponse rate for your sample.

7.4 FROM RESEARCH QUESTION TO DATA COLLECTION: SOME EXAMPLES

This chapter has provided tools for judging the quality of data that have already been gathered, as well as for planning and conducting one's own data collection as part of a statistical study. In the examples presented so far, a well-defined research question has been strongly linked to a particular type of statistical study, either a randomized controlled experiment or a probability sampling. In practice, however, the research question is usually not so well defined at first, and it may not even be obvious which type of statistical study would be best. In this section we will use the tools and ideas presented in this chapter to demonstrate, with simple but realistic examples, how to start with the initial idea of a research question and use it to plan a method of data collection that will answer the question.

Example 7.2

Campus Parking

All research begins with the germ of an idea or question. Suppose you and a few of your friends are students at a college or university, and all of you have been having difficulty finding parking on campus. You complain to each other and wonder what, if anything, you can do about it. Recognizing that there is power in numbers, you realize that you would have a greater likelihood of effecting changes if you could show that a large percentage of the student body are as concerned as you are. This idea can be formulated into a research question: What percentage of students at this university are angry about the parking situation?

To conduct a statistical study to answer this question, you must first make sure that the question is clearly stated. The phrase "are angry about the parking situation" seems too vague. What exactly is it that you are upset about in regard to the parking situation? For simplicity, let us assume that you are angry only about a shortage of parking spaces. In that case, the research question might be rewritten as: What percentage of students at this university believe that there is a need for more parking spaces on campus? Rather than allowing only 'yes' and 'no' answers, you may want to let students respond to a choice of several options. For example, you might ask, "Please answer 'none,' 'small,' 'moderate,' or 'great,' to the following question: How would you describe the need for more parking spaces on this campus?"

Once the question is well defined, you need to determine whether the question will be answered by a randomized controlled experiment or by probability sampling. The question posed here would clearly be best suited to probability sampling.

Then you need to define the population from which the sample will be drawn. You might think at first that the population consists of all students attending the university. But what if most students live on campus and do not drive? You might want to direct the question at only those students who use the campus parking facilities.

Next, you need to determine exactly how the sampling is going to be done. You want to randomly select from the population of students who park on campus. Fortunately, most universities require students who park on campus to register their cars with the campus police. By going through the proper channels, pointing out the value of your study, and signing certain forms, for example, confidentiality of statements, you can probably obtain a list of all the students who are registered for parking on campus. You can then use a random number generator or a list of random numbers from a table in a statistics book to draw a random sample of names. Recall that the use of a random sample ensures that you will avoid selection bias.

Next, you need to decide on the size of the sample. As you will learn in the core chapters on inference (Chapters 8–12), accuracy of the inference improves as the sample increases. Surprisingly, you will learn that the accuracy does not depend much on the size of the population. For example, it requires about the same number of people in the sample to estimate the voter preference of residents in Peoria, Illinois, as in Los Angeles! Let us suppose it is decided that, for purposes of this study, a random sample size of 50 will suffice.

The first step for simple random sampling is simply to have a list of all population members, called a **sampling frame.** Clearly, there are situations where such lists are either too expensive or impossible to obtain. In such cases, other more complex methods of probability sampling, such as the multilevel stratified sampling approach described for the NORC General Social Survey, are required. Let us suppose the list we get from the university includes 943 individuals, which we number 001, 002, ... , 942, 943. Then we choose 50 sets of three-digit triples from a random number table. For example, the first choice might be 249, meaning that the 249th person in the list is selected. Obviously, choices of 000 and 944–999 are simply discarded. In this manner we can choose a simple random sample of 50 parking students. Clearly, as required, each of the 943 students had an equal chance to be in the sample. Thus, the sample of 50 should be representative of the population of 943 parking students.

Finally, to avoid nonresponse bias, you must make a strong effort to obtain responses from 90% to 95% of the students selected for the sample. Perhaps, at first, a simple postcard mailing or e-mail could be used. This could be followed up with either phone or in-person interviews for the remaining selected students who have not responded. Now either you can take the final sample as the number (<50) actually selected, or you can randomly sample to obtain more from the 943 until we actually have 50 respondents; either way is acceptable.

In summary, the steps we went through in this example were as follows:

1. Clearly formulate the research question.
2. Determine whether the question will be answered by doing a randomized controlled experiment or by probability sampling.

If probability sampling is to be used,

3. Define the population from which the sample will be drawn.
4. Determine exactly how the sampling is going to be done (a simple random sampling, if possible).
5. Make a strong effort to obtain responses from 90% to 95% of the selected people in the sample to avoid nonresponse bias.

Example 7.3

Investigation of Bias by Car Dealers

Quite often, buying a new or used car is an uncomfortable experience. One might wonder whether the way one dresses can influence how one is treated by the car dealer. This could be of special interest to students whose fashion styles are on the more casual and sometimes nonconformist side. Suppose a student organization suspects that students who dress casually tend to get a worse deal in buying a new or used car than students who dress more..., well, more like car dealers (all other factors being equal). This idea can also be formulated as a research question. Do students who dress more casually receive a higher final price offer from the dealer in car purchases than students who dress more formally, when all other factors are held equal?

Again we must first make sure that the question is clearly stated. How do the researchers make sure all other factors are held equal? The best way is by randomly assigning prospective car buyers to two groups: one that dresses casually and one that dresses more formally. They can go further and hire actors to pose as people interested in buying cars, and have all the actors use the same script in negotiating for the purchase price of the car they are considering. They can make sure the prospective buyers use the same dealers and negotiate for the same cars. Thus, the question can be more explicitly defined as, "Do students who dress casually receive a higher final price offer than students who dress more formally when they use the same script to deal for the same new car, with the same dealer?"

Once the question is well defined, the researchers need to determine whether the question will be answered by a randomized controlled experiment or by probability sampling. The question posed here is clearly best answered by a randomized controlled experiment, where the variable being manipulated is dress style and the variable perhaps being influenced is car price.

Then the students conducting the experiment need to define the treatment and control groups. The treatment group could be the students who dress casually, in which case the control group would be the students who dress more formally. Pairs of participants from each group would then be taught a script to follow for purchasing a particular new car. A particular set of dealers must also be chosen (who will, of course, be blind to the experiment), and the aforementioned pairs of participants from the treatment and control groups must be assigned to each dealer.

Next, the students need to determine how they will obtain participants for the study. They might use student volunteers. They might recruit actors to pose as students, maybe even student actors. In any case, the participants would be randomly assigned to the treatment and control groups. Recall that this ensures that all other factors will be equalized between the two groups.

Finally, the treatment and control group pairs would perform the experiment and record the price offers they obtain after a reasonable period of negotiation.

In summary, the steps we went through in this example were as follows:

1. Clearly formulate the research question.
2. Determine whether the question will be answered by doing a randomized controlled experiment or by probability sampling.

If a randomized controlled experiment is to be used,

3. Define the treatment and control groups. Define exactly what the treatment will be and how the groups will differ.
4. Determine exactly how the participants for the study will be obtained. Randomly assign the participants to the two groups to avoid confounding the results from unforeseen factors.
5. Perform the experiment and record the data.

Note: As with so many social science experiments, questions of ethics and authenticity can be raised. For example, the students are not really trying to purchase cars, and hence the setting lacks authenticity. Also, it could be seen as unethical to be, in effect, deceiving the car dealers. One remedy for this last problem would be to get several car dealers' permissions to participate, without their knowing the purpose or details of the study.

Example 7.4

Support for After-School Enrichment Activities

Suppose a group of students at a particular high school have a great interest in participating in after-school activities, such as clubs and sports, but the school provides inadequate bus transportation

Section 7.4 From Research Question to Data Collection: Some Examples

to meet their needs. They complain to an administrator at the school, but she is not convinced that there is enough interest among the students in general to warrant the extra cost of providing the additional buses. So this group of students decides that they need to prove to the administration that the demand is indeed present. This idea can be formulated as a research question: What proportion of the student body wants to participate in after-school activities but either cannot or are seriously inconvenienced because of the lack of bus transportation?

Again we must first ask whether the question is clearly stated. Maybe the students should first find out what proportion the administration would require in order to provide additional buses. Then the question would become whether the proportion of interest is equal to or greater than the desired proportion.

Once the question is well defined, the students need to determine whether the question will be answered by means of a randomized controlled experiment or by probability sampling. The question posed here would clearly be best suited to probability sampling.

Then the students need to define the population from which the sample will be drawn. The population is all the students at the high school.

Next, they need to determine exactly how the sampling is going to be done. They want to randomly select from the population of all the students at the high school. There are many ways that this might be done. For example, in some high schools, students are assigned alphabetically to home rooms at the beginning of the school day. In this case, a random sample of the home rooms could first be chosen, resulting in 6 out of 23 homerooms, say. Then one could randomly choose 8 students from each home room. The result would be a probability sample of size 48. Note that this is not a simple random sample but is a multilevel stratified design and that it avoids the need for a list or sampling frame of all the students. The use of a probability sample is important because it ensures that they will have a representative sample and, in particular, will prevent selection bias.

Finally, to avoid nonresponse bias, the students must make an effort to obtain responses from 90% to 95% of the selected students in the sample. The survey might not be considered very important to students who do not plan on participating in after-school events, and the students must assure the administration that such students did indeed fully participate in the survey so that the study does not inflate student interest in after-school transportation.

Example 7.5

Investigation of Employment Discrimination

Acquiring permanent employment after graduation is an important milestone. High school and college graduates take this task very seriously. Students rightly feel that they should be judged primarily on their merits. Unfortunately, generational style differences could sometimes have an unfair impact on the employment process. Suppose, for example, that a certain group of male college graduates suspect that the wearing of an earring may have a negative impact on one's ability to land a job. They may wish to make some sort of formal complaint, but "hard" evidence is needed, so they decide to do a study. The students doing the study wonder whether or not the wearing of an earring has an impact on the perception of the job interviewer. This idea can be formulated as a research question: "Are males with earrings who interview for jobs perceived as having less job potential than males who do not wear earrings, when all other factors are held equal?"

First, the students need to make sure that the question is clearly stated. Again, the best way to ensure that all other factors are held equal is to randomly assign males who will be interviewing for jobs to two groups, one that wears earrings and one which does not. Fortunately, males who wear earrings who are assigned to the non-earring group can simply take their earrings off without this being very evident, and males who do not wear earrings but are assigned to the earring group can wear clip-on earrings that are very hard to distinguish from the real thing. As in Example 7.3, the students can go further here and hire actors to pose as people interviewing for jobs, and have pairs of actors with and without earrings use the same script in the job interview. They can also make sure the résumés and transcripts in the treatment and control groups are of equal quality. Of course, the prospective employees would go to the same employers and interview for the same jobs. Thus, the question can be more explicitly defined as, "Are males with earrings who interview for jobs perceived by their interviewers as having less potential to be hired than males who do not wear earrings, when their credentials are of equal quality, they interview for the same job, and they interview with the same interviewer?"

Once the question is well defined, the student researchers need to determine whether the question will be answered by a randomized controlled experiment or by probability sampling. The question posed here is clearly best suited to a randomized controlled experiment.

Then the students need to define the treatment and control groups. The treatment group could be the males with earrings, in which case the control group would be the males without earrings. Pairs of participants from each group could then be taught a script to follow for the job interview. A particular set of employers must also be chosen (who will, of course, be blind to the experiment), and the aforementioned pairs of participants from the treatment and control groups must be assigned to each employer. To elicit the cooperation of the employers without revealing the purpose of the experiment, the employers could be given a cover story that these particular males are practicing their job interviews and would like the interviewers to evaluate their performance by rating them (perhaps on a 1–5 "Likert" scale) on their potential for being hired if this were an actual interview. (After all the interviews have been conducted, the interviewers would be informed of the true purpose of the experiment.)

Next, the researchers need to determine how they will obtain participants for the study. They might use volunteer male graduates or they might recruit actors to pose as graduates. In any case, the participants would be randomly assigned to the treatment and control groups. Recall that this ensures that all other factors will be equalized between the two groups.

Finally, the treatment and control group pairs would perform the experiment, and the experimenters would record the hiring potential ratings that the interviewers report.

Note: Here again the issues of authenticity and ethics need careful consideration. For example, the design of the experiment and the debriefing procedures would need to be approved by the appropriate university authorities before the experiment could proceed. Also, the interviewers might be considerably more negative towards hiring from the control group if they were actually making job offers to some of the individuals being interviewed.

Section 7.4 Exercises

1. Think of a research question, and use the steps outlined in this chapter to describe how you would collect data to answer the question. Be sure to point out how randomization would be included in your data collection method.

2. Suppose you are in charge of damage control for someone running for your state's House of Representatives. Your candidate needs to produce a statement on a touchy issue. You have to choose between two statements to recommend. State a research question for this predicament, and discuss what factors you feel are the most important to be considered in formulating the question. Determine whether a randomized controlled experiment or probability sampling will be used to help answer the question. Describe the data collection method and where randomization would enter into your method.

3. The principal of a high school believes that standardized test scores are reflective not only of the skills of the test takers, but also of certain psychological factors, especially motivation, confidence, and self-esteem. A psychologist has even recommended that the students of the school be given an emotional boost right before a test to improve their performance. The principal is open to the idea but is reluctant to implement it schoolwide without some proof that it works. State a research question that can be answered by means of a randomized controlled experiment or by probability sampling. Describe the data collection, noting how you would use randomization.
4. Homeopathic medicines are becoming increasingly popular for treating a wide variety of ailments, but many people are skeptical of their efficacy. Come up with a research question that can be answered with statistics and describe the data collection method.
5. A fast-food chain has an idea for a new sandwich. They have many questions that must be answered before they decide whether or not to offer the new sandwich. Determine one question that can be answered with statistics, and describe the data collection process.

CHAPTER 7 SUMMARY

This chapter focused on two major ways to obtain real data: sampling from a population and conducting an experiment.

The key to successful data collection is to use an objective **randomization** mechanism.

When a randomization mechanism is used in conducting an experiment, the experiment is called a **randomized controlled experiment**. A randomized controlled experiment is used to estimate the effect of a **treatment** when it is applied to an **experimental unit**. The effect of the treatment is measured by comparing the **outcomes** produced by units that received the treatment of interest to outcomes produced by units that did not receive the treatment of interest. The units that received the treatment are called the **treatment group**, and the units that did not receive the treatment are called the **control group**. Randomization is used to **randomly assign** the experimental units to the treatment and control groups. Random assignment to groups ensures that the two groups are similar in all respects except for whether or not they receive the treatment of interest.

When a randomization mechanism is not used to assign units to the treatment and control groups in an experiment, the experiment is called an **observational study**. As a result, the two groups might be different in more ways than just whether or not they received the treatment of interest. These other unforeseen differences between the two groups are likely to cause differences in the outcomes between the two groups in addition to differences caused by the treatment effect. This mixing up of the treatment effect with other unforeseen differences is called **confounding**.

The presence of confounding influences in an experiment can nullify or reverse the differences that would be observed if only the treatment effect were present. The observation of conflicting results due to the influence of confounding variables is called **Simpson's paradox**.

One of the more common confounding influences that occurs in observational studies is the **volunteer effect**. When the treatment group contains volunteers and the control group does not and volunteers differ from nonvolunteers in terms of the outcomes of the experiment, then the volunteer effect will cause a confounding error in the experiment.

Another possible confounding influence is the **change in behavior** that can occur when a person knows which group they are in. For experiments where this effect can occur, control group members should be given a fake treatment, called a **placebo**, so that they will be blind as to knowing which group they are in.

Another related confounding influence is the **effect on diagnosis** that can occur when the measurement of the outcome on each unit is done by someone who knows which group the unit is in. This effect can be mitigated by ensuring that those conducting the experiment are blind as to which units are in which group.

A **single-blind study** occurs when the experimental units are blind as to which group they are in. A **double-blind study** occurs when both the experimental units and the experimenters are blind as to which units are in which group.

When a randomization mechanism is used in sampling from a population, the sampling is called **probability sampling**. The goal of probability sampling is to estimate some unknown numerical population characteristic by **randomly** sampling units from the population. The randomization mechanism ensures that all members of the population have an equal probability of being selected for the sample, which means that the sample will be **representative** of the population.

When a randomization mechanism is not used in sampling from a population, the sampling is called **non-probability sampling**. Non-probability sampling is likely to suffer from **selection bias**, which means that the sample is not representative of the population in important ways with respect to the population characteristic that is being estimated.

Another error that can creep into any sampling technique that requires a response from each sampled unit is **nonresponse bias**, which means that the units that do not respond are not representative of the population in important ways.

There are many techniques that can be used for probability sampling. **Simple random sampling** is the simplest technique.

There are also many techniques for non-probability sampling. We reviewed **quota sampling**, **samples of convenience**, and **self-selected samples**. All of these techniques are inferior to probability sampling because they are all liable to unforeseen selection bias.

A **census** is an attempt to sample an entire population.

CHAPTER REVIEW EXERCISES

1. Define the following terms and give a short example of each: observational study, quota sampling, simple random sampling, placebo effect, randomized controlled experiment, treatment group, control group.
2. There has been much emphasis in the past decade on finding a vaccine for HIV, the virus that causes AIDS. Suppose scientists are preparing to test a possible vaccine against the virus on humans. The vaccine is made up of a dead virus, so there is no possibility for a recipient to become infected. From the information in this chapter, design a randomized controlled experiment that would test the effectiveness of the vaccine. In your design, make sure to answer the following questions: How will you choose the sample? How will you reduce the effect of volunteerism in your sample? What problems will you face with your experiment? Are there certain problems unique to HIV transmission and infection that could affect your experiment? How can you control for these problems? (You might need to research characteristics of HIV transmission and infection.)
3. Find a research paper on a medical study and summarize the study and its results. Is this study valid? Which characteristics of the study make the results believable to the scientific community? Which characteristics of the study might make the results unbelievable to the scientific community? Is there anything about the study that you would change?
4. Design a survey of registered U.S. voters for a presidential election. How would you choose the people in the sample? How would you reduce selection bias and nonresponse bias?
5. Research the use of exit polls for predicting the presidential election. (An exit poll is a poll of voters taken on the day of an election after they have voted.) Report on its use in the past 30 years. Have the exit polls ever incorrectly predicted the winner of the presidential election?
6. According to a report by Dr. John Olney in the *Journal of Neuropathology and Experimental Neurology* in November 1998, data from the U.S. government show that brain cancer rates jumped 10% shortly after *NutraSweet* was approved by the Food and Drug Administration for widespread use in 1983.

a. Is this a randomized controlled experiment, or an observational study?
   b. Are there controls?
   c. Is there a placebo?
   d. Does this study give evidence that *NutraSweet* causes cancer? Why or why not?
7. Dr. Jose Manuel Silva and his colleagues at Coimbra University Hospital (in Portugal) conducted a study to see whether fish oil is effective in reducing triglyceride levels in the blood. (It is good for the heart to have low levels of triglycerides.) According to an article by Reuters on December 10, 1996:

   In the study, published in the *International Journal of Cardiology* this month, 40 patients, ages 18 to 70 years, were randomly selected to receive either 12 fish oil capsules a day (3.6 grams of omega 3), or 12 similar appearing soya oil capsules per day for two months. Prior to taking the capsules, all study participants avoided eating any fish for one month. After 8 weeks, triglycerides increased 19.9% among people taking soya oil and decreased 27.8% in the fish oil group (low levels of triglycerides are considered healthy). Total cholesterol levels were not affected significantly.

   a. Is this a randomized controlled experiment or an observational study?
   b. Is there a placebo? If so, what?
   c. Why did the participants avoid eating fish for one month before the study?
   d. Does fish oil appear effective in reducing triglycerides?
8. "Eye-movement desensitization and reprocessing," abbreviated EMDR, is a technique developed by a California therapist, Dr. Shapiro, to help victims of severe psychological trauma. The treatment begins with a therapist passing two fingers rapidly back and forth near the patient's face; the patient recalls the traumatic event while following the fingers' movement with his or her eyes. Shapiro conducted the experiment on 22 patients, using her professional judgment to assign half the patients to a conventional treatment, and half to the conventional treatment plus EMDR. She reported dramatic improvement among those receiving EMDR, but little improvement in those receiving the conventional treatment without EMDR. Shapiro herself provided the treatments to the people in the experiment. Improvement was measured by the patients' own impression of the treatment.
   a. Is this study a randomized controlled experiment or an observational study?
   b. Are there randomized controls?
   c. Is there a placebo?
   d. Is this study double blind?
   e. How convincing is this study? Why?
9. In a college statistics class, people who did well on the homework assignments tended to do about 5–10 points better on the exams than the people who did not do so well on the homework.
   a. Is this study a randomized controlled experiment or observational study?
   b. Is there an association between performance on the homework and performance on the exams?
   c. True or false: These data prove that doing well on the homework causes one to do well on the exams.
10. Gallup conducted a survey in March 1999 to determine support for legalization of marijuana, finding 29% for legalization, 69% against, and 2% don't know. According to Gallup, "The results are based on telephone interviews with a randomly selected national sample of 1,018 adults, 18 years and older."

    Section 7.3 describes the Scotty J. survey, which found that 89% support legalization, and 7% oppose it. That survey is based on 5152 responses.
    a. True or false: The Scotty J. survey is more reliable because it surveyed more people.
    b. Which do you think is a better estimate of the percentage of people in the United States who support legalization: 29% or 89%? Why?
11. Suppose a newspaper is interested in including a new comic strip, but does not know which one of three possible ones to select. Discuss some possible factors that might be considered.

Decide on one particular research question. Describe the data collection method you would use.

12. Teenage drivers pay enormous sums on automobile insurance compared to older drivers. Can you think of any ideas for what can be done to help deserving teenage drivers convince insurance companies that they deserve lower rates? Come up with a research question and describe the data collection method you would use.
13. Suppose you are the member of a club that is having difficulty attracting new members. Consider possible questions that you could discuss with the members to try to find out what is causing the problem. Decide on a question that can be answered with statistics and whose answer could help solve the problem. Describe the data collection method.
14. Teaching methods are often a hot topic of debate. What kinds of questions might we ask about teaching methods that can be answered with either probability sampling or randomized controlled experiments? Decide on one question and describe how you would conduct the data collection.

Statistical Inference: Estimation and Hypothesis Testing

CHAPTER 8 CONFIDENCE INTERVAL ESTIMATION   330

CHAPTER 9 HYPOTHESIS TESTING   394

CHAPTER 10 CHI-SQUARE TESTING   458

CHAPTER 11 INFERENCE ABOUT REGRESSION   510

PROFESSIONAL PROFILE

Dr. Mark C. Otto
Biological Statistician

Population and Habitat Assessment Section
Office of Migratory Bird Management
U.S. Fish and Wildlife Service
Laurel, Maryland

Dr. Mark C. Otto with a survey plane in the background

Dr. Otto's survey work is supported by income from the sale of Migratory Bird Hunting and Conservation stamps, popularly known as "Duck Stamps." Above is the 1998 issue.

Dr. Mark Otto started out diving in Lake Tahoe for periphyton and counting plankton. While waiting to start a masters program in ecology, he did the most practical thing he could think of: He took statistics courses and earned a statistics degree from North Carolina State University. After 14 years with the U.S. Census Bureau, he had the opportunity to go to work for the U.S. Fish and Wildlife Service. Today, Dr. Otto has the job he always wanted; he manages the May Breeding Waterfowl Survey.

Since 1955, this program has been used to estimate the number of ducks, geese, and ponds in the prairie pothole region in Montana, the Dakotas, and throughout much of western Canada up to the Alaskan tundra. Under Mark Otto's direction, the service flies some 50,000 miles of aerial transects, counting birds and ponds all the way. In addition, the U.S. Fish and Wildlife Service and the Canadian Wildlife Services team up to cover about 3000 miles on the ground and in helicopters to verify and correct the air counts.

Dr. Otto analyzes information collected during the flights and ground inspections to determine populations and bird densities in various areas. This information is then used to set hunting regulations and maintain safe bird population numbers.

Dr. Otto is also involved in another high-profile project. He is currently designing a new long-term monitoring project to track bald eagle nest success and population size. With the program he is creating, the Fish and Wildlife service will be able to ensure that our national bird will not become endangered again. Along the way, Dr. Otto is not only using the statistical processes he learned in college but also developing some of his own. He hopes that such new programs will do a better job of protecting the bald eagle and helping our national symbol thrive throughout its range.

8

Confidence Interval Estimation

Everybody believes in the central limit theorem, the experimenters because they think it is mathematical theorem and the mathematicians because they think it is an experimental fact.

G. Lippmann (1845–1921)

OBJECTIVES

After studying this chapter, you will understand the following:

- The statistical estimation problem
- The standard error (SE) as the typical estimation error size
- Confidence interval estimation
- Confidence intervals for the population mean, median, and proportion
- The central limit theorem for $\bar{X}$, $\hat{p}$, $\bar{X} - \bar{Y}$, and $\hat{p}_1 - \hat{p}_2$
- The bootstrap method for estimating SEs
- The SE-based bootstrap method for confidence intervals
- The percentile-based bootstrap method for confidence intervals
- Confidence interval for the difference of two population means
- Confidence interval for the difference of two population proportions
- Confidence interval for the difference of two means in the matched pairs case
- Estimation of the population SD
- Unbiasedness of an estimator

8.1 AN INTRODUCTION TO THE STATISTICAL ESTIMATION PROBLEM 333

8.2 RANDOM SAMPLES, ESTIMATORS, AND STANDARD ERRORS (SEs) OF ESTIMATION 338

8.3 CENTRAL LIMIT THEOREM FOR $\bar{X}$ 346

8.4 LARGE SAMPLE CONFIDENCE INTERVAL FOR THE POPULATION MEAN μ 353

8.5 LARGE SAMPLE CONFIDENCE INTERVAL FOR THE POPULATION PROPORTION p 358

8.6 BOOTSTRAPPING A SAMPLE, BOOTSTRAPPED SEs, AND SE-BASED BOOTSTRAPPED CONFIDENCE INTERVALS 362

8.7 BOOTSTRAPPED PERCENTILE-BASED CONFIDENCE INTERVALS: A UNIVERSALLY APPLICABLE METHOD 369

8.8 CONFIDENCE INTERVAL FOR MEAN μ WHEN THE POPULATION IS NORMAL AND THE SAMPLE SIZE IS SMALL 374

8.9 CONFIDENCE INTERVAL FOR THE DIFFERENCE BETWEEN TWO POPULATION MEANS $\mu_X - \mu_Y$ 379

8.10 CONFIDENCE INTERVAL FOR THE DIFFERENCE BETWEEN TWO POPULATION PROPORTIONS $p_1 - p_2$ 384

8.11 CONFIDENCE INTERVAL FOR THE DIFFERENCE BETWEEN TWO POPULATION MEANS IN MATCHED PAIRS CASE: $\mu_D = \mu_X - \mu_Y$ 385

8.12 POINT ESTIMATE FOR THE POPULATION VARIANCE σ^2 AND SD σ: UNBIASEDNESS OF AN ESTIMATOR 386

CHAPTER REVIEW EXERCISES 390

Key Problem

When a new indoor football stadium was proposed for the St. Louis area, a poll found that 45% of St. Louis County residents opposed the new stadium, 28% supported it, and 27% were undecided. The poll interviewed 301 registered voters in the county. The newspaper article ("Majority Opposes Stadium, Poll Shows," *St. Louis Post Dispatch*, October 12, 1986) states, "The county poll is accurate within plus or minus 5.7 percentage points [giving a 45% ± 5.7% or (39.3%, 50.7%) interval] at a confidence level of 95%. That means if the survey were taken 100 times, the results for the [random] group of respondents would each vary no more than 5.7 percent in either direction [from the true population percentage opposing the stadium about] 95% of these times."

The *St. Louis Post Dispatch* is to be commended for attempting to communicate an important statistical concept: a confidence interval for a population parameter. The Key Problem for this chapter is to understand how to construct confidence intervals such as the 95% interval (39.3%, 50.7%) for the proportion p of members of the population of county registered voters who oppose the stadium construction and to clearly understand why we call such an interval a 95% confidence interval.

Note: It is interesting that in 2002, St. Louis County is again struggling with this issue.

8.1 AN INTRODUCTION TO THE STATISTICAL ESTIMATION PROBLEM

Measurement as Estimation

We sometimes need to measure the height, weight, speed, length, and so forth, of an entity. It is often important that the measurement be **accurate**, that is, be close to its **true** or actual value. Making independent repeated measurements x on the entity and computing the average $\bar{x} = \sum x / n$ allows us to be more accurate than a single measurement would.

For example, one might step on one's bathroom scale three times and estimate one's true weight μ using $\sum x / 3$. Or, one might make several measurements of the height of America's tallest building, the Sears Tower in Chicago, using special surveyors' equipment. Again, the mean $\bar{x}$ of these measurements is an estimate of the Sears Tower height μ.

In Example 2.8 we examined a set of repeated measurements in a famous scientific experiment designed to accurately estimate the true (actual) speed of light μ_{speed}, one of the "fundamental constants" of our physical universe. Each such measurement was the time it took a beam of light to travel 7442 meters. After eliminating one clear outlier (see Figure 2.9), the true time, μ_{time}, for light to travel 7442 meters or 7.442 kilometers (about 4.6 miles) was estimated by $\bar{t} = 24{,}827.29 \times 10^{-9}$ seconds, (using the fact of that 1 nanosecond = 10^{-9} seconds). Then, since *speed = distance / time*, we get

$$\hat{\mu}_{speed} = \frac{7.442}{24{,}827.29 \times 10^{-9}} = 299{,}750.80 \; kilometers / second.$$

Here we have introduced a standard and useful statistical notation. For a parameter such as μ, we denote its estimated value by $\hat{\mu}$ ("mu hat"). Often, but not always, the estimated value of a parameter is the sample average of a set of repeated measurements, such as the sample average $\bar{t}$ above being used as the estimate $\hat{\mu}_{time}$.

Modern and more sophisticated experiments have essentially established that the true speed of light is $\mu_{speed} = 299{,}792.5 \; km/sec$. The **error of estimation** is given by

$$\hat{\mu}_{speed} - \mu_{speed} = -41.7 \; km/sec.$$

The percent error is $-41.7 / 299{,}792.5 \times 100\% = -0.014\%$, a magnitude of only around $1/100$ of 1%. Thus, Simon Newcomb's estimate of the true speed of light around the year 1880 was very close to the true value. Experimental science is dependent on the accurate measurement of unknown parameters, such as the speed of light.

Statisticians see the measurement problem as an **estimation** problem. The true value (one's actual weight, the true height of the Sears Tower, the speed of light, etc.) is a parameter of the distribution of a random variable X. The set of repeated measurements $X_1, X_2, \ldots, X_n$ of this X forms a random sample. The unknown parameter μ being estimated is the expected value of each measurement, that is, $E(X) = \mu$ (assumes there is no measurement bias, as discussed above).

The box model simulation approach of Chapters 4 and 5 helps us to understand concretely how statisticians view the measurement process. Each measurement is modeled as a single random draw X from a box model. The repeated measurement process is simulated by taking n random draws *with replacement* from the box. The probability distribution, specified by the box's numbered balls, models the measurement error. It was the genius of Gauss (1777–1855) to realize that the measurement error follows a normal probability distribution. That is, the box model distribution to use is $N(\mu,\sigma)$ where the mean μ is the true value being measured and σ represents the "typical" measurement error. More precisely, σ must be such that the experimental distribution of the measurements satisfies the 68-95-99.7% rule (recall Figure 2.12) for $\mu \pm \sigma$, $\mu \pm 2\sigma$, $\mu \pm 3\sigma$. We will learn how to estimate σ by S later in this chapter. Clearly, the smaller the value of σ, the more accurately $\overline{X}$ estimates μ for a fixed number of measurements.

Note: A normal distribution cannot be represented by balls in a box because it is a continuous distribution having a density, which cannot be represented by a finite set of numbered balls. Nonetheless, the textbook's accompanying simulation software generates "box-model" normal random variables. We lose the concrete mental image of balls in a box, which has nothing to do with how the computer simulates a box model. Indeed, the simulation software can generate random variables using any of several continuos distributions.

In Section 5.1 we learned that the *law of large numbers* states that when the sample size is large and $X_1, X_2, \ldots, X_n$ are independent repetitions of a random experiment, for example, *repeated measurements*, then $\overline{X} \approx \mu_X$. Thus, the law of large numbers says that, provided we make enough measurements, we can measure the true value μ_X as accurately as we wish.

If this sounds almost too good to be true, it *is*. Measurements have two sources of error: random variability or noise, which is **random measurement error** and **bias**, which is **systematic measurement error**. That is, the fundamental measurement model is

$$\text{measured value} = \text{true value} + \text{bias} + \text{random error}.$$

A simple example may help illustrate the important role of bias. Suppose a bathroom scale has a rotating pointer such that when somebody steps on the scale the pointer rotates from 0 to his measured weight. Suppose a person weighs himself five times, obtaining 161.5, 162, 162, 161, 162.5 (pounds). Then his estimated weight is $\overline{X} = 161.8$ pounds. However, suppose the scale is poorly calibrated and in fact always reads 5 pounds lighter than a person's true weight. Then his true weight is actually fairly close to $\overline{X} + 5 = 166.8$ pounds. This bias of −5 pounds is large and it *persists undiminished* in the computed mean used to estimate one's weight, even if hundreds of measurements are used to compute $\overline{X}$. If we knew the size of the bias we would simply add 5 to $\overline{X}$ and produce a very accurate measurement. The bias is almost always unknown to the people doing the measuring. What is true according to the law of large numbers is that

$$\overline{X} \approx \mu_X = \text{true weight} + \text{bias}$$

when n is large. Then, the true weight is $\mu_X + 5$, whereas the estimation process yields $\overline{X} \approx \mu_X$ as a biased estimator of the true weight $\mu_X + 5$. Thus, bias in the measurement process produces error in the estimated unknown parameter no matter how many repeated measurements we take and what the size of σ is. It was undoubtedly the major source of measurement error for Newcomb's speed of light experiment.

In conclusion, accurate measurement requires enough repeated measurements that $\overline{X} \approx \mu_X$. If σ is small, a relatively small n can suffice to accomplish this. Indeed if a single measurement is very accurate ($\sigma \approx 0$ *on the scale being used*), then $n = 1$ may suffice. But accurate measurement *also requires small bias*, a fact many people simply ignore. For the bias to be small as desired, it is necessary that the true value being measured must approximately equal to μ_X, the expected value $E(X_i)$ of each of the repeated measurements X_i.

It is good statistical practice to choose n large enough to reduce the influence of random error on the measurement process. However, to control systematic measurement error or bias, "good science" is required. For example, scientists and engineers are always calibrating their instruments and equipment and are always thinking about eliminating possible sources of bias. The basic principle for achieving accurate measurement (i.e., estimate $\approx$ true value) is that *both* the bias and the random error must be small.

Estimation of a Characteristic of a Real Population

The process of obtaining repeated measurements is that of random sampling from a **conceptual population** (see Section 7.1). By contrast, many estimation problems involve random sampling from a **real population**, often a population of people.

Section 8.1 Introduction to Statistical Estimation

We learned in Section 4.5 that a random sample of over 100,000 Americans is taken monthly in the *Current Population Survey*. In order to better judge the economic health of the nation, this large sample is used to accurately estimate many parameters, such as the proportion of Americans that are unemployed.

Indeed, the statistical problem of estimating a parameter associated with a real population occurs often. For example, preelection polls, as discussed in Section 7.3, play a major role in election campaigns. Their purpose is usually to estimate for each candidate the population proportion p (a parameter) favoring that candidate.

When random sampling from a real population and recording some quantitative aspect of each sampled member, the parameter usually estimated is the population mean, denoted by μ or μ_X. This population mean is the unknown quantitative characteristic averaged over the entire population. The most common estimator of μ is $\overline{X}$, the mean of a random sample $X_1, X_2, ..., X_n$ from the population. Although it is usually both a practical and a financial impossibility to observe each member of the population (called a **census**), taking a random sample with an empirical distribution like that of the population is both practical and economical. In particular, parameters of the population, such as the mean μ and standard deviation σ should be well estimated by the analogous sample quantities, called **statistics** or **estimators**, namely the sample mean $\overline{X}$ and the sample standard deviation S. Letting n denote the sample size and N the population size, we have that

$$\overline{X} = \sum_1^n X/n \approx \sum_1^N X/N = \mu,$$

where $\overline{X}$ is based on only the sampled members and μ is based on all the members of the population.

For a population survey, the random sampling is *without replacement*. According to *the law of large numbers for population surveys* given in section 5.1, when the sample size n is large, $\overline{X} \approx \mu_X$. The Current Population Survey, with $n > 100,000$ will produce very accurate estimates of the population entities.

Both the repeated measurements process and population survey process require a random sample. Random sampling is done with replacement in the repeated measurements case and without replacement in the population survey case.

Consider next some concrete examples of population surveys (often called polls or sample surveys).

Example 8.1

Twelve hundred people randomly selected from across the country are interviewed in order to determine the popularity of the U.S. president. The sample in this case is the 1200 people, and the population is all the people of voting age in the country. Denote $X = 1$ if a sampled person has a favorable impression of the performance of the President and $X = 0$ if not. Then $\overline{X} = \Sigma X / (1200)$ is a good estimator of p, the proportion of all Americans who have a favorable impression. Clearly 1200 is "large", and hence

$$\overline{X} \approx p$$

by the law of large numbers. If $\overline{X} = \hat{p} = 0.71$, we can be confident that somewhere close to 71% of all Americans have a favorable impression of the President's performance.

Example 8.2

Suppose a sample survey (a random sample without replacement) of 200 Texas married couples is taken where the number of children for each couple is recorded. Suppose

$$\overline{X} = 1.92.$$

By the law of large numbers for population surveys,

$$\overline{X} = \hat{\mu} \approx \mu$$

where μ is the average number of children/family in the population of all Texas families.

Example 8.3

In a medical study of third trimester pregnant women (over 180 days into the pregnancy) in Chicago, a random sample of 50 women is taken and their systolic blood pressures recorded, producing

$$\overline{X} = 132 \text{ mmHG}.$$

This is a good estimate (n is fairly large) of the unknown mean μ_C of the blood pressure of all third trimester pregnant women in Chicago. That is, $\hat{\mu}_C = 132$ mmHg.

Terminology and Notation

Whether the population is conceptual as in measurement or real as in surveying, the estimation problem is viewed as **random sampling from a population.** The quantity being estimated is a **parameter,** sometimes called a **population parameter.** The **estimator** is computed from the random sample according to some formula or algorithm, for example $\overline{X}$ is a common estimator of the population mean μ.

A random variable is usually indicated by an uppercase letter and an observed value of that random variable by a lowercase letter. Thus $P(X = x)$ denotes the probability that the random variable X equals the number x. This notation is carried over to statistics (estimators) obtained from random sampling. For example, $\overline{X}$ and S denote two random variables (or statistics), which can be computed once the random sampling is done. By contrast, the lowercase $\overline{x}$ and s denote the *observed* numerical values of $\overline{X}$ and S that result after sampling. This notational convention for estimators will be used throughout the remainder of the book.

Section 8.1 Exercises

1. For each of the following, identify the sample estimator (statistic), population, and parameter involved. Use the standard notation when appropriate.
    a. In a poll of 64 freshmen at a large college, it was found that 65% were in favor of lowering the legal drinking age from 21 to 18.
    b. Several grains of corn taken at random from a truck unloading corn at a grain elevator were found to have an average moisture content of 35%.
    c. Of 10 frogs randomly taken from a large forest pond, two were found to be deformed. (An increase in deformities in amphibians has been an unresolved environmental quality issue in recent years.)
    d. Twenty-five "one-pound" loaves of bread were randomly sampled from a day's output at a bakery. The average weight of these was 1.07 pounds.
    e. The average cholesterol count of 10 randomly selected long-distance runners from Boston was 157.

2. A poll of 20 randomly sampled people at a shopping center was taken to rate the taste of a new cola drink. The results (1 = "terrible" to 5 "terrific") were as follows:

4	3	3	4	4	3	2	3	5	3
4	2	4	3	3	2	3	4	4	3

What is the mean preference score of the 20 people in the poll? What is the range of preference scores in the poll? What is the sample SD? What parameter could the mean preference score of the poll estimate? What parameter could the sample SD estimate?

3. A random sample of 25 college students at a midwestern university is surveyed about their political views. The students are categorized as liberal, conservative, or neither. Ten of them have liberal views. What parameter could this statistic be used to estimate?

4. In the newspaper account below, what sample and population are involved? What factors could affect how well the sample represents the population?

By measuring annual tree rings in groves scattered across the Western states, scientists here are seeking to define the likelihood of prolonged and extreme drought for each of nine river basins in that region. At least 10 trees are sampled at each site. A hollow drill extracts a pencil-sized core of wood without seriously harming the tree. The score is then studied under magnification to record climate-induced variations in tree ring width back to the time when the tree began growing. Trees on well-drained slopes are preferred since they respond quickly to a drought.

An objective of the project, funded in part by $286,000 from the National Science Foundation, is to determine whether there is evidence for what some hydrologists call a "Noah effect." This would be a weather extreme beyond known precedent, like the 40 days of rain that, according to Genesis, flooded the world in the days of Noah. A variant, known as the "Joseph effect," would be a condition—such as a drought—far more prolonged than any on record. The reference, again from Genesis, is to the seven years of famine predicted by Joseph. It is assumed that, if radical departures from normal behavior have taken place in the past, they may occur again and the frequency of occurrence may be estimated.
(W. Sullivan, "Ancient Tree Rings Tell a Tale of Past and Future Droughts," *The New York Times*, September 2, 1980)

5. Give examples from newspaper or magazine articles in which a sample statistic is used (or assumed to be used) to estimate a population parameter.

6. Give the names and symbols for the parameters and statistics used in estimation of the mean, SD, and proportion.

7. The diameter of a human hair (in micrometers) was repeatedly measured in a laboratory as follows:

49	51	48	48	49	47	50	48	49
49	51	49	48	47	51	50	51	49
48	49	52	49	52	52	53	47	53
47	51	52						

Find the mean and standard deviation of the measurements. Draw a histogram of the data. Do the data appear to be normally distributed?

8. A state inspection office tested 50 gasoline pumps around the state for measurement errors. Five gallons of gasoline (according to the meter) were pumped, and the error was measured. The results, in cubic inches, are as follows (minus signs mean less than 5 gallons):

2	−4	−6	−3	0
−4	−1	0	0	−2
−1	−1	−4	2	−4
−1	−2	−4	−3	1
−2	−4	−2	−1	−4
−2	−6	−2	−1	−5
0	−3	−4	−3	−7
−5	−1	−4	−4	2
−2	−3	−3	1	1
−4	−3	1	−3	−3

Construct a histogram of the errors. Do the data appear to be approximately normally distributed? Find $\bar{x}$. Do you suspect bias in the sense that pumps systematically deliver less than one gallon of gas?

9. The compressive strength of cement is the amount of squeezing pressure it can withstand before crumbling. The following data are values of the compressive strength (in pounds) of the same batch of cement as measured by 50 laboratories. Estimate the true compressive strength of the cement. Does the data appear to be normally distributed?

4466	3914	4084	4135	4084
4154	4123	4030	4120	3626
3771	3889	4180	4141	4524
3810	3777	4072	3871	4181
3948	4081	4334	4046	4130
3676	3888	4126	4128	4058
3589	4135	4018	3675	4251
3842	4409	4134	4017	3888
3922	4300	4447	4228	3907
3825	4172	4005	4241	4339

10. Weigh or measure something repeatedly, say 25 times. Compute the mean (your estimate of the true value) and the standard deviation (your estimate of measurement variability). Do the data appear to be normally distributed? Does using $\bar{x}$ to estimate the unknown true value provide a reasonably accurate measurement, assuming we know there is no bias?

8.2 RANDOM SAMPLES, ESTIMATORS, AND STANDARD ERRORS OF ESTIMATION

Statistical estimation problems start with a **random sample**. Thus we first need to develop a deep and clear understanding of the concept of a random sample that applies in both the random experiment setting and the survey sampling setting.

The Random Sample in the Random Experiment Setting

The measurement and the sample survey settings of Section 8.1 each require a *random sample* from a population, either real or conceptual. For measurements, the random sample consists of *independent* measurements $X_1, X_2, \ldots, X_n$ from the conceptual population of all possible measurements. Each random variable X_i has the same probability distribution, for example, the normal distribution $N(\mu, \sigma)$. This distribution is equivalent to the box model distribution used to simulate repeated measurements by random by sampling with replacement.

Informally, random variables are independent when they don't influence one another. A more formal understanding uses the notion of independent events. Recall that the events A and B are independent if the occurrence of A does not influence the chance of B occurring, that is, $P(B|A) = P(B)$.

> Two random variables X and Y are said to be independent, if knowing the observed value of X does not influence the random behavior of Y, that is, $P(a \leq Y \leq b | X = x) = P(a \leq Y \leq b)$ for all values x of X and all intervals $(a \leq Y \leq b)$ for Y.
>
> By a **repeated trials random experiment** we mean that the individual measurements $X_1, X_2, \ldots, X_n$ are independent random variables, each with the same distribution.

The measurement problem is a special type of repeated trials random experiment. Certain scientific or engineering experiments, such as Newcomb's measurements of the speed of light, consist of repeated, independent trials of a random experiment in which physical conditions, such as the room temperature, are kept constant from trial to trial. The process of carrying out independent trials of a random experiment under the same physical conditions is called **replication**. Replication is a basic design principle in engineering and science experiments and in other areas of inquiry where data is collected.

As stated above, a repeated trials random experiment, or just a **random experiment**, consists of n independent random variables $X_1, X_2, \ldots, X_n$ each with the same probability distribution where n is the number of **replications**. Statisticians say that $X_1, X_2, \ldots, X_n$ are **independent identically distributed** and call this collection of random variables a **random sample**.

Example 8.4

Consider the random experiment of throwing a die five times. Let X_i = number of spots on the ith throw. This is a repeated trials random experiment with each die throw a trial. There are five replications of the basic experiment. $X_1, X_2, \ldots, X_5$ is a random sample and the probability law or distribution of each X_i ($i = 1, 2, \ldots, 5$) is given by

x	1	2	3	4	5	6
$p(x)$	1/6	1/6	1/6	1/6	1/6	1/6

Note that each X_i has this *same* distribution, as required for a random sample. The population being "sampled" is conceptual, namely all conceivable throws of the die. This random experiment can be simulated using five-step simulation to do repeated random draws with replacement from the box.

From the estimation perspective, the random variable $\bar{X} = \sum X / 5$ is an estimator of $\mu = E(X_i)$, the expected value of a die throw. Once the experiment is carried out, $\bar{X}$ becomes a number $\bar{x}$, for example, $\bar{x} = 3.2$.

Example 8.5

Consider the random experiment of Section 5.1, namely observing the time until burnout of n 40-watt electric light bulbs. This can be viewed as random sampling from the conceptual population of all possible 40-watt light bulb burnout times. Let X_i = lifetime (time until burnout) of the ith tested light bulb. This is a repeated trials random experiment with n replications. $X_1, X_2, ..., X_n$ is a random sample of independent identically distributed random variables. A possible probability model for each trial's X_i might be that X_i is $N(\mu,\sigma)$, where μ and σ are unknown. Suppose $n = 50$, $\bar{x} = 1085$, and $s = 310$. Then we can accurately estimate $\hat{\mu} = 1085$, $\hat{\sigma} = 310$ and estimate that the time until failure can be modeled by a $N(1085,310)$ distribution.[1] Note that there is little to no measurement error involved. Instead, the variability in the random X_i variables comes from manufacturing variations in the 40-watt light bulbs.

Other examples of random experiments include a repeated measurement process, tossing a coin repeatedly, randomly drawing playing cards with replacement, playing the lottery day after day, throwing a ball many times and observing the distance of each throw, and so forth. In each case there are natural population quantities (parameters), such as the population mean, and natural estimators of them, such as the sample average, that can be computed from the random sample.

In this textbook, the simulation of a random experiment by repeated random sampling (with replacement) from a box model is emphasized as a "hands-on" way to produce deeper understanding. Sometimes, as with the bootstrap introduced in Section 8.6, box model sampling provides a powerful method to do statistical inference.

The Random Sample in the Real Population Setting (Survey Sampling)

Conducting a poll or, more generally, taking a survey is randomly sampling **without replacement** from a real (actually existing) population. Because the sampling is without replacement, the random variables of the resulting random sampling are *not independent*, as the next example illustrates.

Example 8.6

We can simulate survey sampling by randomly sampling *without replacement* using a box model. First consider a small box model representing sampling without replacement from a small population:

Consider a random sample of size two obtained by randomly sampling without replacement two times (Step 2 of the 5-step method). Note that the conditional probability $P(X_2 = 1 | X_1 = 1) = 0$. By contrast, we compute $P(X_2 = 1)$ using a Section 4.2 probability tree analysis, as shown in Figure 8.1.

---

[1] In actual practice, a better model than the normal would be to use an "exponential distribution." We will not discuss this possibility.

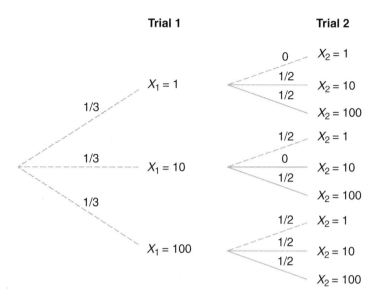

Figure 8.1 Probability tree for a $n = 2$ random sample without replacement.

Using the addition rule and the multiplication rule for probability trees, where the relevant branches are shown by dashed lines, we find that

$$P(X_2 = 1) = \left(\frac{1}{3} \cdot 0\right) + \left(\frac{1}{3} \cdot \frac{1}{2}\right) + \left(\frac{1}{3} \cdot \frac{1}{2}\right) = \frac{1}{3}.$$

Clearly $P(X_1 = 1) = 1/3$. In fact, X_1 and X_2 have the same distribution:

x	0	10	100
$p(x)$	1/3	1/3	1/3

Because $1/3 = P(X_2 = 1) \neq P(X_2 = 1 | X_1 = 1) = 0$, X_1 and X_2 are not independent and indeed are strongly dependent.

Now suppose instead that have we have a large population modeled by a large box of 100,000 ones, 100,000 tens, and 100,000 hundreds representing survey sampling from a large population for which 1, 10, 100 are still equally likely. Thus, the population proportions are the same as for the small box.

100,000 balls   100,000 balls   100,000 balls
    (1)   (10)   (100)

Then, as for the small box, $P(X_1 = 1) = 1/3$ and $P(X_2 = 1) = 1/3$ using a probability tree analysis. Indeed, as for the small box, X_1 and X_2 have identical distributions and have the same distribution as the small box. But $P(X_2 = 1 | X_1 = 1) = (99{,}999)/(299{,}999)$, which is almost equal to $P(X_2 = 1) = 1/3$. This together with the comparison of other analogous probabilities like $P(X_2 = 10 | X_1 = 100{,}000)$ versus $P(X_2 = 10)$, show that X_1 and X_2 are almost independent, and indeed for practical purposes can be considered independent.

342 CHAPTER 8 CONFIDENCE INTERVAL ESTIMATION

We have just observed a special case of an important general rule.

> **Large Population Random Sampling Rule**
>
> When doing real population random sampling and the population size N is large compared to the sample size n, then the random sample $X_1, X_2, \ldots, X_n$ is modeled as being independent and identically distributed. The rule of thumb defining "large" is $N \geq 20n$. That is, the sample size must be at most 5% of the population size.

From the box modeling perspective, this means that when $N \geq 20n$ we can assume random sampling *with replacement*. In our simulation example, because $300{,}000 \geq 20 \times 2$, we can replace sampling without replacement from the $N = 300{,}000$ large box by sampling with replacement from the $N = 300{,}000$ box, knowing that either simulation model will produce almost identical probability sampling behavior. But since box proportions are the same for the $N = 3$ box, we can now replace sampling with replacement from the $N = 300{,}000$ box by sampling with replacement from the $N = 3$ box.

Because real population random sampling usually involves a much larger population size N than the sample size n, the textbook assumes large population random sampling statistical inference problems unless specifically stated otherwise. Hence, the following definition holds for random experiments and survey sampling unless it is specifically stated that the population size is *not* large relative to the sample size:

Definition. A **random sample** $X_1, X_2, \ldots, X_n$ consists of independent identically distributed random variables.

A random sample can always be simulated, sometimes by hand by building a box model and then, perhaps, using a random number table to simulate drawing from a box. A random sample can always be simulated by using the simulation software. One obtains multiple random draws by drawing *with replacement* from a specified box model. This is always at worst a very good approximation to reality because if one is survey sampling from a real population, only large populations are considered, which allows sampling with replacement to be a good approximation to the sampling without replacement.

The Basic Estimation Problem

All estimation problems, whether they have a random experimental setting or a real population setting, start with a random sample of independent identically distributed random valuables. Random sampling can always be simulated by sampling with replacement from an appropriately chosen box model (see Section 5.2). For example, in the Cereal Box Problem of Example 5.9 the expected number (theoretical population mean μ) of cereal boxes required to get at least one of each of six colored pens was estimated, and in Example 5.10 various parameters associated with the airline practice of overbooking flights were estimated.

More commonly, the random sample is a result of observing real-life random phenomenon (polling, random experiment, etc.) rather than the random sample being simulated. However, it does not matter whether the computer or the "real" world generates the data.

In all estimation problems, the goal is to estimate one or more parameters, such as μ, p, or σ, associated with the population (i.e., associated with the distribution of each of the randomly sampled X variables). Estimating the mean μ of randomly sampled X's is the most common estimation problem.

Since estimation problems can involve any parameter of interest to the practitioner, we will denote the parameter being estimated generically by θ. Any estimator $\hat{\theta}$ of the parameter θ

computed from the random sample requires some formula involving the random sample X_1, $X_2, \ldots, X_n$. For example, we can estimate the population SD σ by:

$$S = \sqrt{\frac{\sum(X - \overline{X})^2}{n-1}}.$$

Sampling Distribution of an Estimator

In Section 8.1 we learned that $\overline{X}$ estimates μ accurately if the sample size is large. But, to be useful, one must be able to answer *how accurately* $\overline{X}$ estimates μ. For example, the practitioner desires to know the **typical** estimation error size defined by

$$\overline{X} - \mu.$$

One answers this vital question by obtaining the probability distribution of $\overline{X}$, is called the **sampling distribution** of $\overline{X}$. More generally, we want the sampling distribution of $\hat{\theta}$ (an estimator of θ).

We learned in Chapters 2 and 5, respectively, that the sample SD is a measure of the spread of the data and the theoretical standard deviation is a measure of the spread of (the distribution of) the random variable X modeling each piece of data. In particular, $\pm \sigma_X$ expresses the typical variability of X. More generally, we learned in Section 5.3 that much more can be stated for roughly bell-shaped probability histograms: Random sampling of X is replicated a large number of times,

around 68% of the observed X values will fall the interval $(\mu - \sigma, \mu + \sigma)$;

around 95% of the observed X values will fall in the interval $(\mu - 2\sigma, \mu + 2\sigma)$; and

around 99.7% of the observed X values will fall in the interval $(\mu - 3\sigma, \mu + 3\sigma)$.

Thus, the standard deviation σ_X can tell us *a lot* about the variability of X when replications of X are observed (produced either by actual experimentation, sampling, or simulation). Similarly, the standard deviation of an estimator (i.e., the standard deviation of its sampling distribution) tells us the typical random variability of the estimator and, hence, the typical estimation error. The standard deviation of an estimator is given a special name, the **standard error of the estimator**, denoted by SE($\overline{X}$) in the case of the mean $\overline{X}$ or SE($\hat{\theta}$) more generally for an estimator $\hat{\theta}$.

Note: A standard error is just the standard deviation of the statistic of interest, which is different from the standard deviation of the population (or distribution) producing the random sample from which the statistic of interest is computed.

Example 8.7

Suppose we desire to estimate the average height μ of U.S. male military draftees. Assume a random sample of size 25 yields $\bar{x} = 68.6$. Thus $\hat{\mu} = 68.8$, but we should also report the accuracy of $\hat{\mu}$. If we knew SE($\overline{X}$), then we could express the estimator of μ and its typical error size by

$$\overline{X} \pm \text{SE}(\overline{X}).$$

For example, if SE($\overline{X}$) is approximately $1/2$ inch, we would expect $\bar{x} = 68.6$ to be within ± 1 inch of the true population height. This would be very useful information.

In order to better understand in what sense the standard error of $\overline{X}$ represents the "typical" error, we will use the five-step method to simulate 10,000 values in the case where $X_1, X_2, \ldots, X_n$ is a random sample from a normal population $N(69,3)$, that is, normal with $\mu = 69$, $\sigma = 3$. Based on the data histogram in Figure 2.16 of the heights of 2000 women, it is clearly very reasonable to

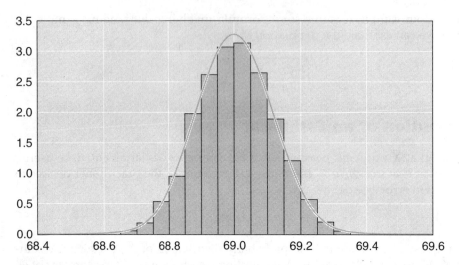

Figure 8.2 The density histogram of 10,000 estimated heights $\bar{x}$ with $n = 25$, $\sigma = 3$, $\mu = 69$, with a $N(69, 0.6)$ density superimposed.

assume that these heights are nicely modeled by a normal distribution. Using $\sigma = 3$ and $n = 25$, it can be shown, as you will see shortly, that $SE(\bar{X}) = 0.6$. The density histogram of the simulated $\bar{x}$ values is given in Figure 8.2.

Note that the density histogram of the $\bar{x}$'s is remarkably bell-shaped. Indeed, it is shaped like the superimposed normal density with mean 69 and standard deviation 0.6. (We will learn shortly that the sampling distribution of $\bar{X}$ is forced to be bell-shaped when n is large.) In particular, the average of the observed $\bar{x}$ values, denoted by $\bar{\bar{x}}$, is 68.9859, which is extremely close to $\mu = 69$. We learned in Chapter 5 that the average $\bar{\bar{x}}$ of a large number of replications of $\bar{x}$ should estimate $E(\bar{X})$ well. Thus because $\bar{\bar{x}} \approx \mu$ has been observed we have evidence that

$$E(\bar{X}) = \mu.$$

This is a mathematical fact that can be shown theoretically. Importantly, this fact tells us that the distribution of $\bar{X}$ is centered around μ; that is, $\bar{X}$ is not a biased estimation of μ.

The observed standard deviation of the simulated $\bar{x}$'s is 0.6030. Note that $\widehat{SE}(\bar{X}) = 0.6030$ is extremely close to the theoretical value $SE(\bar{X}) = 0.6$. This suggests that when $\bar{x}$ values can be simulated, we have an experimental way to accurately estimate $SE(\bar{X})$, an observation that will lead us to bootstrapping.

Because the density histogram is bell-shaped, we can use the 68-95-99.7% rule for data (the $\bar{x}$ values) using $\bar{\bar{x}} = 68.9859$ and $\widehat{SE}(\bar{X}) = 0.6030$. In fact, this rule works extremely well with 68.15% of the $\bar{x}$ values falling in 68.9859±(2 · 0.6030). Such information helps us translate the numerical value of $\widehat{SE}(\bar{X})$ into precise information about the typical size of the variation of $\bar{X}$.

Thus $SE(\bar{X})$ is very informative about the sampling distribution of $\bar{X}$. For example, because the distribution of $\bar{X}$ is bell-shaped, it tells us that about 95% of the time we can expect X to come within +? · $SE(\bar{X})$, here within ±2 · 0.6 = ±1.2 of the true mean μ.

Finally, note that $SD(\bar{X}) = 0.6$ is quite a lot less than the $\sigma = SD(X) = 3$ associated with a single observation. This means that the typical estimation error of $\bar{X}$ is much less than the population SD. In fact, the larger the sample size the smaller $SE(\bar{X})$ will be, as will be seen shortly. Do not confuse the meaning of σ, the typical variation from one observation of X to the next, with $SE(\bar{X})$, the typical variation from one $\bar{X}$ to the next.

Definition. An estimator $\hat{\theta}$ is **unbiased** if $E(\hat{\theta}) = \theta$.

Section 8.2 Random Samples, Estimators, and Standard Errors

Empirically the definition means that if $\hat{\theta}$ is simulated a large number of times, then the average of these $\hat{\theta}$ values is approximately θ. That is,

$$\overline{\hat{\theta}} \approx \theta.$$

Because $E(\overline{X}) = \mu$, as a special case $\overline{X}$ is an unbiased estimator of μ.

Were we just lucky that the density histogram of $\overline{X}$ in Figure 8.2 was shaped approximately like a normal density, thus allowing us to use the 68-95-99.7% rule, and so forth? In Sections 8.3 and 8.5, we will learn the striking and enormously useful fact that when the sample size n is large, both $\overline{X}$ and $\hat{p}$ (the sample proportion) are approximately normally distributed!

Section 8.2 Exercises

1. Suppose $SE(\overline{X}) = 0.2$ is known. Suppose $\bar{x} = 65$ inches for a random sample of 50 heights of female military draftees.
    a. State an estimate of the population height μ and the typical estimation error size.
    b. Supposing evidence that the distribution of $\overline{X}$ is normal, find
        i. $P(|\overline{X} - \mu| \leq 0.4)$
        ii. $P(|\overline{X} - \mu| \leq 0.2)$
        iii. $P(|\overline{X} - \mu| \leq 0.35)$
    Hint: Use the Normal Table E (in Appendix E).

2. Suppose a random sample of size $n = 10$ from a population of $N = 200$ dairy cows.
    a. Can you assume that the random variables are independent as a good approximation?
    b. Is this a real or conceptual population?
    c. If the population size $N = 40$, can the random variables be assumed independent as a good approximation?

3. Consider a random sample of size 10 conducted by each of 1000 statisticians who each find an $\overline{X}$. Suppose $\mu = 6$.
    a. Will Statistician 1 have $\bar{x} \approx 6$? (That is, is $\overline{X}$ guaranteed to have a small estimation error?)
    b. What will be true about the average $\overline{\bar{x}}$ of the 1000 $\bar{x}$ values?

4. Let X_1, X_2 be a random sample of size 2 from a $N(1,2)$ distribution.
    Find $P(X_1 \leq 0, X_2 \leq 2)$
    Hint: Are X_1 and X_2 independent?

5. Define or explain random sample, replication, random experiment, parameter, estimator, real population, conceptual population, sample survey, measurement, estimation, sampling distribution, standard error.

6. Suppose random sampling from a population with SD $\sigma_X = 5$. Suppose $n = 20$. Compare roughly the standard error of $\overline{X}$ with the SD σ_X of one observation. That is, is one a lot larger than the other?
    Hint: You are *not* being asked to solve for $SE(\overline{X})$.

7. Suppose a random sample (1,3,7,5) from a population.
    a. Estimate the population mean μ.
    b. Estimate the population SD σ.

8. Suppose a die is thrown 100 times and a six appears 7 times. Let $p = P(\text{six})$.
    a. Estimate p.
    b. Is the population here real or conceptual? (Be careful to review what conceptual means; clearly the die is "real" of course.)

8.3 CENTRAL LIMIT THEOREM FOR $\bar{X}$

We will take an experimental approach to the question of when $\bar{X}$ is approximately normal, in the process discovering an experimental version of the central limit theorem. We are going to simulate the sampling distribution of $\bar{X}$ when $n = 25$ for three different populations. We classify $n = 25$ as a **large sample** because $n \geq 20$, our rule of thumb for a large sample size. The central limit theorem below is a **large sample theory** result because it only applies when the sample size n is large.

First we state a fundamental result that we will use over and over.

> The standard error of $\bar{X}$, denoted by $SE(\bar{X})$, where $\bar{X}$ is based on $X_1, X_2, \ldots, X_n$, is a random sample with population standard deviation $SE(X) = \sigma$, is given by
>
> $$SE(\bar{X}) = \sigma / \sqrt{n}.$$
>
> Further, $E(\bar{X}) = \mu$.

The first population is a fair die population.

Example 8.8

Throw a die 25 times. Then, $X_1, X_2, \ldots, X_{25}$ is a random sample with each X_i distributed as

x	1	2	3	4	5	6
$p(x)$	1/6	1/6	1/6	1/6	1/6	1/6

For a fair die (see Section 5.3),

$$\mu = \sum xp(x) = \sum_1^6 x\frac{1}{6} = \frac{21}{6} = 3.5,$$

and

$$\sigma = \sqrt{\sum \left(x - \frac{7}{2}\right)^2 \frac{1}{6}} = \sqrt{\frac{35}{12}} = 1.708.$$

By our formula for $SE(\bar{X})$,

$$SE(\bar{X}) = \frac{\sigma}{\sqrt{n}} = \frac{1.708}{5} = 0.342.$$

Thus, the sampling distribution of $\bar{X}$ has a mean of 3.5 and SD of 0.342. We use the simulation software to sample 10,000 $\bar{x}$ values. The resulting density histogram, with the $N(3.5, 0.342)$ normal density superimposed, is displayed in Figure 8.3. Note that $\mu - SE(\bar{X}) = 3.158$ and $\mu + SE(\bar{X}) = 3.842$ are shown.

The following are the main results of the simulation:

1. The average of the observed $\bar{x}$ values, $\bar{\bar{x}} = 3.4988 \approx \mu = 3.5$ and the observed SD based on the $\bar{x}$ values, namely $\widehat{SE}(\bar{X}) = 0.344$, is extremely close to $SE(\bar{X}) = 0.342$.
2. The percentage of the $\bar{x}$ values in the interval $\mu \pm SE(\bar{X})$, namely 3.5 ± 0.342, is 67.7%, a percentage almost identical to the 68% computed by normal theory.
3. The density histogram of the 10,000 $\bar{x}$ values is fit well by the superimposed normal density $N(3.5, 0.342)$.

Section 8.3 Central Limit Theorem for $\bar{X}$

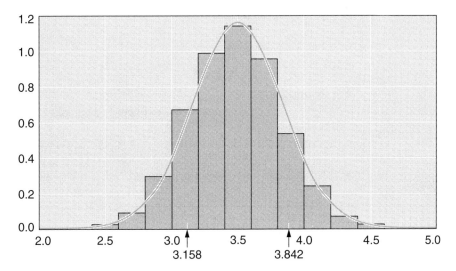

Figure 8.3 Density histogram of 10,000 simulated $\bar{x}$ values from a fair die population, $n = 25$.

Example 8.9

We draw a random sample of size 25 from the following skewed continuous "exponential" density

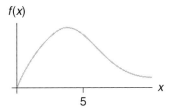

We use the simulation software to sample 10,000 $\bar{x}$ values. The resulting density histogram, with the $N(5,1)$ normal density superimposed, is displayed in Figure 8.4:

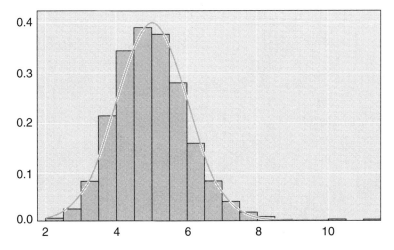

Figure 8.4 Density histogram of 10,000 simulated $\bar{x}$ values from an exponential population with $\mu = 5$, $\sigma = 1$ and $n = 25$.

Note that

$$E(\bar{X}) = \mu = 5, \quad SE(\bar{X}) = \frac{\sigma}{\sqrt{n}} = \frac{5}{5} = 1,$$

which justifies using the $N(5,1)$ density.

We note the following three points as in Example 8.9:
1. The average of the $\bar{x}$ values namely $\bar{\bar{x}} = 5.0028 \approx \mu = 5$ as expected, and $\widehat{SE}(\bar{X}) = 0.9978 \approx SE(\bar{X}) = 1$.
2. The percentage of the $\bar{x}$ values in the interval $\mu \pm SE(\bar{X})$, namely 5 ± 1 is 68.3%, a percentage close to 68%.
3. The density histogram of the 10,000 $\bar{x}$ values is fit by the superimposed $N(5,1)$ normal density with a slight skewness to the right.

The third population is a three-valued skewed discrete distribution.

Example 8.10

We draw a random sample of size 25 from the population distribution given by:

x	1	2	5
$p(x)$	1/6	1/2	1/3

Standard formulas from Chapter 5 show that

$$\mu = \frac{17}{6} = 2.8333, \qquad \sigma = 1.5723.$$

Thus

$$E(\bar{X}) = 2.8333, \qquad SE(\bar{X}) = \sigma/\sqrt{n} = 0.345,$$

leading to a $N(2.8333, 0.345)$ distribution. We note the analogous three points as in Examples 8.8 and 8.9: $\bar{\bar{x}} = 2.8379 \approx 2.8333 = \mu$, $\widehat{SE}(\bar{X}) = 0.3151 \approx SE(\bar{X}) = 0.3145$; about 68% of the $\bar{x}$ values lie in $\mu \pm SE(\bar{X})$; and the experimental sampling distribution obtained graphing the density histogram of the 10,000 $\bar{x}$ values is well fit by the superimposed $N(2.833, 1.573)$ density, as shown in Figure 8.5.

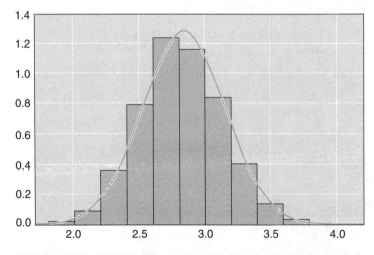

Figure 8.5 Density histogram of 10,000 simulated $\bar{x}$'s from a discrete population with $\mu = 2.8333$, $\sigma = 1.5723$, and $n = 25$.

The previous examples produce powerful empirical evidence for the following result, called the central limit theorem (CLT), which can be stated as an empirical (experimental) fact about observed $\bar{x}$ values or a theoretical fact about the sampling distribution of $\bar{X}$.

Central Limit Theorem as an Experimental Fact

Suppose we take a very large number of random samples, each of size n, from a population with mean μ and SD σ. Then when the sample size n is large ($n \geq 20$), the shape of the density histogram of the sample means (the $\bar{x}$ values) will approximate a normal density whose mean is μ and whose SD is $SE(\bar{X}) = \sigma/\sqrt{n}$. Equivalently, the experimental probability $\hat{P}(\bar{X} \leq x)$ is given for any x by

$$\hat{P}(\bar{X} \leq x) \approx P\left(Z \leq \frac{x - \mu}{\sigma/\sqrt{n}}\right),$$

where Z denotes a standard normal random variable. That is, as we observed in three special cases, the experimental sampling distribution of $\bar{X}$ is approximately $N(\mu, \sigma/\sqrt{n})$ whenever $n \geq 20$.

We now state the central limit theorem as a result about the theoretical sampling distribution of $\bar{X}$.

The Central Limit Theorem as a Mathematical Theorem

Suppose we take a random sample of size n from a population with mean μ and SD σ. Then, when the sample size n is a relatively large ($n \geq 20$), the theoretical probability sampling distribution of $\bar{X}$ will approximate a normal density with mean μ and SD $SE(\bar{X}) = \sigma/\sqrt{n}$. Equivalently, the theoretical probability $P(\bar{X} \leq x)$ is given for any x by

$$P(\bar{X} \leq x) \approx P\left(Z \leq \frac{x - \mu}{\sigma/\sqrt{n}}\right),$$

where Z denotes a standard normal random variable.

We learned in Chapter 4 that experimental probabilities obtained via simulation approximate the corresponding theoretical probabilities when we have many simulations. In the CLT context, this means that for a large number of random samples and for any x,

$$\hat{P}(\bar{X} \leq x) \approx P(\bar{X} \leq x)$$

Thus when n is large

$$P\left(Z \leq \frac{x - \mu}{\sigma/\sqrt{n}}\right),$$

at least approximately.

We have stated both versions of the CLT separately because we will apply both: the experimental version to a large number of simulated $\bar{x}$ values and the theoretical version to the probability sampling distribution of $\bar{X}$. Now we give an example of the application of the theoretical version of the CLT.

Example 8.11

Suppose we have a random sample of size $n = 25$ with population $\mu = 10$ and $\sigma = 7$. Suppose the population is not necessarily normal in shape. Then by the theoretical CLT, noting that $SE(\overline{X}) = \sigma/\sqrt{n} = 7/\sqrt{25}$,

$$P(9.5 \leq \overline{X} \leq 12) = P(\overline{X} \leq 12) - P(\overline{X} \leq 9.5)$$

$$\approx P\left(Z \leq \frac{12-10}{7/\sqrt{25}}\right) - P\left(Z \leq \frac{9.5-10}{7/\sqrt{25}}\right) = P(Z \leq 1.45) - (P(Z \leq -0.36))$$

$$= 0.9236 - 0.3594 = 0.5642$$

Now we give an example of the application of the experimental version of the CLT.

Example 8.12

We constructed a box model of a discrete nonnormal population having a population mean $\mu = 25$ and a population SD $\sigma = 25$. Its probability histogram is shown in Figure 8.6. The probability histogram indicates that the population is very skewed to the right and therefore far from bell-shaped.

Using the five-step simulation, we randomly drew 10,000 samples each of size 4, 12, 24, and 50 *with replacement* from this population, and we calculated the sample mean $\bar{x}$ for each of the 10,000 samples at each sample size. The density histograms of the resulting 10,000 sample means are displayed in Figures 8.7 (a)–(d).

Let's make observations about the results of our sampling as we look at these density histograms resulting from the simulated $\bar{x}$ values for various sample sizes.

1. The "grand" mean of the 10,000 sample means (denoted as above by $\bar{\bar{x}}$) was close to the population mean of $\mu = 25$ in all three cases. For example, the mean of the 1000 sample means of size 4 is $\bar{\bar{x}} = 24.39$ (as computed from the data). Similarly, the other three $\bar{\bar{x}}$'s satisfy $\bar{\bar{x}} \approx 25$.

2. As the sample size increases, the density histograms of the sample means get increasingly closer to the normal density shape. The probability histogram in Figure 8.6, with its very non-normal shape, represents the population distribution. When $n = 4$ (a number much too small for the experimental version of the CLT to apply), the density histogram of the $\bar{x}$ values has only a very crudely normal shape in Figure 8.7(a). As n grows to 12 (a number for which the

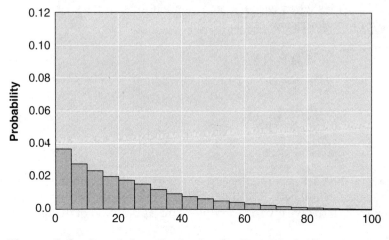

Figure 8.6 Population distribution for Example 8.12.

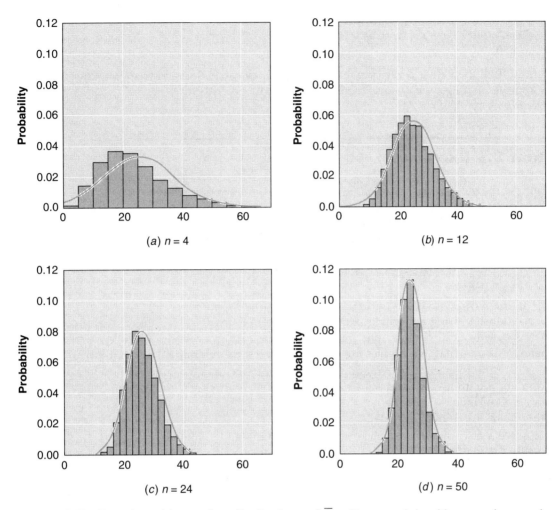

Figure 8.7 Experimental sampling distributions of $\bar{X}$ with normal densities superimposed.

CLT should begin to provide a rough approximation), the density histogram in Figure 8.6(b) appears somewhat more bell-shaped, but still only roughly so. This is what the CLT tells us to expect. In Figure 8.7(c), where $n = 24 > 20$, the density histogram is indeed approximately bell-shaped, although one can still see minor discrepancies. Finally, for the $n = 50$, the density histogram in Figure 8.7(d) shows almost no discrepancies at all from the normal density.

3. As the sample size increases, the variation among the 10,000 sample means decreases. More specifically, as the sample size increases, the simulation-based estimated standard error of $\bar{X}$, namely the sample SD of the 10,000 sample $\bar{x}$ values, decreases as n increases from 4 to 12 to 24 to 50, as shown in Table 8.1. This is precisely the behavior that the theoretical SD of

Table 8.1 Estimated and Calculated Standard Errors of the Mean $\bar{X}$

	n			
	4	12	24	50
SD of observed $\bar{x}$'s	11.39	7.53	5.4	3.58
$SE(\bar{X}) = (\sigma/\sqrt{n})$	12.5	7.72	5.10	3.54

$\overline{X}$, given by $SE(\overline{X}) = \sigma/\sqrt{n}$, predicts. Indeed, the SD of the observed $\overline{x}$ values is quite close to $SE(\overline{X})$ in each case, as estimated from 10,000 simulated $\overline{x}$ values for $n = 4, 14, 24$, and 50.

The central limit theorem is a very powerful tool in statistical analysis. In what statisticians call "large-sample" settings, we will use the CLT to assess the likely error when estimating a population mean μ by $\overline{X}$.

Section 8.3 Exercises

$N(\mu, \tilde{\sigma})$

1. Suppose the mean weight of all male Americans is $\mu = 175$ pounds with SD of $\sigma = 15$ pounds. Give formulas for the estimated theoretical mean and estimated SD of a sample mean from this population based on 50 observations. How is the sample mean distributed?
2. Compute $SE(\overline{X})$, where $\overline{X}$ is the average score on five fair dice. Use the data in the previous exercise to compute parts **a** and **b**.
   a. $\hat{P}(3.5 - SE(\overline{X}) \leq \overline{X} \leq 3.5 + SE(\overline{X}))$
   b. $\hat{P}(3.5 - 2SE(\overline{X}) \leq \overline{X} \leq 3.5 + 2SE(\overline{X}))$
   c. Compare the areas to 0.68 and 0.95, as given by our basic rule.
3. Which density histogram will look more like a normal distribution—the histogram of sample means based on 20 observations per mean or the histogram based on 40 observations per mean? (Assume that the samples are taken from the same nonnormal population).
4. What are the theoretical mean and SD of a sample mean $\overline{X}$ of 40 observations based on a population with mean 0 and SD 1? What is the distribution of the sample mean $\overline{X}$ approximately like? Include information about the theoretical mean and standard error of this sampling distribution of $\overline{X}$.
5. The score X on a fair casino die has mean $\mu = 3.5$ and the standard deviation $\sigma = 1.7$. Let $\overline{X}$ denote the average score on n tosses of a fair die. Use the CLT to estimate $P(3.25 < \overline{X} < 3.75)$.
   a. for $n = 10$ (even though $n \leq 20$, go ahead and do the computation).
   b. for $n = 100$.
   c. for $n = 1000$.
   d. What do you notice when comparing parts a–c?
6. We want to simulate 40 observations of $\overline{X}$, the average score on five fair dice. We shall use the first five-digit column in Table B.2. For the first simulation we use the five digits in the first line, 66533, yielding the average 4.6, and so forth. We get the 40 averages

4.6	3.2	3.2	2.2	4.0	3.0	2.8	3.2	3.6	3.6
3.8	4.0	3.8	3.8	3.6	3.4	3.0	4.2	4.8	4.6
4.0	3.2	3.4	4.0	4.2	3.6	4.0	3.0	2.6	2.6
5.6	4.0	1.6	3.0	3.4	3.8	4.6	3.6	3.2	3.4

   a. Compute the experimental probability $\hat{P}(3.3 < \overline{X} < 3.7)$.
   b. Estimate $P(3.3 < \overline{X} < 3.7)$ using the central limit theorem.
   c. Compare the answers in **a** and **b**.
7. Let X denote the number of heads in 10 tosses with a fair coin. As we have learned, we can write $X = Y_1 + \cdots + Y_{10}$, where $Y_i = 1$ if toss number i results in heads, and 0 if toss number i results in tails. Each Y_i has $\mu = 1/2$ and $\sigma = 1/2$ (recalling that $\mu = np$, $\sigma = \sqrt{np(1-p)}$ and that $n = 1$ for a single Y_i). The proportion of heads in the 10 tosses is

$$X/10 = (Y_1 + Y_2 + \cdots Y_{10})/10 = \overline{Y}.$$

Use the CLT to estimate:
a. $P(0.3 \leq \overline{Y} \leq 0.8)$
b. $P(0.35 \leq \overline{Y} \leq 0.65)$

Since X is binomial with $n = 10$ and $p = 1/2$, use Table G in Appendix G to find the exact probabilities in c and d.

c. $P(0.3 \leq \overline{Y} \leq 0.8) = P(3 \leq X \leq 8) = P(X = 3) + P(X = 4) + \cdots + P(X = 8)$
d. $P(0.35 \leq \overline{Y} \leq 0.65) = P(3.5 < X < 6.5) = P(X = 4) + P(X = 5) + P(X = 6)$

8. We can simulate $Y_1, Y_2, \ldots, Y_{10}$ in Exercise 7 by picking 10 random digits from Table B.1. We shall use the last two five-digit columns to simulate 40 observations of $\overline{Y}$. For the first simulation we use 10001 10110, yielding $X = 5$ and $\overline{Y} = X/10 = 0.5$. We get the 40 observations of $\overline{Y}$:

0.5	0.4	0.6	0.7	0.6	0.5	0.5	0.7	0.6	0.8
0.2	0.4	0.8	0.4	0.4	0.4	0.5	0.5	0.6	0.5
0.3	0.4	0.5	0.5	0.3	0.4	0.6	0.2	0.7	0.7
0.5	0.2	0.6	0.4	0.3	0.8	0.5	0.4	0.6	0.4

Compute the experimental probabilities.
a. $\hat{P}(0.3 \leq \overline{Y} \leq 0.8)$
b. $\hat{P}(0.35 \leq \overline{Y} \leq 0.65) = \hat{P}(\overline{Y} = 0.4) + \hat{P}(\overline{Y} = 0.5) + \hat{P}(\overline{Y} = 0.6)$
c. Compare with the results in Exercise 7.

8.4. LARGE SAMPLE CONFIDENCE INTERVAL FOR THE POPULATION MEAN μ

Estimating SE($\overline{X}$) from a random sample

We have learned that SE($\overline{X}$) is the size of the typical error when using $\overline{X}$ to estimate μ. Thus, it is important to estimate SE($\overline{X}$) accurately. Recall that SE($\overline{X}$) = $\sigma/\sqrt{n}$ where σ is the SD of the population being sampled from. Therefore if σ is unknown (as is usually the case), we will not know the size of the estimation error.

Just as the law of large numbers tells us that $\overline{X} \approx \mu$ when n is large, it is also true that the sample standard deviation S satisfies

$$S \approx \sigma$$

for large n. Thus, when n is large,

$$\text{SE}(\overline{X}) = \frac{\sigma}{\sqrt{n}} \approx \frac{S}{\sqrt{n}}.$$

Thus we have a useful large sample result for accurately estimating the standard error, that is, SE($\overline{X}$) can be estimated by

$$\widehat{\text{SE}}(\overline{X}) = \frac{S}{\sqrt{n}},$$

where S is computed in the usual way from the random sample $X_1, X_2, \cdots, X_n$ when n is large, that is, $n \geq 30$ then S accurately estimates σ and hence $\widehat{\text{SE}}(\overline{X})$ is a good estimation of SE($\overline{X}$).

$$\widehat{\text{SE}}(\overline{X}) \approx \text{SE}(\overline{X})$$

Example 8.13

Recall Newcomb's experiment to measure the speed of light. Excluding the two outliers, the computations in the remaining $n = 64$ measurements show that the average measured time Newcomb's beam of light look to travel 7442 meters was $\overline{T} = 24{,}827.75$ nanoseconds and that the sample SD was $S_T = 5.08$ nanoseconds, where a nanosecond is 1×10^{-9} seconds. Thus,

$$\widehat{SE}(\overline{T}) = \frac{S_T}{\sqrt{n}} = \frac{5.08}{8} = 0.635.$$

Therefore, the estimated time for a beam of light to travel 7442 meters is given by

$$\overline{T} \pm \widehat{SE}(\overline{T}),$$

that is,

$$24{,}827.75 \pm 0.635 \text{ nanoseconds}.$$

Since the standard error is so small, serious measurement error is likely due to bias or systematic error. Indeed, recalling that the true speed of light is now known to be extremely close to 299,792.5 kilometer/second, as contrasted with Newcomb's estimate of 299,750.8 km/sec, the bias in the estimated time $\overline{T}$ (around 3 to 4 nanoseconds) is the major source of the error. Thus, even if Newcomb had the patience to make 1000 observations, his estimate of the time would still be low by 3 to 4 nanoseconds (due to bias) and hence, by using $v = d/\overline{T}$, the estimated speed of light would still have been too slow by 35 to 45 km/sec.

In summary, it is vital to know the magnitude of $SE(\overline{X})$ in order to assess the accuracy of $\overline{X}$ as an estimator of μ. Moreover, $SE(\overline{X})$ is well estimated when n is large using

$$\widehat{SE}(\overline{X}) = S/(\sqrt{n}).$$

However, sometimes the bigger source of estimation error is systematic estimation error (bias) rather than the random error as quantified by $SE(\overline{X})$.

Confidence interval for μ using $\overline{X}$ and $\widehat{SE}(\overline{X})$

Provided a random variable X is approximately normally distributed with mean μ_Y and SD of σ_Y, then the 68-95-99.7% rule applies (recall Figure 6.12). In particular,

$$P(|X - \mu| \leq 1.96\,\sigma) \approx 0.95.$$

This switch from 2 to the more accurate 1.96 is based on the Table E fact that

$$P(-1.96 \leq Z \leq 1.96) = P(|Z| \leq 1.96) = 0.95,$$

where Z is standard normal.

By the CLT, $\overline{X}$ is approximately normal with mean μ and a SD satisfying $\widehat{SE}(\overline{X}) \approx SE(\overline{X})$ when the sample size n is large. Thus, by the CLT,

$$P(|\overline{X} - \mu| \leq 1.96\,\widehat{SE}(\overline{X})) \approx 0.95$$

Section 8.4 Large Sample Confidence Interval for μ

provided n is large. We require $n \geq 30$ instead of $n \geq 20$ so that in addition to the CLT approximation using $SE(\overline{X})$ being accurate, the $\widehat{SE}(\overline{X}) \approx SE(\overline{X})$ approximation will be quite accurate. The event that $|\overline{X} - \mu| \leq 1.96\, \widehat{SE}(\overline{X})$ is identical to the event

$$\overline{X} - 1.96\, \widehat{SE}(\overline{X}) \leq \mu \leq \overline{X} + 1.96\, \widehat{SE}(\overline{X}),$$

where

$$\widehat{SE}(\overline{X}) = S/\sqrt{n}.$$

This random interval is called a 95% **confidence interval (CI)** for μ and has an approximate 0.95 probability of including μ. This is a **large sample** CI because we can only trust the claimed 0.95 probability when $n \geq 30$.

Thus, a 95% CI now replaces the **point estimate** $\overline{X}$ of μ with an **interval estimate**:

$$(\overline{X} - 1.96\widehat{SE}(\overline{X}), \overline{X} - 1.96\, \widehat{SE}(\overline{X})), \quad \text{or} \quad \overline{X} \pm 1.96\, \widehat{SE}(\overline{X}),$$

an interval estimate whose accuracy (0.95 above) we know.

Example 8.14

In Section 2.4 we examined a data set of the heights of 8585 nineteenth century British men for which $\bar{x} = 67.02$ and $s = 2.564$. Thus

$$\widehat{SE}(\overline{X}) = \frac{s}{\sqrt{n}} = \frac{2.564}{\sqrt{8585}} = 0.028.$$

Assuming that the sample is random, a 95% CI for the average height of the population of 19th century British men is given by

$$\overline{X} \pm 1.96\, \widehat{SE}(\overline{X}),$$

that is, by

$$67.02 \pm 1.96 \cdot 0.028 \text{ inches or } (66.97, 67.07).$$

Thus, the resulting CI has a **lower limit** of 66.97 inches and an **upper limit** of 67.07 inches. The claim is that the true μ lies in this interval. This interval is narrow because n is very large.

Interpretation of a confidence interval

The 95% coverage percentage has an intuitive empirical meaning. Certainly 95% suggests that the resulting numerical interval is *correct* in the sense that the confidence limits contain μ. That is, the interval contains μ. Suppose a thousand (say) statisticians each collect a random sample and use it to construct a 95% CI for μ. Then, the law of large numbers implies that the observed proportion of intervals containing μ is approximately 0.95; that is, $\hat{P}(\mu \text{ in interval}) \approx P(\mu \text{ in interval}) = 0.95$. With the textbook's emphasis on simulation, this claim can now be made very concrete.

Example 8.15

Suppose μ is the mean systolic blood pressure of third trimester Chicago pregnant women. Suppose $\mu = 130$ and $\sigma = 9$ for simulation purposes (these are unknown in real-world inference situations) and that $n = 30$. We simulate 1000 statisticians each collecting a random sample of size 30

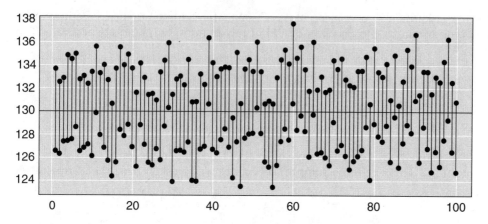

Figure 8.8 100 95% CIs for $\mu = 130$ with sample size 30.

from a $N(130, 9)$ population. A graph of the first 100 CIs in Figure 8.8 shows that 94 are correct in that they contain the unknown $\mu = 130$. Likewise 94.8% of the CIs contain $\mu = 130$, which is very close to the expected rate of 95%. That is,

$$P(\mu \text{ in interval}) = 948/1000.$$

Unlike baseball or softball, the statistician building a CI determines her own batting average. In fact, let's learn how a rate different than 0.95 can be obtained for a CI. If one wants a $(1 - \alpha) \, 100\%$ CI, then one replaces 1.96 with the number z such that

$$P(-z \leq Z \leq z) = 1 - \alpha.$$

Example 8.16

Suppose a 90% CI for μ is desired. Then $\alpha = 0.1$ yields a $(1 - \alpha)100\% = 90\%$ CI. From Appendix E,

$$P(Z \leq 1.645) = 0.95,$$

and hence

$$P(-1.645 \leq Z \leq 1.645) = 0.9.$$

Thus we replace 1.96 with 1.645, yielding the 90% CI:

$$\overline{X} - 1.645 \, \widehat{SE}(\overline{X}) \leq \mu \leq \overline{X} + 1.645 \, \widehat{SE}(\overline{X}).$$

In the blood pressure example, if $\bar{x} = 128.5$, $s = 8.8$, and $n = 30$, the 90% CI for μ is given by

$$128.5 - 1.645 \, \frac{8.8}{\sqrt{30}} \leq \mu \leq 128.5 + 1.645 \, \frac{8.8}{\sqrt{30}},$$

which yields

$$125.86 \leq \mu \leq 131.15.$$

The z value, such as 1.96 or 1.645, providing the desired confidence level is often called a critical value. The general form of a CI is

statistic ± (critical value) · (estimated standard error of the statistic),

where the standard error of the statistic (i.e., estimator) is by definition the SD of the statistic. Symbolically, for a parameter θ with an estimator $\hat{\theta}$, the interval is

$$\hat{\theta} - z\widehat{SE}(\hat{\theta}) \leq \theta \leq \hat{\theta} + z\widehat{SE}(\hat{\theta})$$

with $P(-z \leq Z \leq z) = 1 - \alpha$ used to obtain z. An estimator $\hat{\theta}$ of θ is appropriate to use if $E(\hat{\theta}) \approx \theta$, namely if $\hat{\theta}$ is approximately unbiased, as discussed in Section 8.12.

The almost universally used confidence levels are 95%, 99%, and 90%, with 95% being selected most often. Of course we can obtain *any* confidence level we want by appropriately selecting the critical value.

Confidence interval for μ when σ known

In rare cases, the practitioner may actually know σ from past experience before collecting the data. Then we use the known $SE(\bar{X}) = \sigma/\sqrt{n}$ rather than substituting $\widehat{SE}(\bar{X})$, yielding

$$\bar{X} - z\frac{\sigma}{\sqrt{n}} \leq \mu \leq \bar{X} + z\frac{\sigma}{\sqrt{n}},$$

where z is chosen to satisfy $P(-z \leq Z \leq z) = 1 - \alpha$, with $(1 - \alpha)$ 100% being the desired confidence level provided n is sufficiently large, namely $n \geq 20$. Note that we only need $n \geq 20$ to achieve approximate normality of $\bar{X}$ rather than $n \geq 30$ since σ (and hence $SE(\bar{X}) = \sigma/\sqrt{n}$) is known.

Example 8.17

A laboratory routinely measures red blood hemoglobin levels on a 0–100% scale. Past experiences with this measurement problem allows them to know that $\sigma = 1.2\%$. Standard practice at this lab to take $n = 25$ replications in order to attain high accuracy. If $\bar{x} = 92.5$, find a 99% CI. Solving $P[Z \leq z] = 0.995$ yields $z = 2.575$ and hence that $P[|Z| \leq 2.575] = 0.99$. Since,

$$SE(\bar{X}) = \frac{\sigma}{\sqrt{n}} = \frac{1.2}{5} = 0.24\%,$$

the desired CI is

$$(\bar{X} - z\,SE(\bar{X}), \bar{X} + z\,SE(\bar{X})) = (92.5 - (2.575 \cdot 0.24), 92.5 + (2.575 \cdot 0.24))$$
$$= (91.882, 93.118).$$

Confidence intervals are a widely used statistical technique, and in fact get widely quoted in the media when polling results are reported. Clearly they are much more informative than only reporting a point estimate (e.g., $\hat{\mu} = 128.5$).

Section 8.4 Exercises

1. If possible, calculate a 95% CI for the population mean based on a sample of size 40 with a sample mean of 4.5 and a sample SD of 1.46. If it is not possible, explain your reasons.

2. If possible, calculate a 90% CI for the population mean based on a sample of size 10 with a $\bar{x} = 4.5$ and $s = 1.46$. If it is not possible, explain your reasons.
3. Suppose the population SD for the weight of all male Americans is 15 pounds. Find a 95% CI for the population mean weight based on a sample size of 50 men with a sample mean of 178.9.
4. The popularity of a television show is found by taking a random sample of 100 households around the country. The sample mean rating (percentage of viewers watching) for a particular show was 31, and the sample SD was 4.3. Find a 99% CI for the population mean rating.
5. A set of 64 independent measurements of the length of a certain specimen of microorganism is made. The sample mean value of the measurements is 27.5 micrometers, with a sample SD of 3.2 micrometers. Find a 90% and a 99% CI for the true length of the microorganism. Which is larger?
6. A random sample of 40 students found that the mean study time was $\bar{x} = 19.5$ hours per week with a SD of $s = 4.05$ hours. Find a 90% CI for the population mean study time. Explain the meaning of your CI by imagining 100 students each independently doing such a survey.
7. Thirty cars were tested to determine the fuel economy of a certain model. The mean fuel economy was $\bar{x} = 19.45$ miles per gallon, with SD of $s = 1.14$ miles per gallon. Find a 95% CI for the population mean μ fuel economy of this car model.
8. Explain the meaning of a 99% confidence level.
9. Suppose you want a CI for a population proportion. You want to be as accurate as possible, so you select a 100% confidence level. What would your CI have to be?
10. Which CI, when based on the same data, is wider: an 80% or an 85% CI? Explain.
11. True or false: If you flip a fair coin 100 times and, then calculate a 95% CI for $p = P(\text{heads})$, there is an approximate 0.95 probability that $1/2$ will be in the interval.

8.5 LARGE SAMPLE CONFIDENCE INTERVAL FOR THE POPULATION PROPORTION p

Recall the Key Problem for this chapter. We need to be able to construct a 95% CI for the proportion p of St. Louis County registered voters who oppose the construction of the football stadium. In this section we will learn how to solve this problem

In Section 8.3, with the use of simulation, we discovered in several examples that the experimental distribution of $\bar{X}$ as given by its density histogram is approximately normal when the sample size is large ($n = 25$). We learned by the CLT for *any* population distribution that if n is large, $\bar{X}$ both experimentally (the density histogram of observed $\bar{x}$ values) and theoretically (the theoretical distribution of $\bar{X}$) is approximately $N(\mu, SE(\bar{X}))$, where $SE(\bar{X}) = \sigma/\sqrt{n}$.

Now we follow a similar path for the sampling distribution of $\hat{p}$. Let the sample size be $n = 30$; that is, n is "large." Consider simulating the experimental distribution of $\hat{p}$ when $p = 0.5, 0.3,$ and 0.8. Density histograms of $\hat{p}$ values corresponding to $p = 0.5, 0.3,$ and 0.8, each based on 10,000 simulations, are shown in Figures 8.9, 8.10, and 8.11, respectively. It is clear in all three cases that the density histograms are extremely close to the superimposed normal distributions. Thus, one strongly suspects that the CLT applies for $\hat{p}$, guaranteeing approximate normality. Indeed, the claim that the proportion $\hat{p}$ is approximately normal is really the normal approximation to the binomial (rescaled by $1/n$) given in Section 6.4. This is because the numerator of $\hat{p}$ has a binomial distribution.

We note that the mean of the superimposed normal density is in each case p. The appropriate value of $SE(\hat{p})$, which is the spread of the superimposed normal density, is more complex, but we will learn that $SE(\hat{p}) = \sqrt{p(1-p)}/\sqrt{n}$.

From the CLT we know that the distribution of a sample mean will be approximately normal with mean μ and SD $\sigma/\sqrt{n}$ when $n \geq 20$. But a sample proportion is just the sample mean of data whose only values are 0 and 1 (here, 1 = oppose the stadium, 0 = neutral or favor the stadium).

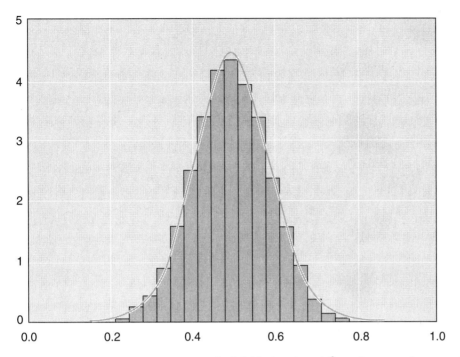

Figure 8.9 Density histogram of 10,000 simulated $\hat{p}$ estimates when $p = 0.5$, $n = 30$.

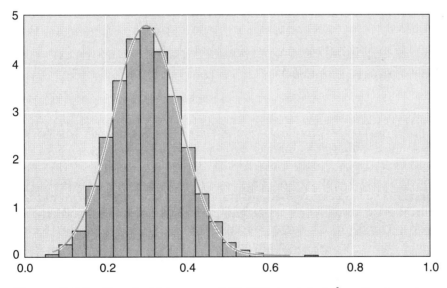

Figure 8.10 Density histogram of 10,000 simulated $\hat{p}$ estimates when $p = 0.3$, $n = 30$.

This trick of recognizing that a sample proportion $\hat{p}$ is really a sample mean $\overline{X}$ is often used in statistics. For example, in the Key Problem, out of the 301 registered voters sampled, 135 opposed the stadium and thus had a value of 1. On the other hand, 166 were not in this category and thus had a value of 0. So the sample proportion $\hat{p} = 135/301$. But

$$\bar{x} = \frac{\sum x}{301} = \frac{(135 \cdot 1) + (166 \cdot 0)}{301} = \frac{135}{301} = \hat{p},$$

as claimed.

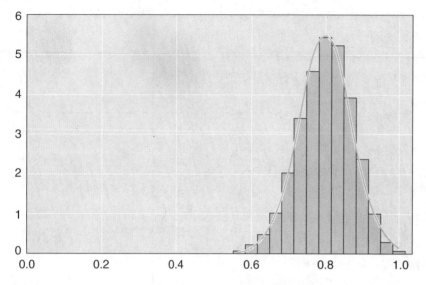

Figure 8.11 Density histogram of 10,000 simulated $\hat{p}$ estimates when $p = 0.8$, $n = 30$.

Since a sample proportion is a special kind of sample mean, we can apply the CLT directly to a sample proportion as well. Therefore, by the CLT, the distribution of the sample proportion, $\hat{p}$, is approximately normal with mean equal to the population mean μ and with SD equal to $SE(\overline{X}) = SE(\hat{p}) = \sigma/\sqrt{n}$, where σ is the population SD. But the population mean μ is just the proportion of 1s in the population, namely p. Further it can be computed that the population SD satisfies

$$\sigma = \sqrt{p(1-p)}.$$

So, replacing $\overline{X}$ with $\hat{p}$ in the previous section, we can find the approximate $100(1-\alpha)\%$ CI for the population proportion p by using the formula

$$\hat{p} \pm z\,\widehat{SE}(\hat{p}),$$

where z is the value from the normal table such that $P(-z \leq Z \leq z) = 1 - \alpha$. However, this is not a useful CI until a reasonable choice for $\widehat{SE}(\hat{p})$ is found.

As learned in Section 6.4, the rule of thumb (for when n is large enough that it is acceptable to use the normal approximation) is that both $np \geq 9$ and $n(1-p) \geq 9$ must hold. Therefore, the closer p is to 0 or 1, the larger n must be to compensate. For example, if $p = 0.5$, we would feel comfortable using the normal approximation for the distribution of the sample proportion $\hat{p}$ as long $n \geq 18$. On the other hand, if $p = 0.10$, then we would need a sample size of $n \geq 90$ before we would feel comfortable using the normal approximation. Of course, we never know p when our goal is to estimate it, so we can only apply the rule in practice by substituting $\hat{p}$ for p: $n\hat{p} \geq 9$ and $n(1-\hat{p}) \geq 9$ are then required to justify our use of the CLT.

Note that for a large sample size, the standard error of a sample proportion is

$$SE(\hat{p}) = \frac{\sigma}{\sqrt{n}} = \sqrt{\frac{p(1-p)}{n}} \approx \sqrt{\frac{\hat{p}(1-\hat{p})}{n}} = \widehat{SE}(\hat{p})$$

because by the law of large numbers $\hat{p} \approx p$ when n is large. So an approximate $100(1-\alpha)\%$ confidence interval for the population proportion p is

$$\hat{p} \pm z\sqrt{\frac{\hat{p}(1-\hat{p})}{n}},$$

where $P(-z \leq Z \leq z) = 1 - \alpha$ for a standard normal Z. Once again, our CI is of the general form:

$$\text{statistic} \pm (\text{critical value}) \cdot \widehat{SE}(\text{statistic}),$$

where the statistic is an estimator of the parameter of the CI.

Example 8.18

Pepsi and Coca-Cola have recently been in competition to secure exclusive rights for their products on various college campuses and in companies across the nation. Suppose a college takes a random sample of 100 students on campus and finds that 60 students prefer Pepsi to Coke. Find a 95% CI for the population proportion p of students who favor Pepsi.

Solution: Because $n\hat{p} = 100 \times 0.6 \geq 9$ and $n(1 - \hat{p}) \geq 9$, we have just learned that one can obtain an approximate 95% CI by using

$$\hat{p} \pm z\widehat{SE}(\hat{p}).$$

Or, more specifically, one can use

$$\hat{p} \pm z\sqrt{\frac{\hat{p}(1-\hat{p})}{n}}$$

where $P(-1.96 \leq Z \leq 1.96) = 0.95$ yields $z = 1.96$. Thus,

$$0.6 \pm 1.96\sqrt{\frac{0.60(0.40)}{100}},$$

and the resulting 95% CI is (0.504, 0.696). We are therefore 95% confident that a majority, indeed at least 50.4% of the student population, prefers Pepsi. Perhaps the college's decision could be influenced by this information.

Section 8.5 Exercises

1. In a random sample of 64 people in a city, 37.5% were in favor of lowering the drunk-driving blood alcohol level from 0.10 to 0.08. Find a 90% confidence interval for the population proportion in favor of lowering the drunk-driving blood alcohol level from 0.10 to 0.08. Is it very likely that a majority of the population is in favor of the change?
2. In a random sample of 2000 U.S. citizens over the age of 18, 53.5% approve of the president. Find a 99% CI for the population proportion who approve of the president. If this is the president's first term and the next election were to occur soon after this poll was taken, would the president be very likely to be reelected?
3. In a random sample of 2000 residents of the populous state of Illinois, 75% believe in the fairness of the jury system. Find a 90% CI for the proportion of people in the state of Illinois who believe the jury system is fair.
4. In a random sample of 2000 residents of the city of Peoria, 75% believe in the fairness of the jury system. Find a 90% CI for the proportion of people in Peoria who believe the jury system is fair.
5. Your answers for Exercises 3 and 4 should be the same. The sizes of the populations are very different, but the sizes of the samples are the same. What can you tentatively conclude about the effect of the size of the population and the size of the sample on the length of the CI?

362 CHAPTER 8 CONFIDENCE INTERVAL ESTIMATION

6. Look at the formula for the standard error of the sample proportion $\hat{p}$, namely $SE(\hat{p}) = \sqrt{p(1-p)}/\sqrt{n}$. What value(s) of p will give the standard error its largest value? What value(s) of p will give the standard error its smallest value? *Hint:* Try various $0 < p < 1$ in the formula.
7. Suppose you believe the value of the population proportion $0 < p < 1$ is very close to 0.5. How large should your random sample be to ensure that the length of a 95% CI for p is no larger than 0.08?
8. In a random sample of 12 college freshmen, 75% were in favor of lowering the drinking age from 21 to 18. If it is possible, explain how you would find a 90% confidence interval for the proportion p of college freshmen on campus who favor lowering the drinking age from 21 to 18. If it is not possible, explain your answer.

8.6 BOOTSTRAPPING A SAMPLE, BOOTSTRAPPED SEs, AND SE-BASED BOOTSTRAPPED CONFIDENCE INTERVALS

Two properties enabled us to find large sample CIs for μ and p. First, the CLT guaranteed that the estimators, $\overline{X}$ and $\hat{p}$ are both approximately normal. Second, both $\overline{X}$ and $\hat{p}$ have simple formulas for their standard errors, namely

$$\widehat{SE}(\overline{X}) = \frac{S}{\sqrt{n}} \quad \text{and} \quad \widehat{SE}(\hat{p}) = \frac{\sqrt{\hat{p}(1-\hat{p})}}{\sqrt{n}}.$$

However an appropriate estimator $\hat{\theta}$ of the parameter of interest θ can satisfy the CLT and, as desired, be approximately unbiased ($E(\hat{\theta}) \approx \theta$) but there may be no formula for finding the estimated standard error of $\hat{\theta}$. Thus, in principle there is an approximate $100(1-\alpha)$% CI for θ, namely

$$\hat{\theta} - z\widehat{SE}(\hat{\theta}) \leq \theta \leq \hat{\theta} + z\widehat{SE}(\hat{\theta})$$

where z satisfies $P(-z \leq Z \leq z) = 1 - \alpha$. But this CI is useless without a formula for $\widehat{SE}(\hat{\theta})$.

However, we learned in Chapters 4 and 5 that when we can't evaluate an unknown quantity associated with a probability distribution because we lack its formula, we can often approximately the answer by simulation. Perhaps we can approximate $SE(\hat{\theta})$ by simulation. Clearly if we could somehow simulate a large number N of $\hat{\theta}$ values, call them $\hat{\theta}*$ values, we could estimate $SE(\hat{\theta})$ by computing the sample standard deviation of the $\hat{\theta}*$s, namely

$$\widehat{SE}(\hat{\theta}) = \sqrt{\frac{\sum_1^N (\hat{\theta}* - \overline{\hat{\theta}*})^2}{N-1}},$$

because the unknown standard error $SE(\hat{\theta})$ is well approximated by the sample SD of the simulated $\hat{\theta}*$ values, provided enough simulations are carried out.

Let's first learn how to use simulations to estimate the standard error of a statistic by random sampling from a box model.

Example 8.19

Toss a fair die 25 times, where X denotes the number of spots appearing on the face of any toss. Estimate $SE(\overline{X})$ by simulation. Actually we have that

$$SE(\overline{X}) = \frac{\sigma}{\sqrt{n}}$$

where elementary calculations show that, for a fair die, $\sigma = \sqrt{35/12} = 1.708$. Hence

$$SE(\overline{X}) = \frac{1.708}{\sqrt{25}} = 0.342.$$

But, interestingly, we could also have obtained this answer approximately, by sampling from the usual fair die box model (Step 1). The remaining steps are

Step 2. Sample randomly with replacement 25 times.
Step 3. Compute $\overline{X}$.
Step 4. Replicate the simulation 10,000 times.
Step 5. Compute the sample SD of the simulated $\bar{x}$ values

$$\widehat{SD}(\overline{X}) = \sqrt{\frac{\sum(\bar{x} - \bar{\bar{x}})^2}{9999}},$$

which, recalling Example 8.8, is 0.344. We notice we have been very successful in that

$$\widehat{SE}(\overline{X}) = 0.344 \approx 0.342 = SE(\overline{X}).$$

The key to estimating $SE(\hat{\theta})$ by simulation is having a box model to repeatedly obtain a random sample $X_1, X_2, \ldots, X_n$. However if a parameter is unknown, the probability distribution of the X_i is not completely specified and hence we cannot build a box model using that distribution.

The genius of Brad Effron, of Stanford, and others who developed the bootstrap method of estimating $SE(\hat{\theta})$, was to realize that although we will never know the population box model in a statistical estimation problem, we can substitute a box model consisting of the observed sample. If the sample size n is moderate to large, this sample-determined box model will approximate the population box model and can be used as the probability model to repeatedly simulate the statistic $\hat{\theta}$. In particular, for large n the error when estimating $SE(\overline{X})$ by the sample standard deviation of the simulated $\bar{x}$ values from the sample-determined box model will be small.

Then we can simulate $\hat{\theta}$ values, which we denote by $\hat{\theta}^*$, the * superscript a reminder that they are simulated using the sample-determined box rather than a box representing the true population distribution. This approach is a way of "lifting ourselves by our bootstraps," that is, from having only the sample data to having useful estimates for both the parameter in which we are interested and the standard error of our estimator of the parameter. Hence this simulation-based method is called **bootstrapping.** Bootstrap sampling is one of the most important and useful contributions to statistics in the last quarter of the twentieth century.

Let us see how the bootstrap approach to estimating the standard error of an estimator works in detail. Consider a large population from which we have taken a random sample of size n and have formed an estimator of some population parameter. We seek an estimate of the standard error of this estimator. We first need to set up the Step 1 box model from which we will obtain our bootstrap samples. We want this box model to be a stand-in for the unknown population. To create a box model that has the same distribution as the random sample and the same size as the population, we replicate the observed sample many times to build up a population sized bootstrap box model. That is, we make a large number of identical copies of the sample to constitute the box model. This large replicated box model is our best estimate of what the unknown large population is like both in size and distributional shape. We can now draw samples *without replacement* in the usual way we sample from a real population, drawing one such bootstrap simulated sample is Step 2. From this simulated sample we compute the statistic of interest (Step 3) $\hat{\theta}^*$ that is our estimator (such as $\overline{X}$). Drawing many such bootstrap simulated samples is Step 4, each time computing our bootstrapped statistic $\hat{\theta}^*$ of interest. This bootstrapping process replaces being able to simulate sampling from the actual population distribution an in Example 8.19, an impossibility in estimation problems.

364 CHAPTER 8 CONFIDENCE INTERVAL ESTIMATION

Because the sample-determined box model will be much like the unknown theoretical population distribution, the experimental distribution of the statistic of interest will be close to its unknown theoretical distribution. Thus the density histogram of the statistic of interest obtained from the bootstrap samples should be close to the true sampling distribution of the statistic.

Example 8.20

Suppose a random sample of size 5 is taken from a population of 200 elderly men and their diastolic (lower) blood pressures are recorded:

$$104 \quad 83 \quad 99 \quad 88 \quad 110$$

We note that for the bootstrap approach to work, most practitioners would want at least a moderate sized n, say $n \geq 10$, so that the shape of the sample distribution (the box model of the sampled observations) really does approximate the population distribution shape, at least roughly.

The sample mean $\bar{x} = 96.8$. Suppose we wish to use $\bar{X}$ as the estimator of the population mean μ. How close is our estimator $\bar{X}$ likely to be to the true parameter μ? Even though we could use $\widehat{SE}(\bar{X}) = S/\sqrt{n}$ to estimate $SE(\bar{X})$, we will instead use our recently-learned bootstrap approach to estimate $SE(\bar{X})$. We perform the five steps:

Step 1. Choice of a Box Model: To form our box model, we "clone" the sample $200/5 = 40$ times to create a bootstrap box model of the same size 200 as the population.

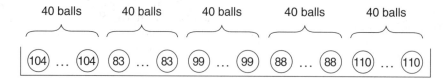

This is our best estimate of what the distribution of diastolic blood pressures in the population of 200 looks like. This box model stands in for the unknown population box model, that is, the unknown population distribution.

Step 2. Definition of One (Bootstrap Sample) Simulation: We choose our first bootstrap sample by drawing five balls at random *without replacement* from the bootstrap box model just as we formed the original real data sample by taking the diastolic blood pressures of five men chosen randomly from the real population without replacement. Suppose the results were

$$(x_1^*, x_2^*, \cdots, x_5^*) = (83, 99, 83, 110, 104).$$

The number "83" (or any number in the sample) can occur two or more times in the simulated sample because of the "cloning" of the sample data.

Step 3. Definition of the (Bootstrapped) Statistic of Interest: Our bootstrapped statistic of interest in this sampling process is the mean $\bar{X}^*$ of the values in the simulated bootstrap sample. The asterisk * in the notation $\bar{X}^*$ indicates that this statistic is a bootstrap-simulated statistic. Do not confuse the actual "real-life" sample $(X_1, X_2, \ldots, X_5)$ and its estimator $\bar{X}$ of μ with a bootstrap sample $(X_1^*, X_2^*, \ldots, X_5^*)$ and its corresponding estimator $\bar{X}^*$.

Step 4. Repetition of the Simulations: To illustrate the process, we perform three simulations (at least 100 is required for valid inferences). The results are shown in Table 8.2. The observed sample mean of these bootstrap-simulated sample means is $\bar{\bar{x}}^* = (95.8 + 96.8 + 94.6)/3 = 95.73$.

Table 8.2 Results of Three Bootstrap Samples

Sample Number	Bootstrap samples					$\bar{x}^*$
1	83	99	83	110	104	95.8
2	99	104	83	110	88	96.8
3	104	104	83	83	99	94.6

Step 5. Estimating the Standard Error of the Statistic of Interest:

$$\widehat{SE}(\bar{X}^*) = \sqrt{\frac{\sum_1^3 (\bar{x}^* - \bar{\bar{x}}^*)^2}{2}} = \sqrt{1.2135} \approx 1.10$$

$\widehat{SE}(\bar{X}^*) = 1.10$ is our bootstrapped estimate of $SE(\bar{X})$. Practically speaking, this will be a very bad estimate because the sample size $n = 5$ is small and the number of bootstrap samples, 3, is totally inadequate. The purpose of the example was only to instruct.

The bootstrap box model is the same size as the real population. But that real population is usually large with respect to the size of the sample. Therefore, as shown in Chapter 4, we can sample *with replacement* here when doing bootstrap sampling, and expect very similar results. When sampling with replacement, it is only the box model *proportions* that determine the box model distribution; for example, a box with 2 zeros and 1 one provides the same bootstrap sampling model as one with 2000 zeros and 1000 ones. The proportions in the bootstrap box model made by "cloning" the original sample are just those of the original sample itself. So the replication of the sample that we did to form a realistic population-sized box model is irrelevant and was a waste of time! We might as well have just drawn random samples *with replacement from a box consisting exactly of the original sample (uncloned)*. Thus we can obtain bootstrap simulated random samples by drawing samples of size n with replacement from a box model that consists only of the original sample itself.

Sampling with replacement is what is done in practice when bootstrap sampling is carried out. Note that in forming the box model we no longer need to know the exact size of the original population either. In summary, the foregoing strategy of cloning the sample to form the box model and then sampling without replacement was presented only to help us understand why bootstrapping is a sensible thing to do. **In practice, bootstrap sampling is always done without cloning of the sample and with replacement.**

When we bootstrap, we draw random samples *with replacement*, many times, from our observations or data. These samples with replacement from our data are called **bootstrap samples** and are the same size as the number of observations, n, originally sampled from the real population. We calculate the bootstrapped statistic of interest from each of the bootstrap samples.

To illustrate how this form of the bootstrap approach works, that is, sampling with replacement to estimate the standard error of a statistic, let us consider a random sample of 25 adult men's heights. The bootstrap approach simply declares these 25 observations that are the observed sample to be our Step 1 population-imitating box model from which (Step 2) 25 heights are randomly drawn *with replacement* to form one bootstrap sample. From each such bootstrap sample we compute the bootstrapped $\bar{X}^* = \sum X^*/25$ as Step 3. Suppose, in Step 4, 10 such bootstrap $\bar{x}^*$ values are:

68.3   68.9   67.4   68.0   67.3   68.2   67.6   68.4   68.8   67.1

For Step 5, we compute the SD of these 10 bootstrapped statistics and obtain $\widehat{SE}(\bar{X}^*) = 0.6$ as our estimate of the standard error of $\bar{X}$. The *key* is that here we have not used simulated data from a computer-generated population (such as a normal population or the fair die population of Example 8.19), but have instead used the actual numbers forming the real sample to create the box model. Thus we do not need to know *anything* about the shape or spread of the population distribution to carry out this process.

This method of bootstrapping to find the standard error of an estimator $\hat{\theta}^*$ is always an option in actual applied problems. It requires only the original sample. This version of the five-step simulation method for estimating $\text{SE}(\hat{\theta})$ is thus *always* open to us.

Let us consider a simple example to illustrate the process in detail.

Example 8.21

Suppose a random sample of four heights (in inches) yields 70, 69, 67, 74. We form a box model consisting of the actual random sample, namely these four numbers. Let us take five bootstrap samples each of size 4, by sampling randomly with replacement. The results are shown in Table 8.3.

Then we obtain the estimated $\text{SE}(\overline{X})$ by finding the sample standard deviation of the five bootstrapped $\bar{x}^*$ values. This yields

$$\widehat{\text{SE}}(\overline{X}^*) \approx 0.48.$$

In actual statistical practice we would require a minimum of 100 bootstrap samples; 1000 is better.

In Section 2.2 we saw that sometimes the median is a better parameter for characterizing the center of a distribution than the mean because it is less affected by skewing and by extreme values. When a population appears to be skewed or to contain outliers, clearly a reasonable estimate of the population median m (defined for the density of a continuous random variable as the point where exactly half the area under the probability density function lies on each side, see Figure 8.12) is the median M of a sample drawn at random from that population.

We do not have a formula for its standard error, which is essential in assessing the accuracy of M as an estimation of m. We can, however, use the bootstrap method to calculate an estimate of the standard error of the sample median. Then, as can be proven theoretically, or suggested empirically using a graphical technique called a probability or Q-Q plot, the sample median obeys the CLT with

Table 8.3 Bootstrap Sample Data for Example 8.21

Bootstrap sample number	Bootstrap samples	Bootstrap statistic $\bar{x}'$
1	67,69,69,74	69.75
2	69,70,67,74	70.0
3	67,74,67,70	69.5
4	70,69,70,74	70.75
5	74,69,69,69	70.25

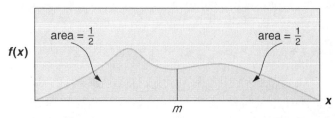

Figure 8.12 Theoretical continuous pdf; *m* denotes its theoretical median.

$E(M) \approx m$. This is a perfect application of the bootstrap method, since nothing in the previous sections allows us to find a CI for the parameter m.

Example 8.22

A random sample of size 21 was drawn from a population whose distribution is not known to the statistician. Because the population is believed to be skewed, it is decided to estimate the population center, the population median m, using the sample medium $M = 9.1$. A 95% CI for m is desired. We now let our observed random sample of size 21 stand in for the unknown population distribution. That is, the sample determines our 21-ball bootstrap box model for Step 1 of the five-step method.

Next we draw samples of size 21 with replacement from our new bootstrap sample-based population. Thus, Step 2 consists of a drawing randomly with replacement one sample of size 21.

The statistic of interest for Step 3 is the bootstrapped sample median M^* of the bootstrapped sample. Here, as usual, the asterisk in M^* serves as a reminder that these sample medians are bootstrapped, as contrasted with the original M computed from the original real-world random sample.

In Step 4 we repeat 100 times this process of selecting a bootstrap sample of size 21, each time obtaining a sample median M^*. This is bootstrap sampling.

Then in Step 5 we compute the standard deviation of the statistic of interest, namely the SD of the 100 bootstrapped sample medians (M^* values), and this sample standard deviation will be our estimate of the standard error, $\widehat{SE}(M)$. The stem-and-leaf plot in Table 8.4 displays the 100 sample medians we obtained via bootstrap sampling.

That is, an estimate of the standard error of the sample median, $SE(M)$, is obtained by calculating the SD of the 100 bootstrap sample medians, which in this case is equal to $\widehat{SE}(M) = 1.005$.

A CI for m: The choice of 100 bootstrap samples is appropriate. In actual practice, 100 is considered enough to produce decent results, although more is better in today's fast computing environment. Thus, we expect our sample median of 9.1 to be in error by around ± 1.005 as an estimate of the population median. An approximate 95% CI for m, recalling that M obeys the CLT, is given by

$$M \pm 1.96 \, \widehat{SE}(M) \quad \text{or} \quad 9.1 \pm 1.970 \quad \text{or} \quad (7.13, 11.07).$$

Table 8.4 Bootstrap Sample Medians (M^* values) for Example 8.22

Stem	Leaf
6	8
7	11444
7	66666
8	000111111111111111122233333
8	666666666888888888888
9	000033
9	666666666888899
10	000000000111133
10	5577
11	1
11	
12	
12	5

Key: "12 5" stands for 12.5

CHAPTER 8 CONFIDENCE INTERVAL ESTIMATION

In summary, we now have a bootstrap-based confidence interval approach for θ using any good estimator $\hat{\theta}$ of θ, provided $\hat{\theta}$ is approximately normal with $E(\hat{\theta}) \approx \theta$ (approximately unbiased). All that is needed is a moderate to large random sample. We do *not* require a formula for $\widehat{SE}(\hat{\theta})$. This is the key to the bootstrap's versatility.

Section 8.6 Exercises

1. When is it necessary to use the bootstrap method to find the standard error of an estimate?
2. Explain how to find an estimate of the standard error of a sample mean for a sample of size 25 using the bootstrap method.
3. Ten people were sampled from a statistics class with 165 people. The weights of the 10 are

   150   185   162   150   117   110   135   110   160   125

   Weight is sometimes skewed to the right. Twenty-five bootstrapped samples of size 10 (with replacement) from these data were taken. The following are the medians of these bootstrapped samples:

133.5	155.0	142.5	150.0	150.0
150.0	126.0	155.0	150.0	130.0
160.0	150.0	137.5	150.0	142.5
150.0	117.0	137.5	137.5	155.0
150.0	147.5	150.0	160.0	150.0

   a. What is the median of the original sample?
   b. Estimate the standard error of the sample median using the 25 bootstrapped sample medians.
   c. Find a 90% CI for m. You may use the fact that M is approximately normal.
4. Three dogs were randomly sampled from a local animal shelter. Their ages in years were given as 1, 2, and 7.
   a. What is the median of the sample?
   b. Ten bootstrap samples of size 3 from the three ages are listed in the following table. Estimate the standard error of the sample median using the bootstrap sample medians.

1	2	1
2	1	1
1	7	2
1	7	7
1	7	7
2	2	7
2	7	2
7	2	2
1	1	2
7	7	7

5. Imagine that we have ten observations from a roughly bell-shaped distribution:

$x_0 = 91$	$x_1 = 151$	$x_2 = 10$	$x_3 = 117$	$x_4 = 101$
$x_5 = 97$	$x_6 = 07$	$x_7 = 95$	$x_8 = 76$	$x_9 = 90$

   a. Compute the mean $\bar{X}$ of the sample.
   b. Find the median M of the sample. (Remember to order the 10 observations first.)

For bell-shaped distributions, both the sample average and the sample median are reasonable estimates of the center of the distribution. To see which estimate is more accurate for this data set, use bootstrapping to evaluate which estimate has the smaller standard error. Use the digits in the last two five-digit columns in Table B.3 of Appendix B to generate five bootstrap samples of size 10 each. (If you are using computer simulation, then you should generate at least 100 bootstrap samples.) For the first bootstrap sample, use the digits 39198 75268 in the first line, corresponding to the bootstrap sample $x_3, x_9, x_1, x_9, x_8, x_7, x_5, x_2, x_6, x_8 = 117, 90, \ldots,$ 76. Find both the average and median of the bootstrap sample. For the second bootstrap sample, use the digits 31151 64726 in the second line, corresponding to the bootstrap sample $x_3, x_1, x_1, x_5, x_1, x_6, x_4, x_7, x_2, x_6$.

c. List each of the five bootstrap samples and the corresponding mean $\bar{x}^*$ values and median M^* values.

d. Estimate the standard error of the sample average.

e. Estimate the standard error of the sample median.

f. Which estimated standard error is smaller? (*Note:* If we generate 100 bootstrap samples rather than five, we can put much more trust in the estimated $\widehat{SE}$s telling us correctly which of the theoretical SEs is smaller.)

6. Consider the following six observations:

$x_1 = 10$	$x_2 = 7$	$x_3 = 12$
$x_4 = 8$	$x_5 = 5$	$x_6 = 9$

a. Find the sample median M.

Use the first six digits in each of the first 10 lines in Table B.2 to generate 10 bootstrap samples of size 6 each. For the first bootstrap sample, use 665334 in the first line corresponding to the bootstrap sample $x_6, x_6, x_5, x_3, x_3, x_4$. For the second bootstrap sample, use 612612 in the second line, corresponding to the bootstrap sample $x_6, x_1, x_2, x_6, x_1, x_2$, and so forth.

b. List each of the 10 bootstrap samples and the corresponding 10 medians M_i^*. Estimate the standard error of the sample median.

8.7 BOOTSTRAPPED PERCENTILE-BASED CONFIDENCE INTERVALS: A UNIVERSALLY APPLICABLE METHOD

The foregoing approach of bootstrapping in order to obtain a standard error-based confidence interval works only if the sampling distribution of the estimator is at least roughly normal. When true, that property probably has likely been established by an application of the central limit theorem.

It turns out that another bootstrapping approach often works even if the sampling distribution of the estimator is quite nonnormal in shape, perhaps heavily skewed or much more heavy tailed than a normal distribution. We call this the percentile-based bootstrap approach to confidence intervals. Let us get its general idea and then consider an example. For an observed random sample $X_1, X_2, \ldots, X_n$, let Y denote an estimator of some population parameter θ (e.g., $Y = \bar{X}$ if $\theta = \mu$). We seek a 95% confidence interval for θ.

Now suppose we bootstrap Y, say, 1000 times. That is, using the observed random sample as the simulation box model, and using appropriate software, such as the instructional software available with the textbook, we select 1000 bootstrap samples and form 1000 bootstrapped estimators:

$$Y_1^*, Y_2^*, \ldots, Y_{1000}^*.$$

Thus, whatever formula was used for computing Y from the real-life sample $X_1, X_2, \ldots, X_n$, we use the same formula for computing each Y^* from its bootstrapped sample $(X_1^*, X_2^*, \ldots, X_n^*)$. It is very important not to confuse Y, which was originally from a real random experiment or an actual random survey, with Y^*, which is computed from a bootstrap sample.

Now, the fundamental idea of bootstrapping is that the experimental probability distribution (density histogram) of the Y^* value will approximate the unknown probability sampling distribution of the estimator Y, whether it is approximately normal or not. If the sample size is moderate to large and the number of bootstrap samples is also large, this approximation will be good. This is really a particular case of the five-step simulation approach that we have been using ever since Chapters 4 and 5. As such, the density histogram or experimental probability distribution, of a bootstrap-simulated estimator Y^* will approximate the theoretical probability distribution of Y as well.

In the case of bootstrapping, when the number of bootstrap samples is reasonably large, the bootstrap experimental distribution formed by the density histogram of experimental y^* values approximates well the true theoretical bootstrap sampling distribution of Y^*. Further, when the sample size n (the number of balls in the box) is large, the box model distribution approximates the true population distribution well, and as a result, the theoretical sampling distribution of the bootstrapped Y^* approximates well the unknown theoretical distribution of the estimator Y. Combining these two approximations, we see that the bootstrap experimental distribution of the Y^*'s found by five-step box model sampling in fact approximates the theoretical distribution of Y well, as desired, provided the sample size n is reasonably large and the number of bootstrapped samples is reasonably large. That is,

$$\text{(bootstrapped estimated distribution of } Y^*) \approx \text{(theoretical distribution of } Y^*)$$
$$\approx \text{(theoretical distribution of } Y).$$

Let $Y_{0.025}^*$ and $Y_{0.975}^*$ denote the 2.5th and 97.5th percentiles of the bootstrapped Y^* values. That is, assuming 1000 bootstrapped samples say, and ranking the y^* values from smallest to largest, $y_{0.025}^*$ is simply the 25th smallest, and $y_{0.975}^*$ is the 975th smallest (that is, it has 25 larger y^* values).

Now, let $p_{0.025}$ and $p_{0.975}$ denote the 2.5th and the 97.5th *theoretical* percentiles of the probability distribution of Y as shown in Figure 8.13. (Note that these percentiles are parameters, like μ and σ, in the same units as the values of the characteristic of interest θ, such as length in centimeters.) That is,

$$P(Y \leq p_{0.025}) = 0.025$$

and

$$P(Y \leq p_{0.975}) = 0.975$$

determine the value of these percentiles.

Since the experimental distribution of the bootstrapped simulated y^* values approximates the theoretical distribution Y well, then, as we need for our 95% CI application,

$$Y_{0.025}^* \approx p_{0.025} \text{ and } Y_{0.975}^* \approx p_{0.975}.$$

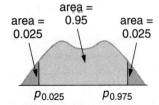

Figure 8.13 The 2.5th and 97.5th percentiles for a continuous random variable.

That is, the observed bootstrapped percentile $Y^*_{0.025}$ estimates $p_{0.025}$ well, and similarly $Y^*_{0.975}$ estimates $p_{0.975}$. A familiar argument leads to

$$P(p_{0.025} \leq Y \leq p_{0.975}) = P(Y \leq p_{0.975}) - P(Y \leq p_{0.025}) = 0.975 - 0.025 = 0.95.$$

Thus, substituting the estimators $Y^*_{0.025}$ of $p_{0.025}$ and $Y^*_{0.975}$ of $p_{0.975}$, we have

$$P(Y^*_{0.025} \leq Y \leq Y^*_{0.975}) \approx 0.95,$$

That is, the interval $(Y^*_{0.025}, Y^*_{0.975})$ is an approximate 0.95 probability interval for the estimator Y. Now, because Y is a good estimator of θ, it seems clear that the above probability should be changed little if we substitute θ for Y. That is, we obtain, reasoning informally,

$$P(Y^*_{0.025} \leq \theta \leq Y^*_{0.975}) \approx 0.95.$$

This is our approximate 95% percentile-based bootstrapped CI for θ.

The following application to finding a CI for a population median nicely illustrates this percentile-based bootstrap approach to CIs.

Example 8.23

We are revisiting Example 8.22, but, from our percentile view point, we will bootstrap the desired 95% confidence interval for the population median, denoted by m. To make calculations a little easier, we generated 1000 bootstrap samples of size 21 and calculated the bootstrap sample medians, $M^*_1, M^*_2, \ldots, M^*_{1000}$ (using a computer, of course). Ranking the M^* values from the smallest to the largest we obtained

$$M^*_{0.25} = 7.28$$
$$M^*_{0.975} = 10.92.$$

Thus our desired 95% percentile-based bootstrapped CI is

$$7.28 \leq m \leq 10.92 \text{ or } (7.28, 10.92).$$

It is interesting to compare this with the normal-distribution-based CI obtained by bootstrapping to estimate the SE(M) in Example 8.22, namely (7.13,11.07). The two intervals are fairly close, as one would hope and expect, because both methods are valid approaches.

Bootstrapping, particularly to estimate standard errors of estimators and to obtain approximate CIs, has been one of the most important and useful advances in statistics in the last 25 years. Entire advanced statistics courses are devoted to the subject! The two confidence interval procedures given in this chapter, namely CLT-based confidence intervals using bootstrapped standard errors and the percentile-based bootstrap CI approach, are discussed in a highly influential textbook by B. Efron and R. Tibshirani, *An Introduction to the Bootstrap* (Chapman and Hall/CRC, London, 1998). Some important refinements of the percentile-based approach perform even better, as discussed in the Efron and Tibshirani textbook. However, the bootstrap percentile approach presented in this section is quite effective in many situations where the normal-distribution SE-based bootstrap approach fails to work well.

Section 8.7 Exercises

1. Ten people were sampled from a statistics class with 165 people. The weights of the 10 are

$$150 \quad 185 \quad 162 \quad 150 \quad 117 \quad 110 \quad 135 \quad 110 \quad 160 \quad 125$$

The median of these 10 weights is 142.5. Twenty-five bootstrapped samples of size 10 were taken with replacement from these data. The following are the medians of these bootstrapped samples:

133.5	155.0	142.5	150.0	150.0
150.0	126.0	155.0	150.0	130.0
160.0	150.0	137.5	150.0	142.5
150.0	117.0	137.5	137.5	155.0
150.0	147.5	150.0	160.0	150.0

   a. Estimate the standard error of the sample median.
   b. Find an approximate 95% CI for the median weight of the entire class, assuming the CLT for the sample median.

2. Three dogs were sampled from a local animal shelter. Their ages in years were given as 1, 2, and 7.

1	2	1
2	1	1
1	7	2
1	7	7
1	7	7
2	2	7
2	7	2
7	2	2
1	1	2
7	7	7

   a. What is the median of the sample?
   b. Ten bootstrap samples taken with replacement from these three are shown in the preceding table. What are the medians of these bootstrapped samples?
   c. Estimate the standard error of the sample median.

3. To estimate the number of hyacinth macaws (a rare, large, dark blue parrot) in a region of the Amazon basin, we capture 60 of them at random, tag them, and release them back into the wild. A week later we capture a random sample of 10 hyacinth macaws in the same region and count the number of tagged birds in the second sample. If, for example, 3 out of the 10 birds in the second sample are tagged, then we estimate that 3 out of every 10 birds in the entire population were tagged. (This obviously will not be quite right, but it is our best guess.) Call the unknown population size N (this is the parameter of interest, of vital importance to those studying ecology and endangered species). Since we know that 60 birds are tagged, we estimate

$$\hat{N} = \frac{10}{3} \times 60 = 200.$$

This is called the **capture/recapture** approach. To evaluate the accuracy of our estimate we need to formalize the procedure. Call the number of tagged birds in the second sample X. Since the second sample is of size 10, the possible values for X are 0, 1, 2,…, 10. Our estimate of the population size is

$$\hat{N} = \frac{10}{X} \times 60 = \frac{600}{X}.$$

Let us represent our second sample of 10 birds by the 10 digits

$$X_0 = 1, X_1 = 1, X_2 = 1, X_3 = 0, X_4 = 0,$$
$$X_5 = 0, X_6 = 0, X_7 = 0, X_8 = 0, X_9 = 0,$$

where a one corresponds to a tagged bird and a zero corresponds to an untagged bird. Using the random digits in the sixth and the seventh five-digit columns in Table B.3, we generate 40 bootstrap samples of size 10 each. For each bootstrap sample we compute X^* and $\hat{N}^*$, thereby obtaining an experimental probability distribution of N^*.

For the first bootstrap sample we use the digits 65394 35595 in the first row, corresponding to the bootstrap sample $X_6, X_5, X_3, X_9, X_4, X_3, X_5, X_5, X_9, X_5 = 0, 0, 0, 0, 0, 0, 0, 0, 0, 0$, yielding $X^* = 0$ and $N^* = 0$. (With 3 tagged birds in the second sample, the population size is unlikely to be much greater than 1000. But it is unclear what value we should assign to $\hat{N}^*$ whenever $X^* = 0$, so we drop this one) For the second bootstrap sample we use the digits 73823 62854 in the second line, corresponding to the bootstrap sample $X_7, X_3, X_8, X_2, X_3, X_6, X_2, X_8, X_5, X_4 = 0, 0, 0, 1, 0, 0, 1, 0, 0, 0$, yielding $X^* = 2$ and $\hat{N}^* = 300$. In other words, X^* is simply the number of zeros, ones, and twos in the group of random digits drawn from Table B.3. We continue to get the 40 bootstrap values of X^*:

0	2	3	6	3	4	2	4	3	3
4	1	5	2	3	3	3	2	1	4
3	3	2	3	2	5	3	1	4	2
3	4	1	6	5	3	2	2	2	5

a. List the 40 bootstrapped values of $\hat{N}^*$.
b. Compute $\hat{N}^*_{0.05}$ and $\hat{N}^*_{0.95}$ (the 5th and the 95th percentiles of the 40 bootstrapped $\hat{N}^*$ values).
c. Give an approximate 90% CI for the population size N.
   Remark: The confidence interval would have been more accurate if the second sample size had been larger than 10—say, at least 50—but then we would have needed a computer to do the bootstrapping.

4. An estimator Y of a parameter θ is known to have a badly skewed sampling distribution. Y is bootstrap sampled 200 times. The smallest three bootstrap sample Y^* values and the largest three Y^* values are

Smallest	99	102	104
Largest	310	340	365

Find a 99% bootstrap CI for θ.

5. An estimator Y of a parameter θ is known to have an approximately normal distribution, but there is no formula for its standard error. Suppose 10 bootstrap samples yield 10 bootstrapped Y^* values

29   34   31   36.5   32   40   28   18   33   27

Find a 95% bootstrap CI for θ.

8.8 CONFIDENCE INTERVALS FOR μ WHEN THE POPULATION IS NORMAL AND THE SAMPLE SIZE IS SMALL

CI for a normal population μ when σ is known.

Our large sample CI approach to estimating μ was based on the fact that, when n is large, $\overline{X}$ is approximately normally distributed with

$$S/\sqrt{n} = \widehat{SE}(\overline{X}) \approx SE(\overline{X}).$$

When n is small, $\overline{X}$ is usually *not* normally distributed and $SE(\overline{X})$ is *poorly approximated* by $\widehat{SE}(\overline{X})$. However, there is one important small sample setting where progress is possible. We have learned that random samples are often normally distributed, for example heights, baseball averages, speed of light measurement errors, and standardized test scores just to name a few. In fact both physiological quantities and measurement errors are often normally distributed. Thus it seems important to explore the sampling distribution of $\overline{X}$ when the population being sampled is normal and n is small.

As we did in the large sample case for $\overline{X}$ and for $\hat{p}$, we explore the sampling distribution of $\overline{X}$ experimentally in the cases $n = 2, 5$, and 20. Random sampling will be simulated from a normal population with $\mu = 0$ and $\sigma = 5$. Thus $E(\overline{X}) = \mu = 0$ and $SD(\overline{X}) = \sigma/\sqrt{n} = 5/\sqrt{n}$. The respective density histograms in each case for 10,000 simulated $\bar{x}$ values are displayed in Figures 8.14, 8.15, and 8.16. A striking fact emerges! The density histograms are all remarkably close to the appropriate superimposed normal density $N(0, 5/\sqrt{n})$, *regardless* of whether n is large or not. The density histogram of $\overline{X}$ for $n = 2$ fits the normal curve just as well as for $n = 20$. It is striking to note how the observed SE of the 10,000 $\bar{x}$ values decreases as n increases from 2 to 20, going from 3.518 to 2.241 to 1.127. Also, as expected, these are all very close to the theoretical SEs of 3.536, 2.236, and 1.118, respectively, using $SE(\overline{X}) = 5/\sqrt{n}$..

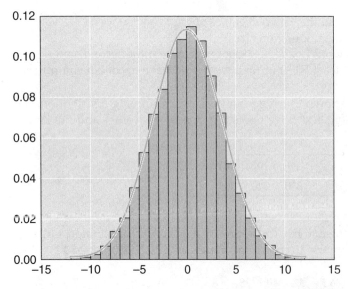

Figure 8.14 10,000 Simulated $\bar{x}$ values when sampling is from a $N(0, 5)$ population, $n = 2$.

In fact, once again, we discover a mathematical theorem experimentally through simulation.

> **The Distribution of $\overline{X}$ when the Population Is Normal**
>
> Let a random sample $X_1, X_2, \ldots, X_n$ be from a $N(\mu, \sigma)$ population. Then, experimentally (the density histogram of sampled $\bar{x}$ values) and theoretically (the theoretical sampling distribution of $\overline{X}$) the distribution of $\overline{X}$ is *exactly* $N(\mu, \text{SE}(\overline{X}))$ where $\text{SE}(\overline{X}) = \sigma/\sqrt{n}$.

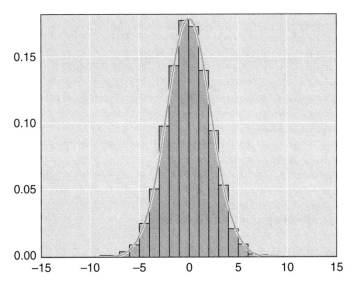

Figure 8.15 10,000 simulated $\bar{x}$ values when sampling is from a $N(0,5)$ population, $n = 5$.

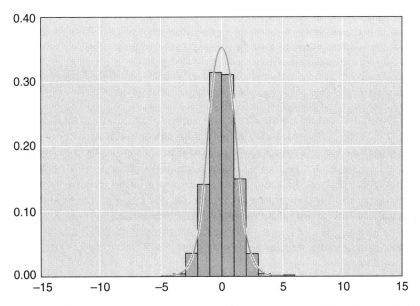

Figure 8.16 10,000 simulated $\bar{x}$ values when sampling from a $N(0,5)$ population, $n = 20$.

Thus, reasoning as in Section 8.4 but without needing the CLT for $\bar{X}$, it is simple to find a CI for μ in the uncommon case where the practitioner knows the value of σ. A $(1 - \alpha)$ 100% CI is given by

$$\bar{X} - z\, SE(\bar{X}) \leq \mu \leq \bar{X} + z\, SE(\bar{X}),$$

where $P(-z \leq Z \leq z) = 1 - \alpha$ determines z and $SE(\bar{X}) = \sigma/\sqrt{n}$.

Example 8.24

Suppose a sample of size 9 from a normal population with known $\sigma = 12$ yields $\bar{x} = 63.4$. We can use the estimator $\bar{X}$ of μ to find a 90% CI for the population mean, μ, using the following formula:

$$\bar{X} \pm z\, SE(\bar{X}),$$

where $P(-z \leq Z \leq z) = 0.9$ determines that $z = 1.645$. Thus, the 90% CI is given by

$$63.4 \pm 1.645 \frac{12}{\sqrt{9}} \quad \text{or} \quad 63.4 \pm 1.645 \cdot 4 \quad \text{or} \quad 63.4 \pm 6.58,$$

yielding the interval (56.82, 69.98). Here we used the fact from Appendix E that $P(Z \leq 1.645) = 0.95$ to obtain the 1.645 value.

CI for a normal population μ when σ is unknown.

Finding a CI for μ in the small sample case where σ is unknown occurs far more often than the σ known case and leads us to one of the classical triumphs of twentieth-century statistics. Recall our general form of a CI:

$$\text{Statistic} \pm \text{critical value} \cdot \widehat{SE}(\text{statistic}).$$

This suggests the following CI for μ:

$$\bar{X} \pm (\text{critical value}) \cdot \widehat{SE}(\bar{X})$$

where $\widehat{SE}(\bar{X}) = S/\sqrt{n}$. However, since $\widehat{SE}(\bar{X})$ *does not approximate* $SE(\bar{X}) = \sigma/\sqrt{n}$ *well* because the sample size is small, we suspect that the critical value cannot come from a normal table as it can when n is large.

In 1908, W. S. Gosset had the brilliant insight (instead of using a normal table z with an increased $\widehat{SE}(\bar{X})$ to guarantee the needed $(1 - \alpha)$100% confidence as others had proposed) of compensating for the inaccuracy of S as an estimator of σ by replacing the critical normal value z by a larger critical value. Armed with this insight, Gosset derived a theoretically justified probability distribution that produces the correct critical value to use in forming a CI for the population mean μ when the sampled population is normal, the population standard deviation σ is not known but estimated by the sample standard deviation S, and the sample size is small. This distribution is known as the **t distribution**.

Carefully stated, Gosset's result is that the standardized statistic

$$T = \frac{\bar{X} - \mu}{S/\sqrt{n}}$$

has a t distribution with $n - 1$ degrees of freedom. Such a t density is shown in Figure 8.17 along with the standard normal density it is replacing.

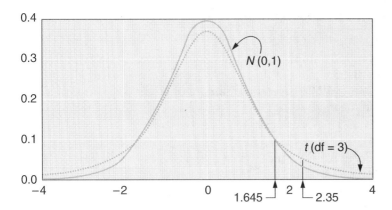

Figure 8.17 Comparison of standard normal and t densities.

We must emphasize that the claim that $T = (\bar{X} - \mu)/(S/\sqrt{n})$ has a t distribution holds *only when our sample comes from a normal population.*

Because Gosset's employer, Guinness Brewery, wished to keep his work secret, he published under the pseudonym "Student." Consequently, the distribution he described is often referred to as **Student's t distribution.** Like the chi-square distribution we will study in Chapter 10, the t distribution is determined by one parameter, the number of **degrees of freedom** (df) of the distribution. Possible df values are the positive integers 1, 2, ... Each different df value corresponds to a different t density, that is, for m degrees of freedom,

$$P(T \geq t(m)) = \beta$$

is the area under the t distribution density to the right of the table value $t(m)$. As an illustration see Figure 8.17, which shows the t density for $m = 3$ (dotted) in comparison to the standard normal density (solid). We see that the t density appears to be bell-shaped like the normal. However, the peak of the t density is lower, and more importantly, the tails of the t distribution are thicker than those of the normal distribution. In other words, there is more probability toward the extremes in the t distribution than in the standard normal distribution. Therefore a t distribution critical value, $t(m)$, is larger than the corresponding standard normal distribution critical value, z. For example, as shown in Figure 8.17, $P(Z \geq 1.645) < P(T \geq 1.645)$, and if $0.05 = P(Z \geq 1.645)$, then $P(T \geq t(m)) = 0.05$ must yield a $t(m) > 1.645$; in fact, $t(3) = 2.35$, as shown. Also shown as shaded is the 0.05 area under the t density to the right of $t(3) = 2.35$.

To summarize Gosset's result as it applies to us, we do obtain a $100(1 - \alpha)\%$ CI provided we use $t(n - 1)$ for the critical value. Thus, a $(1 - \alpha)100\%$ CI can be formed by

$$\bar{X} - t(n-1) \cdot \widehat{SE}(\bar{X}) \leq \mu \leq \bar{X} + t(n-1) \cdot \widehat{SE}(\bar{X}) \quad \text{or} \quad \bar{X} \pm t(n-1) \cdot \widehat{SE}(\bar{X}),$$

where $\widehat{SE}(\bar{X}) = S/\sqrt{n}$ and the tabulated constant $t(n - 1)$ used as our (correct) critical value satisfies

$$P(-t(n-1) \leq T \leq t(n-1)) = 1 - \alpha$$

where T is t distributed with $n - 1$ df. The correct t distribution based CI using $t(n - 1)$ is wider than the corresponding (incorrect) standard normal distribution based CI using the critical value z.

Critical values $t(m)$ are tabulated in Appendix F at the back of this book for various commonly used values of β (such as 0.05 and 0.025) for values of m (the number of degrees of freedom) up to 29 (corresponding to a sample size of 30). As the number of degrees of freedom increases, the t density becomes closer and closer to the standard normal density; for sample sizes greater than 30, the standard normal density can be used with little practical error, as we have seen.

Example 8.25

Suppose the population SD, namely σ, in Example 8.24 is unknown. Suppose, however, that the population is known to be normally distributed and that the observed sample mean is $\bar{x} = 63.4$ and the sample standard deviation is $s = 12.24$ based on a sample of size 9. We can find a 95% confidence interval for the population mean, μ, using

$$\bar{X} \pm t(n-1) \cdot \widehat{SE}(\bar{X}),$$

where $\widehat{SE}(\bar{X}) = S/\sqrt{n}$. This yields, noting from Appendix F for $df = 9 - 1 = 8$ and using the 0.025 column that we obtain $t(df) = t(8) = 2.31$,

$$63.4 \pm 2.31 \frac{12.24}{\sqrt{9}} \text{ or } 63.4 \pm 9.42.$$

Thus the 95% CI is (53.98, 72.82). Note, in comparison to Example 8.23, the interval is wider. This extra width comes from two sources: the level of significance is higher and the critical value was increased from its usual z value to a t distribution value because S is used to estimate σ.

Section 8.8 Exercises

1. A British medical study found the average birth weight of 121 newborn babies born to non-smoking, nondrinking mothers to be $\bar{x} = 7.8$ pounds with a standard deviation of $s = 1.1$ pounds. The birth weights follow a normal curve. Give a 95% CI for the average birth weight of all newborn British babies born to nonsmoking, nondrinking mothers.

2. The following data are from the article "Sensory and Mechanical Assessment of the Quality of Frankfurters" (*Journal of Texture Studies*, 1990, pp. 345–409). It is a random sample of $n = 9$ hot dogs and yielded the following fat content (in percentage):

    25.2 21.3 17.0 29.8 21.0 25.5 16.0 20.9 19.5

    Assume the population is normally distributed. Construct a 95% CI for the population mean fat content μ.

3. Observing the seemingly large numbers of Dutch women on American college volleyball teams, it is remarked that Dutch women seem taller on average than women in general. A random sample of five Dutch women yields heights in inches of 67, 70, 66, 69, 67. Suppose $\sigma = 2.5$ inches is known from other studies of female heights. Find a 95% CI for μ_{DUTCH}.
    Hint: Recall that heights follow a normal distribution.

4. A random sample of 100 women was drawn from the population of all female students at a large university. The heights of the 100 women were plotted, and the density histogram followed a bell-shaped curve. The average height of the 100 women was $\bar{x} = 64.5$ inches, with a sample SD of $s = 2.5$ inches. Give an approximate 95% CI for the average height of all the female students at the university.

5. A random sample of 625 women aged 25 to 34 had average systolic (upper) blood pressure $\bar{x} = 121.5$ mmHG with a sample SD of $s = 12.5$ mmHG. The density histogram for the 625 blood pressure measurements followed a bell curve.
    a. Give a 68% CI for the average systolic blood pressure of all 25- to 34-year-old women. (Note that 0.68 is an unusual choice of confidence level.)
    b. Give a 95% CI for the average systolic blood pressure of all 25- to 34-year-old women.

8.9 CONFIDENCE INTERVAL FOR THE DIFFERENCE BETWEEN TWO POPULATION MEAN $\mu_X - \mu_Y$

Large Sample Case

A common problem is to compare the means of two populations, possibly to know the effects of choosing between two options, such as two treatments for a disease or two teaching methods. This is called the **two-sample problem**, it being understood that mean comparisons are based on two independent random samples. Suppose we want to compare the mean population incomes μ_X and μ_Y for two cities. We can take a random sample of size n from City X and independently a random sample of size m from City Y and calculate the sample means.

Often the goal is to obtain a $(1 - \alpha)100\%$ CI for the mean difference, given by $\mu_X - \mu_Y$. As usual, our general approach is to use

$$\text{statistic} \pm (\text{critical value}) \cdot \widehat{SE}(\text{statistic}).$$

Intuitively it makes sense to use $\overline{X} - \overline{Y}$ to estimate $\mu_X - \mu_Y$. Thus we need to investigate the sampling distribution of $\overline{X} - \overline{Y}$ so that we can both estimate $SE(\overline{X} - \overline{Y})$ and find an appropriate critical value.

By the CLT, both sample means $\overline{X}$ and $\overline{Y}$ are approximately normally distributed with means μ_X and μ_Y, and SDs $SE(\overline{X}) = \sigma_X/\sqrt{n}$ and $SE(\overline{Y}) = \sigma_Y/\sqrt{m}$, respectively, provided that both $n \geq 20$ and $m \geq 20$. In our illustration, σ_X and σ_Y denote the SDs of the incomes of each of the city populations. Since the two random samples are independent of each other, $\overline{X}$ and $\overline{Y}$ are also independent. What we seek is the sampling distribution of $\overline{X} - \overline{Y}$. The hope is that $\overline{X} - \overline{Y}$ is approximately normal by the CLT for n, m both large. We take an experimental approach to this question of whether $\overline{X} - \overline{Y}$ is approximately normal when both n and m are large. For the two-city income comparison, suppose (unknown to the practitioner), $\mu_X = \sigma_X = \$26K$ and $\mu_Y = \sigma_Y = \$29.5K$, K denoting that units are in thousands of dollars. Then we simulate 10,000 $\bar{x}$ values each based on a sample size of $n = 25$ and an exponential population distribution with $\mu_X = 26$ and $\sigma_X = 26$. Likewise we simulate 10,000 $\bar{y}$ values, each based on $n = 25$ from an exponential population distribution with $\mu_Y = 29.5$ and $\sigma_Y = 29.5$. Figure 8.18 shows the pdf of each of these two densities.

The 10,000 simulated $\bar{x} - \bar{y}$ values are graphed in the density histogram which is approximated well by a normal density (see Figure 8.19).

Moreover, the average of the 10,000 $\bar{x}, \bar{y}$, and $(\bar{x} - \bar{y})$ values were computed:

$$\bar{\bar{x}} = \$25.9852K, \bar{\bar{y}} = \$29.4820K, \overline{\bar{x} - \bar{y}} = -\$3.4968K,$$

the latter being the average of all the $(\bar{x} - \bar{y})$ values and approximately equal to $\mu_X - \mu_Y = \$26K - \$29.5K = -\$3.5K$. In fact, according to the law of large numbers $\bar{\bar{x}} \approx E(\overline{X}) = \mu_X$,

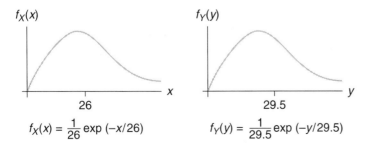

Figure 8.18 The two income distributions sampled from exponential densities.

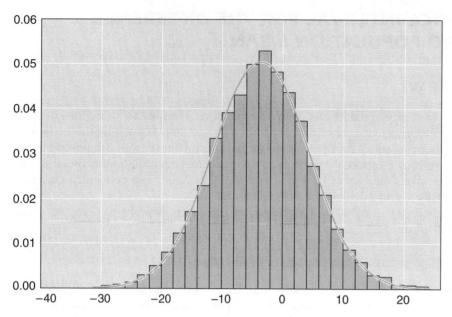

Figure 8.19 Density histogram of 10,000 simulated $(\bar{x}-\bar{y})$ values formed from two exponential populations with $\mu_X = \sigma_X = 26$, $\mu_Y = \sigma_Y = 29.5$.

$\bar{\bar{y}} \approx E(\bar{Y}) = \mu_Y$ and $\overline{(\bar{x}-\bar{y})} \approx E(\bar{X} - \bar{Y})$ must be true because n and m are large. Because $\overline{(\bar{x}-\bar{y})} = \bar{\bar{x}} - \bar{\bar{y}}$, this somewhat tricky argument suggests that

$$E(\bar{X} - \bar{Y}) = E(\bar{X}) - E(\bar{Y}) = \mu_X - \mu_Y,$$

as can be shown mathematically to be true.

Mathematics also shows for independent $\bar{X}$ and $\bar{Y}$ (as must be in the case for two independent samples) that

$$SE^2(\bar{X} - \bar{Y}) = SE^2(\bar{X}) + SE^2(\bar{Y}) = \frac{\sigma_X^2}{n} + \frac{\sigma_Y^2}{m}$$

and hence that

$$SE(\bar{X} - \bar{Y}) = \sqrt{\frac{\sigma_X^2}{n} + \frac{\sigma_Y^2}{m}}.$$

Thus, we can estimate $SE(\bar{X} - \bar{Y})$ by

$$\widehat{SE}(\bar{X} - \bar{Y}) = \sqrt{\frac{S_X^2}{n} + \frac{S_Y^2}{m}},$$

where S_X is the sample SD of X and S_Y is the sample SD of Y. Note that the sample SD for the 10,000 $(\bar{x}-\bar{y})$ values was 7.875. As expected, this is very close to the theoretical SD of $\bar{X} - \bar{Y}$ given by

$$SE(\bar{X} - \bar{Y}) = \sqrt{\frac{\sigma_X^2}{n} + \frac{\sigma_Y^2}{m}} = \sqrt{\frac{(26)^2}{25} + \frac{(29.5)^2}{25}} = 7.864.$$

Thus we have discovered a special case of the CLT for $\bar{X} - \bar{Y}$ when drawing two independent random samples.

CLT for $\bar{X} - \bar{Y}$, the two independent samples case: If n and m are large ($n \geq 20$, $m \geq 20$), then the distribution of $\bar{X} - \bar{Y}$ is, both experimentally (the density histogram of the $(\bar{x} - \bar{y})$ values) and theoretically (the sampling distribution of $\bar{X} - \bar{Y}$), approximately normal with mean $\mu_X - \mu_Y$ and standard error

$$SE(\bar{X} - \bar{Y}) = \sqrt{\frac{\sigma_X^2}{n} + \frac{\sigma_Y^2}{m}}.$$

That is, $\bar{X} - \bar{Y}$ is approximately

$$N\left(\mu_X - \mu_Y, \sqrt{\frac{\sigma_X^2}{n} + \frac{\sigma_Y^2}{m}}\right).$$

Moreover, if $n \geq 30$, $m \geq 30$, $\bar{X} - \bar{Y}$ is approximately

$$N\left(\mu_X - \mu_Y, \sqrt{\frac{S_X^2}{n} + \frac{S_Y^2}{m}}\right).$$

We can now find a CI for $\mu_X - \mu_Y$ in the large sample case using the CLT for $\bar{X} - \bar{Y}$. In particular, if $n \geq 30$, $m \geq 30$ then an approximate $(1 - \alpha) 100\%$ CI for $\mu_X - \mu_Y$ is given by

$$(\bar{X} - \bar{Y}) - z\,\widehat{SE}(\bar{X} - \bar{Y}) \leq (\mu_X - \mu_Y) \leq (\bar{X} - \bar{Y}) + z\,\widehat{SE}(\bar{X} - \bar{Y}),$$

where z satisfies $P(-z \leq Z \leq z) = 1 - \alpha$ and

$$\widehat{SE}(\bar{X} - \bar{Y}) = \sqrt{\frac{S_X^2}{n} + \frac{S_Y^2}{m}}.$$

Example 8.26

A random sample of 100 people in city X have a mean income of \$35,000 with SD of \$2000; an independent random sample of 400 people in city Y have a mean income of \$40,000 with a standard deviation of \$1500. Construct a 95% CI for the difference between the two population income means $(\mu_X - \mu_Y)$.

Using the just derived CI we obtain

$$\$35,000 - \$40,000 \pm 1.96 \sqrt{\frac{(2000)^2}{100} + \frac{(1500)^2}{400}}$$

and, hence, the interval $(-\$5419 \leq \mu_X - \mu_Y \leq -\$4581)$. Thus we are confident that there is a real difference in the population mean incomes favoring city Y by at least \$4581.

Small Sample Normal Populations Case

Comparing two normal distributions by their mean difference $\mu_X - \mu_Y$ when the sample sizes are both small and the population standard deviations are unknown occurs often. Practitioners often use a normal probability plot, called a **Q-Q plot**, to decide when a normal assumption is valid, a topic we won't address. A t distribution–based solution also assumes the two distributions have a common SD, $\sigma = \sigma_X = \sigma_Y$. This is a strong assumption and will often not be valid to assume.

Once again the basic CI approach is

$$\text{Statistic} \pm (\text{critical value}) \cdot \widehat{SE}(\text{statistic}).$$

Here, as in the case of one sample normal population sampling, the t distribution again provides the critical value. The estimated standard error is

$$\widehat{SE}(\overline{X} - \overline{Y}) = S_{\text{POOL}} \sqrt{\frac{1}{n} + \frac{1}{n}}$$

where S_{POOL} (called the pooled standard deviation) estimates the common σ and is defined by

$$S_{\text{POOL}} = \sqrt{\frac{(n-1)S_X^2 + (m-1)S_Y^2}{(n-1) + (m-1)}}$$

which is the square root of the weight average of the sample variances. The correct degrees of freedom is $df = n + m - 2$ and the critical value can be obtained from the t distribution in Appendix F.

Example 8.27 *When to use this?*

Compare the population average amount of calories in a 12-ounce serving of Mountain Dew (X) to that of a 12-ounce serving of Pepsi (Y). Based on random samples of Mountain Dew and Pepsi cans with $n = 6$ and $m = 8$, we obtain, respectively,

$$\overline{x} = 171, \overline{y} = 153, s_X = 3 \text{ and } s_Y = 5.$$

Suppose a 95% CI is desired. Using Appendix F with $df = 6 + 8 - 2 = 12$, $P(T \geq t) = 0.025$ yields $t(12) = 2.18$ as the critical value. Our CI is given by

$$(\overline{X} - \overline{Y}) \pm 2.18 \, \widehat{SE}(\overline{X} - \overline{Y}).$$

Here

$$\overline{x} - \overline{y} = 18, S_{\text{POOL}} = \sqrt{\frac{5.9 + 7.25}{12}} = 4.282 \text{ and } \widehat{SE}(\overline{X} - \overline{Y}) = 4.282 \sqrt{\frac{1}{6} + \frac{1}{8}} = 2.313.$$

Thus the CI is

$$18 \pm (2.18)(2.313),$$

that is,

$$15.69 \text{ calories} \leq \mu_X - \mu_Y \leq 20.31 \text{ calories}.$$

One might ask what procedure to use for finding a CI for $\mu_X - \mu_Y$ when the population distributions are normal but $\sigma_X \neq \sigma_Y$. In this case, a standard procedure is Welsh's approach, the details of which we omit; for an example see Vol. I, p. 206 of *Encyclopedia of Statistical Sciences* (Wiley, New York, 1983).

Occasionally the population SDs are known. Then the above Example 8.27 CI is used with the exception that $\widehat{SE}(\overline{X} - \overline{Y})$ is now replaced with

$$SE(\overline{X} - \overline{Y}) = \sqrt{\frac{\sigma_X^2}{n} + \frac{\sigma_Y^2}{m}},$$

and the critical value t is replaced with a smaller z-value.

Section 8.9 Exercises

1. Find the standard error of the difference of two sample means if the size of both samples is 50 and the population SDs are 10 and 4.
2. In a random sample of size 64, the sample mean was 50.3 with SD 2.5. In another random sample, of size 64 from a second population, the sample mean was 49.25 with SD 3.1. Find the standard error of the difference of the sample means.
3. For Exercise 2, find a 90% CI for the difference of the population means. Is the CI exactly or approximately 90%?
4. Scientists want to study the effect of exercise on the amount of weight loss. One hundred people are randomly divided into two equal groups. Both groups follow the same diet plan, but the first group also completes an exercise program. In the exercise group (X), the mean weight loss over three months was $\bar{x} = 25.2$ pounds with a SD of $s_X = 10$ pounds. In the nonexercise group (Y), the mean weight loss was $\bar{y} = 20.4$ pounds with a SD of $s_Y = 6.3$ pounds. Find a 99% CI for the difference of the ($\mu_X - \mu_Y$) population means.
5. On a college campus, a professor wanted to test the effect that using a computer program to teach calculus has on learning. In the first class of 43 students (X), the professor taught using the computer as an instructional aid. In the second class of 35 students (Y), the professor taught without using the computer. On the final exam the computer class scored a mean of $\bar{x} = 78.32$ with SD of $s_X = 8.07$, but the noncomputer class scored a mean of $\bar{y} = 80.41$ with SD of $s_Y = 8.53$. Find a 99% CI for the difference of the population means. Is it likely that the computer teaching made a difference in student learning?
6. In a random sample of size 75 at one school, the mean SAT verbal score was $\bar{x} = 549.45$ with SD of $s_X = 21.12$. In a random sample of size 75 from another school, the mean SAT verbal score was $\bar{y} = 539.25$ with SD of $s_Y = 20.91$. Find a 90% CI for the difference between the two schools on their population mean verbal SAT scores.
7. Twenty-five independent measurements of the weight of a sample of a chemical compound were made on each of two scales. On the first scale, the mean of the 25 measurements was $\bar{x} = 19.45$ grams with SD of $s_X = 0.49$ grams. On the second scale the mean of the 25 measurements was $\bar{y} = 18.42$ grams with SD of $s_Y = 0.27$ grams. Find a 95% CI for the difference between the mean measurements of the two scales, $\mu_X - \mu_Y$. Assume that the populations are normally distributed with equal variances. Is it likely that the two scales weigh objects the same?
8. In a batch chemical process, two catalysts are being compared for their effect on the output of the process reaction. A sample of 12 batches is prepared using catalyst 1, and a sample of 10 batches is obtained using catalyst 2. The 12 batches for which catalyst 1 was used give an average yield of $\bar{x} = 85$ with a sample SD of $s_X = 4$, whereas the second sample, prepared with catalyst 2, gives an average of $\bar{y} = 81$ and a sample SD of 5. Find a 90% CI for the difference between the population means, $\mu_X - \mu_Y$, assuming that the populations are normally distributed with equal SDs.
9. The tire lives of four randomly selected Goodyear X5W tires driven in Seattle had a mean of $\bar{x} = 48{,}100$ miles with $s_X = 8000$ miles. The tire lives of eight randomly selected X5W tires driven in Chicago had a mean of $\bar{y} = 44{,}000$ miles with SD $s_Y = 7200$ miles. Find a 95% CI for the difference between the population means, $\mu_X - \mu_Y$, assuming that the populations of X5W tire lifetimes in the two cities are normal with equal theoretical SDs.

10. Test scores of men and of women on a statistics final are normally distributed with means μ_Y and μ_X, respectively. Suppose in a small class the eight women score $\bar{x} = 80$ with SD $s_X = 5$, and the six men score $\bar{y} = 82$ with SD $s_Y = 4$.
    a. Find a 95% CI for $\mu_Y - \mu_X$.
    b. Is there any evidence that either group is doing better than the other?

8.10 CONFIDENCE INTERVAL FOR THE DIFFERENCE BETWEEN TWO POPULATION PROPORTION $p_1 - p_2$

Frequently, we want to construct a CI for the difference of two population proportions, $p_1 - p_2$. For example, we might want to compare the cure rates of two medications, the proportions of Native American students graduating at two universities, the proportions of defective units produced on two assembly lines. Let n and m be the random sample sizes associated with these two proportions. Assuming that the two random samples are independent of each other, $\hat{p}_1$ is then unrelated to $\hat{p}_2$. We can argue by the CLT that $\hat{p}_1 - \hat{p}_2$, is approximately normal when $np_1 \geq 9$, $n(1-p_1) \geq 9$, $np_2 \geq 9$, and $n(1-p_2) \geq 9$ all hold. Then the standard form for a $100(1-\alpha)\%$ CI is given by

$$(\hat{p}_1 - \hat{p}_2) \pm z\widehat{SE}(\hat{p}_1 - \hat{p}_2).$$

By noting that $SE(\hat{p}_1) = \sqrt{p_1(1-p_1)/n}$ and $SE(\hat{p}_2) = \sqrt{p_2(1-p_2)/m}$, we obtain

$$(\hat{p}_1 - \hat{p}_2) \pm z\sqrt{\frac{\hat{p}_1(1-\hat{p}_1)}{n} + \frac{\hat{p}_2(1-\hat{p}_2)}{m}},$$

where $P(-z \leq Z \leq z) = 1 - \alpha$ determines z. Here we use the fact that

$$SE^2(\hat{p}_1 - \hat{p}_2) = SE^2(\hat{p}_1) + SE^2(\hat{p}_2) = \frac{p_1(1-p_1)}{n} + \frac{p_2(1-p_2)}{n}$$

and we substitute $\hat{p}_1$ for p_1 and $\hat{p}_2$ for p_2 to obtain $\widehat{SE}(\hat{p}_1 - \hat{p}_2)$.

Example 8.28

A random sample of 100 voters in a city school district produces $\hat{p}_1 = 0.44$ in favor of a tax increase to build a new high school. Worried by this low level of support, the school district pays a public relations (PR) firm to make their case in the community. After the PR firm has carried out its campaign, a second random sample of size 100 yields $\hat{p}_2 = 0.48$.

The PR firm argues that this is a "real" improvement and proposes a further (and expensive) attempt to "bring it above 50%." That is, the firm claims that p_2, the proportion of the population supporting the tax increase after the PR campaign, is greater than p_1. Construct a 95% CI for $p_2 - p_1$ to help evaluate the PR company's claim.

Solution: Note first that we are treating this problem as a difference in proportions between two populations. At first glance, it appears that this is a one-population problem because we are dealing with a single community. However, the issue in question is whether the PR campaign has changed the population proportion. Note that the two random samples are independent of each other, as needed.

We use, noting that $n\hat{p}_1 \geq 9$, $n(1-\hat{p}_1) \geq 9$, $n\hat{p}_2 \geq 9$, and $n(1-\hat{p}_2) \geq 9$ as required, the following CI:

$$(\hat{p}_1 - \hat{p}_2) \pm z\sqrt{\frac{\hat{p}_1(1-\hat{p}_1)}{n} + \frac{\hat{p}_2(1-\hat{p}_2)}{m}},$$

where $z = 1.96$ is determined by $P(-z \leq Z \leq z) = 0.95$. Thus we obtain

$$(0.44 - 4.48) \pm 1.96\sqrt{\frac{0.44(0.56)}{100} + \frac{0.48(0.52)}{100}},$$

which gives

$$-0.04 \pm 0.08$$

or equivalently

$$-0.12 \leq p_1 - p_2 \leq 0.04.$$

Thus this CI for $p_1 - p_2$ allows the possibility that $p_1 - p_2 = 0$, indicating no statistical evidence of an increase in community support. The school district may not wish to spend the money on further PR efforts from this firm!

Section 8.10 Exercises

1. Two detergents were tested for their ability to remove grape juice stains. An inspector determined detergent 1 to be successful on 63 out of 91 independent stain trials and detergent 2 to be successful on 42 out of 79 independent stain trials. Find a 95% CI for the difference $p_1 - p_2$ in success rates of detergent 1 and detergent 2. Is it likely that one detergent is better than the other one for removing grape juice stains?

8.11 CONFIDENCE INTERVAL FOR THE DIFFERENCE OF TWO POPULATION MEANS IN MATCHED-PAIRS CASE: $\mu_D = \mu_X - \mu_Y$

In many applications, two measurements are taken on the same subject (or unit). Often these are "before" and "after" measurements, such as the weight before and after a particular weight-loss program.

Example 8.29

A sample of 24 ninth- and tenth-grade girls were put on an intensive rope-jumping program in an effort to improve their track performance. The time difference for each girl for the 40-yard dash was measured:

$$D_i = (\text{``before program'' time}) - (\text{``after program'' time}),$$

where i indexes the ith person. Note that a positive number indicates an improvement in racing ability. The following data (D_i's) were obtained:

0.38	0.21	0.13	0.33	−0.03	0.17	−0.18	−0.14
−0.33	0.01	0.22	0.29	−0.08	0.23	0.0	0.04
0.20	−0.08	0.09	0.70	0.33	−0.14	0.50	0.36

We will assume that the distribution of a difference $D_i = X_{before,i} - X_{after,i}$ in racetimes for a randomly sampled girl is normally distributed with mean μ_D and SD σ_D. If the conditioning program does not help, then $\mu_D = 0$ will hold. A point estimate for the mean difference (improvement) in race times is the sample mean, $\bar{D} = 0.13$, and the sample SD of the D values for the difference in race times is $s_D = 0.24$.

We now construct a 95% CI for the difference in mean race times. Since n is small but the distribution of differences is assumed to be normal, we can use the formula given by

$$\bar{D} \pm t(n-1) \cdot \widehat{SE}(\bar{D}).$$

Here, from Table F with df = 23, $P(T > 2.07) = 0.025$; that is, $t(23) = 2.07$ only slightly greater than the z value of 1.98. Moreover, $\widehat{SE}(\bar{D}) = S/\sqrt{n} = 0.049$, which gives the interval (0.03, 0.23). This 95% CI contains the true population mean μ_D, with probability 0.95, and moreover the interval is positive (that is, it does not contain zero). Therefore, the statistical evidence supports the effectiveness of the jump-roping program, at the 95% confidence level. Note, if the sample size n is greater than 30, then the tabulated t value is, for practical purposes, the same as the corresponding tabulated z value. For example, if $n = 40$, using a more complete t table than Table F, gives $t(40) = 2.021$ versus the z value of 1.96.

Also because we were able to work with the D values, what at first appeared to be a two-sample problem (X_{before} and X_{after}) can and should be treated as a one-sample problem. Of course, if n is large, then using $D_i = X_i - Y_i$ we have a large sample CI problem for the mean μ_D (Section 8.4).

Section 8.11 Exercises

1. Twelve students participated in a semester-long weight training program. Students bench-pressed their maximum weight at the beginning (X_i) and the end (Y_i) of the program. Let $D_i = Y_i - X_i$ the difference of ("after" − "before") maximum bench press weight for the ith student. The sample mean difference $\bar{D} = 10.2$ pounds, and the sample SD $s_D = 5$ pounds. Assume that the population of (after versus before) D_i's is normally distributed. Calculate a 90% CI for the mean difference in their maximum bench press. Was the program effective? Explain.

8.12 POINT ESTIMATE FOR THE POPULATION VARIANCE σ^2; UNBIASEDNESS OF AN ESTIMATOR

While population means and proportions are two of the most widely used parameters for which we need point estimates and CIs, other parameters are also important. One such parameter that we have already encountered is the population variance (and its square root, the population SD). Information from a sample can be used to estimate the variance of the population from which the sample was taken. If a sample has a large variance, we would expect the variance of the population to be large as well.

We learned that

$$S^2 = \frac{\sum(X - \bar{X})^2}{n - 1}$$

is the usual estimator of σ^2 (and $S = \sqrt{S^2}$ is the corresponding estimator of σ). But why do we divide by $n - 1$ rather than n; that is, why not use

$$(S')^2 = \frac{\sum(X - \bar{X})^2}{n}$$

to estimate σ^2?

In order to understand why most statisticians prefer S^2 to $(S')^2$ as an estimate of σ^2, we have taken 100 samples, each of size 4, from a normal population with $\mu = 50$ and $\sigma^2 = 100$, and we have calculated $(S')^2$ for each sample. The resulting experimental distribution of the $(s')^2$ values is listed in the stem-and-leaf plot in Table 8.5 where the $(s')^2$ values are given in the table to the nearest whole number. The range of these variances is quite large: from 1 to 222. The mean of the 100 $(s')^2$ values is 76.2—a real surprise, because $\sigma^2 = 100$. Clearly, if each $(s')^2$ represents an estimate of $\sigma^2 = 100$, then these 100 statistics are quite low on average!

We repeated the process, but we increased the sample sizes to 16 in Table 8.6 and to 36 in Table 8.7. With a sample size of 16, the range is much smaller than before: from 42 to 148. The mean

Table 8.5 Sample Variances of 100 Samples of Size 4 (Population Values: $\mu = 50.0$, $\sigma^2 = 100.0$)

Stem	Leaf	Frequency
0	1,5,6	3
1	0,0,1,1,2,2,3,3,3,9,9	11
2	0,4,4,6,7,7,7,9	8
3	1,4,4,4,8,9	6
4	0,2,3,4,6,7,7,8,8	9
5	1,1,1,2,5,6	6
6	1,3,4,7,8,9	6
7	0,1,3,6,8,9	6
8	0,1,3,4,4,7,8	7
9	1,2,3,7,9	5
10	1,4,5,5,7,7,7,9	8
11	5,6,7	3
12	1,2,3	3
13	0,0,0,4,4,5,5,6,6	9
14	1,3	2
15	8	1
16	0,0,4	3
17	0,7,7	3
18		0
19		0
20		0
21		0
22	2	1
23		0
24		0
Total		100

Sample values: $n = 4$; mean of sample variances = 76.2

Table 8.6 Sample Variances of 100 Samples of Size 16 (Population Values: $\mu = 50.0$, $\sigma^2 = 100.0$)

Stem	Leaf	Frequency
2		0
3		0
4	2	1
5	1,4,4,5,6,8,8,9	8
6	0,4,7,8,8,9,9,9	8
7	0,1,1,2,3,4,7,8,8,9,9	11
8	1,1,2,2,3,3,3,4,5,5,5,7,7,8,9	14
9	0,1,1,3,3,4,5,6,6,6,9,9,9	13
10	1,1,1,1,1,2,2,2,3,3,3,4,5,5,7,7,8,9,9	19
11	0,0,0,1,2,3,4,4,4,5,5,6,6,6,9	15
12	1,1,4,6	4
13	0,1,3	3
14	1,1,3,8	4
15		0
16		0
Total		100

Sample values: $n = 16$; mean of sample variances = 93.9

Table 8.7 Sample Variances of 100 Samples of Size 36
(Population Values: $\mu = 50.0$, $\sigma^2 = 100.0$)

Stem	Leaf	Frequency
4		0
5		0
6	0,3,6,7,7,7	6
7	3,4,5,7,7,9,9,9	8
8	0,0,1,2,3,4,4,5,5,6,6,7,7,8,8,8,9,9,9	19
9	0,0,0,0,1,2,2,2,2,3,3,3,5,5,5,5,7,8,9,9	20
10	0,1,1,1,1,2,2,2,2,3,3,3,4,4,4,5,5,5,6,6,8,8,8,8,9	25
11	1,1,1,2,2,4,4,4,5,6,8,8,8,9	14
12	0,2,3,3,5,5,7	7
13	5	1
14		0
15		0
Total		100

Sample values: $n = 36$; mean of sample variances $= 97.0$

value of the 100 $(s')^2$ values for a sample size of 16 is 93.9. With a sample size of 36, the range is even smaller—from 60 to 135—and the mean of the 100 $(s')^2$ values is 97.0. The results of the three sets of samples are summarized in Table 8.8.

We see from Table 8.8 that as the sample size increased from 4 to 36, the mean of the naïve sample variance estimates $(s')^2$ approaches $\sigma^2 = 100$. In addition, the range of the estimates decreases from 221 to 75. This is another instance of the fact that increasing the sample size increases the estimation accuracy.

If, on the other hand, we compute the usual $S^2 = \sum (X - \bar{X})^2 / (n - 1)$ for each of the 100 random samples for each sample size as our estimator of σ^2, we get Table 8.9. By comparing Tables 8.8 and 8.9, it is easy to see the advantage of S^2. Clearly S^2 on average is quite close to σ^2, even for small n. This property is called **unbiasedness,** and it is expressed mathematically in terms of expected values by

$$E(S^2) = \sigma^2.$$

By contrast

$$E((S')^2) < \sigma^2$$

by quite a large amount when n is small. As a result of unbiasedness we will use S^2 to estimate σ^2. We note, however, that it can be demonstrated that S is a biased estimator of σ. There are, in fact, other, reasonable arguments that favor $(S')^2$, but we will not discuss these. Following standard statistical practice, this textbook uses S^2 to estimate σ^2 and S to estimate σ.

Table 8.8 Mean and Range of $(s')^2$ as a Function of Sample Size

	Sample size		
	$n = 4$	$n = 16$	$n = 36$
Mean of $(s')^2$	76.2	93.9	97.0
Range of $(s')^2$	221.0	106.0	75.0

Table 8.9 Mean and Range of s^2 as a Function of Sample Size

	$n = 4$	$n = 16$	$n = 36$
Mean of s^2	101.6	100.2	99.8
Range of s^2	295	113.1	77.1

Section 8.12 Point Estimate for σ^2; and Unbiasedness

> **Definition** An estimation $\hat{\theta}$ of θ is unbiased if $E(\hat{\theta}) = \theta$ and is approximately unbiased if $E(\hat{\theta}) \approx \theta$.

Section 8.12 Exercises

1. The following table is a random sample of 15 blood cholesterol levels taken from men at a local hospital. Use these data to find the usual estimate of the population variance and SD. Find also the $(S')^2$ estimate.

164	271	262	287	175
215	247	291	326	202
189	335	305	242	215

2. The following are compressive strengths of random samples from a batch of concrete. Use these data to find an estimate of the population variance and SD.

4466	3914	4084	4135	4084	4125	4123	4030	4120	3626
3771	3889	4180	4141	4524	3810	3777	4072	3871	4181

3. The sample SD of study times for a sample of size $n = 40$ is $S = 4.05$ hours per week of study time. Find a point estimate of the population variance, σ^2.

4. A sample of $n = 3$ men's weights had a variance of $(S')^2 = 225$ when using the estimator that divides by n.
   a. What is the estimate if estimator that divides by $n - 1$ is used? How much difference is there in the two estimates?
   b. Answer the same questions as in part **a**, but with $n = 100$.
   c. For which situation(s) is the difference negligible?

CHAPTER 8 SUMMARY

A **random sample** is a sequence of independent identically distributed random variables. An **estimation problem** can be either a **repeated experiments** (conceptual population) or a **survey sampling** (real population) problem.

Accurate measurement requires both small bias and small variability of the estimator (small SE(estimator)). An estimator is computed from the random sample and estimates a population **parameter**.

The **standard error of an estimator** measures the **typical random variability of the estimator,** and hence the typical random estimation error of an estimator. An estimator $\hat{\theta}$ is an unbiased estimator of θ if $E(\hat{\theta}) = \theta$. If θ is being measured by an unbiased $\hat{\theta}$, then $\hat{\theta}$ has no systematic measurement error.

$E(\overline{X}) = \mu$ and $SE(\overline{X}) = \sigma/\sqrt{n}$ for a random sample from a population with mean μ and SD of σ.

The **Central Limit Theorem (CLT)** says that $\bar{X}$ is approximately $N(\mu, \sigma/\sqrt{n})$ and $\hat{p}$ is approximately $N(p, \sqrt{p(1-p)}/\sqrt{n})$ when the sample size n is large.

All SE-based CIs follow

$$\text{Statistic} \pm (\text{critical value}) \cdot \widehat{SE}(\text{statistic}).$$

The chapter provides CIs for $\mu, p, \mu_X - \mu_Y, p_1 - p_2$ in various settings. The chapter provides three approaches to finding CIs for μ:

  a. small sample normal population sampling based CIs using the t distribution.
  b. large sample CLT SE-based CIs.
  c. bootstrapped based, SE-based CIs.

The chapter provides bootstrap-based, percentile-based CIs. The chapter provides two approaches to finding CIs for $\mu_X - \mu_Y$, including the matched pairs case:

  a. small sample normal population sampling CIs using the t distribution.
  b. large sample CLT SE-based CIs.

Chapter Review Exercises

1. Explain the meaning of a 90% confidence level.
2. Which density histogram will look more like a normal curve: (i) one based on a sample of 10 observations, (ii) one based on a sample of 100 observations, or (iii) neither?
3. Which density histogram will look more like a normal curve: (i) one based on 50 sample means, each of which is based on a sample of 10 observations, (ii) one based on 50 sample means, each of which is based or a sample of 100 observations, or (iii) neither?
4. What are the mean and SDs of a sample mean $\bar{X}$ of size 100 taken from a population with mean value $\mu = 65.25$ and SD $\sigma = 10.52$? What shape does the density histogram of this sample mean have?
5. Explain how to estimate the standard error of a sample variance for a sample of size 30 by using the bootstrap method.
6. In a random sample of 100 people, 70 were regular viewers of the television show *Seinfeld*. Find the estimated standard error of this sample proportion.
7. In a random sample of 100 teenage boys, the mean amount of time spent watching television was $\bar{x} = 3.6$ hours with SD of $s_X = 0.9$ hours. In a random sample of 100 teenage girls, the mean amount of time spent watching television was $\bar{y} = 4.2$ hours with SD of $s_Y = 1.2$ hours. Find the estimated standard error of the difference between the two samples $(\bar{X} - \bar{Y})$.
8. See Exercise 6. Find a 95% CI for the population proportion of people who watched *Seinfeld* on a regular basis.
9. See Exercise 7. Find a 99% CI for the difference in the population mean television hours between boys and girls: $\mu_X - \mu_Y$.
10. Fifty measurements of the length of a bone were made. The mean of the measurements was $\bar{x} = 34.4$ centimeters with SD of $s = 0.25$ centimeters. Find a 90% CI for the true bone length. Is it likely that the bone is 35 centimeters long?
11. In a random sample of 250 residents of a large city, 48% of the people in the sample would favor raising the bar entrance age from 19 to 21. Find a 95% CI for the population proportion of people who would favor raising the age. If a referendum were held, is it likely that more than one-half of the people would vote to raise the bar entrance age?
12. In a random sample of 50 women, the mean height was $\bar{x} = 64.45$ inches with SD of $s = 3.5$ inches. Find a 90% CI for the true mean height of women.
13. In a random sample of 100 women, the mean number of alcoholic drinks consumed per week was $\bar{x} = 3.2$ with SD of $s_X = 1.85$. In a random sample of 100 men, the mean number

of alcoholic drinks consumed per week was $\bar{y} = 5.4$ with SD of $s_Y = 1.25$. Find a 99% CI for the difference between the two population means. Is it likely that in the population men and women consume the same amount of alcoholic drinks per week?

14. Complete the following table, which summarizes the information in this chapter on CIs.

Parameter	Population variance	Standard error	Estimated standard error (if needed)	Confidence interval Lower bound	Upper bound
Mean	Known				
Mean	Unknown				
Proportion	Unknown				
Difference between two means	Known				
Difference between two means	Unknown				

15. The following is a stem-and-leaf plot of bootstrapped sample medians. Find the estimated standard error of the sample median.

Stem	Leaf
232	0
232	
233	0,0,0,0
233	5,5,5,5,5
234	0,0,0,0,0,0,0,0,0,0,0,0,0,0,0,0
234	5,5,5,5,5,5,5,5,5
235	0,0,0,0,0,0,0,0,0,0,0,0,0,0,0,0
235	5,5,5,5,5,5,5,5,5,5,5,5,5,5,5,5,5,5,5,5
236	0,0,0,0,0,0,0,0,0,0,0,0,0,0,0,0,0
236	5,5,5,5,5,5
237	0,0

16. A sample of 25 sapling trees from a large tree farm had a mean height of 6 feet, a median height of $M = 6.3$ feet, and $\sum(X_i - \bar{X})^2 = 97$.
17. Find an estimate of the standard error of the sample mean, if possible.
18. Find an estimate of the standard error of the sample median, if possible.
19. Find an estimate of the population SD.
20. A candidate for president wishes to take a sample to assess support for her. She wants to take a simple random sample and have the standard error of the sample proportion be less than 0.02. Assuming that her support in the population is about $p = 0.4$, what size sample will provide the desired standard error: $n = 100$, $n = 400$, $n = 625$, $n = 900$, $n = 1600$, or $n = 2500$?

21. A random sample of size 6 is observed:

$$12 \quad 15 \quad 10 \quad 13 \quad 13 \quad 18$$

    a. Estimate $SE(\overline{X})$ using six bootstrap samples.
    b. Estimate $SE(\overline{X})$ using the theoretical formula for $SE(\overline{X})$ and s computed from the sample.
    c. Form an approximate 95% CI for μ using the SE-based bootstrapping approach.

22. A random sample of size 6 is observed:

$$12 \quad 15 \quad 10 \quad 14 \quad 13 \quad 17$$

Suppose 100 $\bar{x}^*$ values are bootstrapped using the sample. (For example, if the first bootstrap sample is

$$17 \quad 17 \quad 14 \quad 13 \quad 13 \quad 17$$

then $\bar{x}^* = 15.3$.) Rounding to one digit to the right of the decimal point, suppose the five smallest $\bar{x}^*$ values are

$$11.2 \quad 11.8 \quad 12.0 \quad 12.0 \quad 12.2$$

and the five largest $\bar{x}^*$ values are

$$17.0 \quad 16.0 \quad 15.3 \quad 15.2 \quad 15.3$$

    a. Find a 95% bootstrapped CI for μ.
    b. Find a 90% bootstrapped CI for μ.
    c. Suppose $\widehat{SE}(\overline{X}^*) = 2.52$. Find a 90% CI for μ using the SE-based bootstrapping approach.

PROFESSIONAL PROFILE

A humpback whale dives, showing the identifiable patterns on the ventral side of its tail.

Dr. Sean Todd
Director

Allied Whale

Marine Mammal Research Facility
College of the Atlantic
Bar Harbor, Maine

Dr. Sean Todd

Whales and other large marine mammals were once far more numerous than they are today. Some species of whales have so few left that they are endangered.

Sean Todd is a marine mammal researcher studying these animals in the Gulf of Maine and on the coats of Newfoundland and Labrador. Dr. Todd examines the relationships between variables in the ocean environment and whale behavior. For example, does the surface temperature of the ocean or the amount of chlorophyll available influence the number of whales and where they travel? What impact do these animals have on the marine ecosystem?

Dr. Todd collects large amounts of data on whales, their health, and the sea around them. To get the information they need, researchers may have to identify novel data collection methods. For example, it was not until the late 1970s that scientists discovered that you could identify an individual humpback whale by photographing the underside of the animal's tail when it dove. This simple discovery makes it possible for Dr. Todd to identify and track individual whales as they return each year to their feeding grounds. By following specific individuals year after year, Dr. Todd can learn a great deal about their lives and their families.

Statistical techniques are critical to clarifying the meaning of the data Dr. Todd collects. With statistical techniques, it is possible to judge whether changes in a particular measure are real or just a result of the biological "noise" in the system. Such distinctions are very important to the goal of Dr. Todd's work: understanding what whale populations need for survival and what can be done to stabilize their environment and ensure the future of these creatures.

9

Hypothesis Testing

You don't have to eat the whole ox to know the meat is tough.
Samuel Johnson, quoted by James Boswell

Objectives

After studying this chapter, you will understand the following:

- Formulating a null hypothesis and its alternative
- Using a test statistic to assess the strength of evidence supporting a real-world claim
- Understanding level of significance, P-value, and power
- Testing hypotheses about one and two population proportions
- Testing hypotheses about one and two population means
- Testing hypotheses via normal population theory, central limit theorem, and five-step simulation (including the bootstrap)
- Testing the hypothesis of a treatment effect in a randomized controlled experiment
- Z test versus t test
- Testing the equality of matched pair means
- Significance testing versus acceptance/rejection testing
- Issues involving testing hypotheses about a population standard deviation

---

9.1 THE NULL HYPOTHESIS AND THE ALTERNATIVE HYPOTHESIS 399

9.2 TESTS FOR A POPULATION PROPORTION 402

9.3 TESTS FOR RANDOMIZED CONTROLLED EXPERIMENTS PRODUCING SAMPLE PROPORTIONS 409

9.4 TESTS FOR A POPULATION MEAN 417

9.5 TESTS FOR EQUALITY OF TWO POPULATION PROPORTIONS AND EQUALITY OF TWO POPULATION MEANS 433

9.6 ONE-SIDED AND TWO-SIDED HYPOTHESIS TESTING 441

9.7 SIGNIFICANCE TESTING VERSUS ACCEPTANCE/REJECTION TESTING: CONCEPTS AND METHODS 445

9.8 TESTS FOR A POPULATION STANDARD DEVIATION: ISSUES 452

CHAPTER REVIEW EXERCISES 453

CHAPTER 9 HYPOTHESIS TESTING

Key Problem

An article in the *New York Times* of January 27, 1988, described the results of a study on aspirin's ability to prevent heart attacks. The study focused on a group of 22,071 healthy males at least 40 years old who were followed over 5 years. The men were randomly divided into two groups. The 11,037 men in the treatment group took one children's aspirin tablet (81 mg) every other day for nearly 5 years. The 11,034 men in the control group received placebos that looked and tasted like children's aspirin.

Heart attacks often occur when blood clots form somewhere in the body and travel in the blood stream to the heart where they choke off the heart's blood supply, thereby killing a portion of the heart muscle. Aspirin works as a blood thinner, making the blood less likely to clot. Aspirin may also reduce arterial inflammation, a suspected cause of clogged arteries.

Over the 5 years aspirin users had 104 heart attacks, compared to 189 heart attacks in the placebo group. Could this difference be explained as nothing more than chance variation, or is the difference too big to be explained by chance variation alone? Can we conclude that (at least for healthy middle-aged males) a children's aspirin every other day reduces the risk of a heart attack?

There were 80 strokes in the aspirin group and 70 in the control group. Could this difference be explained as chance variation, or is the difference too big?

In Chapter 4 we learned how to model a random phenomenon by listing the possible outcomes and assigning probabilities to them. We learned that this list of all possible outcomes and their corresponding probabilities is called a probability distribution. We also learned how to set up a theoretical box model for any specified probability distribution. Using such box models we can study chance variation by using the computer to simulate data.

In Chapter 8 we studied the chance variation of estimators. Questions such as "How much is an estimator $\hat{p}$ likely to differ from the parameter p that we are trying to estimate?" were answered by simulating numerous values of $\hat{p}$ and computing the sample SD of the $\hat{p}$ values, denoted by $\widehat{SE}(\hat{p})$. This led us to confidence interval (CI) estimation.

The data in the Key Problem indicates that aspirin reduces a person's risk of a heart attack. We shall use **hypothesis testing**, another important branch of *statistical inference*, to decide whether the evidence provided by the data *"proves"* this claim beyond reasonable doubt. Hypothesis tests are often called **tests of significance** for reasons that will be discussed in Section 9.7. In this chapter we shall use both box model simulation and classical distribution theory (in particular, the central limit theorem for $\hat{p}$ and $\bar{X}$) to perform such tests.

Example 9.1

In the Congressional election of November 8, 1994, 39% of all eligible Americans voted. During the next 2 days Yankelovich Partners, Inc., conducted a telephone poll for TIME/CNN. They interviewed a random sample of 800 Americans of voting age to find out why people had voted the way they had or why they had not voted. A surprising 56% of the persons interviewed told the pollster that they had voted. Can this difference—only 39% of the population voting, but 56% of the sample claiming to have voted—be explained by chance variation, or can we conclude that some of the persons interviewed lied when they claimed that they had voted? Assume the poll was a simple random sample.

Solution: Let p denote the proportion of eligible Americans who voted on November 8, 1994. This is a population parameter, with known value $p = 0.39$. We shall simulate a simple random sample of 800 Americans of voting age and let the estimator $\hat{p}$ denote the proportion of people in the sample who voted. On November 8, 1994, around 74 million Americans voted and around 116 million

Americans of voting age did not vote. This suggests simulating $\hat{p}$ by drawing 800 times at random **without** replacement from the following box model:

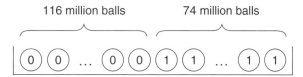

Each American of voting age who did not vote corresponds to a ball with a zero on it, whereas those who voted corresponds to a ball with a one on it. The sum of the 800 numbers drawn corresponds to the number of people in the sample who voted. The sample mean of the numbers on the sampled balls is the proportion $\hat{p}$ of people in the sample who voted.

Since the sample is so small compared to the population, the large population 5% rule is easily met. In fact, it makes virtually no difference whether we draw the 800 balls with or without replacement. We shall therefore draw 800 times from the box *with* replacement. In that case we can replace our original box of 190 million balls with a much smaller box with the same proportions of 0s and 1s. Thus, we shall draw 800 times at random with replacement from the following box:

Note that this sampling scheme involving 95 balls is practical, either drawing by hand from an actual box or by computer, as contrasted with working with 180,000,000 balls. In fact, as a rule, we prefer to draw from a box with as few balls as possible. For each computer simulation of 800 draws we shall compute:

$$\hat{p} = \text{mean of the 800 numbers drawn}.$$

Ten thousand simulations yielded the data in Table 9.1.

Notice that none of the simulated sample proportions even come close to the observed $\hat{p} = 0.56$. In fact, one can show through theoretical calculations that the chance of getting a sample proportion $\hat{p} \geq 0.56$, namely,

$$P(\text{simulated } \hat{p} \geq 0.56) \leq 10^{-12}.$$

In other words, it would take a sampling miracle—the equivalent of a monkey randomly typing Lincoln's Gettysburg Address! So, based on our simulations we conclude that it is out of the question

Table 9.1 Proportion of People in Sample Who Voted

Simulated $\hat{p}$	Experimental Frequency	Simulated $\hat{p}$	Experimental Frequency
0.32	0	0.40	1905
0.33	7	0.41	1222
0.34	32	0.42	537
0.35	180	0.43	165
0.36	513	0.44	42
0.37	1165	0.45	6
0.38	1951	0.46	1
0.39	2274		

that 56% of the people interviewed by Yankelovich actually voted. Some of the people interviewed may have been too embarrassed to admit that they had failed to do their civic duty to vote. Others may have wanted to pretend that they were part of the "Republican Revolution"—even though they had not bothered to go to the polls. All we can say for sure is that our *statistical test proves* beyond any doubt that fewer than 56% of the people interviewed by Yankelovich did in fact vote.

Statisticians have invented **tests of significance** to answer the question: *Was it due to chance, or something else?* In Example 9.1 we provided statistical proof that chance variation could not explain the difference between the 39% of the population who voted and the 56% of the sample who claimed to have voted. In the next example the question is again: Was the observed data due to chance, or something else?

Example 9.2

A labor dispute has arisen concerning the allegedly discriminatory way that 20 laborers at a construction site were given their job assignments. Six of the 20 job assignments were considered highly undesirable, whereas the remaining 14 jobs were considered desirable. The dispute was triggered by the fact that all four minority laborers working on the site were given undesirable job assignments. Could this have happened just by chance, if the jobs were assigned randomly without regard to race? Or do we have evidence of an injustice?

Solution: Imagine that we randomly assign the 20 jobs to the 20 laborers. Let X denote the number of undesirable jobs assigned to the 4 minority workers in this random assignment. We can simulate X by drawing 4 times at random without replacement from the box:

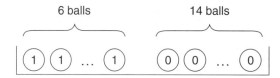

The logic behind how the box was built is as follows: corresponding to each undesirable job assignment there is a ball with a 1 on it; corresponding to each desirable job assignment there is a ball with a 0 on it. Each of the 4 minority workers draws a ball (a job assignment) from the box. This model presumes that job assignments are done totally at random. The labor dispute is over whether this model is correct or incorrect! The sum of the 4 numbers drawn corresponds to the number of undesirable jobs assigned to the 4 minority workers. Ten thousand simulations (each consisting of job assignments to the four minority laborers) yielded the data in Table 9.2.

Our experimental probability of the event of interest is

$$P(4 \text{ undesirable jobs assigned}) = P(X = 4) = 322/10{,}000 = 0.0322$$

Table 9.2 Number of Undesirable Jobs Assigned to the Four Minority Workers

Number of Undesirable Jobs Assigned	Experimental Frequency
0	2055
1	4533
2	2802
3	588
4	322

Although it is not impossible that a random assignment of jobs would give each of the 4 minority workers an undesirable job, it is *highly unlikely*. The odds are very much against it. So we conclude statistically that the jobs are not assigned at random. Note that our statistical test does not tell us how the jobs were nonrandomly assigned. Did seniority play a role? Were the 20 laborers equally qualified? Was discrimination indeed involved? The statistical test does not answer those questions. The test only tells us that the jobs were not assigned at random. It will be up to the courts or a mediation panel to decide whether the nonrandom assignment of the jobs represents an injustice.

In the next sections we shall introduce the rather technical language used in tests of significance: **null hypothesis, alternative hypothesis, test statistic, level of significance,** and *P*-**value**.

9.1 THE NULL HYPOTHESIS AND THE ALTERNATIVE HYPOTHESIS

In the Key Problem the medical **research hypothesis** is that a children's aspirin every other day reduces a middle-aged man's risk of a heart attack. The data seem to support this hypothesis: The men in the treatment group had a lower rate of heart attacks than the men in the control group. Since we are looking at a randomized controlled experiment (see Section 7.1), there are only 2 possible explanations for the different rates of heart attacks:

Research Hypothesis: Aspirin reduces the risk of a heart attack.

Chance Explanation: The observed difference in heart attack rates is due to chance variation alone.

In this set-up, statisticians call the chance explanation the **"null hypothesis."** It nullifies the research hypothesis and is denoted by H_0, the subscript suggesting "null." To prove the research hypothesis, we have to rule out the chance explanation. Following tradition, we call the research hypothesis the **"alternative hypothesis"** and denote it by H_A, the subscript A suggesting "alternative." In summary, in the Key Problem we hope to establish the research claim

H_A: Aspirin reduces the risk of a heart attack,

by ruling out the null hypothesis

H_0: The difference in heart attack rates is due to chance.

In stating these two hypotheses, we use the standard statistical notation of writing "H_A:" and "H_0:" followed by a statement of the corresponding hypothesis.

In statistical inference we use information from samples to draw conclusions about populations, as we saw in Chapter 8 for CI estimation. Therefore, it is important to distinguish between numbers that describe a population and those that describe a sample.

A **parameter** is a number that describes some numerical characteristic of a **population**. It is a fixed number, but we do not know its value unless we examine the entire population. A **statistic** is a number that describes a numerical characteristic of a **sample**. Its value is computed from a sample using an appropriate formula, but its value varies from sample to sample; that is, a statistic is a random variable, such as $\overline{X}$ or $\hat{p}$. We can use a sample statistic to estimate a corresponding unknown population parameter.

Example 9.3

Consider the population of all undergraduates at a large university. The proportion p of the undergraduates who smoke is an example of a parameter. The mean family income μ of all undergraduates is another example of a population parameter. If we take a simple random sample of 100 undergraduates, then the proportion $\hat{p}$ of undergraduates in the sample who smoke is a statistic, and so is the mean family income $\bar{X}$ of the students in the sample.

In tests of significance we often express the null hypothesis and the alternative hypothesis in terms of a population parameter. Further, we often carry out the hypothesis test using a statistic to make our decision, this statistic called the **test statistic**.

Example 9.4

Breast cancer rates are much higher in the affluent West than in the Third World. (See Example 3.1 in Chapter 3.) Heredity no doubt plays a role. But migrant studies suggest that lifestyle and environmental factors also contribute to a person's cancer risk. For example, a medical study found 62 breast cancer cases in a random sample of 10,500 Puerto Rican women who lived in New York City during 1980–86. The breast cancer rate observed in the New York sample, $\hat{p} = 62/10{,}500 = 0.0059$, was a lot higher than the corresponding breast cancer rate for the female Puerto Rican population in Puerto Rico, which was accurately known from past studies to be 0.0035. Can this difference be explained by chance variation alone, or can we conclude that women of Puerto Rican descent living in New York City have a higher breast cancer risk than women living in Puerto Rico?

Let p denote the proportion of the *entire population* of Puerto Rican women living in New York City during 1980–86 who have breast cancer at the time of the study. Using standard statistical notation, our **research hypothesis** is written as:

$$H_A: p > 0.0035.$$

The null hypothesis, which nullifies our research hypothesis, is:

$$H_0: p \leq 0.0035.$$

As often happens, the null hypothesis involves an inequality; thus more than one value of the parameter of interest will satisfy the inequality. For instance, both $p = 0.0025$ and $p = 0.0035$ satisfy the null hypothesis. For any such hypothesis test, this null hypothesis must be replaced by a null hypothesis that specifies a single value of the parameter.

How do we choose the null hypothesis parameter value? Our guiding principle is simple. We cannot accept the research hypothesis made about the population parameter unless we clearly reject the null hypothesis. So, if there are multiple possible values of the parameter under the null hypothesis, we must have strong evidence to reject *all of these values* in order to have strong evidence to reject the null hypothesis. As it turns out, there will always be one value of the parameter that is more difficult to reject than all the other values. If the evidence is strong enough to reject this most-difficult-to-reject null-hypothesis value, then the evidence is *even stronger* that all the other null-hypothesis parameter values should be rejected. This most difficult value is used as the null hypothesis.

In our example, the null hypothesis that will be most difficult to reject in favor of the alternative research hypothesis will be:

$$H_0: p = 0.0035.$$

In general, whenever the null hypothesis is written as an inequality with respect to a certain parameter value, we will simply rewrite it as an equality, noting that this specified value is the most difficult to reject.

Section 9.1 Exercises

1. For each of the following, identify the sample and the population.
   a. A poll of 40 college freshmen is taken to find out the views of the freshman class on lowering the drinking age from 21 to 18.
   b. A few small pieces of rock from a river valley region are analyzed at a laboratory to see how much gold they contain.
   c. Ecologists are worried about a certain parasite infecting tree frogs in northern Minnesota. Twenty frogs are examined for the presence of the parasite.
   d. Scientists are concerned about the salinity (saltiness) of the Gulf of Mexico. Water samples from 30 locations are analyzed.
   e. A building inspector wants to see whether the concrete being used is of sufficient quality. She takes a small sample of mix from a truckload every morning for five days and sends it to the laboratory for analysis.
   f. Lisa is believed to have high blood pressure. Her doctor instructs her to measure her blood pressure at 20 random times during the week.

2. What are the different symbols we use to distinguish a population proportion from a sample proportion?

3. A state's Department of Transportation claims "More than three out of four drivers in our state wear seat belts." To test this claim, a statistician inspects a random sample of 300 drivers.
   a. Identify the population about which the claim is made.
   b. What is the population parameter that is referred to in the claim?
   c. What is the sample statistic that the statistician will calculate?

4. For each of the following, identify the population involved, state the claim in terms of the population parameter (i.e., in terms of the population proportion, p), state the null hypothesis in words, and state the null hypothesis in standard statistical notation.
   a. An organic farming group claims that more than 75% of consumers purchase organic produce on a regular basis.
   b. A polling company claims that more than two-thirds of the public favor reducing the size of the military budget.
   c. A social services organization claims that over 20% of American children are living in poverty.
   d. A public transport advocacy group claims that the majority of Americans living in cities would use public transport if it were available.

5. For each situation in Exercise 4, give an example of a statistic that could be computed from a sample to help determine whether the null hypothesis should be rejected or not.

6. A newspaper claims that over 40% of college undergraduates pull at least one "all-nighter" before they obtain their bachelor's degree.
   a. State the claim in symbolic form, state the null hypothesis in words, and, finally, state the null hypothesis in symbolic form using standard statistical notation.
   b. A statistician then performs a statistical study on a randomly selected sample of recent graduates. In this sample, the proportion who pulled at least one all-nighter during the time they were undergraduates turned out to be 0.45. The statistician determines that the probability of observing a value of 0.45 or more in such a sample is 0.20 when assuming the null hypothesis is true. Discuss whether this constitutes strong evidence for rejecting the null hypothesis.

7. Find two or more examples from newspapers or magazines that involve claims or hypothesis that might be testable by observing a real data sample and testing the null hypothesis. Name the null hypothesis in each case, and give an example of a statistic that could be used in testing it.

9.2 TESTS FOR A POPULATION PROPORTION

As mentioned above, the kind of statistical decision making we are describing is called hypothesis testing. Many statistical problems are hypothesis testing problems. In this section we will learn how to test research hypotheses about the proportion of units in a population (either real or conceptual) that exhibit a particular property or quality.

Claims about population proportions underlie many of the most hotly debated issues that confront our society. Questions that involve single population proportions include:

- What proportion of children in America are living in poverty?
- What proportion of "partial-birth" abortions would result in the mother suffering grievous health consequences if she had carried the child to term?
- What proportion of illegally imported drugs is intercepted by drug enforcement officers?

In testing hypotheses about population proportions, we formulate the null hypothesis by assuming that the "research" claim is false. Recall that these claims are more formerly referred to as "alternative hypotheses." The null hypotheses considered in this section will be of the form

$$H_0: p = p_0,$$

where p_0 is the value of the population proportion assumed by the null hypothesis. (See Example 9.4; keep in mind that H_0 may have initially been stated as an inequality.) We will then assess the probability that a sample proportion generated under the null hypothesis is at least as extreme in the direction of the alternate hypothesis as the value calculated from the real data sample.

For most problems considered in this section, the null-hypothesis probability distribution of the sample proportion is fully specified. In the first part of this section, we will use this distribution to specify a box model. In the second part of this section, via the central limit theorem, we will solve problems having large sample sizes by using a normal distribution to approximate the null hypothesis distribution of the sample proportion. Later in the chapter we will discover that often the null hypothesis does not specify the distribution of the statistic being used to evaluate the strength of evidence against the null hypothesis.

Population Proportion Hypothesis Testing Using Box Models

The distribution of the sample proportion $\hat{p}$ under the null hypothesis H_0 depends on only a single parameter, the population proportion p. The box model for the population is quite simple: The total number of balls in the box is equal to the size of the population, the balls are labeled with either a 1 or a 0, and the proportion of balls labeled with a 1 is equal to the population proportion stated in the null hypothesis. Because the null hypothesis distribution of the sample proportion can be fully specified with a box model, we can apply the box-model-based five-step simulation approach of Chapter 4 to perform hypothesis testing.

Example 9.5

Gun control proponents in a city claim that more than 60% of the 100,000 eligible voters support the mandatory registration of handguns. Let p be the proportion of the population of voters in this city who support the mandatory registration of handguns. This "research hypothesis" becomes the alternative hypothesis, which we write down using the standard notation:

$$H_A: p = 0.6.$$

Suppose also that we have a real data sample of size $n = 20$, based on random sampling without replacement from the population of size 100,000; and it has yielded 17 voters who support the mandatory registration of handguns, resulting in the test statistic $\hat{p}_{OBS} = 0.85$, the proportion in the sample supporting gun control. The question we want to answer is, "Is this value of $\hat{p}$ such an unusually large value that we should conclude that p really is larger than 0.6, or can a $\hat{p}$ this large be reasonably explained by chance variation alone when p really equals 0.6?"

In conducting our hypothesis test, it is helpful to think of the process occurring in three stages. Let's use this example to describe these three stages.

Stage I. Formulation of the Null Hypothesis.

In the first stage, we formulate the null hypothesis. The null hypothesis that nullifies the alternative hypothesis is that p is no bigger than 0.6, which we write formally as:

$$H_0: p \leq 0.6.$$

We rewrite this as:

$$H_0: p = 0.6.$$

Since the null hypothesis must state a single value for the parameter, and that single value must be the one that is closest to the alternative hypothesis (and, thus, the most difficult one to reject). We will typically go directly to this version of the null hypothesis without first writing down the one that contains the inequality.

Stage II. Determine the *P*-Value using a test statistic.

To carry out the second stage, we use the five-step method of Chapter 4 to estimate the probability of occurrence for the event of interest assuming the sample proportion is distributed according to the null hypothesis, that is, $p = 0.6$.

The five-step method proceeds as follows:

Step 1. Choice of a Sampling Box Model: As we stated above, the box model is easy to build for sampling from a population defined by a population proportion. In particular, the box model will represent the population as defined by the null hypothesis.

In this example, the population proportion is equal to 0.6 under the null hypothesis, and the population size is 100,000. The obvious choice of a model is a box of 100,000 balls, with 0.6 of the 100,000 (i.e., 60,000) labeled 1 to represent objects having the attribute of interest and the other 40,000 balls labeled 0 to represent objects lacking the attribute of interest. This model represents the actual population under the null hypothesis. If a simulation consisted of 20 random draws *without replacement*, we would guarantee that the same object would never be sampled twice, as required in survey sampling.

However, recall from Section 4.5 (or recall the Large Population Sample Rule from Section 8.2) that if the population size is large compared to the sample size ($20n \leq N_P$ or $n \leq 0.05 \, N_P$, with N_P denoting population size), we may use sampling *with replacement* instead, with the knowledge that our simulation study is as valid as if we sampled without replacement from the box with 100,000 balls. But if we use sampling with replacement, then the box model needs to contain only as many balls as are necessary to express the null-hypothesis proportion, which equals 3/5 in this example. Therefore, a completely acceptable null-hypothesis box model contains only 5 balls: three balls labeled 1 (where 1 denotes possession of the attribute of interest) and two balls numbered 0.

Step 2. Definition of One Simulation: We want each simulation to draw a random sample from our box model of the same size as the sample that occurred with the real data. In this example, one simulation consists of a random sample of size 20 (drawn with replacement).

Step 3. Definition of the Event of Interest: The event of interest is carefully formulated using the test statistic to assess the strength of evidence against H_0 exhibited by the observed value of the test statistic $\hat{p}_{OBS}$. Let the test statistic $\hat{p}$ be the proportion of 1s in a simulated sample. The event of interest is having a $\hat{p}$ in the simulated sample be at least as extreme (in the direction of evidence supporting the alternate hypothesis) as the numerical value of the proportion $\hat{p}_{OBS}$ that was observed in the real data sample.

Notice that the definition of the event of interest uses the words "at least as extreme," not "equals." Rather than use "simulated $\hat{p}$ exactly equal to the observed real-data $\hat{p}_{OBS}$ value," we specify a range of being at least as extreme as the observed $\hat{p}$ value. The null hypothesis "defends itself" against the original research claim by making the case that typical random variation of the sample statistic can easily produce a value as far or farther away as the observed value. If a simulation produces a value farther away from p_0 than the observed $\hat{p}$, that simulated $\hat{p}$ is evidence against the null hypothesis. Therefore it is included in the event of interest.

So, in this example the event of interest (EOI) is (simulated $\hat{p} \geq 0.85$), 0.85 chosen because $\hat{p}_{OBS} = 0.85$. In this case, that means 17 or more ones being drawn in the sample of 20.

Step 4. Repetition of Simulations: For good accuracy we need at least $N = 400$ repetitions, but 1000 is better. Let N stand for the number of simulations, and N_{EOI} stand for the number of simulations for which the event of interest occurred. In our example, we drew $N = 400$ samples and observed $N_{EOI} = 7$ samples that each contained 17 or more ones and hence produced (simulated $\hat{p}$) ≥ 0.85.

Step 5. Calculation of the Experimental Probability of the EOI (called the ***P*-value**, as explained below): This step is clear. The experimental probability is merely the usual Chapter 4 relative frequency of the EOI based on the simulations. Thus, the five-step method produces an experimental probability calculated from simulated data samples that are all generated from the null-hypothesis box model. If this experimental probability is low, then $\hat{p} \geq 0.85$ is very unlikely to occur in a real-data sample when the null hypothesis is true, so we will decide to reject the null hypothesis (see the Hypothesis Testing Strength of Evidence Principle that follows). Otherwise, we will decide not to reject the null hypothesis.

Using the symbol $\hat{P}(\text{EOI}|H_0)$ to stand for the estimated probability of the event of interest given that the null hypothesis probability distribution is the true distribution for $\hat{p}$, we get

$$\hat{P}(\text{EOI}|H_0) = \hat{P}(\text{simulated } \hat{p} \geq 0.85|H_0) = N_{EOI}/N = 7/400 = 0.0175.$$

Note that in standard statistical nomenclature, $\hat{P}(\text{EOI}|H_0)$ is referred to as the **"*P*-value"** of the hypothesis test. The following principle for assessing the strength of evidence associated with the *P*-value is vital to understand.

Hypothesis Testing Strength of Evidence Principle The smaller the *P*-value, whether computed theoretically or approximated using five-step simulation, the stronger the evidence is against the null hypothesis and, hence, in favor of the alternative research hypothesis. If the evidence is judged strong, we **reject H_0**; otherwise we **fail to reject H_0**.

Stage III. Decide Whether or Not to Reject the Null Hypothesis.

In this stage we use the probability result from Stage II, the *P*-value of the hypothesis test, to decide whether or not the observed proportion $\hat{p}_{OBS}$ in the real-data sample was very unlikely to have occurred if the null hypothesis was true, and thus make a decision to reject (evidence strong) or fail to reject (evidence weak) the null hypothesis.

The standard procedure is to compare the *P*-value to some fixed numerical criterion. This criterion also has a particular name in statistical terminology, called the **"level of significance."** If the *P*-value is less than the level of significance, then the decision is to reject the null hypothesis; otherwise, the decision is not to reject the null hypothesis. The standard choice for the level of significance that is considered strong evidence is 0.05.

In our example, we found that the *P*-value was 0.0175, which is quite small. Using the standard 0.05 level of significance and noticing that $0.0175 \leq 0.05$, we reject H_0. The data thus show strong evidence in support of the alternative hypothesis and prove beyond a reasonable doubt the research hypothesis that more than 60% of the voters in the city support the mandatory registration of handguns.

In summary, when testing a research claim that a population proportion is less than some specified value p_0 (or is greater than some specified value p_0), we employ the following three-stage approach to hypothesis testing:

I. Formulate the null hypothesis ($H_0 : p = p_0$).

II. Estimate the P-value of the hypothesis test, which is the probability of the occurrence of the event of interest under the null hypothesis. The event of interest is the occurrence of a numerical value of $\hat{p}$ under the null hypothesis at least as extreme as the value observed in the real-data sample. In this section the probability of the EOI is estimated by applying the five-step simulation method of Chapter 4.

III. Use the P-value result from Stage II to decide whether or not to reject the null hypothesis by observing whether the P-value is less than the level of significance (usually 0.05).

If the $\hat{p}_{OBS}$ value that occurred in the real-data sample was very unlikely to have occurred if the null hypothesis was true, then the P-value will be less than the level of significance and we will decide to reject the null hypothesis. Otherwise, the decision will be to fail to reject the null hypothesis.

Note: Do not make the mistake that failing to reject H_0 means accepting H_0. For example, the research claim (H_A) that a coin favors heads with 6 heads out of 10 tosses observed would lead to failing to reject $H_0 : p = 0.5$. But this should not lead us to accept the null hypothesis that the coin is fair.

In this section, Stage II is accomplished via the five-step simulation method that was learned in Chapter 4. In particular, the box model represents the population from which the real-data sample would have been drawn if the true population distribution was the one defined by the null hypothesis. We generally consider cases where the sample size is small relative to the population size, thus making our jobs as statisticians simpler. In particular, this enables us to sample with replacement when doing simulations to test hypotheses. Thus, we will be able to use a box that contains only enough balls to represent the values of the proportions, as assumed by the null hypothesis. We do not have to have one ball for each member of the (perhaps huge) population.

The details of the five-step simulation method used in this section are as follows:

Step 1. Choice of the Box Model: Choose a box containing balls marked 1 or 0, and make the proportion of ones correspond to the proportion p_0 in the population that has the attribute in which we are interested, as assumed by the null hypothesis.

Step 2. Definition of One Simulation: Define one simulation to be the random drawing of a sample, with replacement, of the same size as the real data sample.

Step 3. Definition of the EOI: If the claim is that a population proportion is less than some specified value p_0, then

$$\text{EOI} = (\text{simulated } \hat{p} \leq \text{the real data } \hat{p}_{OBS}).$$

If the claim H_A is that a population proportion is greater than some specified value p_0 then

$$\text{EOI} = (\text{simulated } \hat{p} \geq \text{the real data } \hat{p}_{OBS}).$$

Step 4. Repetition of Simulations: For good accuracy we need at least 400 repetitions.

Step 5. Calculation of the Experimental Probability of the EOI: Let $\hat{P}(\text{EOI}|H_0)$ stand for the experimental probability of the EOI when the null hypothesis is true, which is more formally referred to as the estimated P-value of the hypothesis test. Then,

$$\hat{P}(\text{EOI}|H_0) = N_{EOI}/N,$$

where N stands for the number of simulations, and N_{EOI} stand for the number of simulations in which the EOI occurred.

Population Proportion Hypothesis Testing Using the Normal Distribution (Large Sample Case)

Recall from Chapter 8 that the central limit theorem (CLT) showed us that the distribution of a sample mean is approximately normal when we have a random sample (which means with replacement) and the sample size is large enough. If the problem is a sample survey problem involving a real population, note that the population size N_P must be large enough to satisfy the 5% rule, that is, $n \leq 0.05\, N_P$. Because, as discussed in Chapters 8 and 9, the sample proportion is a sample mean, the CLT says that we can approximate its distribution with a normal distribution when the sample size is large. Here the large-sample case is acceptable if n and p are such that both $np \geq 9$ and $n(1-p) \geq 9$. The idea is that for us to be able to apply the CLT to $\hat{p}$, n needs to be large and p needs to be not too close to 0 or 1; and the closer p is to 0 or 1, the larger the sample size must be to compensate. As we saw in Chapter 8, if these requirements for n and p hold, via the CLT, it is accurate to approximate probabilities of events involving a standardized sample proportion using a standard normal distribution.

To conduct our hypothesis tests, we will be able to use the three-stage hypothesis testing approach that was introduced in this section. The only difference will be that Stage II will be carried out by using the CLT for $\hat{p}$ and, hence, using a theoretical probability value of the normal distribution instead of using the five-step simulation method. By bypassing the five-step simulation method, we will be able to carry out our hypothesis testing much more simply.

In short, our three-stage hypothesis testing approach will be:

I. Formulate the null hypothesis ($H_0: p = p_0$).
II. Estimate the P-value of the hypothesis test, which is the probability of the occurrence of the EOI under the null hypothesis. The EOI is the occurrence of a numerical value of $\hat{p}$ at least as extreme as the value observed in the real-data sample ($\hat{p}_{\text{OBS}}$). In this section we will accomplish this by looking up a standard normal probability value in Table E of Appendix E. This will be described in more detail below.
III. Use the P-value from Stage II to decide whether or not to reject the null hypothesis by seeing if the P-value is less than the level of significance.

Let us now consider an example.

Example 9.6

The standard drug treatment is known to produce a cure 60% of the time. A new medication is produced that is claimed to be better. It is tested on 36 people believed to be a random sample from those afflicted with the disease. Of these, 30 are cured producing $\hat{p}_{\text{OBS}} = 30/36$. Can we conclude that the new drug is better? (Note that this is not a randomized controlled experiment since we know the "control" proportion $p = 0.6$. Thus, no control group is needed.)

Let p = cure rate of the new medication. Now let us apply our three-stage approach to hypothesis testing.

Stage I. Formulate the null hypothesis.
The denial of the claim provides the null hypothesis, "The new medication is no better." More precisely and in symbols, this null hypothesis is

$$H_0: p = 0.6,$$

having replaced $\leq$ by $=$.

Stage II. Estimate the P-value of the hypothesis test.

Estimating the P-value for tests about a population proportion was addressed in Example 9.5, but it can be done without simulations by using the CLT. We learned in Section 8.5 that

$$E(\hat{p}) = p \quad \text{and} \quad SE(\hat{p}) = \sqrt{p(1-p)}/\sqrt{n}.$$

According to the CLT, by centering $\hat{p}$ at the true p and dividing by $SE(\hat{p})$, our standardized test statistic of

$$Z = \frac{\hat{p} - p}{\sqrt{p(1-p)}/\sqrt{n}} = \frac{\hat{p} - 0.6}{\sqrt{(0.6)(0.4)}/\sqrt{n}}$$

is approximately standard normal when $np \geq 9$, $n(1-p) \geq 9$ and p is the true parameter value. Under the null hypothesis of $p = 0.6$, we need to check that the sample size is large enough to apply the CLT. Since

$$np = 30(0.6) = 18 \geq 9 \quad \text{and}$$

$$n(1-p) = 30(1-0.6) = 30(0.4) = 12 \geq 9,$$

we can use the CLT. Therefore, using the null hypothesis $p = 0.6$ and $\hat{p}_{OBS} = 30/36$, we have

$$Z = \frac{30/36 - (0.6)}{\sqrt{(0.6)(0.4)/36}} = 2.86.$$

Now we can evaluate the P-value for the EOI using Table E:

$$P\left(\hat{p} \geq \frac{30}{36} \middle| H_0\right) \approx P(Z \geq 2.86 | H_0) = 1 - P(Z < 2.86 | H_0)$$
$$= 1 - 0.9979$$
$$= 0.0021$$

Stage III. Decide whether or not to reject H_0.

Because the approximated P-value of $P(Z \geq 2.86) = 0.0021 < 0.05$ (the standard level of significance), we reject the null hypothesis and conclude that the new medication is better. This hypothesis-testing problem could have been solved by simulation, as was done previously. When it is valid, however, the theoretical approach is preferable, being both easier and usually slightly more accurate.

This example is the first instance of the often occurring situation in which a standardized test statistic will obey the CLT, thus yielding a good method for carrying out hypothesis testing when the sample size is large.

Section 9.2 Exercises

1. Several states are considering banning the use of cellular phones while driving because of the increased risk in traffic accidents associated with such use. State legislators who support such laws might claim that a large majority of their constituents are in agreement with them. Assume that a "large majority" is interpreted as "over 60%."

   Suppose one of these legislators surveys 20 randomly sampled constituents asking whether they think their state government should or should not pass a law making it illegal to use a cellular phone while driving.

The results of the survey showed that 75% of those surveyed did indeed think that their state should pass such a law.

a. In words only, what is the population proportion under consideration in this problem?
b. What is the claim being made about this population proportion?
c. In terms of the population proportion, what is the null hypothesis?
d. Why is the box-model simulation approach preferred over the Z-statistic approach for testing the null hypothesis?
e. To test the null hypothesis, we need to generate simulated samples that conform to the null hypothesis. On each simulation, we will generate a simulated sample having the same size as the original real-data sample. What sample statistic will we measure on each simulated sample that we generate?
f. The EOI for testing the null hypothesis will be based on comparing the sample statistic from the simulated sample to something. To what will we compare it? Why do we make this comparison?
g. The results from 1000 simulations under the null hypothesis are shown in the following table. What are the most frequently occurring values of $\hat{p}$ in the simulated samples? What are the least frequently occurring values in the direction of rejecting the null hypothesis?
h. Is the value of $\hat{p}$ observed in the real-data sample among the least frequently observed simulated $\hat{p}$ values in the direction of rejecting the null hypothesis?
i. Decide whether or not to reject the null hypothesis and explain why you chose that decision.
j. What would your answer have been if 80% of the sample had indicated their state should pass such a law?

$\hat{p}$ observed in simulated sample	Frequency of occurrence	$\hat{p}$ observed in simulated sample	Frequency of occurrence
.00	0	.55	168
.05	0	.60	178
.10	0	.65	182
.15	0	.70	126
.20	1	.75	70
.25	1	.80	24
.30	5	.85	12
.35	7	.90	0
.40	26	.95	0
.45	75	1.00	0
.50	125		

2. In Exercise 1, the simulated samples are generated from a box model. The choice of the box model is required to model the population from which the null hypothesis says the real data sample was drawn.

   Describe the choice for the box model and the type of sampling that should be done. Explain how this results in simulated samples that correspond to the null hypothesis.

3. In a random sample of 64 people in a city, 37.5% were in favor of lowering the drunk-driving blood alcohol level from 0.10 to 0.08.
   a. Test the null hypothesis that the population proportion $p = 0.5$, where prior to data collection it was claimed that $p < 0.5$.
   b. Why is the CLT approach preferred to the box-model approach for testing the null hypothesis in this problem?

4. It is important to travelers that mass transit vehicles exhibit a high percentage of on-time performance. Suppose a regional train system boasts that its trains run on time over 80% of the time.

To check out this claim, an investigative reporter might collect a sample of real data. Suppose a random sample of 40 train arrivals is observed, and 37 of them are observed to be on time.
   a. In words, what is the population proportion in this problem?
   b. What claim is being made about this population proportion?
   c. Why is a box-model simulation approach preferred to a CLT approach in this problem.
   d. What is the null hypothesis?
   e. What is the choice of the box model to be used to generate simulated samples corresponding to the null hypothesis?
   f. What is the definition of one simulation from the box?
   g. Define the EOI and explain why this is of interest in regard to testing the null hypothesis.
   h. In 1000 simulations, 19 of them had a $\hat{p}$ as large as, or larger than 0.925. Explain what this means and whether or not the null hypothesis should be rejected.
5. The President's press secretary claims that a majority of people approve of the President's performance. In a random sample of 2000 U.S. citizens over the age of 18, 53.5% approve of the president's performance.
   a. Test the null hypothesis whether the population proportion is equal to 0.5.
   b. Why is the CLT approach preferred to the box-model approach for testing the null hypothesis in this problem?
6. In a random sample of 2000 residents of Illinois, 75% believe in the fairness of the jury system. Test whether the population proportion is equal to 0.8, as opposed to < 0.8, as has been claimed.
7. It is claimed that 60% of voters favor a proposed constitutional amendment. A telephone poll of 20 voters results in 15 being in favor of the amendment. Should it be concluded, on the basis of this poll, that the community favors the amendment?
8. Records from several years ago show that in a certain community, 60% of the people owned their homes. An advocacy group for affordable housing claims that home ownership has declined since that time. A survey of 20 families, just taken, shows that 9 own their homes. Is this reliable evidence that there has been a decrease in home ownership in the community as claimed by the advocacy group?
9. In a random sample of 60 college freshmen, 75% are in favor of lowering the drinking age from 21 to 18. Test the null hypothesis that the population proportion is equal to 0.7 versus more.
10. In a random sample of 100 people, 70 are regular viewers of the television show *The Simpsons*. Test whether the population proportion is equal to 0.75 or less.

9.3 TESTS FOR RANDOMIZED CONTROLLED EXPERIMENTS PRODUCING SAMPLE PROPORTIONS

Recall that a randomized controlled experiment is used to assess the effect of a treatment (often a medical treatment) by comparing the two possible outcomes (e.g., diseased versus not diseased when testing a vaccine or recovered versus not recovered when testing a drug) that occur on units receiving the treatment to the outcomes that occur on units not receiving the treatment. The difference in the rates of these two sets of outcomes is used to estimate the effectiveness of the treatment. For this experimental method to be successful, the only difference between the experimental units that receive the treatment (the treatment group) and the units that do not receive the treatment (the control group) must be just the presence or absence of the treatment itself. The only way to ensure this principle holds is to randomly assign units to the treatment and control groups. When a large number of units is used, random assignment will ensure that the two groups will be alike in all respects except for whether or not they receive the treatment. Usually the set of units obtained for medical or other experiments is not random. For example, physicians might provide 200 volunteers for a clinical trial of a new drug. The randomization needed occurs in the random assignment of 100 people to the treatment group and 100 people to the control group.

In this section, we will introduce a clever way of using box-model simulations to conduct hypothesis tests for randomized controlled experiments that use proportions to measure treatment versus control effects. The randomized polio vaccine trials of Chapter 7 are an example of a randomized controlled experiment with this type of measurement. The children were the experimental units, the treatment was the polio vaccine, the characteristic of interest was whether a child contracts polio, and the experiment yielded the proportions of children who contracted polio in the treatment and control groups.

Participating children were volunteers and, hence, were *not* a random sample. The most common approach to hypothesis testing in this situation is to compare two sample proportions: the sample proportion of units in the treatment group that exhibit the characteristic of interest (for example, the sample proportion of children in the treatment group who contracted polio) and the sample proportion of units in the control group that exhibit the characteristic of interest. This approach is not conducive to box-model simulations (though it can be handled by theoretical methods under certain conditions). In this section we will take a less well known, but effective, approach to solving this type of problem—an approach that *is* conducive to box-model simulations.

In randomly assigned treatment and control groups, one observes x_T with the characteristic of interest from the treatment (T) units and similarly x_C from the control (C) units for a total of

$$x = x_T + x_C$$

displaying the characteristic of interest.

For example, in the Key Problem, there were $X_T = 104$ heart attacks among the aspirin takers and $X_C = 189$ heart attacks among the placebo takers for a total of

$$x = 104 + 189 = 293$$

heart attacks.

Let p_T denote the proportion of units assigned to the treatment group. This is determined by the investigator and often equals 1/2 or close to it. For example, in the Key Problem when the goal was to choose $p_T = 1/2$ in actuality (perhaps a few dropped out or were removed)

$$p_T = \frac{11037}{22071} = 0.5001.$$

Often the characteristic of interest will be fairly rare, for instance, the polio example of Chapter 7. In the Key Problem, the rate of heart disease (the characteristic of interest) in the n patients of the study was

$$\frac{x}{n} = \frac{293}{22071} = 0.013;$$

that is, heart disease is rare. Later, in Example 9.8, we have a case where the characteristic of interest is not rare.

The null hypothesis in a randomized controlled experiment is that there is no treatment effect. That is, the chances of a unit displaying the characteristic of interest is the same regardless of its random assignment to a group.

There is a clever and simple way to simulate the random process of acquiring the characteristic of interest or not in a randomized controlled experiment *when the null hypothesis is true* using a box-model approach. Consider such an experiment where there are $x_T + x_C = x$ occurrences in the studied units. To build the null hypothesis box model, represent all the units of the experiment by balls in the box, labeling each of the n balls by T or C depending on whether it was assigned to the treatment (T) group or control (C) group. For example, for the Key Problem we would have:

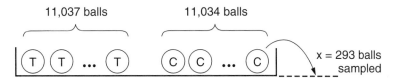

The experiment, under the assumption H_0 is true, is simulated by drawing x balls randomly **without replacement**. Each drawn ball is assigned the characteristic of interest (for example, heart disease) and each ball remaining in the box is assigned to not have the characteristic of interest.

Clearly, whether a unit is assigned the characteristic of interest is not influenced by whether the unit is a T or a C. Thus the null-hypothesis model of no-treatment effect is being simulated. For example, the $x = 293$ heart attacks in the Key Problem can be simulated by taking 293 random draws without replacement from the box displayed above (Step 2 of the five-step model). T and C units are equally likely to have a heart attack (by being drawn), so clearly H_0 (no effect for the aspirin) is true. Thus we do have a probability model of the experiment when there is no treatment effect.

Consider one random sample drawn of size x from the null hypothesis box. Let the statistic of interest defined in Step 3 of the five-step model be

$$\hat{p} = \text{proportion of } T \text{ balls observed in the sample} = \frac{X_T}{x},$$

where X_T is the random number of T balls observed in the sample of size x.

In a randomized controlled experiment the population is the set of units. Thus the balls in the box are the population in the simulation. Recalling that the population (box) proportion of T balls is p_T, the expected value of the statistic of interest $\hat{p}$ can be mathematically shown to satisfy $E(\hat{p}) = p_T$ under the null hypothesis. In general, the treatment may be influential and, hence, the null hypothesis may fail. In fact, when the distribution of $\hat{p}$ is no longer the result of sampling from the null hypothesis box, we expect to have $E(\hat{p}) < p_T$ if the treatment *reduces* the chance of a treated unit acquiring the characteristic of interest. If H_0 fails in the sense that the treatment *increases* the chance of a treated unit acquiring the characteristic of interest, then $E(\hat{p}) > p_T$.

For example, in the Key Problem, the research hypothesis that the aspirin is effective in our box-model simulation amounts to $E(\hat{p})$ satisfying $E(\hat{p}) < p_T = 1/2$, where $\hat{p}$ is the proportion of the sampled heart attack victims that are from the treatment group. By contrast, if the experiment is to test a new drug on a sample of ill people where the characteristic of interest is recovery from the illness, then H_0 being false amounts to $E(\hat{p}) > p_T$. Regardless of whether H_0 or H_A is true, defining the population parameter p by $p = E(\hat{p})$, we then have either

$$H_0: p = p_T \text{ versus } H_A: p < p_T, \quad \text{or}$$

$$H_0: p = p_T \text{ versus } H_A: p > p_T.$$

Thus, just as in Section 9.2, the hypothesis-testing problem for a randomized controlled experiment can be formulated as a test concerning the population parameter p. For example, in the Key Problem we have

$$H_0: p = 0.5001 \text{ versus } H_A: p < 0.5001$$

with the actual experiment producing

$$\hat{p}_{\text{OBS}} = \frac{x_T}{x} = \frac{104}{293} = 0.355.$$

Proceeding as in Section 9.1, we ask whether a simulated $\hat{p} \leq 0.355$ is very unusual under H_0. Because about half of the units were assigned the treatment, but only about 1/3 of all the observed heart attacks come from the treatment group, it seems that the aspirin treatment does have an influence, namely reducing the rate of heart attacks. But we need to carry out a structured hypothesis test to decide if our impression is right.

Let us summarize. As previously stated, the value of p_T is a known constant set by those conducting the experiment. When equal numbers of units are assigned to the treatment and control groups, then $p_T = 0.5$. The experiment results in a data sample that yields $\hat{p}_{OBS}$, the observed proportion of units that received the treatment among all the x units exhibiting the characteristic of interest. By using the null hypothesis to specify a box model as described earlier, we are able to generate simulated samples and, thus, simulated $\hat{p}$ values, to compare to $\hat{p}_{OBS}$. Because the simulated $\hat{p}$ values are generated under the null hypothesis, by comparing them with the real data $\hat{p}_{OBS}$ we will be able to decide whether $\hat{p}$ values as extreme as the $\hat{p}_{OBS}$ value are likely to be generated by the null hypothesis model.

To conduct this hypothesis test, we will use the three-stage hypothesis testing approach that we used in Section 9.2:

  I. Formulate the null hypothesis ($H_0: p = p_T$).
  II. Estimate the P-value of the hypothesis test: Apply the five-step method of Chapter 4 to estimate the P-value, which is the probability of the occurrence of the EOI under the null hypothesis. The EOI is

$$(\text{simulated } \hat{p} \leq \hat{p}_{OBS}).$$

Thus, if N simulations are done and the EOI occurs N_{EOI} times then the P-value is

$$P\text{-value} = \hat{P}(\text{simulated } \hat{p} \leq \hat{p}_{OBS} | H_0) = \frac{N_{EOI}}{N}.$$

  III. Use the P-value result from Stage II to decide whether or not a $\hat{p}$ is at least as extreme as the observed proportion $\hat{p}_{OBS}$ if the null hypothesis is true, and thus make the decision to reject or to fail to reject the null hypothesis.

The generation of simulated data for the null hypothesis is carried out in exactly the same way as in Section 9.2. The use of a real-data sample to define the EOI and the decision-making process are carried out in exactly the same way. Even the choice of whether to use sampling with or without replacement is carried out in exactly the same way!

Now let us look at a simple example to demonstrate hypothesis testing for a randomized controlled experiment that uses proportions to investigate treatment effects.

Example 9.7

A randomized controlled experiment investigated the effect of a new medicine on preventing death caused by heart disease (New England Journal of Medicine, 1996). The treatment group consisted of 4700 subjects who received the medication while the control group consisted of 4300 subjects who received a placebo. The assignment of the 9000 subjects, as treatment or control subjects, was random, as required. Thus, p_T, the proportion of experimental units in the treatment group, is given by

$$p_T = \frac{4700}{9000} = 0.522.$$

The characteristic of interest was whether a subject died of heart disease by the end of the testing period. By the end of the testing period, 658 subjects had died of heart disease. These 658 subjects can be thought of, for statistical purposes, as the people randomly assigned the characteristic of interest. Of these 658, 301 were from the treatment group. Thus, $\hat{p}_{\text{OBS}}$, the proportion of the sample (those who died) who were treatment group members, is given by

$$\hat{p}_{\text{OBS}} = \frac{301}{658} = 0.457.$$

Does this value of $\hat{p}_{\text{OBS}}$ constitute strong evidence that the treatment reduced the number of deaths caused by heart disease?

Stage I. Formulate the null hypothesis.
The null hypothesis denies that the treatment is effective. In particular, the null hypothesis asserts that p, the expected proportion of treatment group subjects in the sample randomly chosen to display the characteristic of interest (i.e., those who died by heart disease) is the same as the proportion of treatment group subjects among all the subjects who took part in the experiment [4700/(4300 + 4700) = 0.522]. Thus, we write

$$H_0: p = 0.522 \text{ versus } H_A: p < 0.522.$$

Stage II. Estimate the P-value of the hypothesis test.
Now we apply the five-step method of Chapter 4 to estimate the probability of occurrence of the event of interest given that the null hypothesis is true. This probability is the P-value of the hypothesis test.

Step 1. Choice of the Box model: Because the sample size (the 658 who died of heart disease) is only slightly greater than 5%, (about 7% of the 9000 subjects in the experiment), we will go ahead and use sampling with replacement, which enables us to use the smallest possible number of balls to represent the proportion of treatment group subjects in the box model.

Thus, we use a box containing a total of 90 balls, 47 of which are marked with a 1 (for the treatment group members) and 43 of which are marked with a 0 (for the control group members).

Step 2. Definition of One Simulation: A simulation consists of randomly drawing a ball from the box $x = 658$ times with replacement to determine each subject who dies of heart disease. The proportion of 1s in the sample is recorded for each simulation. This proportion represents the proportion of those dying from heart disease who were members of the treatment group.

Step 3. Definition of the Event of Interest: The event of interest is defined as the occurrence of a simulated sample proportion $\hat{p} \leq 0.457$, this value being the proportion of subjects that died of heart disease who were from the treatment group ($\hat{p}_{\text{OBS}}$). In other words, if EOI denotes the event of interest, then

$$\text{EOI} = (\text{simulated } \hat{p} \leq \frac{301}{658}),$$

noting that $301/658 = 0.457$.

Step 4. Repetition of Simulations: We then conducted 1000 simulations and counted the number of simulations, N_{EOI}, for which the event of interest occurred.

Step 5. Calculation of the P-value (the Experimental Probability of the EOI): We divide the number of simulations for which the event of interest occurred (N_{EOI}) by the total number of simulations ($N = 1000$), yielding our estimate of the P-value, the probability of the EOI occurring when the null hypothesis is true. When 1000 simulations were done, the EOI happened only once. Thus,

$$P\text{-value} = \hat{P}(\text{EOI}|H_0) = \hat{P}(\text{simulated } \hat{p} \leq 0.457|H_0) = N_{\text{EOI}}/N = 1/1000 = 0.001.$$

Stage III. Decide Whether or Not to Reject the Null Hypothesis.
We reject the null hypothesis because the *P*-value, 0.001, is small, and less than 0.05, the standard level of significance. Our hypothesis test, thus, supports the claim that the treatment reduced the number of deaths by heart disease.

In Section 9.2, we learned that we can use the CLT to test a hypothesis about a population proportion when we have a *large random sample,* which implies sampling is done with replacement. This CLT-based approach is also valid for a randomized controlled experiment, provided the sample size is "large." Our rule of thumb is

$$np_T \geq 9, n(1 - p_T) \geq 9,$$

where p_T denotes the null hypothesis proportion. The further requirement that the random sampling must be with replacement requires that the 5% rule needs to hold, at least approximately, for the sample size n and the population size N_P; that is,

$$\frac{n}{N_P} \leq 0.05.$$

For example, in Example 9.7, the CLT approach produces for $n = 658$ and the null hypothesis value $p_T = 0.522$ that

$$P\left(Z \leq \frac{0.457 - 0.522}{(0.522)(0.478)/\sqrt{658}}\right) = P(Z \leq -3.34) \approx 0.$$

Here $np_T \geq 9$, $n(1 - p_T) \geq 9$ are easily satisfied and the sampling was with replacement because the 5% rule was approximately satisfied. Thus, H_0 would be rejected via the CLT approach as well.

Now let us consider an example where the sample size n is *not* small compared to the population size N_p.

Example 9.8

A randomized controlled experiment investigated whether patients with congestive heart failure would show improvement when given a new medicine (*Heart Lung,* Vol. 19, No. 5). The treatment group consisted of 126 patients who received the medicine, and the control group consisted of 130 patients who received a placebo. Patients were randomly assigned to treatment and control groups. Thus, $p_T = 126/256 = 0.492$. After 4 weeks, 58 patients in the treatment group showed improvement while only 13 patients in the control group showed improvement. Thus, of the 71 patients showing improvement, 58 of them came from the treatment group, which yields

$$\hat{p}_{\text{OBS}} = \frac{58}{71} = 0.817.$$

Does this value of $\hat{p}_{\text{OBS}}$ constitute strong evidence that the treatment is effective? We now apply our three-stage hypothesis-testing approach.

Stage I. Formulate the Null Hypothesis.
The null hypothesis denies that the treatment is effective. If the treatment is ineffective, then the expected proportion, p, of treatment group members in the sample (i.e., in the group of patients who showed improvement) is equal to 0.492, the proportion of treatment group members in the entire experiment. So, the null hypothesis is written:

$$H_0: p = 0.492 \text{ versus } H_A: p > 0.492.$$

Stage II. Estimation of the P-value of the hypothesis test.
Apply the five-step method to estimate the P-value, which is the probability of occurrence of the event of interest under the null hypothesis.

Step 1. Choice of the Box Model: The sample size is 71, which is between a third and a fourth of the population size (the 256 patients in the experiment). The sample is, thus, too large to use sampling with replacement. We must use sampling without replacement from a box model for the population.

The box will contain a total of 256 balls with 126 marked with a 1 (for the treatment group members) and 130 marked with a 0 (for the control group members).

Step 2. Definition of One Simulation: A simulation consists of randomly drawing a ball from the box 71 times, without replacing any of the 71 balls back in the box. The proportion of 1s ($\hat{p}$) in the sample of 71 balls is then recorded. The balls are then all placed back in the box in preparation for the next simulation.

Step 3. Definition of the EOI: The EOI is defined as the occurrence of a simulated sample proportion (a simulated $\hat{p}$) that is $\geq$ the proportion that occurred in the real data sample, which was 0.817. If EOI denotes the event of interest, then

$$\text{EOI} = (\text{simulated } \hat{p} \geq 0.817).$$

Step 4. Repetition of Simulations: We then conducted 1000 simulations and counted the number of simulations, N_{EOI}, for which the event of interest occurred.

Step 5. Calculation of the P-value (the Experimental Probability of the EOI): In the 1000 simulations done, $N_{\text{EOI}} = 0$. We divide the number of simulations on which the EOI occurred (N_{EOI}) by the total number of simulations ($N_P = 1000$), yielding our estimate of the P-value, the probability of the EOI occurring when the null hypothesis is true,

$$\begin{aligned} P\text{-value} &= \hat{P}(\text{EOI}|H_0) \\ &= \hat{P}(\text{simulated } \hat{p} \geq 0.817 | H_0) \\ &= \frac{N_{\text{EOI}}}{N_P} = \frac{0}{1000} = 0. \end{aligned}$$

Stage III. Decide Whether or Not to Reject the Null Hypothesis.
Clearly, we reject the null hypothesis because the EOI has, at most, only a very tiny probability of occurrence when the null hypothesis is true. Our hypothesis test, thus, supports the claim that the treatment helped patients with congestive heart failure to show improvement in their recovery from their illness.

Section 9.3 Exercises

1. In conducting an experiment with a treatment group and a control group, why is it advantageous to randomly assign the experimental units to the two groups whenever possible?
2. In a randomized controlled experiment that uses proportions to measure the treatment effect, describe in words the proportion parameter of interest in the problem. What is the value of this parameter under the null hypothesis?
3. In words, describe the meaning of the null hypothesis in a randomized controlled experiment as described in this section. What does it mean if the null hypothesis is rejected?

4. Describe, in words, the difference in interpretation of the rejection of the null hypothesis in Section 9.3 as compared to Section 9.2.
5. Suppose a nutritional supplement has been invented that is claimed to improve the cognitive performance of rats. The new supplement is tried out in a laboratory with a group of 20 rats. The experimenters assign half of the rats to the treatment group and half to the control group. The treatment group is fed the usual rat diet plus the new supplement. The control group is fed the same diet as the treatment group, but without the new supplement. The two groups of rats run a maze, and the experimenters find that 13 out of 20 were successful on the second try. Of the 13 successes, 10 were by members of the treatment group.
    a. What proportion of all the rats were treatment group members? What proportion of the successful rats were treatment group members?
    b. Of the two proportions in part a, which proportion is a sample statistic and which proportion is a constant whose value was set by the experimenter?
    c. In words, what is the claim that is made about the treatment? How does that claim relate to the proportion parameter mentioned in part b?
    d. In words, what is the null hypothesis that nullifies their claim? Write the null hypothesis using standard statistical notations for the proportion parameters involved.
    e. To test the null hypothesis, we need to generate simulated data samples that conform to the null hypothesis. On each simulation, we will generate a simulated sample that has the same sample size as the original real-data sample. What sample statistic will we measure on the simulated sample that we generate on each simulation?
    f. The EOI will be based on comparing the simulated sample statistic to something. To what will we compare it? Why do we make this comparison?
    g. The results from 1000 simulations are shown in the following table. What are the most frequently occurring values of $\hat{p}$ in the simulated samples? What are the least frequently occurring values of $\hat{p}$ in the simulated samples? What are the least frequently appearing values in the direction that, if observed in a real-data sample, would be evidence for rejecting the null hypothesis?
    h. Is the value of $\hat{p}$ observed in the real-data sample among the most infrequently observed simulated $\hat{p}$ values?
    i. Decide whether or not the null hypothesis should be rejected and explain why you chose that decision.

$\hat{p}$	Frequency of occurrence	$\hat{p}$	Frequency of occurrence
0/13	0	7/13	188
1/13	2	8/13	165
2/13	9	9/13	90
3/13	41	10/13	30
4/13	89	11/13	12
5/13	147	12/13	1
6/13	226	13/13	0

6. In Exercise 5, suppose the proportion $\hat{p}$ observed in the real-data sample had been 9/13. Use the table for Exercise 5 to estimate the probability of observing a simulated proportion that was as large, or larger, than 9/13 under the null hypothesis. Would you reject the null hypothesis in this case?
7. The risk of acquiring Lyme disease is high in many areas of the United States; thus, a safe and effective vaccine is greatly needed. A 1998 article in the *New England Journal of Medicine* reported results from a clinical trial for a new vaccine. For this study, 10,936 subjects who lived in areas

where there is a high risk of acquiring Lyme disease were randomly assigned to two groups of 5468 each. The treatment group received the vaccine and the control group received a placebo. In the first year of the trial, 65 subjects acquired Lyme disease, with 22 of them being from the treatment group. Should we decide from these data that the vaccine is effective in preventing Lyme disease?

8. Children who play violent video games are claimed to display more violent behavior ("violent" behavior must be carefully defined) than children who merely watch violent television programs.

Suppose that a randomized controlled experiment assigns 25 children to a treatment group who play violent video games for one hour and assigns another 25 children to a control group who only watch a violent television program for one hour.

The two groups of children are, then, separately observed at play for 30 minutes immediately afterward. During the 30 minutes of play, a total of 34 children are observed to display violent behavior, with 21 of them being from the treatment group. Should we decide from these data that playing violent video games is more likely to lead to violent behavior in children than merely watching violent television programs?

Note: These numbers are purely hypothetical for this exercise. A real study should take into account *how much* violent behavior the children engage in. An example of a real study done at the University of Missouri at Columbia is reported in the *Journal of Personality and Social Psychology*, vol. 78, no. 4, pp. 772–790.

9.4 TESTS FOR A POPULATION MEAN

In Section 9.2 we looked at sets of data, each characterized by a proportion parameter, and asked whether the data in each set could have arisen from a population in which the population proportion, p, was equal to a certain specified (null-hypothesis) value, p_0. For instance, in Example 9.5, p was the proportion of a population of voters who favored mandatory registration of handguns, and p_0 was 0.6. Gun control proponents in the city hoped for strong statistical evidence to reject the null hypothesis, $H_0 : p = p_0$ in favor of $H_A : p > p_0$ and, thus, to conclude that a strong majority (> 0.6) of the population support the registration of handguns. If the observed sample proportion is so far from p_0 (in the direction of the alternative hypothesis) that it is unlikely to have arisen by chance when $p = p_0$, we reject the null-hypothesis model in favor of the alternative hypothesis. Otherwise we fail to reject the null hypothesis.

The three-stage hypothesis testing approach that we used in Section 9.2 is used for other hypothesis-testing purposes besides the proportion problems of Sections 9.2 and 9.3. In particular, the three-stage hypothesis testing approach can be used for testing about population means that are not proportions. There are many testable hypotheses about population means that are not proportions. Is one of two new drugs more effective than the other at lowering blood cholesterol level, and if so, which? Are the ages of husbands greater than that of their wives on average? Is a particular hypothesized adult blood pressure average correct for the population of healthy active 60-year-olds? We can think of many more examples. This section shows how to assess these types of hypotheses.

The Z Test of the Population Mean

A common and important class of statistical problems involves the central, typical, or expected value of some measured characteristic of a population, such as adult heights. The measure of center is almost always the population mean, μ.

The Large-Sample Case Let us consider the following example.

Example 9.9

You have been taught that 98.6 degrees Fahrenheit is the "normal" adult body temperature. Biologists, however, have come to suspect that the population average temperature for healthy adults is actually less than 98.6 degrees Fahrenheit. Figure 9.1 is a histogram of temperature readings of 130 randomly sampled people from a large population of adults (data are from Allen L. Shoemaker, "What's Normal?—Temperature, Gender, and Heart Rate," *Journal of Statistics Education*, July 1996). Notice that much of the histogram is to the left of 98.6, the supposed normal temperature.

The average of the 130 readings is $\bar{x} = 98.25$, which is less than 98.6. The question is whether this sample average of 98.25 is close enough to 98.6 that the difference could be due to mere chance variation. That is, is such a large difference from 98.6 likely to show up in a new random sample of 130 people drawn from a population whose mean temperature really is 98.6? If so, such a difference is likely to appear by chance, and there is no evidence that the population mean temperature is less than 98.6. Or is the difference of 98.25 – 98.6 = –0.35 so large in size that it is not likely to show up in a new random sample of 130 people drawn from a population whose mean temperature really is 98.6 degrees? If so, such a difference is unlikely by chance, and the evidence is strong that the mean population temperature is indeed less than 98.6. To assist in making this judgment, note that the observed sample SD was 0.73 degrees.

This problem involves a hypothesis or claim about the central or typical value of body temperature of a large population of adults. We have a sample of body temperature readings, which, we will assume, constitute a random sample from the population in question and, hence, are representative of this population. The direct approach to this problem is to frame it as a hypothesis-testing question: whether to believe the "real-world" claim that $\mu < 98.6$.

The first stage, as in the problems in Section 9.2, is to set out a formal null hypothesis (usually the hypothesis of "business as usual" or "nothing of interest here") that denies the claim and then use the data to see whether there is strong evidence that the population average temperature is in fact lower as claimed. Here the claim is that $\mu < 98.6$, and its negation is thus

$$H_0: \mu = 98.6°$$

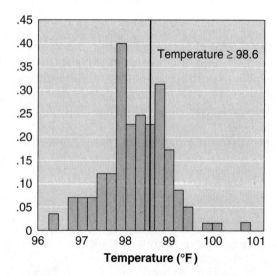

Figure 9.1 Density histogram of 130 body temperatures.

(assuming we know $\mu > 98.6$ is not possible). The burden of proof is on the claim. That is, it is up to the data to yield strong evidence that H_0 is false, thus establishing the claim.

For the temperature problem we have observed the mean of the sample data, $\bar{x} = 98.25$. Then, as in other hypothesis-testing problems, we have two possibilities:

1. The observed sample mean is likely to have come from a null-hypothesis population with a mean of $\mu = 98.6$. In this case we fail to reject the null hypothesis, meaning that there is not enough evidence to say the null hypothesis is false.
2. The observed sample mean is not likely to have come from a null-hypothesis population with a mean of $\mu = 98.6$. In this case we reject the null hypothesis and conclude that the true population average temperature is $\mu < 98.6$.

For the second stage, we need to know whether the sample average is so far below 98.6 that we are compelled to reject the null hypothesis. Specifically, what is the probability that, if the null hypothesis is true, we could obtain a sample mean as low as the observed 98.25 or lower? That is, what is

$$P(\bar{X} \leq 98.25 \mid H_0)?$$

This probability is the *P*-value, and it is read as "the probability the sample mean $\bar{X}$ is less than or equal to 98.25, given that the null hypothesis is true." Its value determines whether we reject H_0 or fail to reject H_0 in the third stage of the hypothesis-testing process.

In Section 9.2, we found the *P*-value using the simulation method in Stage II, that is, by repeatedly simulating the drawing of a random sample from a box model for which H_0 was true. Here, that would be a box model in which the mean box temperature was 98.6. If we used the five-step simulation approach in this problem, the proportion of simulations in which the sample mean was 98.25 or lower would give us $\hat{P}$ (simulated $\bar{X} \leq 98.25 \mid H_0$), an experimental estimate of our desired *P*-value. But because the sample size is large, we can use a theoretical probability approach to find the *P*-value without having to do simulations, just as we did in Section 9.2. In Section 8.4 we constructed a CI for μ using knowledge of the distribution of $\bar{X}$ when the sample size is large. We will take a similar theoretical approach to hypothesis testing. We will assess the strength of evidence against H_0 by finding a *P*-value associated with the **standardized test statistic:**

$$\frac{\text{statistic} - \text{parameter (under } H_0)}{\text{(estimated)standard error of statistic}}$$

Here, as in Section 8.3, we will use the fact that $E(\bar{X}) = \mu$, and that $\mu = \mu_0$ when H_0 is true, and that

$$\widehat{SE}(\bar{X}) = \frac{S}{\sqrt{n}}.$$

The *P*-value we need is the probability that this standardized test statistic, computed for a random sample from a population in which H_0 is true, in particular for $\mu = \mu_0 = 98.6$ in this example, could be at least as extreme as the observed statistic computed from the data. Remember that "as extreme" means "in the direction that provides strong evidence in favor of the real-world claim," that is, in favor of H_A. Thus, when the claim is that $\mu < 98.6$ and $\bar{x} = 98.25$ is observed, "at least as extreme as 98.25" means that the *P*-value event is $(\bar{X} \leq 98.25)$, because low values of $\bar{X}$ provide evidence that $\mu < 98.6$. As in Section 9.2, the *P*-value can often be computed on the basis of theoretical knowledge of the distribution of the standardized statistic when H_0 is true.

In this example, the standardized test statistic involving $\bar{X}$ is

$$T = \frac{\bar{X} - \mu_0}{\widehat{SE}(\bar{X})} = \frac{\bar{X} - \mu_0}{S/\sqrt{n}},$$

where μ is the value of the population mean under H_0. Recall that in this sample $n = 130$ and $s = 0.73$. Thus the observed $\bar{x} = 98.25$ becomes, when standardized,

$$T = \frac{98.25 - 98.6}{0.73/\sqrt{130}} = -5.47.$$

Recall that one consequence of standardizing a statistic is that it produces a random variable with mean of 0 and a SD of 1, and with a distribution that is often approximately standard normal when the sample size is large. Indeed, we learned in Section 8.4 that an enhanced version of the CLT tells us that the standardized statistic

$$T = \frac{\bar{X} - E(\bar{X})}{\widehat{SE}(\bar{X})}$$

is approximately standard normal when $n \geq 30$. We know that $E(\bar{X}) = \mu$ from Section 8.3. Thus, when H_0 is true and hence $\mu = \mu_0$ we have that $E(\bar{X}) = \mu_0$, and

$$T = \frac{\bar{X} - \mu_0}{S/\sqrt{n}}$$

is appropriately standard normal with $E(T) \approx 0$ and $SD(T) \approx 1$. Because T is approximately standard normal, the familiar 68-95-99.7% rule applies to the experimental distribution of T: About 68% of all T values should occur between $E(T) \pm SD(T)$, that is between -1 and $+1$, and about 95% of all T values should occur between $E(T) \pm 2SD(T)$, that is between -2 and $+2$. Thus, $T \leq -5.47$ seems extremely unlikely to occur if in fact H_0 holds.

The statement that the sample mean $\bar{x} = 98.25$ is identical to stating that the test statistic $T = -5.47$ because it is computed by plugging $\bar{x} = 98.25$ into the formula for T. Also, a mean $\bar{X}$ of a random sample will be smaller than the observed mean, 98.25, if and only if it yields a corresponding standardized T smaller than the observed $T = -5.47$. That is, the events $\bar{X} \leq 98.25$ and $T \leq -5.47$ are the same, and thus

$$P(\bar{X} \leq 98.25 | H_0) = P(T \leq -5.47 | H_0).$$

Thus the event $(\bar{X} \leq 98.25)$ is re-expressed as $(T \leq -5.47)$, where T is given by

$$T = \frac{\bar{X} - 98.6}{S/\sqrt{n}}.$$

Therefore, for Stage II of our hypothesis test, we evaluate the P-value, namely $P(T \leq -5.47 | H_0)$, using the CLT to tell us that T is approximately standard normal; that is, we look up the P-value from Table E using $z = -5.47$.

Since Table E runs out of rows at $z = -3$,

$$P(T \leq -4 | H_0) \approx 0$$

to four decimal places.

But $P(T \leq -5.47 | H_0) \leq P(T \leq -4 | H_0)$ and hence $P(T \leq -5.47 | H_0) \approx 0$, which is much less than the "gold standard" level of significance, 0.05. Thus we have strong evidence that the null hypothesis does not appear plausible; that is, the observed sample mean value of $\bar{X}_{OBS} = 98.25$ cannot be ascribed to chance under the null hypothesis. So, for Stage III of our hypothesis test, we reject the null hypothesis, believing the evidence to be very strong that, for the population from which the data were sampled, the average temperature is not the "normal" value of 98.6 degrees but is somewhat lower, as was claimed.

The example we have just examined illustrates the normal distribution-based approach, or Z-test method, of hypothesis testing. It works well when our interest is in making inferences about the mean μ of the population and when the sample size is reasonably large: $n \geq 30$ when σ is unknown and $S \approx \sigma$ is used to form the standardized T.

Normal-Population, σ-Known Case

A second and less commonly occurring setting for which a normal distribution-based Z test is appropriate is that of sampling from a normal (or approximately normal) population with σ (the population SD) known and n possibly small. We illustrate this second case with an example.

Example 9.10

Consider the manufacture of "one-pound" peanut butter jars. Let μ denote the population average weight that results when sampling peanut butter jars as the manufacturing process is currently functioning. When the manufacturing process is "in control," the population average weight of the "one-pound" peanut butter jars is $\mu = 1$. Suppose the manufacturer has reason to think that μ has come to be greater than 1. This is an undesirable state of affairs: the manufacturer is wasting product in the jars now being made, and the manufacturing process may have to be shut down for recalibration. Prior experience indicates that the jar weights are normally distributed with $\sigma = 0.03$. Thus it is reasonable to assume σ known with $\sigma = 0.03$, because even if μ has shifted from 1, σ is likely to have stayed the same. To assess the manufacturer's concern, nine jars are sampled randomly from the production line, with the result that $\bar{x} = 1.02$.

Solution

We follow the familiar three stages of hypothesis testing.

Stage I. The null hypothesis is the negation of the manufacturer's claim that the mean weight has become greater than 1:

$$H_0: \mu = 1 \text{ versus } H_A: \mu > 1.$$

Stage II. To test the null hypothesis, we compute the standardized test statistic from the sample data:

$$Z = \frac{\bar{X} - \mu_0}{\sigma/\sqrt{n}} = \frac{1.02 - 1}{0.03/\sqrt{9}} = \frac{0.02}{0.03/3} = 2,$$

replacing μ_0 by the null-hypothesis value, 1. As we learned at the end of Section 8.8, because the population is normal, $\bar{X}$ is normal, and hence Z is a standard normal random variable when H_0 is true. Thus the P-value is given by Table E:

$$P(\bar{X} \geq 1.02 \mid H_0) = P(Z \geq 2) = 1 - 0.9772 = 0.0228.$$

Stage III. Since $0.0228 \leq 0.05$, we have strong evidence that the null hypothesis is false; that is, the manufacturer is right in suspecting that $\mu > 1$ (excess peanut butter in jars).

Take note that we are not using the CLT here but rather are using the fact that the population being normal implies that $\bar{X}$ is normal. And, note also that this is true whatever the value of n.

Tests for the Mean Difference of Matched Pairs Variables in one Population: Large-Sample Case

A common problem is to test the population mean difference of two random variables measured on each unit of a population. The data are sometimes referred to as **"matched pairs"** because the two variables are paired within each unit in the population. A common example is the comparison of medical subjects before and after a treatment, for example, blood pressure before and after an exercise or a weight-loss program. Because this hypothesis test involves the difference in two variables, it is tempting to automatically think that the test is about the difference between the means of two populations. Therefore, it is important to clearly see that we are actually assessing the mean of a single population, which is the difference of the two variables. This type of problem and its solution are illustrated in the following example.

Example 9.11

Figure 9.2 shows the density histograms of the numbers of years of education of the husbands and of wives of 177 Illinois couples in 1989 (data from the 1989 Current Population Survey). Suppose it is claimed that husbands tend to have more formal education. The distributions look reasonably similar, although it appears that more husbands have at least some college education and, in particular, more than their wives. Also, husbands go through two years of college (14 years of education) more frequently than wives, and more wives have but one year of college (13 years) than husbands. The average for the 177 husbands is $\bar{x} = 12.89$, and that for the 177 wives is $\bar{y} = 12.55$. The difference between the husbands' average attainment and the wives' is $\bar{d} = \bar{x} - \bar{y}$. Is this difference due to chance, or do husbands, on average, have more years of education than their wives in the Illinois population?

This problem is called a matched-pairs comparison of two population means, because the (X,Y) values are paired (a husband, X, is paired with his wife, Y). Thus, the married couples make up the units in the population, and it is the husband/wife difference that is measured on each couple. In other words, the variable we look at is

$$\text{difference} = \text{husband's years of education} - \text{wife's years of education}$$

(i.e., $D_i = X_i - Y_i$ for the ith couple).

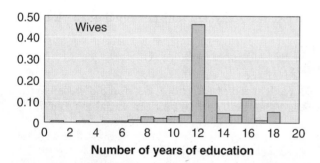

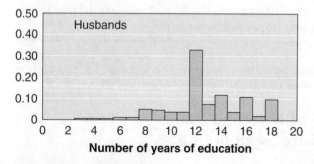

Figure 9.2 Density histograms of years of education for Illinois husbands and wives in 1989.

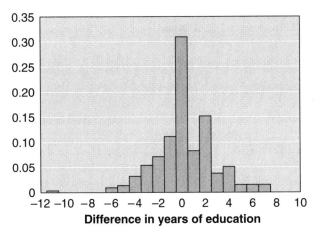

Figure 9.3 Density histogram of the observed differences $d = x - y$ in years of education in Illinois married couples in 1989

Figure 9.3 shows the density histogram of the 177 observed differences. One wife (with 14 years) has 11 more years than her husband (with 3), hence the outlying −11 rectangle. Otherwise, the largest difference is +7 years. The average of these differences is $\bar{d} = 0.24$, and the sample SD of these differences is $s_D = 2.58$.

Is it plausible that these differences could be a sample from a population of differences in which the average difference $\mu_D = 0$; that is, there is no difference on average between the husband's and wife's number of years of education in the population? Note that we have made the problem into a single-population problem where the random variable $D = X - Y$ is being observed. We wish to test the following null hypothesis:

$$H_0: \mu_D = 0 \text{ versus } H_A: \mu_D > 0.$$

We calculate a standardized test statistic Z under the null hypothesis, which we know obeys the CLT (because $n = 177$ is large), namely

$$Z = \frac{\bar{D} - 0}{\widehat{SE}(\bar{D})} = 1.24,$$

where

$$\widehat{SE}(\bar{D}) = \frac{s_D}{\sqrt{n}} = \frac{2.58}{\sqrt{177}} = 0.194$$

and 0 is subtracted from $\bar{D}$ because $\mu_D = 0$ is the null hypothesis value.

Note: In general,

$$Z = \frac{\bar{D} - E(\bar{D})}{\widehat{SE}(\bar{D})} = \frac{\bar{D}}{\widehat{SD}(\bar{D})}$$

is approximately standard normal when H_0 is true (i.e., $E(\bar{D}) = \mu_D = 0$). From Table E, $P(Z \geq 1.24) = 1 - P(Z \leq 1.24) = 1 - 0.8925 = 0.1075$. That is, there is approximately a 0.1 chance of seeing a difference of 1.24 or more just by chance.

The P-value we just obtained, approximately 0.1, is fairly small, but it is certainly larger than our 0.05 gold standard convention for rejection of the null hypothesis. We thus fail to reject the null hypothesis that the average difference in the population is 0. That is, the data do not provide strong evidence that husbands' education exceed their wives' in Illinois.

In Section 9.2, we found that we were able to use the normal density and the corresponding normal table instead of using five-step simulation to get the P-values for hypothesis testing about proportions. Similarly, here, too, we could have used five-step simulation (using a computer to produce random samples from a continuous population model) to carry out a hypothesis test on any of the Z statistics of this section. However, when n is large, we can (and should) bypass five-step simulation and use our knowledge of probability theory (the CLT from Chapter 8) and the corresponding standard normal Table E to conduct hypothesis tests concerning the population mean.

The normal-distribution-based Z test of this section is heavily used in statistics. It applies whenever the null hypothesis concerns one population mean (including the special case of a matched-pairs difference of population means) and the sample size is fairly large ($n > 30$) regardless of whether the population is normal or not. Although used less often, it also applies whenever the assumption of normal population sampling is justified and σ is known, even if n is small.

The *t* Test of a Normal Population Mean

We continue to consider $H_0: \mu = \mu_0$. If the population being randomly sampled is approximately normal, which is often a good assumption in applications (weights, batting averages, repeated measurements, etc.), and S is used as an estimate of the population standard deviation σ, but the sample size n is small, a Z test using the normal table gives inaccurate answers, even if the population distribution is exactly normal. In Section 8.8 we introduced Student's t probability density function for finding confidence intervals for the population mean under these circumstances. For hypothesis testing, an alternative test to the Z test, called the t test is based on the standardized statistic $T = (\bar{X} - \mu_0)/(S/\sqrt{n})$. The t distribution works better than the Z test if the following conditions hold:

1. the distribution of the values in the population being randomly sampled is close to the normal distribution;
2. the sample size is small; and
3. the population standard deviation, σ, is not known to the user.

If the original data do not have a normal shape, at least approximately, and the sample size is small, the t test will work badly and should not be used. (Actually, neither the Z test nor the t test will work well and it is better to adopt a bootstrap simulation hypothesis-testing approach, which will be described later in this chapter.) If the sample size is small, $n < 30$ say, and condition 1 holds, the t test will work better than the Z test. If the sample size is large and hence $S \approx \sigma$, then the t test and the Z test both work well and will give practically the same answer, so either is fine to use. Hence, when the sample size is large, statisticians use the Z test because it is correct and the standard normal table is easier to use than the t table.

Note again the difference between the Z statistic ($Z = (\bar{X} - \mu_0)/(\sigma/\sqrt{n})$) and the T statistic ($T = (\bar{X} - \mu_0)/(S/\sqrt{n})$). A true Z statistic uses the population standard deviation, σ, to obtain the standard error $SE(\bar{X}) = \sigma/\sqrt{n}$. If the sample standard deviation S is used to estimate σ, the statistic is, strictly speaking a T statistic. However, we say "Z test" when we use the standard normal distribution (Table E) for either Z or T and we say "t test" when we use the t distribution for T (Table F).

The Matched Pairs *t* test: Student's Example

The Z test requires that the sample SD be an accurate estimate of the population standard deviation, but the standard deviation of a small sample is not a very accurate estimate. Therefore, in 1908, W. S. Gosset, under the name of "Student" (see Section 8.5), discovered the t distribution and proposed the t test as a refinement of the Z test to adjust for the inaccuracy of the standard deviation estimate from a small sample. Here we shall examine the example Gosset used to illustrate the t test.

Table 9.3 Sleep Gains for Two Medications

Patient	Dextro Sleep Gain	Laevo Sleep Gain	Difference (laevo gain–dextro gain)
A	0.7	1.9	1.2
B	−1.6	0.8	2.4
C	−0.2	1.1	1.3
D	−1.2	0.1	1.3
E	−0.1	−0.1	0.0
F	3.4	4.4	1.0
G	3.7	5.5	1.8
H	0.8	1.6	0.8
I	0.0	4.6	4.6
J	2.0	3.4	1.4

Student, "The Probable Error of a Mean," *Biometrika*, vol. 6 (1908), pp. 1–25.

Two medications intended to increase sleeping time were administered to each of 10 patients, and for each patient the number of additional hours of sleep per night gained after taking each medication was measured. The treatments were dextro hyoscyamine hydrobromide and laevo hyoscyamine hydrobromide (two kinds of molecules that are mirror images of each other).

The data for this study are presented in Table 9.3. Thus, the "Dextro" Sleep Gain column is a random sample of size 10 for a one-population "matched pairs" experiment. The 0.7 in the "Dextro" column for patient A is the number of hours of sleep A had after taking dextro hyoscyamine hydrobromide minus the number of hours of sleep after no treatment. For this patient the Dextro treatment appears to have helped a little. Let D be the value in the "Dextro" column of the table. That is, for the ith patient,

$$D_i = X_{i,\text{ dextro}} - X_{i,\text{ no treatment}}$$

where X_i represents the number of hours of sleep for patient i and the subscripts "dextro" and "no treatment" indicate whether the treatment or control was used.

In the same way, the "Laevo" column provides the number of hours of sleep after taking the laevo medication minus the number of hours of sleep with no treatment. Finally, the "Difference (laevo − dextro)" column contains the number of hours more sleep gained under the laevo medication than when the dextro medication was used, $X_{i,\text{ laevo}} - X_{i,\text{ dextro}}$. These columns will be used here and in the exercises. The values in each column are assumed to be a random sample from a normal distribution.

Three research questions were posed by those doing the study:

Does the dextro treatment increase sleep?

Does the laevo treatment increase sleep?

Does the laevo treatment increase sleep more than dextro?

Each such question is answered by a hypothesis test about the mean of a population based on a matched-pairs random sample of size 10. Here we will address only the first question. For Stage I of our hypothesis test, we form a null hypothesis that denies the research claim of the medical effectiveness of dextro. The null hypothesis is that the dextro treatment does not help increase sleep, that is,

H_0: The population mean increase in sleep due to dextro, in hours, is 0.

or, in symbols

$$H_0: \mu_D = 0.$$

The sample average number of hours of additional sleep when using dextro is $\bar{d} = 0.75$, and the sample SD of the sleep gain using dextro is $S_D = 1.79$. Therefore, the standardized test statistic to use in Stage II is given by

$$T = \frac{\bar{D} - 0}{S_D/\sqrt{n}}.$$

In our example, the observed $\bar{D} = 0.75$, when standardized, becomes

$$T = \frac{0.75 - 0}{1.79/\sqrt{10}} = 1.32.$$

The issue is whether the observed $\bar{D}$ is large relative to the $S_D/\sqrt{n}$ standardizing yardstick. This P-value we need is given by

$$P(\bar{D} \geq 0.75 | H_0) = P(T \geq 1.32 | H_0).$$

Thus the issue is whether $T = 1.32$ is improbably large if H_0 is true. If n had been at least 30, we could have done a Z test using T and looked up the desired probability in Table E, but we cannot obtain a reliable P-value in that way for n as small as 10. Since n is small,

$$T = \frac{\bar{D} - \mu_D}{S_D/\sqrt{n}}$$

has a t distribution with $n - 1$ df. So, T is t-distributed with $10 - 1 = 9$ df under H_0 by Gosset's fundamental result. We can then find the P-value we want by looking into a t distribution probability table, which is given in Appendix F.

We look for $P(T \geq 1.32)$ in the row for df $= 9$ in the t table, but we do not see the number 1.32 in that row. Instead, we do see that for df $= 9$, the number 1.38 is in the column under 0.10. What this means is that $P(T \geq 1.38) = 0.10$ when df $= 9$. The probabilities $P(T \geq t)$ increase as t decreases, so $P(T \geq 1.32) \geq P(T \geq 1.38) = 0.1$ (by a small amount actually).

We now have all we need for our Stage III decision. Under H_0, the probability of obtaining $T \geq 1.38$ slightly exceeds $0.10 > 0.05$, our usual gold standard. Thus we fail to reject the null hypothesis. We are not convinced that treatment with the dextro medication helps increase sleep (it may or may not). Note our failure to reject H_0 did not lead us to make the mistake of concluding that dextro has no influence on sleep. We just don't know based on the statistical evidence we have.

A problem such as the example we have just studied is often called a **matched-pairs** (or **paired-comparisons**) one-sample t test. It contrasts with the matched-pairs one-sample Z test earlier in this section, where the sample size was large and a Z test using the normal distribution was permissible. Like Example 9.11, it is a matched-pairs test because we work with the difference D_i between paired measurements X_i and Y_i. Since each pair of measurements is taken on the same subject, this is a one-sample problem.

The t Test of a Population Mean μ

Note that for any normal-population random-sampling setting (not necessarily matched pairs) where n is small and σ is unknown, that the above approach applies without change to

$$H_0: \mu = \mu_0 \text{ versus } H_A: \mu > \mu_0 \text{ (or } H_A: \mu < \mu_0)$$

using

$$T = \frac{\bar{X} - \mu_0}{S/\sqrt{n}},$$

which has a t distribution with $n - 1$ degrees of freedom when H_0 is true. Its P-value is computed using either $P(T \geq \text{observed } T | H_0)$ or $P(T \leq \text{observed } T | H_0)$ depending on whether we have $H_A: \mu > \mu_0$ or $H_A: \mu < \mu_0$, respectively.

Bootstrap Hypothesis Testing of One Population Mean

Testing $H_0: \mu = \mu_0$ Using a Bootstrap Approach

Let us return to the problem of Example 9.9, where we asked the following question about the population mean: Does the sample average of $\bar{x}_{OBS} = 98.25$ compel us to reject the null hypothesis $H_0: \mu = 98.6$ in favor of the claimed $\mu < 98.6$? To answer that question, we asked what the chance is that, assuming the null hypothesis is true, we observe a sample mean temperature as low as or lower than the observed 98.25. To decide, we calculated a Z statistic, having appealed to the CLT to obtain the sampling distribution of $\bar{X}$ under the null hypothesis. In that case use of the CLT was valid, because $n = 130$ was large.

But in applications where n is small or moderate and the population is not known to be normal, a Z test is not a valid option because n is not large enough for us to use the CLT. For this reason we introduce bootstrap-based hypothesis testing. For convenience we will illustrate the method for the problem of Example 9.9, even though bootstrap hypothesis testing is not needed for this large sample problem. This allows us to compare the two methods in a situation where both are valid and, hence, they should produce similar results.

Bootstrap-based hypothesis testing is a special case of our five-step simulation approach to hypothesis testing. Recall from Section 9.2 that a key step in our simulation approach to statistical hypothesis testing is making a realistic choice of the null-hypothesis box model from which random samples can be drawn. Since we are testing a hypothesis, the specified box model must satisfy the null hypothesis as well as be a realistic model in terms of the shape and spread of the population distribution that produced the observed sample. As we will see, the idea of bootstrapping the observed data to build the null-hypothesis box model required in Step 1 of the five-step simulation is very reasonable.

When we introduced bootstrapping in Section 8.6 as a method for estimating the standard error of an estimate, we learned that the central idea of bootstrapping is to use the sample data as the box model to substitute for the unknown population probability model. That is, the sample data becomes the box model. Because the random sample is representative of the population, it will (approximately) have the shape, spread, and location of the unknown population distribution. Bootstrapping makes statistical inferences using the data alone to produce the model. We do not make the usual assumption of a parametric population probability model (such as the binomial, normal, or Poisson distribution). Such model-free approaches to statistics are called nonparametric statistics. That is, when we assume a population model such as a binomial or normal population, the model is specified by parameters, such as p and μ, on which we then focus our statistical inferences. Nonparametric statistical techniques bypass these distribution-specifying assumptions.

The statistical bootstrap has a valid theoretical justification when sample sizes are large, and indeed the bootstrap often works well in applications. Bootstrap methods are coming into heavy use in modern statistical practice, as was already indicated in Chapter 8. From the viewpoint of this book, bootstrap hypothesis testing is a special version of our five-step method of hypothesis testing. For bootstrap hypothesis testing, the box model of Step 1 is entirely determined by the observed data and the null hypothesis.

Let us return to the setting of Example 9.9 to see how bootstrap hypothesis testing works. Recall the null hypothesis:

$$H_0: \text{The population mean is } 98.6; \text{ that is, } H_0: \mu = 98.6.$$

We now consider the five steps required in Stage II for a simulation-based bootstrap hypothesis test.

Step 1. Choice of a Model (Bootstrap Definition of the Null-Hypothesis Population): We must choose a realistic box model, a population that conforms with the null hypothesis and that approximates the shape and spread of the unknown population distribution.

In our study of hypothesis testing concerning a population proportion p, the null hypothesis alone was enough to determine the null-hypothesis box model. This will hold true again for chi-

428 CHAPTER 9 HYPOTHESIS TESTING

square testing of a fair die, as we will learn in Chapter 10. However, the null hypothesis alone is *not* enough to determine the box model for hypothesis testing of a population mean. Not only must the box-model population have an average of 98.6 as assumed by the null hypothesis; the box-model population must also approximate the distribution of body temperatures of the population of adults from which the actual sample was taken. Many possible populations have a theoretical mean of 98.6. The value at which a population centers is typically independent of its spread about its center and often independent of its general shape (symmetric, skewed, U-shaped, etc.). Thus, usually we do not know what model to assume for the null-hypothesis box model!

When the null hypothesis does not determine the null-hypothesis box model and the CLT is not available to tell us the sampling distribution of the appropriate test statistic, then bootstrap hypothesis testing, as described in this section, becomes a method of choice.

The basic bootstrap sampling presented in Section 8.6 simply takes the observed sample as the box model. The justification for this is that the sample will have approximately the location, spread, and shape of the population, and the approximation will be quite good for large sample sizes and reasonably good for moderate sample sizes. Even for relatively small samples sizes it provides some valid statistical information. The box model that we want to create via bootstrapping for testing a hypothesis concerning a population mean, however, must not only have the population's spread and general shape but must also have as its theoretical mean that which is specified by the null hypothesis ($\mu = 98.6$ in Example 9.9). Taking the sample itself as the box model would approximate the population's spread and shape, as desired, but it would result in a bootstrap sampling box with mean 98.25 (because $\bar{x}_{OBS} = 98.25$) rather than the desired 98.6 specified by $H_0: \mu = 98.6$.

To construct the bootstrap-sampling box model for testing a hypothesis about the population mean, we simply add the difference between the null-hypothesis population mean and the observed sample mean, $98.6 = 98.25 + 0.35$, to each sample value. Then the box model has a new mean of $98.25 + 0.35 = 98.6$, as desired, but its spread and shape stay the same and are approximately those of the larger population being studied, as desired. This is precisely what we wanted as our null-hypothesis box model! That is, we construct a bootstrap box model for $H_0: \mu = \mu_0$ as follows.

Bootstrap Box Model for $H_0: \mu = \mu_0$
Suppose the observed sample $x_1, \ldots, x_n$ produces the sample mean $\bar{x}_{OBS}$.

Basic Principle The box model consists of $x_1^*, x_2^*, \ldots, x_n^*$ defined by

$$x_i^* = x_i + \mu_0 - \bar{x}_{OBS}.$$

Fact $\bar{x}^* = \dfrac{\sum x^*}{n} = \mu_0$

Thus the box model population mean is μ_0 and, hence, H_0 holds for the box model.

The advantage of this nonparametric approach is that we do not need to assume any particular theoretical distribution shape for the population (for example, a bell shape) but let the data alone determine our estimate of the population's shape. The nonparametric approach to statistical inference is very powerful because the user takes no risk of being deceived by assuming an incorrect shape for the population histogram.

Step 2. Definition of One Simulation (One Bootstrap Sample): One simulation consists of randomly sampling, with replacement, 130 readings from the bootstrap null-hypothesis box model population (consisting of the null-hypothesis shifted sample). As we learned in Chapter 8, such random sampling with replacement from the bootstrap population model, based on the actually observed data set, is the proper way to conduct bootstrap sampling.

Step 3. Definition of the Statistic of Interest: The statistic of interest (that is, the test statistic to be used to decide whether or not to reject H_0) is the bootstrap sample mean, denoted $\overline{X}^*$. The asterisk, (*), indicates that $\overline{X}^*$ is a simulated mean of a sample from a bootstrap population model, not to be confused with the original $\overline{X}$ (without an asterisk) often denoted by $\overline{X}_{OBS}$, the observed sample mean.

Step 4. Repetition of Simulations: We perform the bootstrap sampling 100 times, each time obtaining a new mean. The stem-and-leaf plot in Table 9.4 contains these bootstrap sample means (i.e., $\bar{x}^*$ values). The average of the $\bar{x}^*$ values is 98.6048, very close to the null-hypothesis value, as expected from the law of large numbers because the box-model population mean is 98.6. The sample SD of the bootstrap sampled means can be computed: $\widehat{SE}(\overline{X}^*) = 0.0606$.

Step 5. Estimation of the P-Value: The P-value is approximated by $\hat{P}$(bootstrapped $\overline{X}^* \leq 98.25 | H_0^*$), using the 100 bootstrapped $\bar{x}^*$ values. Here we write H_0^* with the asterisk (indicating that it is the sample, shifted to make H_0 true), from which we obtain bootstrapped samples and their corresponding $\bar{x}^*$ values. It turns out that all the simulated sample averages $\bar{x}^*$ values in Step 4 were greater than 98.25: They ranged from 98.45 to 98.75. Thus

$$\hat{P}(\overline{X}^* \leq 98.25 | H_0^*) = 0$$

is much less than the conventional level of significance value of 0.05. We thus reject the null hypothesis, believing the evidence is strong that, for the population from which the data were sampled, the average temperature is less than 98.6 degrees. We note that both the large sample CLT approach and the bootstrapping approach yielded almost identical results, namely P-values of approximately 0.

Example 9.12

We now compute Step 1, the crucial formation of the null-hypothesis-based box model in a simple case. Let us simplify the problem of Example 9.9 by assuming that the sample is of 5 people rather than 130. In this small-sample case we certainly could not justify using a Z test based on the CLT for $\overline{X}$. That is, when we do not know if the population is approximately normal and when n is small, bootstrapping is a good choice for carrying out the hypothesis test.

Suppose the observed temperatures are

$$97.3, 97.5\ 98.4, 98.6, \text{ and } 99.2.$$

The goal is to create a null-hypothesis bootstrap population (from which samples are to be drawn with replacement) with a spread and shape roughly like the observed data but with a mean of

Table 9.4 Sample Mean Temperatures from 100 Bootstrap Samples of Size 130

Stem	Leaf
984	567
985	12222233344444444
985	5555566777777888889999
986	00000000001111112222222333344444
986	55555555666777788899
987	00014
987	55

Key: "984 567" stands for 98.45, 98.46, 98.47 degrees.

98.6. The observed sample mean $\overline{X}$ of the five points is 98.2. To start building the bootstrap box model population, we have to adjust each point so that the mean of the five points is 98.6. By the basic principle for forming a bootstrap box model we have to add $\mu_0 - \bar{x}_{OBS} = 98.6 - 98.2 = 0.4$ to each sample value. We then have the following bootstrap box model (check that these five values have a mean of 98.6, as required) as our null-hypothesis population (box model) from which simulated samples are to be drawn.

Now our plan is to randomly sample five numbers repeatedly with replacement from this population of size 5. That is, we randomly draw one ball, record its value, and then put it back. We then randomly choose another (which could be the same as the first), record its value, and then put it back. We do this five times to obtain each Step 2 bootstrap simulation sample. For example, suppose we draw a size 5 random bootstrap sample with replacement and obtain

$$98.8, 99.6, 98.8, 97.9, \text{ and } 97.7.$$

This first bootstrap sample has a mean of $\bar{x}^* = 98.56$. For accuracy, we would then choose 100 or even 1000 such bootstrap samples to obtain 100 or 1000 $\bar{x}^*$ values, and from those $\bar{x}^*$ values we obtain our estimate of the P-value,

$$\hat{P}(\overline{X}^* \leq 98.2 \mid H_0^*)$$

to decide whether to reject H_0 or not.

Let us illustrate bootstrap hypothesis testing with five bootstrap samples and their bootstrapped $\bar{x}^*$ values:

Sample Number	Bootstrap samples of size 5					$\bar{x}^*$
1	98.8	97.6	98.8	99.9	97.7	98.56
2	97.9	99.0	97.9	99.6	99.6	98.80
3	97.9	97.7	99.6	99.0	98.8	98.60
4	97.7	99.0	97.7	99.6	97.7	98.38
5	98.8	97.9	97.7	98.8	99.0	98.44

Thus,

$$\hat{P}(\overline{X}^* \leq 98.2 \mid H_0^*) = \frac{0}{5} = 0$$

because all bootstrapped $\bar{x}^* > 98.2$.

Our decision in Stage III, based on these five bootstraps (of course we need at least 100 bootstrap samples to trust our answer) is thus to reject H_0 because the estimated P-value is $0 < 0.05$.

In cases in which the population distribution shape is not known and the sample size is too small for valid application of the CLT (i.e., ≤ 30), for example, in Example 9.12, the bootstrap approach is very appealing. But when the sample size used to compute the sample mean is large, as in Example 9.9, the Z test is accurate because the CLT tells us that the distribution of the sample means will be well approximated by the normal distribution.

On the other hand, if the hypothesis test involves a statistic of interest other than $\overline{X}$, bootstrapping can be necessary even if the sample size is large, because not all statistics used in hypothesis testing are normally distributed as a consequence of the CLT. Further, even when a test statistic, Y say, is approximately normal, we may have no formula for SE(Y) and, hence, no way to standardize Y to obtain the needed standardized test statistic. Thus, bootstrap hypothesis testing is a very valuable tool.

Section 9.4 Exercises

1. It is claimed that the average height of men in a particular population exceeds 69 inches. Suppose a sample of 100 heights of men has a mean of $\bar{x} = 70.53$ inches and SD of $s = 3.22$ inches. Use the Z test to test the null hypothesis that the mean height of men in the population is 69 inches.

2. The design of an instrument's controls can determine how easy it is for people to use the instrument. An experiment investigated this by asking 35 right-handed people to turn an instrument knob with their right hand to advance a pointer by screw action. The test device had two knob-and-pointer sets that were identical except that one knob turned clockwise to advance the pointer and the other knob turned counterclockwise. Each of the 35 subjects turned both knobs, and the time in seconds that it took to move each pointer a fixed distance was measured. The observed mean difference of clockwise minus counterclockwise times ($D = X - Y$) was $\overline{D} = 1.4$ seconds, and the SD of the difference was $s_D = 6.62$ seconds. Certain ergonomic engineers believe that right-handed people can turn the counterclockwise knob faster. Test the null hypothesis that negates this claim, namely that the population mean time difference $\mu_D = 0$.

3. For the data in Table 12.1, test the null hypothesis that the mean population increase in hours due to the "laevo" drug is 0. For convenience, the mean $\bar{d}$ of the 10 *laevo* differences is 2.33, and the standard deviation s_D is 1.90.

4. The average length of adult (determined by age) largemouth bass in a certain lake had been 14 inches. A 2-year program was conducted, introducing panfish and building up underwater weed beds in the lake, in the hope of increasing the average adult bass length in the population. It was known from past experience that the length of bass is normally distributed and that the population SD, σ, was 2 inches. Twelve adult bass were randomly sampled at the end of the program, and their mean length was $\bar{x} = 15.5$. State the appropriate null hypothesis and carry out the appropriate hypothesis test.

5. Suppose a sample of 10 heights of men has a mean of $\bar{x} = 70.53$ inches and SD of $s = 3.22$ inches. Explain in detail how to use the bootstrap to test the hypothesis that the mean height of men in the population is 69 inches, where the goal is to assess whether there is strong statistical evidence that the men are taller on average than 69 inches. Be very clear about how to force the box model population to satisfy the null hypothesis.

6. It is claimed that the length of a particular species of microorganism exceeds 25.5 micrometers. A set of 64 independent measurements of the length of the microorganism is made. The mean value of the measurements is $\bar{x} = 27.5$ micrometers with SD of $s = 3.2$ micrometers. Is there strong statistical evidence that the population average length of the microorganism really exceeds 25.5 micrometers as claimed?

7. A large company conducted a study to test the effectiveness of a proposed in-service training program for its sales staff. Forty randomly selected salespersons participated in the program. Their sales records in thousands of dollars for the six-month period prior to attending the program and for the 6-month period following the program were used as the data. The mean difference in "after program" – "before program" $D = X - Y$ sales was $\overline{D} = \overline{X} - \overline{Y} = 5.22$, and the SD of the difference was $s_D = 6.77$. Of course, company management hopes for evidence that the training program improves sales effectiveness. Assuming that the economic situation remained constant during this period, test the null hypothesis that the population mean difference $\mu_D = 0$.

8. Because it has been a very bad year, a farmer believes the corn yield from his land will have a mean of less than 50 bushels per acre, in which case his crop insurance will compensate him for his losses. In a random sample of eight plots, the farmer finds a mean yield of $\bar{x} = 49.25$ bushels per acre with standard deviation of $s = 3.25$ bushels per acre. Test the null hypothesis that $\mu = 50$ that negates the claim that $\mu < 50$. (The data have an approximately normal histogram.)

9. Solve Exercise 8 assuming σ is known and is equal to the sample SD given in the problem.

10. Suppose for Exercise 1 we have 100 bootstrapped means ($\bar{x}^*$ values), each from a bootstrap sample of size 10 (with replacement) taken from the invented box model population with a mean height of 69 inches. The bootstrapped means are recorded in the following stem-and-leaf plot. Is there strong statistical evidence that the mean height of the men in the population is greater than 69 inches?

Stem	Leaf	Stem	Leaf
681	2	690	001233459
682	2	691	000113333799
683	08	692	2223339
684	2	693	1234444478
685	01117	694	1
686	233459	695	
687	01112334444559	696	26
688	03357778899	697	5
689	1122333445555678		

Key: "6975" stands for 69.75 inches.

11. An administrator of the University of Illinois believes that undergraduates study, on average, less than 20 hours per week. State the null hypothesis denying this claim. A random sample of 40 students found the mean study time was $\bar{x} = 19.5$ hours per week with SD of $s = 4.05$ hours. Do you believe the administrator is correct? *Hint:* If needed, reduce the null hypothesis to one specifying a fixed value of μ.

12. A feed company claims that cattle switched to its feed have a mean weight gain greater than 160 pounds. Suppose that weight gain is known to have a normal distribution. A farmer is highly skeptical of the claim. To test whether there is strong evidence against the claim, a farmer gave the feed to 17 head of cattle. They gained the following amounts of weight:

62	83	90	101	104	106
109	109	109	127	143	187
204	205	209	266	277	

Use these data to test $H_0: \mu = 160$, the negation of the company claim.

13. Does aerobic exercise tend to lower the resting heart rate? To answer this question, the resting heart rates of 14 randomly selected female subjects in the 20–30 age range were measured before and after a 3-month intensive aerobic exercise program. The mean difference ("before" − "after" $= X - Y$) is $\bar{D} = \bar{X} - \bar{Y} = 5.0$ beats/minute. The sample SD of the difference is $s_D = 3.40$. Evaluate the claim that there was a significant decrease in resting heart rate after the aerobic exercise program. Suppose it was known that such differences are normally distributed.

14. Suppose 20 small specimens were randomly taken from a certain batch of concrete. The mean compression strength of the specimens was $\bar{x} = 4129.58$ pounds with a SD of $s = 164.12$ pounds. Explain in detail how to use the bootstrap to test the null hypothesis that the mean compression strength of the batch of concrete is 4200 pounds, as opposed to the possibility that the batch is defective and has a lower compression strength. Be very clear how you would force the null hypothesis to be true for the bootstrap sampling box model.

9.5 TESTS FOR TWO POPULATION PROPORTIONS OR MEANS

Until now we have tested hypotheses concerning one population mean. A more common statistical problem is to compare two population means. We have already discussed matched pairs, which we saw was really a one population mean problem in disguise. Alternatively, when sampling from two different populations and observing a common characteristic, the samples are collected independently of each other, that is, each X is independent of each Y. The first part of this section deals with the special case when the two population means are proportions, whereas the second part deals with means of continuous variables.

Testing the Equality of Two Population Proportions (Large-Samples Case)

We tested a hypothesis concerning a single population proportion p in Section 9.2, for example, concerning a hypothesis about the proportion in the population favoring gun control. A more widely occurring statistical problem is that of testing the equality of two population proportions (i.e., $p_X = p_Y$). For example, we could ask whether a new manufacturing method produces a smaller proportion of defectives than the old method. The hypotheses are then written as:

$$H_0: p_X = p_Y \text{ versus } H_A: p_X > p_Y,$$

where X represents the old method and Y the new method.

We illustrate this method of testing with an example, where we use the large sample CLT approach. The basic setting is having two independent binomial experiments.

Example 9.13

A preventative medicine study was designed to study the influence of being married on the likelihood of an individual coming down with the common cold. The research claim is that being married decreases the rate of colds. In the statistical study, 50 married individuals and 50 single individuals are randomly chosen from the community where the study was done. The number of subjects coming down with at least one cold in the 6 months of the study was observed for each group. The number of individuals coming down with a cold in a group is a binomial random variable: Each sample is of fixed size (50), each subject either does or does not come down with a cold, the probability of each person in his/her group catching a cold is assumed to be the same, and one subject's catching a cold is independent of the other subjects, assuming the subjects do not come into close or frequent contact with one another. After a period of 6 months, the researchers interview the individuals to find out whether they have had a cold (Table 9.5).

We are interested in determining whether being married has a positive effect in reducing colds—the research hypothesis. Thus, the null hypothesis is, "being married had no effect in reducing the rate of colds." More specifically, the null hypothesis is $H_0: p_M = p_S$ versus the alternative hypothesis $H_A: p_M < p_S$, where p_M and p_S denote the rates of infection for the two

Table 9.5 Results of Test of Marital Status Effect on Cold Remedies

Group	Number in group	Number catching a cold
Married	50	15
Single	50	25

populations. The null hypothesis can also be written $H_0: p_M - p_S = 0$. How are we going to be able to test whether there is a significant difference between p_M and p_S? The natural statistic to assess whether there is a difference between p_M and p_S is

$\hat{p}_M - \hat{p}_S$ = (married sample proportion catching a cold) − (single sample proportion catching a cold)

$$= \frac{15}{50} - \frac{25}{50} = -0.2,$$

using the data in Table 9.5. Our goal then is to answer the question of whether the difference between $\hat{p}_M - \hat{p}_S$ and $p_M - p_S = 0$ is just due to chance, and thus to determine whether we should reject or fail to reject the null hypothesis.

Note that $\hat{p}_M - \hat{p}_S$ is the difference of two sample proportions obtained from independent random samples. In the process of constructing a CI for the difference of two population proportions in Section 8.10, we learned that the difference of two sample proportions, here $\hat{p}_M - \hat{p}_S$, satisfies the CLT provided the sample sizes, here n_M and n_S, each satisfy our usual rule for being large, namely $n_M p_M \geq 9$, $n_M(1 - p_M) \geq 9$; $n_S p_S \geq 9$, $n_S(1 - p_S) \geq 9$. Thus, when n_M and n_S are large, according to the CLT, the standardized $\hat{p}_M - \hat{p}_S$ is

$$\frac{(\hat{p}_M - \hat{p}_S) - (p_M - p_S)}{\text{SE}(\hat{p}_M - \hat{p}_S)} = \frac{(\hat{p}_M - \hat{p}_S) - (p_M - p_S)}{\sqrt{\frac{p_M(1-p_M)}{n_M} + \frac{p_S(1-p_S)}{n_S}}} = \frac{\hat{p}_M - \hat{p}_S}{\sqrt{p(1-p)\left(\frac{1}{n_M} + \frac{1}{n_S}\right)}},$$

which is approximately standard normal under H_0 (i.e., when $p_M = p_S = p$). As usual, in order to obtain the P-value to assess the strength of evidence against H_0, we seek the distribution of the standard test statistic *when H_0 is true*. However, under H_0, we do not know p. But since $\hat{p}_M$ and $\hat{p}_S$ both estimate p, we can combine or pool them to get

$$\hat{p} = \frac{x_M + x_S}{n_M + n_S} = \frac{15 + 25}{50 + 50} = \frac{40}{100} = 0.4.$$

Thus,

$$\widehat{\text{SE}}(\hat{p}_M - \hat{p}_S) = \sqrt{\hat{p}(1-\hat{p})\left(\frac{1}{n_m} + \frac{1}{s_S}\right)} = \sqrt{(0.4)(0.6)\left(\frac{1}{50} + \frac{1}{50}\right)} = 0.098.$$

Plugging in the estimates, the value for the standardized test statistic is

$$Z = \frac{\hat{p}_M - \hat{p}_S}{\widehat{\text{SE}}(\hat{p}_M - \hat{p}_S)} = \frac{-0.2}{0.098} = -2.04.$$

Thus, the P-value $P(\hat{p}_M - \hat{p}_S \leq -0.2 | H_0) = P(Z \leq -2.04)$. Because Z is approximately standard normal, we can look the probability up in Table E; that is, $P(Z \leq -2.04) = 0.0207$. Note that we choose the EOI to be $(\hat{p}_M - \hat{p}_S \leq 0.2)$ because we are looking for values of $\hat{p}_M - \hat{p}_S$ as extreme or more so than −0.2, in the direction of supporting the research (i.e., alternative) hypothesis.

Thus, we reject the null hypothesis that $p_M - p_S = 0$. That is, this experiment supports the research claim that being married protects one against catching colds. This study is just an example of the general claim that a strong social network protects one against illness, perhaps in part through a strengthened immune system.

Testing for the Equality of Two Population Means

Large-Sample Case: In Section 8.9 we learned that if two random sample sizes are fairly large (each ≥ 30), the difference between the sample means is approximately normally distributed and $S_X \approx \sigma_X$, $S_Y \approx \sigma_Y$. Hence, if the sample sizes are fairly large, we can find a standardized test statistic Z to test a hypothesis involving the equality of two population means, and we can find its P-value from the standard normal table (Table E).

Let us consider an example.

Example 9.14

A beginning statistics class at the University of Illinois at Urbana-Champaign had 104 students with 64 women and 40 men. The average percentage score on homework assignments among the women was $\bar{x} = 78.56$, and that among the men was $\bar{y} = 75.04$. Thus the women did, on average, 3.52 percentage points better than the men. Was this due to chance, or do the sample data provide statistical evidence that women in general perform better on homework in such a beginning statistics class? The standard deviations were $S_Y = 25.08$ for the men and $S_X = 19.54$ for the women.

We first have to decide what populations we are trying to compare. The students in the class are not a random sample from the entire college student population of the United States, nor even of the university. It does seem reasonable to think of these students as a random sample of all the people who take this particular course now or will do so in the near future (a conceptual population, note). We will proceed on that supposition.

The null hypothesis (Stage I of our test) is that the average percentage scores on homework are the same for the populations of women and of men:

H_0: The mean homework score for the population of women equals that of men,

or symbolically,

$$H_0: \mu_X - \mu_Y = 0.$$

This negates the "research claim" (H_A) that $\mu_X - \mu_Y > 0$ (that is, women do better on average). As we did in Section 8.9, we must first be willing to assume that the two random samples are independent of each other. The approximate test statistic is the natural estimator of the difference $\mu_X - \mu_Y$, that is, is the difference of the sample means, $\bar{X} - \bar{Y}$, which we know from Section 8.9 to be approximately normally distributed by the CLT. To obtain a standardized test statistic when H_0 is true, we need to center $\bar{X} - \bar{Y}$ at $\mu_X - \mu_Y = 0$ and divide this difference by $\widehat{SE}(\bar{X} - \bar{Y})$.

As we learned in Section 8.9

$$\widehat{SE}(\bar{X} - \bar{Y}) = \sqrt{\frac{S_X^2}{n} + \frac{S_Y^2}{m}},$$

where n is the number of sampled women and m is the number of sampled men. Thus for our data

$$\widehat{SE}(\bar{X} - \bar{Y}) = \sqrt{\frac{(25.08)^2}{40} + \frac{(19.54)^2}{64}} = \sqrt{21.69} = 4.66.$$

Now we can define the standardized test statistic Z, using the fact that the hypothesized difference $\mu_X - \mu_Y = 0$ under the null hypothesis,

$$Z = \frac{\text{Difference in sample means} - (\text{difference in theoretical means under } H_0)}{\text{Estimated sample SE of difference in means}}$$

$$= \frac{(\bar{X} - \bar{Y}) - 0}{\widehat{SE}(\bar{X} - \bar{Y})} = \frac{\bar{X} - \bar{Y}}{\widehat{SE}(\bar{X} - \bar{Y})} = \frac{3.52}{4.66} = 0.76.$$

To obtain the needed P-value for Stage II of our test, we wish to calculate $P(\overline{X} - \overline{Y} \geq 3.52 | H_0)$. Note that the EOI has "≥" in it because we are looking for evidence supporting the research hypothesis (H_A) that women do better than men on average. The standardizing computations we have just done show that the event given by ($\overline{X} - \overline{Y} \geq 3.52$) is the same as the event that a standard normal $Z \geq 0.76$. So we need the area to the right of $z = 0.76$ in Table E. The area to the left of $z = 0.76$ is 0.7764, from the table, so $P(Z \geq 0.76) = 1 - P(Z \leq 0.76) = 1 - 0.7764 = 0.2236$. (Alternatively, we could simply have looked up the area to the left of $z = -0.76$ to get the same answer.) That is, there is approximately a 0.22 probability of seeing a difference $\overline{X} - \overline{Y} \geq 3.52$ just by chance if $\mu_X - \mu_Y = 0$.

The above calculated P-value of 0.22 is much greater than the usual level of significance of 0.05, showing that it is not unusual to see such a difference when the null hypothesis (that the populations have the same mean) is true. Thus, for Stage III of our test we fail to reject the null hypothesis. That is, we do not have statistical evidence that women outperform men on homework in the introductory statistics course.

Small-Sample Case The Z test we just performed required large samples so that our estimate of the standard error of the difference between the means would be accurate and so that the CLT applies to $\overline{X} - \overline{Y}$. But if we can assume random sampling from two approximately normal populations, we can also carry out a test for small samples, based on the t distribution, analogous to estimating $\mu_X - \mu_Y$ as covered at the end of Section 8.9. We illustrate with an example. *This test also requires the assumption of equal population SD values.*

Example 9.15

A very small study was conducted several years ago to determine whether taking the drug LSD damages one's chromosomes. Four users of LSD and four controls (nonusers) were studied. This is not a randomized controlled experiment; we have not randomly assigned four people to use LSD (highly unethical) but rather randomly sampled four people from each of the population of users and nonusers. A block of tissue cells was taken from each person, and the percentage of cells with chromosomal breakage are given in Table 9.6. For example, the first control had breakage of 3.3% in the studied cells.

We see that the users had a slightly smaller mean breakage, suggesting the bizarre possibility that LSD inhibits rather than increases chromosomal damage! The null hypothesis is that the mean breakages in the control and user populations are equal, that is,

H_0: The mean breakage in the control population = the mean breakage in the user population

or symbolically,

$$H_0: \mu_X - \mu_Y = 0.$$

Although it is bad practice to let the data determine the alternative hypothesis, nonetheless, we will let the fact that $\overline{X} - \overline{Y} > 0$ suggest $H_A: \mu_X - \mu_Y > 0$ to allow us to make the point that one cannot draw a conclusion suggested by the data unless the evidence is strong. Indeed, we strongly suspect that there won't be strong evidence supporting H_A. We have that $\bar{x} - \bar{y} = 5.40 - 4.60 = 0.80$. Is this difference small enough to be compatible with the null hypothesis? What is

Table 9.6 Percentage of Cells with Chromosome Breakage in LSD Users and Controls

Group	Data				Sample Mean	Sample SD
Controls (X)	3.3	4.8	6.4	7.1	5.40	1.47
Users (Y)	0.9	2.6	3.4	11.5	4.60	4.08

the chance of getting a difference that large or if larger if the two population means are the same?

For Stage II, the P-value we seek is $P(\overline{X} - \overline{Y} \geq 0.8 | H_0)$. With sample sizes of only 4, the two-sample Z test will give very poor results. Here, it is assumed that random samples of sizes n and m are taken from two independent normal populations assumed to have identical population standard deviations. Because of these assumptions, a two-sample version of Gosset's t test is available. For independent samples from normally distributed populations with $\sigma_X = \sigma_Y$, we learned in Section 8.9 to substitute the sample standard deviations, denoted by S_X and S_Y, for σ_X and σ_Y, to form a pooled estimate S_P of the common SD(σ) by

$$S_P = \frac{\sqrt{(n-1)S_X^2 + (m-1)S_Y^2}}{n+m-2}$$

$$= \sqrt{\frac{8.66 + 66.74}{4+4-2}} = 3.55$$

Moreover, we learned that

$$SE(\overline{X} - \overline{Y}) = \sigma\sqrt{\frac{1}{n} + \frac{1}{m}},$$

which is estimated by

$$\widehat{SE}(\overline{X} - \overline{Y}) = S_P\sqrt{\frac{1}{n} + \frac{1}{m}}.$$

It can be shown that the standardized $\overline{X} - \overline{Y}$, given by

$$T = \frac{\overline{X} - \overline{Y} - (\mu_X - \mu_Y)}{\widehat{SE}(\overline{X} - \overline{Y})} = \frac{\overline{X} - \overline{Y} - (\mu_X - \mu_Y)}{S_P\sqrt{\frac{1}{n} + \frac{1}{m}}}$$

has a t distribution with $n + m - 2$ degrees of freedom. In particular, when $H_0: \mu_X - \mu_Y = 0$ holds, this simplifies to

$$T = \frac{\overline{X} - \overline{Y}}{S_P\sqrt{\frac{1}{n} + \frac{1}{m}}},$$

a t distribution with $n + m - 2$ degrees of freedom.

Medical experts in this area are convinced that it is reasonable to assume the two populations are normally distributed and that $\sigma_X = \sigma_Y$. Thus T, our standardized test statistic, is t distributed when H_0 is true.

For our data,

$$\widehat{SE}(\overline{X} - \overline{Y}) = S_P\sqrt{\frac{1}{n} + \frac{1}{m}}$$

$$= 3.55\sqrt{\frac{1}{4} + \frac{1}{4}} = 2.51.$$

Now we can compute the standardized testing statistic, T, using the fact that the hypothesized difference $\mu_X - \mu_Y = 0$ under the null hypothesis, that is,

$$T = \frac{(\bar{X} - \bar{Y}) - 0}{\widehat{SE}(\bar{X} - \bar{Y})} = \frac{(\bar{X} - \bar{Y}) - 0}{S_P \sqrt{\frac{1}{n} + \frac{1}{m}}}$$

$$= \frac{0.80 - 0}{2.51} = 0.32.$$

Thus, the P-value is given by $P(\bar{X} - \bar{Y} \geq 0.8 | H_0) = P(T \geq 0.32 | H_0)$, where T has a t distribution with df $= n + m - 2 = 6$ when H_0 is true. So we need to look up 0.32 in the row for df $= 6$ in the t table (Table F). In this row, $P(t \geq 1.58) = 0.10$; that is, the area to the right of 1.58 is 0.1, so the area to the right of 0.32 is much greater than 0.1. There is a large probability of seeing a value of $T \geq 0.32$ (and hence a difference of $\bar{X} - \bar{Y} \geq 0.8$), compared with our usual standard of 0.05.

The P-value we have just calculated, being much greater than 0.1, shows that it is not unusual to see such a difference when the null hypothesis is true (that is, when the populations have the same means). Thus, for Stage III, we fail to reject the null hypothesis. (No surprise!) This shows the value of carefully done inferential statistics: We are sometimes prevented from drawing foolish conclusions!

An interested reader may ask what the statistician would do in the case of small samples if the population SD values are not assumed equal and the populations are assumed normal. We already know from Example 9.14 that, theoretically,

$$Z = \frac{\bar{X} - \bar{Y} - (\mu_X - \mu_Y)}{\sqrt{\frac{\sigma_X^2}{n} + \frac{\sigma_Y^2}{m}}}$$

has a standard normal distribution. So if the SD values are known (whether equal or not), then under $H_0: \mu_X - \mu_Y = 0$,

$$Z = \frac{\bar{X} - \bar{Y}}{\sqrt{\frac{\sigma_X^2}{n} + \frac{\sigma_Y^2}{m}}}$$

is standard normal, and we can use a Z test.

For small samples from normal populations whose SD values are not known and not assumed to be equal, we refer the interested reader to Walsh's approximate t test (see, for example, *Encyclopedia of Statistical Sciences* (New York: Wiley, 1982), Vol. 1, p. 266). This procedure is used often, because the case of $\sigma_X \neq \sigma_Y$ for small samples does arise frequently.

Section 9.5 Exercises

1. Scientists want to study the effect of exercise on weight loss. One hundred people are randomly divided into two equal groups. Both groups follow the same diet plan, but the first group also follows an exercise program. It is suspected that the exercise group may lose more weight on average. In the exercise group the mean weight lost over three months was $\bar{X} = 25.2$ pounds, with SD of $S_X = 10$ pounds. In the nonexercise group the mean weight lost was $\bar{Y} = 20.4$ pounds, with SD of $S_Y = 6.3$ pounds. Test the null hypothesis that the difference between the population means is $\mu_X - \mu_Y = 0$.

2. A study is performed to compare the voting behaviors of women and men during an election. An exit poll was carried out, and the results were as summarized in the following table:

	Total number polled	Number who voted for candidate A
Men	45	21
Women	51	36

Let p_X denote the proportion of men who voted for candidate A, and let p_Y denote the proportion of women who voted for candidate A. It is suspected that $p_Y > p_X$ because of Candidate A's views on various social issues.

   a. What are your estimates of p_X and p_Y?
   b. Using the formula given in the section, determine an estimate of the value of $\text{Var}(\hat{p}_X - \hat{p}_Y)$.
   c. Using the values you found in parts **a** and **b**, find the value of the statistic used to test whether there is a significant difference in p_X versus p_Y?
   d. Is there a significant difference in how the women and men voted in this race?

3. On a college campus, a professor wanted to test the effect on exam scores when she used a computer program to teach calculus. In her first class of 43 students, she used the computer as an instructional aid. In her second class of 35 students, she taught without using the computer. Students were randomly assigned to the two sections. It was claimed that the computer-aided class would perform better. On the final exam the noncomputer class scored a mean of $\overline{X} = 78.32$ with SD of $S_X = 8.07$, whereas the computer-aided class scored a mean of $\overline{Y} = 80.41$ with SD of $S_Y = 8.53$. Test the claim that the difference of the population means was 0.

4. In a study of people who stop to help drivers with disabled cars, some researchers thought that more people would stop if they first saw another driver getting help. In one experiment, 2000 drivers first saw a woman being helped to fix a flat tire, and 2.9% stopped to help a second woman fix a flat tire. Also, out of 2000 drivers who did not see the first helper, 1.75% stopped to help the second woman fix a flat tire. Test whether the two proportions are equal. (From McCarthy, "Help on the Highway," *Psychology Today*, July 1987.)

5. Fifty independent measurements of the weight of a chemical compound were made on each of two scales. There is concern that the second scale is measuring low. On the first scale the mean of the 50 measurements was $\overline{X} = 19.45$ grams with SD of $S_X = 0.49$ gram. On the second scale, the mean of the 50 measurements was $\overline{Y} = 18.42$ grams with SD of $S_Y = 0.27$ gram. Test the claim that the difference between the population means $\mu_X - \mu_Y = 0$.

6. In both 1978 and 1990, a survey was taken of first-year college students, asking whether they opposed the death penalty. The results were as follows:

Year	Number opposing death penalty
1978	165
1980	105

Suppose in both cases the survey polled 500 students. Assess whether there is a significant decrease over time of those opposing the death penalty.

7. One hundred and fifty high school students were randomly assigned to two groups of 75. Group A used a new math text, and group B used the old math text. The sample mean and SD of the SAT math scores for group A were 549.45 and 21.12, respectively, and the sample mean and SD for group B were 539.25 and 20.91, respectively. Test the null hypothesis that the difference between the population means was 0 (as opposed to the hope that the new math text would improve SAT math scores).

CHAPTER 9 HYPOTHESIS TESTING

8. In another survey similar to that in Exercise 4, first-year college students were asked in both 1970 and 1990 what their political orientation was. The following table shows how the percentage of students who replied "liberal/far left" changed:

Year	Number saying they were "liberal" or "far left"
1970	190
1990	125

   Once again, assume that 500 students were randomly surveyed in each case. Is there a significant decrease in the percentages?

9. Twelve independent batches from each of two competing cold medicines are tested for the amount of acetaminophen in milligrams. It is believed the second cold medicine (Brand Y) contains more acetaminophen, on average. For Brand X, $\bar{X} = 494.67$ and $S_X^2 = 233.15$, and for Brand Y, $\bar{Y} = 523.67$, $S_Y^2 = 303.33$. Test the claim that the mean amount of acetaminophen is the same in each brand. (Assume the populations are normally distributed with equal SDs.)

10. Conduct your own poll of students at your school as to whether men and women agree with the death penalty. What claim would you make about p_X (women) and p_Y (men)? Test the null hypothesis $H_0: p_X = p_Y$ against the claim.

11. The World Health Organization air quality monitoring project measures suspended particles in micrograms per cubic meter. Let X and Y equal the concentration of suspended particles in Melbourne and Houston, respectively. The $n = 13$ observations of X from Melbourne resulted in an average of 72.9 with SD of $S_X = 25.6$. The $m = 16$ observations of Y from Houston resulted in an average of $\bar{Y} = 81.7$ and SD of $S_Y = 28.3$. Test the equality of the population means, assuming that the populations are normally distributed with equal SDs. Assume the goal is to detect whether Houston has more suspended particulate matter.

12. To test the effectiveness of the new weight loss medication Redux, a group of 40 women was separated into two groups: Group A, the control group, took a placebo, and group B, the experimental group, took Redux. The amount of weight lost by each member of the groups over a 6-month period is given in the following tables. Test the null hypothesis that the population mean weight loss of the two groups is the same (that is, that Redux is ineffective). Assume that the populations are normally distributed with equal SDs.

Group A: Total weight lost (in pounds) over a 6-month period for each group member				
3	4	5	6	7
10	11	12	15	18
19	20	23	24	25
30	33	38	40	42

Group B: Total weight lost (in pounds) over a 6-month period for each group member				
5	5	7	7	10
10	10	10	15	18
20	24	28	29	38
42	44	49	50	55

9.6 ONE-SIDED AND TWO-SIDED HYPOTHESIS TESTING

Let us first recall two examples with different P-value events in terms of the direction ($\geq$ versus $\leq$) of extreme values. In Example 9.9 (Section 9.2) we considered the normal-distribution-based Z test of

$$H_0: \mu = 98.6 \,°F \text{ versus } H_A: \mu < 98.60 \,°F.$$

We observed $\bar{x} = 98.25$. We based our decision to reject H_0 on the fact that $P(\bar{X} \leq 98.25 | H_0 \approx 0)$ was much less than 0.05, our gold standard of statistical significance. On the other hand, in Example 9.11 we performed a Z test of

$$H_0: \mu_X - \mu_Y = 0 \text{ versus } H_A: \mu_X - \mu_Y > 0,$$

where X is the husband's and Y is the wife's number of years of education. Using $D = X - Y$ we found that

$$P(\bar{D} \geq 0.24 | H_0) = P\left(Z \geq \frac{0.24 - 0}{0.194} \Big| H_0\right) \approx 0.1.$$

Therefore we failed to reject the null hypothesis.

In the first case we computed the probability that $\bar{X}$ was *less than or equal to* the observed sample mean $\bar{x}_{OBS}$; in the second case, we computed the probability that $\bar{D}$ was *greater than or equal to* the observed sample mean $\bar{d}_{OBS}$. It is vital to understand how to choose the correct direction of the inequality sign in the definition of the P-value event.

We need to look closely at the **alternative hypothesis** corresponding to a given null hypothesis. It is this alternative hypothesis that clearly dictates the direction of the inequality sign in the extreme event used to assess the P-value.

In every hypothesis-testing situation, one not only specifies a null hypothesis but also an alternative hypothesis, with which to contrast the null hypothesis. The alternative hypothesis is usually the real-world (research) claim that is negated to produce the null hypothesis. When performing a statistical analysis, we must state the alternative hypothesis explicitly.

Consider the problem of Example 9.9 concerning what body temperature really is "normal." You conjecture, before you ever begin to collect a random sample of body temperatures of healthy adults, that it seems likely that the true population average body temperature is less than 98.6 degrees. Your null hypothesis of

$$H_0: \mu = 98.6$$

in Stage I is contrasted with your alternative hypothesis (the research claim) of

$$H_A: \mu < 98.6.$$

Your view prior to data collection is that either $\mu = 98.6$ (H_0) or $\mu < 98.6$ (H_A) must be true, but you do not know which one. Because of the direction of the inequality ($<$) in the alternative hypothesis, we consider an observed sample average $\bar{x}$ of much less than 98.6 as providing strong evidence that we can reject the null hypothesis in favor of accepting the alternative hypothesis that $\mu < 98.6$. Once we observe a mean $\bar{x}$ in the sample, the null hypothesis "defends itself" against rejection through the size of the probability that a null-hypothesis population could produce a value as low or lower (a more extreme value) than $\bar{x}$ by chance. This probability is the P-value that we will be computing:

$$P(\bar{X} \leq \text{the observed sample mean } \bar{x} | H_0)$$

to test against $H_A: \mu < 98.6$.

The inequalities in the alternative hypothesis and in the definition of the extreme event for which we compute the probability *must* both point in the same direction. This fact provides a mechanical way to decide whether to use $\overline{X} \leq \bar{x}$ or $\overline{X} \geq \bar{x}$ in the P-value computation. This is the same rule that we used in Section 9.2, where we defined "extreme" to mean far away from the null-hypothesis parameter value in the direction of rejecting the null hypothesis, that is, in favor of the original research hypothesis (the real-world claim).

The same rule applies when the alternative hypothesis is that the mean is greater than the assumed null-hypothesis value, as shown in the following variation of the human body temperature example.

Example 9.16

Suppose you have been told that the normal temperatures of young children and infants (age 4 or less) tend to run higher than 98.6 degrees, and you plan to take a random sample of healthy infants to assess this claim. Then you are led to

$$H_0: \mu = 98.6 \text{ versus } H_A: \mu > 98.6.$$

Suppose you now collect a random sample of 36 infants' temperatures and obtain a sample mean of $\bar{x} = 98.8$ and a sample SD of $s = 0.5$. The correct probability to find, under the assumption that H_0 is true, is

$$P(\overline{X} \geq 98.8 | H_0).$$

In preparation for using the CLT,

$$Z = \frac{\overline{X} - 98.6}{S/\sqrt{n}} = \frac{98.8 - 98.6}{0.5/\sqrt{36}} = 2.4.$$

Thus, to test against the alternative hypothesis H_A that the mean is greater than 98.6, we have to find the probability that the Z statistic under the null hypothesis can be greater than or equal to the value obtained from the observed sample. From Table E, $P(Z \geq 2.4) = 0.0082$, less than the usually chosen level of significance, 0.05, so we conclude that the alternative hypothesis holds, and we reject the null hypothesis.

Considering the two variations in the problem of Example 9.9 and in Example 9.16 for assessing the truth or falsity of the null hypothesis that 98.6 is the population mean body temperature, we conclude that there are two possible one-sided alternative hypotheses concerning a null hypothesis of the form H_0: population parameter $= C$, where C is a user-supplied constant. For the population mean μ, in particular, we have

$$H_0: \mu = \mu_0 \text{ versus } H_A: \mu < \mu_0$$

or

$$H_0: \mu = \mu_0 \text{ versus } H_A: \mu > \mu_0$$

as the two possible one-sided hypothesis testing situations. Note that μ_0 is a fixed number such as $\mu_0 = 98.6$.

Section 9.6 One-Sided and Two-Sided Hypothesis Testing

Again, it is the direction of the alternative hypothesis that determines the direction of the inequality in the definition of the event whose probability, or P-value, we compute to assess the strength of statistical evidence against the null hypothesis. This is true even when the observed $\bar{x}$ falls on the "wrong" side of the null-hypothesis parameter value, and thus provides no evidence to reject the null hypothesis in favor of the alternative; for example, consider

$$H_0: \mu = 98.6 \text{ versus } H_A: \mu > 98.6.$$

As in Example 9.16, the $>$ appearing in the alternative hypothesis dictates that we use $\geq$ in $P(\bar{X} \geq \bar{x} \mid H_0)$ when we determine the P-value. Suppose the observed sample mean is $\bar{x} = 98.4$ with an observed sample SD of $s = 0.5$ and a sample size of $n = 36$. Then, assuming H_0 is correct, to get the P-value we compute in the standard way that

$$P(\bar{X} \geq 98.4 \mid H_0) = p\left(Z \geq \frac{98.4 - 98.6}{0.5/6} \middle| H_0\right) = p(Z \geq -2.4) = 0.9918.$$

We note two things about this answer. Remember that small probability values (≤ 0.05) provide strong evidence that we should reject the null hypothesis in favor of the alternative hypothesis. Hence there is a dramatic lack of evidence for rejecting the null hypothesis here. Note also that we would likely have computed the wrong probability here if we had let the observed statistic dictate the direction of the inequality. Namely, we would likely have computed $P(Z \leq -2.4)$, which would not be appropriate for the alternative hypothesis $H_A: \mu > 98.6$.

Perhaps upon reflection, concerning the foregoing discussion, the alternative to $H_0: \mu = \mu_0$ could be a **two-sided alternative hypothesis**, namely

$$H_A: \mu \neq \mu_0.$$

This is most appropriate in situations where we have no prior knowledge or viewpoint that dictates a one-sided alternative. For example, consider the null hypothesis $H_0: \mu = 98.6$. Suppose we wish to test the null hypothesis knowing that both $\mu > 98.6$ and $\mu < 98.6$ are possible alternatives. Then we simply make this our alternative hypothesis, producing

$$H_0: \mu = 98.6 \text{ versus } H_A: \mu \neq 98.6.$$

In this case strong evidence to reject H_0 will be given by an observed $\bar{x}$ that, under the null hypothesis, is *either improbably greater than or improbably less than* 98.6. That is, in order to assess the statistical significance of the observed sample average, the P-value that we must compute is the probability that random variation of $\bar{X}$ around the null-hypothesis value $\mu_0 = 98.6$ can produce a value of $\bar{X}$ that is as far or further from 98.6 in either direction than the observed $\bar{x}$. When the test statistic has a symmetric probability density function than the normal and t distributions do, to obtain the correct P-value, the two-sided alternative hypothesis requires us to double the one-sided alternative P-value. Because the standardized test statistic obeys the CLT and thus is distributed (approximately) as a standard normal, symmetry of the test statistic's distribution is quite common.

To illustrate, suppose we are testing

$$H_0: \mu = 98.6 \text{ versus } H_A: \mu \neq 98.6,$$

and we have an observed sample mean $\bar{x} = 98.8$, with an observed standard deviation $s = 0.64$ and a sample size $n = 36$. First we note that $\bar{x} = 98.8 > 98.6$. If we were testing H_0 only against the alternative hypothesis $H_A: \mu > 98.6$, we would only have to compute the probability that, by chance random variation around $\mu = 98.6$, $\bar{X}$ could go *at least as high as 98.8* in this sample:

$$P(\bar{X} \geq 98.8 \mid H_0) = P\left(Z \geq \frac{98.8 - 98.6}{0.64/6}\right) = P(Z \geq 1.875) = 0.03.$$

444 CHAPTER 9 HYPOTHESIS TESTING

Further, we would reject the null hypothesis, because our P-value would be less than 0.05. However, when we test H_0 against $H_A : \mu \neq 98.6$, the alternative hypothesis has to prove itself against the possibility that the typical random variation of $\bar{x}_{OBS}$ around μ_0 when H_0 is true could produce a sample value at least as far from 98.6 as the observed $\bar{x}_{OBS}$ is, either above or below 98.6. So our P-value becomes

$$P(\bar{X} \text{ at least as far from } 98.6 \text{ as } 98.8 \text{ in either direction} | H_0)$$
$$= P(\bar{X} \geq 98.8 \text{ or } \bar{X} \leq 98.6 - (98.8 - 98.6) | H_0)$$
$$= P(Z \geq 1.875 \text{ or } Z \leq -1.875) = P(Z \geq 1.875) + P(Z \leq -1.875)$$
$$= 2P(Z \geq 1.875) \text{ by the symmetry of the standard normal distribution}$$
$$= 0.06.$$

So this time we fail to reject the null hypothesis, because our P-value exceeds the gold standard of 0.05. We could have represented this probability by:

$$P(|\bar{X} - \mu_0| \geq 98.8 - 98.6) = P(\bar{X} \geq 98.8 \text{ or } \bar{X} \leq 98.4).$$

We have learned that the alternative hypothesis to $H_0 : \mu = C$ can be one-sided or two-sided. Of course, this statement and the principles presented above apply to any other population parameter of interest, such as the population median or population proportion. Moreover, we have learned that our computation of the possible statistical significance of the observed data must take into account which form of alternative hypothesis we have selected. The choice of which of the three alternatives to use is up the person conducting the statistical study and should be made *before* examining the data. It also affects the strength attributed to the evidence.

Section 9.6 Exercises

1. A certain battery manufacturer claims that its batteries will run for an average of 25 hours. You are skeptical and suspect that the batteries likely run for less than 25 hours, so you decide to test the null hypothesis that the batteries will run for 25 hours on average versus less. You test 40 batteries; the sample mean lifetime of the 40 batteries is 23.5 hours with a sample SD of 2.5 hours. Perform a suitable hypothesis test. Make sure to state your null and alternative hypotheses clearly.

2. You work in the quality control department of a light bulb manufacturer. Right now, the light bulbs are supposed to last for an average of 200 hours of use, but you have the impression that the light bulbs are lasting for longer than 200 hours. So you decide to test whether the light bulbs will last for 200 hours on average versus more. You test 100 light bulbs; the sample mean lifetime of the 100 light bulbs is 202.5 hours with a sample SD of 3.5 hours. Perform a suitable hypothesis test. Make sure to state your null and alternative hypotheses clearly.

3. When a car company designs a car, it tries to achieve a target gas mileage for the car. Suppose the target average gas mileage for a new car is 24.5 miles per gallon but the manufacturer does not know whether the true average is better or worse than 24.5. In a test of 200 cars, the average gas mileage is 26.25 miles per gallon with SD of 2.25 miles per gallon. Perform a suitable hypothesis test. Make sure to state your null and alternative hypotheses clearly.

4. It is claimed that the average difference of husband's age minus wife's age has a population mean of 3 years. Suppose you believe that the average difference is less than 3 years. You take a random sample of 100 couples and determine that the mean difference in age between husband and wife (husband's age minus wife's age) for the 100 couples is 2.5 years with a sample SD of 1.3 years. Perform a suitable hypothesis test. Make sure to state your null and alternative hypotheses clearly.

5. The average body temperature of humans is listed as 98.6 degrees Fahrenheit. Suppose you do not believe this number is accurate. To prove your case, you take a sample of 50 people and find that the average body temperature of the 50 people is 98.66 with SD of 0.25.
   a. Perform a test of the null hypothesis against the alternative hypothesis that the average body temperature is not equal to 98.6.
   b. Perform a test of the null hypothesis against the alternative hypothesis that the average body temperature is greater than 98.6.
   c. Perform a test of the null hypothesis against the alternative hypothesis that the average population body temperature is not equal to 98.6 (it could be greater or less than 98.6).
6. An airline reported an on-time arrival rate of 78.4%. Assume that later, a random sample of 750 of its flights results in 630 that are on time. Test the airline's claim that its on-time arrival rate is now higher than 78.4%.
7. Return to the study described in Exercise 4 of Section 9.5 and test the claim that more people would stop if they first saw someone else getting help.

9.7 SIGNIFICANCE TESTING VERSUS ACCEPTANCE/REJECTION TESTING: CONCEPTS AND METHODS

The version of hypothesis testing we have dealt with thus far is known as significance testing. In this type of testing we have two possible decisions available:

1. reject H_0, or
2. fail to reject H_0.

Only one error is possible here from a decision-making viewpoint: rejecting H_0 when H_0 is true. Moreover, our P-value approach of rejecting H_0 only when the P-value is small (typically ≤ 0.05) protects us from making this error often when doing significance testing.

Whenever we fail to reject H_0, this decision means only that the data do not provide strong evidence that H_0 should be rejected! In particular, this is not a decision to accept H_0. The viewpoint when making this decision is that either H_0 or H_A may be true: strong evidence in favor of rejecting H_0 has simply not been found (even though the researcher probably had hoped to reject H_0). We do not make an error when we fail to reject H_0, because we have not decided in favor of either H_0 or H_A.

Sometimes, however, we are in hypothesis testing situations where the setting requires us to decide between two alternatives. Consider the following examples:

- The Federal Reserve Board chairperson looks at economic data and then either raises the prime lending (interest) rate or does not.
- You get the day's weather report ("data") and then either go on a picnic (believing it will not rain) or do not go on a picnic (believing it will rain).
- A manufactured product is judged either to meet product standards or not, based on performance testing data.

From the hypothesis testing viewpoint, the last example gives:

H_0: product meets standards   versus H_A: product does not meet standards

We thus must either accept H_0 (which is the same as rejecting H_A) or accept H_A (which is the same as rejecting H_0). That is, the option of "failing to reject H_0" is not open to us, because the setting requires us to approve the product or not approve it. Real-world risks and costs ride on both possible decisions. If we accept H_0 when H_A is true, the company ships a defective product, may lose customers, and may suffer a lawsuit. If we accept H_A when H_0 is true, the company certainly loses

446 CHAPTER 9 HYPOTHESIS TESTING

Table 9.7 Decisions and Errors in Acceptance/Rejection Testing

Statistical decision based on data (the inference)	Truth about the population (the reality)	
	H_0 true	H_1 true
Reject H_0	Type I error	Correct decision
Accept H_0	Correct decision	Type II error

money, time, and materials on needless rework. We are not allowed the dodge that we do not know whether the product is acceptable or not!

In settings like these, where a decision in favor of H_0 or in favor of H_A must be made, clearly the null hypothesis does not have a special status. Rather, H_0 and H_A are in parallel roles. After analyzing the data statistically, we either embrace H_0 as true or embrace H_A as true. Indeed H_0 and H_A are often equated to two contrasting real-world decisions like raising the prime rate or not. Clearly there are two possible correct decisions and two possible erroneous decisions. These are often expressed in the standard diagram in Table 9.7. Since either H_0 is true or H_A is true, but not both, only one column in this table is active in an actual hypothesis-testing setting. That is, only one of the two errors is possible. For example, if it does rain today, the only possible error is going on the picnic. However, the statistician does not know the state of the real-world (otherwise there would be no need for statistical inference), so which error is possible is not known. The standard terminology of "Type I" and "Type II" errors appearing in Table 9.7 is not very intuitive. The student of statistics simply has to memorize what each term means.

In significance testing, H_0 is rejected only when the computed P-value, P(reject $H_0|H_0$ true) = P(Type I error), is small (≤ 0.05 being the usual gold standard). In this manner, H_0 is rarely rejected when it is true. Since no Type II error is possible, we do not need to consider it.

Now, **acceptance/rejection hypothesis testing**, where we are required either to accept or to reject H_0 (often because we must take one of two possible actions that have costs and risks) we have to consider both $\alpha = P(\text{Type I error}) = P(\text{reject } H_0|H_0 \text{ true})$ and $\beta = P(\text{Type II error}) = P(\text{accept } H_0|H_0 \text{ false})$.

Thus we must also control the probability of making a Type II error. The user who needs an acceptance/rejection hypothesis test carried out would like to turn his data over to the statistician, have her carry out a hypothesis test, and have *both* error probabilities be small. But this desire runs up against the famous (and almost always true) principle: "There is no such thing as a free lunch."

Let us consider how to control the probabilities of Type I and Type II errors when doing acceptance/rejection hypothesis testing. The acceptance/rejection hypothesis-testing approach fixes $\alpha = P(\text{Type I error})$, usually at $\alpha = 0.05$, rather than working with a variable P-value determined by the data, as is often done with significance testing. The chosen fixed level α is called the level of significance. In a testing problem with $n = 30$ and $\alpha = 0.05$, suppose that $\beta = P(\text{Type II error}) = 0.53$, a most unsatisfactory situation! The only way we can reduce β from 0.53 is to allow α to be increased from 0.05. For example, if we allow $\alpha = 0.17$, computation might show that $\beta = 0.23$. Now both errors have about the same amount of probability of occurrence, but most users and statisticians would find $\alpha = 0.17$ and $\beta = 0.23$ both unacceptably high for effective decision making based on hypothesis testing.

Is there anything we can do to remedy a situation like the above illustration? Yes! *We can tell our clients to collect more data!* Further on in this section we will study how collecting more data can make β smaller for a specified small α—making them both acceptably small.

We now study the computational methodology of acceptance/rejection hypothesis testing in some detail. In particular, we will learn three things:

Section 9.7 Significance Testing versus Acceptance/Rejection Testing

1. how to turn a specified level of significance (usually $\alpha = 0.05$) into a hypothesis testing procedure (a rule that looks at the data and chooses either H_0 or H_A);
2. how to compute $\beta = P(\text{Type II error})$ when α is fixed; and
3. how to determine the sample size n needed to meet the client's specified requirements of both low α and low β, if the sample size can be varied.

Specified Level of Significance and the Resulting Critical Region for Acceptance/Rejection Hypothesis Testing The first step in acceptance/rejection hypothesis testing is to select the level of significance. If the consequences of a Type I error are judged by the statistical user to be much more serious than those of a Type II error, then choosing $\alpha = 0.05$ (the gold standard of significance testing), or sometimes $\alpha = 0.01$ (the platinum standard) or 0.1 (the silver standard), is the usual practice, depending on the risk. Otherwise, if the two errors have consequences of approximately equal seriousness, then clearly one should choose α such that it turns out that $\alpha \approx \beta$. It is more common in practice to choose a fixed level of α.

Because our focus has been on significance testing, we have followed the P-value approach to deciding when to reject H_0. For example, for $H_0: \mu = 1$ versus $H_A: \mu > 1$, with $\bar{x} = 2$ observed, we compute the P-value

$$P(\bar{X} \geq 2 \mid H_0)$$

to make our decision. More abstractly, the P-value is the probability, assuming H_0 is true, that a particular test statistic (e.g., $\bar{X}$) is at least as extreme as the observed test statistic (e.g., $\bar{x} = 2$) in the direction that supports H_A being true. In particular, we saw in the previous section that the exact form of the P-value event (e.g., $\bar{X} \geq 2$) depends on the form of the alternative hypothesis. For example, $H_A: \mu > \mu_0$ requires $P(\bar{X} \geq \bar{x})$, and $H_A: \mu < \mu_0$ requires $P(\bar{X} \leq \bar{x})$.

Now, instead of taking a P-value approach, we will choose the level of significance α, say, $\alpha = 0.05$. We will find that here, as with a significance test, the direction of the alternative hypothesis plays the same decisive role in forming the event of interest, which here is called the **critical region**. The critical region is really a rule that tells us whether to accept or reject H_0 for the observed data based on the value of a test statistic.

It is easiest to illustrate this fixed level α approach with an example. Consider (again) the body temperature problem in Example 9.15:

$$H_0: \mu = 98.6 \text{ versus } H_A: \mu > 98.6,$$

with a known population SD of $\sigma = 0.5$, and a sample size of $n = 36$. Under acceptance/rejection hypothesis testing, we decide, prior to data collection, that we want the probability of Type I error, to be 0.05. That is, we will reject the null hypothesis if the observed $\bar{X}$ is sufficiently larger than 98.6, namely if $\bar{X} - 98.6 > C$, where we will determine the unknown *positive* number C by requiring that

$$P(\bar{X} - 98.6 \geq C \mid H_0) = 0.05$$

under the assumption that the null hypothesis is true. Note that choosing C tells us how much greater than the null-hypothesis value $\mu = 98.6$ the observed $\bar{X}$ must be in order to provide strong evidence for rejection of H_0 in favor of accepting H_A. In terms of a number line containing the possible values of $\bar{X}$, we have decided on a region of the line, namely $(\bar{X} - 98.6 \geq C)$, or the region $(\bar{X} \geq 98.6 + C)$, such that, if $\bar{X}$ is observed to be within that region, we will reject the null hypothesis H_0 in favor of the alternative hypothesis H_A. The choice of C is determined by our choice of 0.05 as the level of significance. The region of the line $(\bar{X} - 98.6 \geq C)$ is called the **rejection region** or the **critical region**.

We choose the critical region of the form ($\overline{X} - \mu_0 \geq C$) because H_A is of the form $H_A: \mu > \mu_0$ (in this case, 98.6). Just as in significance testing, we chose the inequality sign in the definition of the critical region event to match the direction of the inequality sign in H_A. Thus, we obtain three types of critical regions to choose from:

1. for $H_A: \mu > \mu_0$, choose critical region: $\overline{X} - \mu_0 \geq C$;
2. for $H_A: \mu < \mu_0$, choose critical region: $\overline{X} - \mu_0 \leq C$; or
3. for $H_A: \mu \neq \mu_0$, choose critical region: either $\overline{X} - \mu_0 \geq C$ or $\overline{X} - \mu_0 \leq -C$ (equivalently: $|\overline{X} - \mu_0| \geq C$).

Returning to the problem of finding the critical region for the temperature problem, let us see how to compute C and thus specify the critical region. To make things simple, suppose we know from past experience with temperature data that the theoretical SD is $\sigma = 0.5$. Recall that $n = 36$. Then $P(\overline{X} - 98.6 \geq C | H_0) = 0.05$ must be used to compute the boundary point C of the critical region. Assuming the null hypothesis to be true, we obtain the standardized test statistic

$$Z = \frac{\overline{X} - 98.6}{\sigma/\sqrt{n}} = \frac{\overline{X} - 98.6}{0.5/6},$$

which is approximately standard normal when H_0 is true (by the CLT). Thus

$$0.05 = P\left(\frac{\overline{X} - 98.6}{0.5/6} \geq \frac{C}{0.5/6}\right) = P\left(Z \geq \frac{C}{0.5/6}\right).$$

(dividing both sides by $\sigma/\sqrt{n} = 0.5/6$). But we know from Table E that $P(Z \geq 1.645) = 0.05$. Hence we obtain

$$1.645 = \frac{C}{0.5/6}.$$

In the latest step we used the obviously true fact that if $P(Z \geq a) = P(Z \geq b)$, then $a = b$. Solving for C yields $C = 0.14$. Thus, our critical region, or rejection region, is the set of all $\overline{X}$ that satisfy $\overline{X} - 98.6 \geq 0.14$; that is, $\overline{X} \geq 98.74$ is so extreme that we reject H_0 and accept H_A.

When we actually do the random experiment, any value of $\overline{X}$ that exceeds 98.74 will provide us with statistically significant (at level 0.05) evidence that we should reject the null hypothesis in favor of the alternative hypothesis that $\mu > 98.6$. For example, if we observe $\bar{x} = 98.8$, we reject the null hypothesis, but if $\bar{x} = 98.7$, we *accept* the null hypothesis that $\mu = 98.6$, because we are doing acceptance/rejection hypothesis testing.

We note that, even if we do acceptance/rejection hypothesis testing and take a fixed level-of-significance approach, we can still compute the P-value once the data are analyzed. Most statisticians would see this additional step as being highly informative and, therefore good, statistical practice.

Power of a Test Once we have used a specified level of significance to determine a critical region, we will want to know how **powerful** the test is. The **power of a test** is defined as how probable it is that the test will reject the null hypothesis for each value of the parameter of interest for which the alternative hypothesis holds. Note that high power amounts to a low probability of a Type II error, which is defined as accepting a false null hypothesis.

The **power of a test** is therefore defined mathematically to be

$$1 - \beta = 1 - P(\text{Type II error})$$
$$= 1 - P(\text{accept } H_0 | H_A \text{ true})$$
$$= P(\text{reject } H_0 | H_A \text{ true}).$$

Thus, we would like the power to be as close to 1 as possible.

Section 9.7 Significance Testing versus Acceptance/Rejection Testing

However, β, and therefore the power of the test, is not the same at all possible values of the parameter of interest for which H_A holds. Furthermore, not all values of the parameter of interest will require high power from the statistical user's viewpoint. For a test to be described as sufficiently powerful, it must have a high probability of rejecting the null hypothesis for all values of the parameter of interest that the researcher considers far *enough* from the null-hypothesis value to be important to detect.

To understand these possibly confusing statements, let us return to the peanut butter jars in Example 9.10. Suppose the peanut butter manufacturer knows that *either* the manufacturing process for "one-pound" jars is "in control" ($H_0: \mu = 1$) or the jars are being overfilled ($H_A: \mu > 1$). If $\mu = 1.0002$, the manufacturer will surely be very comfortable accepting H_0, but if $\mu = 1.2$, the manufacturer will certainly want to reject H_0; otherwise, a sixth of the product is not being paid for. In particular, after a careful consultation between manufacturer and statistician, suppose it is decided that whenever $\mu \geq 1.05$, it is a serious error to accept H_0. Thus, the requirement is then for high power for all $\mu \geq 1.05$, and we make no power requirement for $1 < \mu < 1.05$; that is, the manufacturer doesn't really care whether H_0 or H_A is concluded if $1 < \mu < 1.05$.

Let us again consider the normal body temperature problem in Example 9.15, where the critical region corresponding to level of significance $\alpha = 0.05$ was found to be $\bar{x} \geq 98.74$. Thus H_0 will be rejected if the sample yields $\bar{x} = 98.74$ or larger. Remember that the population is assumed to be normal, with a population SD of 0.5. The medical researcher designing the experiment must make the same kind of choice as the peanut butter maker about how much power to require at what values of the parameter of interest. If μ turns out to be 98.62 degrees, it may not be important to the researcher's work, but a μ of 99.1 would certainly be important to detect.

As will be shown subsequently, any increase in the desired power causes an increase in the required sample size, so the researcher does not wish to demand excessive power for parameter values in H_A too close to the null-hypothesis value to be of practical concern to detect. Suppose the researcher judges that $\mu = 98.8$ is truly important to detect (and, thus, wants to reject H_0 with high probability when H_A holds in the sense that $\mu = 98.8$). That is, the researcher wants a high probability of rejecting that $\mu = 98.8$ is the true population mean.

To evaluate whether the power is adequate in this example, we need to find $P(\text{reject the null hypothesis} | \mu = 98.8)$. Now

$$P(\text{accept the null hypothesis} | \mu = 98.8) = P(\bar{X} < 98.74 | \mu = 98.8)$$

because the acceptance of H_0 region, namely $\bar{X} < 98.74$, is the complement of the critical region. We can compute $P(\text{Type II error})$ for every choice of $\mu \geq 98.6$, that is every μ for which $H_A: \mu \geq 98.6$ holds.

To compute the power $(1 - P(\text{Type II error}))$ when $\mu = 98.8$, we standardize $\bar{X}$ to produce a Z statistic that approximately obeys the standard normal distribution of Table E. Here, the population SD is $\sigma = 0.5$, the sample size is 36, and the H_A value we wish to be relatively sure to detect is $\mu = 98.8$. Thus,

$$Z = \frac{\bar{X} - 98.8}{0.5/6}$$

is, by the CLT, approximately standard normal. Hence, the power at $\mu = 98.8$ is given by

$$P(\text{reject } H_0 | \mu = 98.8) = P(\bar{X} \geq 98.74 | \mu = 98.8)$$
$$= P\left(Z \geq \frac{98.74 - 98.8}{0.5/6}\right) = P(Z \geq -0.72)$$
$$= 0.7642 \approx 0.76.$$

Note that in defining Z we have centered $\overline{X}$ at the temperature of 98.8, *not* at the null-hypothesis 98.6 temperature, as we did in defining Z for significance testing and in defining the critical region. The probability of 0.76 tells us that if $\mu = 98.8$, we will arrive at the correct decision to reject the null hypothesis about 76% of the time. In this precise numerical sense, the test is reasonably powerful against the particular alternative $\mu = 98.8$. Clearly, the power will be higher for any $\mu > 98.8$, such as $\mu = 99.1$, and less for any $\mu < 98.8$, such as $\mu = 98.63$.

Perhaps this capacity to detect that H_A holds 76% of the time when $\mu = 98.8$ is not a sufficiently powerful test from the viewpoint of the person doing the medical study. What if the researcher insisted on a $P(\text{Type II error} | \mu = 98.8) = \beta = 0.05$, namely wanting to detect that H_A holds when $\mu = 98.8$ 95% of the time, while continuing to require $\alpha = 0.05$? Can these two requirements be met? Yes, but only by allowing the sample size, n, to be as large as needed.

Recall that the critical region is given by

$$\overline{X} - 98.6 \geq C$$

and that for $\alpha = 0.05$ we obtained

$$1.645 = \frac{C}{0.5/\sqrt{n}}, \quad (1)$$

where we have replaced $n = 36$ with the now unknown sample size n. Now, the requirement that $\beta = 0.05$ when $\mu = 98.8$ means that $\beta = P(\text{accept } H_0 | \mu = 98.8) = 0.05$. But the event of accepting H_0 is that $\overline{X}$ does not lie in the critical region; in other words, that $\overline{X} - 98.6 \leq C$.

Thus the equation for $\beta = 0.05$ yields

$$0.05 = P(\overline{X} - 98.6 \leq C | \mu = 98.8))$$

But when $\mu = 98.8$, because the population is normal with mean 98.8 and $SE(\overline{X}) = \sigma/\sqrt{n}$, we have by standardizing $\overline{X}$ that

$$Z = \frac{\overline{X} - 98.8}{0.5/\sqrt{n}}$$

is approximately standard normal when $\mu = 98.8$. Thus

$$0.05 = P(\overline{X} - 98.6 < C | \mu = 98.8)$$
$$= P(\overline{X} - 98.6 - 0.2 < C - 0.2 | \mu = 98.8)$$
$$= P(\overline{X} - 98.8 < C - 0.2 | \mu = 98.8)$$
$$= P\left(Z < \frac{C - 0.2}{0.5/\sqrt{n}}\right),$$

by dividing by $0.5/\sqrt{n}$, and recalling from Table E that $P(Z \leq -1.645) = 0.05$, we get

$$\frac{C - 0.2}{0.5/\sqrt{n}} = -1.645. \quad (2)$$

Thus we have two equations, labeled (1) and (2), with two unknowns, namely n and C. Hence we can solve for n and C. Solving yields $C = 0.1$ and $n = 67.65$. But the number of observations n must be an integer and we want the power to be at least 0.95, so by rounding n to the nearest larger integer, our solution is $C = 0.1$ and $n = 68$. The new value of $C = 0.1$ means that our criti-

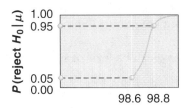

Figure 9.4 Gain in power with increasing true mean temperature for a test with $n = 68$.

cal region is $\overline{X} - 98.6 \geq 0.1$ or $\overline{X} \geq 98.7$. Thus, by increasing n from 36 to 68 and adjusting the critical region to $\overline{X} \geq 98.7$ we have produced a test for which,

$$P(\text{reject } H_0 | H_0 \text{ true}) = P(\overline{X} \geq 98.7 | \mu = 98.6) = 0.05, \text{ which, for } \mu = 98.6,$$

and

$$P(\text{reject } H_0 | H_0 \text{ true}) \leq 0.05,$$

for every $\mu \leq 98.6$. The **power curve**, defined by $P(\text{reject } H_0 | \mu)$, is forced to rise sharply from 0.05 at 98.6 to 0.95 at 98.8. That is,

$$0.05 = \alpha = P(\overline{X} \geq 98.7 | \mu = 98.6) \text{ and}$$

$$0.95 = 1 - \beta = P(\overline{X} \geq 98.7 | \mu = 98.8).$$

This is shown in Figure 9.4.

The procedure we have followed is the essence of good practice in acceptance/rejection testing, namely controlling both $P(\text{Type I error})$ *and* $P(\text{Type II error})$. However, our increased power does not come without a cost. Our sample size was 36; now it is 68. Thus, the budget for research assistants to go around taking healthy children's temperatures nearly doubles. In this case the increased cost might be too great to justify. It would be a different story if, instead, the study concerned the effects of an experimental medication (produced expensively in small quantities) on patients with a rare and swiftly fatal disease. One of the most common problems in academic, government, and industry research is that sample sizes are often too small, because of such resource constraints, to have adequate power.

For example, suppose the current medical treatment for a disease that affects 500,000 people per year produces a cure 60% of the time. The maker of a new treatment wants to conduct a study that would tell, with very high probability, that the new treatment was better if its cure rate were 70% rather than the null-hypothesis rate of 60%. Because of the authorized size of the clinical trial and the level of significance selected ($\alpha = 0.05$, perhaps) the maker may be forced to use a critical region for which $P(\text{reject } H_0 | p = 0.7)$ is not as large as desired. As a result, the treatment may actually fail to show an improvement over the current treatment, even though it would produce 50,000 more cures per year on average if put into general use.

Section 9.7 Exercises

1. What are some common values for the probability of a Type I error?
2. Suppose you lower the probability of a Type I error from 0.05 to 0.01. What happens to the probability of a Type II error?
3. What must you do to decrease the probability of Type I and Type II errors simultaneously?

4. In a court of law in the United States, a conviction for a crime requires "guilt beyond a reasonable doubt." What level of certainty should be required for reasonable doubt? Should you be 99% certain of a person's guilt to convict? Explain your answer.
5. Find the critical region for the hypothesis test in Exercise 1 of Section 9.6 for a level of significance of 0.05.
6. Find the critical region for the hypothesis test in Exercise 2 of Section 9.6 for a level of significance of 0.05.
7. Find the power for the hypothesis test in Exercise 5 above for $\mu = 24$.
8. Find the power for the hypothesis test in Exercise 6 above for $\mu = 201$.
9. A candidate commissions a poll, wishing to reject the null hypothesis that the proportion of the population who support him is $p = 1/2$, in favor of showing that a majority support him. The poll will have $n = 400$ people in it.
   a. What is the alternative hypothesis?
   b. Find C so that $P(\hat{p} \geq C) = 0.05$ when the null hypothesis $H_0: p = 1/2$ is true.
   c. If $\hat{p} = 0.47$, do you accept or reject the null hypothesis?
   d. If $\hat{p} = 0.53$, do you accept or reject the null hypothesis?
   e. If $\hat{p} = 0.61$, do you accept or reject the null hypothesis? (Use level of significance = 0.01.)
10. A random sample of $n = 100$ boxes of cereal was obtained to test whether the population mean weight of the contents is $\mu = 16$ ounces. The alternative is that the population mean weight is greater than 16 ounces. The manufacturer does not want to put more cereal in the boxes than necessary.

    The critical region is to reject the null hypothesis if the sample mean $\overline{X} \geq 16.2$. Assume the population SD is 0.5.
    a. What is $P(\overline{X} \geq 16.2)$ when the null hypothesis is true?
    b. Suppose the population mean is $\mu = 16.1$. What is $P(\overline{X} \geq 16.2)$?
    c. Suppose the population mean is $\mu = 16.2$. What is $P(\overline{X} \geq 16.2)$?
    d. Suppose the population mean is $\mu = 16.3$. What is $P(\overline{X} \geq 16.2)$?
    e. If the population mean is $\mu = 16.1$, is there a high probability one would reject the null hypothesis? What if the population mean is $\mu = 16.3$?

9.8 TEST FOR A POPULATION STANDARD DEVIATION: ISSUES

To test hypotheses about a population SD, the χ^2 distribution can be used whenever the random sample comes from a normal population. We do not discuss this technically interesting but seldom used procedure, because statisticians rarely wish to test hypotheses about the population SD. Moreover, the normal-distribution-based test is not very robust when the population distribution shape departs somewhat from normal, as is often the case. The interested reader can consult any number of standard statistics textbooks to study χ^2 testing concerning the population SD.

Statisticians sometimes do care about the equality of SD values for two or more normal populations, but the procedure usually taught for testing such equality is also quite nonrobust and is therefore not recommended. Even more fundamentally, our practical understanding of the population SD is tightly tied to knowing that the population distributions are approximately normal. For example, the rule that the probability of falling within ±1SD of the population mean is about 0.68 depends on having an approximately normal population. In cases where the shapes of the population are either not known or known to be rather strongly nonnormal, it is much better to use more robust and more easily interpretable statistics to describe the spread in the samples and hence (by the bootstrap principle) in the populations. For example, the interquartile range is effective in this regard.

CHAPTER REVIEW EXERCISES

1. The championship series of the National Basketball Association in June 2000 was played between the Los Angeles Lakers and the Indiana Pacers. The Lakers won the first two games, with Shaquille O'Neal, their star center, leading the way. Because O'Neal makes only 50% of his foul shots on average in regular season games, one of the strategies of Indiana was to foul O'Neal many times, hypothesizing that O'Neal would have an even lower proportion in playoff games.
   When Indiana won the third game, O'Neal made only 3 out of 13 foul shots in the game. Considering this performance as a sample proportion and considering O'Neal's lifetime average of about 0.5 as a population proportion for regular season games, was this sample indicative of O'Neal having a lower population proportion in a championship series game?
   a. In terms of a population proportion, what is the null hypothesis?
   b. To test the null hypothesis, we need to generate simulated samples that conform to the null hypothesis. On each simulation, we will generate a simulated sample having the same size as the original real-data sample. The simulated data samples will be generated from a box model.
   c. What is the choice of the sampling box? What is the definition of one simulation? What sample statistic will we measure on each simulated sample that we generate?
   d. What will be the event of interest for testing the null hypothesis?
   e. The results from 1000 simulations from the null hypothesis box model are shown in the following table. What is the experimental probability of the event of interest?
   f. Decide whether or not to reject the null hypothesis and explain why you chose that decision.

$\hat{P}$ observed in simulated sample	Frequency of occurrence	$\hat{P}$ observed in simulated sample	Frequency of occurrence
0/13	0	7/13	207
1/13	2	8/13	155
2/13	11	9/13	86
3/13	36	10/13	33
4/13	84	11/13	10
5/13	161	12/13	2
6/13	213	13/13	0

2. It is claimed that the mean height of U.S. women is greater than 65 inches. In a random sample of 50 women, the mean height was 64.45 inches with SD of 3.5 inches. Test the null hypothesis that the population mean height of U.S. women is 65 inches.

3. In a random sample of 100 teenage boys, the mean amount of time spent watching television was $\bar{x} = 3.6$ hours with a standard deviation s_X of 0.9 hour. In a random sample of 100 teenage girls, the mean amount of time spent watching television was $\bar{y} = 4.2$ hours with SD of $s_Y = 1.2$ hours. Test whether the difference between the population means is 0 versus the alternative that on average girls spend more time watching television.

4. In a random sample of eight people, the mean amount of sleep per night they got is 7.45 hours with SD of 1.15 hours. Test whether the population mean amount of sleep per night is 8 hours. Assume sampling from a normal population and that we are assessing a claim that sleep time actually averages less than 8 hours.

5. In a random sample of 250 residents of a large city, 48% of the people in the sample would favor raising the bar entrance age from 19 to 21. Test whether the true population proportion is equal to 0.5, versus the claim that it is less.

6. A sample of five body temperatures yields

98.7	98.5	98.6	98.3	99.0

454 CHAPTER 9 HYPOTHESIS TESTING

   a. Find the sample mean.
   b. Create the population for bootstrap testing the null hypothesis that the population mean is 98.6.
7. A study was conducted to investigate the effectiveness of hypnotism in reducing pain. Results for eight randomly selected subjects are given below. The data are based on "An Analysis of Factors that Contribute to the Efficacy of Hypnotic Analgesia," by Price and Barber, *Journal of Abnormal Psychology*, Vol. 96, No. 1. Note this is a (before-after) matched-pairs study.

Before	−5.5	−5.0	−6.6	−9.7	−4.0	−7.0	−7.0	−8.4
After	−1.4	−0.5	0.7	1.0	2.0	0.0	−0.6	−1.8

   Test the claim that the affective responses to pain are the same before and after hypnosis.
8. Between the 1999 and 2000 baseball seasons, Ken Griffey, Jr., was traded from an American League team, the Seattle Seahawks, to a National League team, the Cincinnati Reds. Over Griffey's last 2 years in Seattle, his batting average (that is, the proportion of his at-bats that produced hits) was .285 (353 hits in 1239 at-bats). Because Griffey switched from one league to another, the claim might be made that his batting average is going to be lower, at least for the first season.

   Up to a certain point in the 2000 season, Griffey had 50 hits in 219 tries, a batting average of 0.228. Use the three-stage hypothesis testing approach to decide whether this sample is evidence of a drop in Griffey's batting average after the switch relative to a population proportion defined by his batting average during his last two years in Seattle.
   a. What is the null hypothesis?
   b. What is the choice of sampling box to be used to generate simulated samples corresponding to the null hypothesis?
   c. What is the definition of one simulation from the sampling box?
   d. Define the event of interest and explain why this is of interest in regard to testing the null hypothesis.
   e. In 1000 simulations, 33 of them had a value of as small as, or smaller than, 0.228. Explain what this means and whether or not the null hypothesis should be rejected.
9. A magazine claims to a particular advertiser that the advertiser's ad will be remembered by more than 50% of the readers who see it.

   A statistician tests the claim by surveying a random sample of 25 subscribers. The proportion in the sample who remember the ad is 0.8.

   Apply the three-stage approach of hypothesis testing, using the simulated data for 1000 simulations of a population with $p = 0.5$, in the following table to make a decision about this claim.

$\hat{p}$ observed in simulated sample	Frequency of occurrence	$\hat{p}$ observed in simulated sample	Frequency of occurrence
0.04	0	0.56	111
0.08	0	0.60	79
0.12	0	0.64	74
0.16	1	0.68	53
0.20	2	0.72	14
0.24	5	0.76	9
0.28	11	0.80	2
0.32	19	0.84	2
0.36	69	0.88	0
0.40	87	0.92	0
0.44	140	0.96	0
0.48	178	1.00	0
0.52	144		

10. A double-blind randomized controlled experiment from 1985 to 1986 studied the effectiveness of the drug *enalapril* in healing patients with congestive heart failure. From a convenience sample of 253 patients, 127 were randomly assigned to the treatment group and the remaining 126 were assigned to the control group. The treatment group received *enalapril*, whereas the control group received a placebo. Over 6 months, 33 of the patients in the treatment group died, compared to 55 of the patients in the control group. Could this difference be explained by chance variation alone, or can we conclude that *enalapril* reduces the 6-month mortality rate for patients, with congestive heart failure? Randomized controlled experiment, box model sample without replacement (non-rare disease)

11. Three junior men and three senior men were sampled from a college. Their heights are

Juniors:	68	72	75	Seniors:	70	71	73

   a. Find the sample means of the two groups.
   b. Create the populations for bootstrap testing that the population means are equal.

12. In a random sample of four people, the mean body temperature was 98.25 with SD of 0.73 degrees. Test whether the population mean temperature is 98.6 degrees versus less. Assume that the population distribution is bell-shaped.

13. Fifty measurements of the length of a bone were made. The mean of the measurements was $\bar{x} = 34.4$ centimeters with SD of $s_X = 0.25$ centimeter. Test whether the true length of the bone is $\mu = 35$ centimeters. Suppose there are no claims nor conjectures about which side of 35 μ will be if the null hypothesis is false.

14. Two ambulance services are tested for response times. A sample of 50 response times from the first service produces a mean $\bar{x} = 12.2$ minutes and SD of $s_X = 1.5$ minutes. A sample of 50 responses for the second service produces a mean of $\bar{y} = 14.0$ minutes with SD of $s_Y = 2.1$ minutes. Test the claim that the two services have the same mean response time. It is not known or claimed that either service might be better.

15. In a random sample of 100 men, the mean number of alcoholic drinks consumed per week was $\bar{x} = 5.4$ with SD of $s_X = 1.25$. In a random sample of 100 women, the mean number of alcoholic drinks consumed per week was $\bar{y} = 3.2$ with SD of $s_Y = 1.85$. Test whether the difference between the population means is 0 versus the claim that women drink less on average.

16. In a random sample of 300 people, 58% reported owning at least one cat. Test whether the true population proportion of people who own cats is 0.6. (Test this using $H_1: p < 0.6$.)

17. In a survey of members of the Biopharmaceutical Section of the American Statistical Association, individuals were asked their type of employer and what there current salary was. Out of 188 respondents who said they were working in academia, 41 said that they made more than $91,000 a year. Of the 529 people who worked for pharmaceutical companies, 180 said that they made more than $91,000 a year. Test the null hypothesis that the proportion of people making over $91,000 is the same for both those working in academia and for those working in pharmaceutical companies, versus the alternative hypothesis that the proportion is higher for those working for pharmaceutical companies.

18. A leading candy bar is claimed to have, on average, 170 calories. However, a competitor believes that the candy bar contains more calories on average than 170. The average number of calories in a sample of 40 of the candy bars in question is $\bar{x} = 175.5$, with SD of $s_X = 7.5$ calories. Perform a hypothesis test using a significance level of 0.01. What is the *P*-value?

19. The average height of women in the United States is often listed as 5 feet, 5 inches (65 inches). Suppose you believe this number is incorrect and that women are on average taller than 65 inches. In a sample of 200 women, the average height is $\bar{x} = 65.35$ inches with SD of $s_X = 3.5$ inches.

a. Perform a hypothesis test with the alternative hypothesis that the average height of U.S. women is not 65 inches.

b. Perform a hypothesis test with the alternative hypothesis that the average height of U.S. women is greater than 65 inches.

20. Find the critical region for the hypothesis test in Exercise 18 using a level of significance of 0.05. (Assume $\sigma = 7.5$.)

21. Find the critical region for the test of hypothesis in part b of Exercise 19 using a level of significance of 0.05. (Assume $\sigma = 3.5$.)

22. Find the power of the test in Exercise 21 for $\mu = 66$.

23. In the 1954 polio vaccine field trials, 400,000 children were randomly assigned to equal-sized treatment and control groups. The treatment group received the polio vaccine, and the control group received a placebo. After one year, 199 children in the study had contracted polio, with 57 of the polio cases being in the treatment group. Conduct a hypothesis test to determine if the vaccine was effective in reducing the risk of getting polio.

24. According to the manufacturer, a certain gasoline additive will "significantly" increase gas mileage. Eight cars of the same model were randomly selected and driven by the same professional driver, with and without the additive. The results are given in the table below. Assuming that the population is normally distributed, test whether there was a significant improvement in gas mileage.

Car	Mileage (miles per gallon)	
	without additive	with additive
1	16	17
2	15	15
3	13	16
4	15	14
5	17	16
6	13	16
7	16	17
8	10	12

25. A coin suspected of being biased towards heads is tossed 100 times, yielding 65 heads. Test whether this is a fair coin, that is, whether the population proportion $p = 1/2$.

26. In clinical studies of an allergy medication, 70 of the 781 subjects experienced drowsiness. Evaluate the manufacturer's claim that only 8% of the allergy medication users experience drowsiness, as opposed to a higher percentage.

27. Proponents of alternative medicine might claim that the proportion of people using alternative medicine increased during the 1990's. In a 1990 survey of a sample of 1539 people, 520 had used alternative medicine. In a recent survey of 2055 people, 865 had used alternative medicine. Conduct a hypothesis test to determine whether the claim of increased use is supported.

28. In 1998 an article in the *Journal of Consulting and Clinical Psychology* (vol. 66, pp. 715–730) reported on an evaluation of a new parenting program. It was hoped that the parenting program would result in significant improvements in parenting behavior for high-risk mothers. A total of 235 high risk mothers at nine Head Start Centers were randomly assigned to two groups, 158 to the treatment group and 77 to the control group. The results of the study revealed that 69% of the treatment group showed the desired improvement in parenting behavior as compared to 52% of the control group.

Conduct a hypothesis test to determine whether the difference in percentages can be accounted for by chance variation or whether the difference is so large that we should reject the null hypothesis of equal effects for the two groups.

29. The enrollment of children in preschool programs is claimed to result in long-term societal benefits. To investigate this claim, the Hi/Scope Educational Research Association studied two samples of "at-risk" Michigan students: a sample of 62 from the population of those who attended preschool as children and a sample of 61 from the population of those who did not. For each sample they recorded the proportion who needed social services as adults. The sample statistic was 0.803 for those that did not attend preschool and 0.613 for those who did. Conduct a hypothesis test of the claim that attending preschool reduces the probability of needing social services as an adult.

30. The National Transportation Safety Board has reported that 30% of all accidents involve fatalities. Suppose a claim has been made that the proportion of accidents involving fatalities has recently dropped. If a random sample of 250 accidents reveals that 59 involved fatalities, use hypothesis testing to evaluate if this sample indicates a drop in the population proportion.

PROFESSIONAL PROFILE

Dr. Harold Brooks
Head of the Mesoscale Applications Group
National Severe Storms Laboratory
Norman, Oklahoma

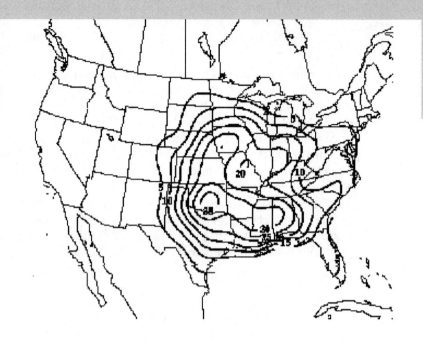

Number of days per century in which a strong or violent tornado touched down within 25 miles of any location in the United States.

Dr. Harold Brooks

Just how likely is it that a severe tornado will strike your community this year?

That's a question Dr. Harold Brooks can answer for you. He studies severe storms around the world, using statistical methods to analyze the history of severe storms and the patterns of their occurrence.

The task is not easy. Over the years, there have been variations in reporting methods and standards. Although a few written reports of tornadoes date back as far as the seventeenth century, the largest body of consistently reported information goes back only to the late 1800s. Since then, there have been many advances in weather reporting and many changes in the methods of data collection and the standards for reporting the data. There are even differences in the ways different nations report storms. For example, France reports only about three tornadoes a year. However, if you apply American standards to France, there are actually about 20 tornadoes a year in that country. Such variations complicate any statistical analysis of global weather data.

The number of severe storms reported each year is going up. What does this mean? Is our weather is getting worse? Or are the numbers getting bigger just because we're doing a more thorough job of reporting storms? "We're still sorting out what's real and what's not," explains Harold Brooks. He is searching for variables that will allow the National Severe Storms Laboratory to look at possible climate change questions and to identify the real weather threats. He also hopes that his work will help community leaders and ordinary people remember that they are vulnerable. "Many years may pass between catastrophic storms," observes Dr. Brooks. "I understand why people become complacent," he says, "but people should be alert and be prepared for that one severe storm in forty years. Their lives could depend on it."

10

Chi-Square Testing

It's Monday again, and I feel sad.
A commonly expressed sentiment.

Objectives

After studying this chapter, you will understand the following:

- The chi-square statistic for assessing the hypothesis of fairness of a many-sided die
- How to use the three-stage five-step simulation approach to test a hypothesis using the chi-square statistic
- How to use a chi-square density in place of five-step simulation
- Chi-square testing in the unequal-expected-frequencies (unfair-die) case
- Why chi-square testing is such a widely used hypothesis-testing procedure

---

10.1 IS THE DIE FAIR?   460

10.2 HOW BIG A DIFFERENCE MAKES A DIFFERENCE?   463

10.3 THE CHI-SQUARE STATISTIC   467

10.4 REAL-LIFE CHI-SQUARE EXAMPLES   473

10.5 THE CHI-SQUARE DENSITY   481

10.6 THE CHI-SQUARE DISTRIBUTION AND ITS USE FOR CHI-SQUARE TESTING   490

10.7 UNEQUAL EXPECTED FREQUENCIES   497

CHAPTER REVIEW EXERCISES   505

Key Problem

The quote at the beginning of this chapter suggests that one's emotional well-being is worse on Mondays than on any other day. Social scientists have studied whether or not certain behaviors related to positive or negative emotional well-being occur more often on particular days of the week. Our Key Problem is an example of such research. Zung and Green (1974) conducted a study to see if the number of suicides in North Carolina tend to occur more frequently on particular days of the week than on others rather than evenly over all days. Out of the population of all suicides in North Carolina, they selected as their sample those that occurred from 1965 to 1971. The following table shows the number of the 3672 suicides in the sample occurring on each day of the week:

Day	Sun	Mon	Tues	Wed	Thurs	Fri	Sat
Number	550	569	527	553	491	496	486

Should we decide from these data that, in the population of all suicides in North Carolina, they occur more frequently on particular days of the week (noting that the most occurred on Monday!) as compared to the null hypothesis that they occur evenly over all the days?

When a hypothesis or claim is made about the proportion of units in either a real or conceptual population that fall into one of two categories, the methods of Chapter 9 can be applied to test such claims. In Chapter 9, the two categories were simply whether or not a unit (coin, die, person, etc.) had a particular quality or property of interest. Thus, the box models had only two types of balls in them (though the proportion of one type was often different from the proportion of the other type). The problems in this chapter, like the one posed in the Key Problem, involve a hypothesis or claim about the proportions of units (people having committed suicide, in this case) in a population that can be classified into more than two categories (in this case, the seven days of the week). To test hypotheses such as these, we will still use the three-stage five-step simulation approach to hypothesis testing introduced in Chapter 9, but our box models will need to contain more than two types of balls—they will need as many types of balls as there are categories into which the units in the population are grouped. In this chapter we will learn how to test such hypotheses using box models that contain three or more different types of balls. Finally we will learn a theoretical approach, using a curve called the chi-square density (or chi-square curve), that bypasses the need for five-step simulation entirely.

10.1 IS THE DIE FAIR?

People use dice to make decisions at play, from the family playing a board game around the kitchen table to the Friday night casino visit. We expect the dice we use to be fair—that is, to turn up each face in equal proportion in the long run. The more money rides on the throw, and the more throws there are, as in casino craps, the more important it is that the dice be absolutely fair. As mentioned in Section 4.1, casinos go to great effort to ensure fairness in their dice. So it is natural to question whether a particular die is fair.

Suppose someone questions the fairness of a particular six-sided die and asks us to conduct a statistical study. We roll the die 60 times and obtain the outcomes in Table 10.1. We see from the table, for example, that 1 was obtained seven times, 2 was obtained nine times, and so on. We want to test the claim that the die may not be fair.

Table 10.1 Sixty Rolls of a Six-Sided Die

Outcome of die	f	$\hat{P}$(outcome)
1	7	$7/60 = 0.12$
2	9	$9/60 = 0.15$
3	14	$14/60 = 0.23$
4	7	$7/60 = 0.12$
5	18	$18/60 = 0.30$
6	5	$5/60 = 0.08$
Total	60	1.00

This is a statistical decision-making situation similar to the ones we studied in Chapter 9. We will use the same three-stage simulation approach to hypothesis testing that was used in the previous chapter.

In Stage I, we formulate the null hypothesis, which in this case is that the data were produced by a fair die.

In Stage II, using the five-step simulation, we estimate the probability of the appropriate event of interest happening under the null hypothesis, thus obtaining a **P-value** that indicates the strength of the evidence that the data provide against the null hypothesis. We will discuss this in more detail below, but essentially, as in Chapter 9, we will be estimating the probability, under the assumption the null hypothesis is true, of obtaining a result as extreme as, or more extreme than, what actually occurred in the real data.

In Stage III, we decide whether or not to reject the null hypothesis based on how small the P-value is that we estimate in the second stage.

Although much of the focus of this chapter is on variations of the fairness-of-a-die question, we will discover that many fascinating and important real-life problems can be solved with the chi-square testing methods described in this chapter for testing die fairness. The methods of this chapter can be used to test whether a given data set is well modeled by a particular many-sided, loaded die; that is, we require neither six categories nor equal probabilities for the categories of the null hypothesis model being tested. Because so many real-world hypothesis testing settings can be modeled as hypothesis that postulate a particular many-sided die, chi-square testing is a method of enormously wide applicability.

Consider two illustrations. The question in the Key Problem, of whether suicides are equally likely on each day of the week, is equivalent to the question of whether a seven-sided die is fair. The fact that a seven-sided fair die cannot be physically built is irrelevant—imagining such a die is all that is needed! We will address this problem in Example 10.3. As a second example, in a famous genetics experiment the proportion of four kinds of peas in 555 pea plants were observed in an attempt to determine whether they occurred with the expected probabilities 1/16, 3/16, 3/16, and 9/16. This hypothesized set of proportions corresponds to a four-sided, loaded die. Statisticians refer to the probability law for the number of times each face occurs in n tosses of a many-sided loaded die as the **multinomial probability law** or **multinomial distribution**. We will analyze Mendel's famous data set in Exercise 2 of Section 10.7. The fundamental issue here is one of **goodness of fit**: how well the null hypothesis model fits the observed data.

Let us return now to the problem of the die whose fairness is the null hypothesis. Think about what sort of outcome (category) frequencies we would expect from a fair die. A natural first question to ask is, "What does 'fair' mean?" Clearly, a fair die is one for which, when it is rolled, *each side has an equal probability, 1/6, of appearing*. This definition is based on theoretical probabilities, as discussed in Chapter 4. From an experimental probability distribution point of view, a fair die is one for which, when it is rolled many times, the experimental probabilities of the faces appearing become closer and closer to 1/6.

In terms of Table 10.1, what does this null-hypothesis model of fairness mean? It means that since we have rolled the die 60 times, the proportion of times we obtained each side (that is, the experimental probability) should each be fairly close to 1/6 (which rounds to 0.167).

In a **frequency table** such as Table 10.1 and often in this chapter, *f* denotes *frequency*, that is, the number of times the outcome occurs. The *f* value on the "1" row in Table 10.1 is 7; that is, we got a 1 seven times in 60 rolls of the die. So

$$\hat{P}(1) = 7/60 = 0.12 \text{ (rounded)}$$

as compared with $P(1) = 1/6 = 0.167$ (under the null-hypothesis fair die model). Similarly, $\hat{P}(2) = 9/60 = 0.15$ can be compared with $P(2) = 1/6 = 0.167$, and so on.

The density histogram (discussed in Section 1.5) in Figure 10.1 presents the observed proportions (experimental probabilities) obtained in the 60 rolls of the die in Table 10.1. Note that the areas of the rectangles do add to 1, as required for a density histogram. The horizontal line is the expected proportion (theoretical probability) of outcomes for a fair die. An off-the-cuff response to the observed frequencies of Table 10.1 might be something like, "We got a one only 7 times, but we got a five 18 times. It seems reasonable to get each outcome about 10 times. Eighteen times seems rather large. So it looks as if the die is *not* fair." But of course, even if the die is fair, these observed frequencies are random and as such are not necessarily equal to their expected values of 10. Just as we do not necessarily get 30 heads and 30 tails when we toss a fair coin 60 times, we also expect some deviation from 10 ones, 10 twos, and so on, when we roll a six-sided die 60 times, *even when the die is absolutely fair*. So the question whether the die of Table 10.1 is fair is still unanswered.

We can start to get a statistical answer to this question by computing a statistical index that measures how much the observed frequencies depart from the expected frequencies for a fair die, that is, we compare the actual outcomes of the 60 rolls of the die with what we would expect from a fair die. In search of such a statistic, let us expand Table 10.1 as shown in Table 10.2 and discuss Table 10.2 column by column:

Outcome of die: Which side fell face up?

Observed frequency (O): This is the observed frequency of each of the given outcomes of the die in 60 rolls—that is, the number of times that each side of the die appeared. The symbol *O* here replaces the symbol *f* used in Table 10.1 to emphasize the fact that this frequency was actually observed.

Expected frequency (E): This is the theoretical or expected frequency (that is, the theoretical expected value of Chapter 5) of the given outcome of the die in 60 rolls—that is, the number of times each side is expected to appear, if the die is fair. For this table, we find *E* using the rule

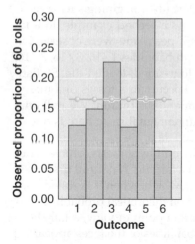

Figure 10.1 Comparison of experimental and theoretical probabilities for a fair die.

that we learned in Chapter 5 for the expected frequency in n independent trials of an event that has probability p:

$$E = n \times p = 60 \times 1/6 = 10.$$

$O - E$: This is the difference, or deviation, between the observed and the expected frequency of the outcome. For example, in row 1 we see that $O - E = 7 - 10 = -3$. This tells us that when the die was tossed 60 times, the 1 appeared three fewer times than expected for a fair die. On the other hand, in row 5 we see that $O - E = 18 - 10 = 8$. That is, 5 was observed eight more times than expected.

$|O - E|$: This is the absolute value of $O - E$. That is, we ignore whether $O - E$ is positive or negative and measure the size of the deviation. For example, for row 6, $O - E = -5$ and $|O - E| = 5$.

Why add a column for absolute values? Look at the total of the $O - E$ column in Table 10.2. The positive and negative values in column $O - E$ cancel each other out, adding up to zero. This sum is thus not very informative about deviations of the data from the fair die model! However, the sum of column $|O - E|$, which we see from the table is equal to 24, is informative. It tells us that in the 60 rolls of the die, which may not be fair, the outcomes differed from what was expected for a fair die by a total difference of 24.

We have found a simple, intuitive statistical index, or statistic, which we will call D and define as follows:

$$D = \sum |O - E|.$$

D is the total of the absolute deviations of the observed from the estimated frequencies and is the sum of the $|O - E|$ entries in Table 10.2. It is intuitively clear that the larger D is, the stronger the evidence is that the die is unfair.

We now need a way to decide whether $D = 24$ is strong *enough* evidence to overcome the null hypothesis assertion that the die is fair and thus lead us to reject the null hypothesis of a fair die. We leave it to the next section to learn how to determine what a "small" or a "large" value of D is with respect to providing strong or weak evidence that the null hypothesis should be rejected.

10.2 HOW BIG A DIFFERENCE MAKES A DIFFERENCE?

At the end of Section 10.1, we were left with the question of how to decide whether a value of the statistic D is large enough to cause us to reject the null hypothesis of a fair die. That is, we need a way of deciding whether our observed value of the D statistic ($D = 24$) is so large that only seldom

Table 10.2 Observed and Expected Outcomes for 60 Rolls of a Six-Sided Die, Assuming the Null Hypothesis that It Is Fair

Outcome of die	Obtained frequency (O)	Expected frequency (E)	$O - E$	$|O - E|$
1	7	10	−3	3
2	9	10	−1	1
3	14	10	4	4
4	7	10	−3	3
5	18	10	8	8
6	5	10	−5	5
Total	60	60	0	24

Table 10.3 Results of 60 Rolls of a Six-Sided Die Known to be Fair

Outcome of die	Obtained frequency (O)	Expected frequency (E)	$O - E$	$|O - E|$
1	4	10	−6	6
2	6	10	−4	4
3	11	10	1	1
4	10	10	0	0
5	15	10	5	5
6	14	10	4	4
Total	60	60	0	20

would a value that large or larger happen by chance for 60 rolls of a die if the die were fair, as assumed by the null hypothesis. In other words, just as in Chapter 9, we need a P-value, in this case $P(D \geq 24 | H_0)$. Our notation "$|H_0$" here is to remind you that the probability is to be computed using the null hypothesis model, in our case that of a fair die.

Suppose we have a six-sided die that we *know* is fair (for example, the manufacturer has carefully constructed it to be a fair die). We roll it 60 times and obtain the results shown in Table 10.3. Let us look at $D = \sum |O - E|$, which is 20. Of the 60 rolls of this fair die, the number of observed outcomes differed from what was expected by a total of 20.

Rather than tediously rolling a real fair die 60 times over and over, we instead use five-step simulation. We repeat these 60 rolls many times for a simulated die that is fair, and each time we calculate a value of D. We will have as a result a table that shows the frequency distribution for the possible values of D obtainable from many simulations of 60 rolls of a fair die. From this frequency table we will be able to determine the value of $\hat{P}$(simulated $D \geq 24 | H_0$). If a D value of 24 or more is unusually large (that is, $\hat{P}$(simulated $D \geq 24 | H_0$) is very small for a Step 1 fair die box model), we would conclude that the real data of Table 10.2, which yielded $D = 24$, indicate a die that is not fair. In summary, we want to find $\hat{P}$(simulated $D \geq 24 | H_0$) for 60 rolls of a fair six-sided die, thus estimating the theoretical probability $P(D \geq 24 | H_0)$, the P-value. This is our five-step simulation-based estimate of the P-value. Typically, as learned in Chapter 9, the P-value is ≤ 0.05, the null hypothesis is rejected and we conclude the die is unfair.

What we have just described is the three-stage hypothesis testing approach of Chapter 9. Let us review in detail how this approach would be applied in this problem:

Stage I. Formulate the null hypothesis. The null hypothesis is that population proportions corresponding to the six sides of the die are equal. That is, the null hypothesis is the following distribution

x	1	2	3	4	5	6
$p(x)$	1/6	1/6	1/6	1/6	1/6	1/6

Stage II. Estimate the P-value using five-step simulation. Use the five-step method to calculate the experimental probability of the event of interest based on constructing a box model determined by the null hypothesis.

Step 1. Choose the null hypothesis Box Model: The box will have six balls in it: one labeled 1, one labeled 2, one labeled 3, one labeled 4, one labeled 5, and one labeled 6. Note that this box is exactly equivalent to a six-sided fair die and to the theoretical probability distribution given in Stage I.

Step 2. Define One Simulation: One simulation is defined as the drawing, with replacement, of 60 balls, the same sample size as occurred in the real data sample. Note that this simulation is exactly equivalent to 60 rolls of a six-sided fair die.

Table 10.4 Frequency Table for D Statistic, 100 Simulations of 60 Fair Die Tosses

D	2	4	6	8	10	12	14	16	18	20	22	24
f	1	3	6	11	15	16	20	9	9	6	3	1

Step 3. Define the Event of Interest: As in Chapter 9, the event of interest will be an occurrence in a simulated data sample of a value of the **test statistic** (in this case D) equal to or more extreme than the value of D that was computed from the real data sample. More extreme values are values (either larger or smaller, depending on the problem) that would provide even stronger evidence toward rejecting the null hypothesis than our observed value $D = 24$ does. In the case of D, "more extreme" always means "larger." If we let "EOI" stand for the "event of interest," then

$$\text{EOI} = (\text{simulated } D \geq 24).$$

Step 4. Repetition of Simulations: We conduct a large number (N) of simulations and keep track of the number of times (N_{EOI}) the event of interest (EOI) occurred. Table 10.4 is a table of the frequencies of each value of D for $N = 100$ for illustrative purposes. If we want reasonable accuracy, we should do 400 or even 1000 or 10,000 such simulations (easy to do quickly with the textbook software).

From Table 10.4 we can obtain N_A by adding up the frequencies (f) of all the columns in which D is 24 or greater. Note that this table contains only even values of D. The D statistic cannot be odd, it turns out.

Step 5. Calculate the Experimental Probability of the Event of Interest: We estimate $P(\text{EOI}|H_0)$—that is, we estimate $P(\text{simulated } D \geq 24 | H_0)$—using the following equation,

$$\hat{P}(\text{EOI}|H_0) = \hat{P}(\text{simulated } D \geq 24 | H_0) = \frac{N_{\text{EOI}}}{N}.$$

According to Table 10.4, there was only one occurrence of a value of D as large as 24 in our 100 simulations. Therefore,

$$\hat{P}(\text{simulated } D \geq 24 | H_0) = \frac{1}{100} = 0.01.$$

Stage III. Decision. We found that, on the basis of 100 simulations, $\hat{P}(\text{simulated } D \geq 24 | H_0) = 0.01$. That is, our estimated P-value is 0.01. Recall from Chapter 9 that the convention (called the "gold standard") is to consider an EOI whose P-value is 0.05 (**specified level of significance**) or less as unusual enough to reject the null hypothesis. Since a P-value of 0.01 is less than our specified level of significance of 0.05, we conclude that it is unlikely that we would obtain $D \geq 24$ if the die were fair. So we conclude that the die of Table 10.2 is not fair. That is, we reject the null hypothesis model of a fair die. Of course, if we want to be able to trust our Step 5 experimental P-value estimate and, hence, our decision, we should do more simulations—at least 400.

Section 10.2 Exercises

Answer these questions using (a) the D statistic and Table 10.4 of simulated D's or (b) the accompanying textbook software.

466 CHAPTER 10 CHI-SQUARE TESTING

1. It is suggested that a particular six-sided die may be loaded, that is, an unfair die. Here are the observed results of rolling the die 60 times. Use the three-stage approach to decide whether the die is fair by answering the following questions.

Outcome of roll	Frequency
1	4
2	17
3	14
4	6
5	18
6	1
Total	60

   a. What is the null hypothesis?
   b. Calculate the D statistic for the real data results.
   c. To determine whether the D statistic for the real data is unusual under the null hypothesis, we must have some idea of the kind of values of D to expect when the null hypothesis is true. We do this by generating *simulated data samples* from a box model based on the null hypothesis. Describe the box model and define one simulation from it.
   d. What is the event of interest used to obtain an estimated P-value?
   e. By using Table 10.4, which simulates D under the null hypothesis of a fair die, or better using a new table created by using the five-step method with many more simulations, estimate the probability of occurrence of the event of interest.
   f. Based on the experimental probability of the event of interest (that is, the estimated P-value), decide whether the die is fair.

2. Suppose we roll a six-sided die that we assume is fair. How many times would we expect each side to occur if we roll the die
   a. 150 times?
   b. 300 times?
   c. 600 times?

3. Needing random digits for a class project, Nancy and Pete go through one page of a telephone book and write down the last digit of each of 50 telephone numbers. Here are their data:

Digit:	0	1	2	3	4	5	6	7	8	9
f	1	6	3	2	5	8	2	10	8	5

   Their instructor is skeptical that their method produces valid random numbers. Apply the three-stage hypothesis testing approach to decide whether this concern is supported by the data by answering the following questions.
   a. Prepare a table of obtained and expected outcomes like Table 10.3, and find the value of D, where the expected outcomes are based on the null hypothesis of equal frequencies for each digit.
   b. The following table gives the results of 30 simulations of 50 random digits (using a box model containing 10 balls marked 0, 1, ..., 9) and their associated D's.

D	4	6	8	10	12	14	16	18	20	22	24
f	1	0	2	1	2	7	7	2	4	3	1

   Define the event of interest and estimate its probability (the P-value).
   c. Decide whether or not the null hypothesis should be rejected. Explain how you arrived at your decision.

4. A breakfast cereal company features a special offer by including one of four differently colored ballpoint pens in a box. In a shopping trip that resulted in 20 boxes of cereal, the following numbers of pens were obtained. What is the value of D? Do you think that the company is distributing the pens in equal numbers of colors, or are some colors more likely to be obtained than others? How would you carry out a simulation to test this hypothesis?

Color	f
Blue	8
Yellow	4
Red	3
White	5
Total	20

5. Explain why large values of D suggest that a die may not be fair.

6. Write down the null hypothesis expected frequencies (the E's) for 60 rolls of a fair die, as given in Table 10.2. Suppose that in reality the six-sided die being tossed is unfair with theoretical probabilities for faces 1, 2, ..., 6 given by $p(1) = 1/20$, $p(2) = 17/60$, $p(3) = p(4) = p(5) = p(6) = 1/6$. Find E for each face if we roll this unfair die 60 times. Now suppose that the observed O's for this unfair die come out exactly equal to their E's for faces 1 and 2 and that $|O - E| = 2$ for faces 3, 4, 5, and 6. That is, the observed frequencies for the unfair die are quite close to their expected values.
   a. What is D for the fair die null hypothesis? *Hint:* Each $E = 10$ for a fair die.
   b. Is the value of D large enough to yield strong evidence the die is unfair, according to the simulations of D given in Table 10.4 (for a six-sided fair die, recall). *Note:* This exercise gives some idea of the power of the D statistic to detect die unfairness. Note that doing 400 simulations would be better.

7. Consider the experiment of 60 rolls of a fair six-sided die. For each trial the D statistic is computed. Use Table 10.4 to estimate the indicated experimental probabilities.
   a. $\hat{P}(\text{simulated } D \geq 16 | H_0)$
   b. $\hat{P}(\text{simulated } D \geq 17 | H_0)$
   c. $\hat{P}(\text{simulated } D \geq 20 | H_0)$
   d. $\hat{P}(\text{simulated } D \geq 8 | H_0)$

10.3 THE CHI-SQUARE STATISTIC

In Section 10.2, we formulated the null hypothesis, calculated the D statistic for a real data sample, and then calculated values of D for a set of simulated data samples generated from the null hypothesis box model using the five-step method. By comparing the value of D from the real data set to the values of D from the simulated data sets, we saw that the D statistic provided us with a convenient yardstick for telling how strong the evidence in the real data is against the fair die null hypothesis.

However, the usual way statisticians measure the strength of evidence against a fair-die null hypothesis is to compute a different statistical index than D called the **chi-square statistic,** written χ^2. The symbol χ is the Greek letter chi (pronounced like "sky," but without the "s"). The χ^2 statistic is very commonly used in practice. It is very important in the life sciences, for example. The kind of testing occurring in this chapter, namely of how well a null hypothesis probability model fits observed data, is sometimes called **goodness-of-fit** testing.

In this section we focus on describing the calculation of this new χ^2 statistic and on its usefulness for goodness-of-fit testing via the simulation approach. First, let's compare how the D statistic and the χ^2 statistic are computed when they are both applied to the real data from the fair die problem. Table 10.5, called a χ^2 frequency table, shows both the D statistic (the total of the

Table 10.5 χ^2 Frequency Table for Die Fairness Data

Outcome of die	Obtained frequency (O)	Expected frequency (E)	O − E	|O − E|	(O − E)²	(O − E)²/E
1	7	10	−3	3	9	0.9
2	9	10	−1	1	1	0.1
3	14	10	4	4	16	1.6
4	7	10	−3	3	9	0.9
5	18	10	8	8	64	6.4
6	5	10	−5	5	25	2.5
Total	60	60	0	24	124	12.4

|O − E| column) and the χ^2 statistic (the total of the (O − E)²/E column) for the same data, 60 rolls of a six-sided die. (In practice, a χ^2 table does not show the value of the D statistic.) The value of χ^2 is given by

$$\chi^2 = \sum \frac{(O-E)^2}{E}.$$

Therefore,

$$\chi^2 = \frac{3^2}{10} + \frac{1^2}{10} + \frac{4^2}{10} + \frac{3^2}{10} + \frac{8^2}{10} + \frac{5^2}{10}$$

$$= \frac{9}{10} + \frac{1}{10} + \frac{16}{10} + \frac{9}{10} + \frac{64}{10} + \frac{25}{10} = 12.4$$

There are similarities, but also important differences, between how we calculate D and how we calculate χ^2. First, instead of taking the absolute value |O − E| (that is, the magnitude of O − E), we square (O − E), obtaining the square of the magnitude of O − E. Like the absolute value, the square of (O − E) is always nonnegative, regardless of the sign of O − E. Second, we then divide each value of (O − E)² by its corresponding expected value E before adding across categories.

Now let's focus on the application of the χ^2 statistic using the three-stage simulation approach to hypothesis testing that was used above with D. The process is identical except D is replaced by χ^2. Indeed the goal here is the same as with the D statistic: We are trying to determine how likely the null hypothesis model is to produce a χ^2 value as large as or larger than the observed χ^2 value, 12.4, that occurred in the real data sample. In other words, we want to estimate $P(\chi^2 \geq 12.4 | H_0)$. To answer this question, in Stage II we again use the five-step method:

Step 1. Choose a Box Model: We use the same box model as we did when estimating $P(D \geq 24 | H_0)$: six balls marked 1, 2, 3, 4, 5, 6.

Step 2. Define One Simulation: As when estimating $P(D \geq 24 | H_0)$, we draw one ball with replacement 60 times and note the frequency of occurrence of each number, O. We then compute χ^2.

Step 3. Define the Event of Interest: The event of interest here is the occurrence of a simulated $\chi^2 \geq 12.4$, the value of χ^2 for the real data sample. "More extreme" means "more likely to lead to rejection of the null hypothesis if that χ^2 value had occurred in the real sample." In this case, "more extreme" clearly means larger, just as with D in Section 10.1.

Step 4. Repeat Simulations: We generate many simulated data samples (400 here), simulating the rolling of a fair die 60 times. We then prepare a frequency table of these null hypothesis χ^2 values from the simulations, shown in Table 10.6, in much the same way as we produced a frequency

Table 10.6 χ^2 Values Obtained from 400 Simulations of Rolling a Fair Six-Sided Die 60 Times

Range	0–0.00	1–1.99	2–2.99	3–3.99	4–4.99	5–5.99	6–6.99	7–7.99	8–8.99
f	5	9	39	64	80	60	46	42	27
Range	9–9.99	10–10.99	11–11.99	12–12.39	12.40–12.99	13–13.99	14–14.99	15–15.99	
f	8	13	4	2	3	5	2	1	

table of null hypothesis D values from the simulations used to estimate $P(D \geq 24 | H_0)$ in the previous section. Note that the 12–12.99 range was split because the observed $\chi^2 = 12.40$.

Step 5. Determine the Experimental Probability of the Event of Interest: From Table 10.6 we see that a χ^2 of 12.4 or larger was obtained in 11 of the 400 simulations. Therefore,

$$\hat{P}(\text{simulated } \chi^2 \geq 12.4 | H_0) = \frac{11}{400} = 0.0275.$$

We thus estimate via five-step simulations that the theoretical $P(\chi^2 \geq 12.4 | H_0)$ (i.e., under the null hypothesis) is small—about 0.0275.

The last stage in using the χ^2 statistic for hypothesis testing is similar to Stage III in previous examples using D. Again, we use our gold standard of 0.05 (level of significance) as the probability of an unusual event. Since our null hypothesis P-value, 0.0275, is less than 0.05, we conclude that the evidence against H_0 is strong and, hence, that the die whose data are shown in Table 10.5 is not fair.

Thus, we have now shown that we can use the same three-stage hypothesis testing approach to evaluate whether our observed χ^2 is unusual under the null hypothesis, just as we did earlier for D. Thus, using either χ^2 or D leads to the same decision.

Section 10.3 Exercises

1. Consider the simulation of 60 rolls of a fair six-sided die. For each simulation the χ^2 statistic is computed. Table 10.6 gives the results of 400 such simulations. Use Table 10.6 to estimate the indicated probabilities:
   a. $P(\chi^2 \geq 6 | H_0)$
   b. $P(\chi^2 \geq 8 | H_0)$
   c. $P(\chi^2 \geq 10 | H_0)$
   d. $P(\chi^2 \geq 14 | H_0)$
   e. $P(\chi^2 \geq 3 | H_0)$

2. A six-sided die was rolled 60 times with the following results:

Outcome of die	Obtained frequency (O)	Expected frequency (E)
1	8	10
2	7	10
3	13	10
4	11	10
5	15	10
6	6	10
Total	60	60

a. Calculate χ^2 for these data. Note that the expected values are based on a null hypothesis that assumes a fair die.

b. Using Table 10.6, estimate the theoretical probability of the event of interest (i.e., the probability of getting a value of χ^2 under the null hypothesis that is as large as, or larger than, the value you obtained with the real data in part **a**). Do you have strong statistical evidence that the die used in part **a** is unfair? Why or why not?

3. In a colored pen offer similar to those in Example 5.9 and Exercise 4 of Section 10.2, this time involving six pen colors, the following numbers of pens were obtained from 90 cereal boxes:

Color of pen	Number obtained
Sky blue	17
Passionate pink	31
Deep purple	7
Burnt orange	10
Boring brown	9
Anemic ash	16
Total	90

a. The null hypothesis is that in the population of all the cereal boxes the manufacturer distributed equal numbers of pens for each color. Explain why the null hypothesis is equivalent to a six-sided fair die.

b. Use the following frequency table of simulated χ^2s under the null hypothesis to help you decide, based on the χ^2 statistic computed from the real data, whether there is strong evidence that the manufacturer distributed more pens of some colors than of others, or not. Note that for good accuracy, one would want 400 or more simulations, rather than the 50 of this exercise.

Stem-and-Leaf Plot for 50 Simulated χ^2s for a Fair Die Rolled 90 Times

Stem	Leaf	f
0	5,7,8	3
1	1,3,5,6,7	5
2	3,5,7,7,8	5
3	1,1,1.2,2,3,5,6,6,6,7,7,9	13
4	1,3,4,4,5,5,6,9	8
5	1,2,3,6,6,6,7	7
6	0,3,3,9	4
7	2,3,5,9	4
8		0
9		0
10	3	1
		50

Key: "1 1" stands for 1.1.

4. During a busy day in a large city, 90 traffic tickets were issued at six locations that each have the same amount of traffic:

Section 10.3 The Chi-Square Statistic   **471**

Location	Number of tickets given
A	12
B	7
C	21
D	15
E	11
F	24
Total	90

Use the frequency table of simulated χ^2s of Exercise 3 and the sample data to help you decide whether, in the population of all tickets, they are being given out with equal probability at each location.

5. The following table gives the outcomes of rolling an actual four-sided die 40 times.

Outcome	Expected number	Obtained number
1	10	12
2	10	7
3	10	14
4	10	7
Total	40	40

a. Calculate χ^2 for these data, under the null hypothesis that the die is fair.

b. A set of 400 null hypothesis simulations was conducted. Each simulation represented 40 rolls of a fair four-sided die. The χ^2 statistic was calculated for each simulation, resulting in 400 χ^2s. The results are shown in the following table. Use this table to decide whether there is evidence that the actually rolled die in part **a** is unfair.

Frequency Table for 400 Simulated χ^2s for Fair Four-Sided Die Rolled 40 Times

Interval	Frequency
0–0.99	55
1–1.99	97
2–2.99	80
3–3.99	72
4–4.99	49
5–5.99	23
6–6.99	11
7–7.99	9
8–8.99	1
9–9.99	2
10–10.99	1
Total	400

6. The following table tells how many of a sample of 40 persons prefer each of four kinds of orange juice. Find D and χ^2 for these data, assuming the null hypothesis is that each juice is equally preferred. Explain why this null hypothesis is equivalent to a four-sided fair die. Then use the five-step simulated χ^2 frequency table (simulated with the null hypothesis true) of Exercise 5 to help

you decide if the sample χ^2 statistic provides convincing evidence that the population of all people prefer some kinds of orange juice over others.

Kind of orange juice	Number of persons preferring
Fresh	13
Freeze-dried	11
Frozen	8
Canned	8
Total	40

7. Suppose that a certain unfair six-sided die has probabilities given by $p(1) = 1/20$, $p(2) = 17/60$, with the remaining probabilities $= 1/6$. Find E for each face if we roll this unfair die 60 times. Now suppose that the observed O's for this unfair die come out exactly equal to their E's for faces 1 and 2 and that $(O - E)^2 = 4$ for faces 3, 4, 5, and 6.
   a. What is χ^2 for the fair die null hypothesis? *Hint:* Each $E = 10$ for a fair die.
   b. Is the value of χ^2 large enough to yield strong evidence the die is unfair, using the fair die χ^2 simulations of Table 10.6?
8. Which produces a larger value of χ^2 for a fair die tossed 60 times: every O differing from E by 2 (in some direction), or two O's differing by 4 from E and the rest each equaling E?
9. Explain the only way that an observed $\chi^2 = 0$ is possible. Is this likely?
10. Could you use a χ^2 test to test whether boys and girls are born equally likely in a certain country? (*Hint:* Using the die analogy, what fair "die" might work here?)
11. Consider the following two sets of data for colors of randomly selected objects in two different situations:

Color	Set A Expected	Observed
Red	100	105
Blue	100	97
Green	100	98

Color	Set B Expected	Observed
Red	5	10
Blue	5	2
Green	5	3

a. By just looking at the data, do you think the evidence is strong that the observed values follow the same distribution as the expected values in the case of data set A? What about data set B?
b. Calculate

$$(\text{Observed red} - \text{expected red})^2 + (\text{Observed blue} - \text{expected blue})^2$$
$$+ (\text{Observed green} - \text{expected green})^2$$

for both data sets A and B.
c. Using your answers from parts **a** and **b**, explain why dividing by the expected number is important in the calculation of the χ^2 statistic.

Table 10.7 χ^2 for Bike Accidents by Day of Week

Day of the week	Expected (E)	Obtained (O)	$(O - E)^2$	$(O - E)^2/E$
Sunday	130	118	$(-12)^2 = 144$	1.11
Monday	130	119	$(-11)^2 = 121$	0.93
Tuesday	130	127	$(-3)^2 = 9$	0.07
Wednesday	130	137	$(7)^2 = 49$	0.38
Thursday	130	130	$(0)^2 = 0$	0
Friday	130	146	$(16)^2 = 256$	1.97
Saturday	130	135	$(5)^2 = 25$	0.19
Total	910	910		4.65

10.4 REAL-LIFE CHI-SQUARE EXAMPLES

Now we provide several applications of χ^2 goodness-of-fit testing to real-life examples, including the Key Problem from the beginning of this chapter. In each case, the real-world setting is analogous to a many-sided fair die null hypothesis.

Example 10.1 Bicycle Accidents

In the course of research on possible ways to reduce bicycle accidents in a large city, it is hypothesized that accidents may occur more frequently on particular days of the week than on others. If this hypothesis were to prove true, better-focused expenditures of time and money could result in reduced accident rates without an increased financial burden on the city. The data for 910 randomly selected bicycle accidents are given in the "Obtained (O)" column of Table 10.7. (Adapted from Insurance Institute for Highway Safety.)

Solution Table 10.7 shows that $\chi^2 = 4.65$.

Stage I. Formulation of the null hypothesis. We want to determine whether the sample provides statistical evidence that bicycle accidents are more likely to happen on particular days of the week than on others. We thus begin by assuming the null hypothesis model that accidents are equally likely to happen on each day. So our box model represents a fair seven-sided die.

Stage II. Estimation of the probability of occurrence of the event of interest. We use the five step method.

> **Step 1. Choice of a Model:** To model the null hypothesis of bicycle accidents equally likely on each day of the week, use a box containing seven balls, each labeled to indicate a different day of the week. With such a model, we *expect* one-seventh of the accidents to happen on each day of the week. That is, the expected number of accidents is $(1/7) \times 910 = 130$.
>
> Now we must determine whether values of χ^2 as large as or larger than the one obtained from the data in Table 10.7 ($\chi^2 = 4.65$) occur rarely or often under the null hypothesis model.
>
> **Step 2. Definition of a Simulation:** A simulation consists of drawing one ball with replacement from the box once for each accident reported in the real data of Table 10.7, for a sample size of 910.
>
> **Step 3. Definition of the Event of Interest:** For each simulated data sample we calculate χ^2. The event of interest is (simulated $\chi^2 \geq 4.65$). The following table gives the results for the first simulation, yielding $\chi^2 = 5.71$. This simulated χ^2 exceeds the observed 4.65, which is an occurrence of the event of interest (check the calculation of this simulated χ^2 by completing the $(O - E)^2/E$ column and totaling the values).

Number of Accidents

Day of the week	Expected (E)	Obtained (O)	$(O - E)^2$	$(O - E)^2/E$
Sunday	130	141	121	
Monday	130	139	81	
Tuesday	130	123	49	
Wednesday	130	119	121	
Thursday	130	145	225	
Friday	130	122	64	
Saturday	130	121	81	
Total	910	910		

Table 10.8 Frequency Table of χ^2 Values for Example 8.1: 1000 Simulations of 910 Rolls of a Fair Seven-Sided Die

χ^2	0–0.99	1–1.99	2–2.99	3–3.99	4–4.99	5–5.99	6–6.99	7–7.99
f	14	68	111	124	133	110	99	95

χ^2	8–8.99	9–9.99	10–10.99	11–11.99	12–12.99	13–13.99	14–14.99	15 or more
f	58	60	37	21	24	15	11	20

Step 4. Repetition of Simulations: We perform a total of 1000 simulations and present the frequency table of χ^2 statistic values in Table 10.8.

Step 5. Estimating the Probability of the Event of Interest: From the original bicycle data in Table 10.7 we obtained $\chi^2 = 4.65$. From Table 10.8 we see that (simulated $\chi^2 \geq 4.65$), the event of interest, was obtained somewhere between 550 (number of simulated $\chi^2 \geq 5$) and 683 (number of simulated $\chi^2 \geq 4$) times in 1000 simulations. So

$$\hat{P}(\text{simulated } \chi^2 \geq 4.65) \geq 0.550.$$

Stage III. Decision: We found in Stage II that (simulated $\chi^2 \geq 4.65$) occurred rather often (over one-half of the time) by chance for the null hypothesis model of equal frequencies for each day of the week. We thus have no evidence to rule out the null hypothesis model. So we fail to reject the null hypothesis: We conclude that there is no statistical evidence that some days of the week have more bicycle accidents than others. The observed differences in occurrence of accidents on the different days of the week are typical of what we might expect because of ordinary chance variation.

Note: You should remind yourself of what decision we would have made if the estimated probability (P-value) we found in step 5 had been small (≤ 0.05, the usual level of significance). In that case, we would have concluded that the obtained statistic happens only rarely by chance for a fair die. So the model we chose in Step 1 would not have been acceptable, because the results we obtained from our original data would not likely have been produced by the model.

Example 10.2 Animal Bites

In 1967, the Chicago Board of Health investigated the hypothesis that during the autumn season, the reported number of animal bites would vary from week to week. A sample of 3 consecutive weeks from 1 year was chosen to be analyzed, and the results were as follows:

Week ending	Number of animal bites
October 26	268
November 2	189
November 9	199
Total	656

Do you think that the weekly variation in the number of bites reported is due to chance alone?

Solution:

Stage I. Formulation of the null hypothesis. We assume a null hypothesis in which each of the 3 weeks is equally likely to have animal bites reported. With such a model, we would expect one-third of the bites to be reported each week (a fair three-sided die, which, of course, cannot be constructed physically). So the expected number of bites is

$$\frac{1}{3} \times 656 \approx 218.7.$$

Now we can compute the χ^2 statistic for these data (see Table 10.9):

$$\chi^2 = \left(\frac{(49.3)^2}{218.7} + \frac{(-29.7)^2}{218.7} + \frac{(-19.7)^2}{218.7} = 16.9 \right).$$

We need to find out whether the chance of getting $\chi^2 \geq 16.9$ is large or small under the assumption that bites are equally possible in each of the 3 weeks.

Stage II. Estimation of the probability of occurrence of the event of interest. We use the five-step method.

> **Step 1. Choice of a Model:** We can use a box model with balls 1, 2, 3 in it, corresponding to the three weeks, drawing with replacement.
>
> **Step 2. Definition of One Simulation:** A simulation would consist of doing 656 draws with replacement from the box, one ball for each bite reported in the real data sample. We would record how many times each of the three balls was drawn. That is, we would find out how many of the 656 simulated bites were reported each week, by chance, in one simulation.
>
> **Step 3. Definition of the Event of Interest:** For each simulated data sample, we calculate the χ^2 statistic. The event of interest is (simulated $\chi^2 \geq 16.9$).
>
> **Step 4. Repetition of simulations:** We do 400 simulations. The results are presented in the frequency table of Table 10.10.

Table 10.9 χ^2 for Data on Animal Bites by Week

Week ending	Expected number (E)*	Observed number (O)	$(O - E)^2$
October 26	218.7	268	$(268–218.7)^2 = (49.3)^2$
November 2	218.7	189	$(189–218.7)^2 = (-29.7)^2$
November 9	218.7	199	$(199–218.7)^2 = (-19.7)^2$

*Expected values do not exactly add to 656 because of rounding.

Table 10.10 Frequency Table for χ^2 for Fair Three-Sided Die

Interval	f
0–0.99	73
1–.99	135
2–2.99	96
3–3.99	55
4–7.99	33
8–16.89	8
≥ 16.9	0
Total	400

Step 5. Estimate the probability of occurrence of the event of interest: We seek to estimate $P(\chi^2 \geq 16.9)$. From Table 10.10, $\hat{P}(\chi^2 \geq 16.9) = 0$.

Stage III. Decision. We found that $\hat{P}$(simulated $\chi^2 \geq 16.9$) = 0. This P-value is much smaller than 0.05, so we conclude that the differences in number of bites reported to the Chicago Board of Health are not due to chance alone. It may be the task of someone to try to find out why the numbers reported are different from week to week. Can you think of any reasons?

Example 10.3 Key Problem: Suicide Rates

As mentioned at the beginning of this chapter, Zung and Green (1974) conducted a study to see whether the number of suicides in North Carolina occur more frequently on particular days of the week. They used recorded suicide data from 1965 to 1971 as their data sample. There is a common belief that people are particularly vulnerable to depressive emotions on Mondays, and Zung and Green wanted to see whether their suicide data would reveal whether any particular day or days of the week, Monday or otherwise, are more prone to suicide than the others. If suicide data reveal significantly more suicides on particular days, then this information might lead to proactive measures to prevent suicides, and it might also shed light on the validity of the theory that Mondays are generally more depressing than other days. The data for 3672 sampled suicides classified by day of the week are as follows:

Day	Sunday	Monday	Tuesday	Wednesday	Thursday	Friday	Saturday
Number	550	569	527	533	491	496	486

We do indeed observe the most suicides on Monday. But, as we are beginning to understand well, this fact *by itself* does not provide strong evidence.

Solution The suicide data are entered in the "Observed number (O)" column of the chi-square table, Table 10.11. If the number of suicides per day does not depend on the day of the week, we expect one-seventh of the suicides to happen on each day. That is, for each day the expected number of suicides (in the "Expected number (E)" column of Table 10.11) is $1/7 \times 3672 = 524.6$, rounded to the nearest tenth. These expected values can now be used to calculate the χ^2 value for the real data:

$$\chi^2 = \frac{645.16}{524.6} + \frac{1971.36}{524.6} + \frac{5.76}{524.6} + \frac{806.56}{524.6} + \frac{1128.96}{524.6} + \frac{817.96}{524.6} + \frac{1489.96}{524.6} = 13.09.$$

Table 10.11 Number of Suicides

Day of the Week	Expected number (E)	Obtained number (O)	$(O - E)^2$
Sunday	524.6	550	$(25.4)^2 = 645.16$
Monday	524.6	569	$(44.4)^2 = 1971.36$
Tuesday	524.6	527	$(2.4)^2 = 5.76$
Wednesday	524.6	553	$(28.4)^2 = 806.56$
Thursday	524.6	491	$(-33.6)^2 = 1128.96$
Friday	524.6	496	$(-28.6)^2 = 817.96$
Saturday	524.6	486	$(-38.6)^2 = 1489.96$
Total	3672.2	3672	

Note: The fact that the E column and the O column do not have exactly the same sum is only due to roundoff error.

Stage I. Formulation of the null hypothesis. We want to determine whether the sample data provide statistical evidence that in the population of all suicides in North Carolina, there is a greater likelihood for them to occur particular days of the week than on other days. We begin by assuming the null hypothesis model that suicides are equally likely to happen on each day. So our box model represents a fair seven-sided die.

1. Next we must determine whether $\chi^2 \geq 13.09$ occurs rarely or often under the null hypothesis model.

Stage II. Estimation of the probability of occurrence of the event of interest. We use the five-step method.

> **Step 1. Choice of a Model:** To model the null hypothesis of suicides being equally likely on each day of the week, use a box containing seven balls, each ball labeled to indicate a different day of the week (a fair seven-sided die), as in Example 10.1.
>
> **Step 2. Definition of a Simulation:** A simulation consists of drawing a ball from the box with replacement 3672 times, once for each of the reported suicides in the real data.
>
> **Step 3. Definition of the Event of Interest:** For each simulated data sample we calculate χ^2. The event of interest is the occurrence of a simulated χ^2 value ≥ 13.09 (the observed χ^2 using the actual North Carolina suicide data).
>
> **Step 4. Repetition of Simulations:** We perform a total of 1000 simulations. For the purposes of this problem, even though the number of trials (3672) is different, it turns out (see Section 10.5) we are allowed to use the same frequency table that was produced for Example 10.1 (with 910 trials), which was presented as Table 10.8.
>
> **Step 5. Estimating the Probability of the Event of Interest:** From our original data we obtained $\chi^2 = 13.09$. The value 13.09 does not appear in the Table 10.8, but values of simulated $\chi^2 \geq 13$ or more occurred on 46 simulations, so we know that values of simulated $\chi^2 \geq 13.09$ could not have occurred any more often than that. Hence,
>
> $$\hat{P}(\text{simulated } \chi^2 \geq 13.09 | H_0) \leq 0.046.$$

Stage III. Decision: We found in Stage II that, under the null hypothesis model of equal frequencies for each day of the week, a simulated χ^2 of 13.09 or more occurred rather infrequently by chance (less than the determining 0.05 rate on average over 1000 simulations). We

478   CHAPTER 10 CHI-SQUARE TESTING

thus have strong statistical evidence to reject the null hypothesis model of equal frequencies among the days of the week in the population of all suicides in North Carolina. Moreover, the one day that is most different from the null hypothesis model is Monday! Thus, because the number of Monday suicides was greater than expected in the sample, the results lend support to the suggestion that the number of suicides on Monday is significantly greater than on any other day of the week. We conclude that the differences between the different days of the week are statistically significant. However, we should also note that the differences between Monday and other days having higher than expected numbers of suicides are not very big.

In the real data sample, 15.5% of the suicides occurred on Mondays, 15.1% on Wednesdays, and 15.0% on Sundays. Furthermore, even though we have strong statistical evidence of unequal suicide rates, even the largest rate (15.5% for Mondays) differed by only 1.2% from the 14.3% rate expected under the null hypothesis. Such a small difference does not seem worthy of further study nor to warrant the targeting of mental health interventions toward Mondays.

Section 10.4 Exercises

1. The following table is a frequency table of χ^2 values based on 1000 simulations of a 200 times tossed 10-sided die, under the null hypothesis of a 10-sided fair die. Use this table to estimate these probabilities:

χ^2 values	Frequency	χ^2 values	Frequency
0–0.99	3	10–10.99	82
1–1.99	6	11–11.99	60
2–2.99	28	12–12.99	46
3–3.99	54	13–13.99	44
4–4.99	82	14–14.99	40
5–5.99	88	15–15.99	23
6–6.99	101	16–16.99	22
7–7.99	104	17–17.99	14
8–8.99	89	18–18.99	10
9–9.99	81	19 or more	23
		Total	1000

   a. $P(\chi^2 \geq 6.0 | H_0)$
   b. $P(\chi^2 \geq 4.8 | H_0)$
   c. $P(\chi^2 \geq 11.2 | H_0)$
   d. $P(\chi^2 \geq 14 | H_0)$
   e. $P(\chi^2 \geq 15.4 | H_0)$
   f. $P(\chi^2 \geq 16 | H_0)$

   Use the three-stage hypothesis-testing approach to solve Exercises 2 through 4.

2. A suspicious-looking octahedral die is tossed 168 times, yielding $\chi^2 = 17.4$. Each of 50 students in a statistics class at Johnson Community College rolled a fair octahedral die 168 times and got the χ^2 statistic results shown in the following table. Use the table to decide whether the suspect die was loaded.

Stem	Leaf	f
0		0
1	5,7,8	3
2	7	1
3	0,4,6,8	4
4	1,2,3,3,3,4,6,6,8	9
5	0,0,1,2,3,8	6
6	1,4,5,5,6,6,9	7
7	0,2,6,9	4
8		0
9	0,0,4,6,9,9	6
10	0,9	2
11	2,6	2
12	7	1
13	5,5,7	3
14		0
15		0
16	2	1
17		0
18	0	1
Total		50

Key: "4 2" stands for 4.2.

3. The statistics class at Buffalo Grove College interviewed 100 students at random and asked them to give their favorite number between 0 and 9. Here are the results. Use the simulations of χ^2s for a fair 10-sided die from the table in Exercise 1 and the survey data below to decide whether among the population of all students, certain numbers are preferred over others.

Outcome	f
0	5
1	3
2	11
3	10
4	19
5	9
6	11
7	15
8	13
9	4
Total	100

4. A clothing store stocks a large number of men's ties that are identical except that they are in four different colors. The records of 40 are shown in the following table. The 40 sales can be considered as a sample of the population of all such sales. Using these sample data and the table of part **b** in Exercise 5 of Section 10.3, decide whether, in the population of all sales, some colors are preferred over others.

Color of tie	Number sold
Amber	7
Blue	9
Orange	14
Maroon	10
Total	40

5. The last digit of each of 100 telephone numbers was taken (in order) from one page of a telephone book. The frequencies were as follows:

Digit	f
0	3
1	8
2	15
3	14
4	10
5	7
6	8
7	9
8	11
9	15
Total	100

This sample can be considered to be a sample of the population of all the last digits of telephone numbers in the entire telephone book. Calculate χ^2 for these sample data. Do you think the telephone book is a good source of random data? Why or why not? (*Hint:* Use the table in Exercise 1.)

6. Using the following stem-and-leaf table based on 50 simulations of 168 rolls each of a fair eight-sided die, find the following experimental probabilities:
   a. $\hat{P}(\text{simulated } \chi^2 \leq 1.7)$
   b. $\hat{P}(\text{simulated } \chi^2 \leq 3.5)$
   c. $\hat{P}(\text{simulated } \chi^2 \leq 13.5)$
   d. $\hat{P}(4.6 \leq \text{simulated } \chi^2 \leq 12.7)$
   e. $\hat{P}(\text{simulated } \chi^2 \geq 14)$

7. Sometimes dishonest researchers and lazy students fake data. When fair die data are faked, there is often a tendency to make the O's too close to the E's to be likely to happen for an actual fair die, which is *very* random in its behavior. John is supposed to toss a six-sided fair die 90 times for homework but decides to fake the data instead. The following are his "observed" data:

Digit	1	2	3	4	5	6
Frequency	14	16	15	13	16	16

   a. Calculate a χ^2 statistic for the null hypothesis of a fair die.
   b. Denote the answer to part **a** by C. Very small χ^2 values will provide strong evidence of this kind of data faking. However, the event of interest is (simulated $\chi^2 \leq$ observed χ^2). Now we'll look at $\hat{P}(\text{simulated } \chi^2 \leq C)$ as our P-value and use the usual 0.05 criterion for rejection of the null hypothesis. What do you conclude, using the table of Exercise 3, Section 10.3, for simulation trials of a fair die tossed 90 times?

8. Refer to Exercise 7. John thinks his data look a little too good to be true and changes the 14 ones to 12 ones and the 16 twos to 18 twos. Now redo Exercise 7.

9. Explain why χ^2 testing is important in real-world problems of interest when all we have learned to test is whether a many-sided die is fair.

10.5 THE CHI-SQUARE DENSITY

We know from Chapters 4 and 5 that the way to obtain good accuracy in our simulation-based probability estimates is to have a large number of simulations in the five-step method. Suppose a chi-square hypothesis-testing problem concerning 90 rolls of a possibly loaded six-sided die produces a statistic $\chi^2 = 9.0$. As we understand from the previous sections, we want to evaluate $P(\chi^2 \geq 9.0 | H_0)$ in order to make a decision about whether a many-sided die is fair. Table 10.12 is a frequency table for the χ^2 values obtained in 100 simulations, each simulation consisting of 90 rolls of a fair six-sided die. Clearly, because of the addition probability rule for mutually exclusive events of Chapter 4,

$$\begin{aligned}\hat{P}(\text{simulated } \chi^2 \geq 9.0) &= \hat{P}(\text{simulated } \chi^2 \text{ in } (9.00, 9.99)) \\ &+ \hat{P}(\text{simulated } \chi^2 \text{ in } (10.00, 10.99)) \\ &+ \hat{P}(\text{simulated } \chi^2 \text{ in } (11.00, 11.99)) \\ &+ \hat{P}(\text{simulated } \chi^2 \text{ in } (12.00, 12.99)) \\ &+ \hat{P}(\text{simulated } \chi^2 \text{ in } (13.00, 13.99)) \\ &+ \hat{P}(\text{simulated } \chi^2 \text{ in } (14.00, 14.99)) \\ &+ \hat{P}(\text{simulated } \chi^2 \text{ in } (15.00, 15.99)) \\ &+ \hat{P}(\text{simulated } \chi^2 \text{ in } (16.00, 16.99)) \\ &= 0.04 + 0.00 + 0.02 + 0.00 + 0.02 + 0.00 + 0.00 + 0.01 \\ &= 0.09.\end{aligned}$$

Table 10.12 Relative Frequency χ^2 Table (100 Simulations of 90 Throws of Fair Die)

χ^2	Frequency f	Relative frequency
0–0.99	0	0
1–1.99	9	0.09
2–2.99	19	0.19
3–3.99	14	0.14
4–4.99	13	0.13
5–5.99	11	0.11
6–6.99	7	0.07
7–7.99	10	0.10
8–8.99	8	0.08
9–9.99	4	0.04
10–10.99	0	0.00
11–11.99	2	0.02
12–12.99	0	0.00
13–13.99	2	0.02
14–14.99	0	0.00
15–15.99	0	0.00
16–16.99	1	0.01

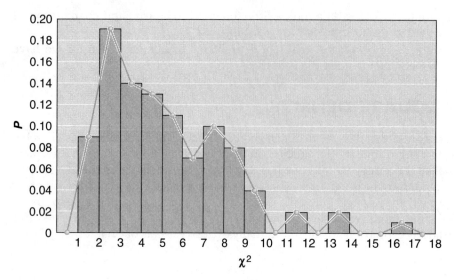

Figure 10.2 Graph of the χ^2 values reported in Table 10.12 (100 trials).

As we learned in Section 1.5, whenever we have a table of frequencies (number of occurrences) for numerical outcomes, we can graph the density histogram of the numerical outcomes. If we then join the midpoints of the tops of the rectangles of the density histogram by straight lines, we get a **density polygon,** which gives the general shape of the distribution of these observed proportions.

Consider $\hat{P}$(simulated $\chi^2 \geq 9.0 | H_0$) as evaluated in Table 10.12. The value of $\hat{P}$(simulated $\chi^2 \geq 9.0 | H_0$) is precisely the area to the right of 9.0 in the density histogram in Figure 10.2, which was constructed from Table 10.12. Indeed, $\hat{P}$(simulated $\chi^2 \geq a | H_0$) for any choice of a can be evaluated as the area to the right of a in the density histogram. Clearly,

$$\hat{P}(\text{simulated } \chi^2 \geq 9.0 | H_0) = 0.04 + 0.02 + 0.02 + 0.01 = 0.09,$$

which is the area of the shaded region to the right of 9.0 in the histogram.

Figure 10.2 also gives the density polygon of the χ^2 values from the 100 trials given in Table 10.12. In each interval of χ^2 values in Table 10.12, we have used the midpoint (for example, 1.5 for the 1.00–1.99 interval) as the horizontal coordinate. This density polygon of χ^2 values gives us a visual idea of which χ^2 values are large (unusual) for 90 rolls of a fair six-sided die.

As we conduct more and more simulations and the experimental probabilities get closer, the density polygon of simulated χ^2 values under H_0 thus becomes closer and closer to a smoothly curved density. To show this very important fact, we have added further columns to Table 10.12, showing the results of 1000 simulations and the results of 10,000 simulations, to produce Table 10.13. In Figure 10.3 the 100-simulation, 1000-simulation, and 10,000-simulation density polygons and the smooth curve that results as the simulations increase to an arbitrary large number are all displayed. Note, in particular, how the 10,000-simulation density polygon is very close to the limiting smooth curve as contrasted with how close the 100-simulation density polygon is to the curve. This smooth curve is called a **chi-square density.**

We have often mentioned that the larger the number of simulations in Step 4, the better that an experimental probability distribution approximates the theoretical distribution. Now we see that as the number of simulations gets very large, the density polygon of the χ^2 statistic approaches the χ^2 density curve. This gives us another way to evaluate $P(\chi^2 \geq 9.0 | H_0)$ approximately, one that avoids simulation and the five-step method *entirely*: We can use the area under the χ^2 density ≥ 9. That is, we can add interval areas ≥ 9 in the last column of Table 10.13 to approximate $P(\chi^2 \geq 9.0)$ very accurately, given in the last column of Table 10.13, instead of a simulated density histogram,

Section 10.5 The Chi-Square Density

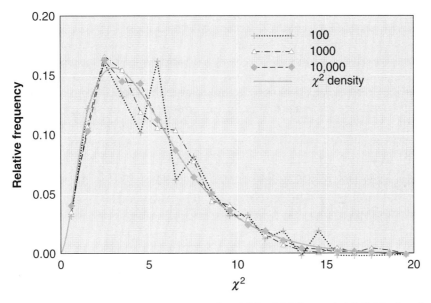

Figure 10.3 Density polygons for 100, 1000, and 10,000 simulations, and χ^2 density.

Table 10.13 χ^2 Values Obtained from Large Numbers of Simulations of 90 Rolls of a Fair Six-Sided Die

Interval	Relative frequency of χ^2 in interval			Interval area probability using chi-square density
	100 simulations	1000 simulations	10,000 simulations	
0–0.99	0.03	0.033	0.0390	0.0366
1–1.99	0.12	0.104	0.1013	0.1120
2–2.99	0.16	0.163	0.1605	0.1476
3–3.99	0.13	0.152	0.1431	0.1491
4–4.99	0.10	0.117	0.1417	0.1323
5–5.99	0.16	0.104	0.1101	0.1087
6–6.99	0.06	0.102	0.0855	0.0848
7–7.99	0.08	0.076	0.0641	0.0639
8–8.99	0.05	0.043	0.0498	0.0468
9–9.99	0.03	0.038	0.0341	0.0335
10–10.99	0.03	0.026	0.0239	0.0237
11–11.99	0.01	0.014	0.0170	0.0165
12–12.99	0.02	0.010	0.0117	0.0113
13–13.99	0.00	0.004	0.0052	0.0077
14–14.99	0.02	0.003	0.0037	0.0052
15–15.99	0.00	0.004	0.0023	0.0035
16–16.99	0.00	0.001	0.0028	0.0023
17–17.99	0.00	0.005	0.0013	0.0015
18–18.99	0.00	0.001	0.0010	0.0010
19–19.99	0.00	0.00	0.0008	0.0007
20–20.99	0.00	0.00	0.0005	0.0004
21–21.99	0.00	0.00	0.0003	0.0003
22–22.99	0.00	0.00	0.0000	0.0002
23–23.99	0.00	0.00	0.0001	0.0001
24–24.99	0.00	0.00	0.0000	0.0001

to estimate the probability of events of interest with the χ^2 statistic. In this way we can obtain an approximate P-value needed to carry out a hypothesis test of goodness of fit, based on the χ^2 statistic. This is useful, because it enables us to avoid doing hundreds or thousands of simulations with the five-step method that we used in Section 10.4! The second stage of our hypothesis-testing approach thus becomes much simpler. To estimate the probability of the event of interest, and thus obtain the P-value approximately, instead of conducting a large number of simulations, we will be able to look up the approximate area in a table of areas under the χ^2 density, provided at the back of our statistics book. For this problem, we seek the area to the right of 9.0 under the χ^2 density, obtained from the last column of Table 10.13:

$$P(\chi^2 \geq 9 | H_0) \approx 0.0335 + 0.0237 + 0.0165 + 0.0113 + 0.0077 + 0.0052 + 0.0035 + 0.0023$$
$$+ 0.0023 + 0.0015 + 0.0010 + 0.0007 + 0.0004 + 0.0003 + 0.0002$$
$$+ 0.0001 + 0.0001$$
$$= 0.1080.$$

Note two notational changes that indicate our change of approach. $\hat{P}$(simulated $\chi^2 \geq 9 | H_0$) that denotes simulations has been replaced by $P(\chi^2 \geq 9.0 | H_0)$ that denotes the exact theoretical probability that we approximate by an area under a χ^2 density. Thus, the simulation estimate $\hat{P}$(simulated $\chi^2 \geq 9.0 | H_0) = 0.09$ is replaced by the χ^2 density approximation $P(\chi^2 \geq 9.0 | H_0) \approx 0.1080$. It is worth noting that when the number of simulations is 10,000 and hence very large, $\hat{P}$(simulated $\chi^2 \geq 9.0 | H_0) = 0.1047$. This value from the next to last column of Table 10.13 is very close to 0.1080, the χ^2 density approximation. Interestingly, some commercial software (such as Cytel's StatXact) can compute probabilities such as $P(\chi^2 \geq 9.0 | H_0)$ for a fair-six-sided die exactly rather than approximately as an area under a χ^2 density. However, such computation is very difficult—and, as we will discover presently, not necessary—when the approximation obtained from the chi-square density is known to be a good one.

Similar to the normal density we have already studied, the χ^2 density is an example of a **continuous distribution**. Figure 10.4 shows this $P(\chi^2 \geq 9.0 | H_0)$ approximation as the area under the χ^2 density to the right of 9.0. You might recognize this as an integration problem in calculus. You will see in the next section that we can consult Table C (in Appendix C), which approximates $P(\chi^2 \geq a | H_0)$ for a large number of choices of a.

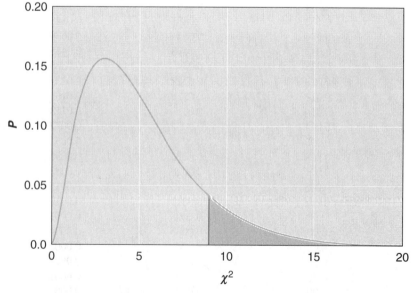

Figure 10.4 Theoretical χ^2 estimate of $P(\chi^2 \geq 9 | H_0)$ for 90 rolls of a six-sided die experiment.

Table 10.14 Experimental χ^2 Frequencies for 100 Simulations of a Fair Six-Sided Die Rolled Many Times

χ^2	30 rolls per trial	60 rolls per trial	90 rolls per trial
0–0.99	2	5	3
1–1.99	13	8	16
2–2.99	20	15	13
3–3.99	12	17	18
4–4.99	11	9	9
5–5.99	8	10	11
6–6.99	17	14	6
7–7.99	3	5	6
8–8.99	5	7	7
9–9.99	3	5	2
10–10.99	3	2	2
11–11.99	2	1	3
12–12.99	0	1	1
13–13.99	1	0	1
14–14.99	0	0	0
15–15.99	0	0	1
16–16.99	0	1	1
Mean χ^2	4.70	4.94	5.01

How good is the approximation of $P(\chi^2 \geq a \mid H_0)$ provided by the χ^2 density? In particular does the number of throws of the die affect the accuracy? Table 10.14 shows the experimental chi-square distribution and the mean χ^2 for 100 simulations of rolling a six-sided die, for 30, 60, and 90 rolls per simulation. Because it is hard to compare the experimental χ^2 distributions from this table, Figure 10.5 shows the three estimated smooth curves that result. These curves have been estimated using least squares nonlinear regression, which is a sophisticated variation of the least squares linear regression method of Chapter 3. The key point to note is that all three curves are *close to each other* and *close* to the *same theoretical χ^2 density*, which is also drawn in Figure 10.5 (the solid line curve). Hence, it is intuitively clear that whether we have 30, 60, or 90

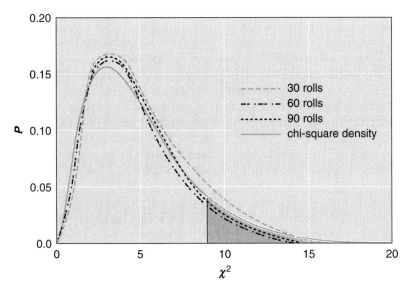

Figure 10.5 Estimated versus theoretical χ^2 densities for a six-sided die experiment.

rolls, we can use areas under the *same* χ^2 density to accurately approximate probabilities of the form $P(\chi^2 \geq a | H_0)$. That is, P-values based on the χ^2 density are not affected much by the total number of rolls of the die involved in calculating the χ^2 statistic. Whether we want to compute, assuming H_0, $P(\chi^2 \geq 9.0;\ 30\ \text{rolls})$, $P(\chi^2 \geq 9.0;\ 60\ \text{rolls})$, or $P(\chi^2 \geq 9.0;\ 90\ \text{rolls})$, the area to the right of 9.0 under the χ^2 density of Figure 10.4 provides a good approximation. That is, we simply use the theoretical χ^2 curve (density) discussed above, regardless of the number of rolls in a trial (as long as there are not too few of them, as we will discover below). This is extremely useful. We will discover that as long as the number of rolls is reasonably large, the χ^2 density approximation is accurate.

The D statistic lacks the crucial property of its distribution not being affected by the number of rolls of the die, as well as other important properties of the χ^2 statistic. Thus no analogous theoretical density approach is available for the D statistic. Of course when using a five-step simulation approach to goodness-of-fit testing, it is *just as easy* to work with D as χ^2. In fact, the textbook software provides both D and χ^2 simulation approaches. Because of its nice approach, statisticians usually prefer to use χ^2 rather than D. We use the fact that areas under a χ^2 density (tabulated in a later section for a few special cases and, more completely, in Table C at the back of this book) can be accurately used to decide whether an observed χ^2 value is strong evidence that a given null hypothesis model is unlikely to have produced the data.

There is one restriction when use of the χ^2 density to evaluate probabilities of the form $P(\chi^2 \geq c | H_0)$ is known to be accurate: Enough data (die throws in this example) must be available. In particular, we require that *every* expected frequency E be ≥ 5 before it is accurate to replace five-step simulation with finding areas under a χ^2 density. For example, this criterion of expected frequencies $E \geq 5$ or greater holds for Table 10.5. Thus we could use the χ^2 density to obtain (approximately) the needed P-value. This $E \geq 5$ rule will be our convention for deciding whether one is allowed to use areas under the χ^2 density instead of the five-step method.

Degrees of Freedom

We have now discovered that we can compute χ^2 probabilities such as $P(\chi^2 \geq 9 | H_0)$, which are useful in hypothesis testing, approximately by using areas under a χ^2 density, namely, that of Figure 10.4. Moreover, we have learned that we can use this curve regardless of the number of die tosses per simulation, provided they are not too few (because of our rule that the expected frequency E must be at least 5). But this was all for a six-sided die. Is the same theoretical χ^2 curve used regardless of the number of sides on the die of the null hypothesis model?

Recall that the null hypothesis model in the animal-bite example was, in effect, a three-sided fair die. Because we have seen that the experimental distributions of the actual χ^2 statistics in simulations are known to be close to the theoretical χ^2 curve (recall Figures 10.3 and 10.5) we can look at χ^2 frequency tables for different numbers of categories (sides in the special case of a die) to find out whether we get the same χ^2 density as the number of simulations gets very large, as we got in Figure 10.3, or whether we need a whole *family* of χ^2 densities depending on the number of categories (and hence a different set of tabulated areas for each different density).

Let us therefore now explore the effect of the number of sides of the fair die on the experimental probability distribution of the χ^2 values. We carry out 50 simulations of rolling a fair die. (We could have done more than 50, but as you will see, 50 is plenty for us to discover what we need to see.) Each simulation represents 60 rolls of a fair die. For the first experiment we use a four-sided die (Table 10.15), for the second experiment we use a six-sided die (Table 10.16), and for the third experiment we use a 10-sided die (Table 10.17). First, notice that as the number of sides on the die increases, so does the mean of the simulated χ^2 values, from 2.7 to 4.7 to 10.4. This is a large and

Table 10.15 Table of Experimental χ^2 Distribution for a Four-Sided Die (50 Simulations of 60 Rolls Each)

Stem	Leaf
0	1,3,4,4,7,7,9,9,9,9
1	1,2,2,2,3,3,3,5,7,7,7,7
2	0,3,3,3,4,5,8,8,8,9
3	1,2,3,3,5,5,6,6
4	1,4,4,7,8
5	7
6	5
7	
8	
9	9
10	3
Mean χ^2	2.7
SD	2.12

Key: "10 3" stands for 10.3.

Table 10.16 Table of Experimental χ^2 Distribution for a Six-Sided Die (50 Simulations of 60 Rolls Each)

Stem	Leaf
0	4,8
1	0,4,6,8
2	0,4,6,8,8
3	0,2,2,2,2,4,4,6,6,8
4	0,0,2,2,4,4,4,6,6,6,6
5	2,4,4,6,6,8,8
6	2,4,4
7	0,4,8
8	2,4,8
9	6
10	
11	2
12	
13	6
Mean χ^2	4.7
SD	2.83

Key: "11 2" stands for 11.2.

Table 10.17 Table of Experimental χ^2 Distribution for a 10-Sided Die (50 Simulations of 60 Rolls Each)

Stem	Leaf
0	
1	7
2	7,7
3	3,3
4	0,0,3,7,7,
5	0,0,3,3
6	3,3,7
7	0,0,0,0,0,3,7,7,7
8	0,0,0,7,7
9	0,7,
10	0,3,3,3,7,7
11	0
12	0,3
13	0,3
14	0,3
15	0
16	3
17	0
18	0
Mean χ^2	8.4
SD	4.04

Key: "10 3" stands for 10.3.

systematic change as the number of faces of the die is increased. Notice also that the sample SD of χ^2 increases as well. That is, as the number of sides of the die increases, the simulated χ^2 values also tend to become more spread out.

Thus, Tables 10.15 through 10.17 strongly suggest that the number of sides of the fair die (that is, the number of possible categories) has an effect on the distribution of the χ^2 statistic. Hence, to be able to use areas under a smooth curve as we did for the six-category case in Figure 10.4, we will need a whole family of curves: one for each number of sides, or categories, of the die.

More precisely, we need one curve for each number of **degrees of freedom** (sometimes abbreviated **df**). The number of degrees of freedom is the number of sides (categories in general) on the die minus 1. This affects the average size and the spread of the χ^2 statistics produced by tables of die-rolling outcomes. Hence, it is important to know how many degrees of freedom are associated with a given χ^2. Therefore, we indicate that a χ^2 has, for example for a four-category experiment or four-sided die, three degrees of freedom by writing

$$\chi_3^2.$$

When we wish to find a probability of a certain χ^2 statistic, *we must first know its number of degrees of freedom*. The number of degrees of freedom will tell us which theoretical χ^2 density to use to find areas. We study this in the next section.

Section 10.5 Exercises

1. Find the following estimated probabilities. You must first decide which table of values to use: Table 10.15, Table 10.16, or Table 10.17.
   a. $\hat{P}(\text{simulated}\,\chi_3^2 \geq 5.7 | H_0)$
   b. $\hat{P}(\text{simulated}\,\chi_3^2 \geq 9.9 | H_0)$

2. Find the following estimated probabilities. You must first decide which table of values to use: Table 10.15, Table 10.16, or Table 10.17.
   a. $\hat{P}(\text{simulated}\,\chi_5^2 \geq 6.0 | H_0)$
   b. $\hat{P}(\text{simulated}\,\chi_2^5 \geq 9.9 | H_0)$
   c. $\hat{P}(\text{simulated}\,\chi_5^2 \geq 11.2 | H_0)$
   d. $\hat{P}(\text{simulated}\,\chi_5^2 \geq 9.6 | H_0)$

3. Find the following estimated probabilities. You must first decide which table of values to use: Table 10.15, Table 10.16, or Table 10.17.
   a. $\hat{P}(\text{simulated}\,\chi_9^2 \geq 10.0 | H_0)$
   b. $\hat{P}(\text{simulated}\,\chi_9^2 \geq 15.0 | H_0)$

4. Find the following estimated probabilities. You must first decide which table of values to use: Table 10.15, Table 10.16, or Table 10.17.
   a. $\hat{P}(\text{simulated}\,\chi_5^2 \geq 2.6 | H_0)$
   b. $\hat{P}(\text{simulated}\,\chi_9^2 \geq 4.6 | H_0)$
   c. $\hat{P}(5.2 \geq \text{simulated}\,\chi_5^2 \geq 11.2 | H_0)$
   d. $\hat{P}(\text{simulated}\,\chi_9^2 \geq 2.7 | H_0)$

5. A six-sided die was rolled 90 times, and the following outcomes were obtained.

Outcome	f
1	17
2	13
3	17
4	10
5	16
6	17
Total	90

   a. Calculate the χ^2 statistic for these data using expected values (E) based on the assumption of a fair six-sided die.
   b. How many degrees of freedom are associated with this χ^2? Explain.
   c. Recall Table 10.14. What is the experimental probability of obtaining a χ^2 as large as or larger than the value you obtained in part **a**, assuming a fair die was used? That is, if the value you obtained in part **a** is called v, what is $\hat{P}(\text{simulated } \chi_r^2 \geq v)$?
   d. Do you believe the die used in this exercise was a fair die? Why or why not?

6. Sue uses her telephone book as a source of random digits. She takes the first 300 telephone numbers in the book, in order, and writes down the last digit of each number. She wants to use a χ^2 statistic on her sample to help her decide if the population of all such digits are randomly distributed across the different possible values. How many degrees of freedom will the χ^2 have? Explain.

7. A record store stocks CDs for the all-time top 20 classic rock bands. The manager of the store wants to determine whether the bands are equally popular (in terms of sales of CDs) or whether some bands are definitely favored over others. If χ^2 testing is used, how many degrees of freedom are there? Explain.

8. Refer to the density histogram for 100 trials of tossing a fair six-sided die 90 times given in Table 10.13 and estimate the following probabilities.
   a. $P(\chi_5^2 \geq 11.0 | H_0)$
   b. $P(\chi_5^2 \geq 13.0 | H_0)$
   c. $P(4.0 \leq \chi_5^2 \leq 9.99 | H_0)$
   d. $P(\chi_5^2 \text{ between 6 and 7} | H_0)$

9. Refer to Table 10.13 and review carefully what is in that table and how to interpret it.
   a. Estimate the probabilities for parts **a**, **b**, and **d** of Exercise 8 using the 10,000-simulation results that are given in Table 10.13.
   b. Estimate the probabilities for parts **a**, **b**, and **d** of Exercise 8 using the areas under the χ^2 density that are given in the last column of Table 10.13.
   c. Draw a sketch of the areas found.
   d. Compare the answers of parts **a** and **b** for the same intervals. Are they quite close or not?

10. Table 10.14 and Figure 10.5 contain a profoundly important fact about the probability law we use to compute a χ^2 statistic when the number of "rolls" is large. (Recall that the actual setting may have nothing to do with rolling a die repeatedly.) Explain this fact and indicate why this is useful.

11. a. Explain what happens to the experimental probability $\hat{P}(\text{simulated } \chi^2 \geq \text{fixed number} | H_0)$ for a fair six-sided die as the number of rolls per simulation gets larger and larger.
    b. Explain what happens to the theoretical probability $\hat{P}(\text{simulated } \chi^2 \geq \text{fixed number} | H_0)$, as approximated using a χ^2 density, as the number of sides of a die changes from 4 to 8 to 20, assuming 90 rolls of the fair die.

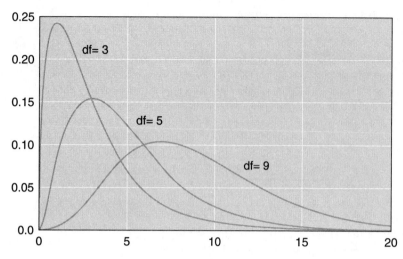

Figure 10.6 χ^2 densities for three, five, and nine degrees of freedom.

10.6 THE CHI-SQUARE DISTRIBUTION AND ITS USE FOR CHI-SQUARE TESTING

All χ^2 probabilities are computed assuming H_0 true in Section 10.6. Hence, for simplicity, we drop "$|H_0$" everywhere. It turns out that the number of degrees of freedom completely determines which χ^2 density we use to compute χ^2 probabilities. The χ^2 density of Figure 10.4 is a χ^2 density having five degrees of freedom, which would be used to compute $P(\chi_5^2 \geq 9.0)$, for example. If a χ^2 problem puts data into 10 categories (that is, the null hypothesis can be viewed as a 10-sided die), we need to use a χ^2 density having $10 - 1 = 9$ degrees of freedom. In Figure 10.6 we display χ^2 densities having 3, 5, and 9 degrees of freedom, corresponding to the χ^2 statistics for the null hypothesis of a 4-sided, a 6-sided, and a 10-sided fair die, respectively. The argument in the preceding section that helped convince us that the χ^2 density with five degrees of freedom works well for testing the hypothesis of a six-sided fair die could just as easily be made for a die of any number of sides. Further, the null hypothesis can even specify a *loaded* die. That is, we can also use χ^2 density areas when the die is hypothesized to be loaded in a specified way, provided we use the right number of degrees of freedom (see Section 10.7). This is a very useful fact!

Provided the amount of data is large, we now have a way that is just as accurate and is faster than using experimental probabilities obtained from the five-step method to find the probability of a χ^2 statistic being as large as or larger than a given value: We approximate the desired χ^2 probability by the appropriate area under a χ^2 density. How can we find such an area? The areas can be computed using advanced integral calculus and numerical methods on a computer, but we do not have to do this, because tables have been produced that provide the correct areas under the theoretical χ^2 densities (that is, the theoretical probabilities). Let us consider an example.

Example 10.4

Find $P(\chi_5^2 \geq 1.6)$.

Solution: Note that no mention of a die is made. But we are dealing with the χ^2 curve that would result from a fair six-sided die, which produces $6 - 1 = 5$ degrees of freedom.

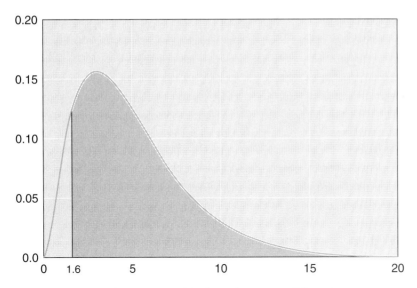

Figure 10.7 χ^2 density with five degrees of freedom.

Table 10.18 Areas under the χ^2 Density

Degrees of freedom	Probability (area) to the right of the interior table entry								
	0.99	0.95	0.90	0.50	0.20	0.10	0.05	0.01	0.001
1	0.00	0.00	0.02	0.45	1.64	2.71	3.84	6.63	10.83
2	0.02	0.10	0.21	1.39	3.22	4.61	5.99	9.21	13.82
3	0.11	0.35	0.58	2.37	4.64	6.25	7.81	11.34	16.27
4	0.30	0.71	1.06	3.36	5.99	7.78	9.49	13.28	18.47
5	0.55	1.15	1.61	4.35	7.29	9.24	11.07	15.09	20.52

We use a table of areas under the χ^2 density to the right of specified numbers. The unique theoretical smooth curve for a χ^2 with five degrees of freedom is shown in Figure 10.7. The shaded area is the theoretical probability in question.

Table 10.18 gives theoretical areas under the χ^2 density. Because different numbers of degrees of freedom produce different distributions of χ^2 values, each row in the table is associated with a problem for a particular number of degrees of freedom. For example, row 5 would be our choice for a χ^2 problem for a fair six-sided die. An expanded version of this table, with many more choices for the number of degrees of freedom, can be found in Table C of Appendix C.

Look at row 5, which gives areas for various regions under the curve in Figure 10.7. Look along row 5 until you find the number closest to 1.6. The number in the table is 1.61, which rounds to 1.6 to the nearest tenth. Now look at the top of the column in which 1.61 is located. You find 0.90, which is the area of the shaded part in Figure 10.7. Thus,

$$P(\chi_5^2 \geq 1.6) \approx 0.90.$$

Example 10.5

Find $P(\chi_5^2 \geq 8.0)$.

Solution: Figure 10.8 shows the area we are trying to find. Again we use Table 10.18 and refer to the row for five degrees of freedom. The value of 8.0 does not appear in this row. The two closest

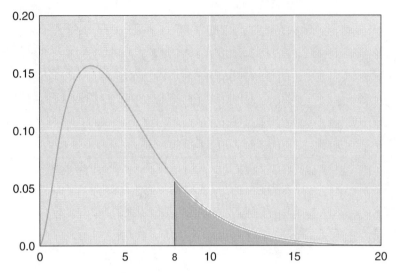

Figure 10.8 $P(\chi_5^2 \geq 8.0)$.

numbers in the row are 7.29 and 9.24. They yield $P(\chi_5^2 \geq 7.29) = 0.20$ and $P(\chi_5^2 \geq 9.24) = 0.10$. So, clearly $P(\chi_5^2 \geq 8.0)$ is between 0.20 and 0.10.

The probability result found by looking at the χ^2 table may be sufficient for our purposes. For example, if we need $P(\chi_5^2 \geq 8.0)$ to test a goodness-of-fit null hypothesis involving six categories, the fact that this probability is greater than 0.10 is enough for us to decide not to reject the null hypothesis. However, we may need a more exact value for the probability.

Here is what we can do to get a closer estimate of this probability. The value 8.0 lies at a certain proportion of the distance between 7.29 to 9.24. The corresponding value of area that we are looking for lies (about) at the same proportion of the distance from 0.20 to 0.10. Before we do any calculating, then, we can estimate that 8.0 is roughly halfway between 7.29 and 9.24. So we estimate that the area (probability) we are seeking is roughly halfway between 2.0 and 1.0, or

$$P(\chi_5^2 \geq 8.0) \approx 0.15.$$

For many purposes this very crudely estimated probability is adequate. However, for a better approximation, we must use the following rule, called **linear interpolation.** It proceeds as follows.

   a. Find the proportion of the distance from 7.29 to 9.24 at which the value 8.0 is located.

$$\text{Distance from 7.29 to 9.24} = 9.24 - 7.29 = 1.95$$
$$\text{Distance from 7.29 to 8.00} = 8.00 - 7.29 = 0.71$$

Calculate the value of the proportion.

$$\frac{\text{Distance between 7.29 to 8.00}}{\text{Distance between 7.29 to 9.24}} = \frac{8.00 - 7.29}{9.24 - 2.29} = \frac{0.71}{1.95} = 0.36$$

   b. Then find the difference. $P(\chi_5^2 \geq 7.29) - \hat{P}(\chi_5^2 \geq 9.24) - 0.20 - 0.10 = 10$
   c. From step **a** we know that the desired value lies at 0.36 of the distance from 0.20 to 0.10. (Because we went from 7.29 to 9.24, we must go from 20, which corresponds to 7.29, to 10, which corresponds to 9.24.) So the area (probability) value is a distance $0.36 \times 0.10 = 0.036$ away from 0.20 toward 0.10. Therefore,

$$\text{desired probability} = 0.20 - 0.036 = 0.164 \approx 0.16$$

or

$$P(\chi_5^2 \geq 8.0) \approx 0.16.$$

Appendix D has further discussion and gives formulas for linear interpolation.

We now return to Example 10.2 to solve it using the theoretical χ^2 area tables rather than the five-step method within Stage II of our hypothesis-testing approach. It will require use of a χ^2 density with a different number of degrees of freedom than 5.

Example 10.6 Animal Bites (Revisited)

Consider Example 10.2, which investigated the weekly variation in the numbers of animal bites during the autumn season in Chicago by studying a sample of 3 weeks of bites reported to the Chicago Board of Health in October and November 1967.

The investigators wanted to investigate the hypothesis that the number of bites reported varies significantly from week to week. The data were given in Example 10.2. We have already solved this problem using the five-step simulation method to estimate the probability of the event of interest. Now let us solve it again using the χ^2 table of areas to theoretically approximate the probability of the event of interest and see whether our answer is close to what we got previously.

Solution:

Stage I. Formulation of the null hypothesis. This is the same as before. The null hypothesis is that the population proportion of reported animal bites is the same for each of the 3 weeks ($p_1 = p_2 = p_3 = 1/3$). Using the expected values ($E = np = 656 \times 1/3 \approx 218.7$) that result from these population proportions we obtain $\chi^2 = 16.9$ for the real data sample.

Stage II. Estimate the probability of occurrence of the event of interest. Because $E = 218.7$ for each category, E being much greater than 5, the use of the χ^2 density will provide an excellent approximation. Because the null hypothesis model is a three-sided-die, the number of degrees of freedom is $3 - 1 = 2$. Recall that $\chi^2 = 16.9$ was obtained for the data. Thus, our interest is in approximating $P(\chi_2^2 \geq 16.9 | H_0)$. This is the P-value on which our hypothesis-testing decisions will be based.

We suspect that we will reject the null hypothesis, because we got $\hat{P}$(simulated $\chi^2 \geq 16.9 | H_0$) = 0 earlier when we used the five-step simulation approach to estimate $P(\chi^2 \geq 16.9 | H_0)$. Now we will use Table 10.21. We go to row 2, because this χ^2 statistic has two degrees of freedom (as we noted at the beginning of the solution). A value of 16.9 does not appear in row 2. The largest value in row 2 is 13.82: $P(\chi_2^2 \geq 13.82) = 0.001$. Because $16.9 > 13.82$, we thus know that P-value, namely $P(\chi_2^2 \geq 16.9)$, is smaller than 0.001. (No wonder we saw 0 reoccurrences of $\chi^2 \geq 16.9$ in 400 simulations.) With a more complete χ^2 table we could estimate what the probability actually is, but we do not need to know this to reach our decision.

Stage III. Decision We found that the P-value is < 0.001. That is, the probability is less than 1 out of 1000 (0.001) that $\chi^2 \geq 16.9$ would be obtained by chance under the null hypothesis. This is a very small probability (much smaller than 0.05), so we strongly reject the null hypothesis of a three-sided fair die. Thus, we conclude that the differences in the numbers of bites reported to the Chicago Board of Health from week to week are not due to chance alone.

Now let us work through a new example.

Example 10.7 The Random Number Generator on a Calculator

In 1988 one of the authors purchased a calculator with a random number generator function that is supposed to produce evenly distributed random numbers between 0.000 and 0.999, each with probability 1/1000. Based on experience over the years, the owner of the calculator has suspected that this random generator is a little off: On average, the numbers have seemed slightly smaller than they should be. To test this claim, the owner generated a sample of 1500 "random" values and counted how many of them fell into each of ten evenly spaced intervals. The results are shown in Table 10.19.

The claim can now be more precisely worded: In the infinite population of numbers that could be generated with the calculator, it is believed that not all of the categories will contain 0.1 of the numbers generated. Does the sample of 1500 data values provide strong statistical evidence that there is a problem with the calculator's random number generator?

Stage I. Formulate the null hypothesis. The null hypothesis is that there is nothing wrong with the random number generator—more precisely, that the numbers it produces are evenly distributed among the intervals. Under the null hypothesis, the expected frequency for each of the 10 intervals is equal: $E(f) = 1/10 \times 1500 = 150$ for all intervals.

Stage II. Using this as the expected frequency for each interval, $\chi^2 = 21.867$. Estimate the probability of occurrence of the event of interest. Is $\chi^2 = 21.867$ an unusually large value for a random number generator that is working properly, which is the null hypothesis model? Thus, we seek $P(\chi^2 \geq 21.867 | H_0)$. We use a χ^2 density table. Since we have 10 categories, the χ^2 statistic has 9 degrees of freedom. Since 150 is much larger than 5, we can use the χ^2 density in Table C in the back of this book to get $P(\chi^2_9 \geq 20.9) = 0.01$. Thus, the P-value satisfies $P(\chi^2_9 \geq 21.867 | H_0) < 0.01$.

Stage III. Decision. Because the observed value of the χ^2 statistic for the calculator's random number generator is highly unlikely to occur with a properly functioning random number generator, we conclude that the calculator's random number generator is not working the way that it is supposed to. Inspect Table 10.21 (in next section) and note whether the observed numbers were off from the expected numbers in such a way to support the owner's observation that the average number is slightly smaller than it should be.

Table 10.19 Distribution of 1500 "Random" Numbers from a Calculator.

Outcome	Frequency
0.000–0.099	166
0.100–0.199	156
0.200–0.299	178
0.300–0.399	153
0.400–0.499	127
0.500–0.599	127
0.600–0.699	164
0.700–0.799	166
0.800–0.899	132
0.900–0.999	131
Total	1500

Section 10.6 Exercises

For Exercises 1 through 4, find the estimated probability (as the capital $\hat{P}$ with a hat indicates) from Tables 10.13, 10.14, or 10.15 and then find the corresponding theoretical probability from Table 10.18.

1. $\hat{P}(\text{simulated } \chi_5^2 \geq 4.4 | H_0)$
2. $\hat{P}(\text{simulated } \chi_5^2 \geq 7.4 | H_0)$
3. $\hat{P}(\text{simulated } \chi_5^2 \geq 9.2 | H_0)$
4. $\hat{P}(\text{simulated } \chi_5^2 \geq 11.2 | H_0)$

For Exercises 5 and 6, find the estimated probability from Table 10.15 and then the corresponding theoretical probability from Table 10.18.

5. $\hat{P}(\text{simulated } \chi_3^2 \geq 4.6 | H_0)$
6. $\hat{P}(\text{simulated } \chi_3^2 \geq 6.5 | H_0)$

7. Find these probabilities. Use the expanded table of χ^2 probabilities in Appendix C.
   a. $P(\chi_4^2 \geq 9.5)$
   b. $P(\chi_4^2 \geq 13.3)$

8. Find these probabilities. Use the expanded table of χ^2 probabilities in Appendix C.
   a. $P(\chi_7^2 \geq 12.0)$
   b. $P(\chi_7^2 \geq 14.1)$

9. Find these probabilities. Use the expanded table of χ^2 probabilities in Appendix C.
   a. $P(\chi_{10}^2 \geq 12.0)$
   b. $P(\chi_{10}^2 \geq 12.0)$

10. Find these probabilities. Use the expanded table of χ^2 areas in Appendix C.
    a. $P(\chi_{20}^2 \geq 18.0)$
    b. $P(\chi_{20}^2 \geq 40.0)$

For Exercises 11 through 14, use the χ^2 table of probabilities in Appendix C to do the χ^2 test. In each case justify using the χ^2 density according to the ≥ 5 rule for the E's

11. The statistics class at Forest View College rolled a six-sided die and obtained the following results:

Outcome	f
1	26
2	7
3	17
4	19
5	14
6	13
Total	96

Calculate the χ^2 statistic for this experiment. How likely are you to get a χ^2 value as large as or larger than the one obtained here, by chance, if the die they used was fair? Do you believe the die used in this question was fair? Explain.

12. The statistics class at Central College interviewed 380 students (at random, we assume) and asked them for their favorite number between 0 and 9. Here are the results:

Outcome	f
0	0
1	12
2	27
3	56
4	37
5	40
6	58
7	103
8	31
9	16
Total	380

Calculate χ^2 for these sampled data. Do you think that the population of all Central students prefers some numbers over others? Why?

13. The number of eighths in the fraction on a closing stock price for a sample of 319 low-priced stocks from the American Stock Exchange is shown below. Assess whether these sampled data indicate some fractions occur more commonly than others in the population of all stock prices.

Fraction	Frequency
0	60
1/8	30
2/8 (1/4)	29
3/8	27
4/8 (1/2)	47
5/8	49
6/8 (3/4)	39
7/8	38
	319

14. In a study of pricing of merchandise a sample of 174 low-priced items were examined and the last digit of the price recorded (for example, $2.99 produces a 9). Here are the results. Would you conclude that, in the infinite population of all priced articles, some last digits are more likely than others?

Last digit	Frequency
0	11
1	10
2	10
3	34
4	7
5	25
6	4
7	18
8	6
9	49
	174

10.7 UNEQUAL EXPECTED FREQUENCIES

The χ^2 statistic can be used in a remarkable variety of ways to solve vital problems for science, medicine, and industry. Indeed, it played a key role in confirming Mendel's theory of genetics, a triumph of modern science. (See Exercise 2 of this section.)

Recall from Section 10.3 that a basic pattern for a χ^2 table is the following:

| Outcome | Expected frequency (E) | Obtained frequency (O) | $(O - E)^2$ |

This applies, for example, to the rolling of a die. If the die is fair, all outcomes are expected to occur an equal number of times, so the expected frequencies of the different outcomes are all equal. So far in this chapter, all but one of our χ^2 problems have involved equal expected frequencies. Now we will see how χ^2 testing can be used to solve problems in which the expected frequencies are unequal, thus greatly widening the scope of problems to which we can apply the χ^2 hypothesis-testing statistic. Note once again that a χ^2 test asks whether the null hypothesis model of a certain die fits the data, in this case a loaded die.

Example 10.8 The Random Number Generator on a Calculator (Revisited)

Consider Example 10.7, which investigated the random number generator for a calculator purchased by one of the authors in 1988. The analysis in Example 10.7 clearly indicated that the random numbers produced by the calculator were not evenly distributed over the 0.000 to 0.999 range as the manufacturer intended.

In the year 2000, the owner of the calculator bought the latest version of that same calculator which again provided the same type of random number generator. The owner was curious as to whether the problem with the 1988 calculator had been fixed in the 2000 calculator. So, again the owner pressed the random number generator key on the calculator 1500 times in a row and kept track of how many of them fell into each of 10 evenly spaced intervals.

The results are shown in Table 10.20.

As in Example 10.7, the owner wants to test the claim that the random numbers are not evenly generated across the intervals against the null hypothesis that denies this claim.

Stage I. Formulate the null hypothesis. As in Example 10.7 the expected frequency under the null hypothesis is $1/10 \times 1500 = 150$ for every interval.

Table 10.20 Distribution of 1500 "Random" Numbers from the Year 2000 Calculator

Outcome	Frequency
0.000–0.099	164
0.100–0.199	178
0.200–0.299	165
0.300–0.399	132
0.400–0.499	142
0.500–0.599	167
0.600–0.699	129
0.700–0.799	164
0.800–0.899	122
0.900–0.999	137

Stage II. Estimate the probability of occurrence of the event of interest. First, we calculate the χ^2 statistic for the given data based on the null hypothesis. This yields $\chi^2 = 23.147$. The event of interest is a χ^2 value greater than or equal to 23.147 under the null hypothesis. Thus, we want to estimate $P(\chi^2 \geq 23.147 | H_0)$. Using the χ^2 density Table C and referring to the row for 9 degrees of freedom, we see that $(P\chi_9^2 \geq 20.9) = 0.01$. Thus, the P-value is less than 0.01.

Stage III. Decision. As we did with the 1988 calculator, we again reject the null hypothesis for the 2000 calculator because the observed χ^2 statistic is highly unlikely to have occurred with a properly functioning random number generator.

So, we conclude that the random number generator on the new calculator also is not working the way it is supposed to. This then leads us to a very interesting question. We know that the true category proportions for the 1988 calculator are not the 0.100 values assumed by the null hypothesis. Let us suppose that we know these true proportion values. Then we could test the random number data generated by the new calculator against a null hypothesis defined by the proportions of the old calculator! This is a natural question, because if the manufacturer is using the same random number generator in the year 2000, the hypothesis should be about the same as those of 1988.

Well, it turns out that over the years, the owner has punched the random number generator key on the 1988 calculator 15,000 times! Thus, we can use the resulting proportions as our true population proportions for the 1988 calculator, which are given in Table 10.21. Now we can use the data from the 2000 calculator to decide whether the new random number generator is the same as the old one.

Stage I. Formulate the null hypothesis. The null hypothesis is that the population proportions for the 2000 calculator are the same as those for the 1988 calculator as shown in Table 10.21, which also gives the expected frequencies for each category

Stage II. Estimate the probability of occurrence of the event of interest. First we calculate the χ^2 statistic for the data generated by the 2000 calculator. This calculation gives us,

$$\chi^2 = \frac{(164-161.2)^2}{161.2} + \frac{(178-166.7)^2}{166.7} + \frac{(165-171.1)^2}{171.1} + \cdots + \frac{(137-132.1)^2}{132.1} = 8.738.$$

The event of interest is $\chi^2 \geq 8.738$ under the null hypothesis. Thus, we want to estimate,

$$P(\chi^2 \geq 8.738 | H_0).$$

Table 10.21 True Population Proportions for the 1988 Calculator Random Number Generator

Outcome	Population proportion p	Expected frequencies for 1500 generated numbers ($1500 \times p$)
0.000–0.099	.10747	161.2
0.100–0.199	.11113	166.7
0.200–0.299	.11407	171.1
0.300–0.399	.09220	138.3
0.400–0.499	.08800	132.0
0.500–0.599	.10060	150.9
0.600–0.699	.10387	155.8
0.700–0.799	.11160	167.4
0.800–0.899	.08347	125.2
0.900–0.999	.08807	132.1

Using the χ^2 table in Appendix C and referring to the row for 9 degrees of freedom, we see that $P(\chi_9^2 \geq 8.738) > 0.25$. Thus, the P-value is greater than 0.25.

Stage III. Decision. Because the P-value is much greater than 0.05, we do *not* reject the null hypothesis. $\chi^2 \geq 8.738$ is fairly likely to occur under the null hypothesis. Thus, the evidence supports the conclusion the population proportions for the 2000 calculator's random number generator are the same as those for the 1988 calculator.

Example 10.9 Effectiveness of Highway Warning Signs

(Adapted from Insurance Institute for Highway Safety.) To reduce highway deaths due to automobiles hitting objects on the sides of highways, highway warning signs were extensively installed to warn drivers of the presence of such objects. After the signs were installed, the question was raised as to whether the signs were more effective for some types of roadside objects than for others. Using data gathered over several years before the signs were installed, eight different types of roadside objects were identified as the ones most frequently hit in fatal accidents, and the proportion involving each type of object was computed for the population of all fatal accidents in those years. The population proportions for the nine categories (the eight types of objects plus the "other" category) are given in Table 10.22.

These are based on so many observations that we can consider this the "true" distribution. Next, after the highway warning sign campaign had been in effect for a while, a sample of 691 fatal accidents was gathered, and the frequencies of accidents involving each of the types of objects hit were recorded. Those data are presented in Table 10.23.

Table 10.22 Distribution of Struck Objects in Fatal Collisions (Prior to Sign Installation)

Type of object struck	P (object struck)
Tree	0.28
Embankment	0.10
Utility pole	0.10
Sign/Post/Fence	0.10
Guardrail	0.09
Ditch	0.07
Curb	0.06
Culvert	0.05
Other	0.15

Table 10.23 Distribution of Struck Objects in Sample of Fatal Collisions (After Sign Installation)

Type of object struck	Frequency of fatal accidents (O)
Tree	178
Embankment	100
Utility pole	107
Sign/Post/Fence	68
Guardrail	57
Ditch	36
Curb	43
Culvert	28
Other	73
Total	691

Use the sample data to decide whether or not the installation of the signs has caused the population proportions of fatal accidents per type of object hit to be different from the population proportions that occurred before the signs were installed.

Solution:

Stage I. Formulation of the null hypothesis. The null hypothesis is that the proportions for the different types of objects hit in the sampled population of fatal collisions after the signs were installed is the same as the proportions for the population of fatal collisions before the signs were installed. That is, H_0 is that the sample is drawn from the (multinomial) distribution (0.28, 0.1, 0.1, 0.1, 0.09, 0.07, 0.06, 0.05, 0.15).

Stage II. Estimation of the probability of the event of interest. Given the null hypothesis, the expected values for the frequencies in the sample of 691 fatal accidents are $0.28 \times 691 = 193.5$ for trees, $0.10 \times 691 = 69.1$ for embankments, and so forth. Using this null hypothesis model, we then set up the χ^2 frequency table of outcomes in the usual way (Table 10.24). We compute χ^2 with the familiar formula:

$$\chi^2 = \sum \frac{(O-E)^2}{E} = 49.723.$$

Now, we want to compute the likelihood of observing a χ^2 value of 49.723 or more (the event of interest) when the null hypothesis is true. So the probability we want to calculate is $P(\chi^2 \geq 49.723 | H_0)$. We will do this by using a χ^2 density table. We need to know the number of degrees of freedom so that we know which row to use in the table. Since we have nine categories, the χ^2 statistic has eight degrees of freedom. Thus, the probability we want to calculate, $P(\chi^2 \geq 49.723 | H_0)$, can be rewritten as $P(\chi^2_8 \geq 49.723)$. By using Table C, we see that $P(\chi^2_8 \geq 27.87) = 0.0005$. Thus, $P(\chi^2_8 \geq 49.723) < 0.0005$. So the difference between the number of accidents per type of object hit that we expected to see, based on the null hypothesis and the number of accidents per type of object hit that was observed in the real data sample, provides such strong evidence (0.0005 is much less than 0.05!) that it is highly unlikely that it would occur by chance under the null hypothesis.

Stage III. Decision. We conclude that the population proportions per type of object hit in fatal collisions has changed since the warning signs were installed. It appears that the signs were more effective for certain roadside objects than for others. Further investigation could lead to improved

Table 10.24 χ^2 Frequency Table for Roadside Collision Data

Outcome (type of object hit)	Expected number of fatal accidents (E)	Obtained number of fatal accidents (O)	O − E
Tree	193.5	179	−14.5
Embankment	69.1	100	30.9
Utility pole	69.1	107	37.9
Sign/Post/Fence	69.1	68	−1.1
Guardrail	62.2	57	−5.2
Ditch	48.4	36	−12.4
Curb	41.5	43	1.5
Culvert	34.6	28	−6.6
Other	103.7	73	−30.7
Total	691.2	691	

highway safety methods that could be used in addition to signs to reduce the proportion of accidents involving, for example, embankments and utility poles.

We have shown in this section how the χ^2 statistic can be used to reach conclusions about many-sided die models, even though the expected numbers of observations of the possible outcomes are not equal. One restriction to keep in mind is that if we are wanting to use the χ^2 density Table C approach, then the expected number of observations of each outcome should not be smaller than about 5, as discussed in Section 10.5.

Let us note how general and powerful the χ^2 approach is. If a problem leads to any hypothesized set of probabilities for a finite number of outcomes or categories (for example, $p(1) = 1/2$, $p(2) = 1/3$, $p(3) = 1/6$), and we can collect frequency data from this experiment by doing enough trials that (number of trials) × (smallest category probability) is at least 5, then we can use the χ^2 testing approach, obtaining probabilities from the χ^2 table, to test the "goodness of fit" of the actual data to the hypothesized model. Because so many applications specify such a finite set of probabilities, χ^2 goodness-of-fit testing has an enormously wide scope of application.

If the smallest expected frequency (that is, number of trials × smallest expected category probability) is less than 5, it is totally legitimate to use five-step simulation to obtain the P-value for the χ^2 statistic. If the number of simulations is large, the five-step method will be very accurate; in particular, it should be much more accurate than a probability inappropriately taken from the χ^2 density table under these conditions.

Section 10.7 Exercises

1. Is it true, as is often assumed, that people prefer the flakiest (lightest) pie crust? An experiment was done at Cornell University to test this assumption. Thirty people were involved. Each person was blindfolded and asked to state the order of preference for three pieces of pastry—one light (L), one medium (M), and one heavy (H)—which were presented in random order.

    There are six different ways in which the order of preference could be given, as shown in the following table. Also given is the number of persons choosing each of the six.

Preference ordering	Number of persons choosing (obtained frequencies)
LMH	10
LHM	3
MLH	3
MHL	8
HLM	1
HML	5
	30

    a. Under the assumption that each of the six orderings is preferred equally, what is the expected number of persons choosing each?
    b. Test the assumption of equal preference using χ^2.

c. A model was developed under the assumption of preference for lighter pastry. According to this model, the probability of each of the six outcomes being chosen is as follows:

Preference ordering	Probability of outcome under assumption of preference for light pastry
LMH	0.23
LHM	0.18
MLH	0.18
MHL	0.15
HLM	0.15
HML	0.11

d. Now what is the expected number of outcomes (that is, expected number of persons preferring each ordering of pie crusts)?

e. Test the assumption of preference for light pastry, using the model given in part c.

2. In the study of genetics, scientists are interested in determining how characteristics of living things are inherited from one generation to another. The fundamental principles of heredity were discovered by Gregor Johann Mendel (1822–1884) when he proved by crossbreeding garden peas that there are definite patterns in the way characteristics such as size, shape, and surface texture are passed on from generation to generation.

One of Mendel's classic experiments involved crossbreeding two kinds of pea plants, one with round yellow seeds and the other with wrinkled green seeds. Here are the results that should have been obtained, according to his law of heredity, and the results actually obtained. Did the seed-growing experiment support or cast doubt on his genetic theory? *Note:* This is one of science's truly famous data sets!

Type of seed	Expected proportion	Expected number	Obtained number
Round and yellow	9/16	312.75	315
Wrinkled and yellow	3/16	104.25	101
Round and green	3/16	104.25	108
Wrinkled and green	1/16	34.75	32
		556	556

3. A biological experiment with flowers yielded the following results. The expected values were based on a genetic theory of inheritance. Did the results support the theory?

Characteristic of flower	Expected f	Observed f
AB	180	164
Ab	60	78
aB	60	65
ab	20	13

4. The results of the election were 60% of the voters favored Mr. Alpha, 30% favored Mr. Beta, and 10% favored Ms. Gamma. A preelection poll produced these "votes":

Section 10.7 Unequal Expected Frequencies   503

Candidate	Number of votes
Alpha	503
Beta	115
Gamma	35

Does it appear that the poll was accurate?

5. A census in a certain city shows the racial composition to be 30% white, 55% black, and 15% others. The city council has the following racial composition (number of persons of each race):

White	20
Black	20
Other	20

Is there evidence that the racial composition of the city is different from the racial composition of the city council? *Hint:* City = population, council = sample.

6. A typist prepared a manuscript of 167 pages, of which 103 contained no errors. Forty-five pages had one typing error each. Sixteen pages had two errors each. Three pages had three errors. None had more than three errors. The table summarizes this information.

Number of typing errors	Number of such pages (observed outcomes)
0	103
1	45
2	16
3 or more	3
Total	167

The probabilistic model called the *Poisson model* (see Chapter 9) has been used to predict the number of typing errors per page. Using this model, the following probabilities are calculated:

Number of typing errors	Probability
0	0.59
1	0.31
2	0.08
3 or more	0.02

a. Calculate the expected number of pages having 0, 1, 2, and so on, errors, according to this model.
b. Use the χ^2 to test whether this model is an appropriate one.

7. A probabilistic model has been used to predict rates of occurrences of heavy rainstorms. The following table shows the number of heavy rainstorms reported by 330 weather stations during a one-year period. (For example, 102 stations reported no heavy rainstorms, 114 reported one heavy rainstorm, and so on.) The third column shows the probabilities of the corresponding numbers of heavy rainstorms according to the model being used. Use χ^2 to test how well the model predicts the occurrence of heavy rainstorms.

Number of heavy rainstorms	Number of stations reporting (O)	Model probability
0	102	0.301
1	114	0.361
2	74	0.216
3	28	0.086
4 or more	12	0.036
Total	330	

8. The composition of pebbles in a stream (e.g., what fraction of the pebbles are quartzite) tells a geologist about the composition of the rock and soil layers over which the stream has flowed. R. Flemal, in 1967, gathered 100 random samples of 10 pebbles each from the Gros Ventre River in Wyoming. In the following table, the results are displayed in the "Observed frequency (O)" column. The right-hand column gives the expected number of samples containing x pebbles according to a binomial probability model (which we discussed in Chapter 6; it assumes the same probability of each pebble being quartzite and independence between pebbles).

Number of quartzite pebbles (x)	Observed frequency (O)	Expected frequency (E)
10	6	9.1
9	25	24.7
8	31	30.0
7	28	21.7
6	9	10.3
5	0	3.3
4 or fewer	1	0.9
Total	100	100.0

   a. Can you use the χ^2 probability table with six degrees of freedom (Table C at the back of this book) to test whether the pebbles can be thought to be distributed according to the model?
   b. How would you do a χ^2 test to see whether the pebbles can reasonably be thought to be distributed according to the model? *Hint:* If you are using the textbook's software package, do the χ^2 test.

CHAPTER 10 SUMMARY

In this chapter we learned how to test hypotheses about the proportion of units in a population that fall into a specified number of categories (greater than two).

The hypothesis testing again followed the **three-stage approach** that was introduced in Chapter 9.

The sample statistic used in the first two sections of the chapter was the D **statistic.**

The preferred sample statistic (used in the rest of the chapter) is the χ^2 **statistic,** which is written as χ^2. But if doing five-step simulation, it is just as easy to use D as it is to use χ^2.

Hypothesis testing with the χ^2 statistic is also sometimes referred to as **goodness-of-fit** testing because we are testing how well a hypothesized set of category proportions fits the observed data.

The estimation of the probability of occurrence of the event of interest was accomplished in two different ways. The first way was to again use the box model simulation approach, which works for any sized data set. The second way was to use a χ^2 **density table** that gives probability distribution information for the χ^2 statistic. This χ^2 density approximation approach is effective only when the minimum $E \geq 5$.

The use of the χ^2 density table requires knowing the number of **degrees of freedom** for the statistic for a specific situation. For the problems in this chapter, the number of degrees of freedom was always equal to one less than the number of categories.

CHAPTER REVIEW EXERCISES

1. At a restaurant there are six choices on the menu. On a certain day, the items are ordered the following numbers of times:

Grilled Chicken Sandwich	12
Patty Melt	14
Steakburger	6
Garden Salad	5
Fish Sandwich	11
Grilled Steak	12

   Were some items preferred over others? Use a χ^2 test.

2. Calculate the following probabilities from Table C:
   a. $P(\chi_5^2 \geq 14.3)$
   b. $P(\chi_5^2 \geq 20.1)$
   c. $P(\chi_5^2 \geq 4.3)$
   d. $P(\chi_5^2 \geq 22.9)$

3. You are told that $P(\chi^2 \geq 9.24) = 0.1$ for a fair die tossed 50 times. Without looking at a χ^2 table, what will this probability be (at least approximately) if the die is tossed 200 times, 1000 times, 10,000 times? Defend your answers.

4. Explain intuitively why the number of degrees of freedom is one less than the number of possible outcomes.

5. A weather model predicts the following for a certain month in your area:

   Fifteen days will have no rain.
   Ten days will have rain, but less than an inch.
   Six days will have more than an inch of rain.

   The rainfall during the month is looked back on after it is over, and the actual results were as follows:

   Thirteen days had no rain.
   Eleven days had rain, but less than an inch.
   Seven days had more than an inch of rain.

   a. How many degrees of freedom are there in this case?
   b. What is the χ^2 statistic?
   c. Did the model predict well what the actual rainfall would be?

6. For each of the following probabilities, give both the estimated value ($\hat{P}$) using the appropriate table from the chapter, and the theoretical value (P) from the tables in the back of the book.

a. $P(\chi_3^2 \geq 4.2)$ (use Table 10.15 for the estimate)
b. $P(\chi_5^2 \geq 5.3)$ (use Table 10.16 for the estimate)
c. $P(\chi_9^2 \geq 15.2)$ (use Table 10.17 for the estimate)

7. Suppose we classify one-half of men and one-half of women as extroverts according to some psychological scale. Assume the rest are classified as introverts. We ask whether being extroverted affects one's choice on marriage partner, and similarly for being introverted. Suppose this personality trait has no influence on the choice of mate. There are four (male, female) combinations: (E, E), (E, I), (I, E), (I, I).
   a. If there is no influence, what is the probability of each combination?
   b. Suppose that the following data have been found for 100 marriages:

Classification	Frequency
(E, E)	35
(E, I)	15
(I, E)	18
(I, I)	32

   c. Test whether one's trait is associated with the trait of one's spouse.

8. In 1000 poker hands, you would expect (based on theoretical probabilities) to see the following hands these numbers of times:

Poker hand	Expected number in 1000 hands
Nothing	502
One pair	423
Two pairs	48
Three of a kind	21
Other (the better hands)	6

   You and your friends play games of poker over an extended period of time, and in 1000 hands you see the following distribution:

Poker hand	Observed number in 1000 hands
Nothing	488
One pair	438
Two pairs	51
Three of a kind	16
Other	7

   a. How many degrees of freedom are there in this case?
   b. Did your results follow the expected distribution? Use the χ^2 test.

9. Explain why we would be more likely to conclude that a die is loaded the larger the χ^2 statistic is, under the fair-die null hypothesis.

10. In 1987 the racial makeup of college campuses was as follows:

White	79%
Hispanic	5%
Native American/Native Alaskan	1%
African American	9%
Asian/Pacific Islander	4%
Nonresident	2%

The total number of bachelor's degrees awarded, broken down by racial categories, was as follows:

Group	Thousands of bachelor's degrees
White	841
Hispanic	27
Native American/Native Alaskan	4
African American	57
Asian/Pacific Islander	33
Nonresident	29

Source: U.S. Department of Education. Cited in QEM, *Education that Works: An Action Plan for the Education of Minorities*, MIT, Cambridge, Mass., 1990.

    a. Out of the total of 991,000 degrees awarded, how many could be expected in each of the racial groups, if you assume they follow the same distribution as the student population?

    b. Do a χ^2 test to determine whether the actual number of bachelor's degrees awarded has the same distribution as the student population.

11. You shuffle a deck of cards forty times and draw out one card.

    a. How many times do you expect each of the four suits (spades, clubs, diamonds, and hearts) to show up?

    b. This test was performed, and the actual results were as follows:

Hearts	12 times
Diamonds	7 times
Clubs	11 times
Spades	10 times

Do a χ^2 test to determine whether the cards were being drawn randomly from a well-shuffled deck so that each suit is equally likely.

    c. Suppose somebody draws 10 cards and gets five hearts, one diamond, no clubs, and four spades. You suspect that the (1/4, 1/4, 1/4, 1/4) model is wrong but realize that you cannot use the χ^2 density (Table C) because $10 \times 1/4 = 2.5 < 5$. There is a way to test this. What would you do?

12. A car dealer believes that each of the five colors of a certain car is equally likely to be chosen by the customer. At the end of the month, these are the results the dealer sees:

Color	Frequency
Red	21
Blue	10
White	15
Black	17
Light taupe	7
Total	70

Use a χ^2 statistic to test whether the dealer is correct that each color is equally likely to be chosen.

13. A clothing store kept track of how often certain sizes of clothing were bought during a day. The results were as follows:

Size	Number of articles sold
Small	32
Medium	40
Large	21
Extra large	15

   a. If you believed that each size is equally likely to be sold, how many of each size would you have expected to be sold?
   b. Using the χ^2 statistic, test the belief that each size is equally likely to be sold.

14. A researcher is interested in determining whether infants are more likely to be born on certain days of the year than on others (ignoring leap years). If the researcher gathered the data for the number of births on every day of the year, what would be the number of degrees of freedom in the χ^2 statistic?

15. a. Calculate the following probabilities using Table C:

$$P(\chi_2^2 \geq 10.2), P(\chi_5^2 \geq 10.2), \text{ and } P(\chi_6^2 \geq 10.2).$$

   b. What trend do you see in your results to part **a**? That is, as the number of degrees of freedom goes up, what happens to the probability of having a χ^2 statistic greater than 10.2?

16. If you were merely using five-step simulation to make your decision in these "dice" problems, would it be as easy or easier to use the D statistic instead of the χ^2 statistic? Explain why people prefer to use the χ^2 statistic when not doing simulations.

17. In a class of 197 students, the professor looks at the first number of every student's social security number. The results are as follows:

Digit:	0	1	2	3	4	5	6	7	8	9
Frequency:	0	2	2	179	7	7	0	0	0	0

Is there evidence the first digit of the social security number is not a good source of random numbers? Perform a χ^2 test (noting that you *know* what your conclusion is going to be!).

18. Recall Exercise 17. This time the professor looks at the last number of every student's social security number. Now the results are as follows:

Digit:	0	1	2	3	4	5	6	7	8	9
Frequency:	25	23	13	16	16	20	26	22	16	20

Is there evidence the last digit of the social security number is not a good source of random numbers? Perform a χ^2 test.

19. In the 1968 presidential election the popular vote was divided as follows:

Richard Nixon	31,770,237 votes
Hubert Humphrey	21,270,533 votes
George Wallace	9,906,141 votes

   a. What percentage of the vote did each candidate win?

b. The actual winner of the election is determined by the votes of the Electoral College. There are 538 members. If the members' votes reflect how the people voted, how many of the electoral votes should each candidate have won?

c. The real Electoral College results were as follows:

Richard Nixon	301 electoral votes
Hubert Humphrey	191 electoral votes
George Wallace	46 electoral votes

Pretending that the electoral college votes are a random sample (they are not), test the expected versus the actual results using the χ^2 test.

PROFESSIONAL PROFILE

Dr. Jian Zhang
Research Scientist

Western &
Intermountain Storms
Hydrometeorology Team
National Severe Storms
Laboratory
Norman, Oklahoma

Jian Zhang studies data in her office at the National Severe Storms Laboratory.

Dr. Jian Zhang

How much rain will fall? How much snow? How quickly? Will there be flash floods? Or avalanches? Will a drought be broken? Will it take all night to plow the streets?

Accurate answers to such questions depend on the ability to estimate the amount of rain or snow that will reach the surface from a particular storm system. Much of that information comes from Dr. Jian Zhang. She analyzes storm information for the National Severe Storm Laboratory.

Dr. Zhang collects data about the precipitation character of storms. She depends heavily on reflectivity, a variable measured by weather radar. In addition, she collects and analyzes data on the temperatures of the tops of clouds using geostationary satellites. Finally, she also collects rain gauge observations from locations beneath storms.

Dr. Zhang uses statistics to find relationships between "true" rainfall amounts, radar reflectivity, and satellite cloud top temperatures. Rain gauges are quite accurate, but there may be only a handful of observations in a large area. "I use rain gauge observations to calibrate, or tune, other measurements of rainfall that cover much larger regions," explains Dr. Zhang. "Statistics help me form regression equations between the different types of observations." The goal is more accurate forecasts of rainfall. "Most people don't realize that floods cause more deaths than tornadoes or lightning strikes," says Dr. Zhang. "Our precipitation estimation techniques provide forecasters with better tools to make more accurate flash flood warnings. I hope that property and lives will be saved."

11

Inference about Regression

Like father, like son (almost).

Objectives

After studying this chapter, you will understand:

- How to compute the standard error of a slope estimate in linear regression
- How to test whether x is useful in predicting Y using a linear model
- How to simulate regression models
- How to obtain a confidence interval for the slope parameter
- How to obtain a prediction interval for Y and a confidence interval for the expected value of Y given x
- How transformations of variables extend the scope of linear regression to nonlinear relationships

---

11.1 INFERENCE ABOUT THE SLOPE 512

11.2 INFERENCE ABOUT THE PREDICTION OF Y AND ESTIMATION OF $E(Y|x)$ 530

11.3 TRANSFORMATION OF VARIABLES 533

CHAPTER REVIEW EXERCISES 537

Key Problem

Can a Child's Age at First Word Measure Overall Development?

The Gesell adaptive score is designed to measure children's behavioral and language development. Some researchers are interested in finding out whether this development can be predicted from the age at which a child utters his or her first word. Table 11.1 gives the Gesell adaptive scores and ages at first word (in months) of 20 randomly selected children. Figure 11.1 is a scatterplot that clearly shows a negative correlation between age at first word (x) and Gesell adaptive score (y). The least squares linear regression equation is

$$\hat{Y} = 109.3 - 1.2x.$$

Because the slope is −1.2, this fitted line predicts that the Gesell adaptive score will be lower by 1.2 units for a child who utters his or her first word one month later.

Suppose that we want to apply this regression line to other children or that we want to do a child development study based on the claimed relationship. Thus we want to know whether such a relationship can be confirmed. In particular, is there strong statistical evidence in the data that the age at which the first word is spoken really helps predict the Gesell adaptive score? If so, how accurate are predictions given by the line? In this regard, we certainly would not expect the age at first word to be more than moderately predictive of something as psychologically complex as the Gesell score.

11.1 INFERENCE ABOUT SLOPE

In Chapter 3, we learned how to analyze (x, y) data, called **bivariate data**, when the scatterplot suggests y is linearly related to x. We considered gas mileage (y) as a function of car weight (x), breast cancer mortality (y) as a function of animal fat consumption (x), age of wife (y) as a function of age of husband (x), and height of son (y) a function of height of father (x), among other linear relationships for which a linear regression analysis was done.

Table 11.1 Age at First Word and Gesell Score for 20 Children

Age at first word (in months)	Gesell score	Age at first word (in months)	Gesell score
15	95	7	113
26	71	9	96
10	83	10	83
9	91	11	84
15	102	11	102
20	87	10	100
18	93	12	105
11	100	42	57
8	104	11	86
20	94	10	100

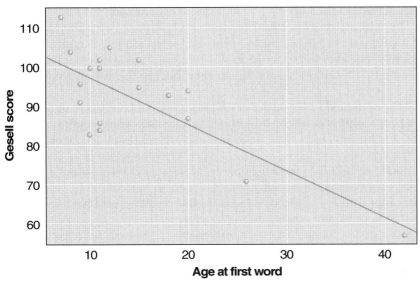

Figure 11.1 *Gesell score versus age (in months) at first word spoken.*

Standard Error of the Slope Estimate

Suppose a scatterplot is roughly elliptical (football-shaped), which suggests a linear relationship between x and Y. From Chapter 3 we know how to compute the slope and intercept of a least squares regression line. The estimated slope parameter is of particular interest. When nonzero, it tells us the average change in the predicted Y, that is, $\hat{Y}$, when x is increased by one unit, as was illustrated in the Key Problem statement. When the estimated slope is close to zero, there is no linear relationship between x and Y and hence when x changes, the value of Y changes only randomly. Knowing whether or not the true slope is zero answers the fundamental question of whether there is a linear relationship between x and Y. The estimated slope helps us infer whether the true slope is 0 or not. Unlike in Chapter 3, we will distinguish between estimated parameters and their true population values. As our notation suggests, in the probability model we develop for linear regression the treatment variable x is considered nonrandom, whereas the predicted response variable Y is considered random.

If x and Y have no linear relationship (and thus the true slope is 0), the slope estimate computed from the data is unlikely to be exactly 0, just as tossing a fair coin 100 times is unlikely to produce exactly 50 heads. Therefore, it is important to know the accuracy of our slope estimate. In particular, as we learned in Chapter 8 for estimators in general, we would like to estimate the *standard error* of the estimated slope. The standard error of the estimated slope enables us to assess how close it is to the true slope. The standard error also tells us how confident we can be that the estimated slope comes within a given range of the true slope. As we learned in our use of standard errors in constructing test statistics in Chapter 9, the standard error of the estimated slope enables us to test hypotheses about the true regression line slope if the distribution of the test statistic is approximately bell shaped. In fact,

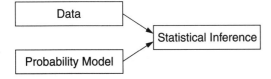

Figure 11.2 *Model-based statistical inference.*

the estimated slope does have a bell-shaped distribution when the sample size is large. Hence, the estimated slope will differ from the true slope by less than twice the estimated standard error about 95% of the time. For example, if the estimated slope is 0.6 but the estimated standard error is 1.3, the true slope could easily be zero, and we would not have strong evidence for claiming that there is a sloped linear relationship useful for predicting Y from x.

Before we can discuss the standard error of the estimator b_1 of the true slope in a regression analysis, we need a probability model for how the bivariate data were generated. This, again, is consistent with our approach to statistical inference throughout the book, and is shown in Figure 11.2.

To form a probability model, we assume for each x that the observed y is the result of a random process, that is, each observed y becomes the realization of a random variable Y. If we have repeated independent observations of Y at a fixed value of x, the resulting observed values, y_1, $y_2, \ldots$, will vary according to the probability distribution of Y given x.

To describe the probability distribution of the random variable Y for each fixed value x, we make three fundamental modeling assumptions:

Regression Model Assumptions

- The Y random variables are all independent of each other. That is, we assume that the data are collected in such a way that the outcome of an observed Y does not depend in any way on the other observed Y values.
- The expected value of Y at x, denoted $E(Y|x)$, is linear in x and can be written

$$E(Y|x) = \beta_0 + \beta_1 x,$$

where β_0 and β_1 are the unknown intercept and slope parameters, that is, we assume that the data can be described by an underlying linear relationship (with a change of notation from $y = mx + b$ in Chapter 3).
- The standard deviation of Y for fixed x, $SD(Y|x)$, is the same at each x, that is, $SD(Y|x)$, denoted by σ, does not change with x.

The second fundamental assumption has a useful empirical interpretation: If we are able to observe Y many times at a fixed x_1, we would obtain an observed mean $\bar{y}_1$. If we graph the point $(x_1, \bar{y}_1)$, then repeat the process at a different x_2 to get $\bar{y}_2$ and graph $(x_2, \bar{y}_2)$, and so on, these points will lie close to a line. However, in a typical regression experiment, there is usually only one observed y for each value of x. But, for some cases, the design of the regression experiment calls for observing multiple y values for each x of the experiment, as shown in Example 11.1.

The second fundamental assumption rules out the possibility that the true underlying relationship between x and Y is nonlinear. Thus, we cannot make this assumption unless we have good reason to believe that the underlying relationship *is* linear!

The quantity σ tells us how close we expect random Y variables to be to the true regression line. From the box modeling perspective, the third assumption states that although for each x a different box generates the random Y values, nonetheless each such box has the *same* standard deviation σ. That is, the typical variation in Y about its regression line at x is the same regardless of the value of x. This assumption often fails to be true for bivariate data; in such a case, the standard regression analysis needs to be replaced by a more sophisticated one carried out by a statistical expert.

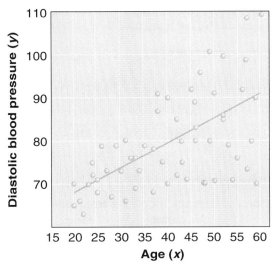

Figure 11.3 Scatterplot of female blood pressure versus age.

Figure 11.3 is a scatterplot that gives the diastolic blood pressure from each of 57 adult women as a function of age. (Diastolic is the pressure when the heart is at rest.) Also displayed is the fitted regression line ($\hat{Y} = 56.1 + 0.59x$). Clearly the observed y values vary much less for x small (e.g., at age 20) than x large (e.g., at age 50). This data set thus violates the assumption that $SD(Y|x)$ is the same regardless of the value of x. Hence, the second fundamental assumption fails and we are not allowed to use the usual formulas we've learned for estimating the regression line.

The standard deviation $\sigma = SE(Y|x)$ of Y for fixed x can be well estimated by what is called the **standard error of the residuals**, which is given by

$$S_E = \sqrt{\frac{\sum(Y - \hat{Y})^2}{n-2}}.$$

Since two parameters, β_0 and β_1, are estimated, the division is by $(n-2)$, that is, the subtracted 2 is described by "two degrees of freedom are lost." When the sample size n is large, a version of the law of large numbers shows that

$$S_E \approx \sigma.$$

As we often do in this textbook, we use simulation in the next example to demonstrate the third bulleted assumption about the standard deviation σ for fixed x and, in particular, to demonstrate the role of σ in controlling the randomness of the Y values for fixed x, in a regression model.

Example 11.1

Using a regression model, namely

$$E(Y|x) = 3x + 2; \quad SD(Y|x) = \sigma = 0.6$$

for every x, seventy-five y values were simulated with 15 at each $x = 0, 1, 2, 3, 4$. In Figure 11.4, the scatterplot together with the true regression line are shown (one cannot see all the points because of overlap). From a box model perspective, we need five boxes to generate the data, one for each x.

We will assume a normal population for each x (hence a computer must do the box model sampling). Thus the five boxes and the normal distributions producing them are:

Box 1 ($x = 0$):	$Y_{0,1}, \ldots, Y_{0,15}$	$N(2, 0.6)$
Box 2 ($x = 1$):	$Y_{1,1}, \ldots, Y_{1,15}$	$N(5, 0.6)$
Box 3 ($x = 2$):	$Y_{2,1}, \ldots, Y_{2,15}$	$N(8, 0.6)$
Box 4 ($x = 3$):	$Y_{3,1}, \ldots, Y_{3,15}$	$N(11, 0.6)$
Box 5 ($x = 4$):	$Y_{4,1}, \ldots, Y_{4,15}$	$N(14, 0.6)$

Here, in each case, $E(Y|x) = 3x + 2$ determines the mean of the box model's normal distribution; for example $E(Y|2) = 8$. Further, in each case $SD(Y|x) = 0.6$; for example $SD(Y|2) = 0.6$.

It is instructive to compare Figure 11.3, where larger x values have larger $SD(Y|x)$ values, with Figure 11.4, where we notice that the spread, or typical variation, of the y values about the line is nearly the same for all x. Thus, the three assumptions, in particular the second one, hold; and, hence, it is okay to use our standard regression formula to solve for the estimated slope b_1.

This simulation is instructive in other ways. From Chapters 5 and 8 we have learned that if the sample size is large, a good estimator, like $\overline{X}$, will be close to the parameter it estimates (μ in the case of $\overline{X}$). Here

$$S_E = 0.65 \approx 0.6 = \sigma,$$

confirming that S_E is at least a moderately good estimator of σ. In fact $SE(S_E) = 0.059$ can be shown theoretically; hence the typical error in estimating σ using S_E is about 0.06 in magnitude.

Finally, $S_E = 0.65$ suggests that the typical distance of an observed y from the line is about 0.65; Figure 11.4 shows this.

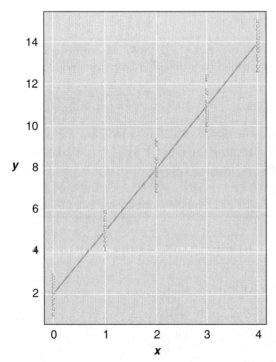

Figure 11.4 Simulation of a scatterplot under regression model assumptions.

We use the symbol b_1, rather than the hat notation $\hat{\beta}_1$, to represent the estimate of β_1 obtained from the data. Recall from Chapter 3 that b_1 is defined by

$$b_1 = r\frac{S_Y}{s_x} = \frac{1}{n-1}\sum\left(\frac{x-\bar{x}}{s_x}\right)\left(\frac{Y-\bar{Y}}{S_Y}\right)\frac{S_Y}{s_x}$$

$$= \frac{1}{(n-1)s_x^2}\sum(x-\bar{x})(Y-\bar{Y})$$

$$= \frac{\sum(x-\bar{x})(Y-\bar{Y})}{\sum(x-\bar{x})^2}.$$

Here S_Y and s_x denote the usual standard deviations for Y and x, respectively. For example,

$$S_Y = \sqrt{\sum(Y-\bar{Y})^2/(n-1)}.$$

Also recall that the computationally convenient formula for the slope estimate b_1 is based on five simple sums computed from the x and y values:

$$b_1 = \frac{n\sum xY - (\sum x)(\sum Y)}{n\sum x^2 - (\sum x)^2}.$$

Note that S_Y^2 is random because it depends on the random Y_i whereas s_x^2 is nonrandom because the x values are nonrandom. Recall from Chapter 3 that $Y - \hat{Y}$ is called the estimation error, or residual, where Y is the observed and $\hat{Y}$ is the predicted value at x. The standard error of the estimator b_1 can be shown to be

$$\text{SE}(b_1) = \frac{\sigma}{\sqrt{\sum(x-\bar{x})^2}} = \frac{\sigma}{s_x\sqrt{n-1}}.$$

We saw in the simulation example above that $S_E \approx \sigma$ when n is large. Thus, substituting S_E for σ in the formulas for $\text{SE}(b_1)$ above, a good estimator of $\text{SE}(b_1)$ is given by

$$\widehat{\text{SE}}(b_1) = \frac{S_E}{\sqrt{\sum(x-\bar{x})^2}} = \frac{S_E}{s_x\sqrt{n-1}}.$$

Note that in Example 11.1,

$$\widehat{\text{SE}}(b_1) = \frac{S_E}{s_x\sqrt{n-1}} = \frac{0.65}{1.42\sqrt{74}} = 0.053,$$

indicating that the estimated slope b_1 should come quite close to the true slope of 3.

Examination of the right-hand expression for the estimated standard error, $\widehat{\text{SE}}(b_1)$, shows the three components that influence its size: the standard deviation of the x values, the sample size n, and the standard error S_E of the residuals. In particular, $\widehat{\text{SE}}(b_1)$ is made smaller by

- increasing the sample size n,
- decreasing the residual errors, that is, S_E;
- increasing the spread of the x values.

We can design a study in which n is large if cost is not prohibitive, and we sometimes have control over the x values and, hence, are able to spread them out widely. But how close the observed data points are to the regression line is never under our control; rather it is a consequence of how capable x is in predicting Y accurately via the linear regression equation. That is, the closeness of the Y values to the line varies depending on the strength of the linear relationship between x and Y. For example, the level of fertilizer x may predict crop yield Y well, whereas age x may predict blood pressure Y less well. Simply put, we do not have control over the size of the model parameter σ that tells us the typical variation of observed Y values at any x.

The standard error of b_1 tells us the size of the typical estimation error $(b_1 - \beta_1)$ due to the inherent randomness in the Y values. The mean (expected value) of b_1 is actually equal to the true slope parameter β_1, that is,

$$E(b_1) = \beta_1.$$

In summary, b_1 is an unbiased estimator of β_1 with $SE(b_1) = \sigma/(s_x \sqrt{n-1})$.

Next we use a simulation study to help learn more about the estimation behavior of b_1.

Example 11.2

In Example 3.12, a regression line was found for predicting a son's height from his father's height, given by

$$\hat{Y} = 0.516x + 33.73.$$

Computation shows that

$$\bar{x} = 67.68, \bar{y} = 68.65, S_E = 2.325, s_x = 2.70, s_Y = 2.71, r = 0.514.$$

Hence, the estimated slope b_1 and intercept b_0 are given by

$$b_1 = r\frac{s_Y}{s_x} = 0.516 \quad \text{and} \quad b_0 = \bar{y} - b_1\bar{x} = 33.73.$$

We will use the estimated regression line to supply the simulation model for each of the three box models used by the computer to simulate the three regression settings:

$$\bar{x} = 67.68, \mu_Y = 68.65, E(Y|x) = 0.516x + 33.73, SE(Y|x) = \sigma = 2.325.$$

We will simulate the estimated regression line repeatedly using three different sample sizes:

$$n = 4, 25, \text{ and } 100.$$

In Figure 11.5, an $n = 100$ scatterplot is shown. A positive association is exhibited but not a strong one. In each of the three cases, 1000 regression line estimations are simulated and, in each case, the variability and the biasedness of b_1, are studied. We expect that the estimator b_1 of $\beta_1 = 0.516$ will perform better as the sample size n is increased. In particular we suspect b_1 should perform poorly when $n = 4$ and well when $n = 100$.

It is desirable that an estimator be unbiased (at least approximately). Here this means

$$E(b_1) \approx \beta_1 = 0.516.$$

We know from Chapter 5 that for a simulated statistic W of interest $\overline{W} \approx E(W)$, if the number of repetitions (replications) of the simulation is large. Thus, in each of our three simulation settings with 1000 replications, we know that $\bar{b}_1 \approx E(b_1)$ holds. Hence for $\beta_1 = 0.516$, if we observe

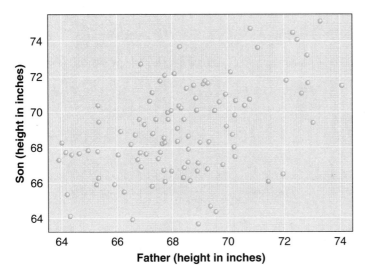

Figure 11.5 Scatterplot of 100 simulated son versus father heights.

$\bar{b}_1 \approx \beta_1$, this will be strong empirical evidence that $E(b_1) \approx \beta_1$ and hence that b_1, at least approximately, is unbiased. We obtained

n	4	25	100	true β_1
$\bar{b}_1$	0.511	0.512	0.516	0.516

Thus, $\bar{b}_1 \approx \beta_1$ in each case, an empirical simulation-based confirmation of the theoretical fact that

$$E(b_1) = \beta_1.$$

We learned above that

$$\text{SE}(b_1) = \frac{\sigma}{s_x \sqrt{n-1}} = \frac{2.325}{2.70 \sqrt{n-1}}.$$

This should be borne out in the simulations. In particular, in each case the sample SD of the 1000 simulated b_1 values should be approximately equal to SE (b_1). This is the case:

	$n = 4$	$n = 25$	$n = 100$
SE(b_1)	0.4972	0.1758	0.0865
sample SD of b_1	0.5296	0.1615	0.0905

This provides an empirical simulation-based confirmation of the above theoretical formula for SE (b_1).

Recall that the true slope $\beta_1 = 0.516$. When $n = 4$, according to the above table, the typical error $(b_1 - \beta_1)$ is given by SE $(b_1) \approx 0.5$. Thus b_1 is clearly a very poor estimator of β_1, even though it is unbiased; it is too variable. One statistician could easily obtain $b_1 \approx 0$ and yet another easily obtain $b_1 \approx 0.9$. Indeed Figure 11.6(a) shows the variability of the 1000 simulated estimated b_1 slopes, that is, the estimates of β_1 by 1000 statisticians with each having only a sample size of four of (x, y) values, with observed b_1 estimates varying from less than −1 to greater than 2.

By contrast, when $n = 25$, SE(b_1) ≈ 0.18, which produces b_1 estimates varying as in Figure 11.6(b). This is better, although not good, in that typical values of b_1 range from $0.516 - 0.18 = 0.336$ to $0.516 + 0.18 = 0.696$. Finally, when $n = 100$, SE(b_1) ≈ 0.09, producing typical variations in b_1 from

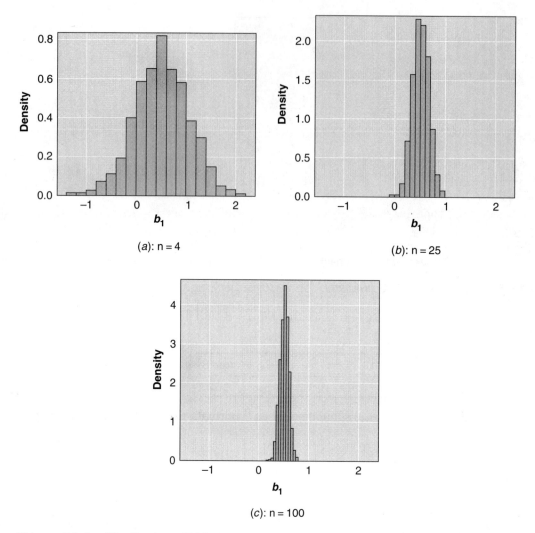

Figure 11.6 Distribution of 100 simulated b_1 values for heights of fathers and sons when $n = 4$, 25, and 100 respectively.

one statistician doing a regression to the next of $0.516 - 0.09 = 0.426$ to $0.516 + 0.09 = 0.606$, which is fairly good, as displayed in Figure 11.6(c). To obtain a typical variation given by $SE(b_1) = 0.04$, say, the statistician would need $n = 464$ observations (a very large sample size).

We now contrast our ability to make accurate inferences about the slope β_1, in Example 11.2, where the correlation was moderate ($r = 0.514$) with a situation where the correlation is close to 1 as in the next example.

Example 11.3

The regression of a wife's age Y as a function of her husband's age x was studied in Examples 3.11 and 3.13. A high correlation ($r = 0.95$) was found between the ages of the roughly 300 husband and wife pairs. The fitted line was

$$\hat{Y} = 0.95x - 0.7.$$

Computation also showed that (unit = year)

$$\bar{x} = 4.06, \bar{y} = 43, s_x = s_Y = 13, S_E = 4.06.$$

We will use these statistics to produce our simulations model:

$$\bar{x} = 4.06, \mu_Y = 43, E(Y|x) = 0.95x - 0.7, \mathrm{SE}(Y|x) = \sigma = 4 \text{ years}.$$

As in Example 11.2, we simulate the regression line estimation process repeatedly (1000 times) for sample sizes $n = 4, 25, 100$. Figure 11.7 shows the scatterplot for n=100.

We already know that the estimator b_1 of $\beta_1 = 0.95$ will perform better as the sample size n increases. We suspect that the accuracy of estimation will be somewhat better for this high-correlation example ($\rho = 0.95$) than it was in Example 11.2.

Further (and strong) confirmation of the unbiasedness of b_1 as an estimator of β_1, namely that $E(b_1) = \beta_1$, comes from the 1000 repetitions of the above simulation for $n = 4, 25, 100$:

n	4	25	100	true β_1
b_1	0.950	0.951	0.949	0.950

Further, we obtained

	$n = 4$	$n = 25$	$n = 100$
$\mathrm{SE}(b_1)$	0.1803	0.0637	0.0314
sample SD of b_1	0.1913	0.0690	0.0314

As expected, the simulation study strongly confirms the theoretical formula $\mathrm{SE}(b_1) = \sigma/(s_x\sqrt{n-1})$, used to obtain the first row of the table. Figure 11.8(a)–(c) again shows the improved capability of b_1 to estimate β_1 accurately as n ranges from 4 to 25 to 100, as was shown in Example 11.2 as well.

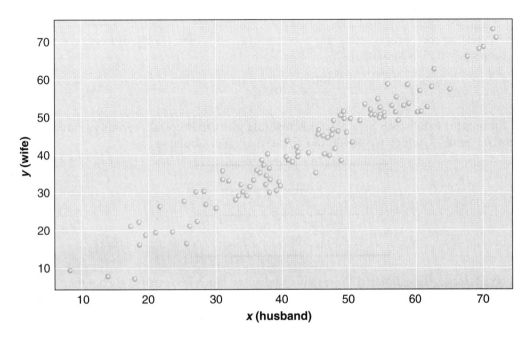

Figure 11.7 Scatterplot of 100 simulated wife versus husband ages.

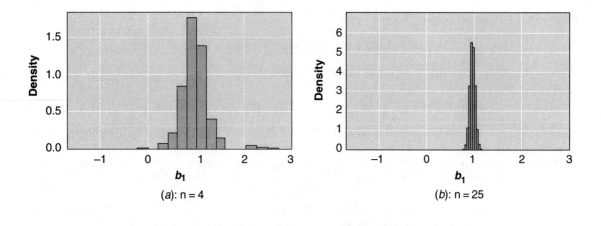

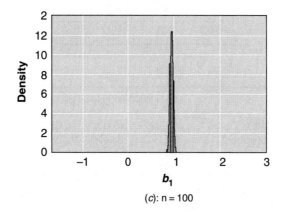

Figure 11.8 Distribution of 1000 simulated b_1 values for ages of husbands and wives when $n = 4, 25,$ and 100, respectively.

Another interesting comparison across the two examples with their widely different correlations ($\rho = 0.514$ versus $\rho = 0.95$) is to compare

$$\frac{\text{SE}(b_1)}{\beta_1} \cdot 100\%$$

values for different n values and ρ values. This formula gives the typical percent variation in b_1, *relative to its true value* β_1, which we would like to be small. It is tabulated as

		n		
		4	25	100
ρ	0.514	96.4%	34.1%	16.8%
	0.95	19%	6.7%	3.3%

Clearly when ρ is large, the estimated slope tends to display much less error at each n than when ρ is moderate. For example, in the $n = 4$ case the 19% error on average is fairly large, but perhaps tolerable. By contrast the 96% error is surely unacceptable!

Clearly a smaller standard error of b_1 means both a more accurate estimate of the true regression line $E(Y|x)$ as x varies, and a more accurate prediction of a new Y for a given x. Indeed, as discussed in Section 3.3, two basic estimation/prediction problems arise in regression: (i) estimate the line *at x*, thus estimating the underlying relationship between Y and x (recall Example 3.12), and (ii) at a particular x predict a new Y that will be observed at a future time (recall Examples 3.10 and 3.11).

We now take a look at our Key Problem.

Example 11.4

Consider the Gesell score data of Table 11.1 of the Key Problem. Recall from Chapter 3 that for ease in computing b, five sums should be computed:

$$\sum x = 285, \; \sum y = 1846, \; \sum x^2 = 5317, \; \sum y^2 = 173{,}514, \; \sum xy = 23{,}807.$$

Recall that the sample size $n = 20$, therefore, by our computational formula for b_1,

$$b_1 = \frac{n \sum xy - (\sum x)(\sum y)}{n \sum x^2 - (\sum x)^2} = -1.2.$$

Also,

$$S_E = \sqrt{\frac{\sum (Y - \hat{Y})^2}{n - 2}} = 8.6 \quad \text{and} \quad s_x = \sqrt{\frac{\sum (x - \bar{x})^2}{n - 1}} = 8.22.$$

Thus the estimated standard error of b_1 is given by

$$\widehat{SE}(b_1) = \frac{S_E}{s_x \sqrt{n-1}} = \frac{8.6}{8.22 \sqrt{19}} = 0.24.$$

Now, for the moderately large $n \equiv 20$, b_1 is approximately normally distributed. This normality can be shown to follow from our three fundamental regression model assumptions. Thus, according to the 95% rule for the standard error of an estimator that is approximately normal, about 95% of the time we would expect b_1 to be within $2 \times 0.24 = 0.48$ of the true slope β_1. Thus, in particular, for $b_1 = -1.2$, $\widehat{SE}(b_1)$ seems to provide strong evidence that the true $\beta_1 \neq 0$. That is, our standard error computation provides strong evidence that the Gesell score is predictable (somewhat) from the age at which a child's first word was spoken. This suggests a formal hypothesis test about β_1, which we do next.

Hypothesis Testing about the Slope β_1

The standard error of b_1 helps in testing whether the true slope $\beta_1 = 0$ or not. Our null hypothesis is

$$H_0 : \beta_1 = 0.$$

This null hypothesis is of central importance because it asserts that there is no information about the value of Y contained in the value of x. In most applications we hope to find sufficient statistical evidence to reject this hypothesis, thus establishing evidence that there is a predictive linear relationship.

To carry out the hypothesis test for the slope, as for any other statistic of interest, we have to find the P-value: the probability that, if the null hypothesis were true, the statistic would be at least as extreme as its observed value. The meaning of "extreme" depends on which alternative hypothesis we have chosen. As was learned in Chapter 9, there are three possible alternative hypotheses:

$$H_{1a} : \beta_1 > 0, \quad H_{1b} : \beta_1 < 0, \text{ and } \quad H_{1c} : \beta_1 \neq 0.$$

Which one is appropriate depends on the investigation. Sometimes the investigator can assert a one-sided alternative because it is known that if x has any influence on Y at all, it causes Y to increase (or to decrease). For example, since increasing the fertilizer level x should not *decrease* the crop yield Y, only H_{1a} needs to be considered. Similarly, the investigator may be interested only in evidence that $\beta_1 < 0$. For example, in the Key Problem for Chapter 3, since it is not likely that heavier cars would get *better* gas mileage, we would choose H_{1b}.

Recall that the standardized test statistic is given by

$$\frac{\text{statistic of interest} - E(\text{Statistic of interest} \mid H_0)}{\text{estimated SE of statistic of interest}}.$$

Thus, the standardized test statistic for the slope is

$$T = \frac{b_1 - 0}{\widehat{SE}(b_1)},$$

where $E(b_1 \mid H_0) = 0$, the hypothesized value.

This standardized T is approximately distributed as a standard normal under H_0 (by an application of the CLT using the regression model assumptions). Therefore, we can use the normal table in Appendix E to find the probability that a normal random variable Z is at least as extreme as the observed value t of the statistic T. The P-value is, then, approximately

$$P(Z > t) \quad \text{for } H_{1a};$$
$$P(Z < t) \quad \text{for } H_{1b}; \text{ or}$$
$$\left.\begin{array}{l} 2P(Z > t) \text{ if } t > 0 \\ 2P(Z < t) \text{ if } t < 0 \end{array}\right\} \quad \text{for } H_{1c}.$$

A decision of whether or not to reject the null hypothesis, H_0, is then made based on the P-value. The test statistic b_1 was standardized by using an estimate of the standard error of b_1. However, this large-sample test is a Z test rather than a t test, because the large sample size tells us that $\widehat{SE}(b_1)$ will be close to $SE(b_1)$.

Example 11.4 (continued)

Continuing our analysis of the Key Problem, we obtain $t = -1.2/0.24 = -5.0$. To test the hypothesis that $\beta_1 = 0$ (or equivalently, $\rho = 0$) versus $\beta_1 < 0$ (or equivalently, $\rho < 0$), we approximate the P-value by

$$P(Z < -5.0) = 0.000,$$

which is much smaller than 0.05, our usual standard of significance. Thus we have strong statistical evidence that there is a predictive relationship between a child's age at first word spoken and Gesell score.

The P-value for the Z test is computed under H_0 assuming T is approximately normally distributed for $n = 20$. This meets our requirement for applying the central limit theorem: $(n \geq 20)$, but it is somewhat less than the sample size of 30 required by our usual convention of standardizing by an estimated SE, so we are cautious about accepting the conclusion.

In many regression problems, it is reasonable to make a fourth assumption, namely that Y, given x, is normally distributed for every x. From our empirical perspective, assuming $SD(Y \mid x) = \sigma$, a constant, for every x amounts to asserting for each x that the centered observations (or residuals) $Y - \hat{Y}$, where $\hat{Y}$ denotes the estimated regression line at x, are sampled from the same normal population with mean 0 and standard deviation σ regardless of the value of x. As a rough check of the validity of this assumption, we can draw a density histogram of the residuals, and visually examine whether the histogram is roughly bell shaped. We must be careful to choose wide enough intervals for this histogram (recall Sections 1.4 and 1.5) so that the true shape of the data is not distorted by too many narrow and unstable intervals. Alternatively, we could do a Q-Q plot analysis.

Assuming random sampling of x values from a normal population, we know from Chapters 8 and 9 that $\bar{X}$ when standardized using its theoretical mean and estimated standard error has a t distribution. A similar analysis based on the four regression model assumptions and the assumption that the null hypothesis $\beta_1 = 0$ is true, shows that the statistic

$$T = \frac{b_1 - \beta_1}{\widehat{SE}(b_1)} = \frac{b_1}{\widehat{SE}(b_1)}$$

has a t distribution, with $n - 2$ degrees of freedom.

Therefore, assuming the four assumptions, we can compute under H_0 the exact P-value according to

$$P\{T > t_0\} \quad \text{for } H_{1a}$$
$$P\{T > t_0\} \quad \text{for } H_{1b}$$
$$\left. \begin{array}{l} 2P\{T > t_0\} \text{ if } t_0 > 0 \\ 2P\{T < t_0\} \text{ if } t_0 < 0 \end{array} \right\} \quad \text{for } H_{1c}$$

where t_0 is the observed value of T and T is t-distributed with $n - 2$ degrees of freedom. For a given level of significance, say 0.05, the critical value of the test can be found from Table F in Appendix F. For example, we have $n = 20$ for our Key Problem. Assuming normal data, the critical value for testing $\beta_1 = 0$ versus $\beta_1 < 0$ at the standard 0.05 level of significance can be found to be $t(18) = -1.734$, based on the t distribution with $n - 2 = 18$ degrees of freedom, shown in Figure 11.9.

That is, we reject the null hypothesis provided that $T < -1.734$ is observed. Since $T = -5.0 < -1.734$, we reject H_0 at the 0.05 level. Indeed, note from Table F for a t-distributed random variable that $P(T < -5.0) < P(T < -2.850) = 0.005$, which provides us much stronger evidence than just noting that we reject at $\alpha = 0.05$! In fact, the P-value is 0.000, rounded to three significant figures, according to more complete tabulations.

When the null hypothesis H_0 is rejected, we say that there is a **regression effect.** That is, we have strong statistical evidence that the regression line slope $\beta_1 \neq 0$. Thus, the hypothesis test tells us that knowledge about x can help predict Y, using the regression line. But it does not tell us *how accurately* we can do such prediction, which we will take up shortly. Measures like r^2 and the standard error (SE) of the residuals help provide some quantitative answers in this regard.

If the null hypothesis H_0 is not rejected, we say that there is no evidence of a regression effect. That is, we are not sure whether the observed positive or negative slope b_1 is merely due to random noise in the data or due to a theoretical $\beta_1 \neq 0$ that we cannot detect.

t Test or *Z* Test for β_1?

Both the Z test and the t test apply to the same null hypothesis $H_0: \beta_1 = 0$ and use the same standardized statistic T, but they differ in the conditions under which we are allowed to apply them. The t test is usable for any sample size n no matter how small, *provided that the rather restrictive fourth assumption of normality of the Y at each x also holds.* If we believe this assumption and $n < 30$,

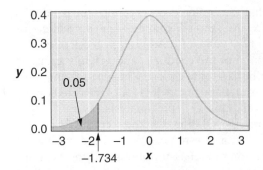

Figure 11.9 *t* distribution with 18 degrees of freedom and showing left tail with area 0.05.

then the *t* test is the better test to use. However, if the assumption of *Y* given *x* having a normal distribution is doubtful (as it often is, and as a histogram or Q-Q plot of the residuals may suggest), we can still use the *Z* test as a good approximation, provided that the sample size *n* is large ($n \geq 30$ is our rule of thumb).

Confidence Interval for β_1

Recall from Chapter 8 that the general form of a confidence interval (CI) for a parameter is

$$\text{statistic} \pm (\text{critical value}) \cdot (\text{estimated SE of the statistic}),$$

where the statistic is a good (hopefully unbiased) estimator of the parameter.

If the true slope is β_1, then, under our assumptions (including the assumption of normality), the standardized statistic

$$\frac{b_1 - \beta_1}{\widehat{SE}(b_1)}$$

is distributed as a *t* distribution with $n - 2$ degrees of freedom. Based on this, analogous to what we learned in Section 8.8 for $(\overline{X} - \mu)/\widehat{SE}(\overline{X})$, we can obtain a $100(1 - \alpha)\%$ CI for β_1:

$$b_1 \pm t(n-2)\,\widehat{SE}(b_1)$$

or

$$b_1 - t(n-2) \cdot \widehat{SE}(b_1) \leq \beta_1 \leq b_1 + t(n-2) \cdot \widehat{SE}(b_1)$$

where $t(m)$ is the upper tail probability point of the *t* distribution with *m* degrees of freedom, given by

$$P(-t(n-2) \leq T \leq t(n-2)) = 1 - \alpha$$

as shown in Figure 11.10 for $\alpha = 0.05$. Note that one then consults the *t* table using $\alpha/2$ rather than α:

$$P(T \geq t(n-2)) = \frac{\alpha}{2}.$$

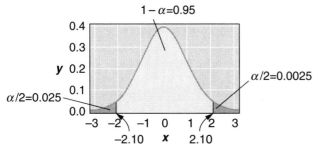

Figure 11.10 *t* distribution with 18 degrees of freedom and showing 95% of the area between −2.10 and 2.10.

Example 11.4 (continued)

Returning to our Key Problem, we can obtain a 95% confidence interval for the slope parameter as

$$-1.2 - (2.10 \cdot 0.24) < \beta_1 < -1.2 + (2.10 \cdot 0.24)$$

or

$$-1.704 < \beta_1 < -0.696.$$

Here we have used the fact that $\alpha/2 = 0.025$ and $P(T > t(18)) = 0.025$ yields $t(18) = 2.10$. Since the 95% confidence interval lies entirely on the negative side of 0, this is another statistical indication that the correlation ρ between the age at first word spoken and the Gesell adaptive score is negative. The negative sign in b_1 is unlikely to be an accident of the natural randomness in the Y random variables.

Inference about the slope parameter is possible even when n is smaller than 30 and when the assumption of normality of the Y random variables is not appropriate. One increasingly used solution is to do bootstrapping to get an empirical estimate of the probability distribution of the statistic b_1 from the data and then to use this to build a percentile-based CI for β_1, as was done in Section 8.7 for a population mean. Suppose that we have a box model containing n balls, each ball being one of the n observed data pairs (x_i, y_i). Each time we perform bootstrap sampling, we randomly draw n pairs of new data from the box by randomly sampling n balls with replacement. We repeat this, say, 1000 times. For each such bootstrap sample, we recalculate a new least squares slope, obtaining 1000 bootstrap-estimated slopes b_1^*.

Then a $100(1 - \alpha)$% CI can then be obtained by taking the $100(\alpha/2)$th and $100(1 - \alpha/2)$th percentiles of the 1000 bootstrapped slopes as the two endpoints. (Recall that the αth percentile of a set of numbers, ordered from smallest to largest, is the smallest number such that at least 100α% of the data lie at or to the left of the number. For example, if the data set is −5, −1, 3, 6, 7, then the 75th percentile is 6, because 60% of the data lie at or to the left of 3 and 80% of the data lie at or to the left of 6.) Fortunately, because the number of bootstrap samples is always large (≥ 200 is reasonable in applications), the αth percentile will have almost exactly 100α% of the observations to its left and the remaining $100(1 - \alpha)$% of the observations to its right, as desired.

The 1000 bootstrapped samples, of course, have to be drawn by a computer. The instructional software accompanying this textbook, or any statistics software package that supports bootstrapping, can perform this task. In a simulation of 1000 bootstrapped samples, we obtained a 95% confidence interval for the slope β_1 in our Key Problem of (−1.52, −0.25). That is, after the 1000 bootstrap sampled b_1^* values were ranked from smallest to largest, the 25th $b_1^* = -1.52$, and the 975th $b_1^* = -0.25$. This is a wider interval than the interval of (−1.704, −0.696) we obtained using our normal-distribution-based approach. But it still does not contain 0, as we might expect.

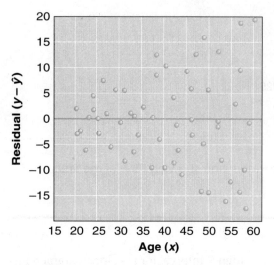

Figure 11.11 Residual plots of female blood pressure versus age.

We recommend using the bootstrap method for finding a confidence interval for β_1 in problems where
  i. $n < 30$ and the normality assumption for the distribution of Y at x is suspect (as would be indicated by a density histogram for the residuals $Y - \hat{Y}$ that is decidedly not bell shaped), or
  ii. our residual plot suggests that the variation of the residuals around 0 increases or decreases considerably with x (indicating that the constant-SD assumption for Y as x varies is wrong). This second point is illustrated by Figure 11.11, which displays the residuals for the blood pressure versus age scatterplot of Figure 11.3).

Section 11.1 Exercises

Assume normal population sampling, as needed in the following exercises.

1. The following table records the time required to cook turkeys of various weights. We are interested in the relationship between the weight of a turkey (x) and required cooking time (Y).

Weight (pounds)	Cooking time (hours)
55	3.5
7	4
10	4.4
14	5.6
18	6.7
22	8.4

   a. Give the least squares prediction equation for $\hat{Y}$ at x.
   b. Give the standard error of the slope estimate b_1.
   c. Give a 90% CI for the slope β_1, and state your assumptions.
   d. Test at the 10% level of significance whether the cooking time increases with the weight of a turkey.
   e. Do part c if the normal population assumption is not justified (assumes software available).

Section 11.2 Inference about Prediction and Estimation

2. Good runners take more steps per second as they speed up. Below are the average numbers of steps per second for a group of women runners at different speeds. The speeds are in feet per second.

Speed (ft/s)	15.8	16.9	17.5	18.7	19.2	20.2	22.5	24.0
Steps per second	3.1	3.2	3.2	3.3	3.4	3.4	3.6	3.8

   a. We want to predict steps per second from running speed for a particular athlete who is not part of this study. Give a prediction equation. If her speed is 23.0 feet per second, predict her number of steps per second.
   b. When the speed increases by 1 foot per second, how many more or fewer steps per second do you expect the runner to take?
   c. Give the standard error of your estimate in part **b**.
   d. Assuming normal population sampling, test at the 5% level of significance whether the model is worthwhile for prediction. *Hint:* It seems safe to assume that $\beta_1 < 0$ cannot occur.

3. The following table gives the math exam scores of twins from eight families. Test at the 5% level of significance to determine whether there is any correlation between the scores of the twins. State your null hypothesis and alternative hypothesis and state your assumptions. *Hint:* Recall that the issue of whether there is correlation is equivalent to whether $\beta_1 \neq 0$.

Score of twin 1	75	86	67	88	76	56	60	92
Score of twin 2	66	75	70	78	82	66	71	88

4. The t statistic for testing whether the slope parameter is 0 or not has another expression based only on the correlation coefficient r between x and Y. It is

$$T = \frac{\sqrt{n-2}}{\sqrt{1-r^2}} r.$$

   Use this formula to recalculate T for the problem in Exercise 3.

5. The data in the following table are the number of push-ups y that could be done by a sample of 12 male instructors at Howard Community College. (The 38-year-old teacher was a track coach.) Assume normal population sampling. Find a 95% confidence interval for the slope parameter β_1 when regressing the number of push-ups on the age of the teachers.

Age	Number of push-ups	Age	Number of push-ups
21	10	22	9
25	8	27	6
22	11	44	4
28	6	48	3
30	7	35	8
38	15	48	5

6. In some cases we may wish to fit a model line to the data (x_i, Y_i) that is known to pass through the origin. The equation of the line then takes the form of $\hat{Y} = \beta_1 x$. The estimated least squares-based slope can be computed by

$$b_1 = \frac{\sum xY}{\sum x^2}$$

and its estimated standard error is

$$\widehat{SE}(b_1) = \frac{S_E}{\sqrt{\sum x^2}}$$

where S_E is defined by $S_E = \sqrt{\sum(Y-\hat{Y})^2/(n-1)}$; note that division is by $n-1$ and not $n-2$ as was done earlier (this is theoretically correct). If the distribution of Y given x is normal with mean $E(Y|x) = \beta_1 x$ and standard deviation σ, it can be shown that

$$\frac{b_1 - \beta_1}{\widehat{SE}(b_1)}$$

has a t distribution with $n-1$ degrees of freedom (note $n-1$, not $n-2$). Give a formula for a $100(1-\alpha)\%$ CI for β_1.

11.2 INFERENCE ABOUT PREDICTION OF Y AND ESTIMATION OF THE LINE $E(Y|x)$

Given an estimated regression line $\hat{Y} = b_0 + b_1 x$, we may attempt to predict a "new" observation Y at x, which we denote as Y_{new} to emphasize that (as yet) it is an unobserved random variable. This variable is one whose unknown value played no part in the computation of b_0 and b_1 used to produce the estimated line. A natural prediction of Y_{new} given x is $\hat{Y}_{\text{new}} = b_0 + b_1 x$, namely to evaluate the estimated regression line at x. This is also the estimate of the line $E(Y|X=x)$, usually denoted by $E(Y|x)$. Thus, throughout this section, our prediction of a new Y at x is the *same* as our estimate of the regression line at x. For convenience, we now denote $E(Y|x)$ by μ_x, emphasizing that the parameter $E(Y|x)$ is a population mean to be estimated and it depends on x. An observed value of the new random variable Y_{new} will have μ_x as its theoretical mean, but it will randomly deviate from μ_x. Thus, even if we know the true regression line, and hence the exact value of μ_x, we still cannot know what Y_{new} at x will be. (Recall how much Y deviated about the true line in Figure 11.1!) Clearly, predicting a Y_{new} at x accurately will be more difficult than estimating μ_x accurately!

Estimating the Mean Value of Y Given x (denoted $E(Y|x)$)

Suppose that in our Key Problem we would like to estimate the average Gesell adaptive score for the *population* of all children who utter their first word at $x = 30$ months. That is, we want to estimate the population parameter $E(Y|x=30) = \mu_{30}$. Based on the least squares line $\hat{Y} = 109.3 - 1.2x$, we estimate μ_{30} by

$$\hat{\mu}_{30} = 109.3 - (1.2 \cdot 30) = 73.3.$$

How accurate is this estimate? We can answer this question by calculating a 90% confidence interval for μ_{30}. We solve this problem under the assumption that the Y random variables (at each x) are normally distributed. A probability argument can establish that the desired confidence interval is based on the t distribution. As usual, it uses our general form of a confidence interval,

(statistic) ± (critical value) × (estimated standard error of the statistic).

Specialized to our regression setting of estimating μ_x this yields

$$\hat{\mu}_x - t\,\widehat{SE}(\hat{\mu}_x) \le \mu_x \le \hat{\mu}_x + t\,\widehat{SE}(\hat{\mu}_x),$$

Section 11.2 Inference about Prediction and Estimation

with the critical value t found in a t table with $n - 2$ degrees of freedom. Here, the estimated standard error of $\hat{\mu}_x$ is given by

$$\widehat{SE}(\hat{\mu}_x) = S_E \sqrt{\frac{1}{n} + \frac{(x - \bar{x})^2}{\sum(x_i - \bar{x})^2}}.$$

As usual, S_E denotes the standard error of the residuals, and n the sample size. Thus, the $100(1 - \alpha)\%$ CI for μ_x is given by

$$\hat{\mu}_x - t(n-2)\sqrt{\frac{1}{n} + \frac{(x-\bar{x})^2}{s_x^2(n-1)}} \leq \mu_x \leq \hat{\mu}_x + t(n-2)\sqrt{\frac{1}{n} + \frac{(x-\bar{x})^2}{s_x^2(n-1)}},$$

where $t(n - 2)$ is the upper $\alpha/2$ tail point obtained from Table F of the t distribution: $P(T \geq t(n - 2)) = \alpha/2$ where T is t-distributed with $n - 2$ degrees of freedom.

For the Key Problem, we have $\hat{\mu}_{30} = 73.3$ at $x = 30$, $\bar{x} = 14.25$, $s_x^2 = 66.09$, $t(18) = 1.734$, and $S_E = 8.6$. Thus, 90% CI for the expected score Y at $x = 30$ is

$$73.3 \pm 1.734 \cdot 8.6 \sqrt{\frac{1}{20} + \frac{(30 - 14.25)^2}{66.09 \times 19}}$$

or the interval $73.3 \pm (1.734 \cdot 4.279)$, namely 73.3 ± 7.41, or $(65.9, 80.7)$. Thus, based on the amount of data we have available (only 20 children), there is a fair amount of uncertainty in estimating μ_{30}. If we had 200 children instead of 20, then, if for simplicity S_E and s_x^2 take the same values, we would have obtained the much more accurate interval 73.3 ± 2.30. This estimate of the line $E(Y|x)$ will be very accurate if n is large (e.g., for $n = 10,000$, we get 73.3 ± 0.33), since $1/n$ approaches 0 for large n.

Predicting an Individual New Observation, Y_{new}

When in the Key Problem we use $\hat{Y} = b_0 + b_1 x$ as a prediction of an individual child's random response Y_{new} at x, there is a larger standard error than there is for an estimate of μ_x. In addition to the variability in the estimation of the regression line, the natural variability of Y about the line at x adds random noise to the observed Y_{new} being predicted. This noise is quantified by the standard error S_E.

An interval used to predict a new individual response Y_{new} is called a **prediction interval**. Note that we use "confidence interval" for the fixed target μ_x, but we use "prediction interval" for the random target Y_{new} at x. A $100(1 - \alpha)\%$ prediction interval for Y at x is also in the form of

$$\hat{Y}_{new} - t\,\text{SE}(\hat{Y}_{new}) \leq \hat{Y}_{new} \leq \hat{Y}_{new} + t\,\text{SE}(\hat{Y}_{new})$$

where

$$\widehat{SE}(Y_{new}) = S_E \sqrt{1 + \frac{1}{n} + \frac{(x-\bar{x})^2}{s_x^2(n-1)}}$$

and t is the same as the critical value in the CI for μ_x.

Comparing this prediction interval with the CI for μ_x, we see that the only difference is that its standard error term has an extra constant 1 added into the expression whose square root is multiplied by S_E, to account for the random variation of Y_{new} around its mean μ_x.

For our Key Problem we have

$$\widehat{SE}(Y_{new}) = 8.6\sqrt{1 + \frac{1}{20} + \frac{(30 - 14.25)^2}{66.09 \times 19}} = 9.606,$$

so a 90% prediction of Y_{new} at $x = 30$ is $73.3 \pm (1.734 \times 9.606)$, namely 73.3 ± 16.66, or (56.64, 89.96). This is quite a bit wider than the 90% CI of μ_{30}, namely (65.9, 80.7). Increasing the number of children studied does little good in improving prediction accuracy. For example, using 10,000 children, with S_E and s_x^2 the same, yields

$$73.3 \pm 14.91,$$

a small gain over the prediction interval 73.3 ± 16.66 we found based on 20 children. That extra "1" makes all the difference!

There is a key lesson here: One can estimate μ_x as accurately as one wants by collecting enough data, but the accuracy of predicting a new observation Y_{new} at x cannot be increased beyond the inherent variability around the line, as given by σ and as estimated by the standard error S_E, regardless of how many data points are collected!

Note that the sizes of both the confidence interval for μ_x and the prediction interval for Y_{new} depend not only on S_E, which is a measure of the inherent variability of Y about the true line, but also on how far the mean $\bar{x}$ is from the x value at which the estimation or prediction is to be done. The farther x is from the center $\bar{x}$ of the observed x values, the less accurate the estimate or prediction will be. This prediction at a value of x near the bulk of the observed data is more accurate than for a value of x near the boundary of the data range.

If we need to make a prediction for x beyond the range of the data (say, for children whose first word comes in less than 7 months or more than 42 months in our Key Problem), we have an **extrapolation** problem. For an extrapolation to be useful, we need sound reasons to believe that a linear relationship between the two variables will continue to hold in the region of x that is of interest. Unless we can justify such an assertion on the basis of substantive knowledge of the subject matter, it is extremely risky to assume linearity outside the range of x values actually observed.

A famous example of faulty extrapolation is found in Mark Twain's *Life on the Mississippi*. The lower Mississippi River weaves back and forth many time across its floodplain, and occasionally the water breaks through the gap between one bend and the next, cutting off the old "oxbow" bend and making the river shorter overall. Twain noted that the river was 242 miles shorter than it had been 176 years ago, and on the basis of this mean rate of shortening, he predicted that in 742 years the lower Mississippi would be only 1.75 miles long. Of course, as the river shortens, it also straightens, so oxbow jumping becomes less and less frequent!

Section 11.2 Exercises

1. Refer to Exercise 1 of Section 11.1. We wish to estimate the cooking time required for turkeys weighing 12 pounds. Assume the normal distribution of Y given x.
    a. Find a 90% CI for the mean cooking time required, as the meat supplier might state on the label supplied with its turkeys.
    b. Find a 90% prediction interval for the cooking time required for a given turkey that weighs 12 pounds, such as one that you are cooking in your oven.
    c. Which of the intervals in parts **a** and **b** is wider? Why?
2. The following table gives the time (in seconds) it took for the decathlon participants in the 1988 World Olympics to finish the 100-meter race (x) and the 110-meter hurdles (y). We wish to use the data to predict the time it takes for the 110-meter hurdles from that for the 100-meter race for top men athletes. Assume normal population sampling.

x	11.25	10.87	11.18	11.02	10.83	11.18	11.33	11.57
y	15.13	14.46	14.8	14.40	14.18	14.39	15.39	16.20

a. For athletes (imagine thousands of them) taking 11 seconds to finish the 100-meter race, how long do you expect them on average to finish the 110-meter hurdles? Give a 90% CI for the mean time.

b. Give a 90% prediction interval for the time it will take to finish the 110-meter hurdles for one athlete who just finished the 100-meter race in 11 seconds.

c. Examine the scatterplot and the residual plot and comment on the validity of your prediction interval. (*Hint:* Do you think that the prediction is less accurate for faster runners in the 100-meter race?)

3. In part c of Exercise 1, if we increase the sample size to 25 but assume the same error variance (that is, the same standard error of the residuals), how wide will the 90% prediction interval be at $x = \bar{x}$?

4. In a linear regression analysis with 16 observations of x and y, we find the five sums as $\sum x_1 = -1.6$, $\sum y_i = 0.6$, $\sum x_i^2 = 14.0$, $\sum y_i^2 = 38.0$, and $\sum x_i y_i = 13.0$.

a. Find the width of a 95% CI for the mean value of Y when $x = -0.1$ (the value of $\bar{x}$).

b. Do part **a** for $x = 2.0$ and compare the two intervals, commenting on the comparison.

c. If the range of the x's is from -2.5 to 2.2, can we trust a prediction of y at $x = 2.0$?

5. A study to find the mail survey response rate among the elderly (ages 60 years to 90 years) estimated the following relationship between age (x) and percentage of people responding (y):

$$\hat{Y} = 90.2 - 0.6x$$

a. What is the predicted percentage responding to a mail survey among 75-year-olds?

b. Find a 95% prediction interval for the percentage responding among the 80-year-olds. Assume that the estimated standard error of the slope is 0.14 and that $n = 200$, $\bar{x} = 68$, $s_x = 5$.

11.3 TRANSFORMATION OF VARIABLES

So far we have learned how to model a linear relationship, make inferences about the slope parameter, and use the resulting linear equation to predict a new Y and to estimate $E(Y|x)$. Not all relationships are linear, as we learned in Chapter 3. However, some nonlinear regression models can be converted into linear regression models by a simple transformation of one or both variables. This requires some new notation for ease of presentation.

Example 11.5

Suppose we have measurements of weights (y) and circumferences (x) of several roundish pebbles (some being flattened, some egg-shaped, and some fairly spherical) collected in one area. To measure the circumference, we take a thread and surround the pebble in the way that makes the needed length of the thread *as long as possible*. For example, an egg-shaped stone would have the thread encircling the egg around its two ends. One can argue that the underlying relationship between these two variables is likely to be in the form of $y = Ax^B$, where, for example, the exponent B is equal to 3 if the stones are spherical. For a set of egg-shaped pebbles, and even more so for a set of flattened pebbles, $B < 3$ seems likely. To try to find a linear model resulting from transforming one or both of the variables, we take the logarithm (that is, the natural logarithm) of y and see what happens to the equation. Then recalling the general rules for logarithms that $\ln(uv) = \ln(u) + \ln(v)$ and that $\ln(u^v) = v \ln(u)$, this yields

$$\ln(y) = \ln(A) + B\ln(x).$$

Thus we get a linear equation by defining $y_1 = A_1 + B_1 x_1$, where $x_1 = \ln(x)$, $y_1 = \ln(y)$, $A_1 = \ln(A)$, and $B_1 = B$. Thus we have succeeded in converting the original nonlinear equation into a linear one, as desired.

Given the data, we can fit the least squares line to the (x_1, y_1) points and thus get an estimate of the intercept A_1 and the slope B_1 using the standard least squares equations of Chapter 3. We can then solve for A and B in the original model. The pebble problem is an interesting data-collection statistics project for you to try.

The natural log transformation is often helpful in producing a linear equation for a positive variable y that increases or decreases exponentially fast with the other variable x. Many situations display so-called exponential growth. A model for such exponential change is

$$y = AB^x$$

with $B > 0$. Try the special case of $y = 2^x$ for $x = 1, 2, 3, 4, \ldots$, to see how truly fast exponential growth can be! (Do not confuse this with $y = x^2$, which grows much more slowly.) Variables such as the early growth (y) in size or weight over time (x) of a living organism having ample nutrients, or of the number (y) of organisms produced over time (x) if conditions are ideal for the breeding of the organism, often display exponential growth. (Charles Darwin once talked about the world being rapidly covered with elephants if such exponential growth were possible.) Another example is the growth of money that increases in value by a fixed percentage every time period, with the profits being reinvested. Examination of the scatterplot is helpful in determining whether the growth in y as x varies seems to display exponential growth.

Consider the exponential growth equation

$$y = AB^x.$$

Do not confuse this with the very similar-looking $y = Ax^B$ that we considered in the preceding pebble example. Let us take the logarithm of y and see what happens to the equation.

$$\ln(y) = \ln(A) + \ln(B)x$$

This result suggests that we can linearize the new equation by: $y_1 = \ln(y)$; $A_1 = \ln(A)$; $x_1 = x$(unchanged); $B_1 = \ln(B)$. This yields

$$y_1 = A_1 + B_1 x_1,$$

which is linear, as desired.

We will use this analysis in the next example.

Example 11.6

The following table gives the annual sales in billions of dollars (x) and return on assets (y) of seven telecommunications companies in the United States for the year 1999.

Company	Sales	Return on assets
AT&T	64.1	2.7%
MCI WorldCom	38.1	5.5%
Bell South	25.7	9.3%
Sprint	17.2	8.7%
GTE	25.6	8.5%
Net2Phone	0.5	35.5%
IDT Corp.	1.0	19.8%

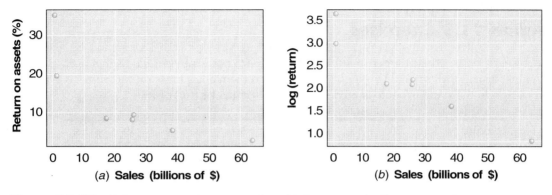

Figure 11.12 Return on assets versus sales of telecommunication companies.

Figure 11.12 gives the scatterplot of the data, indicating higher returns on assets for companies with lower annual sales. The increase in returns for companies as sales get lower is faster than a linear trend would suggest. In fact, it is more like an exponential rate of increase *as sales decrease*. The graph in Figure 11.2 is turned the other way, so we have exponential decay rather than growth as sales (x) increase. The apparent exponential change in y as x increases suggests taking the natural log transformation of the y variable and not transforming the x variable. We do so and examine the scatterplot to see whether it appears linear. The scatter plot of $\ln(\text{return}) = \ln y$ versus sales $= x$ (Figure 11.12) does suggest a linear trend and leads us to the linearized model:

$$y_1 = A_1 + B_1 x_1$$

where $y_1 = \ln(y)$, $A_1 = \ln(A)$, $x_1 = x$, and $B_1 = \ln(B)$.

The standard regression line calculations of Chapter 3 give $\hat{A} = 3.14$ and $\hat{B}_1 = -0.036$ with $r^2 = 0.91$. That is, $100r^2 \% = 91\%$ of the variation in log (return) is explained by the transformed equation. Moreover, the equation proposed, namely $y = AB^x$, gives $\hat{A} = e^{3.14} = 21.10$ and $\hat{B} = e^{-0.036} = 0.965$. Thus our least squares estimated equation becomes

$$\hat{Y} = 3.10 \times (0.965)^x.$$

In addition to the log transformation, the following are some of the nonlinear models that can be converted into linear models with appropriate transformations.

Nonlinear model	Transformation	New Model
$y = 1/(A + Bx)$	$y_1 = 1/y$	$y_1 = A + Bx$
$y = (A + Bx)/x$	$x_1 = 1/x$	$y = B + Ax_1$
$y_3 = Ax^B$	$y_1 = \ln(y), x_1 = \ln(x)$	$y_1 = \ln(A) + Bx_1$
$y^3 = A + Bx$	$y_1 = y^3$	$y_1 = A + Bx$
$y = A + Bx^2$	$x_1 = x^2$	$y = A + Bx_1$

In the last two equations, any fixed power works; 3 and 2 were chosen arbitrarily.

It is important to realize that not all nonlinear models can be linearized by transformations. For example, $y = (A + x)^B$ cannot be linearized by transforming x or y. Even the simple and often occurring

$$y = A + Bx + Cx^2$$

cannot be linearized. However, polynomial models can be solved by using (linear) *multiple* regression, a topic discussed in Chapter 12.

Section 11.3 Exercises

1. A researcher asked a number of subjects to compare notes of various decibel (sound) levels against a standard (set at 80 decibels) and to assign them a loudness rating. Subjects were told to assign to the standard note of 80 decibels a loudness rating (y) of 10.

Stimulus (x)	30	50	60	70	75	80	85	90	95	100
Response (y)	0.2	1.0	3.0	5.0	8.5	10.0	14.0	20.0	29.0	43.0

   a. Is a linear regression appropriate between x and y?
   b. Consider the natural log transformation of y. Is a linear regression appropriate between x and $\ln(y)$?
   c. Give a 95% prediction interval of Y_{new} when $x = 55$. (Hint: You can get a 95% prediction interval for $(\ln(Y))_{new}$ first.)

2. Refer to Exercise 2 of Section 11.2. If we are interested in the speed of the athletes in the 100-meter race (call it x_1) and the 110-meter hurdles (call it y_1), do you think that a linear relationship exists between x_1 and y_1? If not, can you find a transformation that allows you to find a prediction equation of y_1?

3. The following table gives the length of the planetary year in days (y) and the distance from the Sun in hundreds of millions of miles (x) for the nine planets of the Solar System.

x	0.360	0.672	0.929	1.415	4.833	8.867	17.820	27.930	36.640
y	88.00	224.70	365.26	687.00	4332.60	10759.20	30685.40	60189.00	90465.00

   a. Find the least squares estimate for regressing y on x.
   b. Examine the residual plot. Do you see a nonlinear pattern?
   c. Perform linear regression of y on $x^{1.5}$ (i.e., $y = a + bx^{1.5}$) and examine the residual plot.
   d. Compare the standard error of the residuals S of the two regressions.
   e. Compare R^2 of the two regressions.

4. A product is manufactured in batches at a certain factory. The batch size can be varied (within a certain range). A machine manufactures the product at a constant rate. Before the production of a new batch begins, the machine has to be *set up*. The setup time of the machine can be assumed to be constant. A certain number of units of the product were produced each day for the last 14 days. The batch sizes, however, were different on each of these 14 days. The following table gives the details regarding the batch size and the total time (setup time plus production time) taken for production.
   a. Regress total time on batch size.
   b. Regress total time on the reciprocal of batch size.
   c. You will find that the second regression is better than the first. Give reasons why the second regression is superior.

Day	Batch size (number of units)	Total time (in minutes)	Day	Batch size (number of units)	Total time (in minutes)
1	10,000	795	8	24,000	675
2	12,000	760	9	26,000	690
3	14,000	710	10	28,000	675
4	16,000	730	11	30,000	680
5	18,000	680	12	32,000	660
6	20,000	705	13	34,000	660
7	22,000	680	14	36,000	655

5. *Elasticity of demand* is a useful microeconomics concept. Elasticity measures the percentage change in the quantity bought in response to the product. The following table shows the quantity of a certain product at prices. Estimate the elasticity of demand using regression, after applying the variables.

Price	Quantity demanded
25	38
28	31
31	31
33	28
36	27
40	25
42	24
45	23

CHAPTER REVIEW EXERCISES

1. Nine households are surveyed for their monthly expenses on food (y). The following table gives the data together with the number of adults in each household (x).

x	2	3	3	5	6	6	6	6	6
y	220	234	297	304	419	481	569	651	652

   a. Find the least squares line to estimate y from x (either estimating μ_x or predicting Y_{new}). Is the model significant at the 5% level of significance, assuming normal population sampling?
   b. Find a 95% confidence interval for the expected (average over many people) monthly food expenditure for all households with 4 adults, assuming normal population.
   c. Find a 95% prediction interval for the monthly food expenditure for a particular household with 4 adults, assuming normal population sampling.
   d. Comment on the model and prediction.

2. Refer to Exercise 1.
   a. If we randomly change the order of the y values before doing the regression, would we expect the model to be significant?
   b. Use a set of nine random numbers to change the order of the y values randomly. Then perform the hypothesis test that the slope is 0, against the two-sided alternative.

3. The following table gives population (x) and the number of telephones (y) for each of six zones. We wish to see how the population size relates to the number of telephones.

x	4041	2200	30,148	60,324	65,468	30,988
y	1332	690	11,476	18,368	22,044	10,686

   a. Does the scatterplot suggest a linear relationship between x and y?
   b. Even if the scatterplot did not suggest the need for it, consider plotting $\sqrt{y}$ versus $\sqrt{x}$. Fit a linear model ($\sqrt{y} = a + b\sqrt{x}$ using least squares. ($y_1 = \sqrt{y}$, $x_1 = \sqrt{x}$).
   c. Give a 95% confidence interval for the slope parameter of the model in part b.

11: INFERENCE ABOUT REGRESSION

dict the number of telephones for a zone with population size 10,000, using the model in part **b**. Give a 95% prediction interval, assuming normal population sampling.

  e. What would you get in part **d** if you used a linear model of y versus x?

4. In Exercise 3, how would you design a bootstrap method to obtain a 95% confidence interval for the slope parameter β_1, assuming an underlying linear relationship $y = b_0 + b_1 x$?

5. The following table on the left lists the winning times of the 100-meter run event at the Summer Olympics from 1896 to 1996.

Year	Winning time (in seconds)	Year	Winning time (in seconds)
1896	12.0	1956	10.5
1900	11.0	1960	10.2
1904	11.0	1964	10.0
1908	10.8	1968	9.95
1912	10.8	1972	10.14
1920	10.8	1976	10.06
1924	10.6	1980	10.25
1928	10.8	1984	9.99
1932	10.3	1988	9.92
1936	10.3	1992	9.96
1948	10.3	1996	9.84
1952	10.4		

  a. Regress winning times on year. Find the 95% confidence interval for the slope, assuming normal Y's at each x.

6. The following data shows the number of promises made by political candidates and the number of promises fulfilled once they are elected. Fit a linear relationship between promises made and promises fulfilled. (Please do not take the result seriously!)

7. Use the bootstrapping method to find the 95% CI for the slope. (If you do not have access to bootstrap simulation software, just explain in detail how you would carry out finding a bootstrap CI.)

Promises made	Promises fulfilled	Promises made	Promises fulfilled
15	11	39	3
28	8	30	9
22	9	43	4
44	1	21	3
20	8	19	9
35	4	34	5

8. The table at the top of the next page shows the midterm grade and the final grade obtained by each of 15 students.

  a. Regress final grade on midterm grade.
  b. Calculate the standard error of the slope estimate.

Midterm grade	Final grade	Midterm grade	Final grade
90	83	82	84
80	91	99	93
98	97	66	52
72	83	89	75
92	86	37	56
97	80	53	27
56	68	95	81
90	97		

  c. Find the 95% CI for the slope. What are your underlying assumptions?
  d. What is the P-value for $H_0: \beta_1 = 0$ versus $H_{1\alpha}: \beta_1 > 0$?
9. Regression analysis can be used to find out whether the stock market has *momentum* (that is, whether the market is likely to go up/down today because it went up/down yesterday). Do the following exercise: Note the Standard & Poor's 500 stock index for the last month. For each day, calculate the "daily return" as (Today's Index – Yesterday's Index)/Yesterday's Index. For each day's return (call it Today's return) find the corresponding previous day's return (call it Yesterday's return). That is, pair off two consecutive days' returns. Test the hypothesis that today's return is not related to yesterday's return. If this null hypothesis is not rejected, then the stock market displays no statistical evidence of momentum. If the null hypothesis is rejected, then there are two possibilities. A positive relationship between the variables would indicate market momentum. A negative relationship indicates that the market has a tendency to "overreact" one day and correct itself the next day.
10. The rate of soil water evaporation (Y) depends on the speed of the air (x). Suppose the estimated relationship is given by the equation

$$\hat{Y} = 4.5 + 0.0084x,$$

and the standard error of the slope $S_{b_1} = 0.022$. Find a 95% CI for the slope. Assume that the study was based on 11 pairs of data points.
11. Suppose the amount of oxygen consumed (Y) during the physical exercise period (x, in minutes) based on $n = 12$ is estimated as

$$\hat{Y} = 585 + 96.5x.$$

  a. For a 10-minute exercise, what is the prediction of the amount of oxygen consumed by a particular athlete?
  b. If the estimated standard error S_b of the slope is 1.5, find a 95% CI for the average amount of oxygen consumed during a 10-minute workout, assuming normal population sampling, and assuming $\bar{x} = 12$, with $s_x = 4$.
  c. Find a 95% prediction interval for the amount of oxygen consumed by a person during a 15-minute workout, assuming normal population sampling, and assuming $\bar{x} = 12$, with $s_x^2 = 4$.
12. Regression is a very useful method for finding growth rates. The following table shows the population of the United States for different years.
  a. Plot the population of the United States on the y axis and the year on the x axis. What is the natural shape of the graph you observe?

Year	U.S. population
1909	90,490,000
1919	104,514,000
1929	121,767,000
1939	130,879,718
1949	149,188,130
1959	177,829,628
1969	202,676,946
1979	225,055,487
1989	246,819,230
1999	272,690,813

Source: U.S. Census Bureau.

b. If you plot the natural log of population on the y axis and the year on the x axis, what is the shape of the graph you observe?

c. What is the equation you would use in estimating the population growth rate? (Take a hint from part **b**)

Glossary

absolute value The magnitude or size of a number, disregarding whether the number is negative or positive. For positive numbers, the absolute value is simply the value of the number. For negative numbers, the absolute value is the negative of the number (i.e., the number without the minus sign).

acceptance/rejection hypothesis testing Hypothesis testing in which the two possible decisions are to reject the null hypothesis or to accept the null hypothesis. Contrasted with **significance testing** where one never accepts the null hypothesis.

actual (real) population See **population, actual (or real).**

addition rule The fact that $P(A \text{ or } B) = P(A) + P(B)$ when the events are mutually exclusive and $= P(A) + P(B) - P(A \text{ and } B)$ in general.

additive two-way ANOVA model A two-way ANOVA model that assumes that no interaction is present.

alternative hypothesis The hypothesis that specifies the values of the *population parameter* that are possible when the *null hypothesis* fails; can be one-sided or two-sided. Usually corresponds to a real-world research hypothesis. Denoted H_A.

analysis of variance A body of statistical methods for determining the influences of various postulated explanatory categorical variables (often called *factors*) on a continuous *response variable*, such as the influences of wheat variety and soil type on crop yield.

ANOVA Stands for **analysis of variance**.

ANOVA table The standard tabular format for presenting the results of an analysis of variance. Also called *analysis of variance table*.

area fallacy A graph consisting of objects representing frequencies in which the relative areas of the objects are not proportional to the corresponding relative frequencies. Usually caused by making heights of objects proportional to relative frequencies and drawing the objects to scale, with the result that taller objects are also drawn as wider. The resulting distortion of the objects magnifies frequency differences via areas that are visually misleading.

arithmetic average Also called *average*. The sum of a set of numbers divided by the quantity of numbers in the set; same as the **mean** or **sample mean** of the set of numbers.

back-to-back stem-and-leaf plot Compares the stem-and-leaf plots of two data sets by using a common set of stems but two opposite facing sets of leaves.

bar chart A graph consisting of bars (rectangles) that count the number of data points in each of several nonnumerical categories. See **Pareto chart.**

believability The fact that a valid statistical inference can be trusted and relied on, as contrasted with the fact that an invalid statistical inference cannot be trusted.

bell-shaped distribution A density histogram (or probability density function) that is shaped like a Gaussian (or normal) density.

between-samples sum of squares In a one-way ANOVA, the sum of squared deviations, each such deviation being a sample mean minus the overall sample mean of all the observations. Each such squared deviation is multiplied by its sample size, and then these are added. Also called a *treatment sum of squares* and denoted SSTR. Measures variation in means between populations.

between-sample variability In analysis of variance, variability of observations between different samples caused by differences among the population means of the samples. Also called *treatment variability*.

biased estimator An estimator of a population parameter that systematically underestimates or systematically overestimates the parameter. For example, $\sum(X_i - \bar{X})^2/n$ systematically underestimates σ^2. By formula, $\hat{\theta}$ is a biased estimator of θ if $E(\hat{\theta}) \neq \theta$.

bimodal data set A data set with two **clusters** of high concentration, such as durations of Old Faithful geyser eruptions.

binomial distribution The probability model for the number of successes x in a fixed number n of two-outcome ("success," "failure") trials that are independent and are of equal probability p; the distribution of a binomial random variable X is computed as

$$P(X = x) = \binom{n}{x} p^x (1-p)^{n-x} \qquad x = 0, 1, \ldots, n$$

binomial experiment Successive independent trials with a fixed number of trials and with two possible outcomes (designated 0 and 1 for convenience) per trial with $P(1) = p$ the same for all trials. If the statistic of interest is the number of 1s, then this leads to the **binomial distribution.**

bivariate data Data involving two variables, such as height and weight, or amount of smoking and a measure of health; often graphed in a **scatterplot.**

blocking Converting a one-way ANOVA into a two-way ANOVA by introducing a second factor (variable) into the experimental design judged likely to contribute substantially to the observed variability. Levels of the first factor are randomly assigned within homogeneous "blocks" of the second factor. The blocking factor is usually a varying property of the units of the experiment, such as soil type of the various soil plots used in a study of the yields of different varieties of wheat.

Bonferroni method In hypothesis testing of more than two populations, a method to simultaneously compare pairs of population parameters (often population means) when the hypothesis of the joint equality of all the parameters has been rejected. The confidence-interval-based paired comparisons produced by the method satisfy a specified overall statistical confidence percentage of being simultaneously correct.

bootstrap The technique of random resampling from a given sample done in practice with replacement, but done in this textbook first without replacement, using an extensive cloning of the given sample to produce a population-sized box to resample from. This was done to help students understand why bootstrapping works. From each bootstrapped sample drawn, the statistic of interest is computed. Because the given sample should be shaped approximately like the unknown population, the bootstrap-produced experimental distribution of the statistic of interest (as given by its density histogram) should be a good estimate of the unknown theoretical distribution of the statistic of interest. Many estimation and hypothesis testing procedures can be constructed using bootstrapping.

bootstrap confidence interval construction Uses bootstrapping of the sample and then either (1) estimates the standard error of the statistic of interest to form a bootstrapped standard-error-based confidence interval, or (2) estimates the appropriate extreme percentiles for the statistic of interest to obtain a bootstrapped percentile-based confidence interval.

bootstrap hypothesis testing Uses bootstrap resampling of the sample, but with the sample modified so that the null hypothesis holds for it.

bootstrapping to obtain the standard error of a statistic of interest Uses usual bootstrap resampling to generate a large number of bootstrapped copies of the statistic of interest, then computes their SD to estimate SE(statistic of interest).

box modeling (or box model) A method of simulating data by randomly drawing a number from a box repeatedly either with or without replacement between draws, depending on the application. The box may contain any number of real numbers, some or all of which may appear more than once. For example, the box containing 1, 2, 3, 4, 5, and 6 simulates the throwing of a fair die. The box also may be defined by supplying a list of numbers, each therefore having equal probability of being sampled from the box. Finally, one can specify a distribution for the box model, such as normal, binomial, and so forth. In order to carry out bootstrapping, the sample often becomes the box model.

boxplot A graphical way to display data that shows a box with the first and third quartiles as ends, with lines (*whiskers*) extending from the ends to the maximum and minimum values in the data set, and a vertical line through the middle of the box representing the median. Used to show the spread of a data set. Also called a *box-and-whisker plot* or a *five-number boxplot*.

boxplot, modified A boxplot that also displays outliers.

categorical variable or data A variable or data that falls into a finite number of often nonnumerical categories; the emphasis is on the frequency or proportions in each category (an example is the number of men and women in a college class).

causality Causality between bivariate variables x and y holds when it is known that varying x causes a change in the value of y. A correlation, even when large, between x and y does *not* imply causation (although causation is often the reason for the correlation).

census An attempt to sample every member of a population.

central limit theorem (CLT) The empirical and theoretical result that $\overline{X}$ has its experimental distribution (the density histogram of repeated simulations of $\bar{x}$ values) and its theoretical probability distribution, approximately normal with mean μ and standard deviation $\sigma/\sqrt{n}$ when the sample size n is large, where μ and σ are the population mean and SD, respectively, of the random sample that $\overline{X}$ is computed from.

central tendency Measures of central tendency indicate the middle or center of a set of data. The measure of center is also thought of as a typical value for a data set. Important measures of central tendency include the sample mean and the sample median.

chance variation The random variation displayed by a random variable or statistic in a setting where the outcomes are, or seem, random. See **random.**

chi-square density The theoretical curve used to calculate chi-square probabilities via areas under the curve for chi-square hypothesis testing; it is a continuous probability density function (pdf).

chi-square statistic $\chi^2 = \sum[(O - E)^2/E]$, where the sum is over all categories, O is the observed frequency, and E is the expected frequency under the null hypothesis. χ^2 has a chi-square density, approximately, under the null hypothesis when the sample size is large relative to the number of categories.

chi-square table A table that provides numerical values for probabilities involving areas under the chi-square density and that is needed for chi-square testing; provided in Table C of Appendix C.

chi-square test A statistical hypothesis test for situations in which an observed frequency and a theoretical frequency calculated from a null hypothesis are available for the possible categories and the data are from a large number of trials. A chi-square statistic (χ^2) is calculated and the null hypothesis is rejected if the statistic is large. One uses a simulation approach or uses the chi-square pdf depending on the setting.

cluster A portion of high concentration in a data set. Can be of inferential importance, as in the example of an unusually high number of cases of bacterial meningitis in a college town in a data set from many towns, indicating an epidemic in the college town.

coefficient of determination, r^2 The square of the correlation coefficient r of the best-fitting least squares regression line. The quantity $100\,r^2$ is interpreted as the percentage of the total variation in the y's around $\bar{y}$ that is explained by the best fitting regression line.

complementary event Given an event A, the complement of A ("not A") consists of all the outcomes not in A. For example, if A is the event that two of three children are boys, then not A is the event that there are either zero, one, or three boys. The equation $P(\text{not } A) = 1 - P(A)$ is called the complement rule.

conceptual population See **population, conceptual.**

conditional probability The probability that one event will occur given that another has occurred. The conditional probability that event A will occur given that event B has occurred is given by $P(A|B) = P(A \text{ and } B)/P(B)$.

confidence interval (CI) An interval computed from a random sample that is used to accurately estimate the location of some population parameter by an interval. The "confidence" probability that the random confidence interval contains the parameter must be stated along with specifying the confidence interval, such as a 95% confidence interval (95% means a confidence probability of 0.95). Usually of the form: statistic$\pm$(critical value) · (estimated SE of statistic).

confidence level The probability (typically 0.9, 0.95, 0.99) that the statistician's confidence interval contains the true, unknown population parameter. Usually stated as a percentage.

confounding Said of two or more possible causes of an observed effect on the response variable, when the statistical analysis cannot determine which is the cause.

continuous distribution A probability law for a continuous random variable. The probability that the random variable falls within any given interval is given by the area of the interval under a specified curve called a *density* or continuous pdf. Examples include the normal and chi-square distributions.

continuous random variable A random variable whose values are continuous. This means there are no "gaps" between values. Common examples where modeling the situation with a continuous random variable is appropriate are heights of randomly selected people and measurements involving time. The probability of a continuous random variable falling on an interval is given by the area under the random variable's continuous pdf in the interval.

continuous variable or data Data not restricted to a discrete set, such as the integers. Modeled by a continuous random variable.

control group The group of individuals or units not receiving the treatment (perhaps receiving a placebo), allowing for a comparison between the effect of a treatment and that of no treatment.

controlled experiment Experiment in "units" where the treatment is compared with no treatment by having the units split into a treatment group and a control group. Does not necessarily use randomization, although very desirable to do so.

correlation coefficient, r A measure of how close two variables of a scatterplot are to being perfectly linearly related; computed by dividing the sample covariance by the product of the two variables' sample standard deviations. Also called the *Pearson correlation coefficient*.

covariance A nonscaled measure of how closely two variables of a scatterplot are to being perfectly linearly related. A closely related and more easily interpreted measure is the **correlation coefficient.**

critical region Also called the *rejection region*. When doing acceptance/rejection hypothesis testing or a test of significance with a specified level of significance, the set of values for a hypothesis testing statistic for which we decide to reject the null hypothesis. For example we might reject $H_0: \mu = 0$ in favor of $H_1: \mu > 0$ if $\overline{X} > 1.3$. This is the critical region.

critical value In a confidence interval, the distribution-based value (often z- or t-table based) that provides the correct confidence level.

data Numerical information, usually about the real world, obtained by counting or measuring.

decision making See **hypothesis testing.**

degrees of freedom (df) (1) *For the chi-square distribution:* An integer that determines which chi-square density, and, hence, which row of the Table C chi-square table, to use. (2) *For the t distribution:* An integer that determines which t density, and, hence, which row of the Table F, Student's t table, to use. (3) *For the F distribution:* The integer that divides an ANOVA sum of squares to produce a corresponding statistic that is independent of the number of observations and the number of populations being sampled from. For the error sum of squares and the treatment sum of squares, such division allows use of the F distribution to do an ANOVA in the case of sampling from normal populations. The F distribution has a numerator degree of freedom and a denominator degree of freedom that together determine which F density, and hence which entries of the Appendix H F tables, to use.

density See **continuous distribution** and **continuous pdf.**

density histogram A frequency histogram vertically scaled so that the sum of the areas of its rectangles is 1. Used to estimate the population distribution producing the data. See also **experimental probability distribution.**

density polygon A piecewise linear graph obtained by joining the midpoints of the tops of the rectangles of a density histogram. Its shape approximates the probability law producing the data.

descriptive statistics Graphical and numerical techniques for describing or summarizing data that capture the essence of the data. As opposed to **inferential statistics.**

deviation from the (sample) mean The difference between the data value and the sample mean: $(x - \bar{x})$.

discrete probability model Assigns a positive probability to each of a *finite* number or a sequence of outcomes making up the sample space of a model.

discrete random variable A random variable whose values are discrete. This means that values have "gaps" between them. Random variables that involve counting objects (e.g., the number of heads in 10 coin tosses) are common examples of discrete random variables.

discrete variable or data A variable or data restricted to taking on values in a discrete set, such as the integers.

disjoint events Events with no outcomes in common. Also called *mutually exclusive events.*

distribution Also called **probability distribution.** The probability law for a random variable or statistic of interest; for example, the height of a person chosen at random may follow the normal distribution. In particular, in the discrete case, a list of all possible values of a random variable together with their corresponding probabilities: usually denoted as pairs of $(x, p(x))$'s.

dotplot A plot of a data set on a number line where each dot represents a member of the data set. Multiples values that are the same are stacked vertically.

double-blind study A study in which neither the subjects nor the experts conducting the study and evaluating the data know who has received the treatment and who has received the placebo.

draw a sample To produce a sample from a population or from a box model of a population. See **random sample.**

D statistic A sum over all categories of absolute deviations, the $|O - E|$ values. The D statistic could be used in place of the chi-square statistic for measuring the size of the difference between the observed and expected frequencies under a null hypothesis for categorical data. $D = \sum |O - E|$. See **chi-square statistic** and **chi-square test.**

empirical rule for the spread of bell-shaped data States that when the distribution of the data is roughly normal (bell-shaped), approximately 68% of the data values lie within one standard deviation of the sample mean, approximately 95% of the data values lie within two standard deviations of the sample mean, and approximately 99.7% (almost all) of the data lie within three standard deviations of the sample mean (68% within $\pm s$ of $\bar{x}$, 95% within $\pm 2s$ of $\bar{x}$, 99.7% within $\pm 3s$ of $\bar{x}$).

equally likely outcomes A theoretical probability model, such as that for a fair die, in which every outcome has the same probability.

error sum of squares Denoted SSE. In linear regression, SSE is the sum of the squared vertical distances between the observed y values and their corresponding fitted regression line $\hat{y}$ values. It is the unexplained variation. Same as the *residual sum of squares* in regression and the *within-samples sum of squares* in ANOVA.

estimation Using data to make a data-informed inference about the magnitude of a population or probability model parameter.

estimator A sample statistic that is intended to estimate some particular population parameter.

Euler's number Represented by the single letter e, Euler's number is famous in theoretical and applied mathematics. A sufficient approximation for our purposes is $e = 2.718$, although most calculators have much better approximations stored in their memory.

event A set of outcomes of interest in a probability model. For example, the event A that the number showing on a fair die is even is given as $A = \{2,4,6\}$. Usually the goal is to compute or estimate $P(A)$.

event of interest (when hypothesis testing) The event of interest is defined to occur if the null-hypothesis-generated value of the test statistic is as extreme as, or more extreme (in the direction providing evidence that H_A holds) than, the value of the statistic that occurred with the real data sample. Its probability is the **P-value.**

exclusive events Events with no outcomes in common. Also called *mutually exclusive events*. Leads to $P(A \text{ or } B) = P(A) + P(B)$ when A, B exclusive.

expected frequency, E The number of times an outcome is *expected* to occur in a real data sample *based on a chi-square null hypothesis model*. See **obtained frequency, chi-square statistic,** and **chi-square test.**

expected value (experimental) The average value of a random quantity that has been repeatedly observed in replications of an experiment; possibly obtained through the five-step simulation method.

expected value (theoretical) Given a discrete random variable X with probability law $p(x)$, the expected value $E(X)$ is $E(X) = \sum x p(x)$. Also called the theoretical mean of the random variable X and often denoted by μ_X. For a box model of X, μ_X = box average. For a continuous random variable, the summation is replaced by a calculus integration.

experimental probability The estimated probability of an event; obtained by dividing the number of experiments (likely simulated) for which the event of interest occurs by the total number of experiments. It is often obtained by applying the five-step method to obtain simulated data.

experimental probability distribution The density histogram for all possible outcomes or all values of a random variable based on many simulations of the random variable. This distribution estimates the unknown theoretical probability distribution.

explanatory variable In regression, the independent variable that controls or explains the dependent variable.

extreme value An unusually large or small data point.

fair die A many-sided die for which each side has the same chance of occurring. See also **loaded die**.

F distribution The continuous probability law of the ratio of two mean squares (such as the between-samples mean square divided by the within-samples mean square for a one-way ANOVA) used to carry out ANOVAs when sampling from normal populations.

five-step method of simulation Basic method used in this book for carrying out a simulation study. The five steps are the choice of a model, the definition of a simulation, the definition of the event or statistic of interest associated with the simulation, repetitions of the simulation, and the calculation of the experimental probability of the event of interest, experimental distribution of the statistic of interest, or the average of the statistic of interest.

frequency Number of data points, usually in an interval.

frequency histogram See *histogram*.

frequency table A table giving the number of data points in a data set falling in each of a set of given intervals.

F test Standard ANOVA approach that divides treatment (between-samples) mean squares by mean square for error (within-samples) and then judges significance by using the F distribution.

gapping Occurrence of one or more intervals containing no data, caused by the fact that the probability model generating the data assigns a probability of 0 to the interval P. For example, a study comparing tall and short men might randomly sample only men shorter than 5'6" and men taller than 6'0".

geometric distribution A discrete probability distribution useful for calculating the probability of the number of trials before a "success" occurs. If X represents the number of independent trials before the first success, if the outcome of each trial is either "success" or "failure" then

$$P(X = x) = (1 - p)^{x-1} p$$

for $x = 1, 2, \ldots$, where p is the probability of success in a single trial.

goodness-of-fit test A statistical test for testing how well a hypothesized model fits a set of data. For one approach, see **chi-square testing** in Chapter 10. A *goodness of fit statistic* is a statistic on which one bases such an assessment.

H_0 Symbol used to denote the null hypothesis. It is usually followed by a ":" and then the statement of the null hypothesis in either symbols or words. For example, "$H_0: p = 0.50$" indicates that the null hypothesis assumes a population proportion value of 0.50.

histogram A graph whose rectangles count the frequencies of occurrence of data points in specified intervals. Sometimes called a *frequency histogram*.

hypothesis A claim that is made about the value of a population parameter or about the values of several population parameters. The real-world claim of interest or research hypothesis (such as that the experimental drug is better than the traditional treatment) is turned into the alternative hypothesis H_A, while its negation becomes the null hypothesis, H_0.

hypothesis testing Deciding whether or not to reject the null hypothesis H_0 based on the strength of the statistical evidence in a real data sample. The statistical evidence is based on estimating the probability that the null hypothesis model would generate a value of a pertinent sample statistic as extreme as, or more extreme than, its value that occurred in the real data sample. This is called the *P*-value. The *lower* this probability, the *greater* the evidence for rejecting the null hypothesis. See **three-stage significance testing approach.**

independence Property of two events that holds if the occurrence of one does not affect the probability of the occurrence of the other $(P(A|B) = P(A))$; see also **law of theoretical independence.**

independent identically distributed random variables Basis model for a random experiment in a large population survey sample. Called a **random sample.**

inferential statistics Techniques used to draw conclusions from data using probability modeling and statistical techniques.

influence The ability of a single data point to have a major effect on a statistical inference or on the computed value of a statistic like $\bar{x}$. A desirable property of a resistant statistic is that every single data point has, by itself, little influence on the value of the statistic. For example, the sample median is not influenced by the magnitude of very large data points.

interaction In a two-way ANOVA, the influence of the value of one explanatory variable on how varying the values of the other explanatory variable affects the response variable. For example, all varieties of wheat do poorly in extreme drought conditions, whereas with adequate rainfall one variety may be much better than the rest.

interquartile range The difference between the third quartile and the first quartile of a set of data. A resistant measure of spread.

judgment sampling Choosing a sample according to expert knowledge rather than a random mechanism.

large sample approach Use of a statistic to do hypothesis testing, build a confidence interval, or do some other type of inference, where the approximate probability law of that statistic is deduced from the fact that the sample size is large. For example, the CLT may have been applied.

law of large numbers The fact that if the size of a random sample is large, $\bar{X} \approx \mu$. This is the backbone of the five-step simulation in Chapter 5 using $\bar{X}$ to approximately obtain $\mu = E(X)$.

law of theoretical independence The fact that if the events A and B are independent, then $P(A$ and $B) = P(A) P(B)$, where P denotes theoretical probability.

least squares regression The method of locating the best-fitting regression line for a scatterplot by minimizing the sum of the squared vertical distances between the points of the scatterplot and the line.

level of significance User-specified probability of rejecting the null hypothesis when it is true; usually set at 0.05, sometimes at 0.01 or 0.1. Must be used with acceptance/rejection hypothesis testing. May be used with significance testing. In both cases, supplying the *P*-value is recommended.

linear interpolation A technique for estimating the unknown value of a variable (y, say) based on the corresponding known value of a related second variable (x, say). It assumes a linear relationship between two known pairs of (x, y) values, one of which has an x value less than the x for the unknown y and one of which has an x value greater than the x for the unknown y. See Appendix D.

linear relationship A relationship between two variables whose scatterplot is well fit by a straight line; the equation $y = mx + b$ is often used, where m is the slope and b is the vertical axis intercept value.

linear transformation (of a random variable) A linear transformation of a random variable is one in which the random variable is multiplied by a constant value and/or has a constant value added to it.

loaded die A many-sided die for which some of the sides are more likely than others to land face up when the die is rolled. See **fair die**. Often chi-square null hypotheses can be thought of as specifying the probabilities of a fair or perhaps a particular loaded die.

lurking variable A third variable that influences the response variable and is associated with the explanatory variable, thus misleading the user to think the response variable is influenced by the explanatory variable.

matched-pairs problem An apparent two-sample problem that reduces to a one-sample problem because the members of the two samples are naturally paired, such as measurements of the same person before and after treatment.

mean (sample) The sum of a set of numbers divided by the quantity of numbers n in the set; same as the **arithmetic average**, and often defined by $\overline{X} = \sum X / n$.

mean square An ANOVA sum of squares divided by its degrees of freedom.

mean (theoretical) The population or probability distribution mean. In the case of a discrete distribution given by $p(x)$, it equals $\sum x \, p(x)$. Same as the theoretical expected value of a random variable X sampled from the population. Sometimes called the *population mean*. Denoted by μ, μ_X, or $E(X)$.

measurement A special kind of estimation where a random experiment is carried out by repeatedly observing the "amount" of a property such as height, weight, speed, volume, and so forth, modeled by a random sample (independent identically distributed random variables).

median (population) In an actual population, the number such that half of the population values are smaller than this median number and half are greater. For a continuous probability distribution or continuous pdf, the number such that half of the area under the pdf lies on either side of the number. For a discrete pdf, the median is similarly defined.

median (sample) The middle value when a data set is arranged in order from smallest to largest value. When the number of values is even, the median is the average of the two middle values.

missing baseline distortion Occurs when the vertical axis (y) of a graph starts not at 0 but close to where the changes in y are occurring. The effect is to greatly exaggerate the size of the changes occurring.

model See also probability model The value (or values) in a set of data that occurs the most frequently; a seldom used measure of the center of a data set.

model A mathematical set of probability rules, a random physical mechanism, a box model, or a random-number-based simulation for producing data that are as similar as possible to actual real-world data. See also probability model.

Monte Carlo simulation A simulation method for solving probability problems by repeatedly doing an experiment; for example, tossing a coin repeatedly, rolling a die repeatedly, repeatedly drawing from a box model, or repeatedly choosing random digits. The five-step simulation approach in this textbook is, thus, Monte Carlo simulation.

multiple comparisons Simultaneous comparisons of pairs of population parameters in an ANOVA to judge which are distinct, with some measure of overall statistical confidence attached to the results.

multiple correlation coefficient, R^2 A measure of the degree of fit of the best-fitting curve when the regression is nonlinear or linear with multiple explanatory variables; reduces to the ordinary squared correlation or coefficient of determination r^2 when the regression line is $Y = mx + b$.

multiple regression Linear regression where there are at least two explanatory variables influencing the response variable, such as amount of rainfall and amount of fertilizer affecting predicted crop yield.

mutually exclusive events Events with no outcomes in common.

(n choose x) Same as $\binom{n}{x}$; defined by $\binom{n}{x} = \dfrac{n!}{x!(n-x)!}$.

negative relationship A relationship between two variables in which one decreases as the other increases. In the special case of a straight line, a negative relationship means that the slope of the regression line is negative and that the correlation coefficient is negative.

negative slope The slope of a line on which the y value decreases as the x value increases.

nominal data Data falling into named categories (such as hair colors) that are not ordered (the days of the week are ordered). In particular, the data do not represent measurements.

nonlinear regression Regression that seeks the relationship between bivariate (x,Y) pairs when the relationship is not linear, as in the case of $Y = B_0 + B_1 x + B_2 x^2 +$ random error.

nonparametric Said of a statistical procedure that does not require detailed assumptions about the shape of the population distribution. Such detailed assumptions are usually specified by various population parameters.

nonresponse bias Polling bias resulting from the fact that, even if the intended sample is selected by random sampling, the subset of those actually responding to the survey may be quite different from those not responding.

normal distribution Also called the *bell-shaped curve* or the *Gaussian curve*. The most widely used continuous distribution; often used to model biological measurements and errors of measurement. In general, for every choice of theoretical mean and SD there is a different normal density. Probabilities for all normal problems are computed using Table E, which gives standard normal (mean of 0, SD of 1) probabilities.

normally distributed data set A data set whose density histogram is roughly bell-shaped. From a modeling viewpoint, the data is often assumed to have resulted from random sampling from a normal population.

null hypothesis The presumed model (such as that of a fair coin) in hypothesis testing. The data provide a measure of how weak or strong the evidence against this null hypothesis is. The formulation of the null hypothesis is the critical first step in using real data to do hypothesis testing. It usually is the negation of a real-world research hypothesis H_A. It is the first stage of the three-stage hypothesis testing approach, and it is the first step of the five-step simulation method for estimating the probability of occurrence of the event of interest, namely the P-value. Denoted H_0.

observational study A study in which the units of the study were placed in the control and treatment groups according to the natural state of affairs rather than being assigned (hopefully randomly) by the person doing the study. Such data are usually not very useful for statistical inference because there is no control over confounding variables, like age in some examples, differing greatly between the two groups.

observed level of significance Same as *P*-value.

obtained frequency (denoted *O*) The number of times an outcome is observed to have occurred in a real data sample in a statistical study (often a chi-square test). See **expected frequency.**

one-sided hypothesis test A hypothesis test in which the alternative hypothesis is that the population parameter lies to one (specified) side of its null hypothesis value.

one-way ANOVA Analysis of variance in which the population means vary because of the influence of a single variable or factor, as when wheat yields vary because of differing wheat varieties.

ordered categorical Numbers assigned to categories because the categories are ordered.

ordinal data Data placed in ordered categories, like the numbers of accidents on each day of the week.

outcomes The list of the possible ways a random phenomenon can occur.

outlier A data value that is extreme (very large or very small) relative to the rest of the data. Further than ±3 standard deviations from the sample mean is the most common outlier criterion for single-variable data (but boxplots use a different criterion).

p Symbol used to denote "population proportion."

$\hat{p}$ Symbol used to denote the sample proportion, an estimator of the population proportion *p*.

parameter A number describing a characteristic of a population or a probability model, such as the theoretical mean, μ, or the true slope of a regression line.

Pareto chart A bar chart where the bars are arranged from tallest to shortest as one moves to the right.

pie chart A graph for categorical data. The proportion of elements belonging to each category is proportionally represented as a pie-shaped section of a circle.

placebo A nonactive treatment that is administered to the control group so that the human subject cannot tell whether he or she has received the experimental treatment; for example, a shot of saline solution instead of a new vaccine.

point estimate An estimate of a population parameter that is one specific number (as opposed to an interval estimate).

Poisson distribution The discrete probability law of the number of occurrences *X* of a randomly occurring event in a fixed time or fixed area interval when the rate of occurrence is fixed across the interval, separate occurrences are independent, and simultaneous occurrences are precluded. An example is the number of phone calls arriving in a given period of time. The distribution is computed as

$$P(X = x) = \frac{\lambda^x}{x!e^\lambda} \qquad x = 0, 1, \ldots; \qquad \lambda > 0.$$

population The entire collection of objects or people under consideration for statistical study. Often the statistical goal is to use a random sample to make inferences about the population, such as about its center or spread. The population is modeled by a probability distribution. Can be real finite or infinite conceptual. Can be continuous or discrete.

population, actual (or real) A population consisting of actual objects which can, in principle, be measured. An actual population always has finitely many members.

population, conceptual A population consisting of conceptual members. For example, the population consisting of all asthma sufferers or all possible tosses of a die in the future. In practice, this population is not just those currently suffering from asthma, but all future sufferers or potential sufferers as well. Conceptual populations have infinitely many members. Often large finite populations of continuous measurements, such as heights, are modeled as a conceptual population described by a continuous random variable, like the normal.

population, continuous Consists of infinitely many members arbitrarily close together; as such, always a conceptual population.

population, discrete Consists of finitely many members, hence with gaps between members. A real (or actual) population.

population distribution The distribution of one randomly sampled member from a population. See **sampling distribution**.

population parameter See **parameter**.

population proportion A ratio whose numerator is equal to the number of objects in the population that display a particular property or quality of interest, and whose denominator is the total number of objects in the population. It is a type of population parameter. Denoted usually by p.

population size The number of objects or people in a population.

positive relationship A relationship between two variables in which one increases as the other increases. In the special case of a straight line, a positive relationship means that the slope of the regression line is positive and that the correlation coefficient is positive.

positive slope The slope of a line on which the y value increases as the x value increases.

power of a test The probability of rejecting the null hypothesis when it is false. It usually increases as the distance increases between the true value of the parameter of interest and its null hypothesis value.

prediction Inference about what the value of a "new" random variable is likely to be. Different than estimating the value of a (nonrandom) parameter.

prediction interval in regression analysis An interval computed from the data for which a new random observation of Y at x has a high probability of falling within. Contrasted with a confidence interval for $E(Y|X = x)$.

probability See **conditional probability, experimental probability, frequency interpretation of probability, theoretical probability**.

probability density function (pdf) A function that gives the theoretical probabilities associated with the values of a random variable. Discrete pdfs give the probabilities explicitly. Continuous pdfs require that one find the area in a specified range underneath the curve graphed by the pdf in order to find the probability of the random variable being within a specified range.

probability distribution See **distribution**.

probability histogram A histogram made up of rectangles of width 1 where a rectangle centered at integer x has height (and area) equal to $P(X = x)$. This histogram displays the shape of the probability distribution of any integer-valued random variable, like the sum of two dice.

probability model A list of all possible outcomes of a random phenomenon, called a chance experiment, and an assignment of probabilities to these outcomes. Often a physical or computerized mechanism for simulating the outcome of a random phenomenon, such as the simulation of the Chapter 5 cereal box problem. Can be a continuous pdf too.

probability sampling Any sampling scheme where the sample members are selected using a chance mechanism.

probability tree Graphical technique for computing the probability of an event that is the result of a multistage experiment where the probabilities are known for events at each stage given an event from the previous stage.

***P*-value** The probability of observing a value of the test statistic that is as extreme or more extreme than the observed value of the test statistic in a direction suggesting the alternative hypothesis is true. This probability is computed under the assumption that the null hypothesis H_0 is true.

qualitative variable or data Same as categorical variable or data.

quantitative variable or data Having a numerical value like weight or height as opposed to eye color.

quartiles Quartiles divide a data set into four parts. The first quartile is roughly the value that separates the data into the lower 25% and the upper 75% of the data. The third quartile is the value that separates the data into the lower 75% and the upper 25% of the data. The second quartile divides the data into the lower 50% and the upper 50%. The second quartile is the same as the *median*.

quota sampling Choosing a sample by guaranteeing that certain groups (like men and women) are represented in the same proportion in the sample as in the population; otherwise any individuals who fit the quotas can be selected.

r Symbol for **correlation coefficient.**

random Not predictable; occurring by chance. For example, the outcome of tossing a fair coin is random.

random digits Also called *random numbers*. Digits (most commonly 0, 1,..., 9) that occur in equally likely and random fashion, as when produced by using a spinner having 10 equal sectors; often produced by a computer program, in which case the numbers merely look random.

random experiment or replications Repeated observation of a random variable under identical conditions, usually a physical experiment such as tossing a coin or assessing the breaking strength of a steel beam.

randomization Using a totally random mechanism to choose which units receive which treatments in an experiment or which individuals to sample from a population, instead of having experts decide, for example. See also **randomized controlled experiment.**

randomized controlled experiment A controlled experiment where subjects are randomly assigned to the treatment and control groups.

random sample A set of data chosen from a population in such a way that each member of the population has an equal probability of being selected. When sampling is with replacement, the observations are independent random variables and each has the same distribution. When the sampling is without replacement but the population size is large relative to the sample size, the observations are random variables that are approximately independent and each has the same distribution: modeled as a set of independent identically distributed random variables.

random variable A numerical description of the outcome of a random phenomenon, like an automobile accident, a thunderstorm, or a game of roulette. Random variables are denoted by capital letters, usually towards the end of the alphabet, like U, X, and Y. $(X = x)$ denotes the event that the random variable X has an observed numerical value of x. Random variables are represented by a probability distribution function that gives both the values (or ranges of values) and their associated probabilities.

range The measure of spread (variation) that is the difference between the largest value and the smallest value in a set of data. This is not a resistant measure of spread.

real data Data that are obtained from an actual experiment or survey designed to answer a specific research question about the underlying probability model. The data are used to obtain statistical evidence about the probability model underlying the data. Contrasted with **simulated data**.

real population See **population actual (or real)**.

regression equation The equation of the (least squares) regression line.

regression line Same as **regression equation** A straight line used to estimate the relationship between two variables, based on the points of a scatterplot; often determined by a least squares analysis.

rejection of null hypothesis The decision that is made in hypothesis testing when the appropriate event of interest occurs with very low probability (for example, less than 0.05) under the null hypothesis model. See **hypothesis testing** and ***P*-value**.

rejection region Same as **critical region**.

relative frequency interpretation of probability The interpretation of a theoretical probability as approximately predicting the proportion of occurrences of the event of interest (such as the proportion of heads being approximately $1/2$) occurring in a large number of independent (replicated) trials of a random experiment (such as repeated tossing a coin).

relative frequency (of an event) Same as **experimental probability (of an event)**.

repeated measurements See **measurement**.

repeated trials random experiment See **random experiment**.

replication Name for the process of carrying out repeated trials of a random experiment.

resampling methods Any one of several methods, bootstrap statistics being the best known and most emphasized in this textbook, that rely on repeated sampling using a box model that has been somehow determined by the sample or samples of actual collected data.

research hypothesis A real-world claim that becomes the alternative hypothesis H_A and stands against the negation of the claim (the negation being the chance explanation) that became the null hypothesis.

residual In the case of regression, the difference between the actual Y value and the estimated regression line $\hat{Y}$ value: $Y - \hat{Y}$.

residual sum of squares See **error sum of squares**.

resistant A statistic is resistant if it is not sensitive to the influence of individual extreme observations. The median is a resistant measure of the center of a set of data and the interquartile range (IQR) is a resistant measure of the variation (or spread) of a set of data.

response variable In regression, the dependent variable that is being influenced by the independent variable.

S_E See **standard error of the residuals in a regression analysis.**

S_Y The sample standard deviation of a set of data whose values are denoted by Y. Denoted S when no ambiguity is caused by dropping the subscript.

sample proportion A ratio whose numerator is equal to the number of objects in the sample that display a particular property or quality of interest, and whose denominator is the total number of objects in the sample. It is a type of sample statistic and is an estimator of the population proportion. Often denoted $\hat{p}$.

sample size The number of objects or people in a sample.

sample space The set of all possible outcomes of a probability experiment. For example, the probability experiment of throwing a die once has the sample space $S = (1, 2, 3, 4, 5, 6)$.

sample statistic A statistic that has been computed from a (hopefully random) sample.

sample survey A survey of a population, usually human, made by taking a sample judged to be representative of the population. Use of a random mechanism for choosing the sample is essential.

sample variance See **variance (sample).**

sampling distribution The probability distribution of a statistic, such as an estimation hypothesis test statistic that is computed from a sample.

scatterplot A graph of two-variable (bivariate) data in which each point is located by its coordinates (x, y).

SD See **standard deviation.**

selection bias Polling bias resulting from a sampling method that makes some population members more likely to be sampled than others. For example, the *Literary Digest* poll mentioned in Chapter 10 favored the more affluent.

significance, statistical See **statistically significant.**

significance testing Hypothesis testing in which the two decisions are rejecting the null hypothesis and failing to reject the null hypothesis. That is, the null hypothesis is never accepted when doing significance testing. Contrasted with **acceptance/rejection hypothesis testing.**

simple random sample Same as a **random sample.** This expression is only used to distinguish simple random sampling from more complex methods of probability sampling, such as stratified random sampling.

Simpson's paradox Nontechnically, this paradox allows a confounding variable to apparently lead us to the opposite of the correct conclusion about the comparison of a treatment versus a control. In the book's Chapter 7 example, it seemed to suggest that smoking increased life expectancy!

simulation Any method for generating data from a given probability model. Methods used include box models, physical models such as coins or dice, and simulation based on random number generation, often done on the computer. See **five-step method of simulation.**

single-blind study A study in which the subjects do not know who has received the treatment and who has received the placebo.

skewed Said of an asymmetric pdf or data set that is stretched out in one direction. For example, the density of an chi-square distribution is skewed to the right and a histogram of incomes is typically skewed to the right. Causes the mean to be pulled away from the median in the direction of the skew.

slope The rate of change of the y values with respect to the change of the x values in a straight line relationship.

spread (of data) See **variation (of data)**.

standard deviation (sample) The square root of the sample variance. The most widely used measure of the amount of spread (variation) in a set of data. Denoted by s.

standard deviation (theoretical) The standard deviation of the population or distribution, given by the square root of the theoretical variance. Denoted by σ.

standard error of a sample mean $\sigma/\sqrt{n}$, where n is the number of observations used to compute the sample mean and σ is the population SD. Estimated by substituting S for σ.

standard error of a sample proportion $\sqrt{p(1-p)}/\sqrt{n}$, where p is the population proportion of 1s in a population consisting entirely of 1s and 0s; $n =$ the sample size. Estimated by substituting $\hat{p}$ for p.

standard error of a sample statistic For any sample statistic, Y, computed from a random sample, the standard error of Y, written SE(Y), is defined to be the SD of Y, often written SD(Y). This is not to be confused with the population SD, usually denoted by σ. SE(Y) measures the typical estimation error when using Y as an estimator.

standard error of the residuals in a regression analysis $S_E = \sqrt{\sum(Y-\hat{Y})^2/(n-2)}$. Used to estimate σ, the SD of Y given fixed x in a regression analysis.

standard form of a confidence interval Written as: estimate$\pm$(critical value) $\times \widehat{SE}$(estimate), where $\widehat{SE}$(estimate) is often estimated either by bootstrapping or by use of a theoretical equation for SE(estimate).

standardized score Also called *standard score*. Each score x is centered at $\bar{x}$ and this difference is divided by the sample SD, s, producing a standardized score $z = (x - \bar{x})/s$. See **z-score**.

standardized test statistic Of the form: (statistic − parameter under H_0)/(estimated standard error of statistic). Often has a standard normal or t distribution.

standard normal density Same as **standard normal distribution.**

standard normal distribution A normal distribution with a mean of 0 and a variance of 1 (and, hence, an SD of 1).

statistic A piece of numerical information computed from a sample, such as $\bar{X}$. Because the sample is random, the statistic is a random variable and has a sample distribution.

statistical decision making See **hypothesis testing**.

statistical evidence Numerical evidence, usually in the form of a probability or P-value, gathered from a statistical study for use in hypothesis testing. See **hypothesis testing.**

statistical literacy The state of being reasonably informed about the basics of statistical reasoning. Makes one resistant to being deceived by improper data representations or analyses.

statistically significant Said of data behavior that is too unusual to be attributable to chance under the probability model that is presumed to be producing the data under the null hypothesis. For example, if the sample mean is so large that its value cannot be reasonably attributed to chance under the null hypothesis of a population mean of 0, the difference of the sample mean from 0 is said to be statistically significant. See **hypothesis testing.**

statistical regularity The empirical (real-world) fact that the experimental probability becomes closer and closer to a number, called the **theoretical probability**, as the number of trials becomes large; this law is the foundation of statistical reasoning. See **relative frequency interpretation of probability.**

statistics (1) The science of gathering, describing, and drawing conclusions from data; (2) reported numerical information, such as in a newspaper.

stem-and-leaf plot A graphical display of data using certain digits (such as those in the tens place) as *stems* and the remaining digits (such as those in the ones place) as *leaves*. Also called **stemplot.**

stemplot Same as **stem-and-leaf plot.**

straight line equation The slope-intercept form is $y = mx + b$, where m is the slope and b is the y-intercept, namely the point at which the line intersects the y axis.

strength of evidence The degree to which statistical evidence from real data supports rejection of the null hypothesis. The evidence is usually the probability of occurrence of the event of interest under the null hypothesis, called the P-value. The lower this probability is, the stronger the evidence for rejecting the null hypothesis. See **hypothesis testing, event of interest,** and **statistical evidence.**

sum of squares The sum of the squares of observed quantities computed in an ANOVA to estimate the contributions from various sources, such as the effect of one variable in a two-way ANOVA.

survey sampling A study in which a sample is randomly chosen form the population so as to be representative of the population. Then one infers that characteristics of the sample are characteristics of the population. The only truly acceptable way to do this is to use probability sampling to obtain the sample, the simplest being random sampling without replacement, modeled by independent identically distributed random variables or sampling with replacement when sample size is at most 5% of population size.

symmetric distribution of data When the shape of the data to the left of its center is similar to the shape to the right of the center. In this case the sample mean and sample median will be quite close.

test of significance See **significance testing.**

theoretical probability The true probability of an event; it is what the experimental probability will be close to in a very large number of trials. In the special case of an equally likely outcomes probability model, the theoretical probability is obtained by dividing the number of outcomes that produce the event of interest by the total number of possible outcomes.

three-stage significance testing approach Stage I: Form the null hypothesis as a denial of the real-world claim (research hypothesis) of interest. Stage II: Test the null hypothesis using five-step simulation (possibly bootstrapping) or using theoretical results about the hypothesis testing statistic when the null hypothesis is true in order to obtain the P-value. Stage III: In significance testing, reject or fail to reject the null hypothesis based on the strength of evidence computed at Stage II.

totally randomized design Assignment of treatments to units in a completely random manner.

transformation of data Use of x^a, log x, or another transformation to produce a better fitting and simpler model for two-variable (x, y) data. Such transforms can also be applied to y. Sometimes transformation converts a nonlinear regression problem into a linear regression problem solvable by the methods of Chapter 3 and 11.

treatment group The group of individuals or units in an experiment that receive the treatment being studied.

treatment sum of squares Denoted **SSTR**. The ANOVA sum of squares due to differences in the population means. Also described as a between-samples sum of squares.

t statistic A statistic that follows the *t*-distribution under appropriate assumptions. Of the form $(\hat{\theta} - \theta)/\widehat{SE}(\hat{\theta})$. The *t* distribution (or Student's *t* distribution) looks very much like the standard normal distribution. However, its tails are slightly thicker. See **t test**.

t test A special hypothesis test that is usually about the population mean (or about the equality of two population means) used when the population is known to be normally distributed, the sample size is small, and the population standard deviation is unknown. Based on a standardized *t* statistic, obtained by centering at the null hypothesis mean and then dividing by the appropriate *estimated* standard error. Uses the *t* distribution rather than the standard normal distribution. A test about the regression line slope is often a *t* test.

two-sample problem Any inference problem about two populations in which the data consist of two independent random samples, one from each population.

two-sided hypothesis test A hypothesis test in which the alternative hypothesis is that the population parameter may lie on either side of its null hypothesis value.

two-way ANOVA Analysis of variance in which the population means vary because of the influence of two explanatory variables or factors, as when wheat yields vary because of differing wheat varieties and differing soil types.

type I (hypothesis test) error The error of incorrectly rejecting a null hypothesis when it is true.

type II (hypothesis test) error The error of incorrectly accepting a null hypothesis when it is false.

typical variation The standard deviation is interpreted as a measure of the typical amount of variation from one observation to the next.

unbiased estimator If the expected value of a statistic $\hat{\theta}$ used to estimate a parameter θ is equal to the value of the parameter ($E(\hat{\theta}) = \theta$), then that statistic is an unbiased estimator. Put empirically, as the number of statisticians who apply a procedure gets larger and larger, if the average of the estimator over all the statisticians is equal to the value of the parameter that the procedure intends to estimate, then the procedure is unbiased. The sample average is an example of an unbiased estimator of the population mean (assuming the observations are based on random samples).

U-shaped distribution A distribution that is dense in the extremes and sparse in the middle.

validity The correctness of a statistical inference.

variable A quantity that varies, often randomly. For example, the weight of a randomly chosen member of a football team is a variable. Variables are usually represented by letters.

variance (sample) A measure of the variation (spread) in a set of data. The sample variance is almost the arithmetic average of the squared differences of all data values from their mean (one divides by $n - 1$ instead of n; otherwise it would be the arithmetic average). Often an intermediate value in the calculation of the sample standard deviation. Denoted by s^2 or S^2.

variance (theoretical) The variance of the population or distribution. In the case of a discrete distribution given by $p(x)$, it equals $\sum(x - \mu)^2 p(x)$, where μ is the population mean. Denoted by σ^2.

variation (of data) The degree to which data are spread out around their center. Useful measures of spread include the sample SD and the interquartile range.

volunteer effect Bias that arises when response to a survey is voluntary; it happens because even if the survey is sent to a representative random sample, the sample of volunteers is not representative of the population, due to nonresponse bias.

within-samples sum of squares In ANOVA, the sum over all the samples of each sample's sum of squares.

within-sample variability In ANOVA, the variability of observations within a sample that is caused by the population variance.

without replacement Describes random sampling from a set of objects, or from a box, in which each drawn object is not replaced before the next random drawing.

with replacement Describes random sampling from a set of objects, or from a box, in which each drawn object is replaced before the next random drawing.

z-score Also called a *standardized score* or *Z statistic*. Tells how many SDs a data value is above or below the sample mean

$$z\text{-score} = \frac{\text{data value} - \text{sample mean}}{\text{sample standard deviation}} = \frac{x - \bar{x}}{s}.$$

Z test A hypothesis test about the population mean μ (or about the equality of two population means) using either the fact that the population is normal or using the central limit theorem result that $(\bar{X} - \mu)/(\sigma/\sqrt{n})$, where σ is the population SD and n is the sample size, has an approximately standard normal distribution as long as n is large.

A

Computationally Generated Random Digits

Computers (and some calculators) can produce random digits rapidly and in very large quantities. Often it is not practical to require such devices to store long lists of random data. Therefore it is usually better to generate such random digits only as they are needed.

Formulas have been invented to compute random digits. At first, you might think this is impossible, because if you *compute* a number by a formula then you will *know* what the next digit will be, and thus such digits cannot be random. This is true, but although these integers are generated deterministically, they nonetheless appear to be true random digits for all practical purposes. Digits produced by such formulas are therefore sometimes called *pseudo-random*. They are *pseudo*-random because they are not random. For most purposes, such pseudo-random digits work extremely well in that they appear in every observable way to be random and are very convenient to use.

Some such formulas are very complicated; others are surprisingly simple. You can usually find out from a user manual or a consultant the formula that a particular computer program uses for generating its random numbers.

There is one particularly easy method for producing pseudo-random numbers that you can use on a calculator or computer with little or no programming. It has some flaws and hence is not often used, but it shows us how to deterministic approach can produce digits that appear to be random. It is called *mid-square method*, and it was suggested by the mathematician John von Neumann.

Take an arbitrary number of five (or more) digits (you could use the last five digits of your telephone number or social security number, for example.) Suppose we start with 63537. This is not a part of your list of random digits—it is only the starting value for the procedure. We square this number:

$$(63537)^2 = 403\mathbf{6950}369$$

We now take the middle five digits, 69503, shown in boldface in the above equation. These are our first five "random" digits. Next we square these:

$$(69503)^2 = 48\mathbf{30667}009$$

Again, we take the middle five digits and square:

$$(30667)^2 = 9\mathbf{40464}889$$

We repeat this process as long as desired. We then write down the string of random digits. For our example, we get

$$69503 \quad 30667 \quad 04648$$

This string of digits may be regarded as a set of random digits. The advantage is that it has been mechanically produced. In this case, we used an ordinary calculator to do the job.

A caution is necessary about this method: after a while it can sometimes start generating all 0s.

Nobody uses this method for actual production of random number tables. We have presented it because it illustrates the idea that a simple computation done repeatedly can produce digits that appear to be random. In addition, you can certainly use it if you wish.

B

Random Number Tables

Table B.1 Random Number Table for Tossing a Fair Coin: {0,1}

01110	10000	00010	10111	00010	11001	10011	10001	10110
11101	00111	10111	10011	11001	11011	00000	00000	01111
11111	11000	01101	01101	00000	00101	01001	11100	11001
10001	00101	00001	01100	11101	10001	11101	10011	11110
00101	11011	10100	00110	11110	00110	10111	10011	10101
00111	10000	11111	00010	10000	10011	10111	01100	11001
01110	00100	00001	11100	11010	01011	01100	01100	11010
10011	11100	00111	10010	01011	11010	10011	10111	11010
10100	10001	01101	01100	11100	10011	00001	11110	00011
00111	00100	00110	00101	01011	01110	10110	11101	01111
01001	10001	00101	00000	10111	00001	10000	00000	00110
11011	01010	11100	01100	01011	11111	10000	00110	10100
11010	10101	00001	10011	01110	11101	01110	01111	11011
01010	00001	11110	11101	00111	01100	01001	01000	10110
01111	11111	10000	10000	01000	00101	00100	00110	00110
11010	11000	00111	01010	01111	11101	01001	10100	01001
01100	11011	00111	01011	10010	10000	00110	01011	01010
00111	01011	10011	10100	11111	11111	01011	10011	11000
01001	11011	10011	00100	10110	01010	00110	10111	11000
10010	10111	10001	10100	00010	01000	10011	01110	11000
01001	01001	00100	00100	10110	10000	11011	00001	00101
10000	00011	00111	00010	10111	01111	00100	00000	01111
00011	11110	01101	10100	00010	10010	00001	01100	11100
10001	10000	01011	00010	01010	10000	11000	11100	01010
00111	00111	00000	11100	00110	11001	10101	01001	00010
01010	01010	10011	00000	11001	10101	10111	10011	00010
10100	11111	10001	10011	01010	11011	00101	11010	10011
01000	01100	01011	10110	00101	10100	11000	10000	01000
11010	11001	00110	00100	11101	01000	11011	10011	11110
10101	00010	01001	00111	10111	10100	00110	01010	11111
10000	10110	10110	00110	10001	01110	01010	00110	11001
01011	01001	10111	01011	10000	11111	00000	10000	00001
00000	00000	01111	00101	11111	01100	01111	10011	10110
10110	01111	01111	01000	01010	01000	00110	01001	11000
11100	11100	10010	00101	00111	01111	01010	11000	00010
11100	10001	10001	01010	00110	10111	11001	11001	11111
00110	00100	00001	11100	10001	10101	01110	01010	11100
01001	10011	10011	10010	01011	00111	11110	01010	00110
10001	01100	10001	00011	00111	00001	11110	10111	10010
00010	10100	01101	00110	00111	00000	01010	01000	10101

Table B.2 Random Number Table for Rolling a Fair Six-Sided Die: {1,2,3,4,5,6}

66533	45332	24614	22231	26431	35541	12165	62116	16111
61261	22613	26252	14622	32262	33244	34614	13316	41136
61144	46631	56646	24544	36461	14612	21234	23335	16212
21341	66222	53246	24444	13311	44244	41643	54163	21243
15365	46135	23345	53331	46112	54655	65626	24216	11144
51612	21315	16156	56511	31516	26121	23151	66611	64242
31244	26623	33555	53333	24465	14566	54345	25532	43426
41452	65222	25316	44431	33141	64245	62514	45553	24234
25515	41332	25311	65143	61134	65443	45366	61566	63112
35226	33554	55613	54321	21434	22314	11416	64624	26565
41554	23265	43245	46443	26431	23326	51232	45243	54452
24356	54551	52551	64423	13134	25513	43312	43234	12543
56341	24435	44336	62665	26123	25311	55156	46545	45462
26452	16243	13225	66222	31311	41621	51634	63265	22422
41265	26326	23356	62315	32132	55212	26512	56646	42132
13355	53313	42141	56223	61515	22554	44363	42162	52564
22434	61332	13551	43314	45151	43646	15241	55526	61141
62616	22135	55522	62532	43455	14443	61544	55551	34455
53565	42466	22156	21551	13322	54653	62516	66132	36151
55634	65353	51535	66264	61332	11323	15613	44653	36232
64424	24211	63132	25346	41315	66415	33444	16552	56425
34324	62343	35623	22645	41232	35323	25646	52446	24233
44252	16335	35563	33652	12265	25133	56646	14452	45423
55613	35253	13233	26555	56412	23255	61144	36162	51462
34446	33554	16144	16156	23641	24226	52346	26545	55324
16263	13232	52551	32542	41514	52151	43454	32133	34634
34544	15413	14125	62245	13145	53112	52116	63164	42245
23235	36666	61125	44255	62636	53632	56521	26641	56613
32161	66226	66631	33144	12225	42633	61631	14334	45532
13423	54133	12612	15644	41562	36533	25533	22642	34621
66565	24634	23326	42543	55652	25144	12623	46551	42423
52544	45645	23641	64431	41455	21554	23525	26554	66453
21311	46163	14532	63152	12231	34425	41356	64213	62225
14163	41364	32442	55443	54236	45413	43531	12112	45356
63431	26466	35321	56544	25625	66235	56345	15116	65553
66142	14561	25443	53424	45123	32155	41112	45254	54661
56651	14313	51634	65631	45624	34654	32533	63314	11516
61641	44461	14462	35555	26443	61415	62343	22661	31154
25144	62266	63315	51561	26416	26626	46624	55541	64515
26423	66151	45645	44611	16624	56615	55414	36255	25265

Table B.3 Random Number Table for Rolling a Fair 10-Sided Die: {1,2,3,4,5,6,7,8,9,0}*

32236	12683	41949	91807	57883	65394	35595	39198	75268
40336	50658	32089	78007	58644	73823	62854	31151	64726
88795	93736	22189	47004	48304	77410	78871	98387	44647
12807	65194	58586	78232	57097	01430	00304	32036	23671
65929	96713	94452	56211	85446	13656	32155	84455	38125
50339	82178	19650	41283	03944	13736	02627	41929	60613
73840	53838	90804	94332	63639	73187	87067	37557	29635
87062	66298	10731	40629	64955	08081	31443	72112	58006
48038	94580	55223	97799	10105	27952	62493	42176	69615
89830	54426	02692	21233	39553	33483	03141	90919	99219
72234	40065	24052	95658	98335	21125	45364	67989	32451
02833	78254	05112	95160	62546	85982	85567	27427	21436
20565	20846	63664	72162	75338	04022	77166	83339	99021
18090	91089	09799	75883	36480	37067	40933	65634	79883
11519	97203	70899	00697	84864	24470	07933	48202	15392
12732	61573	22852	32281	84871	13331	08947	09023	38248
26823	44530	84507	10396	12240	62603	13396	69378	37173
74622	88768	90819	54769	95306	07685	50369	13763	02205
58535	99062	55182	89858	67701	94838	37317	10432	75653
78551	56329	09024	81507	90137	19241	55198	74006	52851
41477	58940	04016	38081	45519	27559	92403	30967	86797
17004	22782	09508	37331	94994	67305	34040	91360	83009
36925	31844	12940	51503	24822	53594	72930	23342	88646
97569	75612	07237	92264	77989	09054	03863	83891	09041
35122	31549	19982	66024	68615	15959	40347	50052	35312
34358	17573	32838	68335	93497	17412	19850	98965	27357
09285	03384	61410	01932	26797	92577	42580	23354	38677
98363	76867	91821	26538	47181	50938	95676	45306	96725
91769	65764	52386	18551	22196	50282	19985	90730	95175
09393	34982	05654	51208	37731	33916	49063	76700	63094
33202	19891	95374	23650	64877	70661	53330	07223	03469
34111	18376	08231	95163	62837	56995	03022	93618	67560
04410	39026	99536	85083	34607	38979	47259	30921	90092
88209	36990	83284	50578	83549	19006	29501	04565	48865
89257	25109	26253	57523	99297	29901	29472	52817	66611
25581	36013	52215	47684	55094	93140	32969	05603	66922
31742	82956	36361	52786	79761	49819	41375	67628	81707
46012	07959	79667	95325	49142	99596	25691	85964	53568
00174	92988	49499	74089	99209	33816	51757	25031	35862
04861	63886	09763	03265	12748	77513	91010	15062	20270

*This 10-digit case is what is usually referred to when one refers to a *random number table*.

Table B.4 Random Number Table for a Standard Normal Variable

−0.3713	−0.2478	−0.6191	0.0207	−0.3093	−0.6102	−1.5955	−0.4378	0.6066
2.1352	−0.8940	−0.0062	0.6737	−0.6091	−0.3083	−0.0850	1.7532	−1.3563
0.9692	−0.6652	−1.9856	−0.6354	0.5742	−0.7707	0.9028	−0.4957	2.0585
−0.1241	−0.1783	−1.0843	−0.7939	−0.1232	−1.4296	2.7796	0.4968	0.1914
0.0297	0.3037	−1.3847	0.0486	0.2966	0.5970	0.2056	0.1413	−0.1678
−0.3438	1.2274	−1.1718	0.8335	0.7018	0.0750	0.3924	−0.6794	−0.0852
1.1178	1.8089	1.3865	−0.4378	0.6506	1.2432	0.1845	−0.6945	−0.2513
0.7023	−0.1139	0.3233	1.6031	−0.5919	−1.2176	0.7271	−0.0229	−0.1501
0.4092	0.3912	−1.4844	−0.1497	−1.3398	−0.3652	−1.7329	−1.0306	0.9890
0.2007	−2.3895	−0.5677	−1.3207	−1.3434	0.4915	1.5763	0.6146	1.3502
1.0997	−1.4403	−0.6747	−0.3954	−1.5214	1.9471	−0.4529	0.1710	0.1409
−0.8716	−0.9046	0.1550	2.4269	−2.0841	1.2182	−1.5685	−0.0411	1.9245
0.0647	0.3757	0.0207	0.8082	1.3206	0.4904	−0.8063	−1.4134	1.7442
−0.5876	1.0762	0.1382	1.1575	−0.1087	−1.3951	0.0209	−0.3103	−0.1289
0.9306	0.5654	−0.7672	1.4689	0.2111	0.6171	−2.2243	−0.7565	0.8861
0.2412	−0.8951	−0.7605	0.1465	−0.0505	1.0706	−0.0549	−0.3443	0.4225
2.4332	−0.2278	−1.2614	−1.4321	0.1858	−0.2124	0.0483	1.0501	0.3017
−1.3169	−0.9054	−1.7658	0.5712	−0.5247	−0.1075	−1.0862	0.4623	0.0951
1.0249	−1.4323	−0.5404	−1.0173	−0.4266	−1.0423	−0.9953	1.1717	−0.7675
−0.9134	0.3134	−0.9818	0.6456	−1.5494	−1.5576	−1.9423	−1.0402	0.9997
2.3278	0.5612	−1.4350	0.4842	0.2576	0.3703	0.4033	0.1137	0.7600
1.0979	1.4661	1.0422	1.6997	0.3377	1.3285	−0.8753	1.2995	0.6081
−0.0024	−0.1728	0.4132	−0.0027	0.2905	−2.3368	−0.5928	−0.7066	−0.3341
−1.5959	−0.5691	0.6664	1.1476	0.0528	−0.2079	0.4760	−0.3523	−0.6874
−1.2178	0.0393	0.9418	0.9675	−0.0114	−0.3220	−0.9746	−0.0663	0.1718
0.5636	0.8843	1.2293	0.8789	−0.4601	−0.9147	0.6869	−0.7901	−0.9107
−1.8032	1.1091	0.3521	−0.0354	−1.3890	−0.0706	−1.7638	−0.1438	−0.3393
−0.9270	0.2753	−0.3444	1.2687	−0.0046	−1.3528	1.3998	−0.4565	−0.4570
0.8453	0.6641	−0.0013	0.0300	0.0875	0.1401	0.3621	0.9879	−0.5743
−0.1557	−0.1358	−0.1314	−0.6539	−0.9242	0.2399	0.0259	0.0477	−1.0546
−0.4832	−0.2027	1.2441	1.1753	−0.9627	0.6382	1.2822	−0.5381	1.3506
−1.0901	0.2983	−1.3574	−0.7218	−0.2277	0.0415	−0.4500	0.7018	0.4680
0.4859	2.1673	1.3597	−0.3802	0.5927	−2.0448	1.2126	0.9862	−0.0708
0.0313	−0.1177	−0.3527	−0.5585	0.5493	−0.7759	−1.0542	2.7646	1.4251
0.0114	−0.8899	1.1575	1.0503	0.5112	0.9398	0.1870	−1.0656	0.0948
−0.4968	−1.1686	−0.1628	−0.6773	0.8950	1.4502	0.3037	−1.9043	0.8748
2.1473	−2.6982	−0.3968	−0.6915	−1.3053	0.1680	0.9301	−0.1786	−0.2899
−1.6251	1.6827	0.5806	−1.5906	−0.9129	−1.1662	−0.9490	−1.3759	−1.6889
1.1831	−1.1058	−0.7076	0.7401	1.2219	1.1260	−0.6956	−0.2945	−0.6131
1.2256	−0.2552	1.4166	−3.4767	0.3159	−0.1370	0.0577	−1.1569	−0.2177

C
Chi-square Probabilities

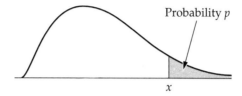

Table entry is the value of x such that $P(\chi^2 \geq x) = p$ for given p (column) and number of degrees of freedom df (row).

Table C.1 Right-Hand-Tail Chi-Square Density Areas

					p				
df	0.25	0.20	0.10	0.05	0.025	0.01	0.005	0.001	0.0005
1	1.32	1.64	2.71	3.84	5.02	6.63	7.88	10.83	12.12
2	2.77	3.22	4.61	5.99	7.38	9.21	10.60	13.82	15.20
3	4.11	4.64	6.25	7.81	9.35	11.34	12.84	16.27	17.73
4	5.39	5.99	7.78	9.49	11.14	13.28	14.86	18.47	20.00
5	6.63	7.29	9.24	11.07	12.83	15.09	16.75	20.51	22.11
6	7.84	8.56	10.64	12.59	14.45	16.81	18.55	22.46	24.10
7	9.04	9.80	12.02	14.07	16.01	18.48	20.28	24.32	26.02
8	10.22	11.03	13.36	15.51	17.53	20.09	21.95	26.12	27.87
9	11.39	12.24	14.68	16.92	19.02	21.67	23.59	27.88	29.67
10	12.55	13.44	15.99	18.31	20.48	23.21	25.19	29.59	31.42
11	13.70	14.63	17.28	19.68	21.92	24.72	26.76	31.26	33.14
12	14.85	15.81	18.55	21.03	23.34	26.22	28.30	32.91	34.82
13	15.98	16.98	19.81	22.36	24.74	27.69	29.82	34.53	36.48
14	17.12	18.15	21.06	23.68	26.12	29.14	31.32	36.12	38.11
15	18.25	19.31	22.31	25.00	27.49	30.58	32.80	37.70	39.72
16	19.37	20.47	23.54	26.30	28.85	32.00	34.27	39.25	41.31
17	20.49	21.61	24.77	27.59	30.19	33.41	35.72	40.79	42.88
18	21.60	22.76	25.99	28.87	31.53	34.81	37.16	42.31	44.43
19	22.72	23.90	27.20	30.14	32.85	36.19	38.58	43.82	45.97
20	23.83	25.04	28.41	31.41	34.17	37.57	40.00	45.31	47.50
21	24.93	26.17	29.62	32.67	35.48	38.93	41.40	46.80	49.01

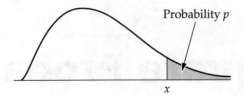

Table entry is the value of x such that $P(\chi^2 \geq x) = p$ for given p (column) and number of degrees of freedom df (row).

Table C.2 (Continued) Right-Hand-Tail Chi-Square Density Areas

df	\multicolumn{9}{c}{p}
	0.25	0.20	0.10	0.05	0.025	0.01	0.005	0.001	0.0005
22	26.04	27.30	30.81	33.92	36.78	40.29	42.80	48.27	50.51
23	27.14	28.43	32.01	35.17	38.08	41.64	44.18	49.73	52.00
24	28.24	29.55	33.20	36.42	39.36	42.98	45.56	51.18	53.48
25	29.34	30.68	34.38	37.65	40.65	44.31	46.93	52.62	54.95
26	30.43	31.79	35.56	38.89	41.92	45.64	48.29	54.05	56.41
27	31.53	32.91	36.74	40.11	43.19	46.96	49.64	55.48	57.86
28	32.62	34.03	37.92	41.34	44.46	48.28	50.99	56.89	59.30
29	33.71	35.14	39.09	42.56	45.72	49.59	52.34	58.30	60.73
30	34.80	36.25	40.26	43.77	46.98	50.89	53.67	59.70	62.16
40	45.62	47.27	51.81	55.76	59.34	63.69	66.77	73.40	76.09
50	56.33	58.16	63.17	67.50	71.42	76.15	79.49	86.66	89.56
60	66.98	68.97	74.40	79.08	83.30	88.38	91.95	99.61	102.7
80	88.13	90.41	96.58	101.9	106.6	112.3	116.3	124.8	128.3

D

Linear Interpolation

Interpolation can be used to help find a value that is not provided in a table. The procedure is called *linear* interpolation, since it assumes that the relations between the values involved are (or are nearly) linear—that is, they are straight-line relationships. This assumption gives very good estimates of desired values for the exercises in this book.

Example 1 Find $P(\chi^2_{10} \geq 20.0)$.

The chi-square table in Appendix C does not have an entry of 20.0 for a chi-square with 10 degrees of freedom. So interpolation must be used.

The chi-square values of 18.31 and 20.48 are entries in the table, with corresponding probabilities (areas) of 0.05 and 0.025, respectively (that is, $P(\chi^2_{10} \geq 18.31) = 0.05$ and $P(\chi^2_{10} \geq 20.48) = 0.025$). So $P(\chi^2_{10} \geq 20.0)$ is somewhere between 0.05 and 0.025. Note that $20.48 - 18.31 = 2.17$ and $20.0 - 18.31 = 1.69$. Now consider the following proportionality argument. The following table shows the proportions we are dealing with.

$$
2.17 \left[\begin{array}{c} 1.69 \left[\begin{array}{cc} \text{Chi-square} & \text{Probability (area)} \\ 18.31 & 0.05 \\ 20.0 & P(\chi^2_{10} \geq 20.0) \end{array} \right] d \\ 20.48 \quad\quad 0.025 \end{array} \right] 0.025
$$

The unknown probability $P(\chi^2_{10} \geq 20.0)$ is proportionally decreased from 0.05 by an amount d according to the proportionality equation

$$\frac{d}{0.025} = \frac{1.69}{2.17}.$$

Thus,

$$d = \frac{1.69}{2.17}(0.025) \approx 0.019.$$

Thus,

$$P(\chi^2_{10} \geq 20.0) \approx 0.05 - 0.019 \approx 0.031.$$

Example 2 Find the value of a in the following:

$$P(\chi^2_{12} \geq a) = 0.03.$$

570 LINEAR INTERPOLATION

The chi-square table in Appendix C does not have a column for a probability of 0.03. However, there are columns for 0.05 and 0.025, so we can use interpolation to find the required chi-square value a with 12 degrees of freedom.

$$2.31 \begin{bmatrix} d \begin{bmatrix} 21.03 & 0.05 \\ a & 0.03 \end{bmatrix} 0.02 \\ 23.34 & 0.025 \end{bmatrix} 0.025$$

Thus the unknown a is proportionally increased from 21.03 by an amount d according to the proportionality equation

$$\frac{d}{2.31} = \frac{0.02}{0.025} = 0.8.$$

So we have

$$d = 0.8(2.31) \approx 1.85.$$

The unknown value a is thus

$$a \approx 21.03 + d \approx 21.03 + 1.85 = 22.88.$$

Therefore,

$$P(\chi^2_{12} \geq 22.88) \approx 0.03.$$

E

Normal Probabilities

Table E.1 Standard Normal Probabilities to Left of Given z

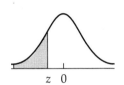

z	0	1	2	3	4	5	6	7	8	9
−3.0*	.0013	.0013	.0013	.0012	.0012	.0011	.0011	.0011	.0010	.0010
−2.9	.0019	.0018	.0017	.0017	.0016	.0016	.0015	.0015	.0014	.0014
−2.8	.0026	.0025	.0024	.0023	.0023	.0022	.0021	.0021	.0020	.0019
−2.7	.0035	.0034	.0033	.0032	.0031	.0030	.0029	.0028	.0027	.0026
−2.6	.0047	.0045	.0044	.0043	.0041	.0040	.0039	.0038	.0037	.0036
−2.5	.0062	.0060	.0059	.0057	.0055	.0054	.0052	.0051	.0049	.0048
−2.4	.0082	.0080	.0078	.0075	.0073	.0071	.0069	.0068	.0066	.0064
−2.3	.0107	.0104	.0102	.0099	.0096	.0094	.0091	.0089	.0087	.0084
−2.2	.0139	.0136	.0132	.0129	.0125	.0122	.0119	.0116	.0113	.0110
−2.1	.0179	.0174	.0170	.0166	.0162	.0158	.0154	.0150	.0146	.0143
−2.0	.0228	.0222	.0217	.0212	.0207	.0202	.0197	.0192	.0188	.0183
−1.9	.0287	.0281	.0274	.0268	.0262	.0256	.0250	.0244	.0239	.0233
−1.8	.0359	.0351	.0344	.0336	.0329	.0322	.0314	.0307	.0301	.0294
−1.7	.0046	.0436	.0427	.0418	.0409	.0401	.0392	.0384	.0375	.0367
−1.6	.0548	.0537	.0526	.0516	.0505	.0495	.0485	.0475	.0465	.0455
−1.5	.0668	.0655	.0643	.0630	.0618	.0606	.0594	.0582	.0571	.0559
−1.4	.0808	.0793	.0778	.0764	.0749	.0735	.0721	.0708	.0694	.0681
−1.3	.0968	.0951	.0934	.0918	.0901	.0885	.0869	.0853	.0838	.0823
−1.2	.1151	.1131	.1112	.1093	.1075	.0156	.1038	.1020	.1003	.0985
−1.1	.1357	.1335	.1314	.1292	.1271	.1251	.1230	.1210	.1190	.1170
−1.0	.1587	.1562	.1539	.1515	.1492	.1469	.1446	.1423	.1401	.1379
−0.9	.1841	.1814	.1788	.1762	.1736	.1711	.1685	.1660	.1635	.1611
−0.8	.2119	.2090	.2061	.2033	.2005	.1977	.1949	.1922	.1894	.1867
−0.7	.2420	.2389	.2358	.2327	.2296	.2266	.2236	.2206	.2177	.2148
−0.6	.2743	.2709	.2676	.2643	.2611	.2578	.2546	.2514	.2483	.2451
−0.5	.3085	.3050	.3015	.2981	.2946	.2912	.2877	.2843	.2810	.2776
−0.4	.3446	.3409	.3372	.3336	.3300	.3264	.3228	.3192	.3156	.3121
−0.3	.3821	.3783	.3745	.3707	.3669	.3632	.3594	.3557	.3520	.3483
−0.2	.4207	.4168	.4129	.4090	.4052	.4013	.3974	.3936	.3897	.3859
−0.1	.4602	.4562	.4522	.4483	.4443	.4404	.4364	.4325	.4286	.4247
−0.0	.5000	.4960	.4920	.4880	.4840	.4801	.4761	.4721	.4681	.4641

*For any $z \leq -3$, the area is approximately 0; indeed, for any $z \leq -4$, the area is 0 to four decimal places.

NORMAL PROBABILITIES

Table E.2 Standard Normal Probabilities to Left of Given z

z	0	1	2	3	4	5	6	7	8	9
0.0	.5000	.5040	.5080	.5120	.5160	.5199	.5239	.5279	.5319	.5359
0.1	.5398	.5438	.5478	.5517	.5557	.5596	.5636	.5675	.5714	.5753
0.2	.5793	.5832	.5871	.5910	.5948	.5987	.6026	.6064	.6103	.6141
0.3	.6179	.6217	.6255	.6293	.6331	.6368	.6406	.6443	.6480	.6517
0.4	.6554	.6591	.6628	.6664	.6700	.6736	.6772	.6808	.6844	.6879
0.5	.6915	.6950	.6985	.7019	.7054	.7088	.7123	.7157	.7190	.7224
0.6	.7257	.7291	.7324	.7357	.7389	.7422	.7454	.7486	.7517	.7549
0.7	.7580	.7611	.7642	.7673	.7704	.7734	.7764	.7794	.7823	.7852
0.8	.7881	.7910	.7939	.7967	.7995	.8023	.8051	.8078	.8106	.8133
0.9	.8159	.8186	.8212	.8238	.8264	.8289	.8315	.8340	.8365	.8389
1.0	.8413	.8438	.8461	.8485	.8508	.8531	.8554	.8577	.8599	.8621
1.1	.8643	.8665	.8686	.8708	.8729	.8749	.8770	.8790	.8810	.8831
1.2	.8849	.8869	.8888	.8907	.8925	.8944	.8962	.8980	.8997	.9015
1.3	.9032	.9049	.9066	.9082	.9099	.9115	.9131	.9147	.9162	.9177
1.4	.9192	.9207	.9222	.9236	.9251	.9265	.9279	.9292	.9306	.9319
1.5	.9332	.9345	.9357	.9370	.9382	.9394	.9406	.9418	.9429	.9441
1.6	.9452	.9463	.9474	.9484	.9495	.9505	.9515	.9525	.9535	.9545
1.7	.9554	.9564	.9573	.9582	.9591	.9599	.9608	.9616	.9625	.9633
1.8	.9641	.9649	.9656	.9664	.9671	.9678	.9686	.9693	.9699	.9706
1.9	.9713	.9717	.9726	.9732	.9738	.9744	.9750	.9756	.9761	.9767
2.0	.9772	.9778	.9783	.9788	.9793	.9798	.9803	.9808	.9812	.9817
2.1	.9821	.9826	.9830	.9834	.9838	.9842	.9846	.9850	.9854	.9857
2.2	.9861	.9864	.9868	.9871	.9875	.9878	.9881	.9884	.9887	.9890
2.3	.9893	.9896	.9898	.9901	.9904	.9906	.9909	.9911	.9913	.9916
2.4	.9918	.9920	.9922	.9925	.9927	.9929	.9931	.9932	.9934	.9936
2.5	.9938	.9940	.9941	.9943	.9945	.9946	.9948	.9949	.9951	.9952
2.6	.9953	.9955	.9956	.9957	.9959	.9960	.9961	.9962	.9963	.9964
2.7	.9965	.9966	.9967	.9968	.9969	.9970	.9971	.9972	.9973	.9974
2.8	.9974	.9975	.9976	.9977	.9977	.9978	.9979	.9979	.9980	.9981
2.9	.9981	.9982	.9982	.9983	.9984	.9984	.9985	.9985	.9986	.9986
3.0*	.9987	.9987	.9987	.9988	.9988	.9989	.9989	.9989	.9990	.9990

*For any $z \geq 3$, the area is approximately 1; indeed, for any $z \geq 4$, the area is 1 to four decimal places.

F

Student's t Probabilities

Table F.1. Right-Hand-Tail t Distribution Probabilities

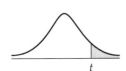

df	0.10	0.05	0.025	0.01	0.005
1	3.08	6.31	12.71	31.82	63.66
2	1.89	2.92	4.30	6.96	9.92
3	1.64	2.35	3.18	4.54	5.84
4	1.53	2.13	2.78	3.75	4.60
5	1.48	2.02	2.57	3.36	4.03
6	1.44	1.94	2.45	3.14	3.71
7	1.42	1.89	2.36	3.00	3.50
8	1.40	1.86	2.31	2.90	3.36
9	1.38	1.83	2.26	2.82	3.25
10	1.37	1.81	2.23	2.76	3.17
11	1.36	1.80	2.20	2.72	3.11
12	1.36	1.78	2.18	2.68	3.05
13	1.35	1.77	2.16	2.65	3.01
14	1.35	1.76	2.14	2.62	2.98
15	1.34	1.75	2.13	2.60	2.95
16	1.34	1.75	2.12	2.58	2.92
17	1.33	1.74	2.11	2.57	2.90
18	1.33	1.73	2.10	2.55	2.88
19	1.33	1.73	2.09	2.54	2.86
20	1.33	1.72	2.09	2.53	2.85
21	1.32	1.72	20.8	2.52	2.83
22	1.32	1.72	2.07	2.51	2.82
23	1.32	1.71	2.07	2.50	2.81
24	1.32	1.71	2.06	2.49	2.80
25	1.32	1.71	2.06	2.49	2.79
26	1.32	1.71	2.06	2.48	2.78
27	1.31	1.70	2.05	2.47	2.77
28	1.31	1.70	2.05	2.47	2.76
29	1.31	1.70	2.05	2.46	2.76
∞	1.28	1.64	1.96	2.33	2.58

Note: The last row of the table (df = ∞) gives values for the area to the right of t using the standard normal density of Table E. For example, the table shows that $P(z > 1.28) = 0.10$ and $P(z > 1.64) = 0.05$, as can also be seen in Table E.

G

Cumulative Binomial Probabilities

The Table Gives the Probability of Obtaining x or Fewer Successes in n Independent Trials, Where p = Probability of Success in a Single Trial.

Table G: Cumulative Binomial Probabilities

						p					
n	x	0.05	0.10	0.15	0.20	0.25	0.30	0.35	0.40	0.45	0.50
2	0	0.9025	0.8100	0.7225	0.6400	0.5625	0.4900	0.4225	0.3600	0.3025	0.2500
	1	0.9975	0.9900	0.9775	0.9600	0.9375	0.9100	0.8775	0.8400	0.7975	0.7500
	2	1.0000	1.0000	1.0000	1.0000	1.0000	1.0000	1.0000	1.0000	1.0000	1.0000
3	0	0.8574	0.7290	0.6141	0.5120	0.4219	0.3430	0.2746	0.2160	0.1664	0.1250
	1	0.9928	0.9720	0.9393	0.8960	0.8438	0.7840	0.7183	0.6480	0.5748	0.5000
	2	0.9999	0.9990	0.9966	0.9920	0.9844	0.9730	0.9571	0.9360	0.9089	0.8750
	3	1.0000	1.0000	1.0000	1.0000	1.0000	1.0000	1.0000	1.0000	1.0000	1.0000
4	0	0.8145	0.6561	0.5220	0.4096	0.3164	0.2401	0.1785	0.1296	0.0915	0.0625
	1	0.9860	0.9477	0.8905	0.8192	0.7383	0.6517	0.5630	0.4752	0.3910	0.3125
	2	0.9995	0.9963	0.9880	0.9728	0.9492	0.9163	0.8735	0.8208	0.7585	0.6875
	3	1.0000	0.9999	0.9995	0.9984	0.9961	0.9919	0.9850	0.9744	0.9590	0.9375
	4	1.0000	1.0000	1.0000	1.0000	1.0000	1.0000	1.0000	1.0000	1.0000	1.0000
5	0	0.7738	0.5905	0.4437	0.3277	0.2373	0.1681	0.1160	0.0778	0.0503	0.0313
	1	0.9774	0.9185	0.8352	0.7373	0.6328	0.5282	0.4284	0.3370	0.2562	0.1875
	2	0.9988	0.9914	0.9734	0.9421	0.8965	0.8369	0.7648	0.6826	0.5931	0.5000
	3	1.0000	0.9995	0.9978	0.9933	0.9844	0.9692	0.9460	0.9130	0.8688	0.8125
	4	1.0000	1.0000	0.9999	0.9997	0.9990	0.9976	0.9947	0.9898	0.9815	0.9688
	5	1.0000	1.0000	1.0000	1.0000	1.0000	1.0000	1.0000	1.0000	1.0000	1.0000
6	0	0.7351	0.5314	0.3771	0.2621	0.1780	0.1176	0.0754	0.0467	0.0277	0.0156
	1	0.9672	0.8857	0.7765	0.6554	0.5339	0.4202	0.3191	0.2333	0.1636	0.1094
	2	0.9978	0.9842	0.9527	0.9011	0.8306	0.7443	0.6471	0.5443	0.4415	0.3438
	3	0.9999	0.9987	0.9941	0.9830	0.9624	0.9295	0.8826	0.8208	0.7447	0.6563
	4	1.0000	0.9999	0.9996	0.9984	0.9954	0.9891	0.9777	0.9590	0.9308	0.8906
	5	1.0000	1.0000	1.0000	0.9999	0.9998	0.9993	0.9982	0.9959	0.9917	0.9844
	6	1.0000	1.0000	1.0000	1.0000	1.0000	1.0000	1.0000	1.0000	1.0000	1.0000

Table G: Cumulative Binomial Probabilities *(Continued)*

						p					
n	x	0.05	0.10	0.15	0.20	0.25	0.30	0.35	0.40	0.45	0.50
7	0	0.6983	0.4783	0.3206	0.2097	0.1335	0.0824	0.0490	0.0280	0.0152	0.0078
	1	0.9556	0.8503	0.7166	0.5767	0.4449	0.3294	0.2338	0.1586	0.1024	0.0625
	2	0.9962	0.9743	0.9262	0.8520	0.7564	0.6471	0.5323	0.4199	0.3164	0.2266
	3	0.9998	0.9973	0.9879	0.9667	0.9294	0.8740	0.8002	0.7102	0.6083	0.5000
	4	1.0000	0.9998	0.9988	0.9953	0.9871	0.9712	0.9444	0.9037	0.8471	0.7734
	5	1.0000	1.0000	0.9999	0.9996	0.9987	0.9962	0.9910	0.9812	0.9643	0.9375
	6	1.0000	1.0000	1.0000	1.0000	0.9999	0.9998	0.9994	0.9984	0.9963	0.9922
	7	1.0000	1.0000	1.0000	1.0000	1.0000	1.0000	1.0000	1.0000	1.0000	1.0000
8	0	0.6634	0.4305	0.2725	0.1678	0.1001	0.0576	0.0319	0.0168	0.0084	0.0039
	1	0.9428	0.8131	0.6572	0.5033	0.3671	0.2553	0.1691	0.1064	0.0632	0.0352
	2	0.9942	0.9619	0.8948	0.7969	0.6785	0.5518	0.4278	0.3154	0.2201	0.1445
	3	0.9996	0.9950	0.9786	0.9437	0.8862	0.8059	0.7064	0.5941	0.4770	0.3633
	4	1.0000	0.9996	0.9971	0.9896	0.9727	0.9420	0.8939	0.8263	0.7396	0.6367
	5	1.0000	1.0000	0.9998	0.9988	0.9958	0.9887	0.9747	0.9502	0.9115	0.8555
	6	1.0000	1.0000	1.0000	0.9999	0.9996	0.9987	0.9964	0.9915	0.9819	0.9648
	7	1.0000	1.0000	1.0000	1.0000	1.0000	0.9999	0.9998	0.9993	0.9983	0.9961
	8	1.0000	1.0000	1.0000	1.0000	1.0000	1.0000	1.0000	1.0000	1.0000	1.0000
9	0	0.6302	0.3874	0.2316	0.1342	0.0751	0.0404	0.0207	0.0101	0.0046	0.0020
	1	0.9288	0.7748	0.5995	0.4362	0.3003	0.1960	0.1211	0.0705	0.0385	0.0195
	2	0.9916	0.9470	0.8591	0.7382	0.6007	0.4628	0.3373	0.2318	0.1495	0.0898
	3	0.9994	0.9917	0.9661	0.9144	0.8343	0.7297	0.6089	0.4826	0.3614	0.2539
	4	1.0000	0.9991	0.9944	0.9804	0.9511	0.9012	0.8283	0.7334	0.6214	0.5000
	5	1.0000	0.9999	0.9994	0.9969	0.9900	0.9747	0.9464	0.9006	0.8342	0.7461
	6	1.0000	1.0000	1.0000	0.9997	0.9987	0.9957	0.9888	0.9750	0.9502	0.9102
	7	1.0000	1.0000	1.0000	1.0000	0.9999	0.9996	0.9986	0.9962	0.9909	0.9805
	8	1.0000	1.0000	1.0000	1.0000	1.0000	1.0000	0.9999	0.9997	0.9992	0.9980
	9	1.0000	1.0000	1.0000	1.0000	1.0000	1.0000	1.0000	1.0000	1.0000	1.0000
10	0	0.5987	0.3487	0.1969	0.1074	0.0563	0.0282	0.0135	0.0060	0.0025	0.0010
	1	0.9139	0.7361	0.5443	0.3758	0.2440	0.1493	0.0860	0.0464	0.0233	0.0107
	2	0.9885	0.9298	0.8202	0.6778	0.5256	0.3828	0.2616	0.1673	0.0996	0.0547
	3	0.9990	0.9872	0.9500	0.8791	0.7759	0.6496	0.5138	0.3823	0.2660	0.1719
	4	0.9999	0.9984	0.9901	0.9672	0.9219	0.8497	0.7515	0.6331	0.5044	0.3770
	5	1.0000	0.9999	0.9986	0.9936	0.9803	0.9527	0.9051	0.8338	0.7384	0.6230
	6	1.0000	1.0000	0.9999	0.9991	0.9965	0.9894	0.9740	0.9452	0.8980	0.8281
	7	1.0000	1.0000	1.0000	0.9999	0.9996	0.9984	0.9952	0.9877	0.9726	0.9453
	8	1.0000	1.0000	1.0000	1.0000	1.0000	0.9999	0.9995	0.9983	0.9955	0.9893
	9	1.0000	1.0000	1.0000	1.0000	1.0000	1.0000	1.0000	0.9999	0.9997	0.9990
	10	1.0000	1.0000	1.0000	1.0000	1.0000	1.0000	1.0000	1.0000	1.0000	1.0000
11	0	0.5688	0.3138	0.1673	0.0859	0.0422	0.0198	0.0088	0.0036	0.0014	0.0005
	1	0.8981	0.6974	0.4922	0.3221	0.1971	0.1130	0.0606	0.0302	0.0139	0.0059
	2	0.9848	0.9104	0.7788	0.6174	0.4552	0.3127	0.2001	0.1189	0.0652	0.0327
	3	0.9984	0.9815	0.9306	0.8389	0.7133	0.5696	0.4256	0.2963	0.1911	0.1133
	4	0.9999	0.9972	0.9841	0.9496	0.8854	0.7897	0.6683	0.5328	0.3971	0.2744
	5	1.0000	0.9997	0.9973	0.9883	0.9657	0.9218	0.8513	0.7535	0.6331	0.5000
	6	1.0000	1.0000	0.9997	0.9980	0.9924	0.9784	0.9499	0.9006	0.8262	0.7256

Table G: Cumulative Binomial Probabilities *(Continued)*

n	x	0.05	0.10	0.15	0.20	0.25	0.30	0.35	0.40	0.45	0.50
11	7	1.0000	1.0000	1.0000	0.9998	0.9988	0.9957	0.9878	0.9707	0.9390	0.8867
	8	1.0000	1.0000	1.0000	1.0000	0.9999	0.9994	0.9980	0.9941	0.9852	0.9673
	9	1.0000	1.0000	1.0000	1.0000	1.0000	1.0000	0.9998	0.9993	0.9978	0.9941
	10	1.0000	1.0000	1.0000	1.0000	1.0000	1.0000	1.0000	1.0000	0.9998	0.9995
	11	1.0000	1.0000	1.0000	1.0000	1.0000	1.0000	1.0000	1.0000	1.0000	1.0000
12	0	0.5404	0.2824	0.1422	0.0687	0.0317	0.0138	0.0057	0.0022	0.0008	0.0002
	1	0.8816	0.6590	0.4435	0.2749	0.1584	0.0850	0.0424	0.0196	0.0083	0.0032
	2	0.9804	0.8891	0.7358	0.5583	0.3907	0.2528	0.1513	0.0834	0.0421	0.0193
	3	0.9978	0.9744	0.9078	0.7946	0.6488	0.4925	0.3467	0.2253	0.1345	0.0730
	4	0.9998	0.9957	0.9761	0.9274	0.8424	0.7237	0.5833	0.4382	0.3044	0.1938
	5	1.0000	0.9995	0.9954	0.9806	0.9456	0.8822	0.7873	0.6652	0.5269	0.3872
	6	1.0000	0.9999	0.9993	0.9961	0.9857	0.9614	0.9154	0.8418	0.7393	0.6128
	7	1.0000	1.0000	0.9999	0.9994	0.9972	0.9905	0.9745	0.9427	0.8883	0.8062
	8	1.0000	1.0000	1.0000	0.9999	0.9996	0.9983	0.9944	0.9847	0.9644	0.9270
	9	1.0000	1.0000	1.0000	1.0000	1.0000	0.9998	0.9992	0.9972	0.9921	0.9807
	10	1.0000	1.0000	1.0000	1.0000	1.0000	1.0000	0.9999	0.9997	0.9989	0.9968
	11	1.0000	1.0000	1.0000	1.0000	1.0000	1.0000	1.0000	1.0000	0.9999	0.9998
	12	1.0000	1.0000	1.0000	1.0000	1.0000	1.0000	1.0000	1.0000	1.0000	1.0000
13	0	0.5133	0.2542	0.1209	0.0550	0.0238	0.0097	0.0037	0.0013	0.0004	0.0001
	1	0.8646	0.6213	0.3983	0.2336	0.1267	0.0637	0.0296	0.0126	0.0049	0.0017
	2	0.9755	0.8661	0.6920	0.5017	0.3326	0.2025	0.1132	0.0579	0.0269	0.0112
	3	0.9969	0.9658	0.8820	0.7473	0.5843	0.4206	0.2783	0.1686	0.0929	0.0461
	4	0.9997	0.9935	0.9658	0.9009	0.7940	0.6543	0.5005	0.3530	0.2279	0.1334
	5	1.0000	0.9991	0.9925	0.9700	0.9198	0.8346	0.7159	0.5744	0.4268	0.2905
	6	1.0000	0.9999	0.9987	0.9930	0.9757	0.9376	0.8705	0.7712	0.6437	0.5000
	7	1.0000	1.0000	0.9998	0.9988	0.9944	0.9818	0.9538	0.9023	0.8212	0.7095
	8	1.0000	1.0000	1.0000	0.9998	0.9990	0.9960	0.9874	0.9679	0.9302	0.8666
	9	1.0000	1.0000	1.0000	1.0000	0.9999	0.9993	0.9975	0.9922	0.9797	0.9539
	10	1.0000	1.0000	1.0000	1.0000	1.0000	0.9999	0.9997	0.9987	0.9959	0.9888
	11	1.0000	1.0000	1.0000	1.0000	1.0000	1.0000	1.0000	0.9999	0.9995	0.9983
	12	1.0000	1.0000	1.0000	1.0000	1.0000	1.0000	1.0000	1.0000	1.0000	0.9999
	13	1.0000	1.0000	1.0000	1.0000	1.0000	1.0000	1.0000	1.0000	1.0000	1.0000
14	0	0.4877	0.2288	0.1028	0.0440	0.0178	0.0068	0.0024	0.0008	0.0002	0.0001
	1	0.8470	0.5846	0.3567	0.1979	0.1010	0.0475	0.0205	0.0081	0.0029	0.0009
	2	0.9699	0.8416	0.6479	0.4481	0.2811	0.1608	0.0839	0.0398	0.0170	0.0065
	3	0.9958	0.9559	0.8535	0.6982	0.5213	0.3552	0.2205	0.1243	0.0632	0.0287
	4	0.9996	0.9908	0.9533	0.8702	0.7415	0.5842	0.4227	0.2793	0.1672	0.0898
	5	1.0000	0.9985	0.9885	0.9561	0.8883	0.7805	0.6405	0.4859	0.3373	0.2120
	6	1.0000	0.9998	0.9978	0.9884	0.9617	0.9067	0.8164	0.6925	0.5461	0.3953
	7	1.0000	1.0000	0.9997	0.9976	0.9897	0.9685	0.9247	0.8499	0.7414	0.6047
	8	1.0000	1.0000	1.0000	0.9996	0.9978	0.9917	0.9757	0.9417	0.8811	0.7880
	9	1.0000	1.0000	1.0000	1.0000	0.9997	0.9983	0.9940	0.9825	0.9574	0.9102
	10	1.0000	1.0000	1.0000	1.0000	1.0000	0.9998	0.9989	0.9961	0.9886	0.9713
	11	1.0000	1.0000	1.0000	1.0000	1.0000	1.0000	0.9999	0.9994	0.9978	0.9935
	12	1.0000	1.0000	1.0000	1.0000	1.0000	1.0000	1.0000	0.9999	0.9997	0.9991

Table G: Cumulative Binomial Probabilities *(Continued)*

						p					
n	x	0.05	0.10	0.15	0.20	0.25	0.30	0.35	0.40	0.45	0.50
14	13	1.0000	1.0000	1.0000	1.0000	1.0000	1.0000	1.0000	1.0000	1.0000	0.9999
	14	1.0000	1.0000	1.0000	1.0000	1.0000	1.0000	1.0000	1.0000	1.0000	1.0000
15	0	0.4633	0.2059	0.0874	0.0352	0.0134	0.0047	0.0016	0.0005	0.0001	0.0000
	1	0.8290	0.5490	0.3186	0.1671	0.0802	0.0353	0.0142	0.0052	0.0017	0.0005
	2	0.9638	0.8159	0.6042	0.3980	0.2361	0.1268	0.0617	0.0271	0.0107	0.0037
	3	0.9945	0.9444	0.8227	0.6482	0.4613	0.2969	0.1727	0.0905	0.0424	0.0176
	4	0.9994	0.9873	0.9383	0.8358	0.6865	0.5155	0.3519	0.2173	0.1204	0.0592
	5	0.9999	0.9978	0.9832	0.9389	0.8516	0.7216	0.5643	0.4032	0.2608	0.1509
	6	1.0000	0.9997	0.9964	0.9819	0.9434	0.8689	0.7548	0.6098	0.4522	0.3036
	7	1.0000	1.0000	0.9994	0.9958	0.9827	0.9500	0.8868	0.7869	0.6535	0.5000
	8	1.0000	1.0000	0.9999	0.9992	0.9958	0.9848	0.9578	0.9050	0.8182	0.6964
	9	1.0000	1.0000	1.0000	0.9999	0.9992	0.9963	0.9876	0.9662	0.9231	0.8491
	10	1.0000	1.0000	1.0000	1.0000	0.9999	0.9993	0.9972	0.9907	0.9745	0.9408
	11	1.0000	1.0000	1.0000	1.0000	1.0000	0.9999	0.9995	0.9981	0.9937	0.9824
	12	1.0000	1.0000	1.0000	1.0000	1.0000	1.0000	0.9999	0.9997	0.9989	0.9963
	13	1.0000	1.0000	1.0000	1.0000	1.0000	1.0000	1.0000	1.0000	0.9999	0.9995
	14	1.0000	1.0000	1.0000	1.0000	1.0000	1.0000	1.0000	1.0000	1.0000	1.0000
	15	1.0000	1.0000	1.0000	1.0000	1.0000	1.0000	1.0000	1.0000	1.0000	1.0000
16	0	0.4401	0.1853	0.0743	0.0281	0.0100	0.0033	0.0010	0.0003	0.0001	0.0000
	1	0.8108	0.5147	0.2839	0.1407	0.0635	0.0261	0.0098	0.0033	0.0010	0.0003
	2	0.9571	0.7892	0.5614	0.3518	0.1971	0.0994	0.0451	0.0183	0.0066	0.0021
	3	0.9930	0.9316	0.7899	0.5981	0.4050	0.2459	0.1339	0.0651	0.0281	0.0106
	4	0.9991	0.9830	0.9209	0.7982	0.6302	0.4499	0.2892	0.1666	0.0853	0.0384
	5	0.9999	0.9967	0.9765	0.9183	0.8103	0.6598	0.4900	0.3288	0.1976	0.1051
	6	1.0000	0.9995	0.9944	0.9733	0.9204	0.8247	0.6881	0.5272	0.3660	0.2272
	7	1.0000	0.9999	0.9989	0.9930	0.9729	0.9256	0.8406	0.7161	0.5629	0.4018
	8	1.0000	1.0000	0.9998	0.9985	0.9925	0.9743	0.9329	0.8577	0.7441	0.5982
	9	1.0000	1.0000	1.0000	0.9998	0.9984	0.9929	0.9771	0.9417	0.8759	0.7728
	10	1.0000	1.0000	1.0000	1.0000	0.9997	0.9984	0.9938	0.9809	0.9514	0.8949
	11	1.0000	1.0000	1.0000	1.0000	1.0000	0.9997	0.9987	0.9951	0.9851	0.9616
	12	1.0000	1.0000	1.0000	1.0000	1.0000	1.0000	0.9998	0.9991	0.9965	0.9894
	13	1.0000	1.0000	1.0000	1.0000	1.0000	1.0000	1.0000	0.9999	0.9994	0.9979
	14	1.0000	1.0000	1.0000	1.0000	1.0000	1.0000	1.0000	1.0000	0.9999	0.9997
	15	1.0000	1.0000	1.0000	1.0000	1.0000	1.0000	1.0000	1.0000	1.0000	1.0000
	16	1.0000	1.0000	1.0000	1.0000	1.0000	1.0000	1.0000	1.0000	1.0000	1.0000
17	0	0.4181	0.1668	0.0631	0.0225	0.0075	0.0023	0.0007	0.0002	0.0000	0.0000
	1	0.7922	0.4818	0.2525	0.1182	0.0501	0.0193	0.0067	0.0021	0.0006	0.0001
	2	0.9497	0.7618	0.5198	0.3096	0.1637	0.0774	0.0327	0.0123	0.0041	0.0012
	3	0.9912	0.9174	0.7556	0.5489	0.3530	0.2019	0.1028	0.0464	0.0184	0.0064
	4	0.9988	0.9779	0.9013	0.7582	0.5739	0.3887	0.2348	0.1260	0.0596	0.0245
	5	0.9999	0.9953	0.9681	0.8943	0.7653	0.5968	0.4197	0.2639	0.1471	0.0717
	6	1.0000	0.9992	0.9917	0.9623	0.8929	0.7752	0.6188	0.4478	0.2902	0.1662
	7	1.0000	0.9999	0.9983	0.9891	0.9598	0.8954	0.7872	0.6405	0.4743	0.3145
	8	1.0000	1.0000	0.9997	0.9974	0.9876	0.9597	0.9006	0.8011	0.6626	0.5000
	9	1.0000	1.0000	1.0000	0.9995	0.9969	0.9873	0.9617	0.9081	0.8166	0.6855
	10	1.0000	1.0000	1.0000	0.9999	0.9994	0.9968	0.9880	0.9652	0.9174	0.8338
	11	1.0000	1.0000	1.0000	1.0000	0.9999	0.9993	0.9970	0.9894	0.9699	0.9283

Table G: Cumulative Binomial Probabilities *(Continued)*

		\multicolumn{10}{c}{p}									
n	x	0.05	0.10	0.15	0.20	0.25	0.30	0.35	0.40	0.45	0.50
17	12	1.0000	1.0000	1.0000	1.0000	1.0000	0.9999	0.9994	0.9975	0.9914	0.9755
	13	1.0000	1.0000	1.0000	1.0000	1.0000	1.0000	0.9999	0.9995	0.9981	0.9936
	14	1.0000	1.0000	1.0000	1.0000	1.0000	1.0000	1.0000	0.9999	0.9997	0.9988
	15	1.0000	1.0000	1.0000	1.0000	1.0000	1.0000	1.0000	1.0000	1.0000	0.9999
	16	1.0000	1.0000	1.0000	1.0000	1.0000	1.0000	1.0000	1.0000	1.0000	1.0000
	17	1.0000	1.0000	1.0000	1.0000	1.0000	1.0000	1.0000	1.0000	1.0000	1.0000
18	0	0.3972	0.1501	0.0536	0.0180	0.0056	0.0016	0.0004	0.0001	0.0000	0.0000
	1	0.7735	0.4503	0.2241	0.0991	0.0395	0.0142	0.0046	0.0013	0.0003	0.0001
	2	0.9419	0.7338	0.4797	0.2713	0.1353	0.0600	0.0236	0.0082	0.0025	0.0007
	3	0.9891	0.9018	0.7202	0.5010	0.3057	0.1646	0.0783	0.0328	0.0120	0.0038
	4	0.9985	0.9718	0.8794	0.7164	0.5187	0.3327	0.1886	0.0942	0.0411	0.0154
	5	0.9998	0.9936	0.9581	0.8671	0.7175	0.5344	0.3550	0.2088	0.1077	0.0481
	6	1.0000	0.9988	0.9882	0.9487	0.8610	0.7217	0.5491	0.3743	0.2258	0.1189
	7	1.0000	0.9998	0.9973	0.9837	0.9431	0.8593	0.7283	0.5634	0.3915	0.2403
	8	1.0000	1.0000	0.9995	0.9957	0.9807	0.9404	0.8609	0.7368	0.5778	0.4073
	9	1.0000	1.0000	0.9999	0.9991	0.9946	0.9790	0.9403	0.8653	0.7473	0.5927
	10	1.0000	1.0000	1.0000	0.9998	0.9988	0.9939	0.9788	0.9424	0.8720	0.7597
	11	1.0000	1.0000	1.0000	1.0000	0.9998	0.9986	0.9938	0.9797	0.9463	0.8811
	12	1.0000	1.0000	1.0000	1.0000	1.0000	0.9997	0.9986	0.9942	0.9817	0.9519
	13	1.0000	1.0000	1.0000	1.0000	1.0000	1.0000	0.9997	0.9987	0.9951	0.9846
	14	1.0000	1.0000	1.0000	1.0000	1.0000	1.0000	1.0000	0.9998	0.9990	0.9962
	15	1.0000	1.0000	1.0000	1.0000	1.0000	1.0000	1.0000	1.0000	0.9999	0.9993
	16	1.0000	1.0000	1.0000	1.0000	1.0000	1.0000	1.0000	1.0000	1.0000	0.9999
	17	1.0000	1.0000	1.0000	1.0000	1.0000	1.0000	1.0000	1.0000	1.0000	1.0000
	18	1.0000	1.0000	1.0000	1.0000	1.0000	1.0000	1.0000	1.0000	1.0000	1.0000
19	0	0.3774	0.1351	0.0456	0.0144	0.0042	0.0011	0.0003	0.0001	0.0000	0.0000
	1	0.7547	0.4203	0.1985	0.0829	0.0310	0.0104	0.0031	0.0008	0.0002	0.0000
	2	0.9335	0.7054	0.4413	0.2369	0.1113	0.0462	0.0170	0.0055	0.0015	0.0004
	3	0.9868	0.8850	0.6841	0.4551	0.2631	0.1332	0.0591	0.0230	0.0077	0.0022
	4	0.9980	0.9648	0.8556	0.6733	0.4654	0.2822	0.1500	0.0696	0.0280	0.0096
	5	0.9998	0.9914	0.9463	0.8369	0.6678	0.4739	0.2968	0.1629	0.0777	0.0318
	6	1.0000	0.9983	0.9837	0.9324	0.8251	0.6655	0.4812	0.3081	0.1727	0.0835
	7	1.0000	0.9997	0.9959	0.9767	0.9225	0.8180	0.6656	0.4878	0.3169	0.1796
	8	1.0000	1.0000	0.9992	0.9933	0.9713	0.9161	0.8145	0.6675	0.4940	0.3238
	9	1.0000	1.0000	0.9999	0.9984	0.9911	0.9674	0.9125	0.8139	0.6710	0.5000
	10	1.0000	1.0000	1.0000	0.9997	0.9977	0.9895	0.9653	0.9115	0.8159	0.6762
	11	1.0000	1.0000	1.0000	1.0000	0.9995	0.9972	0.9886	0.9648	0.9129	0.8204
	12	1.0000	1.0000	1.0000	1.0000	0.9999	0.9994	0.9969	0.9884	0.9658	0.9165
	13	1.0000	1.0000	1.0000	1.0000	1.0000	0.9999	0.9993	0.9969	0.9891	0.9682
	14	1.0000	1.0000	1.0000	1.0000	1.0000	1.0000	0.9999	0.9994	0.9972	0.9904
	15	1.0000	1.0000	1.0000	1.0000	1.0000	1.0000	1.0000	0.9999	0.9995	0.9978
	16	1.0000	1.0000	1.0000	1.0000	1.0000	1.0000	1.0000	1.0000	0.9999	0.9996
	17	1.0000	1.0000	1.0000	1.0000	1.0000	1.0000	1.0000	1.0000	1.0000	1.0000
	18	1.0000	1.0000	1.0000	1.0000	1.0000	1.0000	1.0000	1.0000	1.0000	1.0000
	19	1.0000	1.0000	1.0000	1.0000	1.0000	1.0000	1.0000	1.0000	1.0000	1.0000

Table G: Cumulative Binomial Probabilities *(Continued)*

n	x	p									
		0.05	0.10	0.15	0.20	0.25	0.30	0.35	0.40	0.45	0.50
20	0	0.3585	0.1216	0.0388	0.0115	0.0032	0.0008	0.0002	0.0000	0.0000	0.0000
	1	0.7358	0.3917	0.1756	0.0692	0.0243	0.0076	0.0021	0.0005	0.0001	0.0000
	2	0.9245	0.6769	0.4049	0.2061	0.0913	0.0355	0.0121	0.0036	0.0009	0.0002
	3	0.9841	0.8670	0.6477	0.4114	0.2252	0.1071	0.0444	0.0160	0.0049	0.0013
	4	0.9974	0.9568	0.8298	0.6296	0.4148	0.2375	0.1182	0.0510	0.0189	0.0059
	5	0.9997	0.9887	0.9327	0.8042	0.6172	0.4164	0.2454	0.1256	0.0553	0.0207
	6	1.0000	0.9976	0.9781	0.9133	0.7858	0.6080	0.4166	0.2500	0.1299	0.0577
	7	1.0000	0.9996	0.9941	0.9679	0.8982	0.7723	0.6010	0.4159	0.2520	0.1316
	8	1.0000	0.9999	0.9987	0.9900	0.9591	0.8867	0.7624	0.5956	0.4143	0.2517
	9	1.0000	1.0000	0.9998	0.9974	0.9861	0.9520	0.8782	0.7553	0.5914	0.4119
	10	1.0000	1.0000	1.0000	0.9994	0.9961	0.9829	0.9468	0.8725	0.7507	0.5881
	11	1.0000	1.0000	1.0000	0.9999	0.9991	0.9949	0.9804	0.9435	0.8692	0.7483
	12	1.0000	1.0000	1.0000	1.0000	0.9998	0.9987	0.9940	0.9790	0.9420	0.8684
	13	1.0000	1.0000	1.0000	1.0000	1.0000	0.9997	0.9985	0.9935	0.9786	0.9423
	14	1.0000	1.0000	1.0000	1.0000	1.0000	1.0000	0.9997	0.9984	0.9936	0.9793
	15	1.0000	1.0000	1.0000	1.0000	1.0000	1.0000	1.0000	0.9997	0.9985	0.9941
	16	1.0000	1.0000	1.0000	1.0000	1.0000	1.0000	1.0000	1.0000	0.9997	0.9987
	17	1.0000	1.0000	1.0000	1.0000	1.0000	1.0000	1.0000	1.0000	1.0000	0.9998
	18	1.0000	1.0000	1.0000	1.0000	1.0000	1.0000	1.0000	1.0000	1.0000	1.0000
	19	1.0000	1.0000	1.0000	1.0000	1.0000	1.0000	1.0000	1.0000	1.0000	1.0000
	20	1.0000	1.0000	1.0000	1.0000	1.0000	1.0000	1.0000	1.0000	1.0000	1.0000
21	0	0.3406	0.1094	0.0329	0.0092	0.0024	0.0006	0.0001	0.0000	0.0000	0.0000
	1	0.7170	0.3647	0.1550	0.0576	0.0190	0.0056	0.0014	0.0003	0.0001	0.0000
	2	0.9151	0.6484	0.3705	0.1787	0.0745	0.0271	0.0086	0.0024	0.0006	0.0001
	3	0.9811	0.8480	0.6113	0.3704	0.1917	0.0856	0.0331	0.0110	0.0031	0.0007
	4	0.9968	0.9478	0.8025	0.5860	0.3674	0.1984	0.0924	0.0370	0.0126	0.0036
	5	0.9996	0.9856	0.9173	0.7693	0.5666	0.3627	0.2009	0.0957	0.0389	0.0133
	6	1.0000	0.9967	0.9713	0.8915	0.7436	0.5505	0.3567	0.2002	0.0964	0.0392
	7	1.0000	0.9994	0.9917	0.9569	0.8701	0.7230	0.5365	0.3495	0.1971	0.0946
	8	1.0000	0.9999	0.9980	0.9856	0.9439	0.8523	0.7059	0.5237	0.3413	0.1917
	9	1.0000	1.0000	0.9996	0.9959	0.9794	0.9324	0.8377	0.6914	0.5117	0.3318
	10	1.0000	1.0000	0.9999	0.9990	0.9936	0.9736	0.9228	0.8256	0.6790	0.5000
	11	1.0000	1.0000	1.0000	0.9998	0.9983	0.9913	0.9687	0.9151	0.8159	0.6682
	12	1.0000	1.0000	1.0000	1.0000	0.9996	0.9976	0.9892	0.9648	0.9092	0.8083
	13	1.0000	1.0000	1.0000	1.0000	0.9999	0.9994	0.9969	0.9877	0.9621	0.9054
	14	1.0000	1.0000	1.0000	1.0000	1.0000	0.9999	0.9993	0.9964	0.9868	0.9608
	15	1.0000	1.0000	1.0000	1.0000	1.0000	1.0000	0.9999	0.9992	0.9963	0.9867
	16	1.0000	1.0000	1.0000	1.0000	1.0000	1.0000	1.0000	0.9998	0.9992	0.9964
	17	1.0000	1.0000	1.0000	1.0000	1.0000	1.0000	1.0000	1.0000	0.9999	0.9993
	18	1.0000	1.0000	1.0000	1.0000	1.0000	1.0000	1.0000	1.0000	1.0000	0.9999
	19	1.0000	1.0000	1.0000	1.0000	1.0000	1.0000	1.0000	1.0000	1.0000	1.0000
	20	1.0000	1.0000	1.0000	1.0000	1.0000	1.0000	1.0000	1.0000	1.0000	1.0000
	21	1.0000	1.0000	1.0000	1.0000	1.0000	1.0000	1.0000	1.0000	1.0000	1.0000
22	0	0.3235	0.0985	0.0280	0.0074	0.0018	0.0004	0.0001	0.0000	0.0000	0.0000
	1	0.6982	0.3392	0.1367	0.0480	0.0149	0.0041	0.0010	0.0002	0.0000	0.0000
	2	0.9052	0.6200	0.3382	0.1545	0.0606	0.0207	0.0061	0.0016	0.0003	0.0001
	3	0.9778	0.8281	0.5752	0.3320	0.1624	0.0681	0.0245	0.0076	0.0020	0.0004
	4	0.9960	0.9379	0.7738	0.5429	0.3235	0.1645	0.0716	0.0266	0.0083	0.0022

Table G: Cumulative Binomial Probabilities *(Continued)*

						p					
n	x	0.05	0.10	0.15	0.20	0.25	0.30	0.35	0.40	0.45	0.50
22	5	0.9994	0.9818	0.9001	0.7326	0.5168	0.3134	0.1629	0.0722	0.0271	0.0085
	6	0.9999	0.9956	0.9632	0.8670	0.6994	0.4942	0.3022	0.1584	0.0705	0.0262
	7	1.0000	0.9991	0.9886	0.9439	0.8385	0.6713	0.4736	0.2898	0.1518	0.0669
	8	1.0000	0.9999	0.9970	0.9799	0.9254	0.8135	0.6466	0.4540	0.2764	0.1431
	9	1.0000	1.0000	0.9993	0.9939	0.9705	0.9084	0.7916	0.6244	0.4350	0.2617
	10	1.0000	1.0000	0.9999	0.9984	0.9900	0.9613	0.8930	0.7720	0.6037	0.4159
	11	1.0000	1.0000	1.0000	0.9997	0.9971	0.9860	0.9526	0.8793	0.7543	0.5841
	12	1.0000	1.0000	1.0000	0.9999	0.9993	0.9957	0.9820	0.9449	0.8672	0.7383
	13	1.0000	1.0000	1.0000	1.0000	0.9999	0.9989	0.9942	0.9785	0.9383	0.8569
	14	1.0000	1.0000	1.0000	1.0000	1.0000	0.9998	0.9984	0.9930	0.9757	0.9331
	15	1.0000	1.0000	1.0000	1.0000	1.0000	1.0000	0.9997	0.9981	0.9920	0.9738
	16	1.0000	1.0000	1.0000	1.0000	1.0000	1.0000	0.9999	0.9996	0.9979	0.9915
	17	1.0000	1.0000	1.0000	1.0000	1.0000	1.0000	1.0000	0.9999	0.9995	0.9978
	18	1.0000	1.0000	1.0000	1.0000	1.0000	1.0000	1.0000	1.0000	0.9999	0.9996
	19	1.0000	1.0000	1.0000	1.0000	1.0000	1.0000	1.0000	1.0000	1.0000	0.9999
	20	1.0000	1.0000	1.0000	1.0000	1.0000	1.0000	1.0000	1.0000	1.0000	1.0000
	21	1.0000	1.0000	1.0000	1.0000	1.0000	1.0000	1.0000	1.0000	1.0000	1.0000
	22	1.0000	1.0000	1.0000	1.0000	1.0000	1.0000	1.0000	1.0000	1.0000	1.0000
23	0	0.3074	0.0886	0.0238	0.0059	0.0013	0.0003	0.0000	0.0000	0.0000	0.0000
	1	0.6794	0.3151	0.1204	0.0398	0.0116	0.0030	0.0007	0.0001	0.0000	0.0000
	2	0.8948	0.5920	0.3080	0.1332	0.0492	0.0157	0.0043	0.0010	0.0002	0.0000
	3	0.9742	0.8073	0.5396	0.2965	0.1370	0.0538	0.0181	0.0052	0.0012	0.0002
	4	0.9951	0.9269	0.7440	0.5007	0.2832	0.1356	0.0551	0.0190	0.0055	0.0013
	5	0.9992	0.9774	0.8811	0.6947	0.4685	0.2688	0.1309	0.0540	0.0186	0.0053
	6	0.9999	0.9942	0.9537	0.8402	0.6537	0.4399	0.2534	0.1240	0.0510	0.0173
	7	1.0000	0.9988	0.9848	0.9285	0.8037	0.6181	0.4136	0.2373	0.1152	0.0466
	8	1.0000	0.9998	0.9958	0.9727	0.9037	0.7709	0.5860	0.3884	0.2203	0.1050
	9	1.0000	1.0000	0.9990	0.9911	0.9592	0.8799	0.7408	0.5562	0.3636	0.2024
	10	1.0000	1.0000	0.9998	0.9975	0.9851	0.9454	0.8575	0.7129	0.5278	0.3388
	11	1.0000	1.0000	1.0000	0.9994	0.9954	0.9786	0.9318	0.8364	0.6865	0.5000
	12	1.0000	1.0000	1.0000	0.9999	0.9988	0.9928	0.9717	0.9187	0.8164	0.6612
	13	1.0000	1.0000	1.0000	1.0000	0.9997	0.9979	0.9900	0.9651	0.9063	0.7976
	14	1.0000	1.0000	1.0000	1.0000	0.9999	0.9995	0.9970	0.9872	0.9589	0.8950
	15	1.0000	1.0000	1.0000	1.0000	1.0000	0.9999	0.9992	0.9960	0.9847	0.9534
	16	1.0000	1.0000	1.0000	1.0000	1.0000	1.0000	0.9998	0.9990	0.9952	0.9827
	17	1.0000	1.0000	1.0000	1.0000	1.0000	1.0000	1.0000	0.9998	0.9988	0.9947
	18	1.0000	1.0000	1.0000	1.0000	1.0000	1.0000	1.0000	1.0000	0.9998	0.9987
	19	1.0000	1.0000	1.0000	1.0000	1.0000	1.0000	1.0000	1.0000	1.0000	0.9998
	20	1.0000	1.0000	1.0000	1.0000	1.0000	1.0000	1.0000	1.0000	1.0000	1.0000
	21	1.0000	1.0000	1.0000	1.0000	1.0000	1.0000	1.0000	1.0000	1.0000	1.0000
	22	1.0000	1.0000	1.0000	1.0000	1.0000	1.0000	1.0000	1.0000	1.0000	1.0000
	23	1.0000	1.0000	1.0000	1.0000	1.0000	1.0000	1.0000	1.0000	1.0000	1.0000
24	0	0.2920	0.0798	0.0202	0.0047	0.0010	0.0002	0.0000	0.0000	0.0000	0.0000
	1	0.6608	0.2925	0.1059	0.0331	0.0090	0.0022	0.0005	0.0001	0.0000	0.0000
	2	0.8841	0.5643	0.2798	0.1145	0.0398	0.0119	0.0030	0.0007	0.0001	0.0000
	3	0.9702	0.7857	0.5049	0.2639	0.1150	0.0424	0.0133	0.0035	0.0008	0.0001
	4	0.9940	0.9149	0.7134	0.4599	0.2466	0.1111	0.0422	0.0134	0.0036	0.0008
	5	0.9990	0.9723	0.8606	0.6559	0.4222	0.2288	0.1044	0.0400	0.0127	0.0033

Table G: Cumulative Binomial Probabilities *(Continued)*

						p					
n	x	0.05	0.10	0.15	0.20	0.25	0.30	0.35	0.40	0.45	0.50
24	6	0.9999	0.9925	0.9428	0.8111	0.6074	0.3886	0.2106	0.0960	0.0364	0.0113
	7	1.0000	0.9983	0.9801	0.9108	0.7662	0.5647	0.3575	0.1919	0.0863	0.0320
	8	1.0000	0.9997	0.9941	0.9638	0.8787	0.7250	0.5257	0.3279	0.1730	0.0758
	9	1.0000	0.9999	0.9985	0.9874	0.9453	0.8472	0.6866	0.4891	0.2991	0.1537
	10	1.0000	1.0000	0.9997	0.9962	0.9787	0.9258	0.8167	0.6502	0.4539	0.2706
	11	1.0000	1.0000	0.9999	0.9990	0.9928	0.9686	0.9058	0.7870	0.6151	0.4194
	12	1.0000	1.0000	1.0000	0.9998	0.9979	0.9885	0.9577	0.8857	0.7580	0.5806
	13	1.0000	1.0000	1.0000	1.0000	0.9995	0.9964	0.9836	0.9465	0.8659	0.7294
	14	1.0000	1.0000	1.0000	1.0000	0.9999	0.9990	0.9945	0.9783	0.9352	0.8463
	15	1.0000	1.0000	1.0000	1.0000	1.0000	0.9998	0.9984	0.9925	0.9731	0.9242
	16	1.0000	1.0000	1.0000	1.0000	1.0000	1.0000	0.9996	0.9978	0.9905	0.9680
	17	1.0000	1.0000	1.0000	1.0000	1.0000	1.0000	0.9999	0.9995	0.9972	0.9887
	18	1.0000	1.0000	1.0000	1.0000	1.0000	1.0000	1.0000	0.9999	0.9993	0.9967
	19	1.0000	1.0000	1.0000	1.0000	1.0000	1.0000	1.0000	1.0000	0.9999	0.9992
	20	1.0000	1.0000	1.0000	1.0000	1.0000	1.0000	1.0000	1.0000	1.0000	0.9999
	21	1.0000	1.0000	1.0000	1.0000	1.0000	1.0000	1.0000	1.0000	1.0000	1.0000
	22	1.0000	1.0000	1.0000	1.0000	1.0000	1.0000	1.0000	1.0000	1.0000	1.0000
	23	1.0000	1.0000	1.0000	1.0000	1.0000	1.0000	1.0000	1.0000	1.0000	1.0000
	24	1.0000	1.0000	1.0000	1.0000	1.0000	1.0000	1.0000	1.0000	1.0000	1.0000
25	0	0.2774	0.0718	0.0172	0.0038	0.0008	0.0001	0.0000	0.0000	0.0000	0.0000
	1	0.6424	0.2712	0.0931	0.0274	0.0070	0.0016	0.0003	0.0001	0.0000	0.0000
	2	0.8729	0.5371	0.2537	0.0982	0.0321	0.0090	0.0021	0.0004	0.0001	0.0000
	3	0.9659	0.7636	0.4711	0.2340	0.0962	0.0332	0.0097	0.0024	0.0005	0.0001
	4	0.9928	0.9020	0.6821	0.4207	0.2137	0.0905	0.0320	0.0095	0.0023	0.0005
	5	0.9988	0.9666	0.8385	0.6167	0.3783	0.1935	0.0826	0.0294	0.0086	0.0020
	6	0.9998	0.9905	0.9305	0.7800	0.5611	0.3407	0.1734	0.0736	0.0258	0.0073
	7	1.0000	0.9977	0.9745	0.8909	0.7265	0.5118	0.3061	0.1536	0.0639	0.0216
	8	1.0000	0.9995	0.9920	0.9532	0.8506	0.6769	0.4668	0.2735	0.1340	0.0539
	9	1.0000	0.9999	0.9979	0.9827	0.9287	0.8106	0.6303	0.4246	0.2424	0.1148
	10	1.0000	1.0000	0.9995	0.9944	0.9703	0.9022	0.7712	0.5858	0.3843	0.2122
	11	1.0000	1.0000	0.9999	0.9985	0.9893	0.9558	0.8746	0.7323	0.5426	0.3450
	12	1.0000	1.0000	1.0000	0.9996	0.9966	0.9825	0.9396	0.8462	0.6937	0.5000
	13	1.0000	1.0000	1.0000	0.9999	0.9991	0.9940	0.9745	0.9222	0.8173	0.6550
	14	1.0000	1.0000	1.0000	1.0000	0.9998	0.9982	0.9907	0.9656	0.9040	0.7878
	15	1.0000	1.0000	1.0000	1.0000	1.0000	0.9995	0.9971	0.9868	0.9560	0.8852
	16	1.0000	1.0000	1.0000	1.0000	1.0000	0.9999	0.9992	0.9957	0.9826	0.9461
	17	1.0000	1.0000	1.0000	1.0000	1.0000	1.0000	0.9998	0.9988	0.9942	0.9784
	18	1.0000	1.0000	1.0000	1.0000	1.0000	1.0000	1.0000	0.9997	0.9984	0.9927
	19	1.0000	1.0000	1.0000	1.0000	1.0000	1.0000	1.0000	0.9999	0.9996	0.9980
	20	1.0000	1.0000	1.0000	1.0000	1.0000	1.0000	1.0000	1.0000	0.9999	0.9995
	21	1.0000	1.0000	1.0000	1.0000	1.0000	1.0000	1.0000	1.0000	1.0000	0.9999
	22	1.0000	1.0000	1.0000	1.0000	1.0000	1.0000	1.0000	1.0000	1.0000	1.0000
	23	1.0000	1.0000	1.0000	1.0000	1.0000	1.0000	1.0000	1.0000	1.0000	1.0000
	24	1.0000	1.0000	1.0000	1.0000	1.0000	1.0000	1.0000	1.0000	1.0000	1.0000
	25	1.0000	1.0000	1.0000	1.0000	1.0000	1.0000	1.0000	1.0000	1.0000	1.0000

H

F-Distribution Probabilities

Upper 5% Table

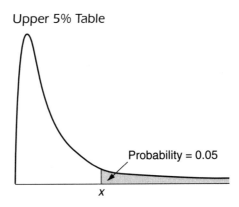

Probability = 0.05

Upper 1% Table

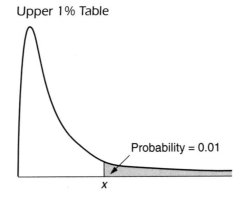

Probability = 0.01

Table H.1 Upper 5% of F Table

Denominator degrees of freedom	Numerator degrees of freedom																				
	1	2	3	4	5	6	7	8	9	10	11	12	13	14	15	16	17	18	19	20	21
1	161.45	199.50	215.71	224.58	230.16	233.99	236.77	238.88	240.54	241.88	242.98	243.90	244.69	245.36	245.95	246.47	246.92	247.32	247.69	248.02	248.31
2	18.51	19.00	19.16	19.25	19.30	19.33	19.35	19.37	19.38	19.40	19.40	19.41	19.42	19.42	19.43	19.43	19.44	19.44	19.44	19.45	19.45
3	10.13	9.55	9.28	9.12	9.01	8.94	8.89	8.85	8.81	8.79	8.76	8.74	8.73	8.71	8.70	8.69	8.68	8.67	8.67	8.66	8.65
4	7.71	6.94	6.59	6.39	6.26	6.16	6.09	6.04	6.00	5.96	5.94	5.91	5.89	5.87	5.86	5.84	5.83	5.82	5.81	5.80	5.79
5	6.61	5.79	5.41	5.19	5.05	4.95	4.88	4.82	4.77	4.74	4.70	4.68	4.66	4.64	4.62	4.60	4.59	4.58	4.57	4.56	4.55
6	5.99	5.14	4.76	4.53	4.39	4.28	4.21	4.15	4.10	4.06	4.03	4.00	3.98	3.96	3.94	3.92	3.91	3.90	3.88	3.87	3.86
7	5.59	4.74	4.35	4.12	3.97	3.87	3.79	3.73	3.68	3.64	3.60	3.57	3.55	3.53	3.51	3.49	3.48	3.47	3.46	3.44	3.43
8	5.32	4.46	4.07	3.84	3.69	3.58	3.50	3.44	3.39	3.35	3.31	3.28	3.26	3.24	3.22	3.20	3.19	3.17	3.16	3.15	3.14
9	5.12	4.26	3.86	3.63	3.48	3.37	3.29	3.23	3.18	3.14	3.10	3.07	3.05	3.03	3.01	2.99	2.97	2.96	2.95	2.94	2.93
10	4.96	4.10	3.71	3.48	3.33	3.22	3.14	3.07	3.02	2.98	2.94	2.91	2.89	2.86	2.85	2.83	2.81	2.80	2.79	2.77	2.76
11	4.84	3.98	3.59	3.36	3.20	3.09	3.01	2.95	2.90	2.85	2.82	2.79	2.76	2.74	2.72	2.70	2.69	2.67	2.66	2.65	2.64
12	4.75	3.89	3.49	3.26	3.11	3.00	2.91	2.85	2.80	2.75	2.72	2.69	2.66	2.64	2.62	2.60	2.58	2.57	2.56	2.54	2.53
13	4.67	3.81	3.41	3.18	3.03	2.92	2.83	2.77	2.71	2.67	2.63	2.60	2.58	2.55	2.53	2.51	2.50	2.48	2.47	2.46	2.45
14	4.60	3.74	3.34	3.11	2.96	2.85	2.76	2.70	2.65	2.60	2.57	2.53	2.51	2.48	2.46	2.44	2.43	2.41	2.40	2.39	2.38
15	4.54	3.68	3.29	3.06	2.90	2.79	2.71	2.64	2.59	2.54	2.51	2.48	2.45	2.42	2.40	2.38	2.37	2.35	2.34	2.33	2.32
16	4.49	3.63	3.24	3.01	2.85	2.74	2.66	2.59	2.54	2.49	2.46	2.42	2.40	2.37	2.35	2.33	2.32	2.30	2.29	2.28	2.26
17	4.45	3.59	3.20	2.96	2.81	2.70	2.61	2.55	2.49	2.45	2.41	2.38	2.35	2.33	2.31	2.29	2.27	2.26	2.24	2.23	2.22
18	4.41	3.55	3.16	2.93	2.77	2.66	2.58	2.51	2.46	2.41	2.37	2.34	2.31	2.29	2.27	2.25	2.23	2.22	2.20	2.19	2.18
19	4.38	3.52	3.13	2.90	2.74	2.63	2.54	2.48	2.42	2.38	2.34	2.31	2.28	2.26	2.23	2.21	2.20	2.18	2.17	2.16	2.14
20	4.35	3.49	3.10	2.87	2.71	2.60	2.51	2.45	2.39	2.35	2.31	2.28	2.25	2.22	2.20	2.18	2.17	2.15	2.14	2.12	2.11
21	4.32	3.47	3.07	2.84	2.68	2.57	2.49	2.42	2.37	2.32	2.28	2.25	2.22	2.20	2.18	2.16	2.14	2.12	2.11	2.10	2.08
22	4.30	3.44	3.05	2.82	2.66	2.55	2.46	2.40	2.34	2.30	2.26	2.23	2.20	2.17	2.15	2.13	2.11	2.10	2.08	2.07	2.06
23	4.28	3.42	3.03	2.80	2.64	2.53	2.44	2.37	2.32	2.27	2.24	2.20	2.18	2.15	2.13	2.11	2.09	2.08	2.06	2.05	2.04
24	4.26	3.40	3.01	2.78	2.62	2.51	2.42	2.36	2.30	2.25	2.22	2.18	2.15	2.13	2.11	2.09	2.07	2.05	2.04	2.03	2.01
25	4.24	3.39	2.99	2.76	2.60	2.49	2.40	2.34	2.28	2.24	2.20	2.16	2.14	2.11	2.09	2.07	2.05	2.04	2.02	2.01	2.00
26	4.23	3.37	2.98	2.74	2.59	2.47	2.39	2.32	2.27	2.22	2.18	2.15	2.12	2.09	2.07	2.05	2.03	2.02	2.00	1.99	1.98
27	4.21	3.35	2.96	2.73	2.57	2.46	2.37	2.31	2.25	2.20	2.17	2.13	2.10	2.08	2.06	2.04	2.02	2.00	1.99	1.97	1.96
28	4.20	3.34	2.95	2.71	2.56	2.45	2.36	2.29	2.24	2.19	2.15	2.12	2.09	2.06	2.04	2.02	2.00	1.99	1.97	1.96	1.95
29	4.18	3.33	2.93	2.70	2.55	2.43	2.35	2.28	2.22	2.18	2.14	2.10	2.08	2.05	2.03	2.01	1.99	1.97	1.96	1.94	1.93
30	4.17	3.32	2.92	2.69	2.53	2.42	2.33	2.27	2.21	2.16	2.13	2.09	2.06	2.04	2.01	1.99	1.98	1.96	1.95	1.93	1.92
31	4.16	3.30	2.91	2.68	2.52	2.41	2.32	2.25	2.20	2.15	2.11	2.08	2.05	2.03	2.00	1.98	1.96	1.95	1.93	1.92	1.91
32	4.15	3.29	2.90	2.67	2.51	2.40	2.31	2.24	2.19	2.14	2.10	2.07	2.04	2.01	1.99	1.97	1.95	1.94	1.92	1.91	1.90
33	4.14	3.28	2.89	2.66	2.50	2.39	2.30	2.23	2.18	2.13	2.09	2.06	2.03	2.00	1.98	1.96	1.94	1.93	1.91	1.90	1.89
34	4.13	3.28	2.88	2.65	2.49	2.38	2.29	2.23	2.17	2.12	2.08	2.05	2.02	1.99	1.97	1.95	1.93	1.92	1.90	1.89	1.88
35	4.12	3.27	2.87	2.64	2.49	2.37	2.29	2.22	2.16	2.11	2.07	2.04	2.01	1.99	1.96	1.94	1.92	1.91	1.89	1.88	1.87
36	4.11	3.26	2.87	2.63	2.48	2.36	2.28	2.21	2.15	2.11	2.07	2.03	2.00	1.98	1.95	1.93	1.92	1.90	1.88	1.87	1.86
37	4.11	3.25	2.86	2.63	2.47	2.36	2.27	2.20	2.14	2.10	2.06	2.02	2.00	1.97	1.95	1.93	1.91	1.89	1.88	1.86	1.85
38	4.10	3.24	2.85	2.62	2.46	2.35	2.26	2.19	2.14	2.09	2.05	2.02	1.99	1.96	1.94	1.92	1.90	1.88	1.87	1.85	1.84
39	4.09	3.24	2.85	2.61	2.46	2.34	2.26	2.19	2.13	2.08	2.04	2.01	1.98	1.95	1.93	1.91	1.89	1.88	1.86	1.85	1.83
40	4.08	3.23	2.84	2.61	2.45	2.34	2.25	2.18	2.12	2.08	2.04	2.00	1.97	1.95	1.92	1.90	1.89	1.87	1.85	1.84	1.83

Table H.2 Upper 1% of F Table

	Numerator degrees of freedom																				
Denominator degrees of freedom	1	2	3	4	5	6	7	8	9	10	11	12	13	14	15	16	17	18	19	20	21
1	4052.2	4999.3	5403.5	5624.3	5764.0	5859.0	5928.3	5981.0	6022.4	6055.9	6083.4	6106.7	6125.8	6143.0	6157.0	6170.0	6181.2	6191.4	6200.7	6208.7	6216.1
2	98.50	99.00	99.16	99.25	99.30	99.33	99.36	99.38	99.39	99.40	99.41	99.42	99.42	99.43	99.43	99.44	99.44	99.44	99.45	99.45	99.45
3	34.12	30.82	29.46	28.71	28.24	27.91	27.67	27.49	27.34	27.23	27.13	27.05	26.98	26.92	26.87	26.83	26.79	26.76	26.72	26.69	26.66
4	21.20	18.00	16.69	15.98	15.52	15.21	14.98	14.80	14.66	14.55	14.45	14.37	14.31	14.25	14.20	14.15	14.11	14.08	14.05	14.02	13.99
5	16.26	13.27	12.06	11.39	10.97	10.67	10.46	10.29	10.16	10.05	9.96	9.89	9.82	9.77	9.72	9.68	9.64	9.61	9.58	9.55	9.53
6	13.75	10.92	9.78	9.15	8.75	8.47	8.26	8.10	7.98	7.87	7.79	7.72	7.66	7.60	7.56	7.52	7.48	7.45	7.42	7.40	7.37
7	12.25	9.55	8.45	7.85	7.46	7.19	6.99	6.84	6.72	6.62	6.54	6.47	6.41	6.36	6.31	6.28	6.24	6.21	6.18	6.16	6.13
8	11.26	8.65	7.59	7.01	6.63	6.37	6.18	6.03	5.91	5.81	5.73	5.67	5.61	5.56	5.52	5.48	5.44	5.41	5.38	5.36	5.34
9	10.56	8.02	6.99	6.42	6.06	5.80	5.61	5.47	5.35	5.26	5.18	5.11	5.05	5.01	4.96	4.92	4.89	4.86	4.83	4.81	4.79
10	10.04	7.56	6.55	5.99	5.64	5.39	5.20	5.06	4.94	4.85	4.77	4.71	4.65	4.60	4.56	4.52	4.49	4.46	4.43	4.41	4.38
11	9.65	7.21	6.22	5.67	5.32	5.07	4.89	4.74	4.63	4.54	4.46	4.40	4.34	4.29	4.25	4.21	4.18	4.15	4.12	4.10	4.08
12	9.33	6.93	5.95	5.41	5.06	4.82	4.64	4.50	4.39	4.30	4.22	4.16	4.10	4.05	4.01	3.97	3.94	3.91	3.88	3.86	3.84
13	9.07	6.70	5.74	5.21	4.86	4.62	4.44	4.30	4.19	4.10	4.02	3.96	3.91	3.86	3.82	3.78	3.75	3.72	3.69	3.66	3.64
14	8.86	6.51	5.56	5.04	4.69	4.46	4.28	4.14	4.03	3.94	3.86	3.80	3.75	3.70	3.66	3.62	3.59	3.56	3.53	3.51	3.48
15	8.68	6.36	5.42	4.89	4.56	4.32	4.14	4.00	3.89	3.80	3.73	3.67	3.61	3.56	3.52	3.49	3.45	3.42	3.40	3.37	3.35
16	8.53	6.23	5.29	4.77	4.44	4.20	4.03	3.89	3.78	3.69	3.62	3.55	3.50	3.45	3.41	3.37	3.34	3.31	3.28	3.26	3.24
17	8.40	6.11	5.19	4.67	4.34	4.10	3.93	3.79	3.68	3.59	3.52	3.46	3.40	3.35	3.31	3.27	3.24	3.21	3.19	3.16	3.14
18	8.29	6.01	5.09	4.58	4.25	4.01	3.84	3.71	3.60	3.51	3.43	3.37	3.32	3.27	3.23	3.19	3.16	3.13	3.10	3.08	3.05
19	8.18	5.93	5.01	4.50	4.17	3.94	3.77	3.63	3.52	3.43	3.36	3.30	3.24	3.19	3.15	3.12	3.08	3.05	3.03	3.00	2.98
20	8.10	5.85	4.94	4.43	4.10	3.87	3.70	3.56	3.46	3.37	3.29	3.23	3.18	3.13	3.09	3.05	3.02	2.99	2.96	2.94	2.92
21	8.02	5.78	4.87	4.37	4.04	3.81	3.64	3.51	3.40	3.31	3.24	3.17	3.12	3.07	3.03	2.99	2.96	2.93	2.90	2.88	2.86
22	7.95	5.72	4.82	4.31	3.99	3.76	3.59	3.45	3.35	3.26	3.18	3.12	3.07	3.02	2.98	2.94	2.91	2.88	2.85	2.83	2.81
23	7.88	5.66	4.76	4.26	3.94	3.71	3.54	3.41	3.30	3.21	3.14	3.07	3.02	2.97	2.93	2.89	2.86	2.83	2.80	2.78	2.76
24	7.82	5.61	4.72	4.22	3.90	3.67	3.50	3.36	3.26	3.17	3.09	3.03	2.98	2.93	2.89	2.85	2.82	2.79	2.76	2.74	2.72
25	7.77	5.57	4.68	4.18	3.85	3.63	3.46	3.32	3.22	3.13	3.06	2.99	2.94	2.89	2.85	2.81	2.78	2.75	2.72	2.70	2.68
26	7.72	5.53	4.64	4.14	3.82	3.59	3.42	3.29	3.18	3.09	3.02	2.96	2.90	2.86	2.81	2.78	2.75	2.72	2.69	2.66	2.64
27	7.68	5.49	4.60	4.11	3.78	3.56	3.39	3.26	3.15	3.06	2.99	2.93	2.87	2.82	2.78	2.75	2.71	2.68	2.66	2.63	2.61
28	7.64	5.45	4.57	4.07	3.75	3.53	3.36	3.23	3.12	3.03	2.96	2.90	2.84	2.79	2.75	2.72	2.68	2.65	2.63	2.60	2.58
29	7.60	5.42	4.54	4.04	3.73	3.50	3.33	3.20	3.09	3.00	2.93	2.87	2.81	2.77	2.73	2.69	2.66	2.63	2.60	2.57	2.55
30	7.56	5.39	4.51	4.02	3.70	3.47	3.30	3.17	3.07	2.98	2.91	2.84	2.79	2.74	2.70	2.66	2.63	2.60	2.57	2.55	2.53
31	7.53	5.36	4.48	3.99	3.67	3.45	3.28	3.15	3.04	2.96	2.88	2.82	2.77	2.72	2.68	2.64	2.61	2.58	2.55	2.52	2.50
32	7.50	5.34	4.46	3.97	3.65	3.43	3.26	3.13	3.02	2.93	2.86	2.80	2.74	2.70	2.65	2.62	2.58	2.55	2.53	2.50	2.48
33	7.47	5.31	4.44	3.95	3.63	3.41	3.24	3.11	3.00	2.91	2.84	2.78	2.72	2.68	2.63	2.60	2.56	2.53	2.51	2.48	2.46
34	7.44	5.29	4.42	3.93	3.61	3.39	3.22	3.09	2.98	2.89	2.82	2.76	2.70	2.66	2.61	2.58	2.54	2.51	2.49	2.46	2.44
35	7.42	5.27	4.40	3.91	3.59	3.37	3.20	3.07	2.96	2.88	2.80	2.74	2.69	2.64	2.60	2.56	2.53	2.50	2.47	2.44	2.42
36	7.40	5.25	4.38	3.89	3.57	3.35	3.18	3.05	2.95	2.86	2.79	2.72	2.67	2.62	2.58	2.54	2.51	2.48	2.45	2.43	2.41
37	7.37	5.23	4.36	3.87	3.56	3.33	3.17	3.04	2.93	2.84	2.77	2.71	2.65	2.61	2.56	2.53	2.49	2.46	2.44	2.41	2.39
38	7.35	5.21	4.34	3.86	3.54	3.32	3.15	3.02	2.92	2.83	2.75	2.69	2.64	2.59	2.55	2.51	2.48	2.45	2.42	2.40	2.37
39	7.33	5.19	4.33	3.84	3.53	3.30	3.14	3.01	2.90	2.81	2.74	2.68	2.62	2.58	2.54	2.50	2.46	2.43	2.41	2.38	2.36
40	7.31	5.18	4.31	3.83	3.51	3.29	3.12	2.99	2.89	2.80	2.73	2.66	2.61	2.56	2.52	2.48	2.45	2.42	2.39	2.37	2.35

Bonferroni Confidence Intervals

Table B.1 Bonferroni Confidence Interval Table

Degrees of freedom	\multicolumn{15}{c}{Number of comparisons}
	1	2	3	4	5	6	7	8	9	10	11	12	13	14	15
1	12.71	25.45	38.19	50.92	63.66	76.39	89.12	101.86	114.59	127.32	140.05	152.79	165.52	178.25	190.98
2	4.30	6.21	7.65	8.86	9.92	10.89	11.77	12.59	13.36	14.09	14.78	15.44	16.08	16.69	17.28
3	3.18	4.18	4.86	5.39	5.84	6.23	6.58	6.90	7.18	7.45	7.70	7.94	8.16	8.37	8.58
4	2.78	3.50	3.96	4.31	4.60	4.85	5.07	5.26	5.44	5.60	5.75	5.89	6.02	6.14	6.25
5	2.57	3.16	3.53	3.81	4.03	4.22	4.38	4.53	4.66	4.77	4.88	4.98	5.08	5.16	5.25
6	2.45	2.97	3.29	3.52	3.71	3.86	4.00	4.12	4.22	4.32	4.40	4.49	4.56	4.63	4.70
7	2.36	2.84	3.13	3.34	3.50	3.64	3.75	3.86	3.95	4.03	4.10	4.17	4.24	4.30	4.36
8	2.31	2.75	3.02	3.21	3.36	3.48	3.58	3.68	3.76	3.83	3.90	3.96	4.02	4.07	4.12
9	2.26	2.69	2.93	3.11	3.25	3.36	3.46	3.55	3.62	3.69	3.75	3.81	3.86	3.91	3.95
10	2.23	2.63	2.87	3.04	3.17	3.28	3.37	3.45	3.52	3.58	3.64	3.69	3.74	3.79	3.83
11	2.20	2.59	2.82	2.98	3.11	3.21	3.29	3.37	3.44	3.50	3.55	3.60	3.65	3.69	3.73
12	2.18	2.56	2.78	2.93	3.05	3.15	3.24	3.31	3.37	3.43	3.48	3.53	3.57	3.61	3.65
13	2.16	2.53	2.75	2.90	3.01	3.11	3.19	3.26	3.32	3.37	3.42	3.47	3.51	3.55	3.58
14	2.14	2.51	2.72	2.86	2.98	3.07	3.15	3.21	3.27	3.33	3.37	3.42	3.46	3.49	3.53
15	2.13	2.49	2.69	2.84	2.95	3.04	3.11	3.18	3.23	3.29	3.33	3.37	3.41	3.45	3.48
16	2.12	2.47	2.67	2.81	2.92	3.01	3.08	3.15	3.20	3.25	3.30	3.34	3.38	3.41	3.44
17	2.11	2.46	2.65	2.79	2.90	2.98	3.06	3.12	3.17	3.22	3.27	3.31	3.34	3.38	3.41
18	2.10	2.45	4.64	2.77	2.88	2.96	3.03	3.09	3.15	3.20	3.24	3.28	3.32	3.35	3.38
19	2.09	2.43	2.63	2.76	2.86	2.94	3.01	3.07	3.13	3.17	3.22	3.25	3.29	3.32	3.35
20	2.09	2.42	2.61	2.74	2.85	2.93	3.00	3.06	3.11	3.15	3.20	3.23	3.27	3.30	3.33
21	2.08	2.41	2.60	2.73	2.83	2.91	2.98	3.04	3.09	3.14	3.18	3.21	3.25	3.28	3.31
22	2.07	2.41	2.59	2.72	2.82	2.90	2.97	3.02	3.07	3.12	3.16	3.20	3.23	3.26	3.29
23	2.07	2.40	2.58	2.71	2.81	2.89	2.95	3.01	3.06	3.10	3.14	3.18	3.21	3.25	3.27
24	2.06	2.39	2.57	2.70	2.80	2.88	2.94	3.00	3.05	3.09	3.13	3.17	3.20	3.23	3.26
25	2.06	2.38	2.57	2.69	2.79	2.86	2.93	2.99	3.03	3.08	3.12	3.15	3.19	3.22	3.24
26	2.06	2.38	2.56	2.68	2.78	2.86	2.92	2.98	3.02	3.07	3.11	3.14	3.17	3.20	3.23
27	2.05	2.37	2.55	2.68	2.77	2.85	2.91	2.97	3.01	3.06	3.10	3.13	3.16	3.19	3.22
28	2.05	2.37	2.55	2.67	2.76	2.84	2.90	2.96	3.00	3.05	3.09	3.12	3.15	3.18	3.21
29	2.05	2.36	2.54	2.66	2.76	2.83	2.89	2.95	3.00	3.04	3.08	3.11	3.14	3.17	3.20
30	2.04	2.36	2.54	2.66	2.75	2.82	2.89	2.94	2.99	3.03	3.07	3.10	3.13	3.16	3.19
>30	1.96	2.24	2.40	2.50	2.58	2.64	2.70	2.74	2.78	2.81	2.84	2.87	2.90	2.92	2.94

J

Cumulative Poisson Probabilities

The table gives the probability of x or fewer events when the expected number of such events is λ.

Table J: Cumulative Poisson Probabilities

	$\lambda = E(X)$									
x	0.1	0.2	0.3	0.4	0.5	0.6	0.7	0.8	0.9	1.0
0	0.905	0.819	0.741	0.670	0.607	0.549	0.497	0.449	0.407	0.368
1	0.995	0.982	0.963	0.938	0.910	0.878	0.844	0.809	0.772	0.736
2	1.000	0.999	0.996	0.992	0.986	0.977	0.966	0.953	0.937	0.920
3	1.000	1.000	1.000	0.999	0.998	0.997	0.994	0.991	0.987	0.981
4	1.000	1.000	1.000	1.000	1.000	1.000	0.999	0.999	0.998	0.996
5	1.000	1.000	1.000	1.000	1.000	1.000	1.000	1.000	1.000	0.999
6	1.000	1.000	1.000	1.000	1.000	1.000	1.000	1.000	1.000	1.000
x	1.1	1.2	1.3	1.4	1.5	1.6	1.7	1.8	1.9	2.0
0	0.333	0.301	0.273	0.247	0.223	0.202	0.183	0.165	0.150	0.135
1	0.699	0.663	0.627	0.592	0.558	0.525	0.493	0.463	0.434	0.406
2	0.900	0.879	0.857	0.833	0.809	0.783	0.757	0.731	0.704	0.677
3	0.974	0.966	0.957	0.946	0.934	0.921	0.907	0.891	0.875	0.857
4	0.995	0.992	0.989	0.986	0.981	0.976	0.970	0.964	0.956	0.947
5	0.999	0.998	0.998	0.997	0.996	0.994	0.992	0.990	0.987	0.983
6	1.000	1.000	1.000	0.999	0.999	0.999	0.998	0.997	0.997	0.995
7	1.000	1.000	1.000	1.000	1.000	1.000	1.000	0.999	0.999	0.999
8	1.000	1.000	1.000	1.000	1.000	1.000	1.000	1.000	1.000	1.000
x	2.2	2.4	2.6	2.8	3.0	3.2	3.4	3.6	3.8	4.0
0	0.111	0.091	0.074	0.061	0.050	0.041	0.033	0.027	0.022	0.018
1	0.355	0.308	0.267	0.231	0.199	0.171	0.147	0.126	0.107	0.092
2	0.623	0.570	0.518	0.469	0.423	0.380	0.340	0.303	0.269	0.238
3	0.819	0.779	0.736	0.692	0.647	0.603	0.558	0.515	0.473	0.433
4	0.928	0.904	0.877	0.848	0.815	0.781	0.744	0.706	0.668	0.629
5	0.975	0.964	0.951	0.935	0.916	0.895	0.871	0.844	0.816	0.785
6	0.993	0.988	0.983	0.976	0.966	0.955	0.942	0.927	0.909	0.889
7	0.998	0.997	0.995	0.992	0.988	0.983	0.977	0.969	0.960	0.949
8	1.000	0.999	0.999	0.998	0.996	0.994	0.992	0.988	0.984	0.979
9	1.000	1.000	1.000	0.999	0.999	0.998	0.997	0.996	0.994	0.992
10	1.000	1.000	1.000	1.000	1.000	1.000	0.999	0.999	0.998	0.997
11	1.000	1.000	1.000	1.000	1.000	1.000	1.000	1.000	0.999	0.999
12	1.000	1.000	1.000	1.000	1.000	1.000	1.000	1.000	1.000	1.000

Table J: Cumulative Poisson Probabilities *(Continued)*

					$\lambda = E(X)$					
x	4.2	4.4	4.6	4.8	5.0	5.2	5.4	5.6	5.8	6.0
0	0.015	0.012	0.010	0.008	0.007	0.006	0.005	0.004	0.003	0.002
1	0.078	0.066	0.056	0.048	0.040	0.034	0.029	0.024	0.021	0.017
2	0.210	0.185	0.163	0.143	0.125	0.109	0.095	0.082	0.072	0.062
3	0.395	0.359	0.326	0.294	0.265	0.238	0.213	0.191	0.170	0.151
4	0.590	0.551	0.513	0.476	0.440	0.406	0.373	0.342	0.313	0.285
5	0.753	0.720	0.686	0.651	0.616	0.581	0.546	0.512	0.478	0.446
6	0.867	0.844	0.818	0.791	0.762	0.732	0.702	0.670	0.638	0.606
7	0.936	0.921	0.905	0.887	0.867	0.845	0.822	0.797	0.771	0.744
8	0.972	0.964	0.955	0.944	0.932	0.918	0.903	0.886	0.867	0.847
9	0.989	0.985	0.980	0.975	0.968	0.960	0.951	0.941	0.929	0.916
10	0.996	0.994	0.992	0.990	0.986	0.982	0.977	0.972	0.965	0.957
11	0.999	0.998	0.997	0.996	0.995	0.993	0.990	0.988	0.984	0.980
12	1.000	0.999	0.999	0.999	0.998	0.997	0.996	0.995	0.993	0.991
13	1.000	1.000	1.000	1.000	0.999	0.999	0.999	0.998	0.997	0.996
14	1.000	1.000	1.000	1.000	1.000	1.000	1.000	0.999	0.999	0.999
15	1.000	1.000	1.000	1.000	1.000	1.000	1.000	1.000	1.000	0.999
16	1.000	1.000	1.000	1.000	1.000	1.000	1.000	1.000	1.000	1.000

x	6.5	7.0	7.5	8.0	8.5	9.0	9.5	10.0	10.5	11.0
0	0.002	0.001	0.001	0.000	0.000	0.000	0.000	0.000	0.000	0.000
1	0.011	0.007	0.005	0.003	0.002	0.001	0.001	0.000	0.000	0.000
2	0.043	0.030	0.020	0.014	0.009	0.006	0.004	0.003	0.002	0.001
3	0.112	0.082	0.059	0.042	0.030	0.021	0.015	0.010	0.007	0.005
4	0.224	0.173	0.132	0.100	0.074	0.055	0.040	0.029	0.021	0.015
5	0.369	0.301	0.241	0.191	0.150	0.116	0.089	0.067	0.050	0.038
6	0.527	0.450	0.378	0.313	0.256	0.207	0.165	0.130	0.102	0.079
7	0.673	0.599	0.525	0.453	0.386	0.324	0.269	0.220	0.179	0.143
8	0.792	0.729	0.662	0.593	0.523	0.456	0.392	0.333	0.279	0.232
9	0.877	0.830	0.776	0.717	0.653	0.587	0.522	0.458	0.397	0.341
10	0.933	0.901	0.862	0.816	0.763	0.706	0.645	0.583	0.521	0.460
11	0.966	0.947	0.921	0.888	0.849	0.803	0.752	0.697	0.639	0.579
12	0.984	0.973	0.957	0.936	0.909	0.876	0.836	0.792	0.742	0.689
13	0.993	0.987	0.978	0.966	0.949	0.926	0.898	0.864	0.825	0.781
14	0.997	0.994	0.990	0.983	0.973	0.959	0.940	0.917	0.888	0.854
15	0.999	0.998	0.995	0.992	0.986	0.978	0.967	0.951	0.932	0.907
16	1.000	0.999	0.998	0.996	0.993	0.989	0.982	0.973	0.960	0.944
17	1.000	1.000	0.999	0.998	0.997	0.995	0.991	0.986	0.978	0.968
18	1.000	1.000	1.000	0.999	0.999	0.998	0.996	0.993	0.988	0.982
19	1.000	1.000	1.000	1.000	0.999	0.999	0.998	0.997	0.994	0.991
20	1.000	1.000	1.000	1.000	1.000	1.000	0.999	0.998	0.997	0.995
21	1.000	1.000	1.000	1.000	1.000	1.000	1.000	0.999	0.999	0.998
22	1.000	1.000	1.000	1.000	1.000	1.000	1.000	1.000	0.999	0.999
23	1.000	1.000	1.000	1.000	1.000	1.000	1.000	1.000	1.000	1.000

INDEX

Material from Chapter 12 (provided on CD) is designated with CD, followed by the page number.

Acceptance/rejection hypothesis testing
   critical region, 447–448
   level of significance, 447
   power of a test, 448–451
   versus significance testing, 445–447
Actuaries, 229
Additional rule for probabilities, 169–173
Additive model, CD 32, CD 33, CD 34, CD 35
   versus interactive model, CD 34–35
Alternative hypothesis, 399, 441
   one-sided, 442–443
   for a slope, 523–524
   two-sided, 443
Analysis of variance (ANOVA), CD 9
   error sum of squares (SSE), CD 8
   and multiple comparisons of means, CD 25–30
   one-way, CD 18–24
   total sum of squares (SSTO), CD 9
   treatment sum of squares (SSTR), CD 7–8
   two-way, CD 31–40
Analysis of variance table, CD 23, CD 52
"And" rule. *See* Multiplication rule for probabilities
ANOVA. *See* Analysis of variance
Area fallacy, 37–38
Arithmetic mean. *See* Mean
Average. *See* Mean

Babe Ruth, 13, 14, 54, 90, 91
Balanced design, CD 39
Bar chart, 21
Believability, 294
Bell-shaped distributions, 28, 69–70, 83
   empirical rule for, 70
Between-samples sum of squares. *See* Treatment sum of squares
Bias, 297, 334
Biased estimate, 297
Bimodal, 103
Binomial coefficient, 250–251

Binomial distribution, 186, 251–258
   histograms of, 254
   probability, 251–254
   probability density function for, 252
   and sampling, 256
   using binomial tables, 255–256
Binomial probability
   normal approximation, 278–283
   Poisson approximation, 264
Binomial probability density function, 252, 257
Bivariate data, 512, 514
Block, CD 38–39
Bonferroni adjustment, CD 27, CD 29
Bonferroni procedure, CD 27–30, 40
   two-way ANOVA, CD 40
Bonferroni table, CD 29
Bonds, Barry, 5, 55
Bootstrapping, 363, CD 3
   comparing means of multiple populations, CD 11–14
   for hypothesis testing, 427–431
   percentile-based confidence intervals, 369–379
   SE-based confidence interval, 362–368
   for a slope parameter, 527–528
   and squared multiple correlation coefficient, CD 47–48
   and two-way ANOVA, CD 39–40
Bootstrap samples, 365
Box-and-whiskers plot. *See* Boxplots
Box mean, 241
Box models, 200–204, 222
   bootstrap, 428
   expected values, 227–229
   formula for expected value of random variable, 225
   formula for standard deviation, 242
   for hypothesis testing, 402–405, 410–411
   variance, 241
Boxplots, 52
   and distribution of data, 95
   five-number summary, 91
   use of, 93
Brooks, Harold, 458

Categorical variable, CD 18
Causation, 148–149
Census, 318, 335
Center, 53, 69, 95
   resistant measure of, 61
Central Limit Theorem (CLT), 390, 406
   estimating P value, 407
   as experimental fact, 349
   as mathematical theorem, 349
   for a sample mean, 346–352
   for difference of two sample means, 379–381
   for difference of two sample proportions, 384
   for sample proportion, 358–361
   standardized statistic and, 420
Chi-square curve. *See* Chi-square density
Chi-square density, 460, 482–486
   theoretical, 490
   table, 509
Chi-square probability law, 5
Chi-square statistic (χ^2), 467, 509
   density polygon and, 482
   formula, 468
   goodness-of-fit testing, 501
   versus D statistic, 468
Chi-square testing, 461
   χ^2 distribution and, 490
   equal expected frequencies, 473–478
   unequal expected frequencies, 497–501
Chodas, Paul W., 156
Clusters, 95
Coefficient of determination, 142
Coefficient table, CD 52
Combination graph, 40
Complement, 169
Complementary event, 169
Completely randomized design, CD 39
Conceptual population, 295, 334, 389
Conditional probability, 176–177
Confidence interval, 355
   Bonferroni adjustment for multiple intervals, CD 27, CD 29
   comparing population means, CD 27

Confidence interval (*continued*)
  for the difference between two populable means, 379–383
  for the difference between two population means in matched-pairs, 385–396
  for the difference between two population proportions, 384–385
  interpretation of a, 355–357
  for μ using $\bar{X}$ and $\widehat{SE}(\bar{X})$, 354–355
  for μ when σ known, 357
  for multiple, CD 28
  for a normal population μ, 374–378
  for p, 358–361
  for a parameter, 526
  percentile-based bootstrap, 369–371
  SE-based bootstrap, 362–368
  for a slope, 526–528
Confounding, 149, 325
Continuous distribution, 484
Continuous random variable, 277
Control, 296
Control group, 295, 300, 325
Correlation
  and causation, 148–149
  strong, weak-to-moderate, or negligible, 120–122
Correlation coefficient, 120, 125–126, 128
  for multiple regression, CD 45– CD 47
Counting variables, 30
Critical region, 447
Critical value, 357
Crockett, Davy, 4–5

Data
  bimodal, 103
  bivariate, 512, 514
  center of, 53
  clusters, 95
  defined, 5
  distribution of, 15–16
  gaps, 95
  ordinal, 220
  overfitting, CD 35
  range, 15
  ranked, 220
  skewness of, 53
  spread of, 53, 70
  underfitting, CD 35
Degrees of freedom (df), 377, 486–488, 509, CD 16–17, CD 19
  determining for ANOVA models, CD 17
  for sums of squares, CD 36, CD 49
Density histograms, 30–34, 186
  and density polygon, 482
Density polygon, 482
Descriptive data analysis, 98–104
Descriptive statistics, 52
  defined, 7
Deviation from the mean, 74–75

Distribution, 4, 15–16
  bell-shaped, 28, 83
  binomial, 251–258
  and boxplots, 95
  continuous, 484
  and density polygon, 482
  geometric, 259–262
  left-skewed, 28
  normal, 82–88, 273–282
  Poisson, 263–270
  right-skewed, 28
  skewed, 16
  symmetric, 63
  uniform, 28, 101
  U-shaped, 28
  for $\bar{X}$ of a normal population, 375
Distribution formula
  for expected value, 225
  for standard deviation, 239–242
Doll, R., 298, 308
Donnellan, Andrea, 2
Dotplot, 11–12
Double-blind experiment, 306, 326
D statistic, 509
  defined, 463
  hypothesis testing, 464–465
  versus χ^2 statistic, 468

Effect, 295
Effect size, 296
Effron, Brad, 363
Empirical rule, 240
  for bell-shaped distributions, 70
Equally likely outcomes
  rule for, 166
  simulation of, 188–193
Error mean square (MSE), CD 17, CD 19, CD 46
  and R^2, CD 45–46
Error of estimation, 135, 333
Error sum of squares (SSE), 136, CD 8, CD 9, CD 17, CD 37
Error types, 446
Error variability, CD 6
  for F distribution, CD 19
Estimated value of y given x, 135
Estimation
  accuracy of, 333
  error, 135, 136
  measurement as, 333–334
  with a real population, 334–336
  of a standard error, 353–354
Estimation problem, 389
Estimator, 335, 336
  sampling distribution of, 343–345
Event, 166
Event of interest (EOI), 403–404
Expected frequency, 473–478
Expected value, 221–229
  box models, 227–229
  and five-step simulation, 232–237
  formulas for, 225

Experiment
  binomial, 252
  designing, CD 23–24
  observational study, 298, 300
  randomized controlled, 298, 303–307, 325. *See also* Randomized controlled experiment
Experimental design, 298
  balanced, CD 39
  completely randomized, CD 39
  randomized block, CD 39
Experimental probability, 159–163, 164, 404
Explained variation, 142
Explanatory variable, 115
Exploratory data analysis (EDA), 13
Exponential growth equation, 534
Extrapolation, 532
Extreme values, 61

Factorial notation, 250
F distribution, CD 16
  and ANOVA problems, CD 19
  justifying use of, CD 18–19
  and one-way ANOVA, CD 21–23
  parameters, CD 19
  tables, CD 20–21
  and two-way ANOVA, CD 36
Fisher, Ronald A., 299, 300, 308
Fitted value, 135
Fitting a line, 133
Five-number boxplot. *See* Five-number summary
Five-number summary, 91
Five-step simulation method, 184–188
  bootstrap approach, 427–429
  bypassing, 424
  and expected (mean) value, 232–237
  for hypothesis testing, 405
  for standard error, 366
Frequency, 24, 462
  expected, 462–463
  observed, 462–463
Frequency histograms, 24–29
  axes of, 24
  defined, 24
  expected, 497–501
  frequency table of, 26
Frequency table, 21, 25, 462
F test
  for ANOVA, CD 19, CD 36–38
  for multiple regression, CD 48–49

Gallup, George, 313, 314, 316, 317
Galton, Sir Francis, 83
Gaps, 95
Gauss, Carl Friedrich, 83, 333
Gaussian distribution. *See* Normal distribution
Geometric distribution, 259–262
  histograms of, 261
  probability density function, 260, 262

Goodness of fit, 461, 467, 509
Gosset, W. S., 376–377, 424
Gould, Stephen Jay, 69, 71
Graphical displays
　bar chart, 21
　density histograms, 30–34
　distortions of, 37–42
　frequency histograms, 24–29
　pie chart, 21

Haphazard samples, 317
Harris, Andrew, 50
Harris Trosper, Jennifer A., 248
Hidden variable, 149. See also Lurking variable
Hill, A. B., 298, 308
Histograms
　density, 30–34, 186. See also Density histograms
　frequency, 24–29. See also Frequency histograms
Hoskin, Alan F., 112
Hoynoski, Bruce W., 292
Hypergeometric distribution, 256
Hypothesis testing, 202, 396
　acceptance/rejection, 445–451
　bootstrapping, 427–431, CD 11
　χ^2 statistic, 467–469
　D statistic, 464–465
　five-step simulation method, 405
　one-sided, 442–443
　population proportion, 402–407
　about the slope, 523–525
　strength of evidence principle, 404
　three stages of, 403–405
　t test, 424–426
　two-sided, 443
　Type I error, 446
　Type II error, 446
　using central limit theorem, 406–407
　Z test, 417–421

Independent events, 175
　product rule for, 260
Independent identically distributed random variables, 339
Independent repetitions, 164
Inference, statistical, 5. See also Statistical inference
Inferential statistics, 7. See also Statistical inference
Influential observations, 61
Integers, 30
Interaction, CD 31–32
　defined, CD 33
　F statistic and, CD 37–38
　ignoring, CD 35–36
　plotting, CD 33–34
Interaction sum of squares (SSAB), CD 36, CD 37

Interactive model, CD 34
　versus additive model, CD 34–35
Interquartile range (IQR), 93
　and significance of outliers, 94
Interval estimate, 355

Judgment sampling, 316

Landefeld, J. Steven, 218
Large population random sampling rule, 342
Large sample, 355
Large sample theory, 346
Law of large numbers, 161, 228, 229, 334
　for independent repetitions, 223
　for population surveys, 227, 335
Law of statistical regularity, 161
Leaf, 13. See also Stem-and-leaf plot
Least square nonlinear regression, 485
Least square regression line, 135–141
Level of significance, 404
Life expectancy, 53
Light, speed of, 64–65
Likert scale, 220
Linear interpolation, 492
Linear models
　converting from nonlinear, 535
　of data, 114
Linear transformations, 77–78
　strength of correlation, 120–122
Lower limit, 355
Lurking variable, 149

Main effect sum of squares, CD 36
Maris, Roger, 13, 14, 54, 55, 61, 63
Matched pairs
　large-sample Z test, 421–424
　one-sample t test, 426
McGwire, Mark, 55, 91
Mean, 53, 69
　deviation from, 74–75
　finding sample, 55
　versus median, 60
　notation, 55–56
　population, 76
　sample, 76
Mean square (MS), CD 16
Median, 52, 53, 366
　defined, 54
　finding, 54
Meta-analysis, 309
Michelson, A. A., 64
Missing baseline, 38
Mode, 53
　defined, 56
　use of, 60
Moivre, Abraham de, 264, 279
Monte Carlo simulation, 183
Mosteller, Frederick, 4
MSE. See Error mean square
MSTO. See Total mean square

MSTR. See Treatment mean square
Multinomial distribution, 461
Multinomial probability law. See Multinomial distribution
Multiple regression
　linear, CD 42–45
　with many explanatory variables, CD 52–54
　regression hypothesis, CD 48–49
　sign of coefficient, CD 52
　squared multiple correlation coefficient (R^2), CD 45–47
Multiplication rule, 169, 302. See also Product rule for independent events
　for multiple-stage experiments, 208
　for probabilities, 174, 177
Mutually exclusive events, 171

National Foundation for Infantile Paralysis (NFIP), 305–307
National Opinion Research Center (NORC), 315
Natural log transformation, 534–535
Negative correlation, 115
Newcomb, Simon, 64, 65, 77, 333, 334, 339, 354
Newton, Sir Isaac, 279
Nominal variable, CD 18
Nonlinear models, 535
Nonparametric statistics, 427
Non-probability sampling, 316–317, 326
　versus probability sampling, 312–319
Nonresponse bias, 313–314, 326
Normal approximation to binomial, 278–283
Normal curve, 241–275. See also Standard normal curve
Normal distributions, 82–88, 273–282
　and random variables, 277
　rule of thumb for binomial approximation, 281–282
Normal population
　confidence interval for μ when σ known, 374–376
　confidence interval for μ when σ unknown, 376
　distribution of $\bar{X}$, 375
　t test of the mean, 424
Normal probability density curve. See Normal curve
n times p formula for the expected number of successes, 225
Null hypothesis
　box model, 403
　defined, 399
　formulating, 403
　F statistic and, CD 37–38
　goodness of fit, 461
　for multiple populations, CD 12

Null hypothesis (*continued*)
  parameter values, 400
  regression effect, 525
  rejecting, 404, 405
  for a slope, 523–525

Observational study, 297, 298, 300, 325
One-way ANOVA, CD 18–24
  for multiple comparisons of means, CD 25–30
  using F distribution, CD 21–23
Ordinal data, 220
"Or" rule. *See* Additional rule for probabilities
Otto, Mark C., 328
Outcomes, 295, 325
  equally likely, 166–168
  not equally likely, 168–169
Outlier, 14, 63, 94, 95
  in boxplots, 91, 95
  discarding, 65–66

Parameter, 336, 389, 399
Pareto chart, 21
Pearson, Karl, 120
Percentile-based bootstrap, 369
Pie charts, 21–22, 42
Placebo, 306, 308–309, 325
Point estimate, 355
  for the population variance, 386–389
Poisson distribution, 263–270
  approximation to binomial, 264
  probability density for, 270
  uses of, 266
Poisson experiment, 267
Poisson, Siméon, 264
Poisson's limit theorem, 266
Pooled standard deviation, 382
Population, 6
  comparing, 16
  conceptual, 295, 334
  mean of, 76
  parameter, 399
  in randomized controlled experiments, 411
  real, 295, 334
  standard deviation of, 75, 452
  variance of, 75, 76
Population mean(s) (μ)
  bootstrapping, 427
  comparing, CD 2–3
  comparing graphically, CD 26
  confidence interval for $\mu_X - \mu_Y$, 379–383
  confidence interval for $\mu_X - \mu_Y$ (matched pairs), 385–386
  confidence interval for, using $\overline{X}$ and $\widehat{SE}(\overline{X})$, 354–355
  confidence interval for, when σ known, 357
  estimating, CD 26

testing for equality of, 435–438
t test of, 426
Z test of, 417–421
Population parameter, 295, 336, 389, 399
Population proportion (p)
  box-model hypothesis, 402–405
  confidence interval for p_1-p_2, 384–385
  grouping, CD 230
  large sample confidence interval for, 358–361
  normal distribution hypothesis testing, 406–407
  testing for equality, 433–434
Population surveys, 225–227, 335
Positive correlation, 115, 116
Potential outlier, 94
Power curve, 451
Power of a test, 448–449
Predicted value, 135
Prediction of accuracy, 295
Prediction interval, 531–532
Probability, 160
  binomial, 251–254
  of an event, 187
  experimental, 159–163
  model, 166–178
  normal, 273–282
  Poisson, 263–270
  of a value, 187
Probability density function (pdf)
  binomial, 257
  for a binomial distribution, 252
  for a geometric distribution, 262
  for Poisson distribution, 264, 270
Probability distribution, 167, 221, 396
Probability histogram, 186
Probability models, 166–178
Probability sampling, 295, 297, 314–316, 326
  versus non-probability sampling, 312–319
Probability trees, 173–175, 301
Product rule for independent events, 175, 260
P-value, 404, 461
  estimating with the central limit theorem, 407
  level of significance and, 404
  standardized test statistic, 419

Q-Q plot, 366, 381
Qualitative variables
  continuous, 7
  discrete, 7
Quantitative variables
  ordered categorical, 7
  ranked, 7
Quartiles, 52, 89–90
  interquartile range (IQR), 93

Quetelet, A., 83
Quota sampling, 316, 326

Random assignment, 296, 325
Random experiment, 159, 339
Randomization, 297, 325
  benefits of, 294–295
Randomized block design, CD 39
Randomized controlled experiments, 149, 295, 298, 303–307, 325
  designing, 296
  hypothesis testing in, 409–415
  null hypothesis in, 410
  population in, 411
Random measurement error, 334
Random phenomena, 158
Random sample, 338, 339, 389
Random sampling, 206–210
  from a population, 336
  of a random experiment, 339–340
  of a real population, 340
  without replacement, 206, 256, 340
  with replacement, 194, 201, 256, 258, 342, 365
  rule for a large population, 342
  simulation via box model, 200–204
Random variable, 186, 221
  continuous, 277
  expected value, 221–229
Range, 15
  defined, 74
Ranked data, 220
Real population, 295, 334, 389
Regression, 114, 132–144
  F test and, CD 48–49
  for many explanatory variables, CD 52
  problems with, 523
  with two explanatory variables, CD 42–45
Regression effect, 525
Regression hypothesis, CD 48
Regression line, 135
Regression model assumptions, 514
Regression sum of squares (SSR), CD 49
Rejection region, 447
Relative frequency, 4, 160
Repeated experiments, 389
Replication, 339
Resampling, CD 48
Research hypothesis. *See* Alternative hypothesis
Residual, 135, 143
  standard error of, 515
Residual sum of squares, 135. *See also* Error sum of squares (SSE)
Resistant measure
  of center, 61
  of spread, 77
  of variation, 93

Response variable, 115, CD 31
  and blocking variables, CD 38
  effect on, CD 37–38
Roosevelt, Franklin Delano, 303–304, 305, 312, 313, 314
Rule of thumb for normal approximation, 360

Salk, Jonas, 304
Salsburg, David, 4–5
Salsburg, Dena, 4–5
Sample, 6, 31, 55
  variance of, 76
Sample mean, 76
Sample of convenience, 317, 326
Sample space, 166
Sample standard deviation, 76
Sample statistic, 295, 399
Sample surveys, 207
Sampling
  bootstrap, 428
  judgment, 316
  quota, 316
  from a real population, 297, 312–319
  without replacement, 403
  with replacement, 403
Sampling distribution
  of an estimator, 343–345
Sampling frame, 321
Scale, change of, 78
Scatterplot, 114, 115–116
Selection bias, 313, 326
Self-selected samples, 317, 326
Sensitivity, 173
Significance, level of, 404. *See also* Hypothesis testing
Significance testing. *See also* Hypothesis testing
  versus acceptance/rejection test, 445–451
Simple random sampling, 256, 314, 326
Simpson's paradox, 303, 325
Simulation, 182–198
  defined, 182
  of equally likely outcomes, 188–193
  five-step method, 184–188
  Monte Carlo, 183
Single-blind experiment, 306, 326
68–95–99.7% rule. *See also* Empirical rule, for bell-shaped distributions
  estimating $\bar{x}$ and s from a frequency histogram, 72–73
Skewness, 53
Slope
  confidence interval, 526
  estimate of, 517
  hypothesis testing, 523–525
  and least squares multiple regression, CD 44

regression, 132
  standardized test statistic, 524, 526
  t test versus Z test, 525–526
Slope-intercept form of a line, 132
Sosa, Sammy, 55
Specified level of significance, 465
Specificity, 173
Spread, 53, 70, 95
Squared multiple correlation coefficient (R^2), CD 45–47
  and bootstrapping, CD 47–48
  and sum of squares, CD 46
SSE. *See* Error sum of squares
SSR/SSTO ratio, CD 54
SSTO. *See* Total sum of squares
SSTR. *See* Treatment sum of squares
Standard deviation, 70
  and interquartile range (IQR), 93
  and measuring spread, 77
  and population, 75
  of a population, 452
  of a random variable, 239–243
  sample, 76
Standard error
  estimation from a random sample, 353–354
  of an estimator, 343, 389
  of residuals, 515
  of a slope estimate, 513–523
Standardized score, 126
Standardized test statistic, 419
  for a slope, 524, 526
Standard normal curve, 85, 275
  area under, 86–88
Standard normal table, 84–88
Standard scores. *See* z-scores
Statistic, 399
  χ^2, 467–469
  D, 463
  nonparametric, 427
  test, 400
Statistical data
  qualitative (categorical) variables, 7
  quantitative variables, 7
Statistical inference, 158
Statistical literacy, 42
Statistical significance, 296, CD 3–4
Statistics. *See also* Estimators
  data, 7
  data analysis, 98
  descriptive, 7
  inferential, 7
  purpose of, 5
Stem, 13
Stem-and-leaf plot, 12
  comparing populations using, 16
  constructing, 13–16
  converting to histogram, 26–27
  defined, 13
Stemplot. *See* Stem-and-leaf plot
Student's t distribution. *See t* distribution

Sum of squares, 75, 76, CD 28
  degrees of freedom, CD 49
  of deviation values, CD 7
  error (SSE), CD 8, CD 9, CD 17, CD 36
  interactions (SSAB), CD 36, CD 37
  regression (SSR), CD 49
  and R^2, CD 46
  total (SSTO), CD 9, CD 17, CD 36
  treatment, CD 7–8, CD 9, CD 17
  variable A (SSA), CD 36
  variable B (SSB), CD 36
Survey sampling, 389
Systematic measurement error, 334

t distribution, 376–377
Test of significance, 396, 398
Test statistic, 400, 465
  for comparing means, CD 6
  standardized, 419. *See also* Standardized test statistic
Theoretical mean. *See* Expected value
Theoretical probabilities, 490
Theoretical probability
  of an outcome, 160
  using probability trees, 301–302
Three-stage approach to hypothesis testing, 509
Todd, Sean, 394
Tolerances, 74
Total deviation, 142
Total mean square (MSTO), CD 17, CD 46
Total sum of squares (SSTO), CD 9, CD 17
Total variation, 142
Transformation of variables, 533–535
Treatment, 295, 325, CD 23–24
Treatment group, 295, 300, 325
Treatment mean square (MSTR), CD 17, CD 19
Treatment sum of squares (SSTR), CD 7–8, CD 9, CD 17
Treatment variability, CD 6
Tree diagram, 167, 172
Trial of an experiment, 159, 251
T statistic, 424
t test, CD 3
  of a normal population mean, 424–426
  for a slope, 525–526
Tufte, Edward R., 38
Tukey, John, 13
Two independent samples case, 381
Two-sample problem, 379
Two-way ANOVA, CD 31–40
  Bonferroni approach to multiple expansions, CD 40
  and bootstrapping, CD 39–40
  experiments with blocks, CD 38–39
  F test, CD 36–38
  and interaction, CD 34
  least squares analysis, CD 40

Type I and Type II errors, 446
Typical random variability of the estimator, 389
Typical variation, 74

Unbiasedness, 388
Unequal expected frequencies, 497–501
Unexplained variation, 142
Uniform distribution, 101
Unit, 295, 325
Upper limit, 355

Validity, 294
Variability
   between sample means, CD 6
   main effects sum of squares, CD 36
   within samples, CD 6, CD 19
Variable A Sum of squares (SSA), CD 36
Variable B Sum of squares (SSA), CD 36

Variables
   blocking, CD 38
   categorical, CD 18
   counting, 30
   explanatory, CD 52
   nominal, CD 18
   plotting three, CD 44
   qualitative, 7
   quantitative, 7
   response, CD 31
   transformation of, 533–535
Variance
   population, 76
   sample, 76
Variation, 141–143. *See also* Spread
   as measure of consistency, 74
   population, 75
   tolerances, 74
   typical, 74

Volunteer effect, 305, 325

Wagner, Peter J., CD 0
Wallace, David, 4
Williams, Ted, 69
Within-samples sum of squares. *See* Error sum of squares

y-intercept (regression), 132

Zhang, Jian, 510
z-score, 84
Z statistic, 424
Z test, CD 3
   for a large sample, 418–421
   for matched pairs, 421–424
   for a normal population, 421
   for a slope, 525–526

CD-ROM Instructions For Chapter 12

To use *Statistics: The Craft of Data Collection, Description, and Inference*, Third Edition Chapter 12 CD-ROM attached to the inside back cover of this book, follow these steps:

Instructions for Windows 95, 98, 2000, and XP

1. Insert the CD into your CD-ROM drive.
2. The CD should start automatically. If it does not, follow the steps below.
3. From the Run dialog box, type

   D:\ch12.pdf

   If your CD-ROM drive is not "D," please replace the "D" with the correct CD drive letter.
4. Follow the instructions presented on the screen.
5. Adobe Acrobat reader is provided for your convenience at the following path: D:\acroread.exe

Instructions for Macintosh

1. Insert the CD into your CD-ROM drive.
2. The CD should start automatically. If it does not, follow the steps below.
3. Double-click on your CD-ROM icon.
4. Double-click on the file titled "Ch12.pdf."
5. Adobe Acrobat reader is provided for your convenience in the same folder. The file is called "acroread."

The material provided on the CD-ROM is distributed with and is considered a part of the book *Statistics: The Craft of Data Collection, Description, and Inference*, Third Edition.

Copyright © 2002 by Möbius Communications, Ltd.
1802 South Duncan Road, Champaign, IL 61822
All rights reserved. No part of the CD-ROM may be reproduced, stored, or transmitted by any means without the prior written permission of the publisher.